Contents

Preface to first edition

The function of a reference book is not always easy to define. My view is that it should be able to provide relatively brief but accurate accounts of a variety of topics under a general heading – in this case audio engineering. Also the assumption has to be made that the reader already knows something of the subject. If not then he or she should instead be looking for a textbook. Furthermore, a reference book should, I feel, be as readable as possible. The last requirement seems to me to be very important. The reader that I have envisaged is a busy person and such people do not want to spend time wading through pages of turgid prose when they require information. They are often going to need information in a hurry, information that can be absorbed with a minimum of effort. There is, of course, a delicate balance to be struck, maintaining accurate information on the one hand with conciseness and readability on the other. I have tried to choose authors who are able to achieve that balance.

The aim, obviously, has been to cover audio engineering as widely as possible. Inevitably any one reader is going to feel that some topic has been missed out. This is unavoidable. A book that contained every potential reader's required subjects would be impossibly big – and expensive! In one or two instances I was unable to track down specialist authors who were available to write on a particular topic – and when that happened I decided that it was better to leave that subject out rather than risk a possibly second-rate contribution. I hope, though, that by covering the range from basic mathematics and formulae to recent digital techniques, there is enough here to fill a significant gap in the currently available literature.

Finally, it gives me particular pleasure to record my gratitude for the never-failing friendliness and helpfulness of three ladies at Butterworth-Heinemann. They are Angela Leopard, Sarah Ward and Margaret Riley. The task of an Editor of a reference book can be at times very difficult and wearisome, but they have consistently smoothed paths for me and given words of cheer and encouragement to make my work enjoyable.

Michael Talbot-Smith

Technical note: Readers may notice an inconsistency in the units of sound pressure measurement made with the A-weighting network; 'dBA' and 'dB(A)' both appear. There seems at present to be no indication, at least in the UK, that one is preferred to the other and I have therefore accepted both, adopting the form which was used in the individual authors' typescripts.

M.T.-S.

Preface to second edition

One of the specifications I gave to contributors to the first edition was that, as far as possible, articles should not be too closely related to current equipment and techniques so that the book wouldn't go out of date quickly.

To a large extent this has worked well, but, of course, there have been major changes in some fields. For example, Perceptual Coding has acquired a much greater importance than it had in 1994 so there has been a need for some new material.

Consequently all the available authors have looked carefully at their original contributions and had the opportunity to revise where necessary. Some of the revisions have been very minor, but others have been quite significant.

A few of the original authors have, for one reason or another, been unobtainable – very sadly in two cases because they have died – but where their contributions were timeless, so to speak, I have retained them without change.

The fact that a second edition was necessary in such a relatively short time suggests that this book has filled a need. I hope that this new edition continues to do so.

Michael Talbot-Smith

Contributors

John L. Andrews, BSc, MIBS, Sales and Marketing Consultant, BCD Audio

Jan Arts, Technical Support Engineer, Sony Service Centre, Europe

Geoff Atkins, MIBS, Consultant Engineer, the Black Mountain Mobile. (Formerly BBC)

Peter Baxandall, BSc(Eng), CEng, FIEE, formerly Electro-Acoustical Consultant (Deceased)

John Borwick, BSc/(Physics) Audio Director, *Gramophone*

Terry Clarke, Technical Director, mc^2 Audio Ltd

David M. Davidson, Chief Engineer, Telephone Systems, Northern Telecom Europe Ltd

Roger Derry, MIBS, Audio and Broadcast Consultant

Steven Dixon, BA(Hons), Department of Physics, University of Warwick

Louis D. Fielder, Principal Staff Engineer, R & D and Business Development, Dolby Laboratories Inc

Douglas Ford, Sky walker Sound, a division of Lucas Digital Ltd

Michael L. Gayford, BSc, MIEE, formerly Consultant (Deceased)

Larry Goodson, formerly Head of Sound, BBC Television

Kenneth Gundry, Principal Staff Engineer, Dolby Laboratories Inc.

Peter Harris, formerly Technical and Managing Director of CTS Studios

Yasmin Hashmi, BSc(Hons), Independent Digital Audio Consultant, SYPHA

John Howell, Senior Sound Supervisor, BBC Television

Peter Jackson, Technical Strategy Adviser, Merlin Communications International Ltd

Jan Maes, Service Operations Manager, Sony Service Centre Europe

Bev Marks, Broadcast Consultant

David Mellor, BMus (Tonmeister) (Hons) ATCL, Media Consultant

Malcolm Nelson, BSc, CPhys, MInstP, AIBS, Head of Operational Training, Centre for Broadcast Skills Training, BBC Wood Norton

Glynne Parry, BSc, CPhys, MInstP, Senior Calibration Engineer, CEL Instruments Ltd

John Ratcliff, FRCDC, MAES, MIBS, AMIOSH, Proprietor, The Television Training Company

Tim Robinson, BSc, Consulting Engineer in Building Services

Francis Rumsey, BMus (Tonmeister), PhD, Senior Lecturer, University of Surrey

Peter Skirrow, BSc, AMIEE, MAES, Designer and Proprietor, Lindos Developments

Michael Talbot-Smith, BSc, CPhys, MInstP, Consultant

Alan Tutton, BEd, MIBS, Lecturer, Audio Operations Training, BBC

Richard Tyler, CEng, MSc, MIEE, MIOA, MSEE, Chief Engineer, CEL Instruments Ltd

Bob Walker, BSc(Eng), CEng, FIDA, MIEE, Senior Research Engineer, BBC

Vivian Weeks, BSc(Eng), ACGI, CEng, MIEE, formerly of BBC Design and Equipment Department

Andy Wilson, BSc(Hons), Studer UK Ltd

J. Patrick Wilson, Lecturer in Electronics, Keele University

Chris Woolf, Technical Consultant, Rycote Microphone Windshields

David C. Yates, CEng, MIEE, Broadcast Transmission Training Consultant

1 Basic principles

John Ratcliff
Proprietor, The Television Training Company

Michael Talbot-Smith
Consultant

J. Patrick Wilson
Lecturer in Electronics, Keele University

Louis D. Fielder
Principal Staff Engineer, R&D and Business Development, Dolby Laboratories Inc

Glynne Parry
Senior Calibration Engineer, CEL Instruments Ltd
Richard Tyler
Chief Engineer, CEL Instruments Ltd

Michael Gayford
Former Consultant (Deceased)

Roger Derry
Audio and Broadcast Consultant

1.1 Mathematics, formulae and equations

1.1.1 SI units

The four fundamental SI units are length, mass, time and electric current.

Length (*L*) The unit of length is the metre (m). One metre is 1 650 763.73 wavelengths in a vacuum of the radiation corresponding to the transition between the levels $2p_{10}$ and $5d_5$ of krypton-86.

Mass (*M*) The unit of mass is the kilogram (kg). One kilogram is the mass of the international prototype held by the Bureau International des Poids et Mesures, Paris.

Time (*T*) The unit of time is the second (s). One second is the duration of 91 926 131 770 periods of radiation corresponding to the transition between two hyperfine levels of the ground state of caesium-133.

Electric current (*I*) The unit of electric current is the ampere (A). One ampere is the current that, flowing in two infinitely long conductors of negligible cross-section spaced apart by one metre in a vacuum, will produce a force of 2×10^{-7} newtons per metre (N/m) between them.

1.1.2 Derived SI units

Velocity (*V*) Velocity is the distance/time in metres/second (m/s).

Acceleration (*a*) Acceleration is the rate of change of velocity. It is given in metres per second (m/s^2).

Force (*F*) The unit of force is the newton (N). One newton will accelerate a mass of 1 kg by 1 m/s^2.

Work or energy (*W*) The unit of work is the joule (J). One joule of work is done when one newton of force moves through one metre in the direction of the force.

Power (*P*) The unit of power is the watt (W). Power is the rate at which work is done: one joule of energy expended in one second requires one watt of power.

Charge (*Q*) The unit of charge is the coulomb (C). One coulomb is the quantity of electricity passing through a conductor when one ampere flows for one second.

Potential Difference (*V*) The unit of potential difference is the volt (V). One volt of potential difference exists between two points if one joule of work is done in transferring one coulomb of charge from one point to the other.

Resistance (*R*) The unit of resistance is the ohm (Ω). One ohm is that resistance in which a current of one amp generates energy at the rate of one joule per second.

Capacitance (*C*) The unit of capacitance is the farad (F). A capacitor of one farad will raise the potential between its plates by one volt with the addition of one coulomb of charge ($Q = CV$).

Magnetic flux (Φ) The unit of magnetic flux is the weber (Wb). When a conductor experiences flux changing at one weber per second a potential difference of one volt is produced across the ends of the conductor.

Flux density (*B*) The unit of flux density is the tesla (Wb/m^2). The flux per square metre at right angles to the field is the flux density. Frequently quoted as webers per square metre: $B = \Phi/a$ (Wb/m^2).

Inductance (*L*) The unit of inductance is the henry (H). A closed circuit has an inductance of one henry when a potential difference of one volt is produced by a rate of change of current of one ampere per second.

Pressure (*P*) Unit of pressure is the pascal (Pa): one pascal is one newton per square metre (N/m^2).

Frequency (*f*) The unit of frequency is the hertz (Hz). Frequency is the inverse of time (1/s). It is defined as the number of waves passing a fixed point in one second.

Wavelength (λ) Wavelength is the distance between successive points (e.g. crests) on successive waves (see *Figure 1.1*). Wavelength, velocity and frequency are related by the formula $v = f\lambda$.

Angle (Θ) Defined as the radian (rad). Two radii of a circle at an angle of one radian will cut an arc equal to the radius of the circle at the circumference. 2π radian $= 360°$. 1 radian $= 57.3°$.

Permeability (μ) The *permeability of space* (μ$_0$) is the ratio of the flux density (*B*) to the magnetizing force (*H*) in the air of a non-magnetic material.

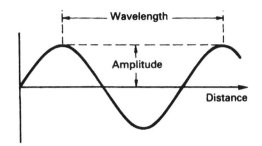

Figure 1.1 *Wavelength is the distance between corresponding points on successive waves, e.g. crests or troughs.*

$$\mu_0 = \frac{B}{H} = 4\pi \times 10^{-7} \text{ (H/m)}.$$

The *relative permeability* (μ_r) of a material is the factor by which the flux density rises for the same magnetizing force.

Flux density in material = Flux density in air = $\mu_0 \cdot \mu_r$. Absolute permeability (μ) = $\mu_0 \cdot \mu_r$

Permittivity (ε) The *permittivity of space* (ε_0) is the ability of space to allow two charges to exert a force on each other at a distance:

$$\text{Force} = Q_1 Q_2 / (4\pi\varepsilon_0 d^2)$$
$$= 8.854 \times 10^{-12} \text{ F/m}$$

The *relative permittivity* (ε_r) of a material is the ratio by which the electric flux increases above its value in air; also known as the *dielectric constant*.

Absolute permittivity (ε) = $\varepsilon_0 \varepsilon_r$

Note: $\mu_0\varepsilon_0 = 1/c^2$, where c is the velocity of light in space (300 000 km/s, to a close approximation).

Thermodynamic temperature (T) The unit of thermodynamic temperature is the kelvin (K). One kelvin is 1/273.16 of the triple point of water.

$0 \text{ K} = -273°\text{C}$.

1.1.3 Powers of ten, prefixes and symbols

	Name	Abbreviation
10^{12}	tera	T
10^{9}	giga	G
10^{6}	mega	M
10^{3}	kilo	k
10^{2}	hecto	h*
10^{1}	deka	da*
10^{-1}	deci	d*
10^{-2}	centi	c*
10^{-3}	milli	m
10^{-6}	micro	μ
10^{-9}	nano	n
10^{-12}	pico	p
10^{-15}	femto	f
10^{-18}	atto	a

*Non-preferred multiples.

1.1.4 Useful relationships

Ratio of circumference to diameter of a circle = π
= 3.142

Velocity of light in freespace = 299 792.458 km/s

Surface area of a sphere $A = 4\pi r^2$

Surface area of a hemisphere $A = 2\pi r^2$

Peak value of a sine wave = $1.414 \times$ r.m.s.

Peak–peak value of a sine wave = $2.828 \times$ r.m.s.

Note: r.m.s. = root mean square.

1.1.5 Inverse square law

If there are no obstructions near a sound source (free field conditions), then the sound pressure wave will travel outwards as an expanding sphere or hemisphere, and the sound energy will be evenly distributed over the surface of this sphere. As the surface area of a sphere is proportional to the square

Figure 1.2 *To calculate the loss due to the inverse-square law, read off the decibel loss for the two distances. The difference between the decibel losses gives the loss due to the inverse-square law. For example, above 3 m read 9.5 dB, above 10 m read 20 dB, 20 − 9.5 = 10.5 dB; so moving from 3 m to 10 m reduces the sound level by 10.5 dB.*

of the radius, the sound energy available over any given area will fall off as the inverse of the (radius).

$$\text{Loss (dB)} = 10\log\left(\frac{r_1}{r_2}\right)^2 = 20\log\frac{r_1}{r_2}$$

Figure 1.2 illustrates a suitable nomogram for calculating the fall off in level (dB) with distance according to the inverse square law.

This relationship only holds when the smaller of the two radii (in practice the smaller of the distances from the sound source being compared) is considerably greater than the dimensions of the sound source. Where the distance from the source is comparable to its size there is little fall off in level.

1.1.6 Trigonometrical relationships

See *Figure 1.3*.

$$c = \sqrt[2]{a+b}$$

$$\sin\theta = a/c$$

$$\sin^2\theta + \cos^2\theta = 1$$

$$\text{cosecant}\,\theta = 1/\sin\theta$$

$$\text{secant}\,\theta = 1/\cos\theta$$

$$\text{cotangent}\,\theta = 1/\tan\theta$$

$$\cos\theta = b/c$$

$$\tan\theta = a/b$$

$$\tan\theta = \sin\theta/\cos\theta$$

$$\sin(\theta \pm \varphi) = \sin\theta\cos\varphi \pm \cos\theta\sin\varphi$$

$$\sin 2\theta = 2\sin\theta\cos\theta$$

$$\cos 2\theta = \cos^2\theta - \sin^2\theta$$

$$\sin\theta\cos\varphi = 0.5[\sin(\theta+\varphi) + \sin(\theta-\varphi)]$$

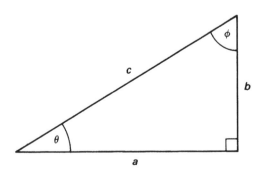

Figure 1.3 *Relationship between the sides of a right-angled triangle with reference to one of the acute angles.*

$$\cos\theta\cos\varphi = 0.5\cos(\theta+\varphi) + 0.5\cos(\theta+\varphi)$$

$$\cos^2\theta = 0.5(1 + \cos 2\theta) = 0.5 + 0.5\cos 2\theta$$

$$\cos\theta + \cos\varphi = 2[\cos 0.5(\theta+\varphi)]$$

$$[\cos 0.5(\theta-\varphi)]$$

1.1.7 Indices

Principles of indices

$$a \times a \times a \times a \times a = a^5$$

$$(a \times a) \times (a \times a \times a) = a^2 \times a^3 = a^5$$

Hence $a^m \times a^n = a^{(m+n)}$

$$\frac{a \times a \times a}{a \times a \times a} = \frac{a^3}{a^3} = a$$

Hence $a^m/a^n = a^{(m-n)}$ and $1/a^n = a^{-n}$

$$a^{1/2} \times a^{1/2} = a^{(1/2+1/2)} = a$$

Hence $\sqrt[2]{a} = a^{1/2}$ and $\sqrt[n]{a} = a^{1/n}$

$$a^3/a^3 = a^{(3-3)} = a^o$$

Hence $a^o = 1$

Manipulation of indices

$$(ab)^n = a^n b^n$$

$$\sqrt[n]{(a/b)} = \frac{\sqrt[n]{a}}{\sqrt[n]{b}}$$

$$\sqrt[n]{\sqrt[m]{a}} = \sqrt[mn]{a}$$

$$(a/b)^n = a^n/b^n$$

$$\sqrt[n]{(ab)} = \sqrt[n]{a}\sqrt[n]{b}$$

1.1.8 Logarithms

Common logarithms

If $x = a^n$ then $n = \log_a x$, e.g.

$$\log_{10} 100 = 2, \text{ or } 10^2 = 100$$

$$\log_{10} 3.17 = 0.5, \text{ or } 10^{0-5} = \sqrt[2]{10} = 3.17$$

$$\log_{10} 1 = 0$$

Natural (naperian) logarithms
e = 2.7183 (to four decimal places)

It is the result of natural growth and equals $\left(1 + \frac{1}{n}\right)^n$

where n approaches infinity:

$$\left(1+\frac{1}{n}\right)^n = 1+1+\frac{1}{1\times 2}+\frac{1}{1\times 2\times 3}$$
$$+\frac{1}{1\times 2\times 3\times 4}+\cdots$$

e is taken from the name of the Swiss mathematician Euler. In general, the term 'log' is taken to mean 'logarithm to the base 10', and 'ln' as 'logarithm to the base e'.

Rules for logarithms

$$\log_a(pq) = \log_a p + \log_a q$$

$$\log_a(p/q) = \log_a p - \log_a q$$

$$\log_a(p^n) = n\log_a p$$

$$\log_a \sqrt[n]{p} = \log_a(p^{1/n}) = \frac{1}{n}\times\log_a p$$

$$\log n_n = 1$$

Conversion between logarithmic bases
To convert between base 10 and base e:

$$\log_e N = 2.30\log_{10} N$$

1.1.9 Solution of quadratic equations

If an equation can be expressed in the form $ax^2 + bx + c = 0$, then it can be solved by using the formula:

$$x = \frac{-b \pm \sqrt[2]{b^2 - 4ac}}{2a}$$

Note that the \pm sign means that there are two possible solutions.

1.1.10 Complex numbers

The operator $j = \sqrt{-1}$. It is regarded as the vertical component of two Cartesian coordinates (see *Figure 1.4*).

Powers of *j*

$$j = \sqrt[2]{-1}$$

$$j^2 = -1$$

$$j^3 = (\sqrt[2]{-1})^2(\sqrt[2]{-1}) = -1(\sqrt[2]{-1}) = j$$

$$j^4 = (j^2)^2 = (-1)^2 = 1$$

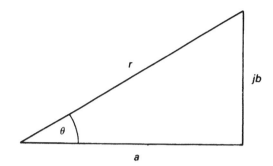

Figure 1.4 *Relationship between Cartesian $(a+jb)$ and polar $(r\angle\theta)$ forms.*

Addition of complex numbers

$$(a+jb)+(m+jn) = (a+m)+j(b+n)$$

Subtraction of complex numbers

$$(a+jb)-(m+jn) = (a-m)+j(b-n)$$

Multiplication of complex numbers

$$(a+jb)(m+jn) = am+mjb+ajn-bn$$
$$= (am-bn)+j(mb+an)$$

Division of complex numbers
Multiply the numerator and the denominator by the conjugate of the denominator:

$$\frac{a+jb}{m+jn} = \frac{(a+jb)(m-jn)}{(m+jn)(m-jn)}$$
$$= \frac{am+bn+j(mb-an)}{m^2-n^2}$$
$$= \frac{am+bn}{m^2-n^2}+\frac{j(mb-an)}{m^2-n^2}$$

Conversion between Cartesian and polar forms
The Cartesian form is expressed as $a+jb$. The polar form is expressed as $r\angle\theta$ (see *Figure 1.4*).

$$a = r\cos\theta;\ jb = r\sin\theta$$

$$a+jb = r(\cos\theta+j\sin\theta)$$

$$r = \sqrt[2]{(a^2+b^2)}$$

$$\theta = \tan(b/a)$$

Multiplication and division of polar forms

$$r_1\angle\theta_1 \times r_2\angle\theta_2 = r_1 r_2(\theta_1+\theta_2)$$

$$r_1\angle\theta_1/r_2\angle\theta_2 = r_1/r_2(\theta_1-\theta_2)$$

1.1.11 Number bases

Base 10 (denary or decimal)

$$237_{10} = (2 \times 10^2) + (3 \times 10^1) + (7 \times 10^0)$$
$$= 200_{10} + 30_{10} + 7_{10} = 237_{10}$$

Base 2 (binary)

$$1011_2 = (1 \times 2^3) + (0 \times 2^2) + (1 \times 2^1) + (1 \times 2^0)$$
$$= 8_{10} + 0_{10} + 2_{10} + 1_{10} = 11_{10}$$

Base 16 (hexadecimal)

Since traditional (denary) arithmetic has no symbol greater than 9, hexadecimal arithmetic uses the symbols A–F to represent the values 10–15. Hexadecimal numbers are conventionally identified by the prefix '&', hence:

$$\&37 = (3 \times 10^1) + (7 \times 16^0) = 55_{10}$$

$$\&A3F_{16} = (10 \times 16^2) + (3 \times 16^1) + (15 \times 16^0)$$
$$= 2560_{10} + 48_{10} + 15_{10} = 2623_{10}$$

Converting between number bases

From base 10 to base 2:
$155_{10} = ?_2$

```
      1 remainder 0
   2)2  remainder 0
   2)4  remainder 1
   2)9  remainder 1
  2)19  remainder 0
  2)38  remainder 1
  2)77  remainder 1
  2)155
```
$155_{10} = 10011011_2$

From base 10 to base 16:
$2879_{10} = ?_{16}$

```
                        11^10 = &B
      11 remainder 3     3_10 = &3
   16)179 remainder 15  15_10 = &F
   16)2879
```
$2879_{10} = \&B3F$

Addition of binary numbers

$1011_2 + 1110_2 (1 + 1 = 0 \text{ carry } 1)$:

```
   1 0 1 1
   1 1 1 0 +
  ---------
   1 1 0 0 1     = 1 1 0 0 1_2
  ---------
   1 1 1
```

Subtraction of binary numbers

$1100101_2 - 11011_2$: Because of the 'borrow across' difficulty, traditional methods are confusing. Use 'reverse' addition, e.g

```
      53
    -17  }  17 + 36 = 53
    ----
      36
```

```
   1 1 0 0 1 0 1
   0 0 1 1 0 1 1 −
  --------------
   1 0 0 1 0 1 0
  --------------
     1 1   1
```

$$0011011_2 + 1001010_2 = 1100101_2$$

Multiplication of binary numbers

$1101_2 \times 101_2$:

```
       1 1 0 1
       1 0 1 ×
     ----------
       1 1 0 1
     0 0 0 0
   1 1 0 1
   ------------
   1 0 0 0 0 0 1 = 1 0 0 0 0 0 1_2
   ------------
   1 1 1 1
```

Division of binary numbers

$$110.1101_2 = \left(4 + 2 + 0 + \frac{1}{2} + \frac{1}{4} + \frac{0}{8} + \frac{1}{16}\right)_{10}$$

$$= 6 + \frac{(8 + 4 + 1)}{16} = 6\frac{13}{16} = 6.8175_{10}$$

Divide 111.01_2 by 10_2

```
                1 1 . 1 0 1
              ----------------
       1 0 ) 1 1 1 . 0 1 0
             1 0
            ----
             1 1
             1 0
            ----
               1 0
               1 0
              ----
                 0   0 1 0
                     1 0
                    ----
                     0 0
```

$$= 11.101_2$$

1.1.12 2s complement arithmetic

Where negative binary numbers have to be processed (e.g. the digital forms of negative half-cycles of audio signals), binary subtraction performed (digital mixing, filtering) or zero voltage crossing points of signals maintained, it is convenient to employ a binary number system that permits these processes with ease. 2s complement is such a system. It is a 'signed' binary system, in which the most significant bit of a negative number indicates its negative polarity, and the remaining bits are arranged such that subtraction (or the addition of negative numbers) can be performed by an addition process (see Section 1.1.11).

To convert a negative binary number to 2s complement, first invert all 0s and 1s, then add 1 to the result:

	1 1 0 0 1 1 0 0
Invert all bits	0 0 1 1 0 0 1 1
Add 1	0 0 1 1 0 1 0 0

To convert a 2s complement number back to normal binary, the process is repeated:

	0 0 1 1 0 1 0 0
Invert all bits	1 1 0 0 1 0 1 1
Add 1	1 1 0 0 1 1 0 0

Note that positive binary numbers are not affected in this system, and that all negative numbers have a '1' as the most significant bit. An 8-bit 2s complement number will represent a maximum positive value of $+127$ and a maximum negative value of -128. If this range is exceeded as a result of the process, incorrect (alias) values will result.

The following example illustrates the use of 2s complement.

Denary value	Unsigned binary	2s complement
$+7$	$+0111$	0111
$+6$	$+0110$	0110
$+5$	$+0101$	0101
$+4$	$+0100$	0100
$+3$	$+0011$	0011
$+2$	$+0010$	0010
$+1$	$+0001$	0001
0	0000	0000
-1	-0001	1111
-2	-0010	1110
-3	-0011	1101

Denary value	Unsigned binary	2s complement
-4	-0100	1100
-5	-0101	1011
-6	-0110	1010
-7	-0111	1001
-8	-1000	1000

Some examples:

$$3 + 2 = 0011 + 0010 = 0101 = 5$$

$$4 - 2 = 0100 + 1110 = (1)0010 = 2$$
(note the leading '1' is discarded)

$$3 - 7 = 0011 + 1001 = 1100 = -4$$

$$-2 - 5 = 1110 + 1011 = (1)1001 = -7$$
(note the leading '1' is discarded)

1.1.13 Calculus

1.1.13.1 Differentiation of a function of the form $y = x^n$

If $y = ax^2$, then

$$\frac{dy}{dx} = 2ax^{(2-1)} = 2ax$$

For example if $y = 3x^{1/3}$, then

$$\frac{dy}{dx} = \frac{3}{3}x^{-2/3} = \frac{1}{\sqrt[3]{x^2}}$$

If $y = ax^n + bx + c = ax^n + bx^1 + cx^0$, then

$$\frac{dy}{dx} = nax^{(n-1)} + b;$$

For example, if $y = 3x^2 + 5x + 4$, then

$$\frac{dy}{dx} = 6x + 5$$

If $y = \frac{1}{ax^n} = \frac{x^{-n}}{a}$, then

$$\frac{dy}{dx} = -\frac{nx^{-(n+1)}}{a} = -\frac{n}{ax^{(n+1)}}$$

For example, if $y = \frac{1}{3x^{1/2}} = \frac{x^{-1/2}}{3}$, then

$$\frac{dy}{dx} = -\frac{1}{2}\frac{x^{-3/2}}{3} = -\frac{1}{6\sqrt[2]{x^3}}$$

If $y = (ax^n)^m = a^m x^{mn}$, then

$$\frac{dy}{dx} = nma^m x^{(nm-1)}$$

For example, if $y = (2x^2)^3 = 8x^6$, then

$$\frac{dy}{dx} = 48x^5$$

1.1.13.2 Differentiation of a function of a function of the form $y = (ax^m + b)^n$

If $y = (ax^n + b)^m$, let $z = ax^n + b$, then $y = z^m$

$$\frac{dz}{dx} = nax^{(n-1)} \text{ and } \frac{dy}{dz} = mz^{(m-1)}$$

$$= m(ax^n + b)^{(m-1)}$$

Thus $\dfrac{dy}{dx} = \dfrac{dy}{dz} \times \dfrac{dz}{dx}$

$$= nax^{(n-1)}m(ax^n + b)^{(m-1)}$$

e.g. if $y = (3x^2 + 5)^6$, let $z = 3x^2 + 5$, then $y = z^6$

$$\frac{dz}{dx} = 6x \text{ and } \frac{dy}{dz} = 6(3x^2 + 5)^5$$

Thus $\dfrac{dy}{dx} = 6(3x^2 + 5)^5 6x = 36x(3x^2 + 5)^5$

1.1.13.3 Differentiation of e^x and $\ln x$

If $y = e^x$, then $\dfrac{dy}{dx} = e^x$

If $y = e^{ax}$, let $z = ax$, then $y = e^z$

$$\frac{dz}{dx} = a \text{ and } \frac{dy}{dz} = e^z$$

Thus $\dfrac{dy}{dz} = ae^{ax}$

e.g. if $y = e^{(2x+3)}$, let $z = 2x + 3$, then $y = e^z$

$$\frac{dz}{dx} = 2 \text{ and } \frac{dy}{dz} = e^z$$

Thus $\dfrac{dy}{dx} = 2e^{(2x-3)}$

If $y = \dfrac{a}{e^x} = ae^{-x}$, *let* $z = -x$, *then* $y = ae^z$

$$\frac{dz}{dx} = -1 \text{ and } \frac{dy}{dz} = ae^z$$

Thus $\dfrac{dy}{dx} = (-1)ae^{-x} = -\dfrac{a}{e^x}$

If $y = \ln(x)$, then $x = e^y$

$$\frac{dx}{dy} = e^y = x$$

Thus $\dfrac{dy}{dx} = \dfrac{1}{x}$

In general, if $y = \ln(ax)$ then $y = \ln(a) + \ln(x)$.

Thus $\dfrac{dy}{dx} = 0 + \dfrac{1}{x}$, since $\ln(a)$ is constant.

In general, if $y = e^{f(x)}$, then

$$\frac{dy}{dx} = f'(x)e^{f(x)}$$

and if $y = \ln[f(x)]$, then

$$\frac{dy}{dx} = \frac{f'(x)}{f(x)},$$

where $f'(x)$ is the differential coefficient of $f(x)$; e.g. if $y = e^{(x^2)}$

$$\frac{dy}{dx} = 2xe^{(x^2)}$$

and if $y = \ln(x^2 + 5x)$

$$\frac{dy}{dx} = \frac{2x + 5}{x^2 + 5x}$$

1.1.13.4 Differentiation of products

(*Note:* $x^n e^x$ is a product, $\ln(\cos x)$ is a function of a function.)

In general, the differentiation of a product is:
(1st factor × Differential coefficient of 2nd factor) + (2nd factor × Differential coefficient of 1st factor)

So, if $y = x^n e^x$

$$\frac{dy}{dx} = x^n e^x + e^x nx^{(n-1)} = e^x(x^n + nx^{(n-1)})$$

and if $y = ax^n \ln(x)$

$$\frac{dy}{dx} = \frac{ax^n}{x} + nax^{(n-1)} \ln(x)$$

$$= ax^{(n-1)} + nax^{(n-1)} \ln(x)$$

$$= ax^{(n-1)}[1 + n \ln(x)]$$

and If $y = e^{ax} \ln(x)$

$$\frac{dy}{dx} = \frac{e^{ax}}{x} + ae^{ax} \ln(x)$$

$$= ae^{ax} \left[\frac{1}{x} + a \ln(x) \right]$$

1.1.13.5 Differentiation of quotients

In general, the differential coefficient of a quotient is:

$$\frac{(\text{Denominator} \times \text{Differential coefficient of numerator}) - (\text{Numerator} \times \text{Differential coefficient of denominator})}{\text{denominator}^2}$$

So, if $y = \dfrac{x^n}{ax^m - b}$, then

$$\frac{dy}{dx} = \frac{(ax^m - b)(nx^{(n-1)}) - (max^{(m-1)})(x^n)}{(ax^m - b)^2}$$

and if $y = e^{ax}/x^n$, then

$$\frac{dy}{dx} = \frac{x^n(ae^{ax}) - e^{ax}(nx^{(n-1)})}{x^{2-n}}$$

$$= \frac{e^{ax}(ax^n - nx^{(n-1)})}{x^{2-n}}$$

and if $y = \ln(x)/e^{ax}$, then

$$\frac{dy}{dx} = \frac{e^{ax}\left(\frac{1}{x}\right) - \ln(x)(ae^{ax})}{e^{2ax}}$$

$$= \frac{e^{ax}\left[\frac{1}{x} - a\ln(x)\right]}{e^{2ax}} = \frac{\frac{1}{x} - a\ln(x)}{e^{ax}}$$

1.1.13.6 Differentiation of trigonometric functions

For the following types, use the function of a function procedure:

If $y =$	$\dfrac{dy}{d\theta} =$
$\sin(a\theta)$	$a\cos(a\theta)$
$\cos(a\theta)$	$-a\sin(a\theta)$
$\sin(a\theta + b)$	$a\cos(a\theta + b)$
$\cos(a\theta + b)$	$-a\sin(a\theta + b)$

For the following types use the product procedure:
If $y = e^x \sin(x)$, then

$$\frac{dy}{dx} = e^x \cos(x) + \sin(x)e^x$$

If $y = 2\sin(\omega t)\cos(\omega t)$, then

$$\frac{dy}{dx} = 2\sin(\omega t)[-\omega\sin(\omega t)] + \cos(\omega t)[2\omega\cos(\omega t)]$$

For the following types use the quotient procedure:
If $y = \tan(\theta) = \sin\theta/\cos\theta$, then

$$\frac{dy}{d\theta} = \frac{\cos(\theta)\cos(\theta) - \sin(\theta)[-\sin(\theta)]}{\cos^2(\theta)}$$

$$= \frac{\cos^2(\theta) + \sin^2(\theta)}{\cos^2(\theta)} = \frac{1}{\cos^2(\theta)} = \sec^2(\theta)$$

If $y = e^x/\sin x$, then

$$\frac{dy}{dx} = \frac{e^x\sin(x) - e^x\cos(x)}{\sin^2(x)} = \frac{e^x[\sin(x) - \cos(x)]}{\sin^2 x}$$

1.1.13.7 Maxima and minima (turning points)

At the turning point of a curve its rate of change (slope) is zero. So, when $dy/dx = 0$ the curve is at a turning point. Solving this equation will give the value(s) of x for maximum or minimum value(s) of y. Putting in the value(s) of x obtained will give the value(s) of y at these turning points.

To calculate whether these turning points are maxima or minima, differentiate the differential coefficient. Put in the value(s) of x obtained earlier in the result, in turn. A positive result indicates a minimum, and a negative result a maximum. For example, if $y = x^3 + 3x^2 - 9x$, then

$$\frac{dy}{dx} = 3x^2 + 6x - 9$$

Let $dy/dx = 0$, then $x = -3$ or 1.

$$\text{If } \frac{dy}{dx} = 3x^2 + 6x - 9, \text{ then } \frac{d^2y}{dx^2} = 6x + 6$$

For $x = -3$, $6x + 6$ is negative; for $x = 1$, $6x + 6$ is positive. Thus at $x = -3$ the curve is at maximum, and at $x = 1$ the curve is at minimum.

Substituting $x = -3$ and $x = 1$ in $y = x^3 + 3x^2 - 9x$ gives values of $y = 27$ and $y = -5$, respectively.

1.1.13.8 Integration of the function ax^n

If $dy/dx = ax^n$, then

$$y = \frac{ax^{(n+1)}}{(n + 1)} + \text{Constant}$$

For example, if $dy/dx = 2x^{1/2}$, then

$$y = \frac{4x^{1/2}}{3} + \text{Constant}$$

This is expressed as:

$$\int dy = \int ax^n \, dx$$

For functions of the type x^n, the process of integration reverses that of differentiation. This is true only for this type of function.
If $\int dW = \int LI\,dI$, then

$$W = \tfrac{1}{2}LI^2 + \text{Constant}$$

If $\int dW = \int CV\,dV$, then

$$W = \tfrac{1}{2}CV^2 + \text{Constant}$$

1.1.13.9 Integration of exponential and logarithmic functions

In general, an integral can be considered as a function which, when differentiated, leads to the given function.

$$\int ke^{kx}dx = e^{kx} + C, \, \& \int e^{kx}dx$$

$$= \frac{1}{k}e^{kx} + \text{Constant}$$

$$\int \frac{1}{ax+b}dx = \frac{1}{a}\ln(ax+b) + \text{Constant}$$

$$\int (1 - e^{-Rt/L})dt = t + \frac{L}{R}e^{-Rt/L} + \text{Constant}$$

$$\int \left(\frac{E}{R}e^{-t/cR}\right)dt = -EC^{-t/cR} + \text{Constant}$$

1.1.13.10 Integration of trigonometrical functions

$\int \cos(\theta)d\theta = \sin(\theta)$

$\int \sin(\theta)d\theta = \cos(\theta)$

To integrate $\int \sin^2(\theta)\,d\theta$, consider $\cos 2(\theta) = 1 - 2\sin^2(\theta)$, then

$$2\sin^2(\theta) = 1 - \cos(2\theta)$$

$$\sin^2(\theta) = \tfrac{1}{2}[1 - \cos(2\theta)]$$

$$\int \sin^2(\theta)d\theta = \frac{1}{2}\left[\theta + \frac{\sin 2(\theta)}{2}\right]$$

Similarly,

$$\int \cos^2(\theta)d\theta = \frac{1}{2}\left[\theta + \frac{\sin 2(\theta)}{2}\right]$$

$$\int \sin^2(\omega t)dt = \frac{1}{2}\left[t - \frac{\sin 2\omega t}{2\omega}\right]$$

$$\int 3\cos^2(\omega t)dt = \frac{3}{2}\left[t + \frac{\sin 2\omega t}{2\omega}\right]$$

1.1.13.11 The area bounded by a curve

To calculate the area bounded by a curve, the x axis, and perpendiculars drawn from the x axis to the curve from $x = a$ and $x = b$:

$$\text{Area} = \int_a^b y\,dx = \int_a^b f(x)\,dx$$

For example, if $y = 4x^2 - 3x$, to find the area bounded by $x = 2$ and $x = 5$

$$\int_2^5 (4x^2 - 3x)dx = \left(\frac{4x^3}{3} - \frac{3x^2}{2}\right)_2^5$$

$$= \left(\frac{4.5^3}{3} - \frac{3.5^2}{2}\right) - \left(\frac{4.2^2}{3} - \frac{3.2^2}{2}\right)$$

$$= 129.1\dot{6} = (-0.6) = 129.8\dot{3} \text{ sq.units}$$

Similarly, if $i = (E/R)(I - e^{-Rt/L})$, to find the area bounded by $t = 0$ to $t = 0.001$ s, given $R = 5\ \Omega$, $E 40$ V and $L = 4$ mH:

$$q = \int dq = \int_0^{0.001} i\,dt = \int_0^{0.001} \frac{E}{R}(1 - e^{-Rt/L})dt$$

$$= \int_0^{0.001} \frac{40}{100}(1 - e^{-5t/4NO^{-3}})dt$$

$$= \int_0^{0.001} 0.4(1 - e^{-1250t})dt$$

$$= 0.4\left[t + \frac{1}{1250}e^{-1250t}\right]_0^{0.001}$$

$$= 0.4\left[0.001 + \frac{1}{1250}e^{-1.25}\right]$$

$$- 0.4\left[0 + \frac{1}{1250}e^0\right]$$

$$= 0.4[1.2292 \times 10^{-3}] - 0.4[8 \times 120^{-4}]$$

$$= 4.9168 \times 10^{-4} - 3.2 \times 10^{-4} = 1.7168 \times 10^{-4}$$

$$= 171.68 \times 10^{-6} \text{ C}$$

1.1.13.12 Simpson's rule for areas under a curve

Divide the x axis into an even number of equal segments h, to start and finish at the limiting values of x, and measure the y ordinates at these points (*Figure 1.5*). The area under the curve is given by:

$$\text{Area} = \frac{h}{3}[\text{Sum of 1st ordinate} + \text{Last ordinate}$$
$$+ 4 \text{ (Sum of even ordinates)}$$
$$+ 2 \text{ (Sum of odd ordinates, excluding}$$
$$\text{1st and last)}]$$

Hence in *Figure 1.5*:

$$\text{Area} = \frac{h}{3}[y_1 + y_9 + 4(y_2 + y_4 + y_6 + y_8)$$
$$+ 2(y_3 + y_5 + y_7)]$$

The result is a good approximation, the accuracy of which improves with the number of segments chosen.

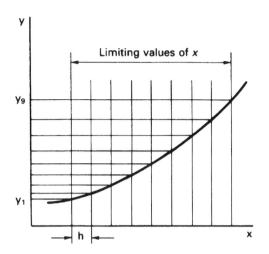

Figure 1.5 *Illustrating the y ordinates for equal segments on x axis.*

To compare this method with that of integration, consider the curve $y = 2\sin(4x)\cos(2x)$, between $x = 0$ and $x = \pi/3$.

By integration,

$$\int_0^{\pi/3} [2\sin(4x)\cos(2x)]\mathrm{d}x$$

$$= \int_0^{\pi/3} [\sin(6x) + \sin(2x)]\mathrm{d}x$$

$$= \left[-\frac{\cos(6x)}{6} - \frac{\cos(2x)}{2} \right]_0^{\pi/3}$$

$$= \left[-\frac{\cos(2\pi)}{6} - \frac{\cos(2\pi/3)}{2} \right]$$

$$- \left[-\frac{cos(0)}{6} - \frac{\cos(0)}{2} \right]$$

$$= \left[-\frac{1}{6} + \frac{1}{4} \right] - \left[-\frac{1}{6} - \frac{1}{2} \right]$$

$$= \frac{3}{4} = 0.75$$

By Simpson's rule (see *Table 1.1*):

$$\text{Area} = \frac{0.1309}{3}[0 + 0.8661 + 4(2.8978)$$

$$+ 2(2.3661)]$$

$$= 0.0436(0.8661 + 11.5912 + 4.7322)$$

$$= 0.7495$$

1.1.14 Conversion between units

From	To	Multiply by
cm	m	0.01
g	kg	0.001
dyne	newton	10^{-5}
erg	joule	10^{-7}
oersted	Magnetizing force (H)	$10^3/4\pi$
maxwell	*weber*	10^{-8}
gauss	*Wb/m²*	10^{-4}
ft	m	0.305
in.	mm	25.4
lb	kg	0.454
ft²	m²	0.093
in.²	mm²	645.2

1.2 Units used in audio engineering

1.2.1 The decibel

1.2.1.1 Decibels as power ratios

The decibel (dB) is basically a method of expressing the *ratio* between two power levels in logarithmic form. Being logarithmic, gains and losses expressed in decibels can be added or subtracted rather than multiplied or divided. The gain (or loss) of a system so expressed is:

$$10\log_{10}\left(\frac{\text{Power out}}{\text{Power in}}\right) = 10\log_{10}\left(\frac{P_{\text{out}}}{P_{\text{in}}}\right)$$

If $P_{\text{out}}/P_{\text{in}} > 1$ the decibel ratio is positive; if $P_{\text{out}}/P_{\text{in}} < 1$ the decibel ratio is negative; if

Table 1.1 Values of x, y, $2\sin(4x)$, $\cos(2x)$ for the curve $y = 2\sin(4x)\cos(2x)$

Ordinate	1	2	3	4	5	6	7	8	9
x	0	0.1309	0.2618	0.3927	0.5236	0.6545	0.7854	0.9163	1.0472
$2\sin(4x)$	0	1.0	1.7321	2.0	1.7321	1.0	0	−1.0	−1.7321
$\cos(2x)$	1.0	0.966	0.866	0.7071	0.5	0.2588	0	−0.2588	−0.5
y	0	0.966	1.5	1.4142	0.8661	0.2588	0	0.2588	0.8661

$P_{out}/P_{in} = 1$ then the decibel ratio is zero. Hence the $+/-$ sign is an integral part of the expression: $+6$ dB is a power gain of $\times 4$; -3 dB is a power loss of $1/2$.

1.2.1.2 Decibels as voltage ratios

In a system where input and output impedances are identical, since Power \propto Voltage2 the gain or loss in decibels can be expressed as:

$$10 \log_{10} \left(\frac{\text{Voltage out}^2}{\text{Voltage in}^2} \right) = 20 \log_{10} \left(\frac{V_{out}}{V_{in}} \right)$$

See *Table 1.2*.

1.2.1.3 Decibels as absolute power levels

For historical reasons, audio communications systems have evolved to use a reference power level of 1 mW, dissipating this in a 600 Ω resistance. Using this reference, decibels can be used to express actual power levels:

$$\text{dBm} = 10 \log \frac{\text{Power}}{1 \text{ mW}}$$

Table 1.2 Decibel values for voltage, current, SPL and power ratios

db	Voltage, current, or SPL ratio	Power ratio
0	1.00	1.00
1	1.12	1.26
2	1.26	1.59
3	1.41	2.00
4	1.59	2.51
5	1.78	3.16
6	2.00	3.98
7	2.24	5.01
8	2.51	6.3
9	2.81	7.94
10	3.16	10.00
20	10.00	100.00
30	31.6	10^3
40	10^2	10^4
50	3.16×10^2	10^5
60	10^3	10^6
70	3.6×10^3	10^7
80	10^4	10^8
90	3.16×10^4	10^9
100	10^5	10^{10}
110	3.16×10^5	10^{11}
120	10^6	10^{12}

SPL = sound pressure level.

The suffix 'm' in dB$_m$ indicates that the quantity referred to is referenced to 1 mW. Hence $+10$ dBm \equiv 10 mW, and -3 dBm \equiv 0.5 mW.

1.2.1.4 Decibles as absolute voltage levels

1 mW dissipated in a 600 Ω resistance results in a potential difference of 0.775 V across the resistor; hence:

$$\text{dBu} = 20 \log \frac{\text{Voltage}}{0.775}$$

The 'u' in dBu indicates that the quantity referred to is referenced to 0.775 V.

Common practice employs dBu when input and output impedances are not equal. In these circumstances the terminating conditions must be clearly understood (e.g. Z(source) $< 10\ \Omega$, Z(load) > 10 kΩ).

Microphone manufacturers (and others) use 1 V as a reference level. In this case the level is quoted relative to an output of 1 V for a sound-pressure level of 1 Pa, and the suffix 'V' is used (e.g. -60 dB$_v$ \equiv 1 mV) and the level quoted is relative to an output of 1 V/Pa. The voltage levels quoted are usually measured under no-load (open-circuit) conditions.

1.2.1.5 Decibels as sound pressure levels

Acousticians and broadcasters use decibels to express sound pressure and intensity levels. These are referenced to a pressure level of 20 μN/m^2 (see *Table 1.3*).

1.2.2 The neper

The neper (Np) is used to express the voltage or current attenuation along a line. The base of the logarithm used is e rather than 10 as used for decibels for power.

$$\text{Neper (Np)} = \ln \frac{V_1}{V_2} \text{ or } \ln \frac{I_1}{I_2}$$

$$\text{Decineper (dNp)} = 10 \ln \frac{V_1}{V_2} \text{ or } 5 \log \frac{P_1}{P_2}$$

To convert nepers to decibels, multiply by 8.686:

$$x \text{ dB} = 8.686 y \text{ Np}$$

1.2.3 Sound level units

1.2.3.1 Sound pressure level

The sound pressure level (SPL) is expressed as the root-mean-square (r.m.s.) pressure in newtons per

Table 1.3 Relationship between SPL and sound pressure

SPL (dB)	Sound pressure (Pa)	Typical sound source
200	2×10^5	Jet engine at take off (atmospheric pressure)
190	6.3×10^4	
180	2×10^4	
170	6.3×10^3	
160	2×10^3	Ram jet
150	6.3×10^2	
140	2×10^2	Propeller engine 30 m overhead
130	63	
120	20	Pain threshold
110	6.3	
100	2	Heavy lorry at kerbside
90	6.3×10^{-1}	Heavy traffic at kerbside
80	2×10^{-1}	Noisy office
70	6.3×10^{-2}	
60	2×10^{-2}	Normal conversation at 1 m
50	6.3×10^{-3}	Average office
40	2×10^{-3}	Quiet house
30	6.3×10^{-4}	Recording studio
20	2×10^{-4}	Gently rustling leaves
10	6.3×10^{-5}	
0	2×10^{-5}	Threshold of hearing

square metre (pascal) and usually given in decibels relative to the 'threshold of hearing' at 1 kHz (20×10^{-6} N/m^2).

$$\text{SPL (dB)} = 20 \log \frac{P_1}{20 \times 10^{-6}}$$

SPL is frequently measured using one of a number of weighting networks originally intended to simulate the changing frequency response of the human ear at different levels, but now used on an empirical basis according to circumstances:

dB(A) Originally used to simulate the response of the ear at 40 phon. It correlates well with the subjective evaluation of traffic noise.

dB(B) Originally used to simulate the response of the ear at 70 phon.

dB(C) Flat from 200 to 1250 Hz with some attenuation outside this band.

dB(D) Used for measuring aircraft noise. It approximates to the 40 noy contour.

See *Figure 1.6.*

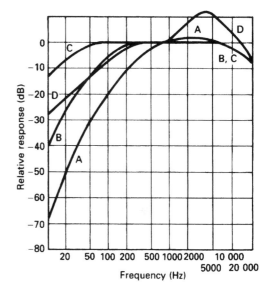

Figure 1.6 *Sound pressure level SPL weighting curves for dB(A), dB(B), dB(C) and dB(D).*

In the UK, the dB(A) weighting curve is specified in the Noise at Work Regulations 1989 for use in ascertaining the exposure of employees to noise (see Section 1.2.3.11).

1.2.3.2 Sound power level
The sound power level (PWL) is a measure of the intensity of a sound.

Intensity $(I) \propto$ pressure $(p)^2$

$$I = \frac{\rho}{\partial^c}$$

where c is the impedance of air; and ∂ is the density of air.

Impedance of air $= \partial \times$ Sound velocity (c)

$= 1.21$ kg/m $\times 343$ m/s at 20°C

$= 415$ rayls (acoustic units; kg m^{-2} s^{-1})

At the threshold of hearing

$p_0 = 20 \times 10^{-6}$ N/m^2.

Hence

$$I_0 = \frac{(20 \times 10^{-6})^2}{415} \simeq 10^{-12} \text{ W/m}^2$$

The PWL is often given in decibels referred to the PWL of the threshold of hearing:

$$\text{PWL (dB)} = 10\log\frac{I}{I_0} = 10\log\frac{I}{1\times10^{-12}}$$

1.2.3.3 Decibel relationship between SPL, intensity and PWL

$$\text{PWL} = 10\log(W/W_0)\ \text{dB re. } W_0$$

$$\text{Intensity level} = 10\log(I/I_0)\ \text{dB re. } I_0$$

$$\text{SPL} = 10\log(p^2/p_0^2)$$

$$= 20\log(p/p_0)\ \text{dB re. } p_0$$

where $W_0 = 1 \times 110^{-12}$ W, $I_0 = 1 \times 10^{-12}$ W/m, and $p_0 = 20 \times 10^{-6}$ Pa $(= 20\ \mu\text{Pa})$.

1.2.3.4 The phon

The phon is a unit of loudness. It is referred to the SPL Level at 1 kHz, and takes account of the changing frequency response of the human ear at different levels (e.g. an SPL of 60 dB re. 20 μPa at 1 kHz will have the same loudness as an SPL of 80 dB at 50 Hz, and 52 dB at 4 kHz).

The line joining points of equal loudness is called the equal loudness contour. The original work was reported by Fletcher and Munsen in 1933 and modified by Robinson and Dadson in 1956. Adopted

by the International Organization for Standardization (ISO) as Recommendation R-226. See *Figure 1.7.*

1.2.3.5 The sone

The sone is a unit of subjective loudness. It relates the logarithmic response of the human ear to its SPL. One sone is defined as the loudness experienced by a person listening to a sound of 40 phon. Two sone will sound twice as loud, 0.25 sones will sound a quarter as loud, etc.

For wide-band noise

$$S_t = S_m + f\left(\sum S - S_m\right)$$

where S_t is the total loudness (sone), S_m is the loudness of the loudest band $f = 0.15$ for 1/3-octave analysis, $f = 0.2$ for 1/2-octave analysis and $f = 0.3$ for whole octave analysis, and ΣS is the sum of the loudnesses in appropriate bands.

See *Figures 1.8 and 1.9.*

1.2.3.6 The noy

The noy is a measurement of the 'noisiness' of a sound. It is defined as the SPL of a band of noise extending from 910 to 1090 Hz that sounds as 'noisy' as the sound being considered. Noys can be expressed as the *perceived noise level* (PNdB) using

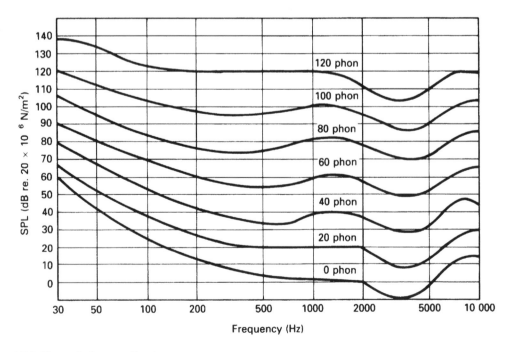

Figure 1.7 *Robinson–Dadson equal loudness contours.*

Figure 1.8 *The relationship between phons and sones.*

Figure 1.9 *Loudness index (sone) versus sound pressure level (SPL).*

the formula:

$$\text{PNdB} = 40 + \frac{100}{3} \log_{10} N$$

See *Figure 1.10*.

1.2.3.7 Relationship between SPL, sones and PNdB

The SPLS produced by a noisy machine are given in *Table 1.4*. From the table:

$$\sum S = 77.5 \text{ and } S_m = 20 \text{ sone}$$

$$S_t = 20 + 0.3(77.5 - 20) = 37.25 \text{ sone} = 92 \text{ phon}$$

$$\sum N = 94.2, N_m = 21$$

$$N_t = 21 + 0.3(94.2 - 21) = 42.96 \text{ noy} = 94 \text{ PNdB}$$

1.2.3.8 Addition of intensities, pressures and decibels

$$\text{Total intensity } (I_t) = I_1 + I_2 + \cdots + I_n$$

$$\text{Total pressure } (P_t) = \sqrt[2]{P_1^2 + P_2^2 + \cdots + P_n^2}$$

Table 1.4 The SPLs produced by a noisy machine

Frequency (Hz) band	SPL (dB)	(sones)	(noys)
20–75	95	20	21.0
75–150	84	12	13.8
150–300	80	11	14.0
300–600	68	7	7.5
600–1200	65	6	6.0
1200–2400	61	6.5	6.1
2400–4800	60	7	9.3
4800–10000	60	8	16.5
Total		77.5	94.2

Decibels cannot be added directly, since they are in logarithmic form. However if $n(\text{dB}) > m(\text{dB})$, and $x = (n - m)$:

$$\text{Total dB} = n(\text{dB}) + [10 \log(1 + 10^{\left(\frac{x}{10}\right)})] - x$$

Figure 1.10 *Contours of perceived noisiness.*

Figure 1.11 illustrates a scale suitable for calculating the result of adding decibel values.

1.2.3.9 Sound intensity at a distance
For a uniformly radiating source:

$$\text{Intensity } (I) = \frac{\text{Sound power } (w)}{4\pi r^2}$$

for a source generating into a hemisphere:

$$I = \frac{W}{2\pi r^2}$$

At a distance r from a sound source radiating in all directions equally:

$$\text{SPL} = \text{PWI} - 20\log_{10} r - 10\log_{10} 4\pi = \text{PWL} - 20$$
$$\log_{10} r - 11$$

and from a hemispherical radiator:

$$\text{SPL} = \text{PWL} - 20\log_{10} r - 10\log_{10} 2\pi = \text{PWL} - 20$$
$$\log_{10} r - 8$$

1.2.3.10 Equivalent continuous sound level (Leq)
The equivalent continuous sound level (Leq) is defined as the level of a steady sound that has, over a given period, the same energy as a fluctuating sound. For audiometric purposes, the integration time is usually taken as 40 h in 1 week.

$$\text{Leq} = 70 + 10\log_{10} \sum E_{\mathrm{i}},$$

where E_{i} is the *partial sound exposure index*;

$$E_{\mathrm{i}} = \frac{\mathrm{d}t}{40} 10^{0.1(L_{\mathrm{i}} - 70)}$$

where $\mathrm{d}t$ is the exposure time (h/week) and L_{i} is the sound level (dB(A))
For example, a 2 h/week exposure to 100 dB(A)

$$E_1 = \frac{2}{40} 10^{0.1(100-70)} = 50$$

and for a 30 h/week exposure to 80 dB(A)

$$E_2 = \frac{30}{40} 10^{0.1(80-70)} = 7.5$$

Thus:

$$\sum E_{\mathrm{i}} = 50 + 7.5 = 57.5$$

and

$$\text{Leq} = 70 + 10\log_{10} 57.5 \simeq 88 \text{ dB}$$

Figure 1.11 *To calculate the result of adding two decibel levels, calculate the difference between the two levels. Read off the correction factor. Add this value to the higher decibel level. For example, to add +8 dB and +11 dB. The difference is 3 dB; the correction factor for 3 dB is 1.7 dB: 1.7 + 11 = 12.7 dB.*

Table 1.5 Correction factor (x in/m^3) at 20°C

Frequency (Hz)	Relative humidity (% $\times 10^{-3}$)					
	30	40	50	60	70	80
1000	3.28	3.28	3.28	3.28	3.28	3.28
2000	11.48	8.2	8.2	6.56	6.56	6.56
4000	39.36	29.52	22.96	19.68	16.4	16.4

1.2.3.11 Personal noise exposure

The UK Noise at Work Regulations 1989 determine how the daily and weekly exposure of employees to noise shall be ascertained. The daily exposure ($L_{EP,d}$ in dB(A)) is given by the formula:

$$L_{EP,d} = 10\log_{10}\left\{\frac{1}{T_0}\int_0^{T_e}\left[\frac{P_A(t)}{P_0}\right]^2 dt\right\}$$

where, t_e is the duration of a person's personal exposure to sound, $T_0 = 8\ h(= 28\ 800\ s)$, $P_0 = 20\ \mu Pa$, and $P_A(t)$ is the time-varying value of A-weighted instantaneous sound pressure (pascal) in the undisturbed field of air at atmospheric pressure to which the person is exposed, or the pressure of a disturbed field adjacent to the person's head adjusted to provide an equivalent sound field pressure.

The weekly average of daily personal noise exposure ($L_{EP,w}$ in dB(A)) is given by the formula:

$$L_{EP,w} = 10\log_{10}\left[\frac{1}{5}\sum_{k=1}^{k=m}10^{0.1(L_{EP,d})k}\right]$$

where $(L_{EP,d})_k$ are the values of $L_{EP,d}$ for each of the m working days in the week being considered.

1.2.4 Reverberation time calculations

1.2.4.1 The Sabine equation

The reverberation time (RT) in seconds is given by:

$$RT = \frac{0.16\ V}{S\alpha}$$

where V is the volume of the room, S is the surface area of the room, and α is the average absorption coefficient:

$$\frac{S_1\alpha_1 + S_2\alpha_2 + \cdots + S_n\alpha_n}{S_1 + S_2 + S_3}$$

Alternatively

$$S\alpha = S_1\alpha_1 + S_2\alpha_2 + \cdots + S_n\alpha_n$$

where S_1, α_1, etc., are the surface areas and absorption coefficients of the various surfaces.

Above 1 kHz a correction factor may be needed to allow for the higher absorption by the air at high frequencies:

$$RT = \frac{0.16\ V}{S\alpha + xV}$$

where x is the absorption per metre of air (see *Table 1.5*).

1.2.4.2 The Norris–Eyring equation

Where the average absorption coefficient is higher than 0.2, the Norris–Eyring formula gives a more accurate value of the reverberation time (RT in s):

$$RT = \frac{0.16\ V}{-S\log_e(1-\bar{\alpha})}$$

where

$$\bar{\alpha} = \left(\frac{S_1\alpha_1 + S_2\alpha_2 + \cdots + S_n\alpha_n}{S_1 + S_2 + \cdots + S_n}\right)$$

Again, a correction factor is applied at high frequency:

$$RT = \frac{0.16\ V}{-S\log_e(1-\bar{\alpha}) + xV}$$

1.2.4.3 Optimum reverberation time

Various empirical methods have been proposed. One formula calculates the optimum reverberation time as:

$$RT = r(0.012\sqrt[3]{V} + 0.107)$$

where V is the volume of the room (m^3); $r = 4$ for speech, $r = 5$ for orchestral music, and $r = 6$ for choral music all with an increase of 40% at low frequencies. *Figure 1.12* illustrates suggested optimum reverberation times for different auditoria.

1.2.4.4 Measurement of absorption coefficients
Reverberation chamber method

RT without material $= t_1(s)$

RT with material $= t_2(s)$

$$d\alpha = 0.16V\left(\frac{1}{t_1} - \frac{1}{t_2}\right)$$

Figure 1.12 *Suggested optimum reverberation time (RT) for different auditoria.*

and

$$\alpha = \frac{d\alpha}{S}$$

where α is the absorption coefficient of material under test, and S is the surface area of material under test. To ensure a diffuse sound field the lowest frequency should be $>125 \sqrt[3]{180/V}$ Hz, where V is the volume of the chamber. The surface area of material should be between 10 and 12 m.

Impedance (Kundt) tube method

Measure the maximum and minimum standing wave amplitudes (see *Figures 1.13* and *1.14*).

$$\alpha = \frac{4}{n + \dfrac{1}{n}}$$

where n is the ratio maximum-amplitude/minimum-amplitude ratio. The maximum diameter (d) of the sample should be $\leq \lambda/2$.

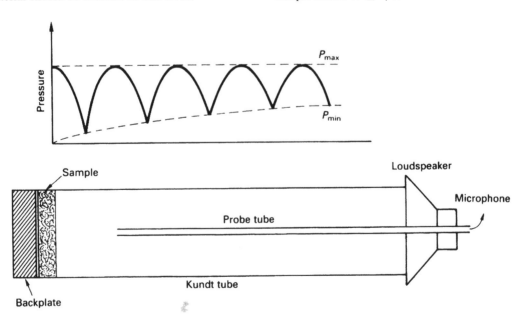

Figure 1.13 *Standing wave tube method for measuring absorption coefficient at an angle of incidence of 90°.*

Figure 1.14 *Graph for converting the standing wave ratio to the absorption coefficient.*

1.2.5 Resonant frequency of resonant absorbers

1.2.5.1 Membrane or panel
The resonant frequency f_0 is given by:

$$f_0 = \frac{60}{\sqrt{Md}}$$

where M is the mass of the panel or membrane (kg/m^2), and d is the depth of the air space in metres.

1.2.5.2 Helmholtz resonator

$$f = \frac{cr}{2\pi}\sqrt[2]{\frac{2\pi}{(2l + \pi r)V}}$$

where c is the velocity of sound in air, r is the radius of the neck. l is the length of neck and V is the volume of the cavity.
If there is no neck

$$f_0 = \frac{c}{2\pi}\sqrt[2]{\frac{2}{rV}}$$

1.2.5.3 Perforated hardboard absorbers

$$f = 5080\sqrt[2]{\frac{P}{1.8l \times D}}$$

where p is the percentage perforation of the hardboard, l is the thickness of the hardboard, and D is the depth of the airspace.

1.2.6 Velocity of sound in solids

The frequency of sound in a solid depends on its elasticity, density and Poisson's ratio (a measure of the deformation characteristics of the solid when stretched), and on whether the sound wave is longitudinal (compression), transverse (bending) or shear.

Velocity of longitudinal waves in a thin rod

$$c_1 = \sqrt{\frac{E}{\rho}}$$

where E is Young's modulus of elasticity, and ρ is the density of material (kg/m).

Velocity of longitudinal waves in a thin plate

$$c_1 = \sqrt{[E/\rho(1 - \mu^2)]}$$

where μ is Poisson's ratio.

Velocity of transverse (bending) waves

$$c_b = \sqrt[2]{(1.8c_1hf)}$$

where h is the thickness of the plate, and f is the frequency.

Velocity of transverse (shear) waves

$$c_s = \sqrt[2]{\left[\frac{E}{2(1 + \mu)\rho}\right]}$$

1.2.7 Simple harmonic motion relationships

$$\text{Displacement } (d) = a\sin 2\pi f t$$

where d is the instantaneous displacement, a is the peak displacement, f is the frequency, and t is time.

$$\text{Velocity } \left(\frac{ds}{dt}\right) = 2\pi f a \cos 2\pi f t (s = 0 \text{ when } t = 0)$$

$$\text{Acceleration } \left(\frac{d^2s}{ds^2}\right) = 4\pi^2 f^2 a \sin 2\pi f t$$

1.2.8 Voltage, current, resistance and power relationships

$$V = IR; \ I = \frac{V}{R}; \ R = \frac{V}{I}$$

$$W = IV = \frac{V^2}{R} = I^2R$$

$$I = \frac{W}{V}; \ V = \frac{W}{I}$$

1.2.9 Inductive, capacitive and resistive networks

Resistor networks

Resistors in series:

$$R_t = R_1 + R_2 + \cdots + R_n$$

Resistors in parallel:

$$\frac{1}{R_t} = \frac{1}{R_1} + \frac{1}{R_2} + \cdots + \frac{1}{R_n}$$

Capacitor networks

Capacitors in series:

$$\frac{1}{C_t} = \frac{1}{C_1} + \frac{1}{C_2} + \cdots + \frac{1}{C_n}$$

Capacitors in parallel:

$$C_t = C_1 + C_2 + \cdots + C_n$$

Inductor networks

Inductor in series:

$$L_t = L_1 + L_2 + \cdots + L_n$$

Inductors in parallel:

$$\frac{1}{L_t} = \frac{1}{L_1} + \frac{1}{L_2} + \cdots + \frac{1}{L_n}$$

Reactance and impedance

It is convenient to use the expression:

$$\omega = 2\pi f$$

Capacitive reactance:

$$X_c = 1/\omega C$$

Inductive reactance:

$$X_L = \omega L$$

Impedance of a capacitive circuit (*Figure 1.15(a)*):

$$Z_C = \sqrt[2]{X_C^2 + R^2}$$

Impedance of an inductive circuit (*Figure 1.15(b)*):

$$Z_L = \sqrt[2]{X_L^2 + R}$$

Resonance

Series resonant frequency (*Figure 1.16(a)*):

$$f_0 = 1/2\pi\sqrt[2]{LC}$$

(a)

(b)

Figure 1.15 *(a) Series capacitive reactance circuit. (b) Series inductive reactance circuit.*

(a)

(b)

Figure 1.16 *(a) Series resonant circuit. (b) Parallel resonant circuit.*

Parallel resonant frequency (*Figure 1.16(b)*):

$$f_0 = \frac{1}{2\pi}\sqrt[2]{\left[\left(\frac{1}{LC}\right) - \left(\frac{R^2}{L^2}\right)\right]}$$

where R is the resistance of the inductor. If $R = 0$, then

$$f_0 = 1/2\pi\sqrt[2]{LC}$$

Circuit magnification at resonance (Q):

$$Q = \frac{\text{Resonant frequency}}{\text{Bandwith}} \quad \text{or} \quad \frac{\text{Reactance}}{\text{Effective resistance}}$$

Thus:

$$Q = \frac{2\pi f L}{R} \quad \text{or} \quad \frac{1}{2\pi f CR} \quad \text{or} \quad \frac{1}{R}\sqrt[2]{\frac{L}{C}}$$

See *Figure 1.17.*

TIME CONSTANTS

For a capacitive circuit (*Figure 1.18*):

$$\text{Time constant} = CR$$

(a)

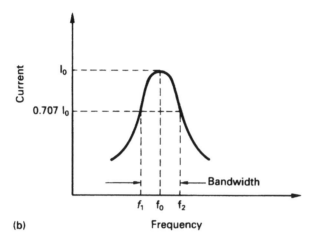

(b)

Figure 1.17 *(a) The Q value of a resonant circuit combines the Q factors of each reactive element. (b) The bandwidth of a resonant circuit is determined by its half-power points, at which the circulating currents are at 0.707 of their peak value.*

For an inductive circuit (*Figure 1.19*):

Time constant $= L/R$

Where an a.c. is applied to a reactive circuit it can be convenient to relate the *turnover frequency* (f_t) to the time constant of the circuit (*Figure 1.20*). When e_0 is -3 dB with respect to e_i, $X_c = R$ and $\frac{1}{2\pi fC} = R$. Thus:

$$CR = \frac{1}{2\pi f_t} \quad \text{and} \quad f_t = \frac{1}{2\pi f_t C}$$

1.2.10 Transformers

See *Figure 1.21*. If $V_1 i_1 = v_2 i_2$,

$$v_2 = \frac{R_2}{R_1} v_1 \quad \text{and} \quad i_2 = \frac{R_1}{R_2} i_1$$

$$Z_1 = \frac{V_1}{I_1}$$

$$Z_2 = \frac{V_2}{I_2} = \left(\frac{n_2}{n_1} v_1\right) \Big/ \left(\frac{n_1}{n_2} i_1\right)$$

$$= \left(\frac{n_2}{n_1}\right)\left(\frac{n_2}{n_1}\right)\left(\frac{v_1}{v_2}\right) = \left(\frac{n_2^2}{n_1^2}\right)\left(\frac{V_1}{I_1}\right)$$

Thus:

$$\frac{Z_2}{Z_1} = \frac{n_2^3}{n_1^2} = \left(\frac{D_1}{n_2}\right)^2$$

Note: a well designed audio transformer will usually introduce a 1 dB loss in power.

1.2.11 Noise

1.2.11.1 Electrical noise
Thermal noise voltage $(e_{\text{R.M.S.}}) = \sqrt[2]{(4KRT\mathrm{d}f)}$, where: K is Boltzmann's constant $(= 1.374 \times 10^{-23}$ J/degree Kelvin), R is the resistance, T is the temperature, $\mathrm{d}f$ is the bandwidth. and $e_{\text{R.M.S.}}(\text{dB}_\text{u}) =$

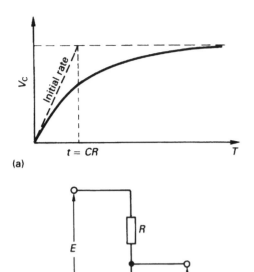

(a)

(b)

Figure 1.18 *(a) A capacitor charges exponentially. If it were to maintain its initial charging rate, it would be fully charged in time* t = CR. *This is known as the time constant. (b) Shows the charging arrangement for calculation of* CR.

$20 \log(e_{R.M.S.}/0.775).$

$$\text{Noise power } (N) = \frac{e^2}{R} = 4KT\mathrm{d}f$$

$$\text{Noise power (dB)} = 10 \log \left(\frac{N}{0.001} \right)$$

$$\text{Noise figure (NF)} = \frac{(S_{in}/N_{in})}{(S_{out}/N_{out})}$$

where S is the signal level, and N is the noise level

$$\text{NF (dB)} = 10 \log (\text{NF})$$

Note: for 600 Ω at 20°C and a bandwidth of 20 kHz, r.m.s. thermal noise (dBu) = −125 dBu.

1.2.11.2 White noise
This is random noise where the energy is constant over constant bandwidths. It is usually band limited, e.g. 20 Hz–20 kHz. When plotted as voltage against a logarithmic frequency scale it will exhibit a 6 dB/octave rise.

1.2.11.3 Pink noise
This is random noise where the energy is constant over successive octaves. It is usually band limited,

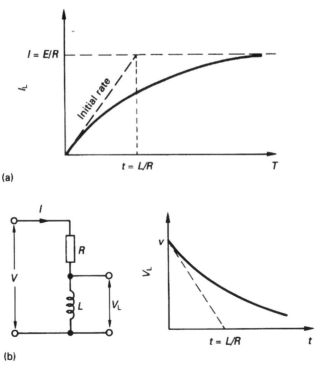

(a)

(b)

Figure 1.19 *(a) If a voltage is applied across an inductor, current through it will rise exponentially. If it were to maintain its initial rate of rise it would reach its final value in* t = L/R. *This is known as the time constant. (b) The relationship between* L, R *and the fall of voltage across* L *with time.*

Figure 1.20 *In a series* CR *circuit there is a 3 dB fall in output voltage with increasing frequency when* $X_c = $ R. *This* −3 dB *point marks the transition from a flat frequency response to one changing at 6 dB/octave.*

Figure 1.21 *The relationship between input and output voltages, currents and impedances, and the turns of wire on the transformer primary and secondary windings.*

e.g. 20 Hz–20 kHz. It can be obtained from white noise by passing it through a 3 dB/octave filter. On a spectrum analyser having octave-based bandwidths, pink noise will show as a flat spectrum.

1.2.12 Resistor networks

'T' attenuators (*Figure 1.22*)

Figure 1.22 *A 'T' attenuator. The attenuator is symmetrical.*

Let the loss be x dB, and $N = 10^{\left(\frac{x}{20}\right)} = $ antilog $(x/20)$, then:

$$R_\text{s} = R_0 \left(\frac{N-1}{N+1}\right), R_\text{p} = R_0 \left(\frac{2N}{N^2-1}\right)$$

Figure 1.23 *A 'π' attenuator. The attenuator is symmetrical.*

'Π' Attenuators (*Figure 1.23*)

Let the loss be x dB, and $N = 10^{\left(\frac{x}{20}\right)} = $ antilog $(x/20)$, then:

$$R_\text{s} = R_0 \left(\frac{N^2-1}{2N}\right), R_\text{p} = R_0 \left(\frac{N+1}{N-1}\right)$$

Star mixers/splitters (*Figure 1.24*)

$$R = R_0 \left(\frac{N-1}{N+1}\right)$$

Figure 1.24 *A star mixer/splitter. Any one of the points shown can be used as an input, the other points being equally weighted outputs. Alternatively, any three of the points shown can provide equally weighted inputs, the remaining point being the output.*

Simple attenuators (non-impedance matching)
Simple attenuators between the 'voltage matched' source and load (e.g. $R_\text{s} < 100 \ \Omega$ and $R_1 > 10$ kΩ) can be made from a couple of resistors. This is not suitable where impedances have to be accurately matched, or where the attenuation must be precise. Nor is it suitable for attenuation ratios of less than 10 dB. It is suitable for level matching 'in the field'.
See *Figure 1.25*; let the attenuation be x dB, and $N = 10$; then:

$$R_1 = (N-1)R_2 \text{ and } R_2 = 100 \ \Omega$$

Attenuators in balanced circuits
All the above calculations assume unbalanced line working. When working with balanced lines the

Figure 1.25 *A simple attenuator. The attenuator is not symmetrical, and can only be used in the configuration shown.*

value of the series resistor should be shared equally between the two legs of the line. So, for instance, in the case of a simple non-impedance matching attenuator designed to give a loss of 20 dB, the value of the series component (900 Ω) will be incorporated as two series resistors of 450 Ω. In practice, two 470-Ω resistors would probably be used.

1.2.13 Resistor colour codes

Most fixed value resistors are colour coded with from three to five coloured bands towards one end (*Figure 1.26(a)*):

First band – 1st digit
Second band – 2nd digit
Third band – denary multiplier
Fourth band – tolerance percentage (if present)
Fifth band – (pink if present) indicates
 high stability

Colour code:
0 Black
1 Brown
2 Red

(a)

(b)

Figure 1.26 *(a) Resistor colour codes are composed of 3–5 coloured bands towards one end of the resistor body. (b) A 4700 Ω, ±5% tolerance resistor.*

3 Orange
4 Yellow
5 Green
6 Blue
7 Violet
8 Grey
9 White

No fourth band – ±20% tolerance
Silver fourth band – ±10% tolerance
Gold fourth band – ±5% tolerance
Red fourth band – ±2% tolerance

The resistor in *Figure 1.26(b)* has a value of 4.7 kΩ, and has a tolerance of ±5%.

Resistors are available in a limited range of values, determined by one of three standard ranges, known as the 'E series'. Resistor values are multiples of 10 of these values. *Table 1.6* gives the three E ranges.

1.2.14 Capacitor standard values

Capacitors come with less variety in values than resistors. Typical values are denary multiples of 1.0, 2.2, 4.7 and 6.8.

Table 1.6 Resistor E-series values

E-24	E-12	E-6
1.0	1.0	1.0
1.1		
1.2	1.2	
1.3		
1.5	1.5	1.5
1.6		
1.8	1.8	
2.0		
2.2	2.2	2.2
2.4		
2.7	2.7	
3.0		
3.3	3.3	3.3
3.6		
3.9	3.9	
4.3		
4.7	4.7	4.7
5.1		
5.6	5.6	
6.2		
6.8	6.8	6.8
7.5		
8.2	8.2	
9.1		

1.2.15 Operational amplifiers

Operational amplifiers can be thought of as 'blocks of gain' requiring only an external power supply and a few components to give predictable and stable amplification. They generally have an 'open-loop' voltage gain in excess of 80 dB, this being modified by external resistors. This open-loop gain falls off above a few kilohertz at 6 dB/octave, depending on the device. Should the 'designed' gain approach the open-loop gain then the approximations that follow will not apply.

1.2.15.1 Inverting amplifiers
In general, inverting amplifiers provide the most useful configuration (*Figure 1.27*). Gain can vary from a fraction of unity to several tens of decibels.

$$\text{Gain} = -\frac{R_1}{R_1}$$

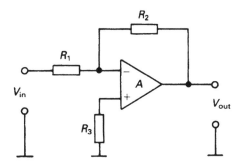

Figure 1.27 *An operational amplifier used as an inverting amplifier (d.c. supply not shown).*

Note the negative sign indicating inversion.

$$\text{Input resistance} = R_1$$

$$\text{Output resistance} = R_0\frac{R_2}{AR_1}$$

where R_0 is the internal output resistance of amplifier, and A is the open-loop gain of amplifier. R_3 should be made equal to R_1 and R_2 in parallel.

1.2.15.2 Non-inverting amplifiers
The gain cannot be less than unity. The input resistance cannot be readily calculated without a specification sheet for the device. See *Figure 1.28*.

$$\text{Gain} = 1 + \frac{R_2}{R_1}$$

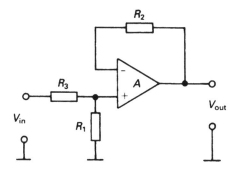

Figure 1.28 *An operational amplifier used as a non-inverting amplifier (d.c. supply not shown).*

$$\text{Input resistance} = R_1\frac{AR_1}{R_2}$$

$$\text{Output resistance} = R_0\frac{R_2}{AR_1}$$

1.2.15.3 Mixing amplifiers
Since the inverting input of an operational amplifier is held very close to ground by virtue of the feedback resistor, the summing point of a mixing amplifier is at virtual earth (*Figure 1.29*). This provides a mixing facility with negligible interaction between the different inputs. The inputs do not have to be equally weighted, the weighting being a function of the various input resistors. Note that it is the inverting amplifier configuration that is most useful for mixing because of its virtual earth facility.

$$V_{\text{out}} = -\left(V_1\frac{R_3}{R_1} + V_2\frac{R_3}{R_2}\right)$$

$$R_4 = R_1//R_2//R_3$$

$$\left(\frac{1}{R_4} = \frac{1}{R_1} + \frac{1}{R_2} + \frac{1}{R_3}\right)$$

Figure 1.29 *An operational amplifier used as a virtual earth mixer (d.c. supply not shown).*

1.2.15.4 The differential amplifier

This amplifier will give an output dependent on the difference between the two inputs. Normally the input resistors would be identical, as illustrated in *Figure 1.30*, in which case both inputs are given equal weighting, with the gain being dependent on the ratio between R_1 and R_2.

$$V_{\text{out}} = \frac{R_2}{R_1}(V_1 - V_2)$$

Figure 1.30 *An operational amplifier used as a differential amplifier (d.c. supply not shown).*

1.2.16 Modulation

All forms of modulation involve the use of non-linear devices. The result is the generation of frequencies other than those fed into the modulation device. These are referred to as 'sidebands'.

The modulation of a carrier by a modulating signal may be undertaken for a variety of reasons other than the transmission of radiofrequency electromagnetic waves. The most common applications for audio include reduction of octave band requirements (audio frequencies from 30 to 16 kHz occupy approximately nine octaves). If this bandwidth can be translated upwards to centre on 100 kHz, the octave requirement reduces to less than one. If a carrier is frequency modulated then it is possible to magnetically record (and replay) representations of d.c. values, and to achieve a process relatively immune to level variations (e.g. dropout).

1.2.16.1 Amplitude modulation

In amplitude modulation, the modulating signal, occupying a range of frequencies commonly referred to as the *baseband*, is used to vary (modulate) the amplitude of a single-frequency signal, the *carrier* *(Figure 1.31)*. As a result of the non-linearity inherent in this process other frequencies, *sidebands*, are generated. The frequency relationship *(Figure 1.32)* is as follows. If F_m is the modulating frequency and F_c is the carrier frequency, the modulated signal comprises:

$$F_m + F_c + (F_c - F_m) + (F_c + F_m)$$

If the carrier is modulated by a range of frequencies, each frequency will generate a pair of sidebands, each spaced in frequency above and below the carrier and spaced from it in frequency by its own value *(Figure 1.32)*. Note that the modulated signal requires a bandwidth twice that of the modulating frequency.

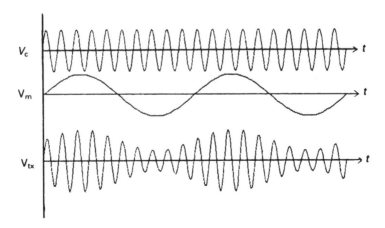

Figure 1.31 *Amplitude modulation: the amplitude of a carrier (V_c) is varied by a modulating signal (V_m). The result is a modulated signal (V_{tx}). Modulation index = 0.5; $F_c/F_m = 10$.*

(a)

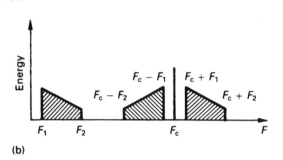

(b)

Figure 1.32 *(a) The amplitude modulation process results in the generation of extra frequencies. (b) For a range of modulating frequencies (the baseband), pairs of sidebands are generated. Note that the total modulated bandwidth is twice that of the baseband.*

The depth of modulation (*modulation index*) is given by (*Figure 1.33*):

$$m = \frac{\text{Max. amplitude} + \text{Min. amplitude}}{\text{Max. amplitude} - \text{Min. amplitude}} = \frac{2B}{2A}$$

Power (in watts) in an amplitude-modulated wave is given by:

$$P_{\text{total}} = P_{\text{carrier}} + P_{\text{sidebands}} = \frac{A^2}{2}\left(1 + \frac{m^2}{2}\right)R$$

where A is the amplitude of the unmodulated carrier current, M is the modulation index, and R is the resistance in which power is dissipated.

The general formula for calculating total modulated power is:

$$P = W\left(1 + \frac{m^2}{2}\right)$$

where W is the power of the unmodulated carrier. From this it can be seen that sideband power increases substantially for modulation indices greater than 0.5.

If the modulation index is increased too much, the point is reached where the carrier is lost (*Figure 1.34*).

1.2.16.2 Suppressed carrier modulation
If the carrier can be suppressed then more effective use can be made of the available power since it will be concentrated in the sidebands (*Figure 1.35*). Since the carrier is missing, it will have to be regenerated at some point before demodulation. In traditional (PAL and NTSC) video systems this has been achieved by incorporating a short burst of unmodulated carrier (the colour burst) at the start of each active (picture-information containing) line.

1.2.16.3 Frequency modulation
In frequency modulation (f.m.), the carrier amplitude remains constant but its frequency is made to vary with the modulating signal (*Figure 1.36*). In this manner a particular d.c. level can be represented by a particular modulated frequency. The analysis of sidebands generated by this process is extremely complex and, in theory, they are infinitely wide, being spaced from the carrier by $\pm f_m$, $\pm 2f_m$, $\pm 3f_m$, etc.

In practice, the higher order components can be neglected at the point where they become very small. This point is a function of the modulation index (m). The degree by which the carrier frequency changes for maximum change in modulating signal amplitude is called the *deviation* (f_D).

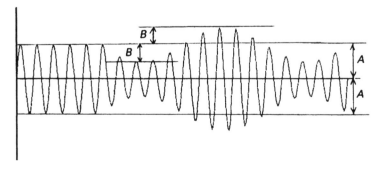

Figure 1.33 *The modulation index is given by* m = B/A.

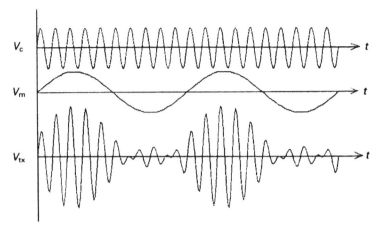

Figure 1.34 *When the modulation index is increased to above unity, the carrier is lost and unwanted harmonics are generated. Modulation index = 1.5; $F_c/F_m = 10$.*

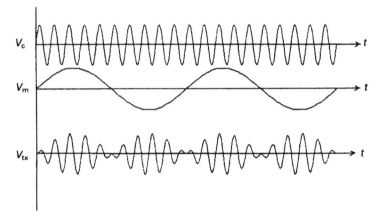

Figure 1.35 *In suppressed carrier modulation the carrier (V_c) is varied by the modulation frequency (V_m). The result (V_{tx}) shows the loss of the carrier, and the phase reversal of the modulated signal when V_m crosses zero. Modulation index = 1; $f_c/f_m = 10$.*

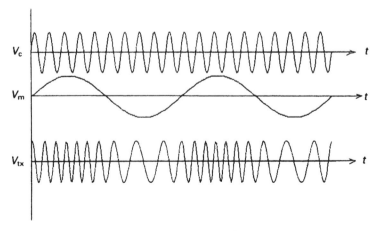

Figure 1.36 *Frequency modulation: the carrier (V_c) is varied in frequency by the modulating signal (V_m). The result is a signal varying in frequency but constant in amplitude. Specific frequencies in the modulated signal relate to specific d.c. levels on the modulating signal. Modulation index = 5; $F_c/F_m = 10$.*

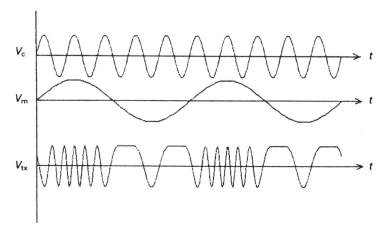

Figure 1.37 *If the f.m. index is too high for the* V_c/V_m *ratio, the result is severe distortion, with the modulated signal having a d.c. component. Modulation index = 10;* $F_c/F_m = 5$.

For an f.m. system,

Modulation index, m

$$= \frac{\text{carrier frequency deviation } (f_D)}{\text{modulating frequency } (f_m)}$$

$$d = \frac{D}{f_m} = \frac{\text{Rated system deviation}}{\text{Modulating frequency}}$$

Too high a modulation index will result in the generation of sidebands that will extend past 0 Hz and 'fold back' into the frequency spectrum as 'aliased' components. The result is severe distortion. The effect is illustrated in *Figure 1.37*.

From a knowledge of the deviation ratio, the bandwidth requirements of an f.m. system can be calculated, e.g. a 15 kHz audio signal is being transmitted on a system the maximum rated deviation of which is 90 kHz. If it is required to transmit three pairs of sidebands, the bandwidth requirement is calculated as:

$$d = \frac{D}{f_m} = \frac{90}{15} = 6$$

$d + 3 = 9$ pairs of sidebands

Total bandwidth = $9 \times 2 \times 15 = 270$ kHz

References

FLETCHER, H. and MUNSON, W.A., Loudness, its definition, measurement and calculation, *Journal of the Acoustical Society of America*, **5**, 82–108 (1933)
ROBINSON, D.W. and DADSON, R.S., A re-determination of the equal loudness relations for pure tones, *British Journal of Applied Physics*, **7**, 166–181 (1956)

1.3 The physics of sound waves

1.3.1 Wave motion, particle displacement, particle velocity and pressure

Sound consists of longitudinal waves in the air. The term 'sound' is also loosely used for corresponding vibrations in other media. The term *longitudinal* means that the mode of transmission is by oscillatory movements of the air particles (in reality molecules of oxygen, nitrogen, etc.), the direction of oscillation being the same as the direction of the wave. Alternatively, sound waves can be regarded as a succession of compressions followed by rarefactions. It is usual, however, to represent sound waves by graphs of convenient parameters. *Particle displacement* is an obvious one but the most useful is normally *pressure*, as shown in *Figure 1.38*.

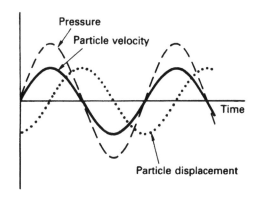

Figure 1.38 *Pressure, particle velocity and particle displacement.*

Particle velocity is another useful parameter for graphical representation and this is also shown in *Figure 1.38*. Note that particle velocity and pressure are in phase since pressure arises from the impact of the moving particles, but pressure and particle displacement are 90° out of phase. In general, we shall refer only to pressure, this being the most important of the parameters as it is the sound-wave pressure which causes movements of microphone diaphragms and the human eardrum.

Various units may be found to express root mean square (r.m.s) sound pressures. The generally preferred one is the *pascal* (*Pa*), which is equivalent to 1 newton per square metre (N/m²). Older literature may be found which uses *dyne/cm²*. The relationship between these units is:

$$1 \text{ Pa} = 1 \text{ N/m}^2 = 10 \text{ dyne/cm}^2$$

Sound-wave pressures fall in the range from 0.00002 Pa to around 200 Pa. (0.00002 Pa corresponds approximately to the pressure required to create an audible sensation in the average human ear at its most sensitive frequency. 200 Pa is around the pain threshold.)

Other units, used mostly for static pressures, are the *bar* and the *torr* (T):

$$1 \text{ bar} = 10^5 \text{ Pa}$$

$$1 \text{ T} = 133.322 \text{ Pa}$$

The bar is frequently used as a measure of water pressure and its subunit the millibar is well known from its appearance in meteorological charts. The use of the torr is generally confined to high vacuum work.

Pressure on its own is a very useful unit, but it gives no indication of the available power or energy in a sound wave. For this the term *intensity* is used, the basic unit for which is watts per square metre (W/m²). However, sound-wave intensities are frequently so small that microwatts per square metre (μW/m²) is usually more convenient. (Note that in everyday speech 'intensity' is often equated with 'loudness'. It is important to realize that these two quantities, although having some relationship, are in fact quite different.)

1.3.1.1 The velocity of sound

The velocity of sound (*c*) depends on the medium and also, in the case of gases, on the temperature (but not, within normal limits, on pressure). Some typical values for *c* are given in *Table 1.7*. The large value of *c* in helium, compared with air, accounts for the raised pitch of speech when an oxygen/helium

Table 1.7 Some typical values of the velocity of sound (*c*)

Substance	c (approx.) (m/s)
Water	1500
Concrete	3400
Steel	5000–6000
Wood	3000–4000
Helium gas	965 (0°C)
Air (dry)	331 (0°C)

mixture is breathed. A general expression for the velocity of sound in gases is:

$$c = (\gamma P / \rho)^{1/2}$$

where γ is the ratio of the specific heats of the gas (1.414 in the case of air), P is the pressure and ρ is the density of the gas. It follows that, provided the gas obeys Boyle's law, increasing the pressure will increase the density in the same proportion, so that under constant-temperature conditions the velocity c is independent of pressure. Increase of temperature, however, causes expansion and thus under constant-pressure conditions an increase in temperature results in a decrease in density. To summarize, an expression showing how c varies with temperature t (°C) is:

$$c(\text{m/s}) = 331 + 0.6t$$

Thus:

$$\text{at} -10°C \; c = 325 \text{ m/s}$$

$$\text{at } 0°C \; c = 331 \text{ m/s}$$

$$\text{at } 10°C \; c = 337 \text{ m/s}$$

$$\text{at } 20°C \; c = 343 \text{ m/s}$$

A fair approximate value at normal temperatures is 340 m/s.

It can generally be assumed that the compressing and rarefying in a sound wave takes place under adiabatic conditions, i.e. there is no heat transfer from high pressure (warmer) regions to low pressure (cooler) regions. The presence of materials that can effect a degree of heat transfer, and even substances such as fibre glass can do this, cause partly isothermal conditions to occur, with a consequent reduction in *c*. A loudspeaker cabinet containing such materials may thus have an acoustic volume which is greater than its physical dimensions.

Table 1.8 Some wavelengths at different frequencies

Frequency (Hz)	Wavelength (λ)
16	21.25
20	17.0
50	6.8
100	3.4
500	0.68
1000	0.34
5000	0.068
10 000	0.034
16 000	0.021

1.3.2 Frequency and wavelength

The frequency of any wave is defined as the number of complete waves emitted or received in one second. The unit is the hertz (Hz). The wavelength is the distance between corresponding points on adjacent cycles.

Frequency f and wavelength λ are related by:

$$c = f\lambda$$

For sound waves, the frequency range which can be detected by the normal human ear is approximately 16 Hz to 16 kHz (higher in children and young persons). *Table 1.8* shows some values of wavelengths in this frequency range.

1.3.3 Plane and spherical waves, the inverse square law

The terms 'plane wave' and 'spherical wave' are used, sometimes to the confusion of the inexperienced, to indicate the apparent 'shape' of a wavefront at a particular point. There is no sharp division between the two. Basically, if waves originate from a distant small source, or a close large plane surface, so that the curvature of the wavefront is insignificant, they may be described as *plane waves*. On the other hand, if the source is small and close to the observer the curvature may be significant and they are then called *spherical waves*.

1.3.1.1 The inverse square law

It is quite obvious that almost any form of wave energy diminishes in intensity as the distance from the source is increased. In practice, this rate of diminution varies greatly with different conditions.

An idealized relationship is given by the *inverse square law*, which states that, for a point source of waves, capable of radiating omnidirectionally and with no obstructions in the vicinity the intensity I decreases with the square of the distance d from the source:

$$I \propto \frac{1}{d^2}$$

In real situations sources of waves are never infinitely small and neither do they radiate omnidirectionally. Furthermore, there are almost always obstacles. Nevertheless, the law can be a useful approximation in conditions where, for example, the source is a small loudspeaker and obstacles and reflecting surfaces are relatively distant so that we are dealing with spherical waves. (Circumstances in which reflecting surfaces and other obstacles are sufficiently far away for their effect to be negligible are often referred to as '*free-field conditions*').

It is important to note that intensity is proportional to the square of the pressure P; the expression thus becomes:

$$P \propto 1/d$$

In free-field conditions, therefore, doubling the distance from the source causes a reduction in pressure by a factor of 2, or 6 dB, and so on. In practical circumstances the reduction with distance may be much less than this because of reflections.

1.3.4 The Doppler effect

This is the apparent change in frequency of a received sound when there is relative movement between the source and the observer. (The Doppler effect applies to all waves, but here we are concerned only with sound.)

Assuming a stationary medium, for a source moving with velocity v_s towards the observer, the apparent sound frequency f' is given by:

$$f' = fc/(c - v_s)$$

Taking c as 340 m/s and a source velocity of 10 m/s (roughly 22 mph), then a sound of true frequency $f = 1$ kHz would be received as:

$$f' = 1000 \times 340/(340 - 10)$$

$$= 340\,000/330$$

$$= 1030.3 \text{ Hz}$$

For an observer moving with velocity v_0 towards a stationary source:

$$f' = f(c + v_0)/c$$

Taking again $c = 340$ m/s and the observer's velocity as v_{10} 10 m/s, a sound of frequency $f = 1$ kHz would be received as:

$$f' = 1000 \times 350/340$$

$$= 350\,000/340$$

$$= 1029.4 \text{ Hz}$$

Note that for the same relative velocities the apparent change in frequency depends on whether it is the source or the observer that is in motion. When the relative velocity approaches c the apparent frequency for a moving source approaches infinity, whereas for a moving observer the frequency approaches $2f$.

1.3.5 Reflection, diffraction and refraction

It may be as well to state that, in general, what follows in this section applies to almost all types of wave capable of propagation in more than one dimension, i.e. to all electromagnetic waves, such as light and radio waves, and surface waves in liquids such as water. For fairly obvious reasons vibrations in rods and strings are excluded.

1.3.5.1 Reflection

This is a well-known phenomenon which is most obvious in sound waves which are reflected from large buildings as an echo. What may be less obvious is that reflection can only occur if the dimensions of the reflecting surface perpendicular to the direction of travel of the wave is relatively large compared with the wavelength. As an approximate guide it may be said that with an obstacle which is less than one-third of wavelength there will be virtually no reflection. When the dimensions of the obstacle are equal to wavelength then about half the incident wave energy is reflected; when the obstacle is three or more times the wavelength then virtually all the incident wave is reflected. By way of an example, if sound waves strike a rigid 1 m × 1 m board perpendicularly, then: frequencies below 100 Hz (wavelength = 3 m approx.) will not be reflected; frequencies of around 300 Hz ($\lambda = 1$ m) will be partially reflected; and frequencies above about 1 kHz ($\lambda = 0.33$ m) will be reflected more or less totally (*Figure 1.39*).

The foregoing is a simplification. In practice, the situation is affected by the rigidity of the obstacle

(a)

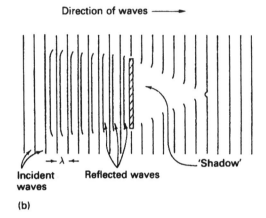

(b)

Figure 1.39 *Reflection from an obstacle. (a) Obstacle much smaller than* λ. *(b) Obstacle much larger than* λ.

and its shape. It should be noted that when reflection takes place, the angle of incidence is equal to the angle of reflection, as with light. (An interesting example in nature is the bat. As is well known, the bat navigates by detecting reflections of its high-pitched sounds from obstacles. Taking a sound frequency of 40 kHz there will be significant reflection from objects less than 1 cm across.)

1.3.5.2 Diffraction

This may be thought of as an effect which is complementary to reflection. Defined broadly as the bending of waves round an obstacle, the same wavelength/dimension relationship applies. Thus, when the wavelength is large and there is little or no reflection it can normally be assumed that the waves are bending round the obstacle. Similarly, with short wavelengths there is marked reflection and little wave energy bends round the obstacle. In the latter case there will be a 'sound shadow'.

Interesting diffraction effects occur with apertures. As *Figure 1.40* shows, when the aperture is large

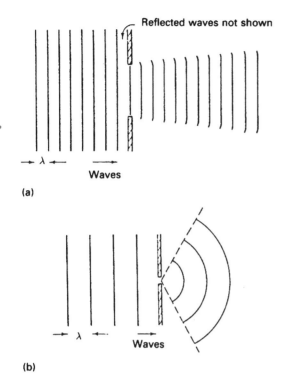

Figure 1.40 *Diffraction at an aperture. (a) Large aperture, small λ. (b) Small aperture, large λ.*

compared with λ the waves pass through it and produce a relatively straight beam. If λ is large, however, there is considerable bending, resulting in a wide beam. These phenomena can often be seen with waves in water.

1.3.5.3 Huygens' Principle
With diffraction and reflection the dependence of the ratio of wavelength to obstacle or aperture size can be more easily understood by the application of a construction due to Huygens (1629–1695). This adopts the notion that a wave progresses by the creation of secondary wavelets at each increment of movement. Thus, the bending of waves round an obstacle can be visualized as resulting from the air particles at the edge of the obstacle, becoming radiators in their own right when set into vibration by an oncoming wave. A very small aperture could be thought of as containing only one secondary source which will clearly radiate over a wide angle. With a large aperture interference effects from the many secondary sources cause cancellation of radiation except along the axis. Similar reasoning can be used to explain other reflection and diffraction effects.

1.3.5.4 Refraction
A common enough phenomenon with light, refraction is seldom significant with sound waves, for the simple reason that we are rarely concerned with sound waves passing into transmitting media – or at least, not in the same way as with light. A good example of refraction of sound can be observed on a cloudless evening late on a warm summer day in the country, when distant sounds can often be heard with surprising clarity. The reason is that on such occasions there is a temperature inversion – the ground has cooled rapidly by radiation and the layers of air a few metres up are warmer than those close to the ground. Refraction occurs because the 'rays' of sound travel more quickly in these higher layers and there is thus a bending affect. *Figure 1.41* illustrates this.

It may be worth mentioning here the effect of wind speed and direction on the transmission of sound waves in the open air. Sounds are said to 'carry well' down-wind. The effect is not due, as is sometimes thought, to the wind speed being added to the velocity of sound, but instead is an effect of a velocity gradient in the air, friction causing the layers of air nearest to the ground to move slower than those higher up, as illustrated in *Figure 1.42*.

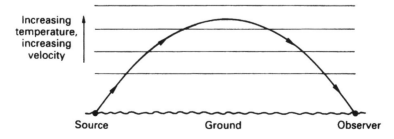

Figure 1.41 *Refraction caused by a temperature inversion.*

Figure 1.42 *Effect of wind on the propagation of sound.*

1.3.6 Components of sounds

A sound is characterized, at least in the human ear-brain system, by various parameters. One parameter, loudness, has been outlined above; two others are *pitch* and *timbre*.

Pitch is generally described, rather unsatisfactorily, as the position of a note in the musical scale, thereby begging at least one question. It is, like loudness, a subjective phenomenon, but in technical terms it is permissible to regard pitch as being very similar to frequency. The difference arises because the pitch of a sound can vary with loudness. If a pure tone from an oscillator is fed into a loudspeaker and the level is abruptly changed by about 20 dB most listeners will experience a change in the pitch, upwards or downwards depending on the original frequency and also depending on the individual person. However, in short, the higher the frequency of the sound the higher the pitch. A portion of a keyboard, as shown in *Figure 1.43*, can be used to aid explanation.

The letters given to different notes correspond, at least ideally, to the frequencies give in *Table 1.9*. The subject of musical scales and pitch is beyond the scope of this book and it must be sufficient here to state that an *octave* – the eight white notes on a keyboard from say, C to C′, contains in fact 12 equally spaced notes. The frequency ratio 2:1 is

an *octave*, as for example A to A (220 to 440 Hz). The frequencies of two adjacent notes, *semitones* in *Table 1.9*, are in the ratio 2^{-12}, or 1.05946. It should be noted that to a trained musical ear the mathematical relationship stated here is not quite correct, so that, for instance, a semitone above F, written as F#, is slightly different from a semitone below G, written Gb. The difference can be achieved with an instrument such as the violin, but keyboard instruments are tuned to a compromise pitch.

Timbre (pronounced 'tarmbre') is the term used to describe the *quality* of a sound, *i.e.* that which distinguishes the sound of, for instance, a trumpet from a clarinet. There are two principal factors that determine the timbre of a sound: the harmonic content and the starting transients.

Harmonics are, of course, exact multiples of a fundamental frequency and the number and relative strengths of the various harmonics goes some way towards giving a sound its individual quality. At least as important are the *starting transients*. These are additional frequencies, not in general related to the fundamental, which last for only a brief time at the start of a sound. It is probably true to say that their duration is typically no more than about 10 cycles of the fundamental. Thus in a sound of basic frequency 400 Hz (period 1/400 = 2.5 ms) the starting transients would probably have died away in

Figure 1.43 *A portion of a keyboard.*

Basic principles

1·35

Table 1.9 The letters given to the notes of the musical scale, and their corresponding frequencies

Note	Frequency (Hz)
A	220.00
A#	233.08
B	246.94
C	261.63
C#	277.18
D	293.66
E	311.13
E#	329.63
F	349.23
F#	369.99
G	392.00
G#	415.30
A*	**440.00**
A#	466.16
B	493.88
C′	523.25

*A = 400 Hz is adopted as a standard and is known as 'International A'.

about 25 ms. The ear – brain attaches great importance to the starting transients, and if they are missing or distorted accurate identification of the source, say a musical instrument, may be difficult or even impossible.

1.3.7 Frequency ranges of typical sounds

Figure 1.44 shows typical frequency ranges of some typical sounds. It should be noted that the *fundamental* frequencies of speech and music only go up to about 4 kHz, but the harmonics may in some circumstances go well into the 10–20 kHz range or maybe even higher.

It is of interest to take the example of a square wave (*Figure 1.45*). This consists of a fundamental together with, theoretically, an infinite number of harmonics of diminishing amplitude:

$$x = A\sin(\omega t) + \frac{A}{3}\sin(3\omega t) + \frac{A}{5}\sin(5\omega t)$$
$$+ \cdots + \frac{A}{n}\sin(n\omega t)$$

where A is the amplitude, $\omega = 2\pi f$, and n is an odd number.

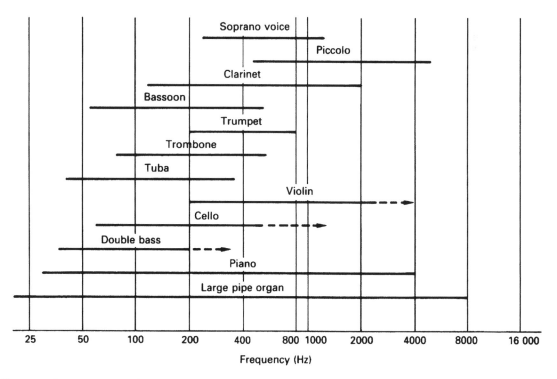

Figure 1.44 *Frequency ranges of some typical sounds.*

Figure 1.45 *A square wave.*

In practice, a good square wave is obtained with some 30 terms. Electrical square-wave signals, or waveforms approaching them, can be a serious source of interference in adjacent circuits because of the likely existence of high frequencies. For example, a square wave of 1 kHz fundamental will contain frequencies of 1, 3, 5 kHz, etc., and the 21st harmonic will be 21 kHz, albeit of low amplitude. Although not strictly square waves, broadly similar waveforms are produced by electronic lighting control systems in television studios and theatres. Where these are used it is almost invariably necessary to use special microphone cables ('star quad') which provide good immunity against the radiated harmonics.

1.3.8 Standing waves

Consider sound waves generated between two parallel reflecting surfaces. If the separation of the surfaces d is such that there is an integral number of half wavelengths contained in d, then the interference between the waves travelling in opposite directions results in *nodes* and *antinodes*, nodes being points of zero displacement and antinodes being maxima. In the simplest case $d = \lambda/2$.

Such a pattern of nodes and antinodes, arising from continuous excitation by the appropriate frequency, is known as a 'standing wave'. It should be noted that the diagrammatic position of nodes and antinodes depends on whether pressure or particle displacement is being represented. At a rigid boundary, longitudinal particle displacement is clearly impossible – air particles cannot burrow into the boundary moving in one direction and they are unlikely to leave a vacuum when moving in the other direction! A pressure antinode or a displacement node therefore occurs at such a boundary. *Figure 1.46* illustrates this.

Standing waves, as presented above, are examples of acoustic resonances and as such are sometimes referred to as 'eigentones'.

1.3.9 Vibrations in pipes and strings

1.3.9.1 Pipes
There are three cases to consider:

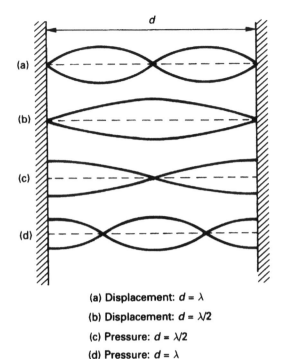

(a) **Displacement:** $d = \lambda$
(b) **Displacement:** $d = \lambda/2$
(c) **Pressure:** $d = \lambda/2$
(d) **Pressure:** $d = \lambda$

Figure 1.46 *Standing waves.*

(1) a pipe open at one end,
(2) a pipe open at both ends, and
(3) a pipe closed at both ends.

Case (3) will be given only the briefest mention for two reasons: sound waves in such a pipe are probably of academic interest only; and the matter has already been outlined on a larger scale in Section 1.3.8.

Pipe open at one end. It is sometimes a matter of surprise that sound waves can be reflected from the open end of a pipe. The reason for this is that the velocity of sound in a large volume of air is slightly different from the velocity when enclosed, because of friction effects at the walls. Thus waves inside a pipe encounter slightly different transmission conditions at the open end; or, put another way, there is a change in the impedance of the medium. Reflection in such a case is far from complete and only a relatively small proportion of energy is reflected, depending on the dimensions of the pipe, etc. However this may be enough to create standing waves. (It is possible to obtain a distinct echo when speaking down pipes of some 300–500 mm diameter and 100 m or more in length.)

As *Figure 1.47* shows, a pipe which is open at one end must have a displacement antinode at the open

Figure 1.47 *A pipe, open at one end.*

end and a displacement node at the closed end. In its simplest mode of vibration the length of the pipe thus approximates to a quarter of a wavelength. The word 'approximately' is used because the acoustic length of the pipe is not quite the same as the physical length. This is because the to-and-fro motion of the air molecules at the open end tends to have a slight streaming effect and it is therefore necessary to apply an 'end correction'. The acoustic length then becomes $l + a$, where l is the physical length and a is the end correction. For a pipe with a significant flange, a is approximately $0.8r$, r being the radius. For an unflanged pipe a is about $0.6r$. From this it follows that the lowest resonant frequency of a pipe open at one end is:

$$f = c/4(l + a)$$

The frequency of the nth harmonic is then

$$f_n = nc/4(l + a).$$

Pipe open at both ends. As shown in *Figure 1.48*, there is a displacement antinode at each open end and the length l approximates to a half wavelength. There will be a correction for both ends so that the acoustic length is now $l + 2a$. Frequencies of resonance are then found from:

$$f = nc/2(l + 2a)$$

n being 1, 2, 3, etc.

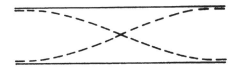

Figure 1.48 *A pipe, open at both ends.*

As a matter of interest the same expression applies to a pipe closed at both ends, except that there are now no end corrections to be applied and the expression becomes:

$$f = nc/2l.$$

1.3.9.2. Vibrations in strings

Full analyses will not be given in the immediately following section as these can be found in a large number of textbooks should they be needed. Instead the broad outlines and equations for the velocities of vibrations are given. We assume that the string or wire is long compared with its thickness and that it is perfectly flexible.

The velocity of a transverse wave is given by:

$$c = (T/m)^{1/2}$$

where T is the tension and m is the mass per unit length. Parallels with sound waves in pipes can be seen when we take the simplest mode of vibration as being that with an antinode at each end so that $\lambda = 2l$. Consequently, the lowest frequency of vibration is given by:

$$f = 1/2l(T/m)^{1/2}$$

In practice, the harmonics produced in a vibrating string depend on the nature and position of excitation. For example, striking a stretched string near to one of its ends may produce a large proportion of the higher harmonics. This fact is made use of by violinists, for example, who can produce very different tonal qualities depending on where a string is bowed. The instruction '*sul ponticello*' (bow near the bridge), results in a much harder tone due to the extra harmonics than when the bow is nearer the middle of the string.

Longitudinal waves may be excited in a string or wire, but they are far less obvious. The velocity of longitudinal waves in a string (or wire or rod) is given by

$$c = (y/\rho)^{1/2}$$

where y is Young's modulus and ρ is the density. Taking steel as an example, for steel $y \simeq 2 \times 10^{11}$ N/m^2 and $\rho \simeq 8 \times 10^3$ kg m/m^3, giving $c = 500$ m/s.

An expression for the lowest frequency of longitudinal waves is

$$f = 1/2l(y/\rho)^{1/2}$$

Substitution of relevant values shows that the frequency of longitudinal vibrations is very much greater than that of transverse vibration. It may also be noted that tension does not enter into the formula for longitudinal vibrations.

1.4 The hearing process

1.4.1 Structure and function of the auditory system

The 'ear' is a term applied to both the sense of hearing and to specific parts of the system. The latter comprises the outer, middle and inner ears (*Figure 1.49*) which convert airborne sound waves into mechanical and fluid vibrations, analyse these into their frequency components and transduce these into electrical and neural signals for transmission to the brain. The neural signals undergo much further processing in the cochlear nucleus, brain-stem nuclei and higher levels of the brain, enabling us to recognize and utilize much of the complex information carried in the original waveform.

1.4.1.1 Outer ear

The outer ear consists of the external ear (ear flap, auricle or pinna) and the ear canal (external auditory meatus). The external ear with its inner cavity, the concha, and the ear canal act together, partly like a horn and partly like an open pipe resonator to

improve sensitivity and impedance matching in the 3–4 kHz region. The complex asymmetrical form of the external ear also provides indirect sound paths to the ear canal by diffraction and reflection from its convolutions at higher frequencies which interfere with the direct sound input. This modifies the incoming sound spectrum in a direction-dependent manner so that, even with one ear alone, the horizontal (azimuth) and vertical (elevation) directions of incidence can be recognized when broadband signals are heard (see Section 1.4.2.13).

1.4.1.2 Middle ear

The ear canal terminates at the ear drum (tympanic membrane) which is a semitransparent conical membrane which can flex in response to sound. Integral with one radius of this is a sliver of bone, the manubrium (handle) of the malleus (hammer), which is attached fairly rigidly to the incus (anvil) which also has a long extension (long process) which is attached more flexibly to the top of the stapes (stirrup). The oval footplate of the stapes forms a piston within the oval window into the cochlea. It is sealed into the bony cochlear wall by a flexible annular ligament. Motion of the ear drum is

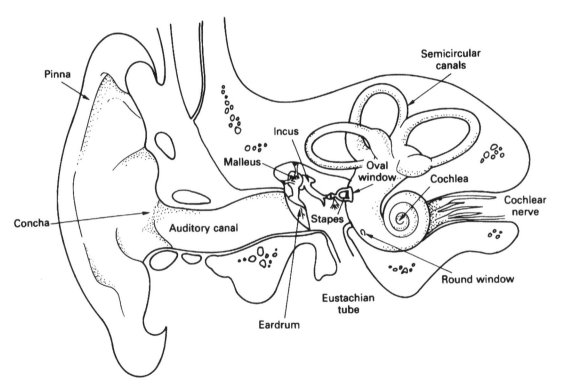

Figure 1.49 *The structure of the external, middle and inner ears. Acoustic pressure on the tympanic membrane causes the stapes to move in and out of the oval window.* (After Kessel and Kardon (1979).)

transmitted by the above three ossicles (middle ear bones) to the cochlea, and in the process the pressure is increased and the volume velocity reduced, thereby matching the low impedance of air to the much higher impedance of the fluid-filled cochlea. Most of this impedance transformation is achieved by the force gathered by the large area of the ear drum (70 mm^2) being concentrated on the much smaller area (3 mm^2) of the stapes footplate, but it is also achieved by the lever ratio of the manubrium to the long process of the incus (1.3:1) and also to some extent by the complex deformation pattern of the tympanic membrane. Most of the mass of the ossicles occurs near their axes of rotation, allowing their moments of inertia to be quite small. At low frequencies their axes of rotation are controlled by various supporting ligaments and at high frequencies by their moments of inertia. The ossicles can be stiffened up by two muscles which are tensed when loud sounds occur and reduce the transmission of sound at low frequencies. The tensor tympani is attached to the manubrium and the stapedius muscle to the side of the stapes. As the ossicles are tightly coupled, the moments of inertia can be combined together to form a single effective mass and the various supporting structures as a single effective stiffness which resonate together at about 1.3 kHz. This resonance is approximately critically damped by resistive components, including the cochlear impedance. The coupling from air to cochlea is most efficient at this resonance frequency where the reactive components cancel.

The middle ear is enclosed in a sealed cavity of about 2 ml to prevent pressure cancellation at low frequencies, but the static internal pressure is equalized with atmospheric pressure from time to time by momentary opening of the eustachian tube during swallowing.

1.4.1.3 Inner ear structure

The inner ear consists of two fluid-filled structures: the cochlea which is the organ of hearing, and the vestibular system which consists of three semicircular canals and the utricle and sacculus which are all concerned with balance and posture.

The cochlea is a snail-shaped bony structure containing a spiral tube divided along its length by two membranes (*Figure 1.50*). The two outer channels (scala vestibuli and scala tympani) are connected by a small opening (the helicotrema), at the apical end of the cochlea and are filled with perilymph which is high in potassium ions and at zero electrical potential. The stapes footplate leads into the scala vestibuli

at the basal end where it also interconnects with the vestibular system. At the basal end of the scala tympani is situated the round window opening covered by a thin membrane exposed to the middle ear cavity. As the cochlear fluids and walls are effectively incompressible, an inward motion of the stapes produces an outward motion of the round window membrane.

The central channel (scala media) is separated from the scala vestibuli by a thin membrane of negligible stiffness, Reissner's membrane, and from the scala tympani by a membrane of graded stiffness (the basilar membrane) which also supports the organ of Corti. Scala media is filled with endolymph which is low in potassium ions and is maintained at a positive potential of +80 mV by the stria vascularis attached to the bony wall of the cochlea and acting as an electrical battery. The organ of Corti contains one row of inner hair cells at its inner edge and, separated by the rods of Corti, three to five rows of outer hair cells situated over the middle of the basilar membrane. The upper ends of all the hair cells are integral with a plate-like structure (the reticular lamina) through which the hairs or stereocilia protrude and make contact or near contact with the tectorial membrane. The latter is a gelatinous substance, attached to the limbus on the inner side of the cochlea, against which the stereocilia are bent when the basilar membrane is displaced by sound.

The inner hair cells are the receptors and transducers. The mechanical input to the stereocilia initiates the sending of nerve impulses to the brain via about 20–30 fibres per cell. These fibres have their cell bodies in the spiral ganglion which lies within the bony shelf (the spiral lamina) supporting the inner edge of the basilar membrane. The axons or output fibres from these cells pass into the centre of the cochlea known as the modiolus forming a twisted rope of fibres getting thicker as it gathers more fibres towards the base. This passes through a hole in the bone of the skull (the internal auditory meatus) into the brain.

1.4.1.4 Inner ear function

Mechanically, the basilar membrane and its associated structures and fluids act like a series of resonators, each tuned to a slightly different frequency. The driving force is provided by the pressure difference between scala vestibuli and scala tympani. High frequencies cause maximal vibration at the basal (stapes input) end of the basilar membrane where it is narrow and thick, whereas low frequencies cause greater effect at the apical end (helicotrema) where

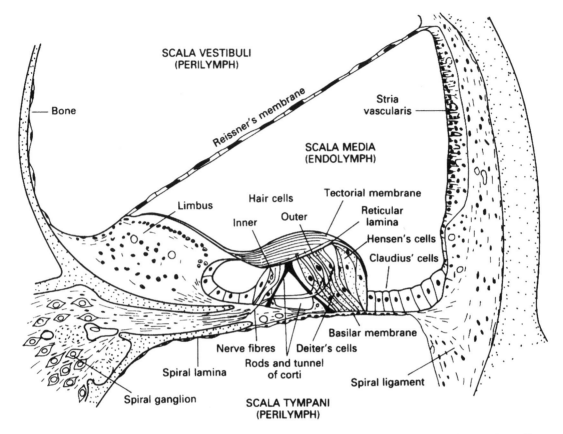

Figure 1.50 *Cross-section of one turn of the cochlea showing the organ of Corti supported on the basilar membrane. Pressure differences between scala vestibuli and scala tympani cause the basilar membrane to move up and down and bend the hairs (stereocilia) of the hair cells against the tectorial membrane. (Courtesy of Davis (1953).)*

it is wide and thin. Because of the high degree of coupling between adjacent segments of the basilar membrane the system acts like a high-order filter and the response to a tone is in the form of a travelling wave (*Figure 1.51*). Older measurements which upset the motile response of the outer hair cells showed the passive mechanics to be poorly tuned. More recent measurements indicate that the basilar membrane response, when sharpened up by the active mechanism of the outer hair cells, probably accounts for the entire tuning properties of the ear.

The function of the outer hair cells has only recently become clear. Earlier studies emphasized their dominant role in the production of the cochlear microphonic, an electrical potential which can be picked up from outside the cochlea by an electrode, preferably placed on the round window membrane. Its waveform provides, like a microphone, a reasonable facsimile of the original sound wave. It is now apparent that the outer hair cells also possess

Figure 1.51 *Travelling wave pattern along the basilar membrane for a 200-Hz tone. The four continuous lines show how the instantaneous displacement pattern moves successively from left (base) to right (apex), building up slowly to a maximum and then falling more rapidly to zero. Higher frequencies would produce a similar pattern but shifted towards the left and lower frequencies towards the right. Modern measurements reveal that active processes make the envelope (dashed lines) much sharper (see text). (Courtesy of Békésy (1960).)*

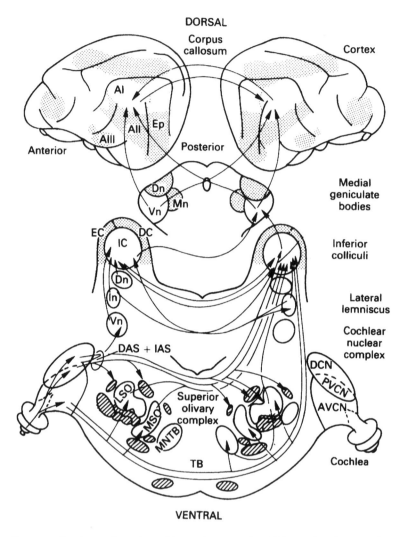

DORSAL

Figure 1.54 *The auditory neural pathways from the cochlea to the cortex. In addition to the afferent pathways shown, there are also many efferent pathways in the opposite direction exercising a controlling influence over the information transmitted as neural impulses.* (Courtesy of Hackney (1987).)

'efferent' (i.e. conducting outwards). However, sensory systems in fact also involve efferent fibres which gate their input and motor systems contain afferent fibres which monitor, for example, extension of muscles. Efferent fibres exist at all levels of the auditory system including many to the outer hair cells.

The auditory pathways (*Figure 1.54*) consist of a number of relay stations or nuclei where many of these complex interactions occur. The first relay station is the cochlear nucleus where the ventral and dorsal parts process information in different ways. The next station is the superior olive, the first place at which interaction between the two ears occurs.

Above this there is much crossing and recrossing of signals.

The highest level of the brain is the cortex, where multiple representations of the input exist and response properties become quite complex.

1.4.2 Psychoacoustics

Psychophysics concerns the relationship between a physical stimulus and the subjective sensation that it evokes. Classical psychophysics treats the observer as a detector, or null instrument, indicating when the strength of signal is just detectable either absolutely,

or above a masking background as an increment, or else just equalling another stimulus. Sensory psychophysics attempts to give a numerical value to the sensory magnitude and involves scaling or fractionation procedures. In classical psychophysics, methods such as the two-interval forced-choice procedure have been designed to avoid problems such as expectation or subject bias. Even such procedures have to be applied with caution, however, to avoid the possibility of some extraneous clue such as different background-noise characteristics providing extraneous information.

1.4.2.1 Hearing range and audiogram

The hearing range is bounded at the lower limit by the threshold curve or audiogram (*Figure 1.55*, dashed curve), which is approximately 0 dB SPL in the 1–4 kHz range and rises to about 70 dB SPL at 20 Hz. The low frequency limit is set at 20 Hz because below this a periodic pulsation is heard without any further reduction in pitch. The upper frequency limit is much steeper and is very age dependent, being about 20 kHz at 15 years and falling roughly by 1 Hz/day so that 2–4 kHz would be quite typical at 70 years. The sensitivity to higher frequencies is dependent upon history of noise exposure as well as age and individual variability – the 4–6 kHz

range is particularly sensitive to noise exposure. For clinical purposes the audiogram is straightened out and inverted using baseline data taken from a large population of normal-hearing subjects. Thus hearing loss is then represented by a downward deviation from the 0 dB hearing level (HL) straight line.

The upper intensity limit of the hearing range is not so well defined and is generally taken as the point where vibration can be felt. At these levels and above (>120–130 dB SPL) quite short exposure durations can result in permanent hearing loss.

1.4.2.2 Pitch perception

The sensation of pitch takes two forms: pitch height and musical pitch.

Pitch height is monotonically related to frequency in a way that suggests a relationship to the place of stimulation on the basilar membrane. A scale in mels has been derived from magnitude estimates and by fractionation techniques.

Musical pitch appears to form a circular or spiral scale on which simple geometrical ratios of frequency, such as the octave (×2) in particular, appear to have greater subjective similarity than smaller ratios. There is a physical basis for this in that a simple ratio of two frequencies will produce a regularly repeating waveform and could also be considered

Figure 1.55 *Frontal free-field audiogram (dashed curve) and equal-loudness contours. These curves are from* Robinson and Dadson (1956) *although such curves were first produced by* Fletcher and Munson (1933).

as different harmonics of the same fundamental frequency. Furthermore, two tones rich in harmonics have many overtones in common when their ratio is simple. This is, of course, a feature of Fourier analysis, where any repeating waveform can be broken down into sinusoidal frequency components that are integral multiples of the reciprocal of the period of the waveform. Although in theory the musical scale is logarithmic, subjectively a slightly stretched scale is preferred. The equal tempered scale is divided equally logarithmically and deviates slightly from simple ratios in order to accommodate many different keys. One intriguing feature of the pitch of harmonically complex tones is that it is not changed by removing the fundamental or lower harmonics. This property goes under various names (false bass, virtual pitch, low pitch, residue pitch) and accounts for the apparent perception of a non-existent fundamental evoked, for example, by a portable radio with a loudspeaker too small to reproduce the lower frequencies. This feature of hearing has led to the concept of pitch mechanisms based on temporal analysis rather than spectral analysis.

In support of this, certain signals, such as chopped noise, have been devised which contain temporal information giving a sensation of pitch without any spectral correlate. There are also many types of binaural (dichotic) stimuli which give a 'central' pitch perception from certain correlated features, even though each ear on its own has no basis for pitch perception in either the temporal or spectral domain. There is also reason to believe from the saturation properties of nerve fibres that spectral patterns which could be resolved at moderate levels may become flattened out at high levels. On the other hand, owing to temporal jitter and the low maximum firing rate of nerve fibres they are unable to convey any temporal information for pure tones above about 4 kHz. Interestingly, this is also approximately the upper limit of musical-type pitch perception. At present it appears that both spectral and temporal information are utilized for the bulk of musical pitch perception.

1.4.2.3 Frequency discrimination
The smallest change in frequency that can be detected for successive stimuli is known as the frequency difference limen (dl). Although it depends upon many factors such as frequency, sound level, stimulus duration and interstimulus interval, it is comparatively constant at a value of about 1 Hz for frequencies below 1–2 kHz, for levels 20 dB or more above threshold and durations above 0.25 s. Surprisingly,

1 Hz at 1 kHz represents a difference of only 0.1% which is about the smallest change that can be detected on an analogue voltmeter! Because music does not normally obey the above constraints a wow and flutter figure of 0.2% is considered acceptable in an analogue tape recorder. At frequencies above 2 kHz the frequency dl tends to increase approximately in proportion to frequency. It also increases quite markedly as threshold is approached and for short-duration stimuli.

1.4.2.4 Frequency resolution and critical bands
Frequency resolution refers to the ability to separate two simultaneously present signals into different auditory channels or more generally to represent the details of a spectral pattern in terms of the pattern of activity across the array of nerve fibres. This facility has been conceptualized in the critical band (CB) or the equivalent rectangular bandwidth (ERB). Signals that appear not to be separated are considered to lie within a critical band. The width of a critical band ranges from about 100 Hz at frequencies up to about 500 Hz and then increases to about 2 kHz at 10 kHz whereas the ERBs are slightly narrower. The whole hearing range can be covered by 24 adjacent non-overlapping critical bands. This situation has been encapsulated in the fixed-channel vocoder for speech processing and is also the basis for the German 'bark' scale of frequency position. Experimental methods to determine critical bands include the change of threshold for complex tones as the spacing of their components change, the masking of a narrow band of noise as a function of the distance to an adjacent tone or tones, loudness summation, the frequency separation of two tones where the border between consonance and dissonance occurs, and the masking patterns of various tone- and band-limited noise combinations. It is also related to the critical ratio which expresses the relative levels of a pure tone and a noise that will just mask it.

1.4.2.5 Masking
Masking is that property of a louder signal which renders a quieter sound inaudible in its presence (*Figure 1.56*). It is greatest when the signals lie within the same critical band, the higher the level of the masker, and for stimuli above the masker frequency rather than below it. For uncorrelated signals within the same band the masker has to be about 4 dB above the target just to mask it. This can alternatively be viewed as an incremental threshold for the masker whose power is increased by 1.5 dB by the target.

Figure 1.56 *Masking curves indicating the degree of threshold elevation as a function of the test frequency for seven different masker levels. The masker is a 90-Hz-wide band of noise centred on 410 Hz presented at the SPLs indicated by the numbers given above each curve. The frequency-specific nature of masking is an indication of the auditory filtering of the inner ear.* (Courtesy of Egan and Hake (1950).)

Masking also occurs when the masker precedes the test signal by a short interval (forward masking) and, more surprisingly, when it follows it (backward masking). These effects are of some theoretical interest because the masking curves thus obtained are narrower than for simultaneous masking.

1.4.2.6 Beats

Two tones of slightly different frequency will be alternately in and out of phase, and when added together will give a single wave the amplitude of which fluctuates at a rate equal to their difference in frequency. This leads to one of three sensations according to this frequency difference. Up to about 10 Hz difference the ear follows the amplitude fluctuations and beats are perceived, for differences between about 10 Hz and 50–500 Hz (depending on the mean frequency) a sensation of roughness is heard, and above this difference the two frequencies can be heard separately. Both the beating sensation and the roughness imply that the two signals are interacting within the same auditory channel, whereas their separate perception implies that they are transmitted by different channels. Beats are not difference tones as such and do not imply nonlinearity. Weak binaural beats can also be heard for lower frequencies when the tones are presented separately to the two ears.

1.4.2.7 Distortion and difference tones

When a system is approaching overload it is common to expect non-linearity and the generation of harmonic and intermodulation distortion. The ear behaves atypically in this respect, however, in that intermodulation distortion occurring within the ear can be heard readily even at low sound levels, whereas harmonic distortion is rarely detected at all. The strongest intermodulation products are the cubic difference tone ($2f_1 - f_2$, where f_1 and f_2 are the lower and upper primary frequencies, respectively) and the quadratic difference tone ($f_2 - f_1$). The strength of both depend upon the frequency ratio (f_2/f_1), being large for ratios of 1.1–1.2 and disappears for ratios greater than 1.5. Higher order difference tones can also be heard, but not the summation or other components occurring above the primary frequencies. The conditions giving the sensation of roughness (Section 1.4.2.6) are those that favour the detection of distortion, and much of this sensation can be attributed to the presence of these products.

In spite of its own inherent non-linearity the ear can be a sensitive detector of distortion occurring within an external system. This ability is, however, strongly dependent upon the conditions. At low sound levels (e.g. less than 20 dB above threshold) even high levels of distortion (e.g. 10%) would be below threshold and, therefore, inaudible, although the shape of the threshold curve (*Figure 1.55*) has to be taken into account. A similar restriction occurs with regard to masking of the distortion by the tone(s) generating it. If, however, the distortion is a difference tone well below the frequencies of high level primaries or is a high harmonic of a low frequency tone, distortion levels of less than 0.1% can be detected. For music, however, where instruments generate their own harmonic distortion as overtones an amplifier producing 3% of predominantly second- and third-harmonic distortion can sound distortionless. Cross-over distortion in class B amplifiers is readily discernible, both because it contains a rich pattern of overtones, many of which are far from the masking effects of the primary, and also because it occurs near the axis of the waveform rather than the peak, it does not disappear at lower amplitudes. For small signals it may represent a higher percentage of the signal than it would near maximum output where amplifier distortion is quoted.

1.4.2.8 Loudness

Loudness is a subjective quality related to sound intensity. Scaling and fractionation methods give a measure in sones, where 1 sone is the loudness of a 1-kHz tone 40 dB above threshold. At mid-frequencies a 10-dB increase gives a doubling of loudness. There is also a loudness level scale which is measured in phons. By definition this is equal

to dB SPL at 1 kHz and the equal-loudness contours (*Figure 1.55*) have been obtained by loudness-matching techniques extended to higher and lower frequencies. At 0 phon this curve approximates the threshold curve and rises towards low frequencies, whereas at 120 phons the curve becomes more nearly flat. This implies a wider dynamic range at mid-frequencies, particularly compared with low frequencies. The standard A-weighting curve was originally designed to compensate for the shape of the equal-loudness contours over the lowest 55 dB of hearing, although it is now common to use it much more generally. Some hi-fi amplifiers have attempted to compensate for this loss of low-frequency loudness, whilst listening at low levels either by a separate switched filter or by a specially tailored volume or loudness control.

The loudness of a complex signal is not the same as that of a pure tone of the same SPL and special techniques have been devised to quantify the loudness of complex signals.

The smallest increment of intensity that can be detected is known as the loudness or intensity difference limen. Weber's law implies that this should be a constant fraction of the stimulus intensity. In fact it decreases from about 1.5 dB within 20 dB of threshold to about 0.5 dB at 80 dB above threshold, more or less independently of frequency. For time-varying signals such as speech and music an incremental threshold of 3 dB is often quoted.

1.4.2.9 Hearing damage

Hearing damage is a complicated topic because of individual differences of susceptibility, the wide range of different possible damaging stimuli, the different time patterns of exposure, and the possible synergistic or protective effects of other influences such as drugs, oxygen level or fatigue. Hearing impairment is defined as an average hearing loss at 0.5. 1 and 2 kHz exceeding 25 dB at which point speech becomes more difficult to understand and medical advice may be sought. For the median individual, 80-dBA exposure over the 8-h working day should be just acceptable without loss, whereas the ear should never be exposed to 140 dBA or above. Hearing loss is usually greatest at half an octave above the exposure ($\times 1.4$) frequency, and a doubling of exposure time necessitates a 5-dB decrease in level. The short-term effects of exposure are known as the 'temporary threshold shift' (TTS) and are generally most marked 2 min after exposure (TTS2) with complete recovery by 36 h. Longer recovery times or TTS2 > 40 dB should be cause for concern,

although permanent threshold shift (PTS) cannot simply be predicted from TTS. When PTS exceeds 40 dB hearing cannot be fully restored with an aid because the frequency filtering and other properties of the ear also become impaired.

An effect known as 'recruitment' means that sounds a little above the elevated threshold may have near-normal loudness, and so when all sounds are amplified these become painful.

1.4.2.10 Habituation, adaptation and after-images

At lower sound levels there are two other effects which superficially might be thought to be related to TTS. Habituation is the effect of attention whereby a constant and repetitive stimulus such as the ticking of a clock can pass unnoticed until attention is directed to it. Adaptation is an effect whereby a steady signal becomes less loud and less detectable and is most noticeable at the upper frequency limit of hearing or in regions of pathological hearing. The effects of adaptation can sometimes be revealed as an afterimage when a neutral stimulus is substituted. A good loudspeaker, for example, may sound inferior when played immediately after one with a deficient response. Similarly, one can be temporarily desensitized to wow and flutter after listening to such a modulated signal.

1.4.2.11 Temporal perception and rhythm

The temporal properties of the ear can be expressed in several ways. Temporal integration implies that some properties of hearing such as loudness or frequency discrimination increase as the duration of the stimulus increases and then become constant and independent of duration. Below this turnover point, at about 250 ms, it would appear that for certain purposes the effects of duration and intensity integrate. This does not mean, however, that a signal of 50 dB SPL and 50 ms duration would be indistinguishable from a 53 dB SPL signal of 25 ms duration. The use of special stimuli, to avoid giving spectral clues to differences, indicates a monaural limit to temporal resolution of about 2 ms. For synthetic speech temporal chunks of 10 ms are considered satisfactory, whereas for interaural comparisons of signals, differences as small as 30 µs can be detected. The incremental threshold for the duration of a signal can be given as a Weber fraction ($\mathrm{d}t/t$) and amounts to about 5%. For the detection of a gap in white noise, simulating a tape drop out, a lower limit of 3 ms is found which corresponds to an integrating time of 25 ms. Reverberations occurring within 30 ms of the stimulus are heard as composite with the source,

whereas reflections later than 60 ms are perceived as separate echoes.

For the perception of rhythm or temporal pattern, intervals have to lie between the limits of 50 ms and 2 s with an optimum range of 1–4 elements/s. There is a tendency to group even isochronous pulse trains in units of 2, 3 or 4 pulses and for more complex patterns the principles of Gestalt psychology can be applied. If the rate of a simple melody is increased, it can break up into two or more separate streams of notes the frequencies of which lie close together in a process known as 'fission'.

1.4.2.12 Binaural hearing

Binaural presentations can be either diotic in which identical (mono) signals are presented to the two ears, or dichotic when differing signals are used. Apart from a centralized, rather than a lateralized image, diotic perception differs little from monotic (monaural) presentation. One of the major benefits of dichotic hearing is in the 'cocktail-party effect', the ability to follow one conversation in the presence of other competing conversations. The main psychoacoustic effect upon which this ability is based is known as the binaural masking level difference (BMLD), although many other factors may also be involved such as language, sex of speaker, topic of conversation, degree of redundancy of message, physical layout of speakers in relation to observer, and amount of reverberation. The BMLD or binaural unmasking is the effect in which a tone which is just masked by white noise if presented monotically or diotically, becomes readily detectable if the same noise, but without the tone, is presented to the opposite ear or if the relative polarity of noise and tone is different in the two ears. As the overall sound levels do not have to be the same in both ears this has led to a hypothesis of equalization and cancellation (EC) for internally subtracting the noise from the tone. For frequencies below 1 kHz improvements in the signal/noise ratio up to 15 dB can be obtained.

Other dichotic effects include central pitch phenomena in which different harmonic components combine centrally to give a percept characteristic of the combination rather than that of the separate ears. Noise signals which include differing phase shifts that would be undetectable monotically can give rise to a central pitch. In the phenomenon of binaural diplacusis a slight frequency-dependent difference of pitch is commonly observed between the ears (±1–2%). Binaural beats have been mentioned in Section 1.4.2.6 above.

1.4.2.13 Sound localization

The strongest clues or cues to sound localization are based on the dichotic presentation inherent in the two ears being situated in different points in space. As a result of this the sounds arrive in general at a slightly different time and a slightly different intensity according to the direction of origin. The time difference can of course also be expressed as a phase difference for sinusoidal signals and this would be more relevant for frequencies and angles for which the path difference exceeds half a wavelength. This situation would lead to ambiguity with sine waves above 1.5 kHz but would be resolved for most naturally encountered sounds owing to their more complex spectra. The time differences are maximal for sounds arriving directly from the sides at ±650 μs which is well above the minimal detectable difference of 30 μs.

Intensity differences between the ears can be due either to the shadowing effect of the head or to the inverse square law for sources close to the observer. The shadowing effect ranges from negligible at low frequencies where the wavelength is much longer than the diameter of the head, up to about 20 dB at 15 kHz. As an interaural difference of 1 dB is detectable, this is clearly a usable clue. Similarly, the inverse square law should give a 1 dB difference for a source 2 m to the side. Unfortunately, these time and intensity clues do not give a unique direction but simply define the surface of a cone of confusion on an axis through the ears. There are two methods by which this ambiguity can be resolved. Motion parallax is a body or head movement, either rotatory or linear, which alters the relative difference between the ears. Pinna clues can also resolve the ambiguity by providing independent evidence of direction. They rely on reflected and diffracted sound paths adding to the direct one (*Figure 1.57*) and introducing peaks and dips into the spectrum – these indirect sound paths are different for different source directions and can give a monotic localization to about 5°–25°. In free-field experiments where all cues are available, minimum audible angles as small as 1° can be found.

There are other clues to the direction and distances of sources or reflecting surfaces, produced by acoustic images, reflection tones (an evenly spaced series of spectral peaks giving a pitch of $1/t$ by interference) and direct to reverberant energy ratios.

It will be seen from the above that the directional information inherent in a stereophonic recording and reproduction system is far from ideally matched to the auditory mechanisms of localization. On the

Figure 1.57 *Perspex model of the pinna showing the direct and indirect sound paths the relative timings of which depend on the direction of the sound source and colour the sound spectrum accordingly by interference.* (Courtesy of Bateau (1961).)

other hand, binaural recording systems using accurate replicas of the ears and proper matching of the earphones, although impressive, usually fail to give fully convincing localization due to the absence of parallax effects.

In a reverberant environment, although the directly received energy is frequently well below that of the reverberant field, the source is usually located correctly. This is not simply due to visual clues, as can be apparent in badly designed sound-reinforcement systems where the sound appears to come from the nearest loudspeaker. This effect is known as the precedence effect or the law of the first wavefront. It is also frequently called the 'Haas effect', although the latter term strictly refers to the greater intensity that an echo can have to be judged equal in loudness to the direct sound, i.e. about 12 dB at 15 ms.

1.4.2.14 Speech perception

The understanding of speech is one of the most important functions of hearing and involves the detection of patterns in the spectral and time domains (*Figure 1.58*). Speech can be characterized by an

Figure 1.58 *Speech sonograms showing the sound spectra (vertical axis) as a function of time for two different filter bandwidths. (a) The broadband (200 Hz) analysis reveals the glottal pulse rate (F_0) as vertical striations and speech formants (F_1–F_4) as horizontal or diagonal bands. (b) The narrowband (30 Hz) analysis tends to be dominated by the harmonics of F_0 as near-horizontal bands, although the formants are also visible. The fricative sound (S) is unvoiced and does not show banding.* (Courtesy of Holmes (1988).)

excitation function, which can be either a pulse train representing vocalized sound or white-noise source representing non-vocalized sounds, followed by filters representing the vocal and nasal cavities. The glottal pulse train is produced by air alternately passing through the open or closed vocal cords and its frequency (F0), ranges from 50 to 200 Hz in males and 100 to 400 Hz in females. Its intonation pattern carries meaning by indicating questions, etc., but is of little relevance in identifying words. Certain bands of harmonics of F0 are enhanced by the vocal tract filters or speech formants (F1–F4). These cover the ranges 0.25–1 kHz for F1, 0.7–2.5 kHz for F2, 1.5–3.5 kHz for F3 and 3–4 kHz for F4, and it is the variations in these, particularly F1 and F2, which carry most of the speech information. These changes are produced by the articulatory motions of the tongue, jaw, lips and velum. The non-vocal or fricative sounds are produced by the passage of air past various constrictions in the vocal tract and, depending on their point of origin, may also be filtered by the vocal cavities. The sudden opening or closing of a pathway produces plosive sounds. Speech perception depends upon the ability to spectrally resolve and follow the speech formants and to identify the temporal pattern of changes. Unlike the analyses shown in *Figure 1.58* the auditory filter bandwidths (critical bands) vary with centre frequency from about 100 Hz at frequencies up to 500 Hz, to 300 Hz at 2 kHz, 700 Hz at 4 kHz and 2 kHz at 10 kHz. Speech recognition is remarkably resistant to spectral distortions such as differentiation or integration. For intelligibility, the telephone band from 0.3 to 3.4 kHz is quite adequate. Infinite clipping of the waveform reduces intelligibility to about 70%, but not if the waveform has been differentiated first. The acceptable signal/noise ratio varies greatly with the vocabulary size from about −12 dB for digits to +12 dB for nonsense syllables.

References

BATEAU, D.W., 'Characteristics of human localisation of sound', *Contract No. 123-(60530)23545A*, Naval Ordnance Test Station (1961)

von BEKESY, G. *Experiments in Hearing*, McGraw-Hill, New York (1960), p. 745

DAVIS, H. 'Acoustic trauma in the guinea pig', *Journal of the Acoustical Society of America*, **25**, 1182 (1953)

EGAN, J.P. and HAKE, H.W., 'On the masking pattern of a simple auditory stimulus', *Journal of the Acoustical Society of America*, **22**, 622–630 (1950)

EVANS, E.F., 'The frequency response and other properties of single fibres in the guinea pig cochlea', *Journal of Physiology*, **226**, 263–287 (1972)

FLETCHER, H. and MUNSON, J.L., 'Loudness, its definition, measurement, and calculation', *Journal of the Acoustical Society of America*, **5,** 82–108 (1933)

HACKNEY, C.M., In *Hearing* (*British Medical Bulletin*, Vol. 43, No. 4) (ed. M.P. Haggard and E.F. Evans), Churchill Livingstone, London (1987), p. 796

HOLMES, J.N., *Speech Synthesis and Recognition*, Van Nostrand Reinhold (1988)

KESSEL, R.G. and KARDON, R.H., *Tissues and Organs*, Freeman & Co., San Francisco (1979)

ROBINSON, D.W. and DADSON, R.S., 'A redetermination of the equal-loudness relations for pure tones', *British Journal of Applied Physics*, **7**, 166–181 (1956)

WILSON, J.P., 'Evidence for a cochlear origin for acoustic re-emissions, threshold fine structure and total tinnitus', *Hearing Research*, **2**, 233–252 (1980)

Bibliography

GREEN, D.M., *An Introduction to Hearing*, Erlbaum, Hillsdale, NY (1976)

HAGGARD, M.P. and EVANS, E.F. (eds), *Hearing (British Medical Bulletin*, Vol. 43 No. 4), Churchill Livingstone. London (1987)

HANDEL, S., *Listening: An Introduction to the Perception of Auditory Events*, MIT Press, Cambridge, Mass. (1989)

MOORE, B.C.J., *Frequency Selectivity in Hearing*, Academic Press, London (1986)

MOORE, B.C.J., *Introduction to the Psychology of Hearing*, Academic Press, London (1986; 4th edn 1997)

PICKLES, J.O., *An Introduction to the Physiology of Hearing*, Academic Press, London (1988)

PLOMP, R., *Aspects of Tone Sensation*, Academic Press, London (1976)

TOBIAS, J.V., *Foundations of Modern Auditory Theory*, Vols. 1 and 2, Academic Press, New York (1970/1972)

1.5 Perceptual audio coding

1.5.1 Introduction

The advent of digital audio coding has allowed for the accurate storage and transmission of music signals. Typical digital audio systems of today possess signal-to-noise and distortion ratios of greater than 90 dB, with audio bandwidths of 20 kHz or greater. Unfortunately, this high performance comes with the price of a greater bandwidth requirement than previous analogue systems. This is true because a 20 kHz bandwidth analogue audio signal is commonly converted to a digital signal composed of 16-bit words sampled at 44.1 or 48 kHz sample rate, resulting in a per channel data rate of 705 or 768 kb/s, respectively.

The fact that digital audio transmission and storage require significantly greater bandwidth than their analogue counterparts has resulted in the need to find methods of reducing this data rate. Unfortunately, simple methods of reducing the word length or sample rate of the digital audio representation provide a limited ability to reduce this bandwidth requirement.

In the quest of greater data-rate reduction, other techniques have used compansion and adaptive filtering to exploit frequency and simple masking characteristics of typical signals. One such system is NICAM or Near Instantaneous Companded Audio Multiplex (e.g. Caine, English and O'Clarey, 1980) using pre-emphasis and compansion (compression on encode and expansion on decode) resulting in a bit rate of 338 kb/s per channel. Another system is AC-1 (e.g. Todd and Gundry, 1984), which combined a variable second-order filter with adaptive delta modulation, a type of companded differential coding, to provide for a data-rate reduction to 220 kb/s. ADPCM or Adaptive Differential Pulse Code Modulation (e.g. Rabiner and Schafer, 1978) has also been used with some success along with pre-filtering via frequency band division filters. The CCITT G722 standard for wideband speech (e.g. CCITT G722 standard 1988) and apt-X coder (e.g. Wylie, 1996) use this method. Unfortunately, the bit rate reduction capabilities of these have been limited to a data-rate reduction ratio of 4:1 with good, but not excellent sound quality.

More powerful audio coding methods successfully exploit a detailed application of the psychoacoustic limitations of the human auditory system. This approach, denoted as perceptual audio coding, has proven to be the most effective method to reduce the data rate required for representation of high-quality audio. Limitations of the human auditory system are much more constant and predictable than the varying characteristics of all possible types of high-quality audio signals. This section contains an examination of the psychoacoustics of masking, a discussion of common tools used to build an effective perceptual audio coder, and a description of the characteristics of the prominent perceptual audio coders.

1.5.2 Psychoacoustics of masking

Application of the psychoacoustics of the human auditory system is the basis of all good rate-reduction music coders. As a result, understanding of the design principles behind perceptual audio coders requires an examination of the psychoacoustics of

masking and detection. It will be shown that the two key assumptions in the design of perceptual audio coders are that the ear exploits a simple frequency division as a first step in the hearing process, and that the masking effect of signals presented to the ear can be understood from a relatively small number of experiments using simple stimuli. This means that the masking characteristics of complex signals can be calculated by extension of the masking characteristics for simple signals, using frequency-division methods to break up the complex signal into a number of simple masking components. Simple superposition of individual masking thresholds is then used to create the combined masking effect of these individual components. This is a bold assumption because typical music signals are often very complex, with many tonal or noise-like components, not at all like the simple stimuli used in the basic psychoacoustic experiments. Despite this, the performance of the better perceptual audio coders can be extremely good over a wide variety of signals.

1.5.2.1 Critical-band model of hearing

Central to the development of a workable model of the auditory system is the critical-band concept and its relationship to the masking characteristics of the ear. The critical-band model of the human auditory system was first developed by Fletcher (1940) to explain why masking experiments showed that signals covering a frequency range narrower than a certain threshold bandwidth produced the same masking and detection properties as other signals with smaller bandwidths. Fletcher's work led to the fundamental approximation that the ear acts as a multichannel real-time analyser with varying sensitivities and bandwidths throughout the audio range. Despite the intrinsic simplicity of the model, it has endured the test of time. Perceptual audio coders rely heavily on this model.

The critical bandwidth represents the minimum frequency bandwidth resolvable for masked signals. For example, the masking of a low-level signal by a larger-level tone nearby in frequency is maximal and continues largely unchanged until the frequency separation between the two exceeds this critical bandwidth. Additionally, detection of a signal component is based on the entire energy within a critical bandwidth, regardless of whether it is tonal in nature, noise-like, or a combination of the two.

This idea has been further examined by others who have refined the critical-band concept, resulting in two groups of bandwidth estimates. These are represented by the earlier Zwicker estimates (e.g. Zwicker

and Terhardt, 1980) which produce larger bandwidths than those of the second group, as represented by Moore (e.g. Moore and Glasberg, 1987). Above 500 Hz there is good agreement between these groups with the later examinations indicating 20–30% smaller bandwidths. Unfortunately, below 500 Hz the disagreement between groups becomes more and more marked; at 20 Hz the critical-bandwidth determinations range from as low as 27 Hz to as large as 100 Hz. This discrepancy in critical bandwidths is the subject of much debate in the literature and still unresolved. *Figure 1.59* shows these two estimates of critical bandwidths compared to 1/5 octave bandwidth.

Examination of this figure shows critical bandwidths as a function of frequency and compares them to 1/5 octave bandwidth. The Zwicker values are generally greater than 1/5 octave over a large part of the audio band, while the Moore critical-band values are generally less. If fixed bands are used to approximate critical bands, this results in 25 or 41 bands for the Zwicker or Moore determinations, respectively.

Typically perceptual audio coders have used the Zwicker bandwidth values, partly due to the fact that these results have been available for a longer time. Despite this fact, often a finer bandwidth than the Zwicker estimates has been used in perceptual audio coders to compensate for the fact that fixed bands have the problem that they do not completely model the adaptive bands of the human auditory system. Unlike a group of fixed filters, the human auditory system process centers a critical-band filter around the appropriate signal component, extending the maximum masking or detection effect ±1/2 a critical bandwidth from the masking component frequency. Fixed bands do not have the capability to centre their frequency of operation and therefore need to have a smaller bandwidth to compensate for the fact that a component may lie at a band edge.

In summary, the critical-band model allows signal detection and masking to be evaluated consistently for both noise and sine wave, or their combination. Most present perceptual audio coders assume the use of the Zwicker critical-bandwidth determination; as a result the following discussion will use the Zwicker values.

1.5.2.2 The use of single-tone masking curves

Information on the masking effect of signal components is available primarily for single tones or bands of noise. Since the most appropriate way to examine masking phenomena is to perform a spectrum analysis based on critical bandwidths, these spectral analyses of masking are used as a basis for the design of the perceptual audio coder filter structures and the methods to reduce the bit rate. Typically masking curves show the single high-level sine-wave component or band of noise that masks the presence of another smaller component (e.g. Fielder, 1987). The masking curve represents the minimum level

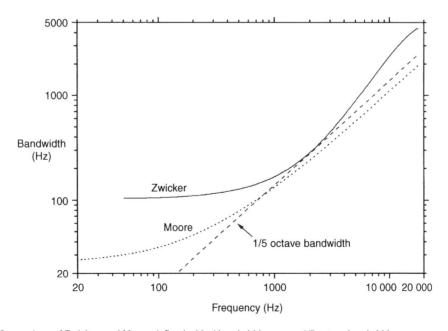

Figure 1.59 *Comparison of Zwicker- and Moore-defined critical bandwidths versus 1/5 octave bandwidths.*

of detectable additional signal when the listener is exposed to a loud sound stimulus, e.g. a 50 Hz sine wave tone at various levels. These curves can be used to derive an upper boundary on the levels of permissible error signals due to the data-rate reduction process because any signal below this minimum level is inaudible.

The masking effect of a signal component varies significantly, depending on its frequency and whether the component is tone-like or noise-like in character. As a result, perceptual audio coders either assume the worst case of a sine-wave (tonal) component masking or separate the input signal into tonal and noise-like components, combining tonal and noise-like masking curves in their operation. *Figures 1.60* and *1.61* are the masking curve families for 50 Hz sine waves at the acoustic sound levels of 110, 100, and 80 dB sound pressure level (SPL) and for 1 kHz sine waves at the acoustic sound levels of 101, 81, 61, and 41 dB SPL. Both figures also include the hearing threshold curve for the typical listener in a quiet environment.

Figure 1.60 represents the typical masking characteristics of low-frequency signals, and shows that the masking effect on the detection of additional sounds varies considerably with level. Note that the largest increase in hearing threshold occurs at frequencies close to the loud stimulus frequency; in these circumstances a significant error can go undetected because of the masking effect of the louder signal. Despite

this fact, low-frequency signals represent a challenge for perceptual coders because of the extreme sensitivity the listener can possess for signals located far from the primary component. In this case a −5 dB SPL signal at 3.5 kHz is audible even though the 110 dB SPL 50 Hz sine wave is present, since the hearing sensitivity above 3 kHz is completely unaffected by the 50 Hz sine waves.

As shown by *Figure 1.60*, in *Figure 1.61* the masking effect varies considerably with acoustic level and is greatest within a critical bandwidth of the masking component. In addition, as the masker's level increases, the masking effect extends upward in frequency considerably, an effect heavily used in low bit-rate coder design. Substantial coding errors are inaudible if they occur higher in frequency and at a lower level than indicated by this masked threshold curve. The masking effect for signals at low acoustic levels (i.e. 64 dB SPL) may extend only an octave upward in frequency while higher-level situations may produce as much as three or more upward octaves of significant masking.

The masking effect for signals lower in frequency than the masker falls off much more quickly. This is particularly evident for component frequencies between 500 and 2000 Hz; in this frequency region the slope can be as steep as 90 dB/octave. This rapid decrease in masking for signals lower in frequency than the primary input frequency component has

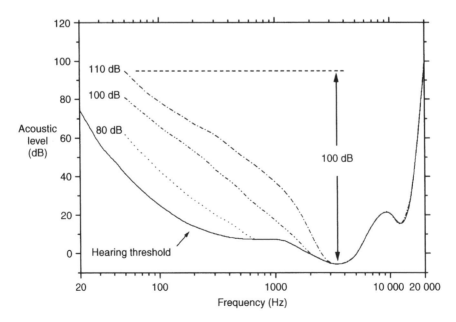

Figure 1.60 *Hearing sensitivity in the presence of quiet and a 50 Hz sine wave at various levels. Levels of 110, 100, and 80 dB sound pressure are shown. 0 dB is defined as 20 μPa.*

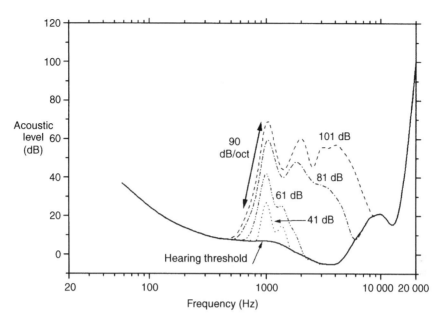

Figure 1.61 *Hearing sensitivity in the presence of quiet and a 1 kHz sine wave at various levels. Levels of 101, 81, 61, and 41 dB SPL are shown. 0 dB SPL is defined as 20 μPa.*

significant consequences in coder design because of the need to implement extremely steep roll-off filters. This has been one of the primary reasons that high-quality perceptual audio coders having data-rate reductions of 4:1 or greater have awaited the availability of powerful DSP architectures capable of implementing the complex filter structures.

The importance of localizing errors nearby in frequency to that of a signal component is demonstrated by a simple experiment where a listener is exposed to a 50 Hz tone at 110 dB SPL, plus the addition of a small amount of noise with differing spectral characteristics. Experimentation shows that if this small noise is band limited between 20 and 110 Hz (within a critical bandwidth) the listener is only able to detect its existence for levels greater than −12 dB with respect to the 50 Hz tone. On the other hand, 20 kHz band limited white noise or a critical band-width band of noise centred at 3.5 kHz is detectable if is greater than −102 or −115 dB with respect to the 50 Hz tone, respectively. This great variation in sensitivity is graphically shown in *Figure 1.62* which is a spectral comparison of the appropriate masking curve with the 50 Hz loud tone and the three examples of noise.

This figure is a spectral comparison of a masking curve and three different types of noise. For clarity, the bands of noise are shown to have rectangular bandpass characteristics and the fact that the overall

level is equal to the critical-band masking curve level is indicated by the fact that the noise band level lies on top of the masking curve. Two of the noise signals are assumed to be exactly a critical band in width and closely follow the hearing threshold curve level within the critical-bandwidth region. The 20 kHz band limited white-noise signal spectrum is measured with critical bandwidths. Using the critical-band model and the appropriate masked threshold curve shows that the noise is audible when at least one critical-band level exceeds the threshold curve. For the 20 kHz band-limited white-noise example, the critical band causing the noise to be audible is centred at 3.5 kHz. Examination of *Figure 1.62* shows that the levels ratios of −12 dB, −102 dB, and −115 dB from the previous experiment result when the noise example's levels just equal the masked threshold over at least one critical band.

In summary, the psychoacoustics of masking indicate that detection of extraneous signals or coding errors is minimized when they are localized to frequency bands smaller than a critical bandwidth and having band-edge roll-offs as rapid as those in the ear. Therefore, when perceptual audio coders reduce the data rate by approximating to the input signal (e.g. via word-length reduction), the errors produced by the process are least audible when they are tightly confined in frequency. In addition, it's also possible to take advantage of the extensive upper-frequency

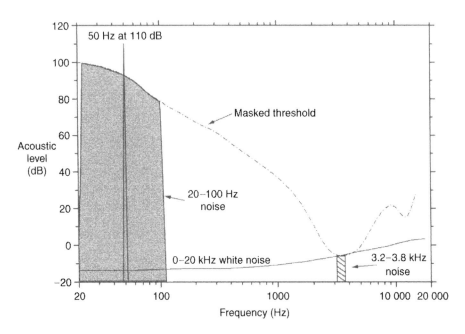

Figure 1.62 *Spectral comparison of a masking tone at 50 Hz, the appropriate masked threshold curve, bands of noise at 20–100 Hz, 3.2–3.8 kHz, and 20 kHz band limited white noise. The 20–100 Hz and 3.2–38 kHz bands of noise are grey shaded and cross-hatched, respectively. 0 dB SPL is defined as 20 μPa.*

masking produced by major signal components by allowing for large errors within the upward frequency masking region of these major components. Coding systems of this type are called multi-frequency band coders and the filter that spectrally confines the coding errors is called a filterbank. Perceptual audio coders are a subset of the larger class of multi-frequency band coders in that the frequency division and quantization are driven by psychoacoustics.

1.5.2.3 Temporal masking and time versus frequency trade-off

Another requirement for audio coders is the accommodation of human auditory characteristics during transient events. Although the frequency resolution of the ear under steady-state conditions is extremely sharp, implying a poor time resolution, the frequency–division characteristics of the auditory system are signal adaptive in nature. For example, the perception of transient events may involve a time resolution on the order of a few milliseconds, much shorter than appropriate for filters modelling human spectral resolution under steady-state signals. If the temporal response of a filterbank is too long, this will cause a dispersion of error signals more widely in time than desirable. For this reason, filterbank design typically involves a trade-off between these conflicting goals.

As in the case of the frequency characteristics of auditory masking under steady-state signal conditions, there is a basic asymmetry in the characteristics of temporal masking. The masking of small signal components occurring in time before a signal causing masking, called backward masking, is substantially less than the forward masking effect in which the same small signals occur after the masker. Noise or coding errors occurring before the signal causing them is called pre-noise. Backward masking remains strong for about 4 milliseconds and disappears for time separations larger than tens of milliseconds, while forward masking lasts approximately ten times as long (e.g. Carterette and Friedman, 1978). The temporal resolution characteristics of a filterbank used for data-rate reduction of music signals should maximize both the time and frequency masking effect, allowing the ear to tolerate the largest error induced by the data-rate reduction. Since a transient event can occur anywhere within the effective time response of a particular filter, this argues strongly for filterbanks with time resolutions less than 4 milliseconds. The need for tight localization of coding errors in frequency on steady-state signals conflicts with the requirement for tight temporal resolution under transient conditions. This has resulted in perceptual audio coders using various strategies.

One strategy employed is to select a single compromise filterbank that allows good spectral resolution and a moderate tendency for pre-noise audibility. This approach is the simplest to implement but has the disadvantage of having neither optimal time nor frequency characteristics.

A second strategy is to employ a filterbank optimized for frequency selectivity but to apply a gain or sensitivity modulation with a finer time resolution than the high-frequency resolution filter. During steady-state conditions this gain or sensitivity modulation is quiescent, allowing the frequency resolution to approach that of the high-frequency resolution filterbank, while transient conditions cause a gain or sensitivity modulation that allows the pre-noise from the high-resolution filterbank to vary with the signal level. This allows the pre-noise to be greatly reduced before transient events and made less audible. Gain or sensitivity modulation is effective but has the limitation that signals that are the combination of transient and steady components are not treated effectively. This is particularly true if the spectral characteristics of the transient and steady-state signals are different. As a result, improved performance is obtained in some coders by pre-filtering the band into separate regions or by using a frequency and time variable method of gain modulation in the basic transform itself (e.g. temporal noise shaping to be discussed later).

The final and most powerful approach is to employ an adaptive filter for the filterbank where the filter characteristics vary between steady-state and transient conditions. Typically this is done by varying the time-frequency trade-off dynamically with the signal. This design philosophy has the advantage of allowing the coding system to directly use several time-frequency trade-offs for the filterband technology employed, allowing for maximum coding efficiency. The perceptual audio coders with the greatest ability to preserve audio quality at low data rates use this method and typically use variable-length transform-based filterbanks.

1.5.3 The multi-frequency band coder concept

It has been shown that the frequency-division and masking characteristics of the human auditory system drive the design of perceptual coders. The previous section has shown that extraneous signals or errors are least audible when they are within a critical band in frequency and the temporal time window of the signal component causing them, or when

their amplitude is such that other louder components mask their presence. This implies that adaptive frequency division and spectral analysis is a central part of perceptual audio coders. Frequency-division is necessary during the encoding to separate each frequency component. This allows perceptually unimportant components to receive a low data allocation, while the important ones get the necessary data. The decoding process also requires a separate filterbank to ensure that errors generated by the approximation implicit in the bit-rate reduction process are tightly confined in frequency around the component.

Another condition for effective reduction of data rate is that the filterbank must be critically sampled or near-critically sampled. This means that the total number of samples at the input and output of the critically sampled filterbank must be the same. For instance, a critically sampled filterbank that divides a 48 kHz sample rate input into ten equal bands must have an output sample rate for each of those bands of 4.8 kHz. This lower sample rate condition is denoted by calling the filterbank outputs subsampled in nature. *Figure 1.63* is a schematic representation of a typical multi-frequency band coder.

The figure shows an example of an audio signal containing three separate frequency components A, B, and C. The basic block diagram is shown in the upper portion of the figure while the lower portion shows spectral representations of the signal at input, output and intermediate stages of the coding process.

The coding process begins as the input filterbank is used to separate the components of the input signal into band numbers 0, 2, and k. This allows a controlled treatment of each component, eliminating the need to allocate data for encoding the other bands without signals. This results in a large data-rate reduction if only a few bands are significant.

Once the various components of the signal have been separated by the input filterbank, the data rate to represent the input signals is reduced by an approximation process via quantization and coding. The quantization and coding errors produce spurious signals or errors that are scattered in frequency, as shown for the intermediate component spectra in the figure. Unless the errors are properly confined in frequency, the perceptual coding system would not be able to take full advantage of the masking properties of the ear. Spectral confinement of the errors is performed by the output filterbank. The output spectrum in the figure shows that the output filterbank causes the signal to be the combination of the original signal with coding errors that are spectrally confined in a controlled manner and minimally audible.

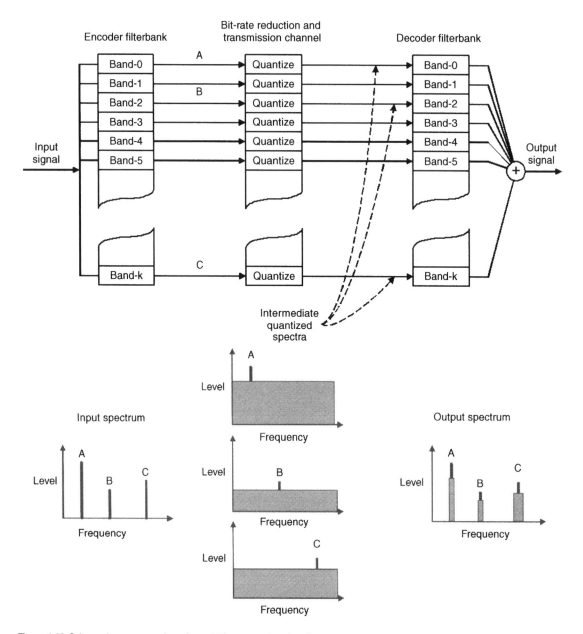

Figure 1.63 *Schematic representation of a multi-frequency band coding system.*

The basic multi-frequency band coding process can also be examined from a block diagram perspective. For this analysis, the perceptual coder is broken down into an encoder, transmission channel (not shown) and decoder.

Examination of *Figure 1.64* shows the input signal is applied to the input filterbank and analysed for its masking properties by a perceptual model. The perceptual model is designed to use the main filterbank

or uses another filterbank as its elementary frequency analysis method. Perceptual audio coders implement auditory modelling in various ways; sometimes the perceptual model is explicitly calculated, other times it is implicit in the bit-allocation algorithm. The perceptual model acts as the controller for the bit-allocation process in the quantization and coding section, where the actual bit-rate reduction occurs. Various techniques, described later, are used

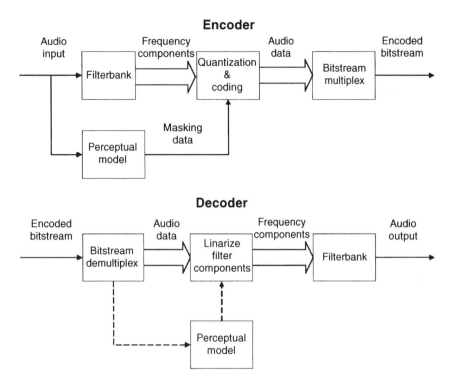

Figure 1.64 *Simplified block diagram of a perceptual encoder–decoder pair.*

to generate a low data-rate approximation to the filterbank signals. Once the quantization and coding have been accomplished, the audio and any necessary control data are then multiplexed into a bitstream that represents the encoded audio signal.

The decoder is basically the reverse of the encoding process. It receives the encoded audio signal and first demultiplexes the bitstream components. This received information contains information on how to relinearize and restore the audio signal. Control of the relinearization and restoration process is done either via bitstream data from the encoder or indirectly through spectral information driving a perceptual model identical to that in the encoder. Perceptual audio decoders use either method or a combination of the two types of relinearization control.

The method of having the encoder sending the complete relinearization information is denoted as a forward adaptive bit-allocation process. This method allows a high degree of flexibility in the design of the perceptual model. Since the relinearization information is sent in the bitstream, decoders can be standardized and widely distributed while the encoder perceptual modelling technology is improved. However, this advantage is offset by the fact that sending

the complete relinearization information generally requires additional data rate.

Another method is the indirect control of the quantization and coding, or relinearization process. This is called a backward adaptive bit-allocation process. Typically, coarse spectral information derived from the filterbank is used to drive identical perceptual models in the encoder and decoder. Since no relinearization control needs to be sent, there is a reduction in the required data rate. This reduction comes at the expense that improvements in perceptual modelling methods require revising of both the encoder and decoder algorithms.

1.5.3.1 Filterbanks used in perceptual coding
The performance and characteristics of the filterbank used in perceptual coding systems play a centrally important role in perceptual audio coder operation. As a result, an overview of the most common filterbank techniques is appropriate for an understanding of perceptual coding. Historically, multifrequency band coding filterbanks have been called subband or transform coders and considered quite different. Usually subband coders have employed a smaller number of bands that have had minimal overlap between bands, while transform filterbanks have had a large

number of highly overlapped bands. Despite these differences, subband and transform techniques are basically the same method, and it is possible to construct subband filterbanks using transform methods and vice versa.

One common similarity that all the effective filterbank techniques possess is a critically sampled or near-critically sampled nature. This characteristic is important because an additional increase in data rate due to the filtering method would reduce the coding efficiency.

The fact that the number of input samples and total filtered samples are equal means that the sample rate of each filterbank output is just sufficient to prevent in-band aliases for passband signals. Unfortunately, this means that the signal content in the transition regions is actually alias products which could create audible problems if not compensated for. Transform filterbanks have the advantage of having an adjacent band alias cancelling property between transform bands, while subband filters must be designed to have this property. Adjacent band alias cancellation means that the aliases in the transition band of one transform band are cancelled by inverse aliases in other bands. Filterbank techniques must be examined to determine the effectiveness of the alias-cancellation process:

Modified discrete cosine transform filterbank
The modified discrete cosine transform or MDCT

is a transform filterbank using time-domain aliasing cancellation (e.g. Princen and Bradley, 1986). This transform filterbank is the combination of a FFT-like transform and a window-overlap process, producing bands that have constant frequency bandwidth. First, the encoder filterbank breaks up an input audio signal into time blocks equal to the transform length with an overlap between succeeding blocks of half the transform length. This overlap means that every input sample is represented by two transforms, but because the frequency-domain subsampling and time domain alias cancellation intrinsic in the MDCT exactly compensates for this 200% overhead, the MDCT filterbank is critically sampled. Each sample in a block of audio data is multiplied by the corresponding value of the window function and then applied to the input of the transform. The decoding process is similar except that it is in reverse and has an extra stage where the successive windowed blocks of audio data are added together.

The function of the window is to reduce the effect of discontinuities at the block edges and to create a constant gain of unity as overlapped blocks in the decoder are added together. If done effectively, this windowing makes the block process implicit in the use of a fixed-length transform seem to be a continuous filtering process. The most commonly used window has the shape of one half-cycle of a sine wave

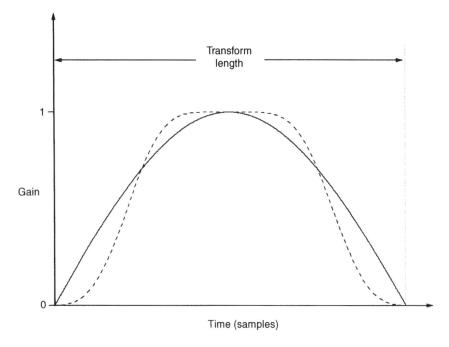

Figure 1.65 *Typical window shapes for use with the MDCT. The solid line is a sine-shaped window and the dashed line is a window with adjustable rejection in the stop band.*

but other types with adjustable stop band rejections have also been developed (e.g. Fielder *et al.*, 1996). *Figure 1.65* shows two typical window shapes with the necessary properties. These windows ensure that the signal level at the transform boundaries is close to zero, have symmetrical gains about their centres, and have smooth transitions from the low gain at frame boundaries to a gain near unity at the centre.

MDCT or related filterbanks are used in many of the audio coding systems because they combine critical sampling, good frequency resolution, and the computational efficiency of a FFT (fast Fourier transform). This type of filterbank also has the property that the time-frequency selectivity trade-off can be varied with transform length and therefore allows for adaptation to transient conditions (e.g. Bosi *et al.*, 1997; Fielder *et al.*, 1996).

Quadrature mirror filter tree-structured filterbanks This type of filterbank splits the time-domain input signal using a cascade of filters, called a quadrature mirror filter or QMF. A QMF divides the input spectrum in two equal, two-to-one sub-sampled bands having the desired alias cancellation property in the transition regions. The method is attractive for perceptual audio coders because it is relatively easy to divide lower-frequency parts of the spectrum more finely by successive division stages. Unfortunately, this method has the disadvantage that it has higher complexity and longer delay than transform techniques. This finer and finer division process gives this filterbank its name of a tree-structured filter due to its similarity to a tree with a large single tree trunk dividing into an increasingly larger number of smaller tree branches.

Polyphase filterbanks Polyphase filterbanks are subband filters of equal bandwidths which combine the filter design flexibility of generalized QMF filterbanks with low computational complexity (e.g. Rothweiler, 1983). This approach has allowed the design of perceptual audio coders with a high ratio of passband frequency to transition region frequency, but has the disadvantage that the alias cancellation property is not inherent in the method. As a result, some practical implementations have only partial alias cancellation.

Hybrid filterbanks Filterbanks which consist of a combination of different types of filterbanks are called hybrid filterbanks. They are used to build filterbanks combining the property of having different frequency resolutions, as with QMF-tree structures and the computational efficiency of transform algorithms. Hybrid filterbanks employing the combination of subband pre-filters and MDCT post-filters

are used in a number of perceptual coders. One disadvantage of this method is that some implementations have impaired alias cancellation in the transition regions even though each separate filter method is intrinsically alias cancelling. This is because the combination of one filter type with another interferes with the alias cancellation properties (e.g. Brandenburg *et al.*, 1992).

1.5.4 Overview of quantization and coding methods

It has been shown that the basic perceptual audio encoder consists of filterbank, perceptual model, quantization/coding, and bitstream multiplex sections, while the decoder is basically the reverse process. This section will examine some of the commonly used methods for quantization, coding, and multichannel coding used in perceptual audio coders. The more important independent channel methods will be first discussed and then the multichannel coding methods examined. Unless otherwise noted, decoding will be considered simply the inverse of encoding and not mentioned.

1.5.4.1 Independent channel coding methods

Linear prediction Linear prediction is a technique that uses previous time samples to predict the value of a new sample. This has the advantage of allowing only a reduced amplitude error signal to be encoded rather than the original signal and minimizes the error generated by the quantization process. A better understanding of linear prediction is obtained by examination of what happens in the spectral domain. Prediction can be looked at as a filtering process that has a frequency-response characteristic that is the inverse of the spectral characteristic of the input signal. When the input signal is treated in this manner its overall level is reduced, its spectrum is flattened and its coding errors are reduced. This process is often called removing the redundancy from the signal because any part of the signal that can be predicted does not need to be transmitted from encoder to decoder, and therefore is redundant.

A practical implementation of linear prediction is called adaptive differential pulse code modulation or ADPCM (e.g. Rabiner and Schafer, 1978). This method continuously uses the prediction and quantization process to adapt the prediction filter to changing signal characteristics. ADPCM can be adaptive in the sense that the sensitivity of the quantizer and/or the frequency characteristics of the prediction filter follow what is needed to minimize the error for the

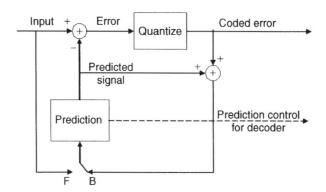

Figure 1.66 *Block diagram for feedforward or feedback implementations of an ADPCM encoder.*

prediction of the input signal. The ADPCM design may be forward adaptive (feedforward topology) or backwards adaptive (feedback topology) in design structure as shown by *Figure 1.66* of a prototypical ADPCM encoder.

Examination of this figure shows that the basic difference between the forward or backward adaptive methods is that the input to the prediction filter either comes from the input signal (switch position F) or from the quantized error signal (switch position B). A forward adaptation process uses the input signal to drive the adaptation process of the prediction filter, sending control information to drive a matching filter in the decoder (indicated by the dashed line). The backward adaptation encoder and decoder predictors are driven from the error signal alone. The figure using the B switch position displays the backwards adaptive configuration if the dotted prediction control line is eliminated.

Both methods are used in perceptual audio coders. Each method has advantages: the backward adaptive method has the advantage that no extra data rate is required to control the prediction filter in the decoder, while the forward adaptive method is faster and more accurate in its ability to adapt to the input signal. Disadvantages of the backward prediction method are that it adapts more slowly to variations in the input signal and requires a very exact match in the prediction process between the encoder and decoder; even very small differences in arithmetic processes can cause the prediction process to diverge in the encoder and decoder. The primary disadvantage of the forward adaptive method is that it requires extra data rate to control the decoder prediction filter.

Block floating point This method allows for the more efficient coding of frequency or time components if they should be grouped together, as in

the case of critical bands. The idea is to represent a block of filterbank output samples (arranged either in time or in frequency) by both a scale factor or exponent value for a band and scaled data values or mantissas within the band (e.g. Fielder *et al.*, 1996). These mantissas are the original values divided by the appropriate scalefactor.

Block floating point is a useful technique in adapting the output from a MDCT filterbank to critical-bandwidth grouping. Since MDCT filterbanks create a multiplicity of constant bandwidth outputs, they must be combined into critical-bandwidth groups for proper treatment in a perceptual audio coder. This is arranged by combining the MDCT outputs into block floating point groups whose bandwidth ranges are critical bandwidth in size. As a result, this technique has found wide application in MDCT-based perceptual coders.

Block floating point works in the following way. Suppose a band has five input components and 40 bits are available for their coding. The most direct approach would simply reduce the word length to 8 bits for each component. Block floating point scales the five values in amplitude by factors of two until the maximum magnitude component is as large as possible without exceeding the maximum amplitude range (in this case -1.0 to 1.0). The scale factor or exponent is the 5-bit binary word representing the number of two-to-one scaling processes (16:1 in this example) while the scaled data or mantissas are represented by binary words 7 bits in length. *Table 1.10* shows the input values and resulting signal to noise ratios or SNRs for conventional word length reduction and use of this example of block floating point.

Note that the SNRs of the block floating point encoded values are much better. As a result, this

Table 1.10 Component levels and signal-to-noise ratios before and after block floating point scaling

	Conventional quantization		Block floating point	
Input signal	*Input SNR (8-bit)*		*Scaled signal*	*Scaled SNR (7-bit)*
−0.02	15.9 dB		−0.32	29.0 dB
0.05	23.9 dB		0.90	38.0 dB
0.01	9.9 dB		0.16	23.0 dB
−0.03	19.4 dB		−0.48	32.6 dB
0.001	−10.1 dB		0.16	3.0 dB

technique is used in many perceptual audio coders due to its simplicity and ability to lower the data rate to represent the input signal. One disadvantage of this method is that it is a less effective scheme if one component of the group is significantly higher in level than the rest of the group because the high-level signal causes the errors in all the others within the group to increase due to the group scaling process.

Huffman coding This method is often used in high-frequency resolution MDCT perceptual audio coders as a means of accurately coding high-level components within a group of lower-level components. It is more complex to implement than block floating point, but avoids the problem that a single high-level component among a group of lower-level components causes increased coding errors. The basic idea is to represent all the possible values to be encoded with different length words, according to their probability of occurrence. The most frequently occurring values have short word-length symbols, while less likely patterns have longer ones

(e.g. Huffman, 1952). If the distribution of spectral amplitudes is uneven, as in the case of tonal signals (those providing the minimum masking), this results in a large reduction in the number of bits needed for coding the filterbank output. In addition, further data-rate savings are obtained by representing pairs, triplets, quads, or larger groups of input components by vector Huffman symbols.

As an example, consider single-component or scalar Huffman coding. Suppose that values between −1.0 and 1.0 in 0.2 steps are to be encoded. If standard PCM encoding is used, this would require 4-bit words (16 values) to represent the 11 possible values. Huffman coding begins by using the fact that the value of zero occurs most frequently, the values of +0.2 and −0.2 somewhat less frequently, and the other values rarely at all, to construct the Huffman code table (*Table 1.11*).

Assume that the sequence of 0.4, 0, −0.2, 0, 0.2, 0, and 1.0 is encoded using *Table 1.11*. This results in the following Huffman sequence, 11000 0 101 0 100 0 11110. As per *Table 1.12*, this

Table 1.11 Huffman encoding table

Input signal	*Huffman code word*	*Input signal*	*Huffman code word*
1.0	11110	−0.2	101
0.8	11100	−0.4	11001
0.6	11010	−0.6	11011
0.4	11000	−0.8	11101
0.2	100	−1.0	11111
0	0		

Table 1.12 Huffman coding example

Input	*0.4*	*0*	*−0.2*	*0*	*0.2*	*0*	*1.0*	*Total*
PCM	0010	0000	1111	0000	0001	0000	0101	28 bits
Huffman	11000	0	101	0	100	0	11110	18 bits

sequence of values via Huffman coding is much more efficient than conventional quantization into linear binary words. The total bit requirement is shown in the far-right column of the table, demonstrating a Huffman coding savings of 10 bits in 28 bits.

Temporal noise shaping This method is used to improve the audible performance of transform filterbank coders for signals which vary more rapidly than the transform filter characteristic response time. It does this by causing a gain modulation in time synchrony with and proportional to the input signal, resulting in a reduction of the audibility of the pre-noise. Temporal noise shaping, or TNS, has a similar effect to the gain modulation method used at the input of an encoder filterbank (e.g., Bosi *et al.*, 1997), except that it is superior in that it is able to adapt to the spectral signal energy of the audio signal instead of being intrinsically broadband in nature.

TNS is a variation of forward adaptive linear prediction techniques and is the time-frequency dual of ADPCM. This means that TNS uses linear prediction on the frequency domain elements of the transform filterbank and sends the error signals forward in a manner similar to that of ADPCM for time signals (e.g., Bosi *et al.*, 1997). In a similar manner to ADPCM, which 'flattens' frequency components by time prediction, TNS uses frequency component prediction to flatten the effective time domain levels within the filterbank. This flattening of levels provides for more efficient coding using block based transform techniques by allowing the noise within the decoder transform block to be modulated by the signal level and be reduced in audible effect. Pre-noise is reduced because the absence of signal before a transient results in reduced levels of filterbank-induced noise. This technique has been used to improve the medium time-resolution properties of MDCT perceptual audio coders with transform lengths equal or greater than 20 milliseconds and is particularly effective for reduction of the pre-noise of speech signals.

Vector quantization Vector quantization or VQ is a method which approximates a set of digital words by a subset of them (e.g., Gray, 1984). This is done by assembling groups of possible input values as coordinates of a single N dimensional vector, where N is the number of input elements. Once this vector is formed, the best approximation output vector from a more limited set of vectors is selected as the vector-quantized value. The coordinates of this vector then form the output quantization values. Since there are fewer output vectors than input vectors, the output data rate is lower.

For instance, a set of ten 3-bit binary words may be replaced with a single vector word of 12 bits. The approximation occurs because the $2^{(10 \times 3)} = 2^{30}$ states represented by the ten words is approximated by the best one out of 2^{12} states represented by the vector. This technique is used in many speech and video coding techniques, but has found limited application in perceptual audio coders.

1.5.4.2 Multichannel coding methods

The following are methods which extend the operation of single-channel perceptual audio coders to applications having more than one reproduction channel. These techniques are useful because most important audio applications require the reproduction for two or more audio channels. Extension to more than one channel creates both opportunities and challenges for the design of perceptual audio coding systems. The extension to more than one channel provides an opportunity for further reduction in the data rate because the ear is not able independently to process many channels at once, but the more complicated and variable playback environment creates additional challenges in the implementation of accurate perceptual models. Two methods of multichannel coding are discussed next. Also methods to convert multichannel programs to other channel configurations during decoding are examined.

Intensity coding or channel coupling The term intensity stereo or channel coupling designates a method which combines multiple channels into a smaller number (e.g. one) during the encode process. This is done by creating combination channel bands and respective spectral control signals for each channel. The spectral control signals allow the decoder to perform approximate restoration of the original discrete channels from the combination channel. This technique works well because the power spectra of multiple channels are often highly correlated at higher frequencies and the inability of the human auditory system to detect phase relationships between channels at higher frequencies. The combination channel or channels are created by taking sums, differences, or phase-shifted sums from the appropriate number of channels. The high-frequency portion of multichannel signals is used to form a single combination channel in the encoder. An example of this encoding process for three channels is given in *Figure 1.67*.

This figure shows that the three-channel combination is performed by adding each channel's two highest frequency bands into 'coupling' or combination channel bands, after an appropriate phase shift

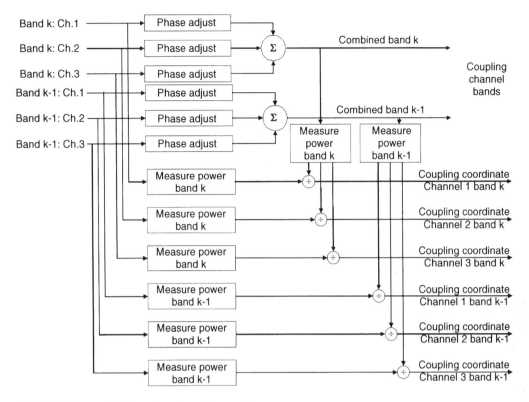

Figure 1.67 *Block diagram of a three-channel coupling encoding process.*

to prevent cancellation effects. The control signals for each channel and band are obtained by taking a band-by-band power ratio of the individual input channels over the combination channel. Even though the lower-frequency bands are passed directly to the quantization and coding sections, a substantial data-rate saving is realized because this method effectively converts the multichannel coding process into a single-channel one for the higher-frequency regions. Bands are often defined as a multiplicity of filterbank outputs grouped together, reducing the data rate for the control signals. *Figure 1.68* shows the inverse process used in the decoder.

The figure illustrates that decoding is done by spectral modulation of the combined channel via control information for each combination channel and combination frequency band. This spectral modulation is performed by multiplication of the two combination bands by the appropriate individual channel control signals to obtain an approximation of the original signal present in each band. These two bands are then combined together with the lower-frequency, individual channel filterbank signals by the output filterbank to create the coded audio signal for each channel. Note that this example does not

perform an inverse phase shift to compensate for the phase shift process in the encoder; often excellent results are obtained without a restoration of phase, due to the ear's insensitivity to phase information at high frequencies.

Sum-difference coding Sum-difference coding is a channel pairwise technique used to code two channels more efficiently. This technique is denoted by several names, such as M/S stereo coding or rematrixing. It involves a matrixing–dematrixing operation similar to the technique used in FM stereo transmission, except that it is done on a frequency band-by-band basis. Instead of transmitting the two channels, normalized sum and difference signals for appropriate frequency bands are quantized and coded (e.g. Bosi *et al.*, 1997; Johnston *et al.*, 1996, or Fielder *et al.*, 1996). These sum and difference signals are often referred to as the middle (M) and the side (S) channels. In situations where more than two channels are coded it is possible to take pairs of channels to perform sum-difference coding to improve the overall performance of the coding system.

Channel format conversion Channel format conversion methods are necessary for practical implementations of multichannel perceptual audio coders.

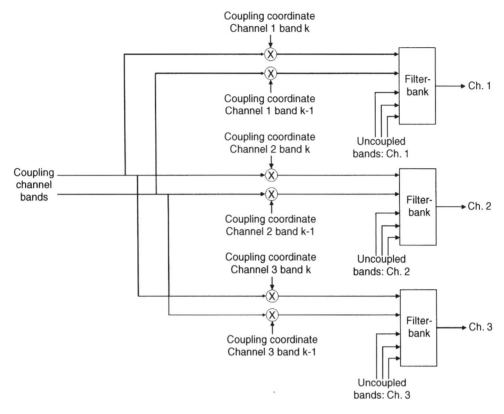

Figure 1.68 *Block diagram of a three-channel coupling decoding process.*

This is true because practical systems must allow for the reconfiguration of the playback conditions at the decoder. For instance, a reasonable rendition of a five-channel programme must be reproduced in a monophonic, stereo, or five-channel format. Two methods designed to address this requirement are matrix encoding and downmixing.

Matrix encoding (e.g. Stoll, 1996) is the extension to the idea of creating a multichannel programme as the simple grouping together of a stereo pair and additional channels. It is more than a simple grouping of channels together because the stereo pair may be a 4–2–4 matrix surround pair, due to the widespread use of matrix surround methods (e.g. Pro Logic surround). Unfortunately, this creates a problem for the proper reproduction of the original multichannel programme since the encoded two channels are no longer two of the original five channels. Therefore, this process creates the matrix surround pair from the five-channel input by adding the necessary components into the stereo pair from the other channels. This creates a five-channel programme that is a matrix surround pair, centre channel, and two surround channels. The advantage of this method is that

two-channel stereo and matrix surround reproduction only requires the partial decoding of the bitstream since only two channels are needed, but has the problem that the original stereo channels in a multichannel programme must be formed by subtraction of some of the components in the other three channels. Signal subtraction is dangerous in a perceptual coder because it alters the signal levels, reduces the masking effect assumed to be present, and causes coding artifacts to become audible.

Another method of addressing the requirements of multiple playback configurations is called downmixing (e.g. Fielder *et al.*, 1996). All the original channels are decoded in their original form, with other channel configurations formed by the summation of channels into the desired output channel configuration. For example, an original five-channel signal may be converted to a matrix surround pair by adding controlled amounts of the two surround and centre channels into the left and right channels in the decoder. This method avoids the subtraction problem of matrix encoding because components are only added together at the decoder but has the disadvantage that the summation of more channels into a

smallar number can cause an increase in signal level that must be carefully handled to avoid overload. Overload is avoided by reducing the overall loudness of the downmixed programme or implementation of peak level compression.

1.5.5 Overview of perceptual coding implementations

The use of perceptual coding systems has resulted in the ability to encode and decode excellent sound quality audio at data rates as low as 64 kb/s per channel and 320 kb/s for five-channel programmes. As a result, the performance of perceptually based audio coders is substantially better than that of non-perceptually based schemes.

This section identifies a number of practical implementations of perceptual audio coding techniques in use today to illustrate the principles so far discussed, and provide an overview of this field. AC-2, AC-3, ATRAC, CAC, the MPEG audio coders: layers 1-3, MPEG-2 AAC, and PAC are examined for their basic approach, important characteristics, and applications where they are used. *Table 1.13* summarizes the characteristics of these systems.

This table delineates some of their important characteristics, such as the filterbank used, frequency resolution at 48 kHz sample rate (44.1 kHz for ATRAC), number of channels supported, the data-rate range, and whether the previously mentioned techniques of linear prediction, block floating point, Huffman coding, temporal noise shaping, vector quantization, and multiple channel coding methods are employed. Coding systems employing the previously mentioned techniques are marked with 'yes' while coders not using these methods are left blank rather than marked 'no'. The designator LFE indicates whether an optional low-frequency effects channel is available.

AC-2 coding The AC-2 designation defines several mono or stereo coding systems with MDCT transform filterbanks having either a fixed block length of 128 or 512 samples, or variable block lengths of 128 and 512 samples (e.g. Fielder *et al.*, 1996). This results in a frequency resolution of 93.75 Hz and 375 Hz for 48 kHz sample rate applications. This coding system family is designed for minimal and equal computational complexity for both the encoder and decoder by use of a moderate length MDCT filterbank, frequency block floating point, simple shift operations during quantization, and a low-complexity bit-allocation process. The equal division of computation between the encoder

and decoder makes it suitable for applications involving point-to-point service and the simple complexity of the decoder has made AC-2 attractive for direct satellite broadcasting applications for stereo sound with video. The AC-2 family of coders operates at data rates between 128 and 384 kb/s for mono or stereo audio at sample rates of 44.1 and 48 kHz.

AC-3 coding The Dolby Digital (AC-3) coder (e.g. Fielder *et al.*, 1996) is a multichannel coding system that uses a MDCT filterbank to perform the frequency-division process and allows for transport of one to five channels of audio, plus a low-frequency effects channel. A distinguishing attribute of AC-3 is that it was the first perceptual audio coding system to employ multichannel coding methods. It processes multiple channels as a single ensemble to improve coding efficiency and is designed for single-point to multipoint consumer applications. It uses sum-difference coding methods for stereo signals and channel coupling techniques to code multichannel signals more efficiently. It also includes provision for sending of user data, i.e. multiple auxiliary data streams, language identifiers, etc.

The bit-allocation control in AC-3 is a hybrid of a backward and forward adaptive process since the perceptual model and bit-allocation process use parametric variables to modify the action of its basically backwards adaptive decoder bit allocation algorithm. AC-3 makes extensive use of downmixing and dynamic level compression to provide a high degree of adaptability to consumer playback conditions. The use of dynamic level compression in the decoder allows for a decoder adjustment of the amount of dynamic level compression employed and the possibility to avoid its use in highest fidelity applications.

The AC-3 algorithm is used in a variety of applications at sample rates of 32, 44.1 and 48 kHz and at data rates of 32–640 kb/s. It is used for point-to-point stereo links over wire and satellite paths, direct satellite broadcasting of sound with video and audio-only programmes, as well as five plus LFE (denoted 5.1) channel sound on LaserDiscs, DVDs, cinema, digital television, and high-definition digital television. In addition, reduced sample rate versions at sample rates of 8, 11.025, 12, 16, 22.05, 24 kHz, with data rates as low as 8 kb/s, are used for transmission of music signals over the Internet.

ATRAC or Adaptive Transform Acoustic Coding The ATRAC coding system (e.g. Tsutsui *et al.*, 1996) is an independent channel coding system used in stereophonic and eight channel applications. It employs a hybrid filterbank which is the combination of a QMF subband filter dividing the signal into

Table 1.13 Overview of common perceptual audio coders

	AC-2	AC-3	ATRAC	CAC	MPEG Layer I
Filterbank type	MDCT	MDCT	Hybrid	Subband	Subband
Frequency resolution (Hz)	93.75 Hz 375 Hz	93.75 Hz 187.5 Hz	86.13* Hz 344.5* Hz 689.1* Hz	1500 Hz	1500 Hz
Channels	1–2	1–5 +LFE	2,8	1–8 +LFE	1–7 +LFE
Data rate (kb/s)	128–384	8–640	292, 1168	32–4096	16–1066
Linear prediction				Yes	
Block floating point	Yes	Yes	Yes		Yes
Huffman coding				Yes	
Temporal noise shaping					
Vector quantization				Yes	
Multiple channel coding method		Coupling M/S		Coupling	Coupling
Channel conversion		Downmix			Matrix

	MPEG Layer II	MPEG Layer III	MPEG AAC	PAC MPAC
Filterbank type	Subband	Hybrid	MDCT	MDCT
Frequency resolution (Hz)	1500 Hz	41.33 Hz 125 Hz	23.44 Hz 187.5 Hz	23.44 Hz 187.5 Hz
Channels	1–7 +LFE	1–7 +LFE	1–48	1–60 +1–3LFE
Data rate (kb/s)	16–1066	16–1066	8–2304**	24–1024
Linear prediction			Yes	
Block floating point	Yes			
Huffman coding		Yes	Yes	Yes
Temporal noise shaping			Yes	
Vector quantization				
Multiple channel coding method	Coupling	Coupling M/S	Coupling M/S	M/S
Channel conversion	Matrix	Matrix	Downmix matrix	

*44.1 kHz sample rate resolution.
**The AAC standard specifies a data rate between 1 b/s to 8389 kb/s. Practical minimum and maximum data rates are 8 kb/s (tested at this rate for mono operation) and 2304 kb/s (assuming 8 channels with a memory limitation of 288 kb/s per channel at 48 kHz), respectively.

three frequency bands of $\frac{1}{4}$, $\frac{1}{4}$, and $\frac{1}{2}$ bandwidths, followed by variable block length MDCT filters with frequency resolutions of 86.13 and 344.5 (lower bands) or 689.1 Hz (upper $\frac{1}{2}$ band only). The bit allocation method is forward adaptive in nature, with the perceptual modelling occurring in the encoder. The reduction of computational complexity is accomplished by the use of frequency block floating point to form bands appropriate for masking calculations and a simple quantization method to perform the bit

rate reduction. ATRAC is used in the MiniDisc stereo playback system and for eight channel cinema playback applications. All applications are 146 kb/s per channel and have a 44.1 kHz sample rate.

CAC or Coherent Acoustic Coding The Coherent Acoustic coder (e.g., Smyth *et al.*, 1996) is a multichannel coding system that uses a polyphase subband filterbank in combination with forward adaptive ADPCM to perform the frequency-division process for sample rates at or below 48 kHz. Higher sampling rates use an additional subband

frequency-division method to divide the spectrum into an audio and ultrasonic band. The audio band is divided into 32 equal bandwidth bands, as in the case for 48 kHz applications, while the ultrasonic band is divided into eight bands.

This system is capable of encoding up to eight channels of audio, plus a low-frequency effects channel. Huffman coding and vector quantization of the ADPCM coefficients are used to improve the efficiency of the coding process and also used to allow for a variable data-rate, lossless coding mode. This mode allows for perfect reconstruction of input signal, but requires high data rates and the use of a slightly different filterbank that has complete frequency alias cancellation properties. Transients are handed by selectively switching off the ADPCM process and modifying the coding process to minimize pre-noise audibility. Multichannel coding efficiency is further improved by the use of channel coupling.

The CAC system is used in LaserDisc and DVD applications. As in the case of AC-3, CAC allows for the transport of time code and other user information, as well as incorporating downmixing and decoder dynamic level compression. This system supports a wide range of sample rates from 8 to 192 kHz and data rates from 32 to 4096 kb/s.

MPEG-1 and MPEG-2 family of coding systems The designations MPEG-1 and MPEG-2 denote algorithms for mono or stereo and multichannel coding, respectively. MPEG-1 and MPEG-2 define three basic perceptual audio coders, designated Layers I, II, and III. In addition, MPEG-2 defines an additional non-backward–compatible coding system that is called Advanced Audio Coding or AAC. For Layers I, II, and III, the MPEG-2 algorithms extend the MPEG-1 characteristics to allow for operation at half sample rates of 16, 22.05, and 24 kHz, in addition to the 32, 44.1, and 48 kHz sample rates supported by the MPEG-1 versions. The MPEG-2 algorithms also extend the MPEG-1 mono and stereo channel operation to seven channels, plus an optional low-frequency effects channel, and increase the maximum data rate to 1066 kb/s. Both the MPEG-1 Layer I and II coders are only specified to have a total of five channels in the MPEG standard but this has been extended to seven channels in some applications. Extension is accomplished by generating an encoded bitstream that is the combination of MPEG-1 and MPEG-2 extension bitstreams in a frame, allowing for backward compatibility to MPEG-1 decoders. Channel conversion to a smaller number of channels is performed using matrix encoding in five channel

configurations or by simulcasting both the stereo pair and the five-channel programme in a seven-channel mode. These coding systems also allow for the carriage of timing and user information in the encoded bitstream.

MPEG Layer I coding This subband coding system is the least complex of the MPEG coding standards. The MPEG Layer I coding system (e.g. Brandenburg *et al.*, 1996 and Stoll, 1996) is a subband coding system that is a simpler version of the Layer II coding system described next. It has lower complexity in that it has a shorter encoded frame length representing 384 input samples, simpler coding of the subband signals, half the frequency resolution in the perceptual modelling process, and a higher data rate of 448 kb/s allowed for in the MPEG-1 bitstream portion. Layer I coding has found application in the Digital Compact Cassette or DCC, and several precursors to the DVD, called CD-i and CD-V. Layer I multichannel coding spans data rates between 32 and 1066 kb/s, sample rates of 16, 22.05, 24, 32, 44.1, or 48 kHz, and allows for the reproduction of one to seven channels of full-band audio, plus an optional low-frequency effects channel through the use of a Layer II multichannel extension bitstream.

MPEG Layer II coding The Layer II coding system (e.g. Brandenburg *et al.*, 1996 and Stoll, 1996) is a multichannel polyphase subband perceptual coding system that divides the audio band into 32 equal bands. It is designed to minimize decoder complexity by having an asymmetrical balance between encoder and decoder complexity by the use of a forward adaptive bit-allocation method. The MPEG-1 portion of the bitstream accommodates data rates of 32–384 kb/s for mono or stereo operation, while the remaining channels utilize the remaining data-rate capacity, to a total of 1066 kb/s. As in the case of Layer I, channel conversion is accomplished by matrix encoding or simulcast techniques. Layer II also makes use of dynamic level compression to provide adaptability to a variety of consumer playback conditions. The use of dynamic level compression in the decoder allows for a decoder adjustment of the amount of dynamic level compression employed, while allowing the carriage of the audio signal with original dynamic levels.

The Layer II system uses an 1152 sample frame to further improve the coding efficiency over the Layer I system. In addition, it uses a perceptual model that has its own FFT-based filterbank of length 1024. The basic coding process involves the splitting of the input signal into 32 bands, the use of a frequency block floating point process, and quantization of the

companded audio data using a frequency variable selection of possible word lengths. The efficiency of the coding in multichannel applications is further increased by the use of intensity stereo to combine the data for the higher-frequency portion of each channel.

The Layer II system is used in a wide variety of applications. They include point-to-point aural links, digital terrestrial radio, direct satellite broadcasting of audio with video, DVD, digital television sound, and Internet audio applications.

MPEG Layer III The Layer III coding system is a multichannel coding system (e.g. Brandenburg *et al.*, 1996 and Stoll, 1996) using a hybrid filterbank composed of a 32-band polyphase subband filterbank followed by MDCT filterbank dividing the audio spectrum up into either 576 or 192 bands. The variable-frequency resolution in the MDCT is used to allow for a better trade-off in time–frequency characteristics of the filterbank for either steady-state or transient signals, respectively.

The Layer III system is optimized more for coding efficiency compared to Layers I and II. The higher-frequency resolution and the ability of the filterbank to adapt to the input signal allows for more efficient coding. A different perceptual model, the use of Huffman coding, and the addition of pairwise M/S multichannel coding further increase the coding efficiency of Layer III. As a consequence, the MPEG-1 portion of the bitstream is only allowed to use 32–320 kb/s of the total 1066 kb/s MPEG-2 data-rate limit.

The improved coding efficiency of Layer III relative to the two related layers has made it suitable for low data-rate point-to-point aural links, digital terrestrial radio, digital satellite radio and Internet audio applications.

MPEG AAC or Advanced Audio Coding
MPEG-2 Advanced Audio Coding is a family of multichannel coding systems designed for maximum coding efficiency and flexibility (e.g. Bosi *et al.*, 1997). These are embodied in three profiles called main, low complexity or LC, and scalable sampling rate or SSR. The main profile provides for the highest possible coding efficiency and makes the most complete use of the possible coding methods. Both the main and low-complexity profiles employ a MDCT-based filterbank having adaptable window shapes with a frequency resolution of 1024 or 128 equal bands, depending on whether steady-state or transient signal conditions occur. The SSR profile uses a hybrid filterbank consisting of polyphase quadrature

filters dividing the band into four equal bands, followed by a MDCT filterbank which further divides each quarter band into 256 or 32 bands. The basic filterbank resolution of all profiles is the same.

The low-complexity profile is a subset of the main profile in that it eliminates the use of the backward adaptive linear prediction of the MDCT values and has a reduced complexity temporal noise shaping. The SSR profile is separate from the other two profiles because of the nature of its filterbank, but is similar to the LC profile because it does not use the backward adaptive linear prediction and makes limited use of temporal noise shaping. Instead, it has a gain modulation system that varies the gain in each of the four subbands to control pre-noise audibility. The division of the audio band into four equal bands allows for reduced complexity decoding if one or more of the four bands of data are not processed.

All profiles use some version of temporal noise shaping to further improve the coding efficiency for signals that are not steady state or completely transient in nature, such as speech signals. Other methods used to improve the coding efficiency are a forward adaptive bit-allocation process driven from the encoder-based perceptual model, Huffman coding of scale factors and audio data, and efficient bitstream packing methods. Multichannel coding efficiency is further improved by a very flexible use of intensity multichannel and M/S pairwise encoding methods. The AAC family of profiles is also very flexible in that it allows 1–48 channels to be encoded, up to three low-frequency effects channels, and a maximum data rate of 288 kb/s per channel at a 48 kHz sample rate. This upper limit is based on an input memory requirement and is 6 kb/s per 1 kHz of sample rate. Operation at a data rate of as low as 8 kb/s has been demonstrated and this represents a practical lower limit. The AAC system operates at sample rates of 8, 11.025, 12, 16, 22.05, 24, 32, 44.1, 48, 64, 88.2, and 96 kHz and employs both matrix encoding for five-channel to stereo applications and a very flexible downmixing method.

The AAC is a relatively new coding system that is designed for very high coding efficiency. Applications in low data-rate point-to-point aural links, digital terrestrial radio, high-definition digital television sound, and Internet audio are being considered.

PAC or Perceptual Audio Coding This coding system is denoted by the generic name of this type of coding system. PAC, sometimes denoted MPAC (e.g. Johnston *et al.*, 1996), is a flexible multichannel coding system allowing for the transport of 1–60 channels, plus one to three optional low-frequency

effects channels. An MDCT filterbank similar to that used by the MPEG-2 AAC main profile coder is used, except that the adaptive window shape function is not present. The PAC system employs a forward adaptive bit-allocation method driven from an encoder-based perceptual model, Huffman coding of the filterband components, and M/S pairwise coding of the multichannel composite. Signals are encoded at data rates between 24 and 1024 kb/s and at a continuum of sample rates. This system has found use in Internet audio applications and is being considered for digital terrestrial radio.

1.5.6 Summary

Perceptual audio coders are the most effective way of transmitting and storing high-quality audio signals. Coders of this type allow bit rate reduction factors of greater than 12:1, with the preservation of the excellent sound quality properties of digital audio sound. The technical success of perceptual audio coders lies in the validity of the primary assumption used in their design, that the human auditory system is effectively modelled by a frequency-division process and masking theory. As a result, perceptual audio coders are being used in many transmission and storage applications.

The design of perceptual audio coders is a relatively new field and much work is being done to improve coding performance. Existing systems are being improved through encoder design innovations, while new systems are also being developed to further improve performance and flexibility. One such effort is the MPEG-4 development of an extremely flexible audio coding system utilizing a combination of music synthesis techniques, speech coding methods, ability to decode useful signals from parts of the bitstream, and the inclusion of further refinements in perceptual audio coder design. The perceptual audio coder part of this evolving standard is an extension of the MPEG-2 AAC system described previously.

References

BOSI, M., BRANDENBURG, K., QUACKENBUSH, S., FIELDER, L., AKAGIRI, K., FUCHS, H., DIETZ, M., HERRE, J., DAVIDSON, G. and OIKAWA, Y., ISO/IEC MPEG-2 advanced audio coding, *J. Audio Eng. Soc.*, **45**, 789–814 (October 1997)

BRANDENBURG, K., EBERLEIN, E., HERRE, J. and EDLER, B., Comparison of filterbanks for high quality audio coding, *IEEE ISCAS Conf., San Diego*, Vol. 3 (May 1992), pp. 1336–1339

BRANDENBURG, K., STOLL, G., DEHERY, Y.F., JOHNSTON, J.D., VAN DER KERKHOF, L. and SCHRODER, E.F., ISO-MPEG-1 Audio: A generic standard for coding of high-quality digital audio, *Collected Papers on Digital Audio Bit-Rate Reduction*, Audio Engineering Society, New York (1996), pp. 31–42

CAINE, C.R., ENGLISH, A.R. and O'CLAREY, W.H., NICAM 3: Near instantaneous companded digital transmission for high-quality sound programs, *Radio Electron. Eng.*, **50**, 519–530 (October 1980)

CARTERETTE, C. and FRIEDMAN, M.P., *Handbook of Perception*, Academic Press, New York (1978), pp. 305–319

CCITT, '7 kHz audio-coding within 64 kbits/s', *General Aspects of Digital Transmission Systems: Terminal Equipment*, Recommendation G.722 (1988)

FIELDER, L.D., Evaluation of the audible distortion and noise produced by digital audio converters, *J. Audio Eng. Soc.*, **35**, 517–534 (July/August 1987)

FIELDER, L.D., BOSI, M., DAVIDSON, G., DAVIS, M. TODD, C. and VERNON, S., AC-2 and AC-3: Low-complexity transform-based audio coding, *Collected Papers on Digital Audio Bit-Rate Reduction*, Audio Engineering Society, New York (1996), pp. 54–72

FLETCHER, H., Auditory patterns, *Reviews of Modern Physics*, **12**, 47–65 (January 1940)

GRAY, H.M., Vector quantization, *IEEE ASSP Magazine (renamed IEEE Signal Processing Magazine)*, **1**, No.2, 4–29 (April 1984)

HUFFMAN, D., A method for the construction of minimum redundancy codes, *Proc. IRE*, **40**, 1098–1101 (September 1952)

JOHNSTON, J.D., SINHA, D., DORWARD, S. and QUACKENBUSH, S.R., AT&T Perceptual Audio Coding (PAC), *Collected Papers on Digital Audio Bit-Rate Reduction*, Audio Engineering Society, New York (1996), pp. 73–82

IWAKAMI, N. and MORIYA, T., Transform-domain weighted interleave vector quantization, *Audio Engineering Society 101st Convention*, Preprint 4377 (November 1996)

MOORE, C.J. and GLASBERG, B.R., Formulae describing frequency selectivity as a function of frequency and level, and their use in calculating excitation patterns, *Hearing Research*, **28**, 209–225 (1987)

PRINCEN, P. and BRADLEY, A.B., Analysis/synthesis filter bank design based on time domain aliasing cancellation, *IEEE Trans. on Acous., Speech, and Sig. Proc.*, **ASSP-34**, No. 5, 1153–1161 (October 1986)

RABINER, L.R. and SCHAFER, R.W., *Digital Processing of Speech Signals*, Prentice Hall, New York (1978), pp. 225–240

ROTHWEILER, J.H., Polyphase quadrature filters, *Proceedings of ICASSP-83*, pp. 1280–1283 (1983)

SMYTH, S.M.F., SMITH, W.P., YAN, M. and JUNG, T., DTS coherent acoustics. Delivering high quality multichannel sound to the consumer, *Audio Engineering Society 100th Convention*, Preprint 4273 (May 1996)

STOLL, G., ISO-MPEG-2 audio: a generic standard for the coding of two-channel and multichannel sound, *Collected*

Papers on Digital Audio Bit-Rate Reduction, Audio Engineering Society, New York (1996), pp. 43–53

TODD, C. and GUNDRY, K., A digital audio system for broadcast and prerecorded media, *Audio Engineering Society 75th Convention*, Preprint 2071 (March 1984)

TSUTSUI, K., SUZUKI, H., SHIMAYOSHI, O., SONO-HARA, M., AKAGIRI, K. and HEDDLE, R.M., ATRAC: Adaptive Transform Acoustic Coding for MiniDisc, *Collected Papers on Digital Audio Bit-Rate Reduction*, Audio Engineering Society, New York (1996), pp. 95–104

WYLIE, F., apt-X100: low-delay, low-bit-rate subband ADPCM digital audio coding, *Collected Papers on Digital Audio Bit-Rate Reduction*, Audio Engineering Society, New York (1996), pp. 83–94

ZWICKER, E. and TERHARDT, E., Analytical expressions for critical-band rate and critical bandwidth as a function of frequency, *J. Acoust. Soc. of Am.*, **68**, 1523–1525 (November 1980)

1.6 Noise measurement

1.6.1 Introduction

Noise is an everyday phenomenon affecting almost everyone. Its effects are numerous and often pass unnoticed, sometimes for many years. However, the result of prolonged exposure to noise can be devastating, producing premature hearing loss to varying degrees. Accurate measurement of noise is therefore essential, but there are other benefits; most people prefer a quiet environment, which seems to improve their general well-being and induces a more relaxed approach to life. The measurement of noise is therefore becoming important and the number of instruments and methods available for doing this is growing continuously. Intelligent use of this equipment often results in noise-reduction programmes which benefit wide sections of the community.

1.6.2 Fundamental properties

The simplest property measured by most noise-measuring instruments is the *sound pressure level* (SPL). Audible pressure fluctuations are measured in pascals (Pa; equivalent to newtons per square metre (N/m^2)) and are usually much less than 1 Pa. Standard atmospheric pressure is 101.3 kPa, so it can be seen that the ear and all other acoustic measuring equipment has to measure extremely small variations about this figure; however, these are vital to everyday life. The *frequency* of the noise is also important: most teenagers can hear frequencies from 20 to 20 000 Hz, although increasing age reduces the

higher figure considerably. Often the lowest frequencies are detected as much by sensing the vibrations they cause rather than their audible effects.

The human ear is not equally sensitive to all frequencies; those from around 300 to 6000 Hz are the best perceived and, therefore, appear loudest, whilst higher and lower frequencies of equal amplitude seem less obvious.

A frequency weighting was introduced in an attempt to measure noise in the same way that the ear perceives it and this is called the *A weighting curve*. Measurements made using this weighting are often indicated as dB(A), but increasingly where measurements of environmental noise are made, the use of A weighting is assumed, and the readings taken are just given in decibels (dB). The response of the A weighting network is shown in *Figure 1.69*.

Other weightings exist; none has the universal usage of the A weighting, although C weighting is specified in some standards, particularly in the USA. Almost every meter measuring SPL will give a reading in dB(A), which for a pure tone of, say, 125 Hz at pressure level of 90 dB will read 73.9 dB(A) (from *Figure 1.69*).

In addition to SPL, many other parameters have been devised in an attempt to produce a single-number result for a particular type of noise. Integration of the SPL with time, to produce the averaged SPL over a given time interval, is one of the most widely used and is known as the *equivalent continuous level* (L_{eq}):

$$L_{eq,T} = \log_{10}\left\{\left[1/T \int_{t_i}^{t_2} p^2(t)dt\right] \Big/ p_0^2\right\} \text{ (dB)} \quad (1.1)$$

where: $L_{eq,T}$ is the equivalent continuous SPL with reference to 20 μPa, determined over a time interval $T = t_2 - t_1$; $p(t)$ is the instantaneous sound pressure; and p_0 is the reference sound pressure of 20 μPa. The frequency weighting may also be incorporated into the symbol, for example $L_{Aeq,T}$.

Because the decibel is a logarithmic unit, higher levels have a much greater effect on the L_{eq} value; but this accords well with the effects of noise and is widely used in assessing the amount of noise that a person is exposed to during, say, a working day.

Another way of assessing noise exposure is to measure the acoustic energy received during a given period. This is referred to as the *sound (or noise) exposure* and can be expressed in units of pascal-squared hours (Pa^2h), or more commonly as a percentage of a maximum allowed exposure. It is frequently referred to as the *noise dose*. An exposure

dB

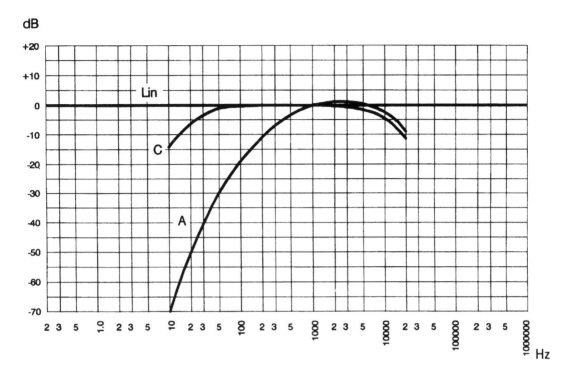

Figure 1.69 *Lin (flat), A and C frequency weightings.*

to a level of 90 dB for 8 h produces a sound exposure of 3.2 Pa^2h, which is commonly taken as 100% of the allowed dose (but see Section 1.6.3 for more details). Both the time interval and the level must be defined, and although the period of 8 h is very common, several different levels can be used for the 100% criterion; 85 and 90 dB are the most common, but it is essential to know these two parameters when making an assessment by this method. Instruments for measuring exposure in this fashion are called *dosemeters, dosimeters* or *personal sound exposure meters* and vary from simple one-answer models to sophisticated units which can give answers for every minute of the exposure time to enable identification of periods of high noise in what might otherwise be a generally quiet environment.

1.6.3 Measuring instruments and techniques

There is no standard measuring instrument or technique suitable for all types of noise. Selecting the most appropriate equipment and method requires experience, but the most popular types of instrument and their uses are described here.

1.6.3.1 Sound-level meters

A sound-level meter consists of a microphone connected directly or via a cable to some electronics which will display a reading of the absolute SPL relative to 20 µPa (deemed to be the quietest sound audible to humans, and used as the reference pressure for acoustics). Thus a pressure of 1 Pa will be displayed as a level of 94 dB.

Four grades of sound-level meter (types 0, 1, 2 and 3) are defined in the International Standards covering these instruments. Type 0 is the highest standard and is rarely encountered, as it is mainly intended for laboratory work. Type 1 is for precision measurements, type 2 for general purposes and type 3 is for sound surveys. In practice, only types 1 and 2 are widely available and the choice of a precision or general-purpose meter reflects the accuracy that is required of the measurement being undertaken.

The most widely recognized standard for sound-level meters is International Electrotechnical Commission (IEC) Standard IEC 651: 1979, which has been renumbered by many national standards bodies (BS EN 60651:1994 in the UK) and is very similar to many national standards, notably ANSI S1.4-1983 (R1994) (American National Standards Institute). Any meter suitable for recognized

measurements to a defined accuracy will quote a standard type (or class). This should be displayed on the meter itself. Failure so to do usually means that the instrument does not conform to any standard and should be used with caution. Instruments which claim compliance only with the obsolete standards IEC 123 and IEC 179 should, likewise, be avoided. The requirements of accuracy are detailed in the standards quoted, but for type 1 meters the overall instrument will be within 0.7 dB and type 2 meters will be within 1 dB. The dynamic span of modern meters is usually around 60–70 dB, with a measuring capability from around 20 dB(A) to 140 dB(A) in a number of ranges. This covers noise from the background levels in the countryside at night to jet engines a few metres away, and suffices for most applications.

In addition to measuring the SPL, many instruments have a maximum-hold facility which retains the maximum level measured during the period; this may be useful in identifying an occasional noise in an otherwise uniform noise climate.

All sound-level meters have a dynamic characteristic or time constant which performs a small degree of averaging on the instantaneous sound pressures. These time constants are universally identified as *fast, slow, impulse* or *peak*. 'Impulse' and 'peak' are self-explanatory and are little used for most noise measurements. 'Fast' averages over a 1/8-s period; whilst 'slow' averages for 1 s. Certain types of measurement stipulate that 'fast' or 'slow' will be used, but for general-purpose use 'fast' is recommended as the meter most closely follows the noise contours and the maximum levels captured are not subject to excessive smoothing.

To measure noise with such a meter, the microphone should be pointed at the perceived source of the noise, keeping the influence of the operator of the meter to a minimum by standing well behind the microphone and, if possible, having at least 1.2 m of clear space in all directions from it, especially with respect to the ground. This is often achieved by placing the meter or microphone on a tripod. Sound-level meters are used for instantaneous readings of fairly short duration, or to sample a noise that is fairly constant.

1.6.3.2 Integrating – averaging sound-level meter

This usually has all the properties of a sound-level meter but, in addition measures L_{eq}, the integrated, averaged SPL. Meters providing this type of measurement are graded in the same fashion as non-integrating meters. but should also conform to IEC

804: 1985 (variously renumbered; BS EN 60804: 1994 in the UK: ANSI equivalent S1.43 in preparation). With the L_{eq} values, the meter can be used to examine the noise climate over longer periods, and to build up a picture of the noise levels experienced throughout the day. This is useful in monitoring environmental noise to assess whether certain sounds constitute a nuisance, or whether they are within the type of noise that the area regularly expects. Some meters can operate and produce answers for set periods, e.g. 15 min, and store each consecutive answer for subsequent analysis. Noise nuisance is covered in Section 1.6.5.

1.6.3.3 Personal sound exposure meters

These instruments, often called *noise dosemeters* or *noise dosimeters*, are designed to be worn during the normal working period, and so measure the noise exposure to which the worker is exposed during his or her working day. Unlike the SPL, this is not an instantaneous measurement but one which combines level and time; the units of sound exposure are pascal-squared hours (Pa^2h), or sometimes pascal-squared seconds (Pa^2s).

The modern sound exposure meter is usually small, lightweight and designed to be worn without impeding the user in any way. It usually complies with national legislative requirements (BS 6402: 1994; ANSI S1.25-1991; IEC 1252: 1993). The microphone is usually mounted on a clip which should be located close to the ear, the recommended distance and mounting method varying from one measurement standard to another; ISO 1999: 1990 recommends 10 cm. Simple instruments give a one-number answer at any time during the measurement period, but more advanced units can store the information in small intervals, typically 1 min, and thus give a profile of the noise exposure directly related to time. These are usually referred to as *profiling* or *logging dosemeters* and often contain sophisticated analysis and report capabilities to enable the acoustician to examine the circumstances of the noise exposure, for example whether it was a steady noise all day or a short period of very high level noise. The noise dose or exposure may give the same answer on a simple instrument, but it could be shown that by preventing the worker from being involved with the high level signal his or her daily exposure would be within permitted limits.

The maximum daily exposure is usually defined nationally, and the amounts vary considerably from country to country. Many of them are not specified in Pa^2h at all, but by a different method entirely. In

(a)

(b)

Figure 1.70 *(a) Comparison of the noise dose in Pa²h(E_A) and percent (D) with the L_ep,d for a period of 8 h and a criterion level of 90 dB. (b) An integrating type-1 sound-level meter, fitted with an octave band filter set, suitable for measuring in accordance with BS4142: 1997.*

order to be related to sound levels (in dB), the sound exposure is defined as a percentage of a given level for a given time, e.g. a continuous level of 90 dB for 8 h may be defined as 100% of permitted exposure. Sound exposure meters therefore calculate correctly by this method only if the defined level, usually called the *criterion level*, is correct for the country or area under consideration. In the UK, 90 dB is usually used, but many countries, including the USA adopt 85 dB, although the reference period is nearly always 8 h. The 8-h normalized SPL is often referred to as $L_{ep,d}$ and the 8-h normalized L_{eq} as $L_{eq,8hn}$.

Figure 1.70(a) shows the relationship between Pa²h, dB, and per cent dose for a 90-dB criterion level and an 8-h period.

It can be seen from *Figure 1.70(a)* that 90 dB for 8 h equals 3.2 Pa²h, and this could be the recommended *maximum daily personal sound exposure*. Allowing a worker to sustain an exposure in excess of this figure is usually an offence, and the employer could be subject to legal action if no remedial action is taken to ensure that the worker is not repeatedly exposed to this noise dose. Remedial action may take the form of reducing the noise level or the time during which the worker is exposed to it. The use of hearing protectors may reduce the level to workers' ears, but the exact effects must be deduced by complex calculations, as a personal sound exposure meter cannot measure accurately inside the protector.

1.6.3.4 Environmental analysers
Instruments of this group are used for medium- or long-term monitoring. Essentially they consist of a sound-level meter with data acquisition and processing capabilities, and often an integral printer to list the data calculated. Often they can be left unattended for periods of 7 days or more, and some have facilities for sending data over telephone lines when required.

Because of the need to be left in all weathers, these instruments are usually found in a protected enclosure, and the measurement microphone is mounted separately on a tripod with sun, wind and rain deflection. If extremes of weather are expected there is often provision for the microphone to be heated to prevent condensation or icing which would impair the accuracy. Most analysers will perform to full specification over temperatures from $-10°C$ to $+50°C$.

Although they can usually measure the SPL, few record this data directly, as the amount of information storage required is huge. The integrated-averaged level now starts to be the main unit of measurement (L_{eq} or L_{avg}) usually for a user-selected period, typically 1 h. Thus all the SPL data for this period are reduced to a single number. Comparison of these numbers for each hour of the day gives an

averaged value for the noise at that location on a manageable basis.

Another parameter usually measured by environmental analysers is the noise distribution. This is presented in a number of different ways, but is always derived from the sampled SPL. The SPL is measured many times per second, and the level is put into a histogram divided into steps, often 0.5 or 1 dB wide, covering the entire dynamic range of the analyser. By knowing the measurement period and the sampling rate, the instrument can calculate the percentage of time for which a given SPL was exceeded, the level in decibels being referred to as L_n, where n is the percentage. Thus L_{10} is the decibel level which was exceeded for 10% of the measurement period. Commonly used numbers are L_1 (near maximum but excluding very occasional high levels), L_{10}, L_{50} (average value, often similar to L_{eq} but not the same, since L_{eq} is a logarithmic average), L_{90} and L_{99} (background noise level). With this information, unusual or interfering noises can often be identified and their effects quantified, either by period or by day. Most analysers provide the means both to perform calculations and simultaneously to save the answers for subsequent analysis.

Environmental analysers can usually perform self-calibration at regular intervals to show that the instrument remains in working order whilst unattended. They may often be set to recognize certain noise events (distinguished by level) and measure these individually. This is particularly useful when known noise sources such as aircraft are being investigated. Many permanent *airport-noise monitoring systems* are based on an environmental analyser set to this mode of operation. The parameter often measured is called the sound equivalent level (SEL previously L_{ax}):

$$\text{SEL} = L_{eq} + 10\log_{10}(T) \qquad (1.2)$$

where T is the duration of the event in seconds.

The frequency weighting may again be added to the symbol. For example, an A-weighted SEL is given by:

$$\text{SEL} = L_{EA,T} = L_{Aeq,t}$$
$$+ 10\log_{10}(T/T_0)(\text{dB}) \qquad (1.3)$$

where T is the time interval and T_0 is 1 s.

This produces a result for each event which is normalized to 1 s, thus giving a comparison between events of different character, e.g. a noisy aircraft flying past quickly will give a similar result to a quieter plane flying more slowly. In order to recognize the

aircraft a threshold is often set, below which all SPL readings are ignored. Once this threshold has been exceeded, usually for a predetermined time of say 5 s, the analyser logs all the data until the SPL falls below the same threshold. The first few seconds of data and their time must be included in the calculation to obtain an accurate value, so a short buffer of SPL data is required in the analyser if it offers this mode of operation.

Over the years, other methods of quantifying noises which annoy people have been suggested, and instruments have been designed to measure them, so environmental analysers often have other selected measurement types available. No method has gained universal acceptance or standardization, however, and these features should therefore be used with caution.

1.6.3.5 Sound-intensity meters

The direct measurement of sound intensity is relatively new. This is partly because the technology required to manufacture such a meter has been difficult to achieve economically, and partly due to a lack of realization of its uses. The advantage of intensity over SPL is that it gives the *direction* of the sound source as well as its level. For simple, single-source environments this is unnecessary, but to evaluate a number of noise nuisances which are close together, and to deal with each one individually, it is invaluable. Most intensity meters are also supplied with frequency-analysis equipment, most commonly dividing the audio spectrum into 1/3 octaves (an octave is a doubling of frequency).

A sound-intensity probe usually consists of two precision microphones, selected for near-identical performance, mounted a fixed distance apart. The intensity meter measures the amplitude and phase response from each microphone and performs calculations to determine both the amplitude and direction of the source. With the frequency analysis divided into bands, different sources (at different frequencies) in varying directions can be quickly identified. Most intensity probes currently available cannot cover the entire audio spectrum in one measurement, as the spacing between the microphones only allows for a limited band of frequencies to be encompassed. This is usually overcome by supplying several microphone spacers, and the user must either fit each in turn to cover the full frequency range, or have some advance knowledge of the area of the spectrum required.

Intensity meters are partly covered by IEC 1043: 1993 (BS EN 61043: 1994). The differences between

alternative probe designs and the corresponding electronics make any but the most general description of these units of little value.

1.6.3.6 Microphone types

Condenser (capacitor) microphones are usually used for the more accurate grades of sound-measuring equipment. Their long-term stability and relatively small temperature coefficient make them the most suitable type.

Lower grade instruments, for example type 2 sound-level meters, are increasingly employing *electret microphones* since these require no polarization voltage circuitry. Long-term stability problems with early electret material have now been largely overcome.

Other types of microphone which have been used include dynamic and piezoceramic designs, although these are becoming less common.

In the UK and throughout most of Europe microphones with a flat *free-field response* are used; in the USA a *random-incidence response* is required. Pressure microphones are generally used only in the laboratory, for example in audiometer calibration.

For further details on microphone types see Bies and Hansen (1988a) and for sound fields see Bies and Hansen (1988b).

1.6.4 Industrial noise

The phenomenon of *temporary threshold shift* is well known most people have experienced the transient 'deafness' caused by loud sounds in industrial complexes, pop concerts, etc. However, exposure to noise can also produce a permanent threshold shift, i.e. hearing damage.

1.6.4.1 Hearing damage

Hearing loss has many possible causes, not all related to noise problems. These include perforations, calcification (joining together of the bones in the ear), fluid in the ear, or simply an obstruction of the ear canal. *Presbycusis*, a natural effect increasing with age, causes loss of high frequencies. It affects men more than women, and there is some evidence that it may be ethnically dependent. Failure of the hair cells, by far the most common type of damage caused by excessive noise, may also be caused by common ototoxic chemicals.

The diagnosis of work-related deafness is further complicated by leisure activities which expose people to a considerable portion of the maximum allowable noise dose. It is thought that there is now a

generation in which a majority have some hearing damage due to rock concerts and, more particularly, to the wearing of headphones with the volume set excessively high.

Occupational deafness is regrettably common, particularly amongst older people who were subjected to high noise levels before regulations were enforced. Failure of the hair cells is very difficult, if not impossible, to treat because conventional hearing aids are of little use. They merely amplify the sound, but if the patient has no hearing mechanism the result is still unacceptable.

Apart from simple hearing loss, noise-induced impairment may take several other forms, the two most common being tinnitis and recruitment. *Tinnitis*, usually caused either by a very loud impulsive sound or a blow to the ear, has the symptom of continuous noise in the affected ear. This is often a 'ringing' sound but may also be a buzz or a wind-like sound, and often varies in apparent loudness from day to day. In severe cases it may not only be a general nuisance, but can also prevent the sufferer from sleeping. Often the only relief is provided by masking with a more pleasurable sound such as music. Unfortunately, in severe cases the level required may be so high that there is a risk of further hearing damage. *Recruitment* is another otological problem, the symptom of which is a small range of levels between the inaudible and the unbearable. This can make life extremely difficult for the sufferer and those around him, or her.

1.6.4.2 Stress

The effects of noise are not confined to hearing damage. It can also trigger medical problems normally associated with other forms of stress, and severely affect concentration. If the latter interferes with the work process, this can in turn lead to further stress-related problems due to the difficulty in keeping up productivity. However, since these problems have so far not been quantified, the noise regulations relate purely to hearing damage.

Away from the workplace, similar effects can be found. The recognition that noise is not only a nuisance but also has a detrimental effect on the quality of life has resulted in, for example, grants being given for sound insulation in homes near airports and motorways.

1.6.4.3 Regulations relating to noise in the workplace

Most industrialized nations now have legal requirements stipulating the maximum noise exposure to which employees may be subjected. These commonly

set both a maximum SPL and a maximum daily sound exposure, usually normalized to an 8-h day. This reflects the fact that hearing can be damaged both by a very loud sound of even short duration, and by lower sound levels that occur over prolonged periods. The maximum SPL allowed in the UK is 140 dB (unweighted), which is equivalent to 200 Pa. At present the sound exposure limit is equivalent to an L_{eq} of 90 dB(A) over an 8-h day. However, since it has been found that exposure to this maximum may result in approximately a quarter of subjects sustaining some hearing loss, some countries such as the USA have introduced a limit of 85 dB(A). It should be noted that maintaining sound levels below the maximum limit does not necessarily confer immunity from legal liability if an employee still suffers hearing loss. This shows the need for a system of regular audiometric testing where the noise levels even approach the limit.

Recommended measurement methods vary widely from one country to another, but there is now some consensus within Europe, at least, with the publication of ISO 1999: 1990. This gives measurement details and methods for predicting the *noise-induced permanent threshold shift* (NIPTS). It also gives details of the 'average' otologically normal person used as a norm for audiometric testing, and of the age-related corrections for male and female subjects.

In the UK the regulating body is the Health and Safety Executive (HSE); in the USA this function is performed by the Occupational Safety and Health Administration (OSHA).

1.6.4.4 Weighting curves

Since the only reason for measuring sound is related to its effect on people, it is logical to attempt to make measuring instruments respond in a similar manner to the human ear. As explained in Section 1.4, the ear is an extremely complex device. Perceived loudness, as opposed to the actual SPL, is dependent on many factors, including the frequency spectrum of the sound. Hearing loss is, to a great extent, dependent on the same weighting factors as perception. A sound-level meter with linear weighting (called 'flat' in the USA) does not, therefore, mirror human hearing, and is not generally used for measurements relating to hearing loss.

A weighting is generally used, although C weighting is specified in some standards (see Section 1.6.2).

1.6.4.5 Hearing-conservation programmes

A hearing-conservation programme may be instigated and should contain at least the following six points:

assessment; reduction of noise at source; reduction of exposure; regular monitoring of noise levels; audiometric testing of employees; and an employee-education programme.

Assessment Whilst it is technically quite correct to begin the noise assessment using highly sophisticated equipment and techniques, in practice it is often better to start with simpler measurements. It is, however, vital that the equipment is correctly designed, calibrated and used. For a description of the various types of instrumentation available, see Section 1.6.3.

Consideration must first be given to the nature of the problem and the circumstances under which the employees are exposed to noise. Is the noise relatively constant, widely varying, cyclic, or impulsive? Does the employee remain in the same place during the day? Does the sound source move past the employee (e.g. vehicles passing)?

If both the sound source and the employee are stationary, an initial survey using a simple (but accurate) sound-level meter measuring dB(A) is often used to determine whether a problem is likely to be present. Firstly, it will indicate (if not already obvious to the listener) whether the noise is of a reasonably steady nature or is widely varying. Secondly, if even the highest momentary noise levels indicated do not exceed, say, 85 dB(A) then the average SPL cannot exceed a limit of 90 dB(A). No further measurements would then be necessary.

If such a simple survey indicates that a problem may exist, then the equivalent continuous noise level L_{eq}, must be determined by using an integrating sound-level meter. Averaging must be carried out over a period long enough to obtain a steady reading over a number of duty cycles of the equipment causing the noise.

Where the noise source and/or employee is mobile, a personal sound exposure meter should be used to record sound exposure, either directly or as a percentage of a maximum allowable dose. Sound exposure E depends on both the SPL and the exposure time, and is a measure of the sound energy received:

$$E = \int_{t_1}^{t_2} p^2(t)\mathrm{d}t \qquad (1.4)$$

$$E = (p_0{}^2 T)(10^{0.1 \times L_{eq}}) \qquad (1.5)$$

where: $p^2(t)$ is the square of the instantaneous sound pressure as a function of time t for an integration period starting at t_1 and ending at t_2; and T is the averaging time.

Reduction of noise at source It is better to reduce the noise at source, using all reasonable methods, than to rely on ear protectors. Noise reduction will generally be the province of a specialist consultant, but can often be achieved by reducing vibration, fitting sound insulation or baffles, moving noisy plant to an unmanned location, or fitting sound-absorption panels to the room. Changing the process or the duty cycle of the machinery may also help.

Reduction of exposure The sound exposure experienced by a machine operator, for example, is directly proportional to the time for which he or she is exposed to the noisy environment. It may therefore be possible to alternate two employees between tasks in noisy and quiet areas, perhaps changing over at lunchtime; the daily sound exposure for each may then be below the legal maximum.

Failing this, *personal ear protectors* may be prescribed, although they should be a last resort in the noise-reduction effort. There are several problems which make their use undesirable. First, their efficiency depends not only on their design but also on the way they fit to the ear. Since no two people have the same shaped ears, it must follow that some people will receive less protection than the average figure quoted. Second, people are generally reluctant to wear ear defenders due to their discomfort and the loss of 'wanted' sound such as speech. Headbands may be adjusted by the employee to minimize discomfort, thereby rendering the ear protectors partially or totally ineffective; or they may simply be removed for much of the time. Third, there is some evidence that the protracted wearing of these and similar devices such as headphones, particularly in warm environments, may cause medical problems such as bacterial ear infections.

Two basic types of ear protector are used: circumaural devices (fitting around the ear) usually called 'ear defenders' or 'muffs'; and ear inserts (commonly called 'earplugs'). They are specified in BS EN 352–1: 1993 and BS EN 352–2: 1993 respectively. Circumaural devices generally provide the greatest attenuation and rely on a good seal around the ears, with the ear cup being pressed against the head by a sprung headband. They are specified in BS 6344: 1984. Ear inserts rely on a good seal into the ear canal. Both types of protector must be correctly fitted for the particular employee concerned, and trained personnel are needed to do this. Standard 'one-size' earplugs from a local 'do-it-yourself' store cannot be expected to perform adequately for employees who are habitually exposed to high sound levels.

Regular monitoring of noise levels The methods used for regular monitoring are the same as those described for the initial assessment.

Audiometric testing of employees This can only be carried out sensibly by a trained audiologist. The tests are used to compare the subject's hearing with that of an 'average' otologically normal person, and to monitor changes in their hearing. An initial measurement, known as the 'baseline audiogram', is therefore taken at the earliest opportunity and later results compared with it. Where the working environment is known to be noisy the baseline *audiogram* should be produced before employment in the noisy area commences. In all cases the subject should not have been exposed to high noise levels for a period immediately before the test, so that temporary threshold shift is eliminated. This can usually be achieved by testing on arrival at work in the morning - provided that the employee has not been using a personal stereo on the way to work with the volume turned up.

Audiology, the scientific assessment of hearing, requires that repeatable tests may be carried out over long periods of time and over a wide selection of human subjects. However, this is not an easy subject in which to perform objective experiments. Any test of hearing must be subjective, and people vary widely in their perception of given sounds. Furthermore, it is unacceptable to perform experiments on live subjects in which, for example, a sound level is increased until hearing loss occurs. A certain amount of knowledge about the geometry and effects of the ear has been gained from experiments using ears from corpses, but tests of hearing are confined to those who have already suffered impairment. Much work has been done to determine the 'average' response of the human ear, and this is generally used as a baseline.

Several forms of hearing measurement have been developed, the most common of which is pure-tone audiometry. Pure (sine wave) tones of different frequencies are played to the subject over headphones at known SPLs, and the lowest level which can be heard is recorded. The tones may also be presented via a bone vibrator held against the mastoid bone, this method being referred to as 'bone-conduction audiometry'. Pure-tone audiometers are specified in IEC 645–1: 1992 (BS EN 60645–1: 1995). Relevant audiometry methods are specified in ISO 6189: 1983 (equivalent to BS 6655: 1986 (1993)) and ISO 8253–1: 1989 (rejected by the UK as inconsistent with recommended practice).

Sound-field audiometry attempts to produce a more realistic situation, in which the subject is

exposed to free-field rather than pressure situations. The subject is seated in a room which has suitable acoustic properties, surrounded by loudspeakers. These are often in a tetrahedral arrangement with the subject at the centre. The appropriate tones (sinusoidal or narrow band random noise) are then played over the loudspeakers, and again the minimum audible level is recorded. ISO 8253–2: 1992 (BS 7636: 1993) covers this type of audiometry.

Speech audiometry, whilst relatively new, is currently being researched as another useful tool for the audiologist. It uses 'standard' recordings of word lists, replayed in standard acoustic rooms at set levels. Since the intelligibility of speech depends also on such parameters as local dialect and accent, the results may vary from area to area, and it is necessary to compile sufficient data to calculate a local norm. Note that speech audiometry is impossible if the subject does not speak the language in which the recording has been made. IEC 645–2: 1993 (BS EN 60645–2: 1997) covers this field as will ISO 8253–3.

Employee-education programme This is a frequently overlooked part of a hearing-conservation programme. Noise and its effect on people are not generally well understood, and people are frequently unaware that their hearing can be irreparably damaged. The importance of using ear defenders should be explained. The different types, their suitability and, perhaps most important of all, how they should be used must be made clear. There is a natural reluctance to place anything over, much less in, the ears and this must be overcome if a hearing-conservation programme is to succeed. Note that in some real-life situations it is possible that an employee who removes his or her ear defenders for 5 min during each day can receive a noise exposure far in excess of the permitted maximum.

1.6.5 Noise measurement and nuisance assessment

Noise measurement has enjoyed considerable attention from both equipment designers and acousticians for the last 50 years or so, with ever-increasing accuracy and detail for all types of noise. However, the assessment of the nuisance value has proved more difficult to quantify. Research programmes have been conducted in an attempt to specify a measurable index for nuisance or its effects, but none has been universally successful. One of the pioneering documents in this field has been BS 4142, first published

in 1967. This was titled *Method for Rating Industrial Noise Affecting Mixed Residential and Industrial Areas* and introduced the principle of adjusting actual sound-level readings by given amounts depending on the type of noise being created and when it was occurring, e.g. a factory operating at 6 a.m. could have 5 dB added to all readings made of its noise due to the time of day and its effects on sleeping residents nearby. This standard has been used, with some minor variations, until very recently, and was also adopted in the main by other Standards Bodies.

In 1990 and 1997, major revisions of BS 4142 were published under the same title and made significant improvements. Accurate and repeatable calibration of all equipment was made mandatory, and L_{eq} became the principal noise descriptor.

The microphone position is specified as 1.2–1.5 m above the ground and 3.5 m from walls and other facades. If measuring at heights significantly above ground level, then 1 m from the facade is acceptable. These distances are almost universally accepted as satisfactory conditions for the measurement of noise in the open air and can be applied to most types of environmental-noise measurement.

Details of measurement methods specified in BS 4142: 1997 are too lengthy for inclusion here, but seek to define what the background level would be *without* the obvious noise source, and then determine the level including the source. Corruption of the measured value of the specific noise (that from the source being investigated) when its level is close to the measured background is also addressed. In certain circumstances alternative measurement locations are seen as a solution, substituting a background measurement at a similar location without the source.

External influences, such as weather conditions and electrical interference, must be documented. Ground cover and other variables that affect sound propagation, and hence the noise arriving at the facade, must be considered. Options are also given to ensure that the measured value for the specific noise is not corrupted by high background noise. These include corrections to be applied to the specific noise level which mirror the corrections applied when decibel levels occur together.

The background level with which the specific noise source is to be compared may only be determined by measurement, hence the use of alternative locations, but it is necessary to provide a rationale for the choice of an equivalent site.

The specific noise level having been determined, it is corrected to give the rating level. The assessment

for complaint or nuisance purposes is based on the arithmetic difference between the rating level and the background level.

Most of BS 4142: 1997 is relevant to any outdoor noise measurement and can be recommended as a safe procedure in many applications, notably road and entertainment noise, not just those listed in its title.

In the UK the Noise Act 1996, with its 1997 amendment, applies to noise from and within dwellings at night. It creates a new night noise criminal offence with a £100 fixed penalty and allows for seizure of the equipment causing the disturbance. Persistent or gross offenders may face prosecution and fines of up to £1000.

1.6.5.1 Airport noise

One of the major items of noise nuisance today is that produced by jet aircraft taking off or landing. Most airports have monitoring equipment strategically placed to measure noise in locations which give rise to the best average readings. This can be misleading, as it assumes that all aircraft follow exactly the same flightpath and that the weather has no variable effect, neither of these assumptions being true. Elaborate computer programs have been written to offset these problems, even trying to track the flight with radar and correcting for deviations from a constant flightpath. However the principal factor remains not with the airport but with the manufacturer of the aircraft and the pilot's handling of it.

Aircraft-noise measurement is usually performed using an unattended device which is left to attempt to recognize aircraft amongst the other prevailing noises. Often this is straightforward, but the less obvious the noise against the background, the harder

it is to obtain accurate identification, although it could be argued that in this case it is not necessary, since the aircraft is not intruding significantly. This is where instruments providing an SEL measurement parameter come into their own and provide the most useful measure of the noise. This enables automatic aircraft recognition if (referring to *Figure 1.71*) the lower threshold is set to a value somewhat above the average background level of the measurement site, but lower than most of the aircraft passing by (a few hours of initial observation will give this value). The meter will be set to ignore all readings below this threshold and to monitor only when it is exceeded.

By introducing a delay of a few seconds into the recognition of an 'event', short noises such as thunder, car horns, etc., can be eliminated and only those which remain above this threshold for this period of time (3–7 s are typical values) can be considered as possible aircraft. A second threshold may be set at a level which the airport considers excessive, and thus aircraft overflights can be divided into 'acceptable' and 'unacceptable' for the location in question. When the level falls below the first threshold, usually also for a few seconds to ensure the entire 'event' has been measured, the system ceases measuring and calculates the SEL for all the time and energy that were above this threshold level. Thus a single number is used to quantify each overflight. In addition to this number, the maximum value recorded during the event, and its exact time and duration are often logged in order to relate the noise to a particular aircrafts. As an additional identifier of aircraft, a maximum duration time, based on the average time it takes an aircraft to pass the monitor, may be included. Noises that continue above the first threshold for significantly longer than this time are unlikely to be aircraft (with the possible

Figure 1.71 *Noise-event definition.*

exception of hovering helicopters) and can also be ignored in aircraft-recognition terms.

Some airports impose financial penalties on aircraft that exceed their specified second threshold, as this is often adjusted to equate to a noise level which brings complaints from local residents, but no standardization exists for this. Aircraft manufacturers spend large amounts of time and money both measuring and attempting to quieten their aircraft, and it is here that the solution to a quieter environment lies.

1.6.5.2 Vehicle and domestic equipment noise

In some countries, standards exist for measuring the noise emitted by a vehicle when driven past a test site under given conditions. Other countries test with the vehicle stationary by running the engine at a fraction of the maximum r.p.m., usually 75%, and noting the maximum level obtained at a fixed position relative to the exhaust pipe. Equipment for performing these tests is available but often only to meet a particular country's requirements. In the UK, BS 3539: 1986 specifies the requirements for measuring instruments, but not how to use them. Permitted noise levels for different types of vehicle are stipulated, usually by national legislation, and an enforcement agency tests each vehicle type for compliance. Few countries have regular retesting and usually individual vehicles are checked only if they are causing concern. It is likely that acoustic emission will be added to the tests performed on vehicles as part of their roadworthiness certification, and Norway and Italy have made this mandatory.

Standards for the permitted noise levels made by domestic equipment are gradually being produced. These are usually by class of product, e.g. lawnmowers, and define 'acceptable' levels when measured with a given type of equipment in a prescribed manner. Most of these are unknown to the general public and it is only the desire of the manufactures not to be 'too noisy' which makes them comply. Most of these levels are specified as a single number, broad-brand measurement and take little if any account of the frequencies being emitted. Thus it is still possible to have two items which give identical levels, but cause significant difference in nuisance value due to, say, one having mainly low-frequency noise and the other a single midfrequency output.

1.6.5.3 Masking noises and antinoise

Everyone is familiar with the phenomenon that it is difficult to hear clearly and understand speech in the presence of a background of some other loud noise,

such as the roar of an engine. The human ear has a unique and remarkable ability to discriminate in both the frequency and time domains, outclassing electronic analysis equipment. However, it is not a perfect filter, and suffers from problems in resolving closely spaced frequencies. Research has shown that one sound *masks* another if it has a significant amount of its energy around the frequency to be masked. Hence a pure tone of, say, 1200 Hz could be masked by a band of noise of, say, 1100–1300 Hz, but not by a band of 2300–2500 Hz. Furthermore, it is found that if the bandwidth is increased and the amplitude reduced such that the energy level remains constant, masking remains. The effect continues until a certain point is reached. This is known as the *critical bandwidth of the ear* and is found to approximate to one-third of an octave at mid- to high-frequencies, with larger bandwidths at low frequencies.

It can be seen that any attempt to quantify noise nuisance must, therefore, take into account both the amplitude and frequency of the noise source. The climate of the noise is also important as a person always moving within a fairly high background level will not notice other individual noise sources of similar level as much as someone who is usually in a low background level and encounters the same noise sources. Sometimes a way of disguising this is to permanently introduce some form of masking noise, e.g. an air conditioning system, into the location affected. Another much newer method is the concept of *antinoise*, where the offending noise signal in a given area is measured, has its phase inverted 180°, and is replayed via loudspeakers into the same area, thus causing a cancellation of the SPL. This system has been shown to give reductions of up to 20 dB at frequencies below 1000 Hz. Several car manufacturers are pursuing this as a method of making vehicle interiors quieter. It has no effect on the noise observed externally. Above 1000 Hz, scatter effects reduce the attenuation rapidly, and other more conventional methods, e.g. absorbent materials are more successful in reducing unwanted noise.

1.6.6 Calibration of acoustic devices

For acoustic measurements to be accurate, it is essential that instrumentation is both used and calibrated correctly. Instruments should generally undergo thorough laboratory calibration at least every 2 years.

1.6.6.1 Traceability and quality

There is an increasing requirement for accuracy in acoustic measurements, especially where litigation and compensation are involved. In order to ensure that the equipment has definable accuracy, calibrations should be traceable to a defined acoustic standard. This is normally derived from basic units at the national standards laboratory of the country concerned; e.g. the National Physical Laboratory (NPL) in the UK, the National Institute of Standards & Technology (NIST) in the USA, and the Physikalisch Technische Bundesanstalt (PTB) in Germany.

Traceability, as defined in the International Standards, requires only that there is an unbroken chain of recorded calibrations leading back to the appropriate Standard. Unfortunately, there is no requirement relating to the quality of the calibrations performed. Traceability in itself is meaningless unless every calibration in the chain is technically sound. This means that the staff, equipment, environment and measurement procedures involved must be adequate for the task. It is therefore desirable to have periodic calibrations performed by a laboratory accredited by a recognized quality-assurance scheme. In the UK, the United Kingdom Accreditation Service (UKAS) assesses, accredits and monitors calibration and testing laboratories under the National Accreditation of Measurement and Sampling (NAMAS) scheme. Similar bodies operate in many other countries.

The calibration of instruments can be divided into three basic types: field calibration with an acoustic calibrator; field calibration without a sound source; and laboratory verification.

1.6.6.2 Field calibration

This is a check performed on the instrument before, and also preferably after, it is used; it is a simple calibration carried out by the user. For sound-level meters, dosimeters and environmental-noise analysers this usually consists of the application of a single-frequency sound source to the microphone and appropriate adjustment of the instrument's gain control. The sound source should be periodically calibrated by a suitable (preferably accredited) laboratory. Care should be taken in the choice of sound source, which is usually called an *acoustic calibrator*.

Acoustic calibrators This device is used for the field calibration of acoustic instrumentation, and generally falls into one of two categories: electronic or mechanical. New calibrators should conform to IEC 942:1988 (under revision; equivalent to BS 7189: 1989) or ANSI S1.40–1984 (R1994). Older calibrators may be suitable, but no International Electrotechnical Commission (IEC) standard existed prior to 1988.

Electronic calibrators (*Figure 1.72*) generally employ a stabilized high-quality oscillator connected to a transducer in an acoustic cavity. The microphone is sealed into another opening in the cavity and a known acoustic pressure is generated by the transducer. The accuracy of this type of calibrator varies markedly from one design to another, but those meeting IEC 942: 1988 and many earlier devices from reputable manufacturers are suitable for everyday applications.

Mechanical calibrators have historically taken various forms, but the only readily available one worthy of consideration is the dual-piston *pistonphone*. In this device a rotating cam with a sinusoidal profile causes a matched pair of pistons to vibrate. The pistons in turn move the air inside the cavity of the calibrator. The change in pressure, and hence the SPL, may be calculated from the diameter and stroke of the pistons. Pistonphones are generally

Figure 1.72 *An electronic acoustic calibrator meeting IEC 942: 1988 class 1.*

more accurate than electronic calibrators, but more expensive due to the precision engineering required. It has so far proved impossible to produce a pistonphone which operates at the standard reference frequency of 1 kHz. The most common frequency is 250 Hz.

The SPL generated by most acoustic calibrators, electronic or mechanical, may vary considerably due to several influences. Calibrators must therefore be used in an informed manner if the results are to be meaningful. The main influences are: temperature, air pressure, configuration, and type of microphone.

Electronic calibrators are generally most susceptible to change with temperature: for example a calibrator meeting IEC 942: 1988 class 2 is allowed to change by ±0.5 dB over the range 5–35°C. These temperatures are attainable if, for example, the instrument is left in a car in the extremes of even UK climatic conditions. The frequency of electronic calibrators may also change with temperature, but this is usually less significant.

Air pressure can change with weather patterns or altitude. Most calibrators increase in SPL with increasing ambient pressure. As an example, those meeting IEC 942: 1988 class 2 may change by ±0.5 dB. Many calibrators, particularly pistonphones, are supplied with barometers so that pressure corrections may be applied.

Calibrators are frequently supplied with couplers for use with a variety of microphones. The sound pressure level generally varies when different couplers are used due to the different cavity volumes so produced. The common practice of stacking couplers to obtain the correct diameter should, therefore, be avoided unless specifically recommended by the manufacturer. Likewise, couplers from one calibrator should not be used in another. The conversion of calibrators to accept different microphones is not just a matter of simple plumbing.

Due to different microphone effective volumes and acoustic impedances, the SPL generated by a calibrator may depend slightly on the actual microphone type in use. The manufacturer's instructions should be followed in order to obtain accurate levels.

Field calibration without a sound source
Installations such as airport-noise monitoring systems present calibration problems. It is impractical to visit all microphone sites at frequent intervals and to use a sound source to check the system, particularly since microphones are frequently situated several metres above ground level. They are therefore generally fitted with an electrostatic actuator system which simulates a sound field. This is activated at regular intervals by the controlling computer. The accuracy achieved is usually lower than with a sound calibrator, but is a convenient compromise.

There are some cheap and/or old sound-level meters on the market that rely on an internal oscillator for field calibration, there being no acoustic calibrator available which fits the microphone. They rely on the stability of the oscillator, which is frequently suspect, and check only the gain of the electronic circuitry, ignoring the microphone. Since microphones, particularly cheaper ones, can vary considerably in their sensitivity (for example, across the range of temperatures encountered) the calibration of the complete instrument is likely to be inaccurate. These meters should therefore be avoided.

1.6.6.3 Periodic verification or calibration
This procedure comprises a more thorough laboratory calibration intended to ensure that an instrument continues to perform correctly throughout its life. This generally consists of a subset of the tests required for the type approval. Requirements for verification vary from country to country; in Germany, for example, instruments may be used for legal metrology only if they have undergone type approval and are regularly verified by state laboratories. In the UK there has, until recently, been very little requirement for calibration of any variety, it being only a recommendation; however recent legislation and standards have required periodic verification either annually or every 2 years. The main UK standard for the verification of sound level meters is BS 7580: Part 1: 1997, with a reduced set of tests for Type 2 meters given in Part 2. Requirements specifying the quality of the calibration are also being imposed (see Section 1.6.6.1).

1.6.6.4 Type (or pattern) approval
This is the most thorough test of all, in which the design of a particular model of acoustic instrument is verified. This usually takes the form of verification against the national or International Standard which the instrument is claimed to meet, and should be performed by reputable manufacturers as part of the product development. The final type approval is normally undertaken by the primary standards laboratory of the country concerned, as in Germany for example. There is at present no such independent type approval for acoustic intrumentation in the UK, except for acoustic calibrators.

1.6.6.5 Laboratory calibration methods for periodic verification and type approval

Pattern evaluation and periodic verification are generally split into two types of test, acoustic and electrical. The latter is used to check the main functions of the instrument, and simulates the electrical signal generated by the microphone. It would be extremely difficult, if not impossible, to generate suitable acoustic signals reproducibly.

Electrical calibration The main functions tested are: the linearity of the instrument; its frequency response; the time constants; its r.m.s. characteristics; and its integrating capability, if applicable. On more complex instruments other parameters may also be tested, though in modern instruments these are almost exclusively software calculations from the raw data.

Acoustic calibration Acoustic testing will usually be markedly different between pattern evaluation and periodic verification tests. Periodic verification will generally consist of a few simple checks to ensure that the microphone has not been damaged; an example is given in BS 3539: 1986. Here the instrument is calibrated acoustically using a traceable 1 kHz calibrator and a 125-Hz pistonphone. The 125-Hz pistonphone would show up a perforated microphone diaphragm, since this type of damage tends to cause a marked drop in sensitivity at low frequency. This is a common cause of microphone failure, and is brought about not only by careless handling but also by particles blown in through the protective grid. Use of a foam windshield, which stops most particles, is advisable in dusty atmospheres.

Pattern evaluation calls for a much more thorough test of the instrument's acoustic performance. This includes not only a full frequency response of the complete instrument but also measurements on its directional response. Both are affected by the basic response of the microphone and by the shape of the instrument case, which may cause sound reflections. This is why most sound-level meters are of a broadly similar shape, with a microphone mounted on a stalk attached to a tapered body. There are still some meters available which consist of a rectangular box with a microphone mounted flush in one end. These are unlikely to meet the directionality requirements of IEC 651: 1979 (BS EN 60651: 1994) and equivalent standards.

The directional response of a sound-level meter is stipulated in the national or International Standard to which it was designed. There are two basic types; in the UK and the rest of Europe, most meters have a 'free-field' response, whereas in the USA 'random-incidence' meters are preferred. This difference makes transatlantic comparison of measurements difficult. For a definition and explanation of free and random fields, see Bies and Hansen (1988b).

1.6.6.6 Calibration of pure-tone audiometers

The output from audiometer headphones is calibrated by coupling the transducer and ear cushion to a microphone attached to a measuring system. Two basic designs of coupler are used for supra-aural earphones; the design used for a particular calibration is dictated by the earphone type and the ear cushion fitted. The earlier design, called the 'reference coupler', is specified in IEC 303: 1970 (equivalent to BS 4668: 1971 (1988)). A relatively simple coupler, it is almost identical to the NBS9A coupler (US former National Bureau of Standards). The appropriate headphones are specified in BS 2497: 1992 (ISO 389: 1991) as are the *reference equivalent threshold sound pressure levels* (RETSPLs) for each combination.

The second coupler is considerably more complicated, being designed to have acoustical characteristics which simulate the human ear. Known as the 'IEC artificial ear', it is specified in IEC 318: 1970 (equivalent to BS 4669: 1971 (1988)). The appropriate head-phones and RETSPLs are also given in BS 2497: 1992 (ISO 389: 1991).

Much simplified, the method used is to apply each earphone in turn to a calibrated measuring system, comprising a reference microphone inside the appropriate coupler, attached to a sound-level meter. The earphone is driven by the audiometer circuitry at each frequency in turn, and adjustments made to obtain the correct level. A set of reference headphones is kept to cross-check the equipment.

1.6.6.7 Microphone calibration

There are several methods for the laboratory calibration of precision microphones. They are fairly complex and involve greater accuracy than is necessary for everyday noise measurement, but are mentioned here for completeness.

Laser pistonphone Not yet commercially available, a refinement of the pistonphone which allows accurate calculation of the SPL generated in a cavity at low frequencies. The piston displacement is measured by laser interferometry and the consequent pressure change is calculated.

Insert voltage calibration An accurate method for determining the open-circuit sensivity of a condenser microphone. For details see ANSI S1.10–1996 (R1986).

Reciprocity The most accurate means of determining the sensitivity of a condenser microphone, utilizing its ability to act as a loudspeaker by driving it with a signal. For details see ANSI SI. 10–1966 (R1986), Beranek (1986) or Kinsler *et al.* (1982).

References

BERANEK, L.L., *Acoustics*, American Institute of Physics, New York, p. 377 (1986)
BIES, D.A. and HANSEN, C.H., *Engineering Noise Control*, Unwin Hyman, London, p. 47 (1988a)
BIES, D.A. and HANSEN, C.H., *Engineering Noise Control*, Unwin Hyman, London, pp. 134 and 163 (1988b)
KINSLER, L.E., FREY, A.R., COPPENS, A.B. and SANDERS, J.V., *Fundamentals of Acoustics*, 3rd ed. Wiley, New York, p. 390 (1982)

National and International Standards

In the following list, equivalent standards are given in bold type, in parentheses after the British Standard numbers.

ANSI American National Standards Institute
BS British Standards Institution, London, UK
IEC International Electrotechnical Commission
ISO International Standardization Organization, Geneva, Switzerland

BS 2497: 1992 **(ISO 389: 1991)**
 Standard reference zero for the calibration of pure tone air conduction audiometers
BS 3539: 1986
 Sound level meters for the measurement of noise emitted by motor vehicles
BS 4142: 1997
 Method for rating industrial noise affecting mixed residential and industrial areas
BS 4668: 1971(1988) **(IEC 303: 1970)**
 Specification for an acoustic coupler (IEC reference type) for calibration of earphones used in audiometry
BS 4669: 1971(1988) **(IEC 318: 1970)**
 Specification for an artificial ear of the wide band type for the calibration of earphones used in audiometry
BS 6402: 1994 **(IEC 1252: 1993)**
 Personal sound exposure meters
BS 6655: 1986(1993) **(ISO 6189: 1983)**
 Specification for pure tone air conduction threshold audiometry for hearing conservation purposes
BS 7189: 1989 **(IEC 942: 1988)**
 Sound calibrators **(Note: revision of IEC 942 due to be published Dec 1997)**
BS 7580: Part 1: 1997
 The verification of sound level meters. Part 1: Comprehensive procedure
BS 7580: Part 2: 1997
 Part 2: Shortened procedure for type 2 sound level meters

BS 7636: 1993 **(ISO 8253–2: 1992)**
 Method of determination of thresholds of hearing using sound field audiometry with pure tone and narrow-band test signals
BS EN 352–1: 1993
 Hearing protectors. Safety requirements and testing. Part 1: Ear muffs
BS EN 352–2: 1993
 Part 2: Ear plugs
BS EN 60645–1: 1995 **(IEC 645–1: 1992)**
 Audiometers. Part 1: pure tone audiometers
BS EN 60645–2: 1997 **(IEC 645–2: 1993)**
 Part 2: Equipment for speech audiometry
BS EN 60651: 1994 **(IEC 651: 1979)**
 Specification for sound level meters
BS EN 60804: 1994 **(IEC 804: 1985)**
 Integrating-averaging sound level meters
BS EN 61043: 1994 **(IEC 1043: 1993)**
 Instruments for the measurement of sound intensity with pairs of pressure sensing microphones
 (ISO 1999: 1990)
 Acoustics – determination of occupational noise exposure and estimation of noise-induced hearing impairment
 (ISO 8253–1: 1989)
 Acoustics – audiometric test methods. Part 1: basic pure tone air and bone conduction threshold audiometry
ANSI S1.4–1983(1994)
 Specification for sound level meters
ANSI S1.4A–1985
 Amendment to S1.4–1983
ANSI S1.10–1966(R1986)
 Specification for calibration of microphones
ANSI S1.25–1991
 Specification for personal noise dosimeters
ANSI S1.40–1984(R1994)
 Specification for acoustical calibrators

1.7 Electromechanical analogies

1.7.1 Introduction

Newton's laws of motion lead to the equations describing the displacement of mechanical particles under various forces or constraints, whilst the equations describing the relationships between electrical potentials, charges, conduction and dissipation have been derived by Ampere, Faraday, Ohm and other early masters. The French mathematicians in the early 1900s pointed out the similarity of the linear equations describing the motions of mechanical bodies and electrical charges in different systems. Wegel is credited with being the first to show symmetrical equations for the coupled electrical and mechanical structures of telephone receivers.

An early example of the use of electromechanical analogues as the basis of the complete design of a radically improved disc recording cutter head was described by Maxfield and Harrison of Western Electric in the early 1920s, the system being realized as a mechanical filter equivalent to an electrical bandpass filter, designed by the conventional electrical filter theory then in use (Hunt, 1954). Since then it has become common to use electromechanical analogues in the design, and also to improve the understanding of the principles of operation of electromechanical acoustical instruments.

There are necessary limitations on the scope and validity of these analogues. It is obvious from the initial concepts of discrete mechanical particles and electrical components that an assembly of 'lumped constants' has to be assumed. We know that, generally, this assumption is only valid over a limited frequency range. Any mechanical object will generally exhibit higher modes of vibration or harmonics in addition to its fundamental mode. Similarly, electrical systems and acoustical systems may exhibit distributed properties, harmonics and standing waves at higher frequencies.

Thus for larger objects such as loudspeakers, analogies may only be strictly valid at lower frequencies. For microphones, much of the structure is relatively small, as are the associated holes, passages, slits and air cavities involved in the structure, thus extending the validity of the analogies. For instance, if the volume of an enclosed space has dimensions less than 1/10 to 1/20 of the wavelength of the sound, the fundamental analogy is fairly accurate. For extended narrow pipes, one may consider transmission line theory for an analogy in place of a single component. However, for a volume extending in more than one dimension, transmission lines in three dimensions would be called for, which is clearly not practical.

1.7.2 Basic analogue concepts

In considering analogues, initially one considers the basic quantities involved in the various systems concerned: mechanical, acoustical and electrical. For mechanical elements, the displacement is x, velocity is dx/dt (or \dot{x}), and acceleration is d^2x/dt^2 (or \ddot{x}). The force (Force $=$ Mass $M\ddot{x}$) represents the kinetic energy stored, as the mass (M) is the quantity of matter in the element. Resistive dissipation is represented by $R\dot{x}$ and the potential energy is stored by stiffness multiplied by the displacement x. (*Table 1.14*).

Conversion to other systems such as electrical and acoustical is greatly facilitated by the use of the SI system of units, in that energy is then measured in joules. *Table 1.14* gives the overall units and dimensions. *Table 1.15* is a more complete table and includes magnetic units, etc. The basis of the direct or impedance analogies thus becomes easy to grasp.

A mass suspended on a spring and subject to some energy dissipation such as friction, air loading or viscosity, is represented by the equation:

$$F = jWM\dot{x} + R\dot{x} + \left(\frac{1}{jWc}\right)x$$

The mechanical equivalent circuit is seen to be of the same form as the electrical series resonant circuit.

For the acoustical case, a simple Helmholtz resonator (see *Figure 1.73*) may be considered. The mass of the air in the opening is considered to move as a 'slug' against the compliance or stiffness of the air in the resonator volume. Air friction in the neck and sound radiation adds dissipation. For acoustical relationships other than those occuring in completely free-air conditions, it is necessary to use the acoustic excess pressure as the driving force and the volume velocity of the air displaced in order to take account of the different areas of the passages that the sound wave has to traverse. The equation of motion is shown in *Figure 1.73*. We see that the analogue

Mass on spring

$$F = j\omega M\dot{x} + R\dot{x} + (1/j\omega C)\dot{x}$$

Acoustic resonator

$$P = j\omega M\dot{x}A + R\dot{x}A + 1/j\omega C$$

Electrical series resonant circuit

$$V = j\omega Li + Ri + (1/j\omega C)i$$

Figure 1.73 *Simple resonant systems.*

Table 1.14 Summary of analogue quantities, units, symbols and relationships

Mechanical	Acoustical	Electrical
Mass, M (kg)	Mass M_a, Inertance	Inductance, H (Henry)
Force, F (N)	Sound pressure, p	Volts (V)
Displacement, x (m)	–	Charge Q (C)
Velocity, \dot{x} (m/s)	Volume velocity	Current (Q/s, A), i
Energy, work (J)	joules (J)	joules (J)
Power (W, J/s)	watts (W)	watts (W)
Stiffness, F/x	–	–
Compliance x/F	Compliance, C_a	Capacitance, C
Frequency, f (1/s, Hz)	–	–
Resistance R_m	Acoustic resistance, R_a	Resistance, R
Temperature (K)	–	Magnetic flux density, B
Length ℓ (m)	–	–

Impedance relationships

Mechanical impedance, $Z_m = F/\dot{x} = \dfrac{PA}{\dot{x}}$

Specific acoustic impedance, $Z_s = p/\dot{x}$
Analogous acoustic impedance, $Z_a = p/\dot{x}A$
Where A is the area of the element, and p is sound pressure

Sound pressure level, SPL (Pa) referred to threshold of hearing
$= 2 \times 10^{-5}$ Pa 0 dB.
1 pascal (Pa) \equiv 1 newton per square metre (N/m^{-2}).

Admittance $= 1/Z$
Transformations: $F = B\ell i$; $v = B\ell\dot{x}$
Newton's laws: $F = M\ddot{x}$
Kinetic energy $= 1/2\ M\ddot{x}^2$
Potential energy $= 1/2\ sx^2$

of the resonator is an electrical series resonant circuit. This is the direct impedance analogy of the above systems and it is easy to grasp the idea of voltage being equivalent to force, current to velocity and resistance to dissipation. With the SI system of units it is convenient to carry out exact transformations from the mechanical and acoustical systems into the equivalent electrically analogous circuit, the impedance, current and voltage of which may be set up as actual electrical circuits for measurement, or may be calculated by using well-established circuit theory. Mechanical or acoustic systems may be optimized by applying electrical filter theory to the analogues. This was done in the original case of the disc cutter head and has been more recently applied to microphone design and the realization of optimum bass roll-off of closed box loud-speaker cabinets (the drive unit constants and box volume are made equivalent to the nth-order filter roll-off, with a consequent improvement in bass efficiency and flatness, etc.) (Thackeray, 1988).

However, in some cases transformation must be introduced between the electrical and mechanical parts of a system, and thus a direct analogy cannot be followed. For example, the electromagnetic interaction which occurs to the drive current to a moving coil loudspeaker produces a force on the voice coil and the system reflects as a parallel resonant circuit to the electrical input, because it presents an electrical impedance maximum, despite the fact that the simple mechanical system of diaphragm and coil on the suspension in isolation would be represented by a series resonant circuit (Ford, 1988; Kelly, 1988) (see *Figure 1.73.*).

So far, we have only considered a simple resonant structure described as having one degree of freedom. Most structures which one desires to analyse have several identifiable masses coupled with compliances and subject to dissipations. Several modes of vibration are possible and the system is said to have an 'equivalent number of degrees of freedom' or 'resonant modes'. When the system is excited

Table 1.15 (After Hunt)

Analogue quantity	Symbol	Units in which quantity is expressed
Energy		joules \equiv newton metre
		\equiv coulomb volts
Power	W	watts $=$ joule seconds
		$=$ ampere volts
Charge	Q	coulomb metre
Mass	M	kilograms $=$ newton seconds2/metre
Potential	V	volts
Current	I	amperes $=$ coulombs/second
Transduction coefficient	T	newtons/ampere $=$ volt seconds/metre
Force factor		newtons/ampere $=$ volt seconds/metre
Magnetic flux	φ	webers $=$ volt seconds
Magnetic flux density	B	webers/metre
Magnetic field vector	H	amperes/metre
Magnetization	M	amperes/metre
Magnetomotive force	F	ampere turns $=$ coulombs/second
Magnetic reluctance	R	ampere turns/weber $=$ coulombs/
		volt seconds2
Inductance	L	henries $=$ webers/ampere
		$=$ volt seconds2/coulomb
Susceptibility	χ	
Electric impedance	Z	ohms
Electric resistance	R	$=$ volts/ampere
Electric reactance	X	$=$ volts seconds/coulomb
Force	F	newtons $=$ joules/metre
		$=$ coulomb volts/metre
Pressure	P	Pascals $=$ newtons/metre2
		$=$ joules/metre3
Density	ρ	kilograms/metre3 $=$ newton seconds2/metre4
Stiffness	S	newtons/metre $=$ coulomb volts/metre
Compliance	C_m	metres/newton $=$ metres2/coulomb volt
Specific acoustic impedance	ρ_c	kilogram/metre4 second
		$=$ coulomb volts/metre4
		$=$ newton seconds/metre3
Mechanical impedance	z_m	newton seconds/metre
		$=$ kilograms/second
Acoustic impedance	Z_A	newton seconds/metre5
		$=$ kilograms/metre4 second
		$=$ ohms (acoustic)
Permittivity of free space	ϵ_0	$8.854 \times 10^{-12} = \dfrac{1}{36\pi} \times 10^{-9}$ farads/metre
Electric field	E	volts/metre
Electric displacement		coulombs/metre2
Electric polarization		coulombs/metre2
Capacitance	C	farads $=$ coulombs/volt

Figure 1.74 *The acoustic elements, electrical equivalent circuits and R, L and C expressed in terms of the most readily measurable parameters: l, b, d (m), S (m², V (m³), frequency f (Hz). l = effective length including end correction l'. Viscosity coefficient η = 1.86 × 10⁻⁵ kg/(m/s). Air density ρ = 1.2 kg/m³. Speed of sound c₀ = 340 m/s. Electrical analogy (SI units) p, q system: p (Pa), g (m³/s) equivalent to U (V), I (A) respectively. 1 Ω = 1 kg/(m⁴ S), 1 H = 1 kg/m, 1 F = 1 m⁴/(s² kg). (Courtesy of Poldy (1988).)*

in a specified manner, the motion of the various masses (assumed to be linear) is described by a series of simultaneous linear differential equations, from which the equivalent electrical analogous circuit may be derived (Olson, 1991). *Figure 1.75* shows examples of systems with two degrees of freedom. A driven mass on a spring suspension has an auxiliary mass suspended from the main mass by a separate spring. This is a well-known type of structure which may take the form of an equivalent electrical filter. It can also represent a loudspeaker unit in a vented cabinet, the response of which may again be optimized.

It obviously becomes difficult to derive the various modes of vibration and degrees of freedom of more complicated structures with many masses and their associated compliances and dissipations. If the system can be drawn in the form of an energy-flow diagram in which a coordinate can be assigned to each discrete mass, use of the Lagrange equation enables the various equations of motion relating to the meshes of the equivalent circuit to be identified and the circuit drawn (Gayford, 1970). A good example of the application of Lagrangian analysis is provided by the analysis of a moving coil microphone

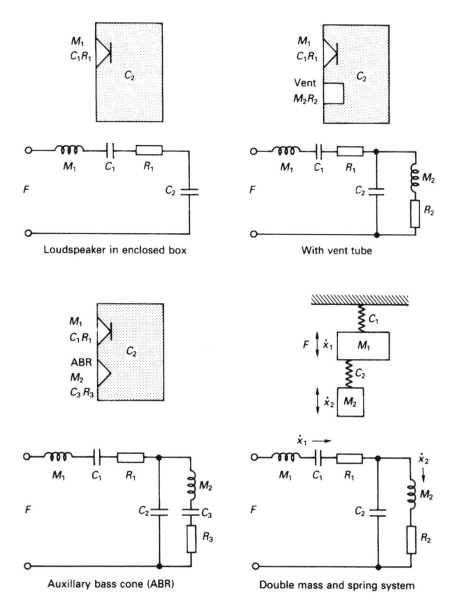

Loudspeaker in enclosed box

With vent tube

Auxillary bass cone (ABR)

Double mass and spring system

Figure 1.75 *Systems with two degrees of freedom.*

(see Section 1.7.8). Various examples of practical systems are considered below.

1.7.3 A vibration isolating mounting for a microphone or other instrument

Figure 1.76 shows a microphone suspension for use in situations where mechanical vibrations have to be isolated, e.g. in mobile boom mountings. The microphone is suspended by a system of soft rubber thongs which are designed to provide the correct mounting stiffness in the appropriate planes, with enough mechanical resistance to control the resonance of the system (Robertson, 1963). *Figure 1.77* shows the theory, and the isolation provided against mechanical vibrations from the boom, etc. The system is designed to, as much as is possible, have a single degree of freedom, the resonant frequency being placed as low as possible, preferably at a subaudible frequency. If no resistance is present there will be maximum transmission at the resonant frequency

Figure 1.76 *Microphone suspension.* (Courtesy of AKG)

f_r and maximum isolation at frequencies f above this point, as shown by the curve $R_m = 0$. If an intermediate damping value is obtained by using an appropriate grade of rubber or plastic for the thongs, so as to give slightly under-critical damping, the basic resonance is damped sufficiently to prevent the microphone being shock-excited and bounce about when being moved, etc. The excitation is shown as a vertical excursion but, in practice, the thong geometry is arranged to provide isolation and damping for horizontal and intermediate directions of excitation. Assigning values for the mass of the microphone and mounting, the spring stiffness and the resistance, will enable the analogous mechanical circuit to be set up for measurement or calculation.

It is good practice to design the thongs and supporting framework to be as slender as possible so as to avoid any appreciable sound reflections which might affect the microphone response. This design is an example of a simple mechanical filter and analogy. Many examples of more complex mechanical filters occur in electrical, mechanical and civil engineering which are amenable to analysis by the use of analogues.

1.7.4 Analogue circuits and interfaces

In audio applications, analogues are largely concerned with transducers such as loudspeaker, earphones and microphones. These all have electrical, mechanical and acoustical attributes which act together to give the desired response to particular inputs. It is usually difficult to analyse the functions in terms of one 'domain' only and the analogues require the mechanical and acoustical constants to be represented in a single analogous electrical circuit. Having established the lumped constants, one has to decide on the class of circuit to be adopted. If voltages and currents represent the actual drive currents from an amplifier into a loudspeaker, electrical analogues will represent the true electrical loading on the amplifier at the interface. If, however, the analogue components represent mechanical or acoustic values, then forces, velocities and mechanoacoustic impedances and values are represented. The transducing element associated with the diaphragm system is a crucial factor in the analogue circuit as it represents an interface between two different regimes.

1.7.5 The moving coil loudspeaker and headphone

The input current in a conductor of length L in a magnetic gap of flux density B gives rise to a force $F = BL_i$ driving the diaphragm. In the analogy this force must be represented by a mathematical gyrator rather than a straight transformer because it converts an input *current* into a *force* on the diaphragm. This can cause difficulties in the analogy because the diaphragm and suspension system, which mechanically is a simple series resonant circuit, when viewed from the amplifier side must appear as a parallel resonance, as it produces a back e.m.f. rising to a maximum at resonance. The mechanical impedance direct analogy is thus in effect transformed to an inverse admittance or 'mobility' analogue.

In order to produce one coherent analogous circuit capable of straightforward calculation, conversion to one form of analogy is necessary throughout. *Figure 1.77* shows the initial complete analogue circuit of a moving-coil loudspeaker radiating into free air from the front and rear of the diaphragm when driven from a constant-voltage amplifier of low source impedance, which is usual nowadays. The BL gyrator requires the loudspeaker mechanical constants for the coil and cone (M, C and R) to be represented by the mobility parallel resonant circuit, whilst the air loads for the front and rear of the

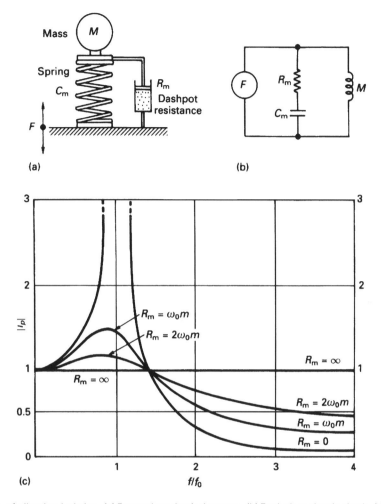

Figure 1.77 *Analysis of vibration isolation. (a) Damped mechanical system. (b) Equivalent electric circuit. (c) Isolation factor (ip) plotted against f/f₀ for various degrees of damping. It can be seen that the larger the damping the greater is the reduction in the resonance peak but the less effective is the mounting as an isolating device. Damping is achieved by replacing the spring by a resilient material which is not loss-free. The most commonly used materials of this type are rubber and certain plastics. Studio boom microphones are often suspended on soft rubber-thong mountings, whilst stand units are mounted on concentric bonded-rubber shear-type mountings.*

diaphragm are also in mobility form, connected via a transformer of ratio S_d (the area of the diaphragm). Thus we have the electrical section of the analogue circuit, with the mechanical circuit and the acoustic section in mobility form (*Figure 1.78(a)*).

Unless one is very familiar with the mobility representation of mechanical structures, it is difficult to see the actual physical form when shown a mobility circuit. A direct impedance analogous circuit is easier to comprehend as one naturally associates voltage with force, current with velocity, compliance with reciprocal stiffness, etc. Thus in explaining the action of moving-coil loudspeakers in various practical cabinets, etc., it is helpful

to ignore the BL current drive and its inversion and to represent the system in terms of mechanical impedances, with the diaphragm system shown as a series resonant circuit. *Table 1.14* and *Figures 1.73* and *1.75* show the direct mechanical analogies, electrical, mechanical and acoustic elements and simple resonators. *Figure 1.75* shows the mechanical analogous circuits for moving-coil loudspeakers in closed (infinite baffle) cabinets, vented cabinets with a pipe vent, and cabinets with a passive acoustically driven diaphragm as the vent element. The passive diaphragm type of vent has the advantage that it may, with careful design, be possible to extend the low bass response significantly below the bass

resonance of the loudspeaker unit in a well-controlled fashion.

When it is desired to study and to calculate the exact performance of such loudspeakers, including the damping effects of the amplifier source impedance and the air loading, one can start with a full general circuit such as that shown in *Figure 1.78(a)*. If the transformers are eliminated, the circuit can appear as *Figure 1.78(b)* which refers everything to the input and would enable the final output to be calculated, albeit in a possibly inconvenient form due to the shunt connections. A further rearrangement enables a simpler series impedance calculation to be carried out (Ford, 1988). More complicated analogue circuits may be evolved on the above lines to enable the calculations on loudspeakers in various enclosures to be made, particularly with regard to the low-frequency performance, which can be more accurately represented by the lumped constant analogies.

The case of moving-coil headphones is similar to that of the moving-coil loudspeaker, except that the output feeds a semiclosed cavity in the form of the ear cavity and earcap or circumaural ear seal, etc. In some cases, leakage into free air around the earcap or from designed rear vents from the diaphragm system may have to be allowed for (Poldy, 1988) (see *Figure 1.79*).

1.7.6 Impedance analogies

We may note that, in the case of moving-coil microphones, the position of analogues is simplified because the electromagnetic function $V = Bl\dot{x}$ is accommodated without difficulty. The system is designed to give constant diaphragm velocity for constant sound pressure (p_a) input. The resulting constant voltage output is required to drive the preamplifer, which is generally arranged to have an input impedance one order of magnitude or more greater than that of the microphone. Thus a straightforward impedance analogy without gyrators is used. If a pure mechanical analogy is used throughout, transformers of ratio A_o/A must be used to allow for different areas of the various passages in the microphone structure that the sound traverses. The convention of using the volume velocity $\dot{x}A$ in place of the mechanical particle velocity in the wave eliminates these transformers. Ideal transformers may still have to be used when different parts of the diaphragm feed different passages, e.g. in the case of the diaphragm dome and surround (see Section 1.7.7).

All microphones require some communication via a passage to the rear of the diaphragm in order to equalize the barometric pressure on each side of the diaphragm. For pressure-operated omnidirectional microphones the leak is through a tube or

Figure 1.78 *Analogue circuits for a moving-coil loudspeaker. (a) The initial analogue circuit with the BL transformation and the Sp transformer covering the diaphragm area to airload. (b) The circuit referred to the electrical input terminals. The transformers have been omitted. The air-load constants depend on the baffle or cabinet and may be in series or parallel form.*

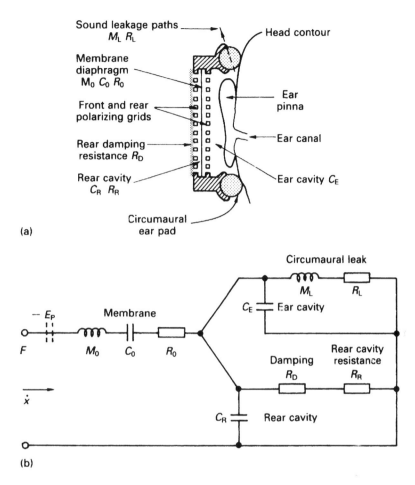

Figure 1.79 *An earphone on the ear. (a) A membrane-type electrostatic earphone with circumaural pads. The rear acoustic resistance fabric damps the system resonances. Some rear radiation is said to reduce the in-head localization effect. Circumaural leaks reduce bass reproduction. (b) Direct mechanical analogue circuit of an earphone. The electrostatic polarizing voltage appears as a negative compliance. The electrical analogue circuit is of identical form due to common transduction components.*

hole of high acoustical impedance which blocks the passage of audio frequencies, except at very low frequencies. However, directional moving-coil microphones and ribbon microphones must be responsive to the pressure gradient in the sound wave, the rear opening then having a specified phase shift (Gayford, 1970). The same types of analogous circuits, as described above may be derived, but phase-displaced voltages must be applied to the parts of the circuit representing the front and rear openings. It is not always easy to derive these analogue voltages at all frequencies, because the case geometry and diffraction effects cause complications. Nevertheless, the analogous circuits are still valuable in the explanation of the operation of the internal phase shift networks of directional gradient-operated microphones (Gayford, 1970). *Figure 1.82* (see below) shows examples of the direct impedance analogous circuits for such microphones.

1.7.7 Electrostatic devices

Electrostatic loudspeakers, earphones and microphones also lend themselves to study and analysis by the use of analogous circuits. Their structures are usually less complicated than those of electromagnetic transducers, although in the case of electrostatic loudspeakers the diaphragm systems are much larger and more distributed. In such cases, single-sided and push–pull double-sided types are used. These have been described and analysed extensively (Hunt, 1984). Unlike electromagnetic transducers, where the drive and generation coefficients Bl_i and Bl_v differ,

the corresponding transduction coefficients are identical for drive and generation in the electrostatic cases, which conveniently gives identical form to both the electrical and the mechanical analogous circuits. These are related by the common transduction coefficient (Baxandall, 1988).

Figure 1.80 shows the basic structural form of both single-sided and double-sided electrostatic loudspeakers and microphones, and *Figure 1.81* shows the basic electrical and mechanical analogous circuits

Figure 1.80 *A double-sided electrostatic transducer. A single-sided transducer has only one plate.*

that exemplify the above points. One subtlety to note is that the necessary constant-charge polarization which has to be applied to the diaphragm appears as

a negative force displacing the diaphragm. This force is a factor which has to be considered in the design, particularly that of single-sided transducers. In modern electrostatic loudspeakers, constant-charge operation is achieved by applying a thin high-resistance conductive coating to the surface of the diaphragm, whereas in electrostatic (condenser) microphones the diaphragm is uniformly conductive and is polarized via a very high electrical resistance, or by a permanent electret charge on the diaphragm or backplate. This is normally a more solid structure which is generally designed to add air damping and stiffness to the diaphragm movement, whereas for electrostatic loudspeakers the electrodes need to be substantially acoustically transparent (Baxendall, 1988).

1.7.8 Moving-coil microphone

The systematic detailed derivation of the analogous circuit of a moving-coil microphone can be taken as an example for the analysis of a fairly complex mechanoacoustic system (Gayford, 1970). A detailed sectional drawing of such a microphone is shown in *Figure 1.82*. From inspection of the mechanical layout, key mechanical and acoustic masses are identified that are suitable for lumped constant representation. Each mass will oscillate about a mean position

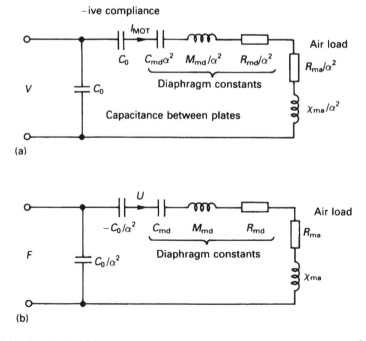

Figure 1.81 *Electrical (a) and mechanical (b) analogous circuits. α, transduction coefficient; $\alpha = \varepsilon_0 A V_{POL}/d^2 = F/V = I_{mot}/U$, where U is the velocity, $-C_0$ is the negative compliance due to polarization.*

and will represent a resonant mode or 'degree of freedom' in the overall motion of the structure. If the section of the microphone is laid out across the page, then vertical lines through the masses represent the degrees of freedom. It is convenient to take the zero degree of freedom to represent the motion of the diaphragm and coil mass, the velocity of which is proportional to microphone output, given by $BL\dot{x}$, where B is the radial flux density in the coil gap, L is the length of the coil conductor, and \dot{x} is the velocity of the coil perpendicular to the flux.

The air in the holes of the front protective grille act as masses in the -1 degree of freedom and form a resonator with the air volume before the diaphragm. Air masses are also identified in the coil slots and in other paths in the microphone case and rear access tube or apertures, giving other degrees of freedom (1, 2, etc.). The application of the Lagrange equation (Gayford, 1970) enables the equations of motion of the various masses to be derived as follows.

First a functional diagram is drawn from a knowledge of the mechanical construction of the microphone. The various parts acting as masses are identified by reason of their ability to store kinetic energy, and their vibrational mode displacements are chosen as independent coordinates. The various constants in the form of lumped masses, resistances, and stiffnesses are identified, and the appropriate areas and linear velocities are designated. From Lagrange's equation:

$$\frac{\mathrm{d}}{\mathrm{d}t}\left(\frac{\partial T}{\partial \dot{x}_n}\right) + \left(\frac{\partial F}{\partial \dot{x}_n}\right) + \left(\frac{\partial V}{\partial x_n}\right) = e_n(t)$$

where: $e_n(t)$ is the applied force for each coordinate; \dot{x}_n are the linear coordinate velocities; x_n are the linear coordinate displacements; $n = -1, 0, 1, \ldots$; T is the kinetic-energy function; F is the dissipational-energy function; and V is the potential-energy function.

The kinetic-energy function is derived by inspection of *Figures 1.82(b) and 1.82(c)*.

$$T = \tfrac{1}{2}M_{-1}A_{-1}^2\dot{x}_{-1}^2 + \tfrac{1}{2}M_0(A_0 + A_0')^2\dot{x}_0^2 + \tfrac{1}{2}M_1A_1^2\dot{x}_1^2$$
$$+ \tfrac{1}{2}M_2A_2^2\dot{x}_2^2 + \tfrac{1}{2}M_3A_3^2\dot{x}_3^2 + \tfrac{1}{2}M_4A_4^2\dot{x}_4^2$$

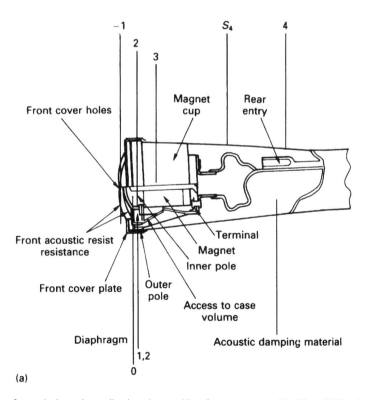

(a)

Figure 1.82 *(a) Section of a typical moving coil microphone with reference masses identified. (b) The functional diagram of a moving-coil microphone. The various mechanical and acoustical masses and compliances are identified in order to facilitate the derivation of the analogous electric circuit. (c) The analogous circuit for the acoustical and mechanical elements of the microphone shown in (a). The current through the diaphragm element M_0l/S_0R_0 represents the response of the microphone.*

(b)

(c)

Figure 1.82 (*continued*)

Similarly, for the dissipational-energy function:

$$F = \tfrac{1}{2}R_{-1}A^2_{-1}\dot{x}^2_{-1} + \tfrac{1}{2}R_0(A_0 + A'_0)^2\dot{x}_0^2 + \tfrac{1}{2}R_1A_1{}^2\dot{x}_1{}^2$$
$$+ \tfrac{1}{2}R_2A_2{}^2\dot{x}_2{}^2 + \tfrac{1}{2}R_3A_3{}^2\dot{x}_3{}^2 + \tfrac{1}{2}R_4A_4{}^2\dot{x}_4{}^2$$

and for the potential-energy function:

$$V = \tfrac{1}{2}S_{-1}[A_{-1}\dot{x}_{-1} - (A_0$$
$$+ A'_0)\dot{x}_0]^2 + \tfrac{1}{2}S_0(A_0 + A'_0)^2\dot{x}_0{}^2$$
$$+ \tfrac{1}{2}S_1(A_0x_0 - A_1x_1)^2 + \tfrac{1}{2}S_2(A'_0x_0$$
$$- A_2x_2)^2 + \tfrac{1}{2}S_3(A_1x_1 + A_2x_2 - A_3x_3)^2$$
$$+ \tfrac{1}{2}S_4(A_3x_3 + A_2x_4)^2$$

The constants M_{-1}, M_0, $M_1 \ldots$, R_{-1}, R_0, $R_1 \ldots$, and S_{-1}, S_0, S_1, \ldots, are assumed to be in analogous acoustic impedance units and thus are multiplied by the square of the area concerned in order to give the equivalent mechanical constants.

Taking the x_{-1} coordinate differentiating the equations, and with respect to \dot{x}_{-1}, x_{-1} and t as indicated and substituting in the appropriate equation, we obtain:

$$j\omega M_{-1}A_{-1}\dot{x}_{-1} + R_{-1}A_{-1}\dot{x}_{-1}$$
$$+ (S_{-1}/j\omega)[A_{-1}\dot{x}_{-1} - (A_0 + A'_0)\dot{x}_0] = p$$

Similarly, taking the x_0 coordinate, we obtain:

$$j\omega M_0(A_0 + A'_0)\dot{x}_0 + R_0(A_0 + A'_0)\dot{x}_0$$
$$+ (S_{-1}/j\omega)[(A_0 + A'_0)\dot{x}_0 - A_{-1}\dot{x}_{-1}]$$
$$+ (S_0/j\omega)(A_0 + A'_0)\dot{x}_0$$

$$+ (S_1/j\omega)A_0/A_0 + A_0'(A_0\dot{x}_0 - A_1\dot{x}_1)$$

$$+ (S_2/j\omega)A_0'/A_0 + A_0'(A_0'\dot{x}_0 - A_2\dot{x}_2) = 0$$

Taking the x_1 coordinate:

$$(S_1/j\omega)(A_1\dot{x}_1 - A_0\dot{x}_0) + j\omega M_1 A_1\dot{x}_1 + R_1 A_1\dot{x}_1$$

$$+ (S_3/j\omega)(A_1\dot{x}_1 + A_2\dot{x}_2 - A_3\dot{x}_3) = 0$$

Taking the x_2 coordinate:

$$(S_2/j\omega)(A_2\dot{x}_2 - A_0'\dot{x}_0) + j\omega M_2 A_2\dot{x}_2 + R_2 A_2\dot{x}_2$$

$$+ (S_3/j\omega)(A_2\dot{x}_2 + A_1\dot{x}_1 - A_3\dot{x}_3) = 0$$

Taking the x_3 coordinate:

$$j\omega M_3 A_3\dot{x} + R_3 A_3\dot{x}_3 + (S_3/j\omega)(A_3\dot{x}_3 - A_1\dot{x}_1$$

$$- A_2\dot{x}_2) + (S_4/j\omega)(A_3\dot{x}_3 + A_4\dot{x}_4) = 0$$

(a)

(b)

(c)

Figure 1.83 *(a) Cross-sectional view of a unidirectional moving-coil microphone. (b) The analogous electrical circuit: p₁, the acoustic pressure on the front openings; p₂ the acoustic pressure on the case-side openings; p′₂, the acoustic pressure on the bass equalizer pipe opening. T₁ and T₂ are ideal transformers. (c) Free-field response of a microphone. (Courtesy of Gayford (1970).)*

Taking the x_4 coordinate:

$$j\omega M_4 A_4 \dot{x}_4 + R_4 A_4 \dot{x}_4 + (S_4/j\omega)(A_4 \dot{x}_4 + A_3 \dot{x}_3) = p$$

The equations are seen to be the mesh equations of the analogous electrical circuit given in *Figure 1.82(c)*. They show that the two separately acoustically loaded areas, represented by the rear surface of the diaphragm dome and the rear surface of the diaphragm surround, leads to the two ideal transformers, T_1 and T_2, which have turn ratios of $A_0/(A_0 + A_0')$ and $A_0'/(A_0 + A_0')$, respectively.

The input voltage applied to the network represents the sound pressure applied to the microphone front openings, whilst the current through M_0, the inductance representing the diaphragm and speech coil moving mass, is proportional to the e.m.f. generated or the response of the microphone. In this analysis, wave effects are assumed to be absent, and thus it is legitimate to assume that the same sound pressure is applied to the microphone front openings M_{-1} and to the rear M_4 openings, without any phase displacement, and also that the microphone main case volume can be represented by a single stiffness S_4. In practice, the fact that the leak tube is operative at frequencies below 100 Hz (3.5 m), and the employment of distributed damping in the microphone case volume, mean that the approximation is valid for omnidirectional microphones.

It will be noted that the analogous circuit normally gives the pressure response rather than the free-field response of the microphone, as the effects of diffraction around the case cause a modification of the constant input voltage condition for the circuit.

1.7.9 Pressure-gradient microphones

We note that the generalized construction and layout shown above can apply to both omnidirectional pressure-operated moving-coil microphones and to pressure-gradient operated microphones. For the omnidirectional microphones, the rear leak path $M_4 R_4$ is usually made in the form of a long narrow tube with a high acoustic mass and resistance which effectively blocks sound entry, except at very low frequencies, where its mass resonates with the main case volume and gives a useful boost to the low bass response in a manner similar to the action of the vent in a vented enclosure (reflex) loudspeaker cabinet. (Some leak has always to be provided to equalize barometric pressure.)

For directional microphones which are pressure-gradient operated, the rear entry path is given quite different constants, and may even be of a distributed nature, as it functions as a phase-shifting network which is designed to give the proper delay to the sound pressure p. The system thus becomes pressure-gradient operated and the other microphone constants such as the diaphragm stiffness and the controlling resistances are arranged to make the overall system substantially mass controlled instead of resistance controlled, as for the omnidirectional pressure operated microphone (Gayford, 1970).

We also note that, for small narrow passages such as where the edge of the diaphragm dome comes close to the inner pole piece, it may be difficult to decide whether this area should be assigned as an additional mass added to the inner magnet gap or whether it should simply form part of the small volume of air under the diaphragm dome. We also note that two ideal transformers of ratios shown have to be introduced to couple the inner and outer diaphragm areas into the appropriate sides of the coil slot. Subtleties of this sort may reduce the absolute accuracy of the necessary lumped constant analogy, but do aid the understanding of the system.

References

BAXANDALL, P.J., *Loudspeaker and Headphone Handbook*, Butterworths, Oxford, p. 106 (1988)

FORD, R.D. *Loudspeaker and Headphone Handbook*, Butterworths, Oxford, p. 20 (1988)

GAYFORD, M.L., *Electroacoustics*, Newnes-Butterworths, Oxford, p. 142 (1970)

HUNT, F.V., *Electroacoustics*, Wiley, New York (1954) (Reprinted by Acoustic Society of America 1984)

JORDAN, E.J., *Loudspeakers*, Focal Press, London (1963)

KELLY, S., *Loudspeaker and Headphone Handbook*, Butterworths, Oxford, p. 36 (1988)

MORENO, J.N., Measurement of loudspeaker parameters, *Journal of the Audio Engineering Society*, **39**, 243 (1991)

OLSON, H.F., *Acoustical Engineering*, Professional and Audio Journals, Philadelphia, PA (1991)

ROBERTSON, A.E., *Microphones*, Iliffe, London (1963)

THACKERAY, D., *Loudspeaker and Headphone Handbook*, Butterworths, Oxford, p. 249 (1988)

POLDY, C.A., *Loudspeaker and Headphone Handbook*, Butterworths, Oxford, p. 468 (1988)

1.8 Introduction to digital audio

1.8.1 Why digital?

Analogue audio signals consist of the variation of sound pressure level with time being mimicked by

the analogous change in strength of another phenomenon. This might be an electrical voltage, a magnetic field or the deviation of a groove.

The principle of digital audio is very simple and that is to represent the sound pressure level variation by a stream of numbers. These numbers are represented by pulses. The major advantage of using pulses is that the system merely has to distinguish from pulse and no-pulse states. Any noise will be ignored provided it is not sufficient to prevent that distinction. Wow and flutter are eliminated from recording systems along with analogue artefacts such as frequency response and level changes and transient distortion due to tape crushing. If the audio is copied as digital data then it may be cloned many times without any degradation. This is unlike analogue systems where generation loss is an important factor in deciding operational logistics. These aim to reduce the number of copy generations between the original and the final broadcast programme.

Digital circuits are not subject to phase error and one of the earliest digital systems introduced in broadcasting was the BBC's PCM system used to distribute radio networks from studio centres to transmitters around the UK. The requirement was to convey stereo signals that could also be combined in mono receivers to provide a satisfactory compatible signal. Analogue lines suffer from phase errors that would have led to high frequency loss or even comb filter effects when the left and right signals are combined.

Analogue apparatus suffers from component drift and circuits usually have to be designed assuming component tolerance of 1%, 5% or 10%, depending on cost. Similar functions using digital technology can emulate components that are precisely accurate. To design an analogue filter with 2000 active elements would be hopelessly impractical. In the digital domain this is quite feasible.

1.8.1.1 Disadvantages
At first digital techniques had the disadvantage of expense, but the use of very large scale integrated (VLSI) chips overcomes this. They are expensive to design and to produce the first few samples. But thereafter they can be stamped out at low cost. Digital audio is often riding on the back of developments in the computer industry, adopting components and techniques as they become cost effective. Analogue audio switches, variable resistors and potentiometers are required to be of the highest grade to avoid contact noise. Controls for digital

devices can be very cheap as contact noise can be eliminated electronically.

Operationally, it is simultaneously an advantage and a disadvantage that a digital system can conceal or eliminate interference. The characteristic of most digital systems is that they tend either to behave apparently perfectly or not at all. This is because with a series of absolute values being sent they are either decoded successfully or they are not. Some delta modulation systems like Dolby AC-1 used for satellite broadcasting in Australia, exhibit a gentle degradation. The advantage for broadcast engineers of soft degradation is the simple one that it is possible to hear a problem (and possibly do something about it) while it remains usable and before a feed fails. Normally there is a relatively narrow area where deterioration is heard as spits, bangs and mutes.

By far the greatest disadvantage is delay. Most digital audio systems have delays in them. This could be while a buffer fills or while a series of mathematical operations are carried out on the data. For the reproduction of recordings this can be merely annoying especially if it is inconsistent as with the compact disc where the delay will vary depending where the data are on the disc. If the start button is pressed just after the required section has passed then the disc has to revolve a whole revolution before it is available. On the other hand the start might be virtually instantaneous if the data is approaching the compact disc's laser when the button is pressed. Some professional machines have a buffer on the output so that instant starts are both possible and reliable.

On live transmissions the delays due to tandemmed systems can make off-air cue problematic.

1.8.2 What is digital audio?

Pulse and digital systems are well established and can be considered to have started in Victorian times. The best known pulse system is the Morse code. This code uses short and long pulses combined with short, medium and long spaces between pulses to convey the information. The original intention was to use mechanical devices to decode the signal. In practice it was found that human operators could decode by ear faster.

Modern digital systems run far too fast for human decoding and adopt simple techniques that can be dealt with by microprocessors with rather less intelligence than a telegraph operator. Most systems use two states that can be thought of as on or off, short or long, dot or dash. Some line transmission systems

are bipolar and use three states: positive, off, negative. The rest of this introduction considers two-state systems.

1.8.2.1 Binary arithmetic

With two-state systems the convention is to describe the two states as '0' and '1'. This is because binary arithmetic is used. We are used to decimal mathematics using ten symbols. Numbers larger than nine are represented by two or more symbols in a line. The right-hand symbol is the 'least significant' as it conveys its absolute value. The character to the left represents the number of tens; the next, the number of hundreds; and so on.

Binary arithmetic is similar except that only the two symbols '0' and '1' are used. The right-hand digit represents one or nought. The digit to its left represents the number of twos; the next the number of fours, etc. Unlike human use of decimals, machines are happier with a fixed-length number. The length of this number is fixed by the designer. This is known as a digital word. (Confusingly when text is transmitted digitally a single digital word may contain the code for a single letter.)

In the early 20th centrury, the word length often adopted was one of five digits, the best known implementation being the Creed teleprinter system. An early railway signalling system used by London. Transport also used a five binary digit word. The number of combinations that are possible using five digits is 32 including zero. The value zero will be represented as:

00000

The value 31 will be represented as:

11111

This is equivalent to:

$(1 \times 16) + (1 \times 8) + (1 \times 4) + (1 \times 2) + (1 \times 1)$

or, put in another way:

$2^4 + 2^3 + 2^2 + 2^1 + 2^0$

Because the least significant bit (LSB) corresponds to 2^0 it is usually known as 'bit 0'. Each other bit is also called by the power of 2 that it represents. The most significant bit of an 8-bit word will therefore be called 'bit 7'. This can be thought of as equivalent to the British convention of describing floors in a building. In the UK the first floor is the first floor above ground level, floor 'zero' being the ground floor. .

Some binary values are listed below:

0 = 0	1 = 1	10 = 2	11 = 3
100 = 4	101 = 5	110 = 6	111 = 7
1000 = 8	1001 = 9	1010 = 10	1011 = 11
1100 = 12	1101 = 13	1110 = 14	1111 = 15
10000 = 16	10001 = 17	10010 = 18	10011 = 19

The length of this binary word controls how many combinations there are and the highest value that can be conveyed. Each 0 or 1 value is known as a bit (BInary digiT). The binary digits on the left of the word are known as the most significant bits (MSBs). Those on the right are known as the least significant bits (LBSs). Early domestic computers used data words of 8 bits. This grouping became known as a *Byte* (allegedly a contraction of 'by eight').

Home computers developed in the mid-1980s used a data word of 16 bits with 32 bits and 64 bits being used from the early 1990s. The byte remains the fundamental unit to measure data – in particular memory which is usually measured in Kbytes. The large 'K' indicates that the 'kilo' is not a metric kilo, meaning 1000, but a binary kilo which is actually 2^{10} or 1024. Similarly, 1 megabyte (Mbyte) is actually 2^{20} or 1 048 576.

Binary is a difficult system for humans to work with and a nomenclature known as 'hexadecimal' or 'Hex' is often used. In this system, 16 symbols are used rather than the two of binary or the ten of decimal. The values 0 to 16 are written:

0, 1, 2, 3, 4, 5, 6, 7, 8, 9, A, B, C, D, E, F, 10

This means that the 256 values that can be represented by 1 byte range from 00 to FF. A 16-bit datum value can be represented by 0000 to FFFF. It is much easier for a human to notice the difference between B19C and 599C than between 1011000110011100 and 101100110011100.

The four bits represented by a single hex digit are usually known as a 'nibble' (by punning reverse-derivation from byte).

1 bit = 2 levels	13 bit = 8192 levels
2 bit = 4	14 bit = 16384
3 bit = 8	15 bit = 32768
4 bit = 16	16 bit = 65536
5 bit = 32	17 bit = 131071
6 bit = 64	18 bit = 262143
7 bit = 128	19 bit = 524287
8 bit = 256	20 bit = 1048575
9 bit = 512	21 bit = 2097151
10 bit = 1024	22 bit = 4194303
11 bit = 2048	23 bit = 8388607
12 bit = 4096	24 bit = 16777215

The word length becomes a measure of the resolution of the system and, with digital audio, the fidelity of the reproduction.

How much resolution is needed for digital audio? CDs use 16-bit words giving 65 536 states. The original stereo distribution devised by the BBC in the 1970s used 13-bit words (8192 states). A central audio switching system designed and installed by NBC in the USA around the same time used 12-bits (4096 states). What is the effect of reducing the resolution of the digital audio system? It means that the value reproduced at the output of the system may differ from the input by a value that will vary randomly. This difference could be as high as the absolute value represented by the difference in level between adjacent states. The effect is to etch a 'staircase' onto the smooth edge of a sine wave. This is because the levels produced in digital-to-analogue conversions are a finite number of discrete voltages controlled by the resolution of the system. This manifests itself as programme-modulated noise known as quantizing noise. If the resolution is low then the audio takes on a granular quality. The difference between the actual input voltage and the voltage output from the system is the 'quantizing error'.

A rule of thumb is that for every extra bit available in the data word there is an improvement of 6 dB in the signal/noise ratio. The compact disc uses 16 bits and has a theoretical signal/noise ratio of about 96 dB.

For a sine wave, and ignoring filtering, the actual equation for N bits is:

$$10 \log_{10} \tfrac{3}{2}(2^{2N})(\text{dB})$$

which is equivalent to:

$$6N + 1.75(\text{dB})$$

This reveals 16 bits theoretically to produce a 97.75 dB signal/noise ratio. In practice the operational noise figures are worse than this and the 6 dB/bit formula is sufficient for normal use, although it fails for a small number of bits. The first reason for this is that, unlike analogue systems, digital systems have no overload margin. Once the maximum value that can be represented by a given word length has been reached there is nowhere to go and hard clipping takes place (this is at least preferable to the value overflowing to 0000 and producing a peak level click when the decoded analogue signal leaps from maximum level to zero). Modulation has to be reduced to provide headroom for the system; 10–12 dB headroom is common. Most meter systems

designed for analogue systems like traditional peak programme meters (PPMs) were actually designed to under-read short transients, as analogue systems could stand overloads of such short duration. Digital systems need to be protected from this.

An objectionable aspect of quantizing noise is that it is programme modulated. If the input is less than that needed to activate the LSB then digital silence reigns. An imaginary 4-bit system would have a signal/noise ratio of 25.75 dB. Yet if a 4-bit and a 16-bit system were both fed with speech the 4-bit system would be quieter in the pauses. This is because the room noise would be insufficient to activate the LSB, whilst the wide dynamic range 16-bit system would reproduce it without difficulty. However, unless special techniques are used, the noise floor switching on and off would make the 4-bit system unusable.

1.8.2.2 Pre-emphasis
The signal/noise ratio can be improved further by using high-frequency pre-emphasis. The choice of the pre-emphasis characteristic must be made with care. If too much high-frequency boost is applied this will cause the sound balancer to reduce the general modulation level, which would be counter-productive.

The curves for pre-emphasis are designed making assumptions about the programme content's spectrum. For example, the 50 μs pre-emphasis used for FM radio in Europe was chosen at a time when single mic balancing of music was the norm. Modern mic techniques and electronic instruments like synthesizers have considerably increased the high-frequency content of programme material causing transmitter limiters to operate on much modern music.

Modern design choices tend to roll off the rate of increase of boost at high frequencies. A typical pre-emphasis curve is the J17 equalization used by NICAM (near instantaneously companded audio multiplex) stereo broadcasts (*Figure 1.84*). Compact disc and DAT, among other systems, allow the user a choice of whether to use pre-emphasis or not. A typical decision might be to use pre-emphasis for classical orchestra recording but not to use it for pop music. For the first, a wide dynamic range places a premium on low noise at low modulation levels. For pop recordings there is a perceived requirement to make them sound as loud as possible requiring the highest possible modulation level. The commonplace use of electrical and synthesizer instruments or very close mic balances of acoustic instruments means that the high-frequency content is usually very

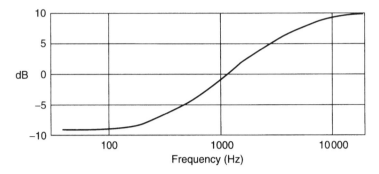

Figure 1.84 *The J17 equalization pre-emphasis curve used by NICAM stereo broadcasts.*

high. The systems use one bit of each block of digital data to act as a flag to show whether pre-emphasis is being used.

1.8.2.3 Dither
At low modulation levels only a small number of states is available. This can lead to audible distortion on some programme material, most notably the die-away of a piano. The quantizing error possible for each sample remains constant regardless of level and therefore becomes a greater proportion of the low level signal. A half quantizing level error will be swamped when the modulation is measured in thousands of quantizing levels but is very significant if only eight levels are being used as might be the case in a pianissimo passage. A very low level sine wave that had a peak amplitude of only one quantizing level would at best be reproduced as a square wave. If its amplitude were to vary only slightly then the mark/space ratio would change introducing unnatural harmonics that the ear can find particularly offensive.

It has been found that adding a small amount of random noise significantly improves the perceived quality. This noise 'dithers' the LSB and can be regarded as the digital equivalent of bias in magnetic recording. (This is due to the nature of the digital signal not of the recording medium.) Dither allows the perception of sounds that are quieter than the theoretical resolution of the system as the audio 'rides' on the noise. When a digital editing device alters levels, care must be taken to re-dither the audio if it is attenuated. A 6 dB reduction in level will lose the bottom 'dithered' bit.

1.8.2.4 How many bits?
For conventional systems the minimum number of bits that can be used is eight and this must use what is known as 'non-linear coding'. This means that more

'numbers' are allocated to the low level range than to the high level range. This reduces quantizing noise and distortion for quiet passages.

In *Figure 1.85* the principle is shown for a 4-bit system. In this imaginary system, quarter level (-12 dB) would be allocated a quantizing value of 10, rather than 4 as with the linear system. The artefacts caused by the reduced resolution for high-level sounds are masked by the sound itself. For linear coding systems using a companding system 10 bits is the minimum thought to be needed (NICAM). For straight linear coding using no special techniques then 12 or 13 bits is an operational minimum. 13 bits were used by the BBC to distribute its programmes from the early 1970s. The early Sony PCM-F1 system used 14 bits as an option but also offered 16 bits as used for compact discs. Professional users who are having to handle uncontrolled or only partially controlled signals need to adjust levels in postproduction and need an even higher resolution, 18 or

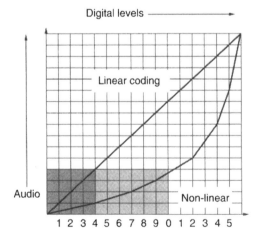

Figure 1.85 *Principle of non-linear coding.*

20 bits. The ultimate aim has been to achieve 24 bits or better although, here, the law of diminishing returns determines whether this is practical and economical. However, the AES/EBU standard for digital connection allows for 24-bit audio. With a 120 dB dynamic range this could code the range from the threshold of hearing to the threshold of pain. Digital devices that perform mathematical manipulation on audio already captured may have internal resolutions of 32 bits so as not to produce rounding errors during intermediate calculations.

1.8.3 Positive and negative

In digital audio, minus values have to be dealt with by a convention that may be different between different systems.

1.8.3.1 Offset binary
Here an offset value equal to half the numeric range is added to the coded value. For an 8-bit system a peak negative excursion gives 00000000 and peak positive excursion gives 11111111. Zero modulation is represented by 10000000, minus one quantizing level by 01111111, and plus one quantizing level by 10000001.

1.8.3.2 Sign and magnitude coding
The signal is described as being 'positive' or 'negative' by a single bit. For an 8-bit system a peak negative excursion gives 11111111 and peak positive excursion gives 01111111. Zero modulation is represented by 10000000 or 00000000 (the sign bit is indeterminate), minus one quantizing level by 10000001, and plus one quantizing level by 00000001. This has the advantage that the magnitude can be dealt with separately from the sign.

1.8.3.3 Twos complement
Again the signal is described as being positive or negative by a single bit. For an 8-bit system a peak negative excursion gives 10000000 and peak positive excursion gives 01111111. Zero modulation is represented by 00000000, minus one quantizing level by 11111111, and plus one quantizing level by 00000001. This has the advantage for computer processing that minus numbers can be added rather than a separate subtraction process being needed. This is used inside most computers and NICAM-728.

This relies on the fact that the number is always a fixed length. For example, imagine a number system based on the decimal system with each number always being four digits long. 1, 2, 3, etc.,

become 0001, 0002, 0003, etc., and −1, −2, −3, etc., become 9999, 9998, 9997, etc. The maximum number that can be represented is 4999; minus 4999 is 5001; 4999 + 5001 = 10000. We lose the '1' that has fallen off the end and the correct answer of 0000 is obtained. Thus 3−1 becomes 0003 + 9999, which gives (1)0002. In binary, 0001 + 1111 (1−1) gives (1)0000 and 0011 + 1111 (3−1) gives (1)0010.

1.8.3.4 Floating point
These systems can be extended to represent an extremely wide range of values by using a floating point convention. How this is implemented varies between systems but essentially works by representing the number as two parts; one that gives its value, the other its range. This is the binary equivalent of the scientific notation that describes 0.002 as 2×10^{-3} and 20000 as 2×10^4. The value is called the 'mantissa' and the multiplying value the 'exponent'.

1.8.4 Sampling rate

A major design parameter of a digital system is how often the analogue quantity needs to be measured to give accurate results. The changing quantity is sampled and measured at a defined rate. If the sampling rate is too slow then significant events may be missed. Audio should be sampled at a rate that is at least twice that of the highest frequency that needs to be produced. This is called the 'Nyquist limit' after Harry Nyquist of Bell Telephone Laboratories who developed the theory. A practical figure is 2.2 times the highest frequency.

1.8.4.1 Aliasing
It is also important that the sampling process is protected from out-of-band frequencies. These beat with the sampling frequency and produce spurious frequencies that not only represent distortion but because of their non-musical relationship to the intended signal, also represent a particularly objectionable form of distortion. These spurious frequencies are called 'alias frequencies' and the filters 'antialiasing filters'. It is these alias frequencies that lead to the Nyquist limit.

Beating between two frequencies is a well-known phenomenon. Musicians use it to tune instruments to a tuning fork. Radios use it to tune to stations (there it is called 'superheterodyne'). Suppose we take a sampling frequency of 10 kHz. With an input of 2 kHz the system would produce not only the 2 kHz but also its 'alias' frequency of 10 − 2 = 8 kHz. As the

input frequency increases this alias frequency would become lower and lower until the input frequency was exactly half the sampling frequency when they would coincide ($10 - 5 = 5$ kHz). Until this point the alias frequency is 'out of band' and can be filtered. With a frequency of, say, 7 kHz there will be a $10 - 7$ kHz $= 3$ kHz beat. This would be well within band and unfilterable. In practice, because antialiasing filters cannot be perfectly sharp, sampling should be about 2.2 times the highest required frequency of audio. The design of these filters is critical as they can adversely affect the audio. They have to be extremely steep as we need to achieve 90 dB of attenuation in less than an octave.

The outputs of digital devices need to be filtered to remove harmonics generated by the switching nature of the signal. Although these harmonics are inaudible they could, if passed onto the next piece of equipment, cause intermodulation distortion, reduced headroom and other artefects in the analogue circuits. They could even beat with the bias of an analogue tape recorder. In an extreme case it might even cause tweeters to burn out. Again, sharp well-designed filters that will not affect the quality of the audio are needed.

1.8.4.2 Oversampling

A common technique to reduce the burden on the filters is oversampling. By reading the data two, four or eight times the spurious frequencies are raised 1, 2 or 4 octaves. They therefore need much less severe filters to attenuate them to insignificant proportions. This can be done at the input as well. There is also a significant advantage as far as noise is concerned. Because a high sampling rate is being used the noise and distortion due to quantizing error remains the same, but is spread over the wider bandwidth. The noise is 'spread more thinly'. When that noise spectrum is filtered so that only the audible range remains, most of the quantizing noise is filtered out as well. Because the audio input has been bandwidth limited nothing is removed. This, with proper design, can have the effect of increasing the resolution of the system. Oversampling of $\times 4$ can make a 14-bit converter behave as if it were 16 bit.

1.8.5 Bit-rate reduction

The 1990s have seen the introduction of bit-rate reduction systems that can achieve remarkable results even at 2-bit resolution. There are economic pressures to do this. It allows more audio to be recorded on a given piece of recording medium, and it reduces

the cost of line rental (charged by the number of bits per second that the circuit can handle). Bit reduction also allows more audio to be processed in a given time. Halving the audio bit rate on a hard-disk system can double the number of audio signals that can be simultaneously output from it.

1.8.5.1 Digital compression

It is important to distinguish between two techniques both of which are sometimes called 'digital compression'. For many years there have existed techniques used to transfer computer data in compressed form. These compression techniques were developed for conveying financial and other precise information and are transparent: the original data are reconstituted at the other end. Because the transfer does not have to take place in real time it is possible to have a variable data rate. This is why fax machines vary the speed of the paper during transmission depending on the complexity of the image.

A digitally conveyed image is represented as a matrix of 'dots' individually known as 'pixels'. Where there is a large length of the same colour, an obvious technique is to replace 1000 data words that all say 'white pixel' with 4 or 5 data words that say that the next 1000 pixels are white. An interesting early example was the Creed teleprinter which was able to transmit more than 32 symbols using only 5 bits as it had a pair of shift codes which changed the set of characters sent.

Audio bit-rate reduction techniques go a stage further and are designed to exploit the imperfections of the ear and omit data that will not be audible. The code that is sent is often one that describes the difference between the new sample and the previous sample (delta modulation). Some systems use an algorithm that predicts what the waveform will be and send a signal describing the difference between the prediction and reality. Others attempt to send data that describe the spectrum rather than an instantaneous level. The precise algorithms are complex and are optimized to take advantage of what the ear can and cannot distinguish.

1.8.5.2 Delta modulation

Delta modulation is a system where the information that is sent is the amount the present sample differs from the previous one. Such a system is the Dolby AC-1 adaptive delta modulation bit reduction system that gives about a 2:1 saving on the number of bits per second needed to be stored or broadcast.

Delta modulation can be summed up very briefly. Whereas the PCM digital system used by the CD

sends a series of absolute values for each sample, delta modulation sends only one bit of information. Is the sample higher or lower than the previous one? In other words what is received is a succession of 'up a bit, up a bit, up a bit, down a bit, down a bit', etc., and so there is no theoretical maximum level as with PCM. This is obviously not the whole story because it begs the question of how often one must send this information (the sampling rate) and is the step up or down? If the step height is too small the system will not be able to handle transients. If it is too big then noise and distortion is introduced. For low bandwidth telephone systems delta modulation works well but for wideband audio what is needed is a way of changing the step height – to make it adaptive hence 'adaptive delta modulation' (ADM). AC-1 includes in the digital signal not only the 'audio' channel but also a separate one that relays the step-height information. This need only have a very low data rate (about 50-bit/s) for good results. An effect of changing the step size is to change the coarseness of the quantization and change the level and character of the quantizing noise potentially causing programme modulated noise. A second low data rate channel is used to control high frequency emphasis to counter this.

This form of delta modulation has the advantage that each sample represents a small increment, so if it is lost this will not be too dramatic a disruption to the audio. With PCM systems if an MSB is wrongly decoded then a gross level change is introduced causing a loud click. Sampling rate is usually 10 or 12 times the highest audio frequency to be conveyed, but as it is effectively a 1-bit system low data rates can be achieved. There are also multibit systems that convey the step height as a value.

1.8.5.3 Companding
As with analogue systems it is possible to reduce noise by compressing the audio before transmission and then restoring the correct level on reception. In the digital domain it is immaterial whether the audio signal is negative or positive because that information is carried in the MSB. It is possible to code the signal so that only a low-resolution signal is sent backed up by 'shift' data to represent the absolute loudness of the signal. NICAM achieves this by sending 10-bit samples in groups of 32. With each group a 3-bit datum is sent describing the absolute level. The system uses 32 kHz sampling so 32 samples represent 1 ms, and hence the term 'near instantaneously companded audio multiplex' (NICAM).

1.8.6 Using digital audio

No audio system is of any use unless there are ways of transferring it from one place to another either down a wire or as a recording. There are many methods for doing this.

1.8.6.1 Handshaking
Most computer systems have some form of handshaking. Here a return circuit allows the receiving machine to request corrupted data to be re-sent or for the sending machine to wait, while the receiver processes information already received, before sending any more. Digital audio has to work in real time and cannot afford this luxury. The exception to this is when audio is recorded on hard disk. Audio manufacturers have found no need to 'reinvent the wheel' so the techniques of recording onto hard disk are identical to those used by the computer industry. The hard disk is spinning sufficiently fast to allow the operating system time to request a re-reading of the data if they are found to be in error. The data are output to a memory buffer. Data on and off of a hard disk or tape streamer will be error free or will fail completely. Once audio data exist as a computer file they may be transferred using standard computer techniques, but not necessarily in real time.

1.8.6.2 Parallel
If parallel circuits are used then for every data bit in the word there is a separate circuit. Stationary head tape recording systems such as digital compact cassette (DCC) use a multiplicity of tracks to reduce the linear speed of the tape required.

1.8.6.3 Serial
Only one circuit is used for data, although other circuits may be used for control information and to convey the clock. Most digital audio systems are serial as this cuts down wiring inside the studio. Outside the studio the use of multiple circuits is impractical due to the expense of line hire and recording equipment.

The bits of the data word are sent one after the other. Multiple circuits can be emulated by sending data faster and sequencing through data from two or more sources. This is known as 'time division multiplexing'. The compact disc has two audio channels plus several data channels. The data channels contain information for the control of the disc as well as having plenty of capacity for text and graphics.

Time division multiplexing is sometimes thought of as having fast rotating switches at each end of

Figure 1.86 *Multiplexing. The rotary switch analogy.*

the circuit feeding each circuit in sequence. Each complete sequence of data items (usually known as a *frame*) must include a framing word to keep the flow in strict order (see *Figure 1.86*).

1.8.7 Coding

To state the obvious, it is essential that when coding the audio for sending from one place to another it must be sent in a form that is readily decodable at the other end. Not only must the data be sent, but also some indication of which section of the data is being sent. The receiving end needs to 'know' when one bit of data ends and the next begins. Within the sample word it needs to be able to distinguish which bit is being sent. When dealing with audio the data is almost invariably multiplexed so that there will be two or more audio channels and usually several channels of control information and data. The receiving end also needs to 'know' which channel is being sent at any instant.

Because the data are conveyed by 'on' and 'off' pulses, any recording medium has only to deal with two quantities, so no bias is required by a magnetic recorder. An erase head is not needed either because the recording head will take the tape to just short of saturation in either polarity. This is sufficient to erase the original data. Magnetic recording media for digital use usually have recording characteristics with much more hysteresis than an analogue recorder. With the compact disc an optical system using a laser is used that detects the presence or absence of pits.

1.8.7.1 The clock
Consider a period of digital silence. This is where the audio data are zero for many samples or even seconds or minutes. Suddenly a '1' bit arrives. What is its significance? Like the percussionist in an orchestra

counting bars until the next tubular bell entry, a digital system needs a 'conductor' to convey the beat. It is therefore vital that what is fed to a recording head or down a circuit contains a guaranteed number of 0/1 transitions to allow the distant end to 'keep its place'.

The data are synchronized by using a 'clock'. This clock is a frequency corresponding to the number of bits per second. This is usually constant but need not do so, as in longitudinal SMPTE/EBU time code used in TV editing systems and automated mix-down systems. This has to be readable over a wide range of tape speeds when the material is being spooled.

Within a specific piece of equipment or even within a studio area it is economical to send a feed of the clock signal on a separate wire. This is impractible for recording systems and for transmission down lines or for broadcast. This would lead to doubling of line costs and perhaps halving of the capacity of the recording medium.

One solution is to have short and long pulses with spaces between pulses. This is known as a 'return to zero' (RTZ) system (*Figure 1.87*). RTZ systems contain the clock information in the leading edge of each pulse being equally spaced. The alternative is

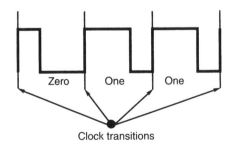

Figure 1.87 *Return-to-zero data.*

to use fixed-length data words each bit being either on or off.

Non-RTZ (NRZ) systems can send twice as much data within a given bandwidth. This is because the 'worst case' of data in the form of 1010101010101010 requires a waveform equivalent to a frequency of half the clock frequency (*Figure 1.88*).

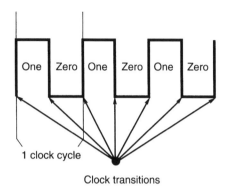

Figure 1.88 *Non-return-to-zero data.*

1.8.7.2 Maximum run length

It is not always necessary to send a continuous stream of clock information if it cannot be ensured that enough transitions will occur to keep a local clock generator in sync with the incoming feed. With modern technology the clock generator is going to be very accurate and will only need the transitions to correct the phase of the clock often enough to prevent it drifting one clock pulse out of sync. A design criterion of a system is therefore the maximum run length. The system needs to prevent a long run of 1s or 0s as this is equivalent to trying to record

d.c. Some of NRZ's extra capacity is traded off by dividing data into blocks and having a header to each block that is arranged to have a unique sequence of pulses. These are used to synchronize a clock generator. The generator acts as a tuning fork that is 'struck' often enough by the header blocks to 'sound' continuously. There is a limit to the period between 'strikes' as recording media will have an element of jitter and the system needs to be impervious to this. While some systems will be crystal controlled and will start from approximately the right value, others will not. The compact disc changes its speed of rotation to maintain a constant linear speed under the laser pick-up. It is the detection of the data bitstream that controls the rotational speed (*Figure 1.89*).

Data are usually sent in blocks or frames. Provided the start of the block can be detected then the order of data within the block is known as it has been designed in. A common technique is to start the block with a sequence of 0s and 1s that, either cannot occur within the data, or are extremely unlikely to occur. This is also designed to ensure that enough transitions are introduced to give the receiving end enough information. In bandwidth terms what is being done is to reduce the frequency range that the system has to cope with. A long period of 1s or 0s corresponds to d.c. on an NRZ system. This would require the system to have a bandwidth of an 'infinite' number of octaves. By ensuring that d.c. is never transmitted then the octave range is brought to practical levels.

1.8.7.3 Biphase/FM

The use of biphase modulation can combine the advantages of both the above systems and has the advantage that a phase inversion does not affect

Figure 1.89 *Example of a typical audio dataframe (AES/EBU standard).*

data reception. (This system is also used with SMPTE/EBU time code.) It is effectively a frequency modulation system and, therefore, is often called 'FM' (*Figure 1.90*). Biphase/FM is a robust system with a narrow bandwidth, has plenty of clock edges and is d.c. free. But it is inefficient and can be regarded as sending a stream of 1s separated by data bits.

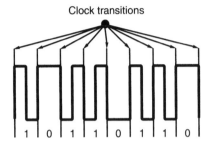

Clock transitions

| 1 | 0 | 1 | 1 | 0 | 1 | 1 | 0 |

Figure 1.90 *Biphase or FM data.*

Modified FM (MFM) (*Figure 1.91*) or Miller code (after its devisor) is similar to phase modulation and doubles the amount of data that can be sent down the same bandwidth or on the same amount of recording medium. The system relies on a phase lock loop being used at the receiving end to recover the clock. This works like a flywheel (our continuously struck tuning fork). MFM works by causing a transition between adjacent 0s and in the middle of the clock period for a 1. A disadvantage of this system is that when an even number of 1s is encoded a d.c. component is introduced.

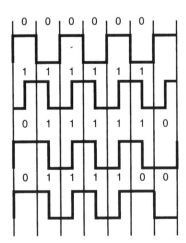

Figure 1.91 *Modified FM data.*

By one of those coincidences that are doomed to cause confusion, the Miller code was modified by another researcher with the name Miller and rather than call it the 'Miller Miller code' it was dubbed 'Miller Squared code'. It has nothing to do with mathematical squaring. This introduces another rule that suppresses the transition in the middle of the last 1 of a sequence of an even number of 1s. The decoder takes account of this and the data are correctly retrieved.

1.8.7.4 Block codes

Another approach to the problem of ensuring enough transitions take place reliably to drive the distant end clock is to use what is called a 'block code'. Compact discs use a block of 14 bits into which are encoded 8 data bits. This, for obvious reasons, is called 'eight to fourteen modulation'. Fourteen bits give 16384 possible combinations to convey only 256 states. A look-up table can be created to decode the valid codes – any other codes are, by definition, errors. The rules used for compact discs are that the word must only have single 1s, at least two consecutive 0s, no more than 10 consecutive 0s and must not begin or end with more than eight consecutive 0s. It turns out that these rules give 257 valid codes. One of these is not used.

1.8.8 Processing digits

Once audio has been converted to numbers it can be processed by any number of mathematical means.

1.8.8.1 Mixing

The digital equivalent of mixing is to add the numbers from the two sources together. If the two numbers are identical the resulting value will be twice the original. This is equivalent to the 6 dB increase in level when two identical audio sources are mixed together. Two equal but opposite-sign numbers cancel, like out-of-phase audio.

1.8.8.2 Gain

The digital equivalent of gain is multiplication. To obtain 6 dB gain the original number is multiplied by 2, and to obtain 12 dB gain the multiplication factor is 4. Attenuation is achieved by multiplying by values smaller than 1: for −6 dB gain the multiplication factor is 0.5.

As well as to addition, subtraction, multiplication and division, various logical manipulations can be undertaken using what is known as 'Boolean algebra'. Devised by the Reverend Boole in Victorian

times it was only with the arrival of digital electronics that it found a major role. In Boolean algebra 1 is known as true and 0 as false. Practical devices usually 'pull' the logic .voltage to earth rather than 'pushing' a voltage to logic 1. Because of this negative logic is often used. This effectively makes AND equivalent to OR. Similarly, OR becomes equivalent to AND. EOR remains functionally the same:

AND	OR
1 and 1 = 1	1 or 1 = 1
1 and 0 = 0	1 or 0 = 1
0 and 1 = 0	0 or 1 = 1
0 and 0 = 0	0 or 0 = 0

Exclusive Or (EOR or XOR)

1 EOR 1 = 1
1 EOR 0 = 0
0 EOR 1 = 0
0 EOR 0 = 1

NOT inverts
NOT AND and NOT OR are usually written as NAND and NOR

1.8.9 Basic devices

1.8.9.1 Memory
These logical tests can be performed by logic gates. Different combinations of these gates are the basic building blocks of any digital system. They can be combined to form latches, shift registers and counters. A collection of latches that can be switched to 1 or 0, and then hold that setting until another switching pulse is detected, form the basis of a form of digital memory called 'static random access memory'. 'Random access' refers to the fact that any bit of memory can be accessed directly by the computer system.

Another form of random access memory (RAM) is called 'dynamic memory'. Here the status is held in a storage capacitor. This will discharge itself unless it is refreshed at frequent intervals by having its data read and then rewritten to recharge the capacitor.

Most forms of memory are volatile. Unless some form of battery backup is applied the memory contents will be lost when the power is removed. The basic programs and data needed to start up a system are held in read only memory (ROM). This cannot be altered by the software and survives switch off. An intermediate form of memory is the erasable programmable read only memory (EPROM). This is a chip that behaves like ROM but is programmed

in a special machine. The chip has a window in it and about half an hour's exposure to hard ultraviolet light (not a sun bed!) will erase it ready for reprogramming.

Traditionally the mechanical parts of a computer system have been known as hardware. The program that is loaded to tell the machine what to do is known by reverse derivation as 'software'. The operating system programs stored in ROM and EPROM is often called 'firmware'.

1.8.9.2 Counters
A different combination of latches can be formed to make a binary counter. When pulses are input the LSB output toggles on the falling edge of every pulse. The next output toggles every time the first bit has a 1 to 0 transition, and so on. Counters have reset inputs so that they can be made to count any number by depending on the setting of logic gates.

For example, a divide-by-10 counter can be made from a 4-bit counter by using a logic circuit monitoring for the presence of 1 on the outputs of both bits 3 and 1. When both are 1 then a reset is forced. Because there is an instant where the counter output is an illegal 1010 there is a short 'glitch' and the output is not valid. Similarly, when the counter goes from 111 to 1000 there will be a finite time while the carry travels from one bit to the next. Any practical system will use extra circuitry to ensure that data are not used until time has been allowed for them to stabilize (settling time).

1.8.9.3 Shift registers
Shift registers are devices consisting of a series of memories (or registers) that are connected in tandem. Data are presented to the first memory which latches the data on the edge of a clock pulse. On each clock pulse the datum in memory one is passed to memory 2, 2 to 3, etc. Some shift registers can also be loaded or read 'from the side' in parallel form and can therefore be used to convert from parallel to serial or serial to parallel. They also form the basis of a first-in first-out (FIFO) memory. They can be made reversible and can be used as a stack or last-in first-out (LIFO) memory.

1.8.10 Some practical devices

The introduction of digital technology has allowed the development of devices that can achieve effects that were either never before possible or only achievable in the analogue domain using expensive, bulky and usually unreliable mechanical means.

1.8.10.1 Delay lines

The first digital audio devices to become generally used within the professional audio industry were delay lines. The digital audio is fed into a shift register. If the clock pulse is 32 kHz then with 32 000 registers in tandem, the signal will emerge 1 s after it entered. The delay can be varied by changing the tapping point or changing the clock frequency. Many commercial delay lines have a times-two mode that uses half the clock frequency and hence gives half the audio bandwidth (because of the Nyquist limit).

1.8.10.2 Echo

If a portion of the output is fed back to the input there will be a flutter echo. If there is feedback of a multiplicity of taps off the delay line of registers the device emulates the multiple reflections of a room. A practical system uses a microprocessor to control a digital signal-processing chip allowing software simulation of many different acoustics.

1.8.10.3 Pitch change

If the shift register is in a ring and it is arranged that the audio data can be input at any point and output at any point independently we have the basis of a pitch changer. If the data are fed in at a faster rate than it is removed, the pitch will be reduced. This much is easy until, of course, the ring becomes full, represented by the input pointer 'lapping' the output pointer. If nothing is done there will be a very audible short perturbation known as a glitch. Different manufacturers have devised different software strategies (or algorithms) to deal with this. Equally pitch shift upwards is possible – data are effectively output more than once. The glitches are again disguised by software. A combination of techniques can be used to change the speed of a performance without changing the pitch.

1.8.10.4 Equalizing a number

Once audio is in digital form it may be processed and manipulated. Gain can be changed. Mixing may take place. Equalization and filtering can be achieved. Usually this is done using specialized digital signal processing (DSP) chips.

How can one equalize a number? In an analogue circuit, filtering and equalization is achieved using capacitors and, perhaps, inductors. They operate by delaying the signal through them, allowing frequency-dependent operations to be performed by combination with the original signal. The same principle applies to digital signals. The use of software allows 'circuits' of great complexity to be constructed that would be hopelessly expensive to produce using analogue components. The digital software equivalent does not suffer from component tolerances and can be configured into a totally different setting in an instant.

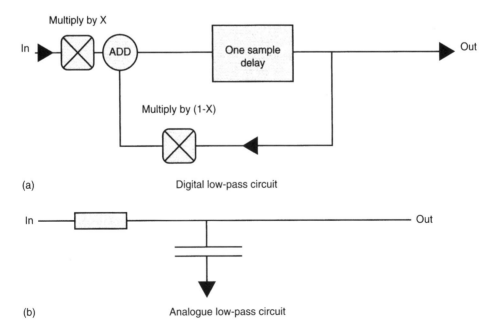

Figure 1.92 *(a) A digital low-pass circuit. (b) An analogue low-pass circuit.*

Figure 1.92 shows the basic building block of digital equalization systems, namely a low-pass filter and its analogue equivalent. As the input frequency increases so the feedback through the delay changes phase until it subtracts rather than adds to the incoming signal.

1.8.11 Digital-to-analogue conversion

Turning numbers into an analogue voltage is relatively straightforward using a ramp generator. This is a binary counter with each of its outputs feeding via a resistor to a common junction with a pull-up resistor. This forms a variable potential divider. As the counter pulls down different combinations of resistors so the output from the potential divider changes. Different value resistors give different strengths to each digital bit. Instead of a counter the resistors can be fed by the output of recovered data.

1.8.11.1 Aperture circuit
In practice an aperture circuit is required. This is because it takes a finite time for the digital data to settle to their final values. An analogue electronic switch is opened only for the settled period. The resulting output is a series of pulses at the sampling rate. The amplitude of the pulses is proportional to the analogue audio value. Low-pass filters remove the pulses and leave the audio. (See *Figure 1.93*).

It is possible to economize by using a single digital-to-analogue converter (DAC) and switch it between channels. However if this is not also done at the beginning with the analogue-to-digital converter (ADC) process then one channel will be delayed with respect to another. Some compact disc players do this. For stereo reproduction there is not a problem as the delay is equivalent to the listener's head turning and moving the ears by a couple of centimetres. It can be a problem for broadcasters as a mono output produced by summing the left and right outputs will

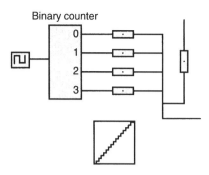

Figure 1.93 *A ramp generator.*

lose high frequencies from about 11 kHz for centre images.

1.8.12 Analogue-to-digital conversion

Analogue-to-digital conversion is performed by sampling the analogue voltage and then generating a voltage using a digital-to-analogue circuit that matches it. The number used to generate that voltage is sent as data. The changing voltage is first sampled using a sample-and-hold circuit (*Figure 1.94*). A switch is momentarily closed, connecting a capacitor to the changing voltage for long enough for it to charge or discharge to the same voltage. The switch is then opened and the sampled voltage measured. The most obvious method involves a counter starting from zero feeding a ramp generator. A comparison circuit compares the two voltages and when the ramp generator equals or exceeds the sampled voltage the counter is stopped at the resulting binary number read. If this is fed into a shift register then the data can be clocked out serially. If the capacitor is at all leaky then the sampled voltage will decrease, introducing what is known as a 'droop factor'. The chances are that the voltage generated by the ramp generator is not exactly equal to the sampled voltage and this is a source of quantization distortion.

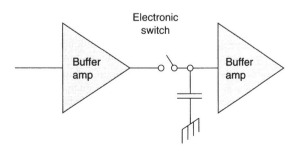

Figure 1.94 *Sample-and-hold circuit.*

For 16-bit resolution conversion it is unnecessarily complex to have to generate up to 65 536 discrete levels to allocate a number to a voltage. It also requires a clock generator that is much faster than the sampling clock (65 536 times faster for 16-bit data). A much better way is to use successive approximation. Each bit is tested in turn, starting with the most significant one. If the voltage from the MSB is higher than the sampled voltage then the test is continued with the next digit. If the sampled voltage is higher then the MSB is added to the next significant bit for the test.

For a 4-bit system, ignoring negative values, with a range of values from 0 to 15 units, the bits have the values

$$\text{Bit 3} \quad 2^3 = 8$$
$$\text{Bit 2} \quad 2^2 = 4$$
$$\text{Bit 1} \quad 2^1 = 2$$
$$\text{Bit 0} \quad 2^0 = 1$$

The testing sequence runs thus. Let us say the sampled voltage is $5\frac{1}{2}$ units.

> Set bit 3 high and compare the generated voltage with the sampled voltage:
> $8 > 5\frac{1}{2}$
> too high, therefore bit 3 = 0.
> Set bit 2 high and compare the generated voltage with the sampled voltage:
> $4 < 5\frac{1}{2}$
> too low, therefore bit 2 = 1.
> Set bits 2 and 1 high and compare the generated voltage with the sampled voltage:
> $(4 + 2) > 5\frac{1}{2}$
> too high, therefore bit 1 = 0.
> Set bits 2 and 0 high and compare the generated voltage with the sampled voltage:
> $(4 + 1) < 5\frac{1}{2}$
> too low, therefore bit 0 = 1.

We have now run out of resolution and the remaining $\frac{1}{2}$ unit represents the quantizing error.

1.8.12.1 Buffers

It is not necessary for data to be continuous. It is possible to design systems where a memory buffer is filled with bursts of data. Provided that data input keeps pace with the data output a continuous reproduction can be achieved. When a memory buffer is being used the output clock rate can be crystal controlled. This means that for recording systems wow and flutter are eliminated as audible effects. The mechanical rotating parts still have to be made

to high standards so that the raw data can be decoded. In digital terminology the effects of wow and flutter are described as 'jitter'.

The most common use of this is to multiplex many channels into one data stream. This is a process in which separate blocks of information from several different sources can be fed in sequence onto a single transmission channel; it works on the conveyer belt principle. If the blocks are always in the same order they can be removed at the far end and packaged appropriately.

The compact disc multiplexes two high-quality audio channels to provide stereo. (Both compact disc and DAT have within their specification the ability to contain four audio tracks with a corresponding decrease in recording time capacity.) In addition, extra data channels are included. In compact discs two of these are used for internal housekeeping for the system. This includes time and track information and whether pre-emphasis is being used or not. The data channels are sent as an 8-bit word. Each bit is called by a letter namely p, q, r, s, t, u, v or w. Each bit is modulated to provide a serial channel to provide control information, text and graphics.

Another use of the ability to send data in short bursts is in sound-in-syncs TV sound distribution. The basic television waveform was specified in the mid 30s and a considerable portion of the waveform is devoted to vertical and horizontal synchronizing pulses. This allows economies to be made in the design of domestic television sets. Because professional equipment does not need such large sync pulses it is possible to insert bursts of data within the synchronizing pulse. With the PAL 625 line system this gives a basic sampling frequency of only 625 lines × 25 frames = 15 625 kHZ. However, that is assuming that only one sample is inserted. Practical systems insert two samples for a mono sound-in-sync system and four samples for a stereo system. The data are filtered out before transmission to prevent domestic sets from failing to lock to the picture. A sound-in-syncs signal has the advantage for distribution purposes of being tied to the picture. This prevents the wrong sound feed being used. It prevents different routings being used for sound and picture which can lead to a loss of synchronization. It also saves money as a separate line does not have to be booked. (For international hook ups like football matches, etc., the sound-in-syncs feed is used for sound effects which are distributed internationally with the picture while each country will book separate audio circuits for its native language commentaries.)

The Dolby digital sound system for the cinema also makes use of digital audio's ability to be fed in bursts. The system is able to add digital audio to release tracks without removing existing analogue tracks by placing the data for the six digital audio tracks between the film sprockets.

In a practical professional digital audio installation that is expecting to transfer audio from one area to another and in and out of the building the audio data will be locked to a common crystal reference. There are going to be occasions when audio is fed from outside the building where there is a finite difference in frequency between the house clock and the incoming clock. This need not always be the case as there are nationally available frequency references broadcast. When there is a difference then, unless something is done about it there will be periodic loss of synchronization causing audio drop out. If a memory buffer is interposed then this problem will only occur if the buffer becomes full or empty. Before this happens an intelligent device will be able to detect silences and dump or repeat data to maintain the buffer at a half full state.

1.8.13 Errors

This could be the end of the story if perfect transmission and recording media were in use. A practical digital audio system has to cope with the introduction of errors due to noise and mechanical imperfections of recording and transmission media. Errors, especially to most significant bits, can cause distortion, clicks, bangs and drop outs.

A measure of the robustness of the system is the number of errors per second that it can deal with. Digital systems incorporate extra 'redundant' bits. This redundancy is used to provide extra information to allow the system to detect, conceal or even correct the errors.

1.8.13.1 Interleaving
Most errors tend to come in bursts. This may be in the form of recording medium drop-out, or impulse interference for line and broadcast systems. Because of this the data are usually sent out of order so that a burst of errors is effectively spread out along the data stream. This technique is known as 'interleaving'. If the data from different words are interleaved then the burst of errors will appear as single bit errors in several words. These words will contain redundant information that can be used to correct errors. With recording systems some stationary head tape systems like ProDigi (PD) format and digital compact cassette

also use a multiplicity of narrow tracks to spread the data across the width of the tape as well as along it.

1.8.13.2 Burst correction length
Error correction systems fail if a burst of errors lasts longer than the burst correction length. With compact disc this has been thought of as the size of the hole that can be drilled into it before the correction fails. Because microprocessors are able to apply arithmetic to the data it is possible to use coding systems that can detect errors. Computer systems for text, etc., can use handshaking protocols to request that corrupted data are re-sent. This is not possible with digital audio as it has to work in real time and the medium is often only one way. However, audio is not like accountancy and the occasional error can be accepted as it may not be audible in a well-designed system. Having detected the error the system may be able to:

(1) correct the error.
(2) conceal the error – this is usually done by either sending the last correctly received sample (replacement) or by interpolation where an intermediate value is calculated; or
(3) Mute the error – a mute is usually preferable to a click.

1.8.13.3 Checksums
Data can be divided into blocks. As each datum passes it can be used to contribute to a mathematical operation. For example, each datum can be added to provide a total that is sent at the end of the block forming a checksum. This does not necessarily have to be the full number: any overflow can be lost. For example, the maximum value that can be represented by an 8-bit word is 255 (FF hex). If we add 1 to it the result should be 256 or 100 hex, but by losing the overflow we come up with a result of 0 (00 hex). This can then be sent at the end of the block. On reception provided the receiving device follows the same rules it can detect whether a block is in error as the chances are that this rule would lead to the least most significant 8 bits being different. Rather than send the 'answer' at the end of the block redundant data can be placed in the bit stream which is given values such that the 'answer' is zero. A common strategy is to use division rather than addition. The method used in compact disc is due to Reed and Solomon. It is then arranged for the division to have no remainder if the transmission is error free. Furthermore, the non-zero answer can be used to calculate where the error is. Simple digital systems use look-up tables but these can quickly become unwieldy. Often 'cyclic redundancy'

is used. Here the data results are fed back through a small number of auxiliary registers. Practical error-correction systems usually use a battery of techniques implemented in commercially available large scale integration chips.

1.8.13.4 Parity

This is the major weapon in the armoury of error detection and correction. At its simplest this consists of adding an extra bit to each data word. This bit signals whether the number of 1s in the binary data is odd or even. Both odd and even parity conventions are used. With an even-parity convention the parity bit is set so that the number of 1s is an even number (zero is an even number). With odd parity the extra bit is set to make the number of 1s always odd. Odd parity has the advantage that data is never all 0s.

Received data can be checked during decoding. If the signal is encoded with odd parity and arrives at the decoder with even parity then it is assumed that the signal has been corrupted. A single parity bit can only detect an odd number of errors. Quite complicated parity schemes can be arranged to allow identification of which bit is in error and for correction to be applied automatically. Remember that the parity bit itself can be affected by noise. *Figure 1.95*

is a method of parity checking a 16-bit word. The data bits are shown as 'b' the parity bits are shown as 'P'. The parity is assessed both 'vertically' and 'horizontally'. The data are sent in the normal way with the data and parity bits intermingled. This is called a 'Hamming code'. If the bit in column 2 row 2 were in error its two associated parity bits would indicate this. As there are only two states then if the bit is shown to be in error then reversing its state corrects the error.

1.8.13.5 Majority voting

Here the data are sent an odd number of times: 3, 5, 7, 9, etc. Because of the enormous increase in data overhead this is rarely used although it is a feature of part of the NICAM 728 stereo sound television transmission system. Here an ingenious method is used called 'signalling in parity'. NICAM divides audio into 1 ms blocks consisting of 32 samples. To save data each audio block is divided into one of 5 level ranges spaced 6 dB apart. The level range is indicated by dividing the first 27 samples into three groups of nine. The 3 bits of data that carry the level information are signalled by setting each group to use odd or even parity according to the value of the bit that they represent. On reception

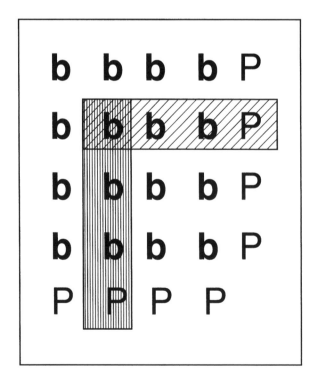

Figure 1.95 *An example of Hamming code.*

the decoder reconstitutes the data by looking at the incoming parity. With perfect reception there will be no doubt as each block of nine will decode as 100% odd or 100% even parity. With poor reception the data can still be reconstituted by choosing the parity used by the majority of data words. This means that five have to be in error before the data bit is wrongly decoded.

1.8.13.6 Cyclic redundancy checks

Complex parity checks could need equally complex arrays of gates or large look-up tables to be implemented. However, the fact that the data are serial can be exploited. By having a series of registers in a loop it is possible to cycle back data so that it can interact mathematically with later data. The cycle is reset from time to time, usually at the end of each block.

1.8.13.7 Editing

Convolutional codes take this concept a stage further and generate very long sequences. Editing convolutional codes that cross frame boundaries can be very difficult. The normal technique requires one head to decode the sequence and another head to rewrite it with a new sequence. This second head has to be critically placed so that data are re-recorded on the same point of the tape. Early digital recording systems were based on video recorders. Some systems were designed so that edits could be made at the junction of video frames (not to be confused with digital frames – compact disc has 75 frames/s but may be mastered from a video recorder using 30 video frames/s). To edit audio to one simple resolution it must be decoded and loaded into memory and

Figure 1.96 *An eye pattern.*

then re-recorded. This is not only to remove the error correction codes but also to allow manipulation such as using a search wheel to 'scrub' back and forth to locate an edit point.

1.8.13.8 Eye pattern height

It is possible to arrange a display on an oscilloscope where all the 0s and 1s are superimposed giving an 'eye' pattern, as illustrated in *Figure 1.96*. This is a good way of assessing how good digital reception is. If the 'eye' is open then reception is likely to be reliable. It is closed if there is a high error-rate. The term 'eye height' is used. The effect of noise is to blur the trace in the vertical direction. The effect of jitter is to blur the trace in the horizontal direction.

2 Acoustics and acoustic devices

Bob Walker
Senior Research Engineer, BBC

Michael Talbot-Smith
Consultant

Chris Woolf
Technical Consultant, Rycote Microphone Windshields

John Borwick
Audio Director, *Gramophone*

Francis Rumsey
Senior Lecturer, University of Surrey

John L. Andrews
Sales and Marketing Consultant, BCD Audio

Peter Baxandall
Former Electro-Acoustical Consultant (Deceased)

Alan Tutton
Lecturer, Audio Operations Training, BBC

2.1 Room acoustics

2.1.1 Introduction

The acoustics of a room are usually thought of in terms of the subjective perception of sound within the room or, if the room is being used for recording, of the resulting output, via microphones and loudspeakers. For brevity, the word 'room' will be used throughout to denote all types of space in which some aspect of the sound quality is important. This includes technical areas such as studios, control rooms, listening rooms, editing suites as well as live-performance venues, domestic environments, offices, public spaces, etc. This field is too wide for a general treatment. For the present purposes, the subject will be limited to small and medium-sized rooms (mainly up to about 400 m^3) used for the purposes of recording and listening to speech and music. The general principles may be extended to other spaces with some modification but, often, with very specialized additional considerations.

In fact, the subject of room acoustics has three aspects which are closely interrelated.

(1) Perceived sound quality – quantified by a number of objective parameters such as 'reverberance' (the musician's 'resonance'), rise time, early and lateral energy ratios and many others.
(2) The continuous internal noise levels in the room – generally quantified by the 'background noise level'.
(3) The sound isolation required around the room in order to reduce the external sounds to acceptable levels within it.

The present section is confined to discussion of the first of these, leaving the significantly interdependent considerations of noise levels and sound isolation to Section 2.2

2.1.2 What is sound and how is it perceived?

Sound energy consists of fluctuations of some physical attribute of a compressible medium, for example, the position of an elementary point in a solid or of the local instantaneous pressure in a fluid. It propagates in the form of waves, by virtue of the mass and elasticity of the medium. Such waves falling within a certain range of frequencies produce the sensation of 'hearing'. For human beings, the range of frequencies which are audible is usually taken to be about

30–20 000 Hz, although these limits depend significantly on the sound level and the age and otological history of the individual.

Because air is a fluid medium, airborne sound energy can only travel as compression waves, with the pressure at any point being alternately above and below the static atmospheric pressure. These waves propagate at a speed of about 340 m/s in straight lines with very little inherent energy loss with distance (at least at medium and low frequencies). Close to a small source the wavefront, that is the line joining points of equal pressure, is spherical and the sound power emitted by the source is spread over a progressively increasing area, leading to a four-fold reduction in intensity for each doubling of distance. It is this 'spreading' effect which is responsible for most of the apparent reduction in sound levels with distance in the vicinity of small sources.

2.1.3 Sound energy in rooms

In practice, and particularly inside rooms, the sound energy emitted by a source does not travel for very long before its path is obstructed by some solid object. Even in a large room such as a television studio or auditorium, this will occur within about 30 ms (10 m). In a small room, this first interaction may well be within 1–2 ms. What happens then is invariably very complicated. Because of the wave properties of sound, the effects of obstacles depend on the relative sizes of the obstacle and the wavelength. Just as in the analogous situation with light, very large objects (relative to the wavelength) will completely obstruct the wave and possibly reflect some or all of it in one or more different directions, the remainder being absorbed. Smaller objects will distort the wavefront, this diffraction causing some of the sound energy to be diverted into different directions. Very small objects will not affect the wave at all.

Unfortunately for the acoustic designer, the range of wavelengths of normal sound is between 20 mm and 10 m. This nicely encompasses the sizes of features often contained within rooms, including the size of the room itself. Thus, over the whole sound frequency range, all ratios of wavelength to object dimension are encountered. This means that the room design must deal with the entire range of effects from total specular reflection to insignificance, through all scales of diffraction.

Additional complications are introduced by the acoustic properties of the various surfaces within

the room, with which the sound field might interact. Some surfaces are intrinsically hard and acoustically reflecting but others may absorb some of the sound energy or modify the shape of the wavefront by virtue of a reactive component of acoustic impedance. Furthermore, in typical sizes of rooms, the sound waves might travel many times across the room interacting with all the objects and surfaces encountered, before becoming insignificant. In many rooms the sound might travel a distance of 20 or 30 times the largest dimension of the room in the time taken to fall in level by 60 dB, which is approximately the range taken for a reasonably loud sound to decay into inaudibility.

Because of this very complicated behaviour over a long period of time, a complete analytic approach to the description of the sound field in even a simple real room is not feasible. The problem may, however, be divided into three distinct frequency ranges: low, middle and high. The basis of the division is essentially that of the relationship between wavelength and object dimensions. At low frequencies the wavelengths are so long that only the principal dimensions of the room are relevant. The detail of all other objects is irrelevant. At high frequencies, many objects are large compared with the wavelength and cause both specular reflection and selective absorption of some parts of the wavefronts. Both of these factors lead to discrete reflections in the short term and to highly diffuse spaces in the long term, that is, spaces in which the sound energy is evenly dispersed throughout the room and, at any instant, is travelling in a highly randomized pattern of directions. Between these two frequency bands is a region of medium frequencies where some or all of these factors are operative.

2.1.4 The three frequency ranges in detail

Throughout this chapter, the phrase 'rectangular room' will be used to denote a three dimensional space in which all of the boundary internal angles are right angles. Strictly speaking, the description should be 'rectangular parallelopiped'.

2.1.4.1 Low frequencies
If one considers a sound wave approaching a room surface in the direction of the normal, then the incident and reflected waves will be coincident (but travelling in opposite directions). In a rectangular room, such a reflected wave will eventually be reflected again from the opposing surface to complete the round trip. If the wavelength happens to be simply

related to the room dimension, then the reflections will be synchronized. Two such waves travelling in opposite directions will establish a standing-wave pattern, or mode, in which the local sound pressure variations are consistently higher in some parts than in others. This will occur at those frequencies for which the room dimension is an integer multiple of one-half wavelength. For example, in a room of 6 m length, this will occur at a lowest frequency of about 28 Hz and then at all multiples thereof. Similar patterns of standing waves will also be present for the other two pairs of surfaces. Furthermore, this triple, infinite subset of 'axial' modes is only one of three types of mode, the other two being 'tangential', involving reflections from four surfaces in turn, and 'oblique', involving reflections from all six surfaces.

In general, a mode will exist at a frequency given by:

$$f_m = c/2 \sqrt[2]{(n_L/L)^2 + (n_W/W)^2 + (n_H/H)^2} \quad (2.1)$$

where n_L, n_W and n_H are positive integers (including 0), L, W and H are the room dimensions, and c is the sound-wave velocity (approximately 340 m/s) (Rayleigh, 1929).

These room modes have the attributes of resonant systems, that is, they have characteristic natural resonance frequencies, bandwidths dependent on their individual loss (damping) factors and amplification factors, also dependent on the damping, by virtue of an energy accumulation and storage mechanism. As in any other resonant system, the energy storage takes the form of repetitive interchange between kinetic and potential energies. A way of visualizing at least a fundamental axial mode is to consider the volume of the air within a room being divided into two parts, one acting as a mass and oscillating from end to end and being resisted by the stiffness of the other part of the air volume, acting as a spring. In the middle, no pressure is generated and all the energy exists as kinetic energy. Conversely, at the walls, no velocity components may exist and the energy of the field is entirely potential energy.

In a control room, listening room or small studio, typical values for the fundamental frequencies of the axial modes are 28, 35 and 45 Hz. Thus, in such rooms, the lowest part of the acoustically important frequency range, up to about 120 Hz, is characterized by a relatively small number of discrete room modes. At or near to each of these frequencies, sound energy will be amplified (to a degree dependent on the damping factor) by the resonance effect. At higher frequencies, above some critical lower limit, there

are very many modes present in a typical room and this amplification is the norm for listening to sound inside rooms. At frequencies between the modes this amplification will not occur. This will be perceived as a relative attenuation. Thus, the response will be irregular. In practice, nothing can be done to eliminate the fundamental modal structure of a room but if the proportions of the room are chosen carefully the coincidence of several modes can be avoided. Rooms whose dimensions are simply related are particularly subject to coincident modes. Rooms which are square or, worse still, half of a cube, are poor indeed. If the modal density is fairly

high it may be approximated by:

$$N \approx 4\pi f^3 V/3c^3 + \pi f^2 S/4^2 + fL/8c \quad (2.2)$$

where N is the number of modes below the frequency f, $V = LWH$ is the volume. $S = 2(LW + WH + HL)$ is the total surface area, and $L = 4(L + W + H)$ is the sum of the edge lengths (Richardson, 1953).

Equation (2.2) shows that the number of modes in any frequency interval increases very rapidly with frequency. At some frequency, the average modal density will be so high that the mode resonances will overlap significantly. This frequency marks the upper limit of the low frequency region of a room.

Figure 2.1 *Distribution map of root-mean-square mode spacing for a 100 m³ room up to 120 Hz.*

The usually accepted criterion for overlap is that the spacing between modes is equal to one third of the bandwidth B given by:

$$B = 2.2/T_{60} \qquad (2.3)$$

where T_{60} is the reverberation time. For typical studios, control rooms and listening rooms of about 100–200 m³, this limit is about 100–120 Hz.

It is often suggested (Olson, 1957) that a pair of 'golden ratios' may be used to select the room proportions and, indeed, the use of such non-simple ratios is helpful, but the essential problem remains. Much work has been done, by this author (Walker, 1992a,b, 1993a) and others (Bolt, 1947; Bolt and Roop, 1950: Louden, 1971; Bonello, 1981; Milner and Bernhard, 1989), on the selection of such ratios. Using some criterion of modal spacing, such as the root-mean-square (r.m.s.) distance between adjacent modes for frequencies up to the critical limit, distribution maps showing the 'quality' may be generated for ranges of room proportions. *Figure 2.1* shows such a distribution contour map for a rectangular room of 200 m³, using the simple r.m.s. mode spacing criterion for all mode frequencies up to 120 Hz. The axes are the length and width as multiples of the room height. For nearly square rooms, ratios near to 1.37 and 1.15 times the room height are optimum. For rooms in which the width and length are both about twice the height, proportions near 2.20 and 1.85 are optimum.

The irregularity of the frequency response as a result of the listener's location within the complex arrangement of standing waves is only one half the problem – the other is the location of the sound source within the same pattern. The listener is usually free to move in order to arrive at some kind of mental average impression of the response. The source, especially if it is a loudspeaker in a control room, will usually be immovable. It is also subject to an additional effect of the modal pattern – that of modification of its actual sound power output by the sound field (Allison, 1974; Salava, 1988). Many sound sources, including most loudspeakers, are actually generators of volume velocity rather than sound power. The delivered acoustic power arises as a result of the cross-product of volume velocity and instantaneous pressure. If the source is located at a pressure node, then the acoustic pressure (radiation impedance) is low for one direction of radiation and the output power in that direction is correspondingly low. This results in reduced actual output power, which is perceived throughout the room as a dip in the frequency response, in addition to the

variations in the modal coupling as a result of the listener's position. *Figure 2.2* shows a typical calculated response for a particular room and positions of source and receiver. *Figure 2.3* shows some of the 200 individual modal responses which contributed to that overall response. This may be compared with the actual measured response for the same conditions, shown in *Figure 2.4.*

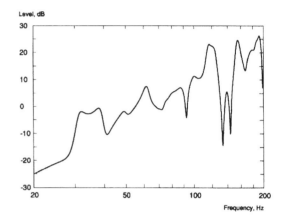

Figure 2.2 *Calculated low-frequency room response.*

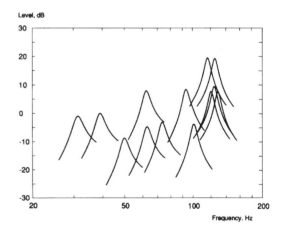

Figure 2.3 *Some of the 120 room modes which make up the overall response shown in* Figure 2.2.

It may be noted that many control rooms are about 6 m wide, with the stereophonic loudspeaker pairs located carefully at the points of symmetry 1.5 m in from each wall. This is precisely at the pressure nodes of the second harmonic of the room width, which occurs at 57 Hz. Many of these rooms also show significant dips in their measured

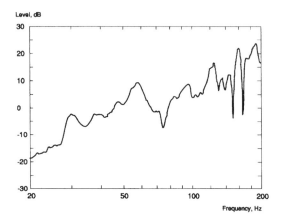

Figure 2.4 *Measured overall response corresponding to* Figures 2.1 *and* 2.2.

frequency response in the 63 Hz one-third octave band.

Two factors may be employed to reduce the magnitude of these effects. One is to increase the damping factor of all the modes by suitable disposition of acoustic absorption which is effective at these low frequencies. This is generally done in any case as part of the routine overall design, to control the measured 'reverberation time' and thereby avoid excessively 'boomy' conditions. In fact, at low frequencies in reasonably small rooms, the reverberation time strictly cannot be either defined or measured because it is a statistical property of completely diffuse spaces. Much better control of the low-frequency response can be obtained by consideration of the lowest modal frequencies individually and then selecting and placing acoustic treatment to deal with each one. By increasing the damping factor, the maximum amplitude of the resonance is reduced and the range of frequencies over which it acts is widened. The latter reduces both the effective mode spacing and lowest frequency at which the modes may be considered to be overlapping, that is the critical low-frequency limit.

It is often said that a possible solution is to make the room non-rectangular. Before discussing that, it may be useful to consider how it is that the surfaces of a room can cause low-frequency specular reflections when the wavelength is comparable with or even greater than the extent of the surface. In an otherwise free space, such a reflection would be anything but specular. Modes can only exist because the reflections are more or less specular. However, if one considers a single boundary surface of a rectangular room, the virtual images of that surface, formed by reflection in the adjoining surfaces, effectively extend the real surface to infinity. Thus, acoustically, a rectangular room consists of three sets of interlocking infinite planes. Of course, in practice, any boundary absorption eventually truncates the planes, but the effect is to produce large reflectors in each principal direction. The effective reflection may therefore be specular, even though the physical surfaces are small.

If one then considers a deviation from parallel of one of the pairs of surfaces, for example a tapering of the width of the room towards one end, the images of that taper will alternate in angle, producing effective surfaces which have a 'ripple-like' structure. Unless the peak-to-peak magnitude of the ripple is a significant fraction of the wavelength, this will have no effect except to shift the effective reflector to the mean position, that is, the room width at the midpoint of the side walls. To have any other effect at, say, 60 Hz the extent of the taper would have to be of the order of 3 m per side!

Many modern rooms are significantly non-rectangular, generally for reasons other than the low-frequency modal behaviour. In these cases, the mode structure cannot be derived by simple equations or by inspection (although approximate methods based on mean dimensions may give adequately accurate results for most purposes). In such cases, mathematical modelling techniques, for example, finite element analysis (Gladwell, 1965; Schenk, 1968; Goransson and Davidsson, 1987) can give arbitrarily accurate answers. However, the modal behaviour is no less pronounced than in an equivalent rectangular room – just different.

Another factor often employed in control rooms, again usually for other reasons, is the incorporation of loudspeakers into the walls. Theoretically, all modes have pressure antinodes at their reflection points. Thus a hard wall surface is a pressure antinode for one-third of the axial modes, two-thirds of the tangential modes and all of the oblique modes. Placing the sound source there ensures that it is better coupled to more of the room modes. Such considerations can help to reduce the response irregularities, at least for the coupling of the loudspeaker to the room. However, it is generally unrealistic to consider loudspeakers located in the corners of the room and such an approach does nothing whatever about the irregularities resulting from the listener's position with respect to the room modes.

One arrangement in which the loudspeakers can be located in the extreme corners of the room without detriment to the higher frequency image localization

is the use of 'subwoofers'. Because the problem is one of low-frequency response and there can be no image localizing information at frequencies of the order of 100 Hz (consideration of time–frequency resolution limitations simply does not permit it), the extreme low-frequency part of the spectrum could be handled by loudspeakers specially located to achieve an even response. This arrangement may also have side benefits in loudspeaker design and power-handling capabilities.

Much of the above discussion has been related to the reproduction of sound in a control or listening room. All of it is also relevant to studios and live performance spaces – they also have sources and listeners (or microphones) subject to precisely the same considerations. However, many of these spaces are larger, some very much larger, so that the critical frequency limit may be very low indeed. In concert halls, the modal density at all reasonable frequencies is high and these low-frequency problems do not exist at all for the wanted sound. However, some cases of very low-frequency resonances causing problems have been recorded (Knudsen and Harris, 1950; Yamada *et al.*, 1983; Hubbard and Shephard, 1986).

2.1.4.2 High frequencies

The high-frequency range of internal acoustic design is characterized by objects and surfaces being large compared with the wavelength. This results in generally specular reflection of that part of the sound wave which is not absorbed although some objects do produce diffuse reflections, despite their size. Because of the way human beings perceive sound the acoustic design at these frequencies falls naturally into two distinct time zones: long-term effects and early effects.

Long-term effects The two processes of absorption and diffusion dominate the long-term region of the time-scale. All materials absorb some fraction of the incident sound, converting the acoustic energy to heat in different ways. Some materials absorb so little that they may generally be neglected, but most are acoustically significant – even those that have no intended acoustic effect! For calculation of the overall, average effect within a room a useful concept is the mean surface absorption coefficient, usually expressed as $\bar{\alpha}$. This is given by

$$\bar{\alpha} = \frac{\sum A_n \alpha_n}{\sum A_n} \qquad (2.4)$$

where A_n is the area and α_n is the absorption coefficient of an individual surface.

The other characteristic of the sound field in this time zone is that of diffusion. A fully diffuse space is theoretically defined as one in which the statistical energy density is uniform throughout the space and in which there is an equal probability of a sound wave travelling in any direction. It is a theoretical ideal which cannot, in practice, be verified. However, it is a useful concept and many diffuse spaces approach the ideal to an adequate degree. Seriously non-diffuse spaces are much easier to identify. Diffuse sound fields also take some time to become established in most rooms.

The concept of the diffuse field is particularly important in the calculation of absorption. In practice, one difficulty with absorption coefficients are that they are often dependent on the angle of incidence of the sound. However, in general, it is not possible to know the angle of incidence of a sound wave on a material in a room. Also, after many reflections, a sound wave may even impinge on a particular material several times from a number of different directions. Therefore, materials are generally tested and room design calculations carried out on the assumption that the fields are diffuse. Whilst this is convenient and quick, it can and frequently does lead to significant discrepancies between calculated and achieved results.

Beyond the time interval of about 50–80 ms after the arrival of the direct sound, the human perception of sound is of an integrated whole, gradually decaying in level as the sound field energy is absorbed by the structure, furnishings and fittings. This gradual, regular decay was one of the first aspects of perceived sound to receive formal study. There is some clear evidence that the ancient Greeks understood the concept and introduced means of controlling it in their theatres. Vitruvius, writing about Roman theatre (which was based on the Greek) says:

> ... By the adoption of this plan the voice which issues from the *scena* expanding as from a centre and striking against the cavity of each vase will sound with increased clearness and harmony from its unison with one or the other of them. (Translation by Granger (1931).)

The objective in these open-air theatres was probably to increase the resonance rather than to absorb the sound, but the concepts are clearly evident. For

other early work, Bagenal and Wood (1931) and Hunt (1978) give many sources and translations.

In more modern times, a degree of measurement has been applied, to derive descriptions more numerical than philosophical. This began with Sabine (1927) at Harvard in the later part of the 19th century. When asked about the poor acoustics of his University's large hall, he began to make measurements of the time taken for a sound to decay by a given amount under different conditions. He used the sources of test signals which were readily available to him – organ pipes, whistles, horns, etc., and a stopwatch. The range of levels over which he could generally make these measurements lay between the loudest source and the local ambient noise – a range of about 60 dB. Thus, his measure of decay time, and one which is still used today, is the 'reverberation time', i.e. the time taken for a sound to diminish to a level 60 dB below the initial level.

The reverberation time is a statistical description of the general properties of the acoustic field inside a large enclosure. It is unrelated to any of the fine detail of the sound field. It is also, by definition, an average property both of the whole space and over the whole time interval. It can only be measured by averaging the individual results for a large number of positions and can give no information about the form of the decay. Despite these limitations, it has an important acoustic significance. It approximates to what musicians call 'resonance' and is also closely connected with the musician's 'warmth'. It is the only numerical description of a room's acoustics about which there is much consensus. Yet it plainly does not give a complete description, otherwise a small bathroom and a large television studio would sound the same (they might easily have the same reverberation time). Having the appropriate reverberation time could fairly be described as a necessary but not sufficient condition for 'good' acoustics.

In practice, the reverberation time as precisely defined is rarely used. Almost invariably, the number of positions used for the measurement is less than the statistically sufficient number, especially at low frequencies. Indeed, in some areas, the 'reverberation time' at one position may be discussed with some real meaning. Equally commonly, the measurement is rarely made over the necessary 60 dB range of levels. What is more frequently measured is the time taken for the sound to fall over some other range and the results expressed as though that rate of decay applied throughout. In practical rooms, this is actually much more useful because it can reveal significant departures from the ideal and, thereby,

provide an additional degree of specification or analysis. Rather in the manner of group delay (the rate of change of phase with frequency), the characteristic of decay rate as a function of time can be specified and measured.

Because it is a generalized measure, independent of the detailed sound energy pattern in the room, its relationship to the other generalized properties of the room can be quantified. The average acoustic absorption coefficient of the surfaces within the room, together with the size of the room will determine the reverberation time.

The link between the size of the room, its contents and the reverberation time was originally expressed by Sabine as:

$$T_{60} = 0.161V/S\overline{\alpha} \tag{2.5}$$

where T is the reverberation time (s). V is the volume (m^3), S is the total surface area (m^2), and $\overline{\alpha}$ is the average absorption coefficient of the surfaces.

This is approximately true for rooms with small amounts of absorption. For rooms with larger amounts, like most studios and control rooms, the more general expression first derived by Eyring and Norris (Eyring, 1930) is:

$$T_{60} = 0.161V/[-S\ln(1-\overline{\alpha})] \tag{2.6}$$

In practice, in all but the smallest rooms, the average surface absorption coefficient must have added to it a component representing the absorption of the air itself. This is usually small at low frequencies but can become dominant at higher frequencies.

Tables of absorption coefficients covering a wide range of materials are readily available (Gilford, 1972b; Parkin *et al.*, 1979). Some of these are very extensive. *Table 2.1* gives, as examples only, a short-list of some common materials. One of the most serious difficulties of acoustic design in this frequency and time region is the correct identification of all of the materials within a room. It is commonly believed (by non-acousticians) that materials which are included for other than acoustic reasons have no acoustic effect – the materials themselves are generally not aware of that restriction!

Acoustic defects which can arise in this region, apart from simple errors in the achieved reverberation time, are mainly related to serious departures from a diffuse space. Two relatively highly reflecting surfaces facing each other may confine sound energy to a repetitive double-reflection process between them. This process may well have a much lower decay rate than the rest of the room so that, eventually, this becomes the dominant signal present in

Table 2.1 Typical acoustic absorption coefficients of common building and furnishing materials

Material	Octave frequency band centre (Hz)						
	63	125	250	500	1000	2000	4000
225-mm brickwork (structural)	0.05	0.05	0.04	0.02	0.01	0	0
113-mm brickwork (structural)	0.10	0.08	0.05	0.02	0	0	0
75-mm breeze block (structural)	0.09	0.13	0.16	0.03	0	0	0
Timber stud wall (structural)	0.27	0.24	0.12	0.06	0.02	0	0
Board on joist floor (structural)	0.10	0.10	0.07	0.01	0	0	0
12-mm wood panels on 25-mm battens	0.33	0.31	0.33	0.14	0.10	0.10	0.12
Glass (>0.25-in. thick)	0.03	0.03	0.03	0.03	0.03	0.03	0.03
3-mm hardboard on 25-mm battens	0.30	0.32	0.43	0.12	0.07	0.07	0.11
Brick (surface)	0.02	0.02	0.02	0.03	0.04	0.05	0.07
Rough concrete (surface)	0.01	0.01	0.02	0.04	0.06	0.08	0.10
Smooth plaster, painted	0.01	0.01	0.01	0.02	0.02	0.02	0.02
Wood (surface)	0.05	0.06	0.07	0.09	0.10	0.10	0.12
Lino	0.02	0.02	0.02	0.03	0.03	0.04	0.04
Rubber flooring	0.01	0.02	0.03	0.04	0.04	0.02	0.02
Audience, units/person	0.15	0.33	0.40	0.44	0.45	0.45	0.45
Orchestra, units/person including instruments	0.20	0.40	0.85	1.15	1.39	1.30	1.20
Orchestral rostrum (per m^2)	0.44	0.36	0.11	0.10	0.15	0.23	0.15
Haircord carpet on underfelt	0.05	0.13	0.17	0.24	0.29	0.30	0.30
Wilton carpet on underfelt	0.04	0.08	0.22	0.51	0.64	0.69	0.71
Typical carpet tiles	0.01	0.02	0.11	0.11	0.39	0.45	0.55
Curtains, drama, sailcloth, draped	0.03	0.03	0.04	0.10	0.17	0.18	0.15
Curtains, velour, draped	0.05	0.06	0.31	0.44	0.80	0.75	0.65
Lightweight fabric over 50-mm airspace	0	0.04	0.10	0.20	0.50	0.60	0.50
Fabric cover for low-frequency absorbers	0	0	0	0	0.03	0.10	0.40
Ceiling tiles over 25-mm airspace	0.02	0.11	0.33	0.68	0.72	0.51	0.47
Ceiling tiles over 600-mm airspace	0.66	0.74	0.75	0.67	0.66	0.62	0.68
25-mm mineral wool, 5% perf. hardboard cover	0.03	0.09	0.47	1.12	0.90	0.57	0.31
25-mm mineral wool, 25-mm airspace, 5% cover	0.04	0.14	0.65	1.18	0.90	0.57	0.31
25-mm mineral wool, 175-mm airspace, 5% cover	0.35	0.51	0.89	0.99	1.02	0.82	0.44
50-mm mineral wool, 5% perf. cover	0.10	0.31	1.10	1.20	0.90	0.57	0.31
50-mm mineral wool over 150-mm airspace	0.60	0.90	0.95	0.80	0.81	0.83	0.85
A1 Modular absorbers	0.46	1.11	1.27	1.23	1.20	1.18	1.15
A2 Modular absorbers	0.25	1.07	0.69	0.59	0.38	0.24	0.22
A3 Modular absorbers	0.40	0.88	0.94	1.12	1.00	1.00	0.91
A8 Modular absorbers (BBC, 1990)	0.45	1.17	1.11	1.28	1.15	0.98	0.87
A9 Modular absorber (BBC, 1990)	0.42	1.13	0.91	1.30	1.25	1.17	1.25
D2 measured in 106 m^3	1.25	0.81	0.87	0.43	0.30	0.20	0.13
D2 measured in 8151 m^3	1.13	1.23	1.18	0.62	0.40	0.20	0.03

the room. The decay rate would then generally be two-valued, with a higher early rate followed by a slower rate. This is generally disturbing, subjectively described as a 'flutter echo' for a fast repetition rate of high frequencies or a 'honk' if the frequency band affected is somewhat lower. Double decays may also occur in situations with two or more interconnected spaces with widely different mean absorption coefficients, such as the orchestra pit and the main auditorium in a theatre or opera house or as a result of objects, such as fire-extinguishers, acting as resonators.

Early effects In the time immediately following the emission of a sound wave from a source, the wave will propagate as though it was in an open space. Each radial vector from the point of emission can be treated somewhat like a light ray. An entire branch of acoustic design methods has been developed to study such ray-like propagation. At its simplest, each elemental ray may be followed through its interactions with surfaces and objects as they are encountered. Many of these methods follow these hypothetical rays until they vanish into insignificance. In reality, the accuracy of representation of the interactions becomes inadequate for such extended multiplication and the system becomes more or less chaotic. However, in the short term, for a few interactions, the representation may be fairly realistic. On this scale, the particular properties of human hearing become relevant. Many workers, notably Haas (1951), have established that within a period of about 5 ms a reflection cannot be distinguished from the direct sound (provided that its level is reasonable). In the region of 10–20 ms, reflections are capable of causing confusion about the apparent direction of a sound source. Within a period of about 50–80 ms no sense of space or reverberance is conveyed. Beyond that time the sensation of reverberance and blending of the sound into a whole is important. For the acoustic designer, these figures mean that some control of the pattern of reflections reaching the listener is essential.

One effect of the very early reflection pattern which is generally detectable is the interaction of two sound waves separated by a small time difference to form a regular interference pattern. For example, a reflection delayed by 1 ms will introduce some degree of cancellation at 500 Hz and reinforcement at 1000 Hz (and similarly at all multiples of these frequencies). This regular distortion of the frequency response, usually called a 'comb-filter' because of its appearance in the frequency domain, is very disturbing for time delays in the range 5–0.5 ms. Fortunately, most occurrences are relatively easy to avoid. Two common situations are the reflection from a table-top to a microphone above the table and a glancing reflection from the top of a studio mixing desk. The former requires the use of a specially designed, non-reflecting table-top and the latter careful design of the desk and the layout of the control room. Ceiling reflections may also fall in this time range. The effect is only serious if the amplitude of the reflection is comparable with the direct sound. A reflection 6–10 dB down is likely to be quite innocuous.

It has been known for many years that reflections in the time zone from about 5 to 20 ms are very critical, especially for stereophonic listening or recording. Even fairly low levels of reflection amplitude can change the apparent image position. In the past, it was common practice, especially in control rooms, to attempt to absorb all such reflections by the extensive use of acoustic treatment. This has led to progressively more anechoic rooms, to the point where they have become rather oppressive environments. More recently, attempts have been made to control the direction of the reflections and to provide an area within a room which, in this time zone, is relatively free from significant reflections (Walker, 1993a, 1994a,b, 1995 a–c). This method is only applicable in control rooms and listening rooms for very small numbers of people. It is not generally relevant to studios or live-performance spaces.

In practice, there is also an upper time limit to the extent to which this freedom from significant reflections may be achieved. Whilst the first reflections are relatively simple to direct and some of earliest second reflections may also have to be controlled individually, it is generally impractical to consider higher orders. In average size rooms, this limit is about 20–30 ms. Above that, up to about 80 ms, the best that can be achieved is that the reflection pattern should be free from isolated, strong reflections. This means that, by that time, the acoustic field within the room should be diffuse (see later).

Other measures of acoustic quality There have been many attempts, over a period of about 60 years, to find correlations between the subjective acoustic quality in a room and any kind of objectively measurable parameter. Clearly, the first and still the most important is the reverberation time. No one has ever designed an acoustically successful room of any kind with a reverberation time which was grossly different from the normal expectation. More than that, many of the recognized failures of acoustic design are largely due to incorrect reverberation time

and, with only a few exceptions, the retrospectively fitted 'cures' have sought to modify only the overall, mean sound decay characteristic.

However, if there was nothing but the idealized reverberation time there would be no difficulties. Given the correct value, all rooms would sound satisfactory and uniform. Every room would be perfectly diffuse and the decay curve would be perfectly regular. This is clearly far from either the requirement or the achievement. Thus much of the subjective quality, good or bad, lies in departures from the ideal diffuse space and regular reverberation decay.

These other factors, which determine the perceived sound quality in a room (those which do distinguish between a small bathroom and a large television studio), have been studied by many people, particularly in relation to concert halls (Barron, 1971; Hawkes and Douglas, 1971; Schroeder *et al.*, 1974). This is mainly because of the relatively great cost of such rooms and the public and media awareness of the outcome.

In the concert hall the principal requirement is to provide some idealized acoustic experience over as wide a fraction of the audience area as possible. Acoustic measures such as the rate at which the sound field builds up to its steady-state value, the decay rate during the early part of the reverberation phase, the ratio of sound energies arriving from different directions and in different time intervals and many others have been studied. Links to subjective experience have been found and many of the most modern halls built to these criteria are regarded as excellent.

In contrast, the study of measurement and criteria for smaller rooms has been essentially neglected. Indeed, one study at least (Walker, 1987) suggests that there are good reasons why many of the concert hall criteria at least may be inapplicable to small rooms. This is principally because of the differences in time-scales between the large hall and a small room with, probably, a short reverberation time. Because of the inherent uncertainty in the resolution of time and frequency. it is not easy to make measurements of short time-scale events in any worthwhile frequency band (but see Section 1.2.8.4).

2.1.4.3 Midfrequencies
As in many other fields covering a range of dimensions from the very small to the very large, the mid-frequency range in room acoustic design is a mixture of effects. Some of these resemble the low-frequency range and others the high-frequency range. In general, the wavelengths are comparable with the sizes

of objects within the room, but rather smaller than the room itself. The modal density is therefore too high (and too complex) for calculation by summation of modes, but the propagation can by no means be treated as ray-like. Most objects will cause significant diffraction and the wavefronts will rapidly become chaotic.

In principle, numerical methods such as finite element analysis could be employed to derive the distribution of the sound field at each frequency (Gladwell, 1965; Schenck, 1968; Goransson and Davidsson, 1987). However, even today, the necessary computation would be considered extensive. For example, a simple rectangular room of 7 m × 5 m × 3.5 m with no contents would require a minimum of about 670 points/m^3 at 500 Hz, equivalent to a total of 82 000 points. The computer would be required to solve an 82 000 square matrix of simultaneous equations for continuity of pressure and volume velocity in three dimensions. This is readily achieved in other fields but beyond the scope of the type of computer generally available for commercial acoustic work!

The practical acoustic designer is thus limited to applying overall statistical considerations, such as reverberation time and average absorption coefficient, relying on, and even deliberately enhancing, the fact that the sound field is at least fairly diffuse.

Diffusion is one of the unquantifiable aspects of a studio design. Theoretically, it is a measure of the 'evenness' of the reverberant sound field, but the numerical value has little subjective significance and there is no practicable method of 'designing' to a given level of diffusion. In practice, the room is made to be as diffuse as possible by intermingling the different types of acoustic treatment in small patches (Olsen, 1957; Gilford, 1972a), so that diffraction effects distort the otherwise uniform wavefronts. It used to be common to angle some or all of the room surfaces or to provide geometric shapes on the walls to aid diffusion but this is constructionally expensive and does not improve upon the alternative method of dividing up the acoustic treatment.

2.1.5 Measurement of reverberation time

The principle of the measurement of reverberation time is unchanged since the days of Sabine. Some means of producing a suitable excitation signal and of recording the resultant time response are required. Usually, the measurement is required to be carried out in narrow frequency bands rather than the whole frequency range at once, so that some form of filtering is also needed. In practice, a loudspeaker

driven by an electronic signal source is used, usually producing suitably filtered random noise and under the control of an automatic sequencer for starting and stopping. The recording device may be a chart recorder but is now more likely to be some form of computer acquisition, storage and analysis device. In many cases, the whole instrument is self-contained (Walker, 1982).

The instrumental limitations are few. Firstly, the dynamic range, from the steady-state test signal to the room ambient noise level must be sufficient to produce a reasonable fraction of the decay. This is sometimes a non-trival limitation. In particular, those types of real-time audio spectrum analysers which can carry out this measurement, using a single pulse to measure all frequency bands simultaneously, are almost always deficient at low frequencies. The second limitation is that the analysis response time must be capable of resolving the decay. In small, heavily treated studios or control rooms, a response rate of 1000 dB/s may be required. At low frequencies, even the filters used to define the measurement bandwidth may not respond at that rate, especially if they are one-third octave filters.

The analysis of the decays presents more problems than the instrumentation. Most recordings of individual decays are somewhat irregular. It is rarely adequate simply to time the decay between upper and lower threshold levels. The uncertainties in the crossing points would be very large. Most instruments carry out some form of straight-line, least-squares fit to the whole data within the range decided on for the test (often −5 to −35 dB). It is also common practice to average several decays at each frequency. One method (Schroeder, 1965) uses reverse integration of one decay to produce the theoretical equivalent to the average of a infinite number of decays. That curve is then used in place of the actual decay for the straight-line fitting. The range of signal levels over which the decay rate is measured is often set as an arbitrary fixed range, but some instruments are capable of using whatever range is available and others may use two ranges to obtain some measure of variations from a constant rate. All results are usually normalized to the standard 60-dB range, that is:

$$T = 60 \times 1/\text{Decay rate}$$

Figure 2.5(a) shows a typical single decay at 100 Hz in a small room, *Figure 2.5(b)* shows that the average of five such decays at the same location, and *Figure 2.5(c)* shows the same average with the least-squares error straight line.

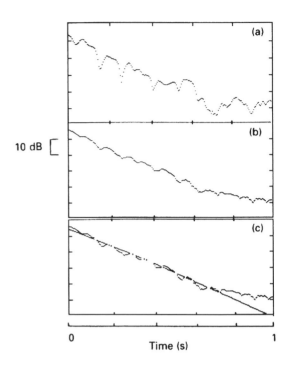

Figure 2.5 *Measurement of reverberation time. (a) Single reverberation decay at 100 Hz. (b) Average of five reverberation decays at one position. (c) Best-fit straight-line showing a measured 1.09 s reverberation time.*

2.1.6 Sound absorption

2.1.6.1 Acoustic treatment

In the design of a room it is very rare to find that the basic structure and furniture provides just the right degree of control of the internal acoustics. The introduction into a room of materials the sole or main purpose of which is the efficient absorption of airborne sound energy (acoustic treatment) is the principal means of control. Thus, the study of sound absorption is fundamental. Essentially, some fraction of the total sound energy is converted into heat energy when sound waves traverse any transition between two media. In general, the two media will both have characteristic acoustic impedances, each with real and reactive components.

From the maximum power transfer theorem, the ratio of the energy absorbed to the maximum which could be absorbed by a perfectly matched load is given by:

$$\alpha = 4R_2^2/[(R_1 + R_2)^2 + (X_1 + X_2)^2] \qquad (2.7)$$

where R_1 and R_2 are the real parts of the impedances of medium 1 and medium 2, and X_1 and X_2 are the reactive parts.

In many cases of sound absorption, the first medium will be air and the wavefront will be essentially plane. This makes the impedance of the first medium purely real and of magnitude about 410 acoustic units (rayls). This value is derived from the product of the density of the medium, ρ_0, (1.18 kg/m^3) and the wave velocity. c. (340 m/s). It is usually denoted by $\rho_0 c$. Thus, the conditions for high values of sound absorption are that the reactive impedance of the absorbing material should be small and the real part should be roughly equal to $\rho_0 c$, in which case the absorption coefficient will be close to 1.

There are two main principles of acoustic absorption. The most common uses a dense mass of fibrous material which has many open pores, typically glass wool or mineral wool. This material behaves as a (very) lossy transmission line. Sound energy which enters the pores is quickly dissipated as heat by the viscous friction as a result of the relative motion between the air molecules and the material fibres. Provided that the material thickness is great, practically all of the sound energy will be absorbed before reaching the far side. Thus the material absorbs practically all of the incident sound, that is, it has a high absorption coefficient. The main problem with such an absorber, apart from its excessive thickness, is the reflection from the front surface which, by definition, cannot have the same characteristic impedance as the air. For the most demanding applications, such as laboratory measurement rooms, the material is made in the form of tapered wedges to provide a gradual transition from air to the material.

For thinner sections of the same kind of material with a solid backing, the sound energy is still significant at the rear surface. This remaining energy is reflected and re-emerges after another transit through the material. If the material is very thin compared with the wavelength there will be very little relative velocity between the air molecules and the material and, consequently, little loss, because of the pressure maximum (and velocity minimum) near to the hard surface. As the material thickness increases so does the absorption coefficient until at about one-quarter wavelength the absorption is at a maximum. Depending on the loss factor of the material, the absorption may fall somewhat at about one-half wavelength and thereafter show a diminishing cyclic characteristic. *Figure 2.6* illustrates an idealized version of such behaviour.

Such absorption mechanisms are clearly unsuitable for low frequencies – even an unreasonably expensive and space-consuming 0.5-m thickness would

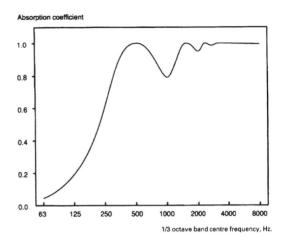

Figure 2.6 *Idealized absorption characteristic of spaced mineral wool blanket, with a 175-mm airspace.*

only absorb effectively down to about 150 Hz. Some saving can be made from the fact that the parts of the material nearest to the hard surface are progressively less effective and can be omitted without much detriment to the overall performance. An absorber consisting of, say, 50-mm mineral wool over a 0.5-m airspace would be acoustically nearly as effective as the full thickness and only unreasonable in terms of the space occupied.

For moderately low frequencies, down to about 500 Hz, this form of absorption, in the form of 30–100-mm mineral wool over a 150–200-mm airspace, is the most cost effective and is widely used (BBC, 1990). The same thickness of mineral wool with a hard surface backing can form simple high-frequency absorbers for frequencies above about 2 kHz.

For low frequencies, the second main method of sound energy absorption can be used. This is based on some form of resonant system, usually a panel of material which is free to vibrate as a membrane. By including a damping component, the vibrational energy can be dissipated, thus absorbing energy from the sound field. This type of absorber is effective because, over a limited range of frequencies around the resonance frequency, the reactive impedance is small. With a suitably adjusted damping factor, the real part of the impedance can be made to be effectively similar to $\rho_0 c$. As an example, a sheet of ordinary 3-mm hardboard fixed over a 180-mm deep airspace has a resonant frequency of about 80–100 Hz. Adjustment of the damping factor is not trivial – one solution used in the past, but with reliability problems, was the attachment of a layer of

proprietary damping compound to the panel. Other solutions include layers of mineral wool and auxiliary damping circuits (Walker and Legate, 1982).

Because they are resonant systems, these types of absorbers work only in a relatively narrow band of frequencies. They are therefore only used to supplement the first type and, even then, several different types might be needed simultaneously to cover just the low-frequency end of the spectrum. However, they are also of vital importance for a quite different reason – resonating panels are a frequently encountered 'accidental' feature of many rooms (for example, decorative finishes, lightweight partitions, the walls of ventilation trunking or of equipment bays). Many occurrences of unwanted sound absorption have ultimately been traced to such panels, both in studios (Walker, 1984) and in public concert halls.

An alternative form for the low-frequency absorber is a Helmholtz resonator. In this arrangement, a small volume of air forms the mass component and an enclosed space forms the compliance component. These are generally less satisfactory than the panel absorbers, principally because their performance is more critically dependant on the location in the room. Nevertheless, they can be effective in the control of individual room modes if accurately tuned and positioned.

One feature of acoustic absorption in isolated patches which is frequently encountered is diffraction, especially at low frequencies in sparsely treated rooms. A single patch of efficient absorber on an otherwise reflecting surface will have an effective absorption 'cross-section' determined by the relationship between its size and the wavelength of the sound. A 100-Hz absorber panel of 60 mm × 600 mm is likely to have an effective area equal to at least twice the actual surface. Thus, apparent absorption coefficients up to 2 are common and 4 has been recorded in an experimental chamber (Spring *et al.*, 1971). In essence, because of distortion of the wavefronts, the panel absorbs more sound energy than would have fallen on the surface in its absence. At higher densities of acoustic treatment, the 'share' of the energy available for each panel is reduced and the excess absorption is less. This is one of several ways in which the apparent absorption coefficient of a material can depend on the other contents of the room.

Acoustic treatment of all types is often provided in the form of 'boxes' of about 600 mm × 600 mm (Burd, 1971). This modular construction allows the architectural design to proceed without requiring the details of the acoustic treatment and permits easy

changing of the treatment for final tuning, if necessary. In principle it also allows for the acoustic treatment to be removed and reused. Because it is modular and usually constructed in a factory, away from the building site and by a specialist manufacturer, it is also relatively cheap and reliable.

However, for large television studios with the large amount of treatment required to produce short reverberation times, the modular approach would need a very large number of boxes. It is generally more economical to provide that amount of material as part of the main contract, at the time of construction.

2.1.6.2 Incidental (accidental?) absorption

All materials within a room have some effect on the sound fields. Some of these effects will be so small as to be negligible, except sometimes in very large spaces or rooms, such as reverberation chambers, where the intention is to minimize the absorption. The acoustic designer is well advised to check the significance of every single element, beginning with the exterior shell. Any panel of material which is not rigidly supported by a relatively massive structure will produce some degree of low-frequency absorption – these are likely to include stud partitions, floor and ceiling structures, as well as decorative wall coverings (often used to disguise ventilation ductwork).

High frequency 'accidental' absorption may result from carpets and ceiling tiles (both of which are relatively obvious and often included directly in the acoustic design), curtains and seating (the latter is of the greatest importance in large auditoria (Beranek, 1969)). The contemporary trend for fabric finishers to wall surfaces is virtually impossible to accommodate in an acoustic design. The stretched fabric over the inevitable airspace behaves in just the same way as the mineral wool over an airspace of the intentional acoustic treatment. It is difficult to achieve an airspace much less than about 10 mm in practice, giving a peak of acoustic absorption at about 8 kHz. Even with the lightest fabric which will be optically satisfactory, the maximum absorption is likely to be about 0.3. It is impossible to compensate for this in the other acoustic treatment, especially as the wall surfaces are such a large fraction of the total. Almost inevitably, areas so finished will show severely diminishing reverberation times above about 4 kHz.

2.1.7 The design of a speech studio

In this section, as a demonstration of the design procedure and method, an acoustic design for a small

studio for speech or discussion, of about 200 m³, will be discussed in detail. Many of the principles are directly applicable to other types of areas.

2.1.7.1 Shell

The acoustic design of the interior of a studio begins with the shell of the room, as determined by other considerations, for example, intended usage, location or finance. Acoustic problems at this stage are limited to those of room proportions and the existence of room modes. For such a volume, the room is likely to be about 3.7 m high. If the choice is available, a plan size of 2.2 × 1.85 times the height (= 8.1 × 6.9 m) would give close to the optimum distribution of room modes. The fundamental axial modes are 21, 25 and 46 Hz. From equation (2.1), the first few modes can be calculated; some are shown in *Table 2.2*, together with their order. The three below 42 Hz are unlikely to present significant difficulties for a design covering the normal audio frequency range. For a reverberation time of about 0.3 s, the bandwidth is about 7 Hz. Thus, the mode spacing is less than

Table 2.2 Room mode frequencies

Order			Frequency
Length	Width	Height	(Hz)
1	0	0	21.2
0	1	0	24.9
1	1	0	32.7
2	0	0	42.5
0	0	1	46.5
2	1	0	49.2
0	2	0	49.9
1	0	1	51.1
0	1	1	52.7
1	2	0	54.2
.	.	.	.
.	.	.	.
.	.	.	.
4	0	0	84.9
2	3	0	86.0
0	3	1	88.1
4	1	0	88.5
1	3	1	90.6
0	0	2	93.0
3	2	1	93.3
1	0	2	95.4
.	.	.	.
.	.	.	.
.	.	.	.

one third of the bandwidth for frequencies above 90 Hz and much less at frequencies above about 100 Hz. The low-frequency region, dominated by isolated room modes lies, therefore, between 40 and 100 Hz. In this frequency range the room response will be uneven and, at best, the irregularities can only be moderated by the selective application of low-frequency, resonant absorption. For example, three of the four modes around 50 Hz have a 'room-width' component. These can all be absorbed more effectively by treatment on the side walls. If such treatment is additionally placed near to the corners with either the floor/ceiling or the front/rear walls then the mode of order 101 will also be absorbed more efficiently.

2.1.7.2 Internal design

For the middle and high frequency ranges the sole objective parameter is generally the reverberation time. From the size of the room and the intended purpose, a choice of an appropriate average reverberation time can be made from a more-or-less limited range of possibilities. Simple graphs (*Figure 2.7*) are widely used and represent the sum of many years of accumulated experience with rooms of all types (Fry, 1988). However, as with many other things, fashions change and the recent tendency has been for shorter reverberation time than the graph in *Figure 2.7* shows. The specified reverberation time is often independent of frequency but may, for special purposes, be chosen to be a particular function of frequency. In some cases, in order to obtain a reduction in the cost, relaxations in the specification may make the reverberation time non-uniform (Spring and Randal, 1970). For the example talks/discussion studio

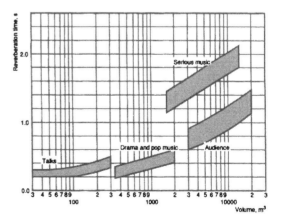

Figure 2.7 *Recommended reverberation times for BBC sound areas.*

of 200 m^3, a specified average reverberation time of 0.30 s will be assumed.

The acoustic effects of structural and other incidental materials and furnishings can be estimated. Then the necessary type and quantity of additional acoustic treatment can be calculated. In this context, a number of different computer programs are used to simplify the arithmetic, chiefly as private, custom-written programs, but a general-purpose 'spread-sheet' program is just as satisfactory. A reference database of commonly encountered materials need not be very large (the author's own database, used mainly for studio design, has about 200 materials, most of only historic value). Most modern spread-sheets allow the use of programmed commands (macros) so that the effort can be reduced to the entry of dimensions and the selection of material reference numbers and areas. *Table 2.3* shows the design for the hypothetical studio, with half stud-framed lightweight walls, half masonry walls, a wooden floor with carpet and underlay and draped curtains. The 'acoustic treatment' is in the form of two types

of standard modular absorbers (Walker and Legate, 1982; BBC, 1990). The wall surfaces are finished with stretched fabric over a 10-mm airspace. As can be seen in this design, a substantial bass rise has been allowed. This may be permissible in speech studios, in order to reduce the costs of acoustic treatment, but is unlikely to be acceptable in a music studio. In order to control the three lowest one-third octave bands, a significant amount of very low-frequency absorption would be required, in addition to the other treatment already included. This is rarely justifiable in speech-only studios. It may also be noted that the commonly specified suspended 'acoustic tile' ceiling is not included. Had it been so, the combination of fabric wall covering and ceiling tiles would have resulted in a calculated high frequency reverberation time (above about 2 kHz) of less than 0.2 s. The room then would probably have given a rather oppressive acoustic impression.

Once the quantity of treatment is known, the disposition must be planned. In this example, the only material with any degree of freedom in its location is

Table 2.3 Room acoustic design spreadsheet

Room data ———			Length:	8.1	Freq.	Old RT	New RT
			Width:	6.9	50		0.80
			Height:	3.7	63		0.80
		Relative humidities (%), old:		50.0	80		0.68
			new:	50.0	100		0.36
			Volume:	206.8	125		0.34
Ref.	Title			Area	160		0.35
7	9-IN. BRICKWORK (STRUCTURAL)			55.5	200		0.36
10	CAMDEN WALLING (STRUCTURAL)			55.5	250		0.38
11	BOARD ON JOIST FLOOR (STRUCTURAL)			55.9	315		0.35
21	SMOOTH PLASTER PAINTED			54.6	400		0.32
24	GLASS > 0.25-IN. THICK			6.0	500		0.30
37	WILTON CARPET ON UNDERFELT			55.9	630		0.30
45	CURTAINS VELOUR DRAPED			6.0	800		0.30
49	Fabric over A2			32.4	1k		0.31
50	Fabric over A3			28.8	1.25k		0.31
75	A2			32.4	1.6k		0.31
77	A3			28.8	2k		0.31
					2.5k		0.30
					3.15k		0.29
					4k		0.29
					5k		0.29
					6.3k		0.29
					8k		0.29
					10k		0.29
					Average		0.30

the acoustic treatment itself, in the form of 90 low-frequency absorbers and 80 high-frequency modules (each of 600 mm × 600 mm). The total surface area of the treatment is thus nominally 61.2 m². The total surface area of the walls and ceiling is 166.9 m², from which 6 m² must be subtracted for the windows, 6 m² for the doors and inevitable cupboards, and probably another 6 m² for the light fittings, 10 m² for the ventilation trunking and 3 m² for technical patresses, etc.

This leaves 135.9 m² for acoustic treatment. The low-frequency absorbers are, by definition, high frequency reflectors and must not be positioned at either the height of the sound source (a seated or standing person) or the microphone. They should also be located in the corner between at least two boundary planes for maximum effect. The perimeter of the studio is about 26 m (excluding doors and other obstructions). Thus, a single row close to the floor would accommodate about 9 m². A similar row near to the ceiling would accommodate about 10 m² and the border of the ceiling about another 10 m². Thus 29 m² of the low-frequency absorbers can be more-or-less optimally located in the various corners. Most of the remaining space on walls and ceiling will be occupied by the high frequency absorbers, with a small amount of intermixing of the remaining low-frequency absorption. Once the structure of this basic layout is established, some small variations can be made to intersperse some of the unbroken bands of low-frequency treatment with some high frequency absorption, to assist in providing additional diffusion. The displaced low-frequency absorbers should be relocated so that they are not near to likely source or microphone positions, not opposite doors or windows and not opposite to each other. This design should ensure good low-frequency control, probably with an achieved reverberation time significantly less than the predicted bass rise, together with freedom from distracting reflections in the studio working zone (between about 1 and 2.5 m from the floor).

2.1.8 The design of other types of rooms

The example above was for the design of one specific type of studio. The general principles for the design of other rooms are essentially identical, but usually with significant additional requirements and constraints.

2.1.8.1 Music studios
Music studios are needed in a wide variety of sizes and functions, from the large orchestral music studio to a modern 'pop group' recording venue. The former

is generally the place where a resident orchestra has its base. Because of the demands of much of the classical romantic repertoire and because, usually, there are provisions for a small audience (perhaps up to 500–600 people), the studio has much of the character of a concert hall. The volume is likely to be of the order of 5000–10 000 m³ and the reverberation time should be fairly long (around 1.5–2.0 s) to give the necessary 'blending' of the sound. The orchestra space should be free from discrete, long-delayed reflections but fairly rich in short-term (<50 ms) reflections to facilitate ensemble. Unfortunately, the space is not usually as large as a real concert hall, which generally means that the orchestra occupies a significant fraction of the floor space and cannot reasonably be enclosed by a canopy. These limitations make it virtually impossible to maintain a long reverberation time whilst at the same time, avoiding significant long-term discrete reflections. A common solution is to provide an acoustically reflecting canopy at about 8–9 m above much of the orchestra. The considerations normally applied to the design of real concert halls are generally inapplicable here because the limited size makes it impractical.

Some more modern orchestral work requires little reverberance for blending and greater separation between instruments. Such works are more suited by a much shorter reverberation time, perhaps below 1 s. Other classical music studios, for example for small groups, may be smaller, perhaps 500–1500 m³. The reverberation time requirement for, say, classical quartets is about 1 s.

'Light music', for example, dance bands, generally require good definition of instruments and good separation, rather than acoustic blending. Such studios are therefore large enough to accommodate a large band comfortably, at least 500 m³, but with heavy acoustic treatment giving a low reverberation time, for the size of the room.

The isolation of individual instruments is carried to the extreme in the case of 'pop music' recording studios. The acoustic design of these rooms is at the discretion of individual acoustic designers, each of whom will have his or her own particular approach. However, much of the modern instrumentation is not acoustic at all – it is not unusual to find everyone in the 'control room' except, perhaps, the vocalist in the studio and the drummer and drum kit in another, completely isolated room.

2.1.8.2 Drama studios
General-purpose drama studios require a range of acoustic environments to simulate different locations.

The main part of the studio is frequently divided into two parts – one with much acoustic treatment, a carpet and curtains, etc., and the other more sparsely treated, probably with a wood or vinyl floor. The division is frequently in the form of two sets of curtains, one made of a hard fabric such as sailcloth and the other of conventional heavy curtaining material. In some cases, the acoustic division is made more positive by the use of a solid sliding wall, finished on one side with a hard surface and on the other the same acoustic treatment as on the walls. The design average reverberation times for these areas usually follow the recommendations in *Figure 2.7*. Curtains provide very little isolation at low frequencies, so that the separation which can be achieved by curtains is limited to the higher frequencies.

Ancillary areas required are usually a 'dead room' with as near total absorption as possible, a 'live' room with little absorption and special localities, such as balconies and stairs.

2.1.8.3 Television studios
Because most large television studios are general purpose, the design simply attempts to produce as acoustically 'dead' a space as possible. With any other solution the 'live' acoustic would intrude on some types of production. Reverberation for effects must be applied artificially. The acoustic treatment is therefore designed to be as efficient as possible and to cover all available surfaces. Even then, because of the large size of the studios, reverberation times of less than 0.5 s are not generally feasible and values approaching 1 s may be the practicable limit in very large studios. The quantity of technical equipment and ventilation systems usually limit the area available for acoustic treatment, frequently leaving only about 50% of the total internal surface area. *Figure 2.8* shows recommended reverberation times for television studios of different sizes. The lower limit of the recommended range represents effectively the lowest practicable value.

2.1.8.4 Control and monitoring rooms
The objective, in a control/monitoring room, is to evaluate the sound from a recording or a live performance. In principle, the acoustic environment should reflect that of the eventual target audience. In the past, control rooms were designed, at least to some extent, to represent the domestic environment. Many studies have been carried out to determine the 'typical' acoustic characteristics of domestic rooms (Jackson and Levethall, 1972; Burgess and Utley, 1985). These results have never been fully incorporated into

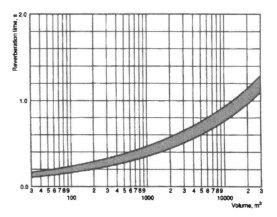

Figure 2.8 *Recommended reverberation times for BBC television areas.*

the design of control rooms but, until fairly recently, there has been a token element of this, with a design midband reverberation time of about 0.3 s.

Technical considerations about the need for a more 'analytic' acoustic and for the better control of unwanted reflections has, over the past 10–20 years, led to the design of much more heavily treated rooms than the domestic environment. This has culminated in a typical design which is more or less anechoic (reverberation times of much less than 0.2 s). Such rooms are oppressive and uncomfortable to work in. They make it difficult for the occupants to converse, unless they are face to face and, also, are quite expensive to construct because of the very great quantity of acoustic treatment.

Some acoustic designers began to question this approach and have developed alternatives, initially from the concept of making one end acoustically 'live' and the other end 'dead' (Davis and Davis, 1979, 1980). By implication if nothing else, the reverberation time would be longer, not least because only part of the surfaces was available for acoustic treatment. Further developments have resulted in more attention being paid to controlling reflections rather than simply absorbing them (D'Antonio and Konnert, 1984; Muncy, 1986; Walker, 1993; Walker, 1994), in particular eliminating 'early' reflections from the operator's position. The intended benefits of this approach are precise stereo imaging and much less effect of the room on the perceived sound quality and the stereo localization. The acoustic absorption can be reduced, giving a room which is more comfortable and easier to work in. One of the penalties is that the stereo imaging illusion works over a rather narrower range of lateral displacement from the centre position. In practice, in reasonable sized rooms,

only about the first 20–30 ms of reflected sound can be deliberately redirected in this way. This appears to be sufficient to achieve most of the intended benefits.

2.1.8.5 Listening rooms

Listening rooms present the same acoustic requirements as control rooms, in principle. However, for any reasonable size of audience, the area over which good sound quality is required is larger. For up to about five persons in an average size of room (about 150 m³), a controlled-reflection room design would be feasible. Much beyond that, the size of the audience exceeds the maximum achievable size of the controlled zone and the design can only be based on the principle of absorbing unwanted reflections. The design discussed above for a 200 m³ studio would be satisfactory, although more effort would be necessary to control the bass rise if the room was intended for music.

2.1.8.6 Concert halls and theatres

The design of concert halls and theatres is a very specialized subject, open only to a small, select band of acoustic designers. It would be presumptuous, in a general reference such as this to give more than the briefest outline of the contemporary philosophy. First, for the conventional large classical concert hall it is vitally important to maintain the sound energy. Any accidental acoustic absorption is a serious matter. Many of the well-documented failures of concert hall design are due, to some extent, to an inadequate reverberation time, particularly at low frequencies. The audience absorption is a significant factor and the average space allocated to each seat should be so large that the reverberation time is adequate even in the presence of the audience. It is also desirable that the absorption of the seating should be the same when empty as when occupied.

These rooms are so large that the question of room modes and the optimum shape does not arise. The predominant contemporary considerations are the maintenance of powerful early reflections from the side walls and the minimizing of later discrete reflections, especially from above.

The reverberation time in a theatre must be short to maintain intelligibility of the speech. At the same time, it is generally important to arrange reflectors near to the stage area in order to project the relatively small amount of acoustic energy which is available into the audience area. The absorption of the seating should also be independent of occupation.

2.1.8.7 Multi-purpose spaces

Spaces intended for multi-function use are notoriously difficult to design. To be entirely satisfactory for different types of function, the acoustic properties must be changed by large amounts. Some successful variable spaces have been constructed using essentially room acoustic considerations (Jordan, 1980), but these generally involve massive constructional difficulties in moving structures (whole walls and ceilings) and large quantities of adjustable acoustic treatment. More frequently, electroacoustic means are used to provide reinforcement and artificial acoustic environments in an otherwise acoustically 'dead' space.

References

ALLISON, R.F., The influence of room boundaries on loudspeaker output power, *Journal of the Audio Engineering Society*, **22**, 314–320 (1974)

BAGENAL, H. and WOOD, A., *Planning for Good Acoustics*, Methuen, London, p. 338ff. (1931)

BARRON, M., The subjective effect of first reflections in concert halls, *Journal of Sound and Vibration*, **15**, 475–494 (1971)

BBC (British Broadcasting Corporation), *BBC Guide to Acoustic Practice*, 2nd edn, BBC Building Design Services, London, 91–103 (1990)

BERANEK, L.L., Audience and chair absorption in large halls. *Journal of the Acoustical Society of America*, **45**, 13–19 (1969)

BOLT, R.H., Normal frequency spacing statistics, *Journal of the Acoustical Society of America*, **19**, 79–90 (1947)

BOLT, R.H. and ROOP, R.W., Frequency response fluctuations in rooms, *Journal of the Acoustical Society of America*, **22**, 280–289 (1950)

BONELLO, O.J., A new criterion for the distribution of normal room modes, *Journal of the Audio Engineering Society*, **29**, 597–604 (1981)

BURD, A.N. Low Frequency Sound Absorbers (*BBC Research Department Report 1971/15*), BBC, London (1971)

BURGESS, M.A. and UTLEY, W.A., Reverberation times in British living rooms, *Applied Acoustics*, **18**, 369–380 (1985)

DAVIS, C. and DAVIS, D. (LEDE) Live-end-dead-end control room acoustics, time-delay spectrometry (TDS), and pressure-zone microphones (PZM), *Recording Engineering/Production*, **10**, 41ff (1979)

DAVIS, D. and DAVIS, C., The LEDE concept for the control of acoustic and psychoacoustic parameters in recording control rooms, *Journal of the Audio Engineering Society*, **28**, 585–595 (1980)

D'ANTONIO, P. and KONNERT, J.H., The RFZ/RPG approach to control room monitoring. In: *76th Audio Engineering Society Convention*, New York, October 1984, preprint No. 2157 (1984)

EYRING, C.F. A derivation of the reverberation formula, *Journal of the Acoustical Society of America*, **1**, 217 (1930)

FRY, A. (ed.), *Noise Control in Building Services*, Pergammon Press, Oxford, 95ff (1988)

GILFORD, C.L.S., *Acoustics for Radio and Television Studios*, Peter Perigrinus. London, 178 (1972a)

GILFORD, C.L.S., *Acoustics for Radio and Television Studios*, Peter Perigrinus. London, 245–247 (1972b)

GLADWELL, G.M.L., A finite element method for acoustics. In: *Proceedings of the 5th ICA*, Liege, Paper L-33 (1965)

GORANSSON, J.P.E. and DAVIDSSON, C.F., A three-dimensional infinite element for wave propagation, *Journal of Sound and Vibration*, **115**, 556–559 (1987)

GRANGER, F., Translation of *De Architectura*, Book 5, Chapter 8, by Marcos Vitruvius Pollio. Loeb Classical Library, Harvard University Press, Cambridge, Mass. **1** (1945)

HAAS, H. The influence of a single echo on the audibility of speech, *Acustica*, **1**, 49ff (1951)

HAWKES, R.J. and DOUGLAS, H., Subjective experience in concert auditoria, *Acustica*, **28**, 235–250 (1971)

HUBBARD, H.H. and SHEPHARD, K.P. The Helmholtz behaviour of single and multiple tones. *NASA CR-178173*, NASA, Washington (1986)

HUNT, V.M., *Origins in Acoustics*, Yale University Press, New Haven, CT (1978)

JACKSON, G.M. and LEVETHALL, J.G., The acoustics of domestic rooms, *Applied Acoustics*, **5**, 265–277 (1972)

JORDAN, J.C., *Acoustical Design of Concert Halls and Theatres*, Applied Science, London, 151–157 (1980)

KNUDSEN, V.O. and HARRIS, C.M., *Acoustical Designing in Architecture*, Wiley, New York (1950)

LOUDEN, M.M., Dimension ratios of rectangular rooms with good distribution of eigentones, *Acustica*, **24**, 101–104 (1971)

MILNER, J.R. and BERNHARD, R.J., An investigation of the modal characteristics of nonrectangular reverberation rooms, *Journal of the Acoustical Society of America*, **85**, 772–779 (1989)

MUNCY, N.A., Applying the reflection-free zone RFZ concept in control room design, *dB The Sound Engineering Magazine*, **20**, 35–39 (1986)

OLSON, H.F., *Acoustical Engineering*, Van Nostrand, Princeton, NJ, 544 (1957)

PARKIN, P.H., HUMPHRIES, H.R. and COWELL, J.R., Acoustics, noise and buildings, Faber & Faber, London, 279–283 (1979)

RAYLEIGH, Lord, *Theory of Sound*, Vol. II, Macmillan, London (1929)

RICHARDSON, E.G., *Technical Aspects of Sound*, Elsevier, London, 133 (1953)

SABINE, W.C., *Collected Papers on Acoustics*, Harvard University Press, Cambridge, MA, 43–45 (1927)

SALAVA, T., Acoustic load and transfer functions in rooms at low frequencies, *Journal of the Audio Engineering Society*, **36**, 763–775 (1988)

SCHENCK, H.A., Improved integral formulation for acoustic radiation problems, *Journal of the Acoustical Society of America*, **44**, 41–58 (1968)

SCHROEDER, M.R., A new method of measuring reverberation time, *Journal of the Acoustical Society of America*, **37**, 409–412 (1965)

SCHROEDER, M.R., GOTTLOB, D. and SIEBRASSE, K.F., Comparative study of European concert halls: correlation of subjective preferences with geometric and acoustic parameters. *Journal of the Acoustical Society of America*. **56**, 1195–1201 (1974)

SPRING, N.F. and RANDALL, K.E. *Permissible Bass Rise in Talks Studios (BBC Research Department Report 1970/10)*, BBC, London (1970)

SPRING, N.F., RANDALL, K.E. and SMITH, M.K.E., *Acoustic Scaling: A One-Eight Scale Model Reverberation Room (BBC Research Department Report 1971/3)*, BBC, London (1971)

WALKER, R., Microprocessor-controlled equipment for the measurement of reverberation time. In: *IOA Spring Conference*, Guildford, Paper C.2.2.1 (1982)

WALKER, R. *Acoustic Scaling: A Re-evaluation of the Acoustic Model of Manchester, Studio 7 (BBC Research Department Report 1984/18)*, BBC, London (1984)

WALKER, R. Alternative measurements in small rooms, In: *Reproduced Sound 3 Conference*, Windemere (*Proceedings of first IOA*), 195–204 (1987)

WALKER, R., *Low-frequency Room Responses: Part 1, Background and Qualitative Considerations. (BBC Research Department Report No. BBC RD 1992/8)*, BBC, London (1992a)

WALKER, R., *Low-frequency Room Responses: Part 2, Calculation Methods and Experimental Results (BBC Research Department Report No. BBC RD 1992/9)*, BBC, London (1992b)

WALKER, R., *Optimum Dimension Ratios for Studios, Control Rooms and Listening Rooms (BBC Research Department Report RD 1993/8)*, BBC, London (1993a)

WALKER, R., A new approach to the design of control room acoustics for stereophony, Paper G1-1, 94th AES Convention, Berlin, March 1993 (preprint no. 3543). (1993b)

WALKER, R., Early reflections in studio control rooms: the results from the first controlled Image Design installations, 96th AES Convention, Amsterdam, 1994, Paper P12-6 (preprint no. 3853) (1994a)

WALKER, R., *High Frequency Room Responses: Acoustic Design and the Control of Stereophonic Image Quality (BBC Research Department Report No. RD1994/11)*, BBC, London (1994b)

WALKER, R., *Controlled Image Design: the Management of Stereophonic Image Quality (BBC Research Department Report No. RD1995/4)*, BBC, London (1995a)

WALKER, R., *Controlled Image Design: the Results from the First Installations (BBC Research Department Report No. RD1995/5)*, BBC, London (1995b)

WALKER, R., *Controlled Image Designs: The Measurement of Time-Frequency Responses (BBC Research Department Report No. BBC RD 1995/3)*, BBC, London (1995c)

WALKER, R. and LEGATE, P.H.C., *The Design of a New Wide-band Modular Absorber (BBC Research Department Report 1982/8)*, BBC, London (1982)

YAMADA, S., ICHINOSE, K., KAMIYA, H., WATANABE, T. and KOSAKA, T., Occurrence and effects of low frequency noise. In: *Proceedings of the 1983 Conference on Noise Control Engineering*, Vol. 2, Institute of Ploughkeepsie, New York, 859–862 (1983)

Bibliography

Books

BERANEK, L.L., *Music, Acoustics and Architecture*, Wiley, New York (1962)
GILFORD, C.L.S., *Acoustics for Radio and Television Studios*, Peter Perigrinus, London (1972)
KNUDSON, V.O., *Architectural Acoustics*, Wiley, New York (1932)
OLSON, H.F. and MASSA, F. *Applied Acoustics*, 2nd edn, P. Blaikiston's Son & Co., Philadelphia, PA (1939)
SABINE, P.E., *Acoustics and Architecture*, McGraw-Hill, New York (1932)
WATSON, F.R., *Acoustics of Buildings*, Wiley, New York (1932)

Journals

Acustica
Applied Acoustics
Journal of the Acoustical Society of America
Journal of the Audio Engineering Society
Journal of Sound and Vibration

2.2 Sound isolation and noise control

2.2.1 Introduction

Three closely inter-related aspects of room acoustics are:

(1) perceived sound quality – quantified by a number of objective parameters such as 'reverberance' (the musician's 'resonance'), rise time, early and lateral energy ratios and many others;
(2) the continuous internal noise levels in the room – generally quantified by the 'background noise level'; and
(3) the sound isolation required around the room in order to reduce external sounds to acceptable levels within it.

Section 2.1 dealt with the first of these. This present section will deal with the significantly inter-dependent considerations of noise levels and sound isolation. For similar reasons, the complexity of the sound fields in ordinary rooms, the analytic approach to background noise and sound isolation is equally impractical. These must also be treated statistically, with general rather than explicit expressions of the relationships.

2.2.2 Noise masking and the perception of extraneous noise

For noise control in many industrial, commercial and health-and-safety applications, noise limits are generally expressed in terms of absolute levels. The necessary standards are generally reasonably clearly set out by various authorities, in terms of International Organization of Standardization (ISO) noise curves (ISO, 1973) or single-figure safety limits. The achievement of the standards, in principle, is unambiguously indicated by some form of measuring instrument. The necessary degree of noise control, again, in principle, is indicated by the difference between the noise levels in the absence of the control measures and the target controlled levels.

In the specialized field of audio engineering applied to the production of entertainment material or for the entertainment event itself, it is more likely that the criteria for noise control will be those of 'perception' rather than absolute level. In such cases, the criteria are derived from highly variable, subjective judgements, as well as economic constraints.

One psychoacoustic factor is important in the study of noise control at the limits of perception. It is self-evident that under some conditions one sound will not be perceptible in the presence of another, much louder sound – that the latter will 'mask' the former. This masking relationship is, in all its detail, an extremely complex one which has been extensively studied for many years (Fletcher, 1940; Egan and Hake, 1950; Beranek 1954a: 393; Schafer *et al.*, 1950). However, some of the main features of this process, as far as they affect the acoustic design of rooms and noise control, may be fairly simply summarized – the mammalian ear behaves in some respects like a series of sharply tuned resonators, covering the whole audible range from 30 Hz to 20 kHz. Within each 'band', sounds at different frequencies may mask each other. The term 'critical bandwidth' used to describe this effect (proposed by Fletcher) is an important concept in the perception of sound, and especially for the derivation of sound isolation criteria. Other masking effects, such as those of sounds displaced in time ('forwards' and 'backwards' masking) and of sounds

widely spaced in the frequency domain ('remote' masking) are much less relevant to the present discussion.

The masking effect allows a perceptibility criterion to be fairly easily, if rather roughly, stated as 'an interfering sound will be just audible if it is equal in level to the continuous, wide-band noise within the same critical band'. Robinson (1978) has shown this to be reasonably true for low sound levels, appropriate to the design of critical areas such as studios, control and listening rooms, etc. Furthermore, he also showed that at 5 dB above that level the sound is judged to be clearly audible and at 10 dB 'disturbing'. As a further rough approximation, and one that is frequently implicitly made in much room-acoustics work, the critical bands can be approximated by one-third octaves. This is true at about 500 Hz and again at about 4–5 kHz. Between those frequencies the critical bands are somewhat narrower (down to about one-sixth octave), and below 500 Hz significantly wider (much more than one octave at 100 Hz) (Beranek, 1954a: 394). It is commonly believed that sounds such as discrete tones are audible when they are very much lower in level than a continuous, wideband noise. However, if both measurements are carried out in a critical bandwidth it is usually found that the audible discrete tone is not, in fact, at a lower level.

The requirements for the control of extraneous noise are, therefore, that the control measures should reduce the maximum interference from an external source to a level equal to that of the room's own internal noise in each one-third octave band. The result then will be just at the threshold of perception. A relaxation of this requirement by 5 dB will result in interference which is, at worst, clearly audible but probably not disturbing. A relaxation by 10 dB will result in occasional disruption because of disturbing interference levels (Robinson, 1978).

These masking considerations are so significant that electronically generated masking noise has been introduced into a studio where the achieved ventilation noise levels were significantly below the requirement. It is also the reason that noise is deliberately introduced into many open-plan office spaces. The objective there being to provide a degree of acoustic privacy and confidentiality by the same masking effect, although in those cases not at levels close to the threshold of hearing.

Thus, in the quest for 'adequate' noise control, the continuous, internal background noise levels must first be determined.

2.2.3 Criteria and standards for noise control

2.2.3.1 Background noise criteria
Noise criteria for industrial, commercial and health-and-safety applications are usually given as some form of single-figure, overall value (HMSO, 1989). These contain no frequency-dependent information and are almost completely useless for the design of noise-control measures. Critical audio areas are often specified using standard curves (for example the ISO noise rating (NR) (ISO 1973) (*Figure 2.9*), and noise criteria (NC) (Beranek, 1954b: 519) (*Figure 2.10*)) or, sometimes, special criteria (BBC, 1990: 24; IBA, 1980). This noise will normally be generated by the ventilation system and the criterion serves as part of the specification for the ventilation system designer (see Section 5.8). *Figure 2.11* shows the background noise criteria curves used by the BBC for the design of areas of all types.

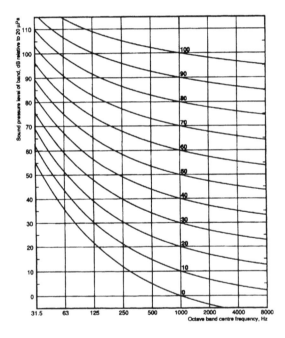

Figure 2.9 *ISO noise rating (NR) curves.*

Simplistically, it might be supposed that the levels ought to be as low as is technically feasible. However, this would be an uneconomical approach, both financially and in the space occupied in the building by the ventilation plant, in addition to the increased demands on the sound-insulation measures. The criterion will generally be chosen to be the highest tolerable level for the intended area function. It is

Figure 2.10 *Noise criteria (NC) curves.*

Figure 2.11 *Background noise criteria for BBC studios and control rooms.* (Courtesy of BBC (1990).) *(i) Relaxed criterion; (ii) normal criterion; (iii) radio drama studios.*

also implicit that if the achieved ventilation noise levels are below the specification then interference from adjacent areas might be audible, even if the sound-insulation systems actually meet their design specifications.

In deriving appropriate criteria for different types of audio areas, consideration must be given to the acoustic signal/noise ratio, the programme dynamic range and the typical microphone directivities as well as the costs of ventilation plant and of providing sound isolation. Based on the intended usage of the room, an appropriate criterion for the background noise level can be chosen (Beranek, 1954b: 523; Gilford, 1972: 38–40; Meares, 1980). Because of the

ear's changing sensitivity as a function of frequency, these criteria approximately follow the hearing sensitivity curves for the appropriate sound level.

It is fortunate that ventilation systems typically produce noise spectra close to those which are required. It is difficult (and therefore expensive) to obtain low-frequency attenuation in ventilation systems. The low-frequency noise levels produced are consequently fairly high, even in supposedly low-noise areas. However, low frequency sound isolation is also difficult and expensive to achieve – therefore high levels of ventilation noise are required for masking of the potential interference. At higher frequencies, the low ventilation noise levels and high sound isolation values are also compatible. If these factors were not naturally so then critical areas of all types would be even more difficult to design and expensive to construct than they are.

2.2.3.2 Airborne sound isolation

The criteria for sound isolation can, in principle, also be simply stated – the sound insulation should exclude all noise at all times from whatever source. This is unrealistic. Sound isolation, especially at low frequencies, can be obtained only from large and massive structures. At its simplest, for each doubling of the mass of a partition an extra 6 dB of sound isolation is (theoretically) obtained. In practice, it is not usually as much as this, and thus overspecifying the sound isolation can be very expensive indeed. The proper derivation of criteria for sound isolation seeks to minimize the isolation requirement, whilst simultaneously achieving a separation which is adequate for the purpose.

The requirement for 'adequate' airborne sound isolation has been stated above. A criterion can be derived (in decibels) for any pair of areas simply by subtracting the design internal background noise level from the level of the maximum potential external source noise in each one-third octave band. The numerical difference is the required degree of airborne sound isolation. In many cases, the direction of potential interference will be clear. In others, there might be interference in either direction. The criterion then is the higher of the two differences at each frequency.

One of the difficulties is determining, before construction, the actual level of the potential interfering sound. Many standard tables of sound levels have been produced and published (Beranek, 1954a: 332–344; Gilford, 1972: 55). In other cases, for example the protection against existing traffic noise, a survey may be carried out to determine the level.

The table of differences as a function of frequency is not a practicable design criterion because it is usually irregular and bears no relation to the performance of practical structures. The requirement can be approximated, without significant overspecification, by fitting an appropriate line or lines representing the idealized sound isolation values for different types of barrier/partition, or by using the ISO or American Society for Testing and Materials (ASTM) standardized partition characterization curves (Northwood, 1962; BSI, 1984), although using such a rigidly defined shape is bound to lead to greater degree of overspecification.

One comprehensive set of airborne sound isolation criteria for a broadcasting studio centre, covering the requirements for rooms of many types (Walker, 1990), describes the derivation of a set of sound isolation criteria from the raw data. The first step was to obtain typical sound levels for the different activities which might cause interference. A survey of the typical types of programme which might occur in a broadcasting centre was carried out (Meares, 1973). Figures for road traffic, aircraft and many other types of general noise source were also obtained and/or were measured. The result of this work was a database giving not only the maximum levels of noise against which any sensitive area had to be protected, but also much information of the statistical distribution of the sound pressure levels for each type of source. To represent the majority of cases encountered in a broadcasting studio centre, 30 representative types of area were taken. For each pair of areas, the required sound-level difference was calculated in both directions at each of eight octave band centre frequencies. To the composite requirement, made up from the larger of the two values at each frequency band, a composite criterion, made up from the idealized performances of different types of partitions, was fitted. A chart was prepared (BBC, 1990: 60; Walker, 1990) showing the required isolation for about 400 combinations of area types, including building exteriors and public spaces, as well as acoustically critical areas.

2.2.3.3 Structure-borne noise and vibration control

The requirements for the control of structure-borne noise can, in principle, be calculated in the same way as for airborne noise. However, the potential sources are so variable in type and intensity that no significant categorization can be implemented. In practically every case, a survey of the actual installation (or of a very similar one) will have to be carried out. The performance of vibration-control devices and structures are also significantly dependent on the details of the installation. It is, therefore, not practicable to derive generally useful criteria for the control of structure-borne noise.

2.2.4 Paths for the transmission of unwanted noise and vibration

2.2.4.1 Airborne paths

'Airborne noise transmission' is generally thus called, despite the fact that at least part of the path is usually within a solid material. These paths typically involve airborne transmission from a source to a materially solid barrier, which is then caused to vibrate as a result of the fluctuating sound pressure acting on its surface. This vibration will be conducted to the air on the far side through the structure of the barrier. The relative amplitudes of these exchanges will be controlled by the relative impedances of the media. In general, the barrier will be much heavier than the air and there will be a significant reduction in magnitude at the first interface. The ratio will also be greater for heavier barriers than for lighter ones and for higher frequencies than for lower ones, according to Newton's second law. At the second interface, the impedance ratios are in the opposite sense and the barrier will transmit roughly the same amplitude of vibration to the air on the far side. Thus, the essential transmission loss of such path is determined by the mass of the barrier and the frequency of the sound, according to:

$$p_t/p_s = \rho_0 c/\pi f m \qquad (2.8)$$

where p_s and p_r are the sound pressure levels on the source and receive sides of the barrier, respectively, $\rho_0 c$ is the characteristic impedance of the air, m is the mass of the barrier per m^2, and f is the frequency. (This approximation is only valid when $\pi f m \gg \rho_0 c$ and is for the idealized condition of normal incidence. It is also 6 dB less than might at first be thought because of the pressure doubling which takes place at interfaces with large impedance mismatches.

This idealized relationship is commonly called the 'mass law'. It shows that the pressure reduction is inversely proportional to both the mass of the barrier and the frequency of the sound. However, it takes no account of the very significant amount of resonances and other effects which always control the performance of real partitions (Beranek, 1960: 280–360; Fry, 1988). At 500 Hz for a partition of

250 kg/m^2 (125 mm dense concrete), equation (2.8) gives about 54 dB reduction.

The rate of increase of this function with the mass shows that very high values of isolation cannot reasonably be achieved by massive, single-leaf partitions. More effective use of materials can be obtained by the use of multiple leaves. This introduces additional air–solid interfaces at which the majority of the actual attenuation takes place.

2.2.4.2 Structural paths

The transmission paths for structure-borne sound may be defined as having the majority of the path within solid materials. Most frequently, the paths lie entirely within the structure until the final, airborne element which is perceived as noise. Unlike airborne paths, structural paths are frequently confined by the boundaries of the elements so that there is little or no spreading loss with distance. Furthermore, the transmission media, especially in modern buildings, are formed of materials with little inherent energy loss (steel, reinforced concrete and, to a lesser extent, brickwork and blockwork). Practically the only energy dissipation mechanisms occur at joints, where the vibration energy is shared between the meeting members – even then, because of the low loss, some of the energy will travel to the receiving area by different paths. Thus, structure-borne noise is characterized by low path losses and multiple routes. It can travel great distances, compared with airborne sound, and can arrive from unexpected directions.

With the continuing pressure to reduce building costs and especially building services costs, trends in the design of buildings have tended to worsen the vibration problems. In relatively lightweight buildings, sometimes with the heavy services plant on upper floor levels, it is difficult and expensive to provide effective vibration isolation.

All these factors make the control of structure-borne noise more complicated. The sources of structure-borne vibration are usually machines containing moving parts. By far the best method of control is to prevent the excitation at source by the use of a suitable isolation system. In many cases, the machines are fixed and reasonably compact, making the fitting of control measures simple in principle. Sources which are dispersed over a large area or which move, such as road and rail transport, are more difficult to control at source, although many successful examples of control of these at source have been described (Clayden, 1967; Stubbs, 1986), and are better controlled by isolation of the receiver.

2.2.5 Airborne noise control

2.2.5.1 Sources of airborne noise

Sources of airborne noise which usually have to be considered in the design of critical areas fall into two main categories – those inside the building and those outside. For the first, examples of which are adjacent rooms, corridors, other public spaces, building services machinery and office machinery (printers and photocopiers), some degree of control might be applicable to limit their potential for interference.

External sources are mainly transportation vehicles and systems – road, rail, aircraft, etc. None of these can be controlled to any significant degree by the owner/occupier of an individual building. Source sound levels can be very high indeed, although the once-threatening problem of shock waves from supersonic aircraft no longer seems to be a problem, at least in urban areas. Other external sources, from nearby buildings and installations, are subject to statutory controls that can be enforced but may still leave a significant residue of noise, especially during the normal working day.

2.2.5.2 Planning to reduce the noise

The most effective and low-cost form of airborne noise control is planning to reduce the problem. Potential sources of interference can often be located away from sensitive areas. Less critical areas, such as store rooms and even offices, can be used to form a shield. Even the location of an entire building may be chosen with some recognition of boundary noise levels and potential shielding by other structures. Building services plant rooms should not be located in the centre of the building, particularly on floors above ground level, without proper consideration of the consequential acoustic difficulties.

2.2.5.3 Prevention

The second most effective form of airborne noise control is the prevention of noise. Apart from being a statutory requirement, at least for high sound levels, the prevention of noise helps to provide a more comfortable and productive environment. Many external sources of airborne noise, with a few notable exceptions, are already constrained by law and, in any case, are usually outside the control of the acoustic consultant. Internal noise levels can be controlled by selection of plant, for example rotary rather than reciprocating compressors, low-speed fans and pumps and the use of effective silencers. Interior noise levels can be reduced by the selection of surface finishes, for example acoustically absorbing walls and ceilings and resilient floor coverings.

2.2.5.4 Reduction

Despite the preventative measures which might be taken, the main form of airborne noise control is inevitably the use of barriers or partitions. The noise reducing properties of partitions have been the subject of much research and development for many years (Kurtze and Watters, 1959; Watters, 1959). The principles remain as outlined for the mass law above, but the variety of departures from that simple theory are legion. Over the ordinary range of acoustic frequencies, no arrangement of solid materials remains inert and totally 'mass like'. Instead, the mass and natural stiffnesses of the material form resonant systems which act to reduce the isolation at some frequencies.

One of these resonances is particularly important – the 'coincidence effect' (Beranek, 1960: 290), which depends on the propagation characteristics of bending waves in the partition. These waves travel at a speed which is related to their frequency so that, for some angle of incidence, an airborne sound wave will have the same projected wavelength on the surface as the bending wave within it (*Figure 2.12*). At this frequency and angle of incidence, the resonance caused by the coincidence of the wavelengths will cause a large reduction in effective sound isolation. In a random incidence field in a real room, one angle of incidence and one frequency are not very significant, but the aggregate effect of many such coincidences are sufficient to cause sound isolations which are significantly less than would otherwise be the case over bands of frequencies. The effect is illustrated in *Figure 2.13*.

In multiple-leaf partitions, another resonance is also important – that of the fundamental resonance of the mass of the leaves of the partition with the stiffness of the enclosed airspace. The effect is to reduce the effective isolation of the partition to a very low value at the resonance frequency. For most heavy masonry walls this effect occurs at a frequency too low to cause significant problems. However,

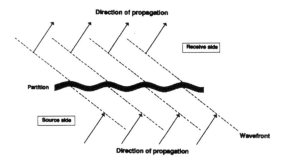

Figure 2.12 *Diagrammatic illustration of coincidence effect in partitions.*

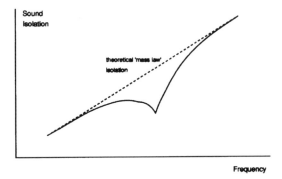

Figure 2.13 *Effect on airborne sound insulation of coincidence effect.*

for light-weight partitions, this frequency might be around 100 Hz.

Table 2.4 gives the expected airborne isolations for a small selection of the types of partition commonly encountered in audio facilities and studio centres. Note that these are not sound reduction indices but the actual sound level differences for cases where the partition forms the entire wall (typically about 15–20 m^2) between two areas of typical size (typically 80–100 m^3) and with reverberation times around 0.3 s. These values are taken directly

Table 2.4 Typical sound insulation values

Partition type:	Frequency (Hz)						
	63	*125*	*250*	*500*	*1000*	*2000*	*4000*
Single-leaf stud partition	19	24	33	37	40	47	51
Double-leaf stud partition	26	39	43	52	61	65	70
112–112 brick, 50 cavity	31	37	42	52	60	63	68
112–225–112 brick, 50 cavity	48	53	56	62	69	78	83
100–100 blocks, 50 cavity	28	33	40	48	59	64	68
100–100–100 blocks, 50 cavity	33	40	50	58	64	73	82

from the *BBC Guide to Acoustic Practice* (BBC, 1990: 64–82).

A common misconception about sound isolation is the belief that adding absorption to one face of a partition can increase the sound isolation. In comparison with the mass of, say, a 225-mm thick masonry wall, a layer of acoustic tiles is negligible. Even adding half as much mass again to the wall will only make a just noticeable difference to the interference. The absorption may have a small effect on the actual noise level in the source because of the reduced reverberation time, but its effect on the noise reduction properties of the partition will be minute.

In the laboratory, the sound isolation of a partition is measured as the difference in levels between a source room and an adjoining receive room, with the test partition in the common surface. To make comparisons possible between different laboratories, the effects of the area of the partition and the acoustic conditions of the rooms are removed numerically. The resulting normalized value for the sound isolation is called the 'sound reduction index' (SRI). For practical design to a criterion calculated as above, on the required sound level differences (SLD) the corrections appropriate to the particular design must be reapplied to the SRI figures to obtain the actual differences likely to be achieved in practice.

Another specification trap for the unwary are single-figure indices of sound isolation performance. Two of these are in common use: R_w in Europe, and the sound transmission class (STC) in the USA (Northwood, 1962; BSI, 1984). These are convenient descriptors of the achieved performance of a partition, in terms of a standardized shape of response (*Figure 2.14*). However, both of these specification methods permit some shortfalls from the quoted curve value, as much as 8 dB at a single frequency, and an average of 2 dB over all (specified) frequencies. Due allowances should be made in critical designs for these possible 'shortfalls'.

Many of the isolation requirements for audio purposes are very stringent, requiring constructions greatly in excess of those required simply for structural purposes. They also fall outside the experience of most non-specialist building designers. For example, the least structure which is likely to be satisfactory between two TV studios is a double-leaf wall, each leaf being made of 350-mm concrete (or the equivalent) and with a 225-mm airspace. The most substantial partition known to the author comprises four leaves, one of 228-mm brickwork and three of 325-mm brickwork with three airspaces of 228 mm, giving a total thickness of 1.9 m. This partition meets

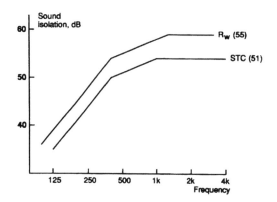

Figure 2.14 *Single-figure sound-insulation specifications: R_w and sound transmission class (STC).*

the appropriate criterion, between a 'pop' music studio and a radio drama studio, except at the lowest frequencies.

With such stringent design criteria the greatest care must be taken in the construction and in the provision of throughways for ventilation ductwork and other services. Generally, these must take indirect routes if the performance of the partition is not to be compromised. The design and construction of any windows must also be carried out carefully (BBC, 1990; 64–82).

The provision of access between areas almost always involves a two-door system with a sound lobby. Even so, the doors may be very heavy. The large doors required for scenery access to TV studios are very special items, costing many thousands of pounds (at 1997 prices) and, in some cases, these are also arranged as a two-door system with a lobby. For personnel access, even with two doors and a sound lobby, the doors will be required to be of a special design rather than ordinary building products.

Randall *et al.* (1986) give a collection of many field measurements, made over a period of about 15 years, of airborne sound isolations appropriate to audio facilities. For most types of partition, it includes a sufficient number of measurements to allow mean performances and standard deviations to be extracted.

2.2.5.5 The selection of partitions for airborne noise control

Assuming that an appropriate criterion or specification for the degree of airborne noise reduction has been obtained, either from direct measurements or from the considerations given in Section 2.2.3.2, the remaining problem is one of selecting a suitable type of construction to satisfy that need. This requirement

will usually be in the form of a sound-level difference as a function of frequency (see *Figure 2.15*).

Figure 2.15 *Typical airborne sound insulations: (a) hypothetical requirement; (b) suitable single-leaf partition, 5 dB octave; (c) suitable double-leaf partition, 10 dB octave; (d) suitable triple-leaf partition, 15 dB octave.*

Most actual requirements will be of the same general form as *Figure 2.15(a)*, with a relatively low isolation at low frequencies because of the high receive-area background noise levels and the ear's insensitivity to low-frequency noise. The requirement increases with frequency, mainly because of the lower midfrequency levels required in the receive area. At some frequency, typically around 1 kHz, the receive area requirement becomes fairly constant but the potential source sound level is reducing (Meares, 1973). Thus the isolation requirement becomes level or actually decreases slightly. The first principle to be recognized in the selection of a real partition is that very little benefit can be obtained from any local minima in the requirement or of the overall reduction at high frequencies. Partitions always show a generally increasing isolation from low to high frequencies. Attempting to specify or fit local deviations at a few frequencies would require correspondingly local deviations from the basic character of the partition. It is not generally feasible to improve the isolation at individual frequencies and it is pointless to try to degrade the performance locally. Thus, a partition must be selected for isolation properties which just satisfy the requirement at the points of closest approach.

The first major constraint on the selection is that of the load-bearing properties of the site. Unless it was planned for from the beginning, this will generally limit the possible options. Usually, the best available option is just not adequate to meet the specification

with certainty. Compromises have then to be made. The low-frequency isolation, up to about 80–100 Hz for typical structures, is set essentially by the total mass, irrespective of the number of leaves. It is frequently this low-frequency region where the likely shortfalls will occur, because of the weight limitations. Above the purely mass-controlled frequency range, the isolation of the partition depends on the number and the spacing of the leaves. Simplistically, a double-leaf wall uses twice as much material as one of its leaves but, at high frequencies, gives double the number of decibels instead of the +6 dB of a single-leaf wall of the same total mass. In reality, this is an extreme simplification, but the average slope of a typical cavity wall is 8–10 dB/octave rather than about 5 dB/octave for a single leaf wall. Triple-leaf walls, widely used in studio environments, typically have slopes of 15 dB/octave.

Using these approximations, any particular requirement could be met with any of these walls. *Figure 2.15.* curves (b)–(d) show some possible solutions. Of course, those walls with shallower slopes have higher isolations at the low-frequency end of the spectrum. This is equivalent to 'heavier'. Taken with the maximum permissible structural loading, this forms a method of selecting both the minimum total weight of the partition and its essential construction details. It can then form the basis of a subsequent search for suitable real partitions from databases of known and measured sound isolation performances (Parkin *et al.*, 1979; Randall *et al.*, 1986). Possible differences in the conditions of the two cases, for example SRI versus SLD, or differences in areas, perimeter fixings, reverberation times, etc., must be taken into account.

Properly constructed, based on reliable data and in the absence of flanking paths, this design method should lead to the expected values of low- and midfrequency sound isolation. However, the new construction may have incorporated parts of existing structures, especially floors and ceilings. Flanking paths through those parts may limit the achieved isolation, despite prior estimates that they should have been satisfactory – it is generally impossible to carry out tests to verify this before the construction of the new partition.

The high frequency isolation is ultimately limited by flanking paths and leakage. Leakage will tend to cause the achieved sound isolation to reach a plateau at a frequency where the increasing attenuation characteristic of the partition becomes equal to the high-value, but less frequency-dependent, attenuation of the hole. Provided that this limitation is not less than

the required high-frequency isolation, no corrective measures need be taken. This is, however, frequently not the case and remedial work on window frames, service penetrations, door seals or ventilation system will be required.

2.2.6 Structure-borne noise control

2.2.6.1 Sources of structure-borne noise
Most sources of vibrational energy involve some kind of mechanical motion. Actual or residual imbalance in these moving parts causes reaction forces in the bearing surfaces, required to constrain the motion to its proper path. These forces, modified by and transmitted through the structure of the machine or equipment, in turn generate reaction forces at the supports. Additional internal forces may also arise as a direct consequence of the machine's function, for example gas pressures on piston crowns and cylinder heads in compressors or internal combustion engines. The supporting structure is always flexible to some degree and therefore moves under the influence of the forces and their reactions.

Obvious local sources of structure-borne vibration are building services plant and distribution systems, including air-conditioning equipment, ventilation fans, compressors, pumps, lifts (elevators), transformers and water pipes. Less obvious but still common sources are occupational activities such as door slams, trolleys and office machinery. Obscure sources are some types of fixtures and fittings, for example fluorescent-light fittings, toilet flushing and wallmounted service facilities such as towel-roll dispensers.

External sources are mainly surface-transport related, e.g. road and rail vehicles, although aircraft, particularly low-flying helicopters, may induce structure-borne vibration in addition to the more obvious airborne noise.

In most cases where the source of vibration is local and fairly compact, it is preferable to isolate the source itself from the remainder of the building. This is usually the most economical and successful option and reduces the noise problems in all other areas simultaneously. In other cases, principally those where the source is not local, such as external traffic, the option to treat the source is not available and the only means of control is to isolate the critical room(s). This option is always expensive in both direct monetary and space costs but has the advantage of reducing the noise in one area from all sources simultaneously, including the otherwise very difficult problem of internal occupational noise

from door slams, trolleys, etc. In principle, these two options can be dealt with by the same theoretical approach. In some of the following sections, the bias is towards the isolation of a room because that is often the most demanding, but the considerations are exactly reciprocal.

2.2.6.2 Planning to reduce structure-borne noise problems
As in the case of airborne noise, planning can help to reduce the potential problems. However, because of the lower loss transmission characteristics of structure-borne noise, remotely locating potential sources or receivers may not be as beneficial.

The coupling of vibration energy into a building structure is almost always less if the part of the building used for support is heavier. Ideally, large and heavy sources of vibration should be located at or below ground level, preferably on a separate set of foundations. Even much lighter sources are better mounted on solid walls rather than on lightweight partitions.

2.2.6.3 Prevention
The scope for control of vibration by prevention is also more limited than in the case of airborne noise. Many of the principal sources of vibration, especially building services plant, are constrained by performance, size, weight and running-cost criteria rather than any acoustic considerations. Some items, for example circulation water pumps, are available in different grades of balance but, even then, it would be doubtful whether such minor improvements could be maintained for the lifetime of the plant.

As in the case of airborne noise, the difficulties of isolation of the vibration from all types of areas should be a factor in the selection of types of plant. For example, most reciprocating plant operates at speeds equivalent to 25 or 50 Hz (30 and 60 Hz in places with 60 Hz public electricity supply). This range of fundamental frequencies is difficult and expensive to isolate. Some types of compressors, for example, operate at much higher speeds, perhaps up to the equivalent of 250 Hz, and are also inherently better balanced. The control of higher frequency vibration is usually simpler and more effective.

2.2.6.4 Reduction
Vibration isolation When all other options have been considered or incorporated, there remains only the provision of some form of vibration isolation system. A typical machine, if it could be operated freely suspended in space, would exhibit

a finite degree of movement at its mounting points. This movement would be determined by the relationship between the mass and acceleration of the moving parts and the dynamic impedance of the stationary parts. There would, clearly, be no forces acting on the mounting points. If the same machine were fixed to an infinitely massive or stiff foundation there would be no movement at the mounting points, but there would be large forces. Thus, as a source of vibration, a machine exhibits the same properties as any other type of source, that is it has both magnitude and impedance. For real machines, the magnitude has a spectrum which is usually composed of a relatively small number of harmonically-related component frequencies. The impedance is usually a complex vector sum of many resonances.

As a receiver of vibrational energy, the supporting structure also has a characteristic impedance, controlled by the mass, stiffness and damping factors of the material.

The objective of most 'vibration isolation' systems is to permit the movements of the plant or machine, whilst transferring a minimum of varying forces to the supporting structure. (A small class of an alternative type, usually known as 'active' systems, attempts to control the vibrations by predicting and producing equal and opposite forces to cancel those produced by the machine.)

Vibration isolators Springs or other compliant elements have been used to reduce the forces transmitted by machines or other vibration sources for a time probably extending back into prehistory. The same technique was applied only much later to isolate buildings or parts of buildings from ground vibration, but is at present rapidly becoming more widespread. The first use of intentional compliance (in the form of rubber mat) to isolate a BBC studio dates back to the late 1950s, and was recognized even than as providing less isolation than expected.

The concept of controlling vibrational energy flow to or from a structure by supporting it on 'antivibration mountings' probably arose from the intuitive picture of a mass supported upon a spring. Such a system has a resonant frequency at which it oscillates if disturbed. Losses in the system can be represented by a third element, i.e. resistance. The resultant three-element system is shown in *Figure 2.16*. By using the circuit analogue (Olson, 1958; Beranek, 1985) (*Figure 2.17*), the behaviour of the system is easier to envisage, and well-developed network theory can be used for the analysis of this or more complex arrangements. *Figure 2.18* shows the main features of the isolation achieved by this idealized

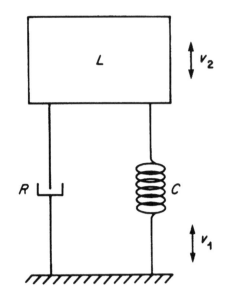

Figure 2.16 *Arrangement of mass supported by an antivibration mounting.*

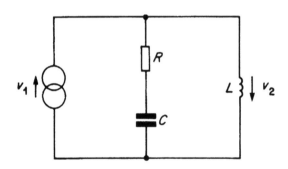

Figure 2.17 *Circuit analogue of a mass supported by an antivibration mounting.*

system. The Q-factor of the system at resonance depends on the type of antivibration mounting and may vary from 20 or less for a rubber pad to 200 or more for a steel spring. The isolation (expressed in decibels) is zero at zero frequency, negative at all frequencies up to about 1.4 times the fundamental resonance, and then increases at +12 dB/octave above that frequency. At Q times the resonance frequency, the slope changes to +6 dB/octave and remains at this value. The peak value of the amplification at resonance is about Q times.

Any practical investigation of a resiliently supported structure shows that it deviates in three respects from the ideal three-element model shown in *Figure 2.16*. These are the modal behaviour of the supported structure, the modal behaviour of the

Figure 2.18 *Ideal response of the circuit shown in* Figure 2.17.

antivibration mounting, and the airborne flanking (the transfer in parallel with the antivibration mounting of airbone energy to or from the supported structure). The relative magnitudes of these effects for a typical building structure have been the subject of detailed investigations (Mathers and Walker, 1985; Smith *et al.*, 1986; Walker and Mathers, 1989a, 1991).

Non-ideal antivibration mountings The simple theory of vibration isolation assumes that the mountings can be accurately represented by a single compliance with some form of damping, usually assumed to be viscous (proportional to velocity). In practice, the construction of the mounting inevitably involves finite dimensions and mass, in addition to its principal property of compliance. The combination means that the mounting can itself be considered as a transmission line rather than as a single, pure component. Alternatively, the mounting itself may consist of a series of masses and compliances. (Rubber-based isolators are often of this form, with metal sheets within the body of the rubber. The object is to obtain the appropriate degrees of stiffness and load-bearing, but a side-effect is to add several distinct resonances to the mount impedance.) Helical metal springs have a number of theoretically infinite series of resonance modes. In all these cases, the equivalent transfer impedance, for a lossless antivibration mounting, can take all values of reactive impedance from $+j\infty$ to $-j\infty$ over a range of frequencies.

Structural resonance modes The mass which is supported, i.e. the studio or the machine, may also

have non-ideal behaviour. Above some limiting frequency the supported mass will cease to behave as a rigid body, moving as a whole, and will begin to move differently at different points. The frequency at which this will begin to happen will depend on the size of the body, its structure and the material of which it is made. At such frequencies the effective source impedance for a machine or load impedance for a room will not be the mass of the supported object and the achieved isolation will be different from that predicted by the simple theory, even with ideal mountings. In practice, real bodies have resonance modes just like those of the antivibration mountings described above. In the absence of any loss components the impedance with which the mount is loaded can also take all possible reactive values from $+j\infty$ to $-j\infty$.

For the isolation of large machines or entire rooms, especially large rooms such as drama studios, the highest frequency at which the structure may be considered as a simple mass is well within or even below the audio frequency band. The behaviour is similar in kind to that obtained from non-ideal mountings with ideal loads. Both types of imperfections cause the same shortfalls from the ideal performance at some frequencies and, in general, both types will be present in any practical construction. There are, of course, other frequencies where the isolation is improved with respect to the theoretical ideal. For the isolation of discrete, harmonic, machine vibrations some advantage can be taken of this, or at least no disadvantage, by choosing that the adverse modes do not correspond with any of the frequency components present. For the isolation of structures from continuous spectrum noise this choice is not available and the worst-case available isolation will always be that governed by the adverse resonances in either the mounts or the structure. This marks one difference between the two vibration isolation requirements.

Acoustic coupling One further means by which vibration energy may be coupled from a source to a receiver is acoustically, via the intervening airspace. Consider a structure comprising two plane, parallel, limp surfaces of infinite extent separated by a large airspace. If one of the surfaces is vibrating with a fixed magnitude then the vibration velocities in the adjacent air layers and the surface will be equal. If this sound energy propagates as a plane wave, as it will if the vibration amplitude is uniform all over the surface, then it will arrive at the opposite surface with the same magnitude, ignoring any energy loss due to propagation (which will be very small over short distances and at moderate frequencies).

The mechanical force exerted on the second surface by the sound wave will depend on the acoustic pressure which is related to the acoustic velocity by the acoustic impedance. From Newton's laws of motion, this force variation will then induce mechanical vibrations in the second surface. It can easily be shown that, ideally, the ratio between the vibrations in the two surfaces, i.e. the effective coupling c, of the vibrations, is given by:

$$c = Z\rho_0 c/\omega m \qquad (2.9)$$

where $\rho_0 c$ is the plane wave acoustic impedance, m is the mass of the second surface per m^2, and ω is the frequency (rad/s).

At a frequency of 100 Hz and for a wall of mass 625 kg/m^2, this expression has a value of −53 dB. Therefore, if this arrangement were part of a vibration isolation system, there would be a theoretical limit of 53 dB for the isolation for this factor alone. The assumption of plane, parallel walls and of planewave propagation mean that this limit is theoretically independent of the spacing between the walls. In practice, the walls are not infinite in extent or plane or parallel and, therefore, the propagation of the sound energy within the cavity is not so simple. But at low frequencies, with the typical dimensions of ordinary cavity walls, the idealized model is a reasonable approximation. For relatively small machines, the sound energy propagation will be spherical, with the usual spreading loss of 6 dB per doubling of distance. However, it is self-evident (and frequently encountered in practice) that the airborne noise in a plant room can excite vibrations, especially in relatively lightweight structures of large area such as ventilation ductwork. In both these cases, if the limiting transmission path is airborne then no amount of attention to vibration isolators will make any improvement.

Also in practice, the vibration modes of the structure have an effect on the magnitude of this airborne flanking path. At some frequencies and at some positions on the surfaces there will be an amplification of the vibrations, both of the transmitting surface and the receiving surface. If the modal Q factors of the surfaces are about 20 then the effective acoustic flanking path can be increased, at some frequencies and positions, from the −53 dB calculated above to −26 dB, even if there is no coincidence between the modes of the two surfaces. The effect of this modal amplification averaged over the whole structure is a reduction in the isolation limit imposed by the acoustic coupling. For full-size structures, where the modal density is high over most of the audio-frequency range, this effect will occur at all significant frequencies. Measured values for this limit of about 15–25 dB are frequently encountered (Walker and Mathers, 1989a).

Figures 2.19 and *2.20* are typical of vibration isolation values obtained in practical systems, i.e. a 10-t slab of concrete (approximating to the floor slab of a half-scale model studio) supported on rubber compression mountings and helical steel springs, respectively. In both cases, the theoretical isolation of the equivalent ideal three-element system is indicated by the dotted line. The behaviour is close to the ideal only up to about 30 Hz, which was the frequency of the first main vibration mode of the slab. Above that frequency, the isolation is limited by modal behaviour of the systems and by airborne coupling across the 100–400 mm airspace.

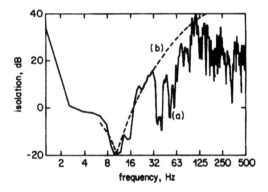

Figure 2.19 *Example of a real antivibration-mounting response, a rubber compression' mount.* (Courtesy of Walker and Mathers (1989a).) *(a) Measured; (b) predicted from the simple model of* Figure 2.17.

Figure 2.20 *Example of a real antivibration-mounting response; a helical steel spring mount.* (Courtesy of Walker and Mathers (1989a).) *(a) Measured; (b) predicted from the simple model of* Figure 2.17.

2.2.6.5 The design of vibration control systems

Machine isolation For the proper design of a machine isolation system it is necessary to know the degree of isolation required. For an existing case of unacceptable interference, this is all too obvious. However this information is rarely available for new installations because of the large uncertainties involved. Firstly, the machine itself as a source is likely to be somewhat unknown, although measurements can be made using identical or similar machines already in service. Secondly, the mechanical properties of the supporting structure, if not yet built, are at best subject to large tolerances. Thirdly, the properties of the transmission path to the remote critical area are likely to be completely unknown, whether the building exists or not.

In an existing building, where it is proposed to install new or replacement plant, some of these properties can be measured and a prediction made. Walker and Mathers (1989b) have described a methodology for measuring the source vibration of an identical sample already operating and measuring the transfer function of force at the proposed machine location to sound level in the critical area. From these measurements and the desired final noise levels, the requirements for the antivibration-mounting system can be calculated. This is a complicated and time-consuming process, which uses elaborate instrumentation and computer processing to deal with the large quantity of complex data. It is only justifiable in critical cases.

Most cases of so-called 'design' of antivibration-mounting systems simply seek to install the maximum reasonably achievable isolation. The data generally available on the performance of isolation devices is, in any case, not specific enough to permit a much more rigorous approach.

The design of a vibration isolation system is also more complicated than the airborne equivalent because of the usually significant interactions between the component parts. The vibration of the source itself will be affected to some degree by the restraining effects of the mountings. The mountings behave in a far from ideal manner except at very low frequencies because of their inherent modal responses, which in turn may be modified by their terminations. The supporting structure may be very far from rigid. In fact, if a machine is supported directly on a steel frame with a significant static deflection it is already effectively on one set of antivibration mountings. The additional isolation system would then have to interact with the steel framework. The result of all these factors is likely to be very different to the 90%, 99% or 99.9% isolation usually presented in data sheets for vibration isolation components.

One rigorously correct way to deal with all of these interactions is to represent each element as a transfer matrix describing the relationship between the input and output forces and velocities. In this representation, the matrices may be concatenated, the output of one becoming the input of the next for the entire system. The overall behaviour of the system is then the product of all of the matrices. To do this for any real situation, incorporating several, if not all degrees of freedom, together with the variations with frequency, would be a major undertaking.

However, some simple rules may be used to give guidance for the installation in less critical situations. The fundamental resonant frequency of a vibration isolation system, f_0 in *Figure 2.18*, is simply related to the static deflection of the mount produced by the mass of the machine by:

$$f_0 = 1/2\pi\sqrt{(g/d)} \qquad (2.10)$$

where g is the gravitational acceleration (9.81 m/s^2), and d is the deflection (m).

It should be noted that d is the deflection that would occur if the mass were directly supported by the mount. It does not mean that the deflection has actually to be present (for example, the antivibration mounting may be horizontal and not supporting the machine mass) or the machine may have other, non-gravitational forces acting on it (for example loads transferred through pipework to a compressor).

The idealized isolation characteristic (as illustrated in *Figure 2.18*) is obtained up to a frequency corresponding to the first modal resonance of the entire system, provided that it is well above f_0.

The design fundamental resonance must, therefore, be significantly lower than any of the vibration frequencies which require isolation. For example a machine running at 1500 r.p.m. will produce vibration components at 25, 50, 75 Hz, etc. An isolation system with a natural resonance of 6.25 Hz would ideally give 24 dB of attenuation at 25 Hz. that is just under 20-fold reduction in the force transmitted to the supporting structure. Such a system would have a static defection of about 6.5 mm, and all of the machine services and connections would have to accommodate movements on that scale, especially during start-up, when the machine speed passes through the resonant frequency.

Machine vibrations and supports are often viewed as purely vertical effects. In most cases, the rotation of the machine parts generates force vectors which rotate

in space and cause laterally, as well as vertically, acting forces. A free body has six degrees of freedom due to translation and rotation on three orthogonal axes. It is important that the vibration isolation system is considered in an adequate number of degrees of freedom – for a simple, symmetrical machine two directions of translation and one of rotation would probably suffice. Many real machines will require all six degrees to be considered. If the natural resonance frequency of any of these degrees of freedom is significantly different from the others then difficulties will be encountered. If higher, the expected degree of isolation will not be achieved; if lower, that mode might correspond with the actual running speed and large amplitude oscillations would result.

One method of reducing the magnitude of the vibration forces and of making the supported machine more resistant to externally imposed forces, such as the pressure reaction from a compressor or ventilation fan, is to add mass in the form of an inertial base. If it is added directly to the machine, such a mass reduces the external effects of the internal forces by increasing the ratio between the machine's 'moving' and 'fixed' masses. Alternatively, a second set of antivibration mountings may be included which, if correctly designed, allows the whole system to act as a two-stage isolator with correspondingly greater rate of increase of isolation with frequency.

Room isolation To isolate a whole room from vibrations in the remainder of the building, it is common practice to build the room as a completely separate box, entirely contained within the structure of the remainder of the building, but with no rigid contact between them. Some form of support against gravity must be included; this usually takes the form of large antivibration mountings.

Many of the considerations of machine isolation also apply to the isolation of a room from the rest of a building. However, only in very few cases are there large amplitude movements between the two parts. On the contrary, the extremely small movements involved, typically less than the diameter of a hydrogen atom, mean that very small amounts of Coulomb friction can completely bypass the antivibration mountings (Crede, 1951). It requires a total force of less than 50 kgf to move the mass of a medium size studio (450 t) by an amount sufficient to generate within it a noise equal to the maximum permitted background noise for such a studio. Such proportionally small forces are difficult to avoid and require the greatest care in the installation of the services.

One extensive investigation (Smith *et al.*, 1986; Mathers and Walker, 1985; Walker and Mathers, 1989) into the isolation of tube-train vibrations from possible studio locations included an approximately half-size model of a radio drama room. It was found that the limit of the performance of such a system was about 20 dB, set mainly by the acoustic coupling across the cavity and the structural resonance modes of the isolated structure.

For many areas, this approach to the isolation of external vibrations can be exceedingly expensive. An alternative which might be cheaper, but one which does nothing to isolate the internally generated vibrations, is to build antivibration mountings into the building's main supports or the pile-caps. This is becoming increasingly common in urban areas. In London in 1991 there were said to be more than 40 such buildings, mainly for isolation from the vibration caused by transport systems. In areas subject to earthquake activity, such as the Pacific Rim, their numbers are very large (hundreds if not thousands).

2.2.7 Measurements and investigations

2.2.7.1 General

Compared with, say, fault finding in electronic equipment, the investigation of noise problems in buildings is complicated by the fixed and heavy nature of the objects involved. It is not generally practicable to demolish and rebuild parts simply in order to identify transmission paths. Much literature exists on theoretical work on the prediction of transmission paths (Cremer, 1953; Plunkett, 1958; Craik, 1988; Moorhouse, 1988). In particular, the concept of statistical energy analysis (SEA) (Lyon, 1975), allows the general properties of a complex building to be represented in a way that allows some identification of the relative magnitudes of possible different paths. It involves extensive measurements of airborne and structureborne energy levels and measurement or calculation of modal densities in each element of the building. From consideration of the general principle that there will be a nett flow of energy from high energy density regions to lower ones, matrix equations can be derived and solved to give the energy flow patterns even in the presence of multiple flanking paths. It is more suited to the analysis and prediction of average properties (for example, the average sound level in a room), than to specific values at particular points.

Sometimes, simpler techniques can be useful. For example, a common problem is to identify whether

airborne noise in a room from some nearby source is transmitted mainly by the structure or as airborne noise. In the first case it would be necessary to improve the vibration isolation of the source whilst in the second, improvements to the partitions between the areas or a reduction in the source airborne emission would be required. Two measurements of airborne noise level difference between the source and receive areas are needed – one with the source operating, and one with the source stopped and replaced by a generator of purely airborne noise, such as a loudspeaker, producing a similar level in the source area. If the two measured differences are similar then the transmission is predominantly airborne, caused by the airborne noise generated by the source. If the measurement with the loudspeaker source gives higher differences then the problem is caused by inadequate vibration isolation of the source. This technique makes several assumptions, for example that the airborne noise generator does not cause significant vibrations in the (non-functioning) source, but these are usually reasonable.

2.2.7.2 Airborne isolation

The field measurement of airborne noise isolation is covered by a number of specifications, for example ISO 140–1978: Parts 4 and 5 (ISO, 1978), which require that the average noise levels in the source and receive areas are adequately assessed. This usually means measurements at a number of microphone positions or with a continuously moving microphone. In some cases, it will be necessary also to carry out a reverberation time measurement in the receive area, in order that the sound level difference can be converted to sound reduction index, if the specification was written in those terms.

Airborne noise problems are either due to inadequate specification, in which case the solution is either to tolerate the problem or to construct additional partitions, or due to the specified construction failing to meet the expectations.

The most common cause of failure is inadequate control of minor details, such as the fitting of doors and windows and the sealing of penetrations for services. In a substantial partition, of the kind likely to be required in studio and other critical areas, the performance can be greatly compromised by even very small gaps. In one example, a shrinkage crack about 0.5 mm wide and 1 m long through a 225-mm masonry wall over a lintel caused a reduction of more than 20 dB in the isolation of high frequencies. Windows and door frames must be bedded throughout the thickness of the partition on non-hardening

mastic. Service ducts should take indirect routes and be thoroughly 'pugged' after installation of the services. If the ducts have to take direct routes then they must be pugged with material similar in density to the partition. A common weakness in brickwork is inadequate filling of the 'frogs' leading to large holes within the structure. Bricks must be laid 'frog-side up' and fully filled. All these factors are expensive and must also be allowed for in the 'bill of quantities'. Good site control is required to ensure that the specified precautions are actually included.

2.2.7.3 Vibration

The most common problem encountered in the isolation of vibration is that the installed system apparently does not function as it was supposed to. Of the many possible reasons for this the most common is that the expectations were optimistic. Even properly installed vibration control equipment has only a finite potential for improvement. It may not have been even theoretically capable of providing the necessary isolation and the only solution in these cases, apart from relocating the sensitive areas or some other way of accepting the situation, is to redesign the system correctly.

Of the remaining problems, the possible reasons for failure are unexpected modes in the supporting structure or the vibration isolators and airborne coupling. Some measurements can be carried out to determine the likely cause. The method of measuring sound level differences with and without the plant running, outlined in Section 2.2.7.1 above, can usually identify airborne limitations. Measurements of the vibration levels at both ends of the antivibration mounting, i.e. on the machine 'foot' and on the supporting structure at the base of the antivibration mounting immediately beneath can easily be carried out. The 'machine' levels are likely to be much higher than those of the 'floor' in all cases, but differences of less than about 40–50 dB are likely to indicate problems with modes of the supporting structure. Sometimes, and especially with helical steel mounts, the vibration levels at both ends of the mount are much less than those in the middle. Cases where the steel spring coils are just a 'blur' are commonly encountered. This indicates modal behaviour of the antivibration mounting itself.

For more rigorous investigations, the methods and measurements of Mathers and Walker (1985), or similar methods, can be applied retrospectively to quantify the behaviour of the successive elements of the transmission path.

References

BBC, *BBC Guide to Acoustic Practice*, 2nd edn, British Broadcasting Corporation. Building Design Services, London (1990)

BERANEK, L.L., *Acoustics*, McGraw-Hill, New York (1954a)

BERANEK, L.L. (ed.), *Noise Control*, McGraw-Hill, New York (1954b)

BERANEK, L.L., *Noise Reduction*, McGraw-Hill, New York (1960)

BERANEK, L.L., Some remarks on electro-mechano-acoustical circuits, *Journal of the Acoustical Society of America*, **77**, 1309–1313 (1985)

BSI (BRITISH STANDARDS INSTITUTION), *BS 5821: Part 1 Rating the Sound Insulation in Buildings and of Building Elements*, Milton Keynes (1984)

CLAYDEN, K.J.C., Noise and vibration attenuation in the Barbican scheme. *Use of Rubber in Engineering*, MacClaren, London, p. 204 (1967)

CRAIK, R.M.J., Sound transmission paths through large structures, *Proceedings of the IOA*, **10**, 747–652 (1988)

CREDE, C.E., *Vibration and Shock Isolation*, Wiley, New York, p. 193 (1951)

CREMER, L. Calculation of sound propagation in structures, *Acustica*, **3**, 317–335 (1953)

EGAN, J.P. and HAKE, H.W., On the masking pattern of a simple auditory stimulus, *Journal of the Acoustical Society of America*, **22**, 622ff (1950)

FLETCHER, H., Auditory patterns, *Reviews of Modern Physics*, **12**, 47–65 (1940)

FRY, A. (ed.), *Noise Control in Building Services*,' Pergamon Press, Oxford, pp. 101–129, (1988)

GILFORD, C.L.S., *Acoustics for Radio and Television Studios*, Peter Perigrinus, London (1972)

IBA, *Independent Broadcasting Authority Code of Practice for Independent Local Radio Studio and Outside Broadcast Performance*, London (1980)

ISO, *ISO-140-1978: Part 4 Field Measurements of Airborne Sound Insulation Between Rooms: Part 5 Field Measurements of Airborne Sound Isolation of Facade Elements and Facades*, Geneva (1978)

ISO, *International Standards Organisation Recommendation R1996–1973*, Geneva (1973)

KURTZE, G. and WATTERS, B.G., New wall design for high transmission loss or high damping, *Journal of the Acoustical Society of America*, **31**, 739–748 (1959)

LYON, R.H., *Statistical Energy Analysis of Dynamical Systems: Theory and Applications*, MIT Press, Cambridge, MA, (1975)

MATHERS, C.D. and WALKER, R., Some practical aspects of vibration isolation. *Proceedings of the IOA*. **7**,3ff (1985)

MEARES, D.J., *Statistics of Typical Programme Sound Pressure Levels in Sound Rooms and their Control Rooms (BBC Research Department Report No. 1973/37)*, BBC, London (1973)

MEARES, D.J., *Revised Background Noise Criteria for Broadcasting Studios (BBC Research Department Report No. 1980/81)*, BBC, London (1980)

MOORHOUSE, A.T., Determination of the propagation path of structure-borne sound, *Proceedings of the IOA*, **10**, 759–766 (1988)

NORTHWOOD, T.M., Sound insulation ratings and the new ASTM Sound Transmission Class, *Journal of the Acoustical Society of America*, **34**, 493–501 (1962)

OLSON, H.F., *Dynamical Analogies*, 2nd edn, van Nostrand, New York (1958)

PARKIN, P.H., HUMPHRIES, H.R. and COWELL, J.R., *Acoustics, Noise and Buildings*, Faber & Faber, London, pp. 286–287 (1979)

PLUNKETT, R., The interaction between a vibrating machine and its foundations, *Noise Control*, **4**, 18–22 (1958)

RANDALL, K.E., MEARES, D.J. and ROSE, K.A., *Sound Insulation of Partitions in Broadcasting Room Centres: Field Measurement Data*, BBC Engineering Division, London (1986)

ROBINSON, A.P., *The Masking Effect of Background Noise (BBC Research Department Report No. BBC RD 1978/3)*, BBC, London (1978)

SCHAFER, T.H., GALES, R.S., SHEWMAKER, C.A. and THOMPSON, P.O., Frequency selectivity of the ear as determined by masking experiments, *Journal of the Acoustical Society of America*, **22**, 490–496 (1950)

SMITH, T.J.B., WALKER, R. and MATHERS, C.D., Studies on a vibration-isolated radio studio mock-up. In: *12th International Congress on Acoustics*, Toronto, Paper D1–5 (1986)

STUBBS, G., Noise and vibration from underground railways, *Noise and Vibration Control Worldwide*, **17**, 293–295 (1986)

WALKER, R., *A Review of the Airborne Sound Insulation Requirements for Broadcasting Studios and Ancillary Areas (BBC Research Department Report No. 1990/10)*, BBC, London (1990)

WALKER, R. and MATHERS, C.D., *The Control of the Audible Effects of Ground Vibrations in Building Structures. (BBC Research Department Report No. RD1989/2)*, BBC, London (1989a)

WALKER, R. and MATHERS, C.D., The prediction of noise from machine vibration. In: *13th ICA*, Belgrade, Paper 3.H.3 (1989b)

WALKER, R. and MATHERS C.D., The vibration isolation of building structures. In: *Plastics, Rubber and Composites Processing and Applications*, **18**, 1, Elsevier, London (1992)

WATTERS, B.G., The transmission loss of some masonry walls, *Journal of the Acoustical Society of America*, **31**, 4, 898–911, (1959)

Bibliography

Books

BBC, *BBC Guide to Acoustic Practice*, 2nd edn, British Broadcasting Corporation, Engineering Division (1990)

BERANEK, L.L., *Acoustics*, McGraw-Hill, London (1954)

BISHOP, R.E.D. and JOHNSON, D.C., *Mechanics of Vibration*, Cambridge University Press, Cambridge (1979)

CRANDALL, S.H. and MARK, W.D., *Random Vibrations in Mechanical Systems*, Academic Press, New York (1963)

CREDE, C.E., *Vibration and Shock Isolation*, Wiley, New York (1951)

DEN HARTOG, J.P., *Mechanical Vibrations*, McGraw-Hill. New York (1961)

FRY, A.T. (ed.), *Noise Control in Building Services*, Pergamon Press. Oxford, (1988)

GILFORD, C.L.S., *Acoustics for Radio and Television Studios*, Peter Perigrinus. London (1972)

HARRIS, C.M. (ed.), *Handbook of Noise Control*, 2nd ed, McGraw-Hill, New York (1979)

MORROW, C.T., *Shock and Vibration Engineering*, Wiley, New York (1963)

MORSE, P.M., *Vibration and Sound*, McGraw-Hill, New York (1948)

RAYLEIGH, Lord, *Theory of Sound*, Macmillan, London (1928)

REYNOLDS, D., *Engineering Principles of Acoustics: Noise and Vibration Control*, Allyn & Bacon, Boston, MA (1981)

SNOWDON, J.C., *Isolation and Absorption of Machine Vibration (US Navy Report TM 766188)*, US Navy Department, Washington (1976)

SNOWDON, J.C., *Vibration and Shock in Damped Mechanical Systems*, Wiley, New York (1968)

TIMOSHENKO, S., *Vibration Problems in Engineering*, Wiley, New York (1974)

Journals

Acustica
Applied Acoustics
Journal of the Acoustical Society of America
Journal of the Audio Engineering Society
Journal of Sound and Vibration

JONES, R.S., Noise and vibration control in buildings, *Sound and Vibration (USA)*, **20**, 16–22 (1986)

2.3 Microphones

2.3.1 Requirements of professional microphones

As a device for producing an electrical voltage which is, as far as possible, a replica of the sound waves striking the sensitive part of the microphone, it may first be useful to consider some of the characteristics of a microphone which are important in a professional unit. These can be listed as follows, but it should be noted that some of the requirements may be mutually conflicting so that compromises are needed.

(1) The frequency response should be as flat as possible over the frequency range 20 Hz to 20 kHz. In practice, ±2 dB over much of this range with slightly less tight tolerances at the high and low frequency ends may be acceptable. Note, however, that for some specialized microphones there may be some advantages in having a frequency response that is not flat. A reduced bass response may, for example, be desirable in microphones used for public address.

(2) The microphone's response to starting transients is important and this response may not always be apparent from the frequency-response graph. Essentially it is a matter of the diaphragm's ability to respond rapidly to the almost violent changes in sound wave pressure which can sometimes occur at the start of a sound.

(3) Sensitivity – the electrical output should be as high as possible so that the audio signal produced by the microphone is high relative to system noise produced in, for instance, mixing consoles. Typical sensitivities are quoted later (see Section 2.3.6).

(4) The self-generated electrical noise should be low. This is often quoted as the equivalent *acoustic* noise (in dBA) which would be present even if a perfect, noiseless microphone were used.

(5) The polar diagram (directivity pattern) should vary as little as possible with frequency.

(6) Environmental influences should have negligible effects on the microphone's output. Such influences include temperature and humidity, and alternating magnetic and electrostatic fields.

(7) In all reasonable practical situations a microphone should be mechanically as robust as possible; repair and maintenance services should be quick and reliable. Accessories such as clips and windshields should be readily available. Finally, the cost may be an important factor, especially in a broadcast organization where large numbers of microphones may be purchased.

2.3.2 The basic microphone

Any microphone can be regarded as consisting of three main components.

(1) The diaphragm, a thin light membrane on which the incident sound waves impinge so that its movements follow accurately the sound wave pressure variations.

(2) The transducer, which is coupled to the diaphragm (in some cases the diaphragm may

be an integral part of the transducer). The function of the transducer is to produce a voltage which is a precise analogue of the diaphragm movements.

(3) The casing which encloses, or partly encloses, the transducer and diaphragm. The significance of the casing in the present context is that its nature determines the directional response of the microphone.

The diaphragm and transducer will be considered together because, as stated above, one may be an integral part of the other and even if this is not the case they are so closely coupled mechanically that it may be misleading to treat them separately.

2.3.3 Transducer systems

A number of types exist. The first two described below are virtually obsolete, but are included for completeness.

2.3.3.1 Carbon granules

A quantity of small granules of carbon is in contact with the diaphragm so that the pressure on the granules varies with the movements of the diaphragm. A characteristic of such an arrangement is that the electrical resistance varies with the pressure on the granules, with the result that a d.c. current flowing through them is modulated.

The carbon microphone has the merits of being cheap and giving a high output. Its drawbacks are that there is a considerable background 'hiss' and control of the polar diagram is not easy. For many years this type of microphone was used in telephones.

2.3.3.2 Piezo-electric ('crystal')

Slices of certain crystals such as quartz and barium titanate create an e.m.f. when they are deformed. By coupling such a crystal slice to a suitable diaphragm, a small, lightweight and inexpensive microphone can be made. The disadvantage of this system is that the impedance of the crystal is high and capacitive and this means that the microphone cable must be short (no more than perhaps 1 m) to reduce the effects of induction from adjacent circuits. In addition, the input impedance of the load must be high. If longer cable lengths are needed, a preamplifier must be incorporated in the microphone and if this is done there is no financial advantage over electrostatic microphones (see below).

2.3.3.3 Moving-coil ('dynamic') microphone

These are widely used. They depend for their operation on the production of an e.m.f. in a small light-weight coil attached to a diaphragm. Movements of the latter as a result of sound wave pressures cause alternating voltages to be produced. Such voltages are small – typically of the order of 0.5–1 mV for speech at a distance of 0.5 m. The coil is often made of aluminium, for lightness, and the impedance is commonly about 30 Ω. Moving-coil microphones have the advantage that they are generally robust and are unaffected by humidity. They tend to be relatively expensive as skilled assembly is needed and, although designed to be very light, the coil/diaphragm assembly is heavier than the diaphragm of an electrostatic microphone and this is apt to affect adversely the transient response.

The basic moving-coil microphone (*Figure 2.21*) is resistance controlled, i.e., the natural resonant frequency of the diaphragm is in the middle of the audio range, at around 1 kHz, and this is damped acoustically to give a suitably broadband response.

Figure 2.21 *A moving coil microphone.*

2.3.3.4 Ribbon microphones

The diaphragm consists of a very light strip of corrugated aluminium ribbon, typically some 20 mm long, 5 mm wide and 0.0006 mm thick. This is mounted between the poles of a permanent magnet. An e.m.f. is induced in it when it moves as a result of sound wave pressures. The ribbon is sometimes corrugated lengthways, to give stiffness, with transverse corrugations at the ends to allow the stiffened portion to move in the magnetic field. More commonly, through, the whole length is corrugated, the corrugations being perpendicular to the length. The e.m.f. generated is very small because of the short length of conductor and also because of the fact that the relatively wide gap results in a lower than ideal

flux density. The ribbon impedance is small, usually less than 1 Ω, but a voltage step-up transformer is often incorporated to give an output voltage which approaches that of a moving-coil microphone and also to give an output impedance of ≥30 Ω.

Ribbon microphones (*Figure 2.22*) lend themselves to pressure gradient operation (see below) and this requires the ribbon to be mass-controlled – the mass of the ribbon is the principal factor compared with resistance or compliance so that the mechanical characteristics of the ribbon are such that it moves relatively easily at low frequencies, and less easily at high frequencies.

Magnet pole pieces

Ribbon or corrugated aluminium, typically 2.5 cm long, 0.6 cm wide and 0.00006 cm thick

Magnet

Pole pieces

Ribbon

Plan view

Figure 2.22 *A basic ribbon microphone.*

2.3.3.5 Electrostatic microphones

In their simplest form these consist of a conducting diaphragm (for example, plastic coated with a metallic deposit) positioned near to a rigid metal plate so that the two form a capacitor. The combination of diaphragm and back plate is referred to as the *capsule*. The spacing between the two is about 0.02 mm, resulting in a capacitance of the order of 20 pF. If a charge Q exists on the capacitor so formed, then any variation in the value C of the capacitor will result in a change in the voltage V across it, from the expression $Q = CV$, so that:

$$\delta V = Q/\delta C$$

Capacitance changes occur when the diaphragm moves as a result of sound wave pressures acting on it. There are two methods of putting a charge on the capacitor.

The first is to connect the two halves of the device to a d.c. source of (commonly) about 50 V through a high-value resistor. The latter and the microphone's capacitance have a time constant which is sufficiently long for the charge Q to be effectively constant (*Figure 2.23*).

To 'pre-amp'

Capsule

R (high value)

d.c.

Figure 2.23 *An electrostatic microphone with a d.c. voltage supply.*

The second method of putting charge on the capacitor is to use *electrets*. These are substances which can be given a nearly permanent electric charge by raising the material to a fairly high temperature while it is in a strong electrostatic field and then allowing it to cool while still in the field. While the charge is not totally permanent, the leakage rate is estimated to be such that the sensitivity of the microphone reduces at a rate of less than 1 dB per decade. The use of electrets for either the diaphragm or the backplate is currently very widespread because of the obvious simplification in not having to provide a d.c. source of voltage.

An advantage in having the electret form of the diaphragm is that low-cost production is possible. On the other hand, electret materials do not generally have the mechanical characteristics suitable for diaphragms, so that an electret layer on the backplate may result in a better microphone frequency response, albeit at a higher cost.

The high impedances associated with electrostatic microphones of the types just mentioned mean that it is essential for some form of impedance-matching amplifier (usually termed *the pre-amp*) to be installed close to the capsule. In most microphones this is fitted inside the microphone case, but an acceptable alternative is to have the amplifier within a distance of about 1 m, as for example where the capsule is on the end of an extension tube. The high impedance (the series resistor shown in *Figure 2.23* has a value of several hundred megohms) means that good insulation is vitally important and, as a consequence, electrostatic microphones are liable to be severely affected by moisture, albeit only temporarily. Bringing this type of microphone from a cold environment into a warm one is apt to cause condensation and the microphone's output is likely to be seriously affected by 'frying' or similar noises until the moisture has evaporated.

A third form of electrostatic microphone is the so-called radio-frequency electrostatic microphone. This is represented in simplified form in *Figure 2.24*.

Figure 2.24 *A simplified radio-frequency (rf) electrostatic microphone. fm, frequency modulated.*

The radio-frequency oscillator has high-frequency stability and operates, typically, at about 8 MHz. Its output is fed into a frequency discriminator the tuning of which is controlled by the circuit consisting of the microphone capsule and an inductor. The output of the discriminator thus follows the capacitance variations of the capsule.

An advantage of the radio-frequency electrostatic microphone over the more conventional types is that it is largely unaffected by humidity as the impedances in the radio-frequency circuitry are relatively low. Its cost, though, tends to be high.

2.3.4 Directivity patterns ('polar diagrams')

An important characteristic of a microphone is its response to sound arriving from different directions, and for the user this is generally more important than the type of transducer. This directivity pattern or polar diagram depends basically on the way in which sound waves strike the diaphragm.

2.3.4.1 Pressure operation
In acoustic terms the simplest type of microphone is that shown in *Figure 2.25*. It can be seen from this figure that sound waves can strike only the

Figure 2.25 *A simplified pressure-operated microphone.*

front of the diaphragm, so that the movements of the latter are determined by the sound wave pressure and not some derivative of it, as is the case in the other arrangements dealt with later. Not shown in the diagram is a small aperture which serves to equalize slow changes in air pressure. It can be taken that this aperture is too small to allow significant ingress of sound waves.

The shape of the directivity pattern of a pressure-operated microphone depends on the frequency of the sound waves or, more precisely, the relationship between wavelength and the diameter of the microphone. At frequencies where the wavelength is large compared with the microphone's diameter it can be assumed that sounds incident from any direction will be capable of striking the diaphragm, those incident at 180° (the rear) reaching the diaphragm by diffraction. The microphone is thus omnidirectional.

At higher frequencies, when wavelength is comparable with the diameter of the diaphragm, sounds incident at 90° may show phase cancellations. For example, a complete wavelength across the diaphragm will give little output because the high-pressure and low-pressure regions will effectively balance out (*Figure 2.26*). Also, at this kind of frequency diffraction of 180° sounds will be likely to be incomplete.

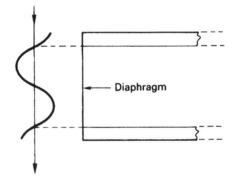

Figure 2.26 *Phase cancellation.*

At higher frequencies still, when wavelength is appreciably smaller than the microphone's diameter, diffraction processes will virtually cease and the microphone will be markedly insensitive to sounds from the rear. *Figure 2.27* shows a typical set of directivity patterns for a pressure-operated microphone.

An alternative, and sometimes more useful, method of representing the directional properties of a microphone is shown in *Figure 2.28*. It can be seen from this figure that the microphone's frequency response is virtually flat up to the frequency f',

Figure 2.27 *Directivity patterns for a pressure-operated microphone.*

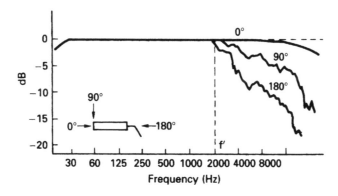

Figure 2.28 *Frequency-response graphs for a pressure-operated microphone.*

above which the sensitivity falls off for non-axial sounds. As a rough rule of thumb, the frequency f' is usually about two to three octaves below the frequency for which wavelength equals microphone diameter. Thus for a microphone having a diameter of 20 mm, the sound wave frequency for which the wavelength equals 20 mm is about 17 kHz. Two octaves below 17 kHz is roughly 4 kHz; three octaves is around 2 kHz. In practice. $f' = 3$ kHz will be found to be a reasonable approximation for a microphone of this diameter.

It is common practice to refer to pressure-operated microphones as 'omnidirectional' ('omni' for short) as, for many purposes, the tendency for a directional response at the higher frequencies is not signifi-cant. This is especially true for very small 'personal' microphones, which can be effectively omnidirec-tional up to several kilohertz.

2.3.4.2 Pressure-gradient operation

The essential feature of this type of microphone is that sound waves have equal access to both sides

Figure 2.29 *A basic pressure-gradient microphone.*

of the diaphragm, as shown in *Figure 2.29*. It can be seen from this figure that the microphone is symmetrical front to rear, so that it will give, at least in principle, equal outputs for sounds from both 0° and 180°. Sounds approaching from 90° will produce equal pressures on both sides of the diaphragm so that there will be no resulting force on the diaphragm and no output. A little thought shows that the output of the microphone will, other things being equal, be proportional to $\cos\phi$, where ϕ is the angle of sound incidence. $\phi = 0°$ for on-axis sounds.

A plot of $r = \cos\phi$ results in a figure-of-eight pattern, as shown in *Figure 2.30*. With a well-designed

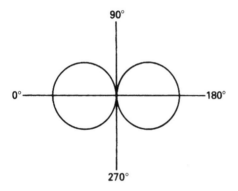

Figure 2.30 *The directivity pattern of a pressure-gradient microphone.*

microphone the departure with frequency from the figure-of-eight pattern shown is likely to be small.

The force acting on the diaphragm is a function of the *pressure gradient* (pressure difference) between the two sides. This is illustrated graphically in *Figure 2.31*. An alternative method of representation and one which lends itself better to quantitative results is shown in *Figure 2.32*, where $0F_\mathrm{f}$ and $0F_\mathrm{r}$ represent the forces on the front and rear of the diaphragm, respectively. The angle β between $0F_\mathrm{f}$ and $0F_\mathrm{f}$ is found from the effective front-to-rear path for sound waves and the frequency f.

Figure 2.32 *Method of determining the pressure gradient.*

If d is the length of the front-to-rear path, then β is equal to $d/\lambda \times 360$, where λ is the wavelength. Since $\lambda = c/f$, c being the velocity of sound, then:

$$\beta = df/c \times 360$$

The force corresponding to the pressure gradient F_G is represented by the line $F_\mathrm{f}F_\mathrm{r}$, and this is given by:

$$F_\mathrm{G}^2 = F_\mathrm{f}^2 + F_\mathrm{r}^2 - 2F_\mathrm{f}F_\mathrm{r}\cos\beta$$

If F_t and F_t are equal, as will be the case when the source of sound waves is relatively distant so that inverse-square-law effects can be neglected, then the expression reduces to:

$$F_\mathrm{G}^2 = 2F_\mathrm{f}^2(1 - \cos\beta)$$

Figure 2.31 *Pressure gradient.*

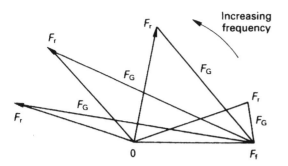

Figure 2.33 *The variation of the pressure gradient with frequency.*

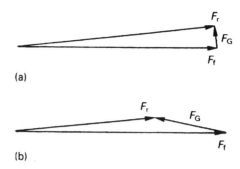

Figure 2.34 *Proximity effect.*

Figure 2.33 shows how the pressure gradient varies with frequency. It can be seen that the pressure gradient is very small at low frequencies, rises to a maximum when $\beta = 180°$ and is then expected to decrease. In fact the latter decrease is not normally observed as the wavelengths are then, in practice, likely to be smaller than the dimensions of the microphone and diffraction round to the rear of the diaphragm does not occur. In these circumstances the microphone is pressure operated. (It still retains an approximately figure-of-eight directivity pattern because of the front-rear symmetry.)

From the foregoing it is clear that the pressure gradient in a typical microphone is very frequency dependent, being extremely small at low frequencies. A flat overall frequency response is obtained by making the diaphragm mass controlled. This means that the diaphragm is given a very low resonant frequency. If this frequency is correctly chosen and the damping (or Q) is appropriate then the mechanical frequency response of the diaphragm can be a mirror image of the pressure gradient curve, resulting in a flat frequency response of the microphone. It may be mentioned that ribbon microphones (Section 2.3.3.4) lend themselves well to pressure-gradient operation.

A further and often important point about pressure-gradient microphones is that they exhibit a phenomenon known as proximity effect or bass tip-up. This means that there is an exaggerated output at low frequencies when the source of sound is close to the microphone. *Figure 2.34* shows why this occurs. For distant sound sources it can be assumed that F_r and F_f are equal. *Figure 2.34(a)* illustrates this at a low frequency so that the angle β is small. The pressure gradient F_G is small. In *Figure 2.34(b)* the conditions for a close sound source are shown. Here inverse-square-law effects result in the sound pressure on the rear of the diaphragm being rather smaller than that on the front because of the extra distance that

the rear waves have had to travel. F_r is then shorter than F_f and for the same frequency as in (a) the pressure gradient F_G is clearly greater. A similar analysis at a higher frequency when β is larger (approaching 180°, for example) shows that the pressure gradient for close sound sources may even be slightly smaller than for distant ones. In practice this effect at high frequencies is virtually impossible to detect. The rise in bass frequencies for close sources may, however, be very marked. The proximity effect can be detected with typical pressure-gradient (figure-of-eight) microphones when the sound source is less than about 30–50 cm.

The effect is turned to good use in certain so-called 'noise cancelling' microphones. In such a microphone the pressure-gradient unit is positioned close to the speaker's mouth and the resulting rise in the low frequency output is corrected by a simple electrical equalizing circuit. Distant sounds which would not have an increased low-frequency content are, nevertheless, also processed by the equalizer and are thus reduced in level. Microphones of this type are widely used in broadcasting for purposes such as sports commentaries.

2.3.4.3 Cardioid directivity patterns

A plot of $r = 1 + \cos\phi$ produces the cardioid shape shown in *Figure 2.35*. This is a particularly useful directivity pattern as it means that the microphone is very insensitive to sounds arriving from the rear. Almost all cardioid microphones make use of a phase-shift principle, as shown in simplified form in *Figure 2.36*.

The function of the acoustic labyrinth is to introduce a slight delay into sound waves which enter through the apertures. The labyrinth is designed so that the time of travel of sound waves entering the apertures to the diaphragm (the internal path) is equal to the external path, i.e. the route from the apertures

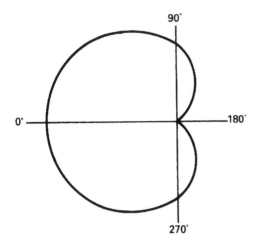

Figure 2.35 *A cardioid directivity pattern.*

Figure 2.36 *Simplified phase-shift principle.*

round the outside of the casing to the diaphragm. Sounds approaching the microphone from the 0° angle simply have a pressure-gradient operation as a result of sound wave pressures on both sides of the diaphragm. However, for sound waves arriving from the rear (180°) those entering the apertures arrive on the back of the diaphragm at the same time as those travelling round the outside and there is no resultant force on the diaphragm. Care

has to be taken to ensure that as far as possible the internal and external paths are equal at all frequencies and also that the amounts of sound wave energy in two paths are equal. Both requirements are in practice impossible to achieve totally, but very close approximation to the ideal cardioid can be made over fairly wide frequency ranges. *Figure 2.37* gives frequency-response graphs for a typical cardioid microphone.

Because a cardioid microphone is basically a pressure-gradient device it exhibits a degree of proximity effect, the amount depending on the individual microphone's design. In some cases it may be very small or negligible unless the sound source (person's mouth) is within a very few centimetres of the diaphragm; in other instances the effect is detectable at source distances of up to 30–50 cm.

2.3.4.4 Hypercardioid pattern

A hypercardioid pattern (*Figure 2.38*) can be thought of as intermediate between a figure-of-eight and a cardioid pattern. Such microphones are produced by, for example, allowing a greater proportion of sound wave ingress to the aperture of a cardioid microphone so that the microphone becomes closer to the pure pressure gradient outlined in Section 2.3.4.2.

For the user the main advantage of a hypercardioid microphone is that its frontal angle of acceptance is narrower than that of a cardioid, so that it is in effect more directional, despite the existence of the rear lobe.

2.3.4.5 Variable directivity microphones

Microphones are available with a variable directivity pattern facility, the directivity pattern being selected by a switch which may be on the body of the microphone or on a separate control unit. In such microphones there are, in effect, two electrostatic cardioid capsules back-to-back. In practice, both units share

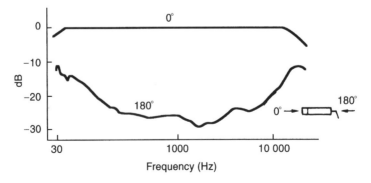

Figure 2.37 *Frequency-response graphs for a typical cardioid microphone.*

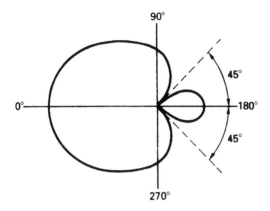

Figure 2.38 *A hypercardioid directivity pattern.*

a common 'backplate' which contains the acoustic labyrinth needed to give cardioid characteristics.

The electrical circuit is shown in somewhat simplified form in *Figure 2.39*. The front diaphragm is maintained at earth potential. The 'backplate' carries a potential *V* (typically about 50 V) The rear diaphragm is switchable from 0 V to approximately 2 V in steps. When the rear diaphragm is at *V* (the same as the backplate) it is inoperative as there is no potential difference between it and the plate. Thus the front cardioid only is effective. When the rear diaphragm is at earth potential then both cardioids are working and the effective output is the addition of their outputs. This is given by:

$$r = (1 + \cos\phi) + (1 - \cos\phi)$$

Figure 2.39 *A variable directivity pattern microphone.*

The second term in brackets has a minus sign to take account of the fact that the rear cardioid is acting in the opposite direction to the front cardioid. The sum of $(1 + \cos\phi) + (1 - \cos\phi)$ is 2, indicating that the microphone's output is independent of angle, i.e. it is omnidirectional.

When the rear diaphragm is at a potential of 2 V the expression becomes

$$r = (1 + \cos\phi) - (1 - \cos\phi)$$

which is $2\cos\phi$, implying a figure-of-eight pattern. Hypercardioid, or indeed other directivity patterns can be obtained by a suitable choice of resistors in the potential divider chain.

2.3.4.6 Highly directional microphones

This term is used for microphones with directivity patterns the main characteristic of which is a frontal lobe which is significantly narrower than any of the directivity patterns described above. A total angle of $\pm20°$ or less would generally be regarded as being highly directional. There are two principal methods of achieving such a pattern.

(a) The use of a parabolic reflector with the microphone at the focus. Very narrow-angle lobes can be obtained (e.g. $\pm3°$), but only when the sound wavelengths are appreciably smaller than the diameter of the parabolic dish. For most practical applications the maximum diameter of such a dish is likely to be about 50 cm and then the system will be more-or-less omnidirectional below about 150 Hz and will only be really highly directional above about 2 kHz. The gain to on-axis sounds will be considerable in the latter frequency range.

(b) The use of interference tubes (so-called 'gun' or 'rifle' microphones) is shown in simplified form in *Figure 2.40*. Sounds that arrive off the axis of the tube and enter the slots will arrive at the diaphragm at different times, depending on which part of the tube they entered. There will thus be a tendency for phase cancellation, the amount of cancellation depending on the angle of incidence and the wavelength. In short, cancellation occurs only to a significant degree if the sound wavelength is less than the

Figure 2.40 *An interference-tube microphone.*

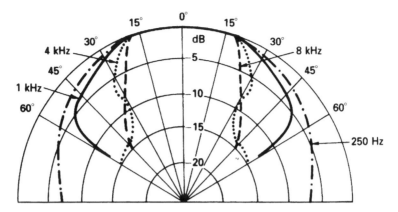

Figure 2.41 *The response of a typical interference-tube microphone.*

acoustic length of the tube. The majority of practical interference tube microphones have an effective length of about 50 cm, so that they are basically not particularly directional below about 500 Hz. However, an improvement in this can be effected by making the basic capsule have a cardioid directivity pattern. Directivity patterns for a typical interference-tube microphone are shown in *Figure 2.41*. The on-axis frequency response of this type of microphone is relatively flat, as cancellation occurs for off-axis sounds, unlike the parabolic type of microphone where gain is given to on-axis sounds, but, as has been pointed out, this is negligible at lower frequencies.

The interference-tube microphone is widely used, especially in news-gathering work in radio and TV as the sound quality of its pick-up can be good and also it can be held and pointed relatively easily by an operator.

2.3.4.7 Boundary layer ('pressure zone') microphones

A microphone placed close to a hard reflective surface with a large area will be in a region of 'pressure doubling'. In other words the reflection of sounds incident at 90° to the surface results in a pressure antinode and the output of the microphone is higher by 6 dB than when it is away from the reflecting surface. Sounds incident at less than 90° will not produce this pressure-doubling effect and at grazing incidence (0°) the surface has no effect at all. A conventional microphone placed close to a hard surface may experience this boundary layer or pressure-zone effect, but because of its physical size there may be appreciable unwanted phenomena in that the distance from the diaphragm to the surface may be of the same order as the wavelength for the higher sound

wave frequencies, so that interference between the incident and reflected sounds can occur, giving rise to a comb-filter response for these frequencies.

The use of very small transducers mounted close to the boundary plate, or inserting the transducer actually in the reflecting surface so that its (suitably protected) diaphragm is effectively in the plane of the surface, gives the advantages of the pressure doubling without the comb-filter characteristics. Microphones of this type have become widely employed in recent years where they can be used to give a good omnidirectional pickup for discussions, etc.

2.3.5 Phantom power

This is a commonly adopted system of providing electrical power for electrostatic microphones using standard three-core microphone cable. There are two systems in use: 48-V standard or T-powering; and A–B powering.

2.3.5.1 48 V standard or T-powering

A 48-Volt d.c. power supply is connected, usually to the microphone sockets on a sound desk, so that the 48 V positive side is 'shared' between the programme leads. The 0 V side is connected to the earth of the socket (*Figure 2.42*). The voltages in the cable itself are then as shown in the figure. It is important to notice that there is no d.c. voltage difference between the wires carrying the programme signals. Thus any non-powered microphone such as a moving coil can be connected without risk of damage provided it is a professional quality one with a balanced output. In any case the resistors R in *Figure 2.42* limit the current which could flow to about 15–20 mA and this means that a short

Figure 2.42 *Phantom powering.*

circuit at one microphone socket does not affect other microphones fed from the same supply.

The microphone end of the phantom-powered system is shown in *Figure 2.43*. The centre-tapped transformer delivers the +48 V to the microphone amplifier and capsule. It also acts as an output transformer for the microphone's audio signal.

Figure 2.43 *Phantom power – microphone end.*

2.3.5.2 A–B Powering

This system is almost entirely confined to providing a low voltage (9–12 V) to individual microphones. It is often used with microphones for portable single-camera (PSC) units and in location film work. *Figure 2.44* illustrates the principle. A drawback of this system is that if there is a phase reversal in the microphone cable then the polarity of the power supply to the microphone is reversed and the microphone will not work.

2.3.6 Microphone sensitivities

Different manufacturers use a variety of systems of units to indicate the sensitivities of their products. An approximate conversion between the different systems is given in *Table 2.5*.

Typical sensitivity ranges are given in *Table 2.6*, although it should be noted that it is possible to find exceptions. For example, some small 'personal' or

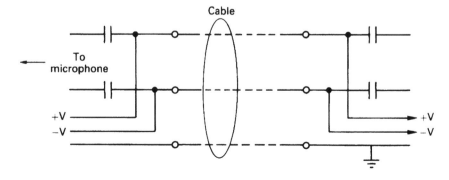

Figure 2.44 *A–B powering.*

Table 2.5 Conversion table for different systems of indicating microphone sensitivities

dB relative to 1 V/Pa	dB relative to 1 V/dyne/cm²	mV/μbar	mV/10 μbar = mV/Pa
−20	−40	9.5	95
−25	−45	5.5	55
−30	−50	3.0	30
−35	−55	1.8	18
−40	−60	1.0	10
−45	−65	0.55	5.5
−50	−70	0.30	3.0
−55	−75	0.18	1.8
−60	−80	0.10	1.0

Table 2.6 Typical microphone sensitivity ranges

Type of microphone	Sensitivity (dB relative to 1 V/Pa)
Electrostatic	−35 to −45 dB rel. to 1 V/pa.
Moving-coil	−55 to −60 dB rel. to 1 V/pa.
Ribbon	−60 to −65 dB rel. to 1 V/pa.

'tie-clip' microphones, although having electrostatic transducers, are less sensitive than would be indicated by the range of figures given in the table.

2.3.7 Windshielding

A microphone works by detecting the minute changes in air pressure that a sound produces. If the surrounding air was otherwise completely still the electrical output would be a faithful and noise-free analogue of that sound. In practice, though, air is never still. It is a turbulent fluid full of convection currents and vortices, and many of the pressure fluctuations that the microphone senses have nothing at all to do with the wanted sound source. They are, by definition, noise and although the source of these disturbances is not always wind in the conventional sense they are usually lumped together as 'wind-noise'.

The inevitability of wind-noise leads automatically to the requirement for windscreening. For microphones in very carefully controlled conditions the level of screening required may be minor – no more than the minimum level that the manufacturer builds into the capsule structure. For microphones that are used out of doors in rugged conditions something very much more effective will be needed. It is quite impossible for the shielding not to have some effect

on the acoustic properties of the microphone but the severity of this and the effectiveness of the reduction in wind-noise can vary enormously. An understanding of the principles involved allows optimal use of the techniques available.

2.3.7.1 Shelter

The usual action of a person caught in a gale is to find shelter behind a hedge or a building where the wind-speed is reduced. Similarly someone sitting in a thermally induced draught will try to block the flow of air in some way to make things more comfortable. The same technique should be the first-line approach for microphones. Positioning them out of the fiercest airstreams, in the lee of any windbreak, away from fans or gaps in doors may well produce an improvement that is more significant than any add-on shielding.

2.3.7.2 Pop-shields

The flat, circular pop-shield is, in many respects, just a portable shelter for a microphone. It is a screening disk of about 100 mm diameter that is used to prevent popping sounds when an artist is speaking close to a microphone. Certain consonants, for instance P, are accompanied by a blast of air – try holding your hand in front of your face and saying 'Pop'. The puff – a rare unidirectional wind – is a by-product of the way the sound is produced, not an essential part of it. It is normally inaudible and its energy is dissipated into the surrounding air but if the direct air-stream hits a microphone diaphragm it will cause a massive deflection and gross distortion. Less obviously, if the stream hits any part of the rest of the microphone, particularly its rough grille, the turbulence created will cause mechanical vibration and thus generate acoustic noise. It

is an uncomfortable truth that the acoustic source of wind-noise is always uncomfortably close to the microphone.

For sounds where the wavelength is significantly greater than the size of the pop-shield there would be virtually no effect on the sound quality even if the shield were completely solid. For higher frequencies, though, it would start to become an obstacle that 'shadows' the microphone. If a way could be found of making the disk transparent to the wanted audio but not to the wind this would solve the problem. Bleazey (1961), who wrote one of the most quoted articles on windscreens, dealt with the theory of how this is done. He used the Bernoulli gas equations to calculate the behaviour of sound and wind waves. He calculated the particle velocity of sound in air, which is a factor of its frequency and intensity, to be, for most practical purposes, less than a few cm/s. Wind, on the other hand, has particle velocities that roughly match the wind-speed itself – perhaps two orders of magnitude higher – and also has associated turbulence. Bleazey then calculated the flow resistance of a thin silk cloth – we would substitute loose-woven polyester now. For low particle-velocity sound waves it is very small. The attenuation is no more than that of a 100 mm layer of air. For higher particle velocities the flow resistance increases and for fully turbulent wind it would increase by the first power of the velocity. He calculated a theoretical attenuation of 39 dB for a brisk 20 mph (32 km/h) wind and a single layer of silk mesh. He also noted that it did not appear possible to achieve this figure in practice.

If the pop-shield is therefore made from a wire ring with a fine mesh stretched over it it will produce the transparency to audio that is needed while still attenuating the wind-noise. The fact that the shield appears 'opaque' to wind means that it can be buffeted by it and in doing so will act as a source of sound itself. To minimize this the fabric needs to be made as stiff as possible, perhaps by adding a coarser supporting wire mesh.

For less controlled conditions the separate screen is too limiting and the cylindrical or spherical pop-screen fitted as part of the microphone has to be adequate. This will have a similar structure – a fine mesh of sintered bronze, wire or fabric with whatever support is needed to make a stiff and mechanically strong shield. Its size is also important. On a hand-held microphone it cannot be very large but Bleazey's experimental work on spherical windshields has proved that the larger the diameter of the shield, the more effective it is (see *Figure 2.45*). The curve is described by his equation dB (attenuation) = $\log V + 10.4$ where V is the free air volume in cubic inches – for convenience I have translated the scaling to show shield radius in millimetres. Bleazey was sure that since the windshield also acted as a noise source, making it larger would improve matters. In general terms, moving the noise source (the sphere's surface) further away from the microphone reduces the noise level by the square of the distance. Practically the case is less clear-cut because the larger the sphere becomes, the harder it is to keep it rigid and the greater the level of turbulence it creates in a complex airstream.

Obviously a design of windshield that reduces this turbulence is a sensible idea and the aerodynamic design of both the shape and the surface are

Figure 2.45 *Windshield radius versus attenatuation of wind-noise.*

important. Reference microphones in wind-tunnels often use carefully shaped nose-cones but since real wind is chaotic and unpredictable in direction simpler spheres or round-ended cylinders are the norm. The metal mesh grilles of most microphones owe more to strength than to aerodynamic refinement but the addition of a cover that can slow the wind-speed and reduce its turbulence is quite simple. The other requirements of this cover are that it should not be able to generate any noise itself and that it should be transparent to low-particle velocity audio waves. One such material is open-cell plastic foam.

2.3.7.3 Foam

A good foam windshield forms an extensive labyrinth of interconnecting cells. The material is both stiff enough to hold its shape without any supporting structure and will damp turbulence-induced vibration. The long, indirect path length through the labyrinth will slow the wind and its energy will be dissipated. The ratio of radius to effectiveness still applies – larger shields give better wind-noise reduction but the audio signal must now traverse a much greater depth of material than a single fabric mesh, so foam shields will cause a slight, but usually equalizable, attenuation of HF audio.

All these forms of shielding apply equally to pressure and pressure-gradient microphones and take no account of the spectral distribution of the wind-noise. Wuttke (1991) deals with both these factors in an excellent paper. The spectrum of wind-noise is biased strongly to LF (see *Figure 2.46*). A highpass filter could be used to remove much of this noise and indeed dynamic microphones that often have a poor bass response will be significantly less troubled by wind. However, although some LF roll-off is almost essential for any microphone a corner frequency high enough to provide a substantial reduction in

wind-noise would have the disadvantage of losing some wanted sound. *Figure 2.46* also highlights the difference in wind sensitivity between pressure and pressure-gradient microphones.

Wuttke's second diagram (*Figure 2.47*) shows interfering wind signals in vector terms as an addition to the wanted signal for both types of microphone. The wind-noise amplitude is expressed as a vector inside the circle at the end of the signal vector. In the case of the pressure microphone (*Figure 2.47(a)*) the device responds to absolute pressure and its output is the vector sum of the wanted sound plus the interfering wind-noise. The noise is, at worst, only a small component of the combined signal. However, the pressure-gradient microphone (*Figure 2.47(b)*) responds to the *difference* in amplitude of the signals at its two ports – a difference that may be very small at LF. It is this that allows it to discriminate between sounds from different directions and thus be the basis for all directional microphones. *Figure 2.47* shows how the pressure-gradient – the difference – between the two independent wind-noise vectors may be much larger than the gradient due to the wanted sound source.

A solution to this is to arrange that the two wind-noise vectors are always identical. If they can be correlated in this way they will have no net effect on the microphone at all. This can be done by forming a pressure chamber around the microphone. For any wavelength of sound that is much greater than the size of the chamber the momentary pressure inside will be the same at all points – there will be no pressure difference between the two inlet ports to produce an output signal. The consequences of this are a significant reduction in wind-noise but also a corresponding bass roll-off and loss of directivity at LF since the correlation applies equally to the wanted LF audio. The pressure chamber is conventionally

Figure 2.46 *Typical spectra for wind-noise – pressure and pressure-gradient microphones.*

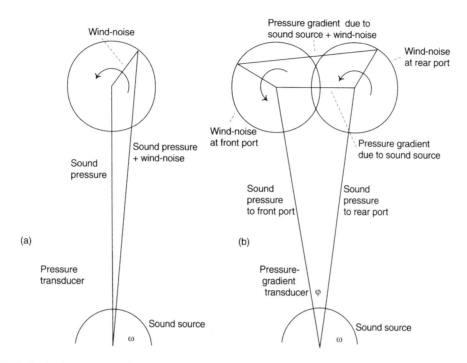

Figure 2.47 *Wind-noise in vector terms for pressure and pressure-gradient microphones.*

formed by a basket structure around the microphone. The 'skin' of the basket is supported on a frame-work in such a way that it is, in momentary terms, rigid and air tight. There are also some foam wind-shields which emulate baskets by having cavities around the capsule zone. These are more effective at reducing wind-noise than ordinary foam shields on pressure-gradient microphones but less so than bas-kets because of their poorer rigidity. They do have the advantage of generating less coloration though since they do not need the structural components of a basket that inevitably cause some reflection or scat-tering of sound.

2.3.7.4 Fur
Just as it is possible to add a foam layer to a microphone to make it less prone to turbulence-generated sound so it is possible to coat a basket with material such as fur for the same purpose. The fur may be integral to the basket or a close-fitting sleeve but its benefit is that it can absorb large amounts of turbulence energy in complete silence. The fur also conforms to the underlying shape of the basket but smoothes any small roughness that would generate noise. It is important that the matrix of fibres and the backing material maintain adequate transparency to audio and it is usual to find a small degree of

HF loss due to the density of the fur material. As with the foam windshields, this normally takes the form of a gradual roll-off which can be satisfactorily equalized.

Clearly many of these wind-noise reduction tech-niques can be used in combination – a pressure-gradient microphone fitted with mild LF filtering can be held in a sheltered position within a basket windshield equipped with a fur windjammer to very good effect. However, some combinations cannot work – the use of a foam windshield inside a bas-ket will cause decorrelation of LF sounds and hence a poorer attenuation of wind-noise. Multi-layered basket assemblies are possible and have been used although the coloration due to the complex struc-tural elements necessary and the degree of HF loss through multiple thicknesses of fabric detract from their benefits.

2.3.7.5 Measurements
Perhaps the last point to mention is the measure-ment of the effectiveness of windshields. The greatest difficulty is defining and reproducing a standard-ized wind. Various techniques with whirling arms, wind tunnels and multiheaded fans have been tried but though useful for some development work they do not provide anything comparable with the wide

range of 'real' wind. It is therefore very difficult to obtain comparisons between different windshields that translate from the test laboratory into real use. A microphone must also be used within a suspension system which will have a marked effect on the mechanical transmission of sound from the windy exterior to the sheltered interior of the shield. In addition, the microphone often needs to be considered as part of a dynamic system – it may be fixed to a boom or pole, also subject to wind and thus able to generate wind-noise, and its orientation and position within the shield will be varying greatly over time. The best that can be said is that windshields can be very effective, producing improvements as high as 40 dB in the reduction of wind-noise – but any comparative claims should be proved on an individual basis.

Bibliography

BLEAZEY, J.C., Experimental determination of the effectiveness of microphone wind screens, *Journal of the Audio Engineering Society* (January 1961)

BORWICK, J., *Microphones; Technology and Technique*, Focal Press, London (1990)

GAYFORD, M.L., *Electroacoustics, Microphones, Earphones and Loudspeakers*. Newnes-Butterworths, Oxford (1970)

NESBITT, A., *The Use of Microphones*, 4th edn, Focal Press. Oxford (1994)

WUTTKE, J., Microphones and wind, *Journal of the Audio Engineering Society* (October 1991)

ROBERTSON, A.E., *Microphones*, Iliffe, London (1963)

SANK, J.R., Microphones, *Journal of the Audio Engineering Society*, **33**, 514–591 (1985)

2.4 Loudspeakers and headphones

2.4.1 Introduction

The basic function of a loudspeaker is simply stated: it is to respond to the incoming audio frequency electrical signals representing music or speech by performing physical vibrations, which in turn radiate sound waves corresponding as closely as possible to the original signals (see *Figure 2.48*).

In practice, things are not quite so simple, and considerable design efforts are needed if the air wave reaching the listener is to match the original even approximately. The difficulties include making the diaphragm and its suspension capable of: (a) responding evenly to the full frequency range, nominally from 20 Hz to 20 kHz; (b) handling the full dynamic range, say 80 dB, i.e. a range of displacements in the ratio of 10 000:1; and (c) radiating the sound waves over a sufficiently wide dispersion angle at all frequencies to reproduce the original frequency balance at the listener's ears. Set such formidable tasks, it is little wonder that the loudspeaker remains the weakest link in the whole audio recording and reproduction chain.

Three basic configurations are:

(1) the transducer motor system is coupled at the centre of a radiating diaphragm or cone which in turn is built into a baffle or cabinet as a means of reinforcing the efficiency at sound wavelengths which are large compared with the diaphragm diameter;

(2) the diaphragm is set at the apex of an acoustic transformer or horn to provide greatly enhanced coupling to the air load: or

Figure 2.48 *A loudspeaker converts electrical energy into acoustic energy.*

(3) the diaphragm is simultaneously driven over its whole surface.

These three basic systems are found, for example, in:

(1) the electromagnetic moving-coil cone loud-speaker (by far the most common);
(2) the horn-loaded moving-coil or ribbon loud-speaker; and
(3) the electrostatic membrane loudspeaker.

2.4.2 Moving-coil drive units

Figure 2.49 shows the elements of a moving-coil drive unit, the type of transducer used in nearly all of today's loudspeakers. The voice coil is wound on a cylindrical former and suspended in the radial field of the magnet. Passing an a.c. through the coil will produce an electrodynamic force due to interaction between the fixed field and that set up by the current. This force will act at right angles to the fixed field and cause the coil to oscillate back and forth at the frequency of the applied current.

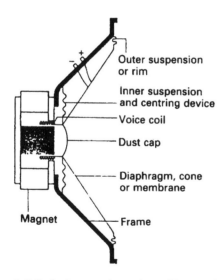

Figure 2.49 *Basic elements of a moving-coil loudspeaker.*

The magnitude of the force is given by:

$$F = BlI \text{ (N)} \tag{2.11}$$

where B is the flux density (Wb/m^2), l is the coil length (m), and I is the current (A). Note that the force increases in proportion to any increase in flux density, current of effective coil length (i.e. that portion of the coil situated within the magnetic field).

Since coil motion would not of itself give rise to significant air movement, the coil former is attached at the apex of a lightweight diaphragm or cone better able to act as a radiator of sound waves. At rest, the system is held in position by a centering device (spider) at the cone centre and a flexible surround at the rim which between them encourage fore-and-aft piston-like motion and prevent lateral or rocking motion.

The velocity of coil/cone motion is given by:

$$v = F/Z \text{ (m/s)} \tag{2.12}$$

where F is the force (N), and Z is the mechanical impedance (mech.-Ω).

The mechanical impedance is a complex function of the total moving mass (i.e. coil, former, diaphragm, air loading, etc.), resistance due to friction losses in the suspension and radiation, and the total stiffness (the reciprocal of compliance) of the suspension. By analogy with electrical theory, the system as a whole will exhibit mechanical resonance at the frequency for which the mass (inductance) reactance and compliance (capacitance) reactance are equal, i.e. mechanical impedance is a minimum and entirely resistive.

In practice, this main resonance occurs at the lowest frequency down to which we can expect a linear response. *Figure 2.50* shows the theoretical power response, f_0 being the principal resonant frequency just mentioned, at which the total mass resonates with the restoring force of the system. Below f_0 the system is stiffness controlled and the power output is proportional to f^4, falling by 12 dB/octave. At frequencies above f_0 the system is mass controlled and provides a flat power response to f_1, the point at which the wavelength and diaphragm diameter are roughly equal. For higher frequencies, the total power output should fall at 12 dB/octave due to the mass of the system. However, two factors help to maintain useful output. First, there is the onset of cone flexure, and the finite velocity of transverse vibrations out from the centre of the cone, which

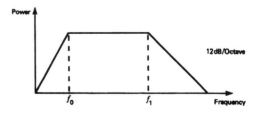

Figure 2.50 *Theoretical power response of a loudspeaker diaphragm behaving as a rigid piston.*

together cause a steady reduction in the effective radiating area of the real cone. Second, there is the progressive narrowing of the radiated beam which at least restores the power level on axis, though at the expense of a more rapid fall in the off-axis response.

2.4.2.1 The voice coil

Typical voice coils comprise two or four layers of round insulation-coated copper or aluminium wire. Copper has higher density and lower electrical resistance so that, for coils of equal size and d.c. resistance, a copper coil will usually have more turns and greater inductance. For maximum efficiency, coil mass should equal the cone and air-load mass with both being as low as possible. In general, copper provides the better compromise but, where high-frequency response is important, the lower mass and inductance of aluminium may be preferred. Copper-clad aluminium eliminates some of the manufacturing difficulties of pure aluminium and annealing (heat treatment) improves mechanical strength. Efficiency is notoriously low in loudspeakers, typically only a few per cent, and may be halved as signals approach the maximum power handling capacity due to the temperature rise in the coil, which increases resistance by about 0.4%/°C.

Amongst measures taken to control temperature rise at high signal levels, particularly in high-frequency drivers, is the introduction of a colloidal ferrofluid into the magnet gap. This magnetic fluid is held in the gap by the magnet's field and helps to dissipate heat from the coil to the main mass of the magnet, acting as a heat sink. Its viscosity also adds to the mechanical damping of the system.

Rectangular or flat wire is sometimes used instead of round wire and is usually wound edgewise in a single layer. This can pack about 40% more conductor material into the available gap area and increase the motor efficiency by about 2 dB, despite the increase in coil mass.

The voice coil is generally wound on a cylindrical tube or former attached at the apex of the diaphragm or cone and centering spider. The early use of Kraft paper formers soon gave way to synthetic nylon constructions having greater rigidity and heat resistance, raising the maximum allowable temperature from about 100°C to 150°C. In high-power loudspeakers a further move to polyimides such as Kapton or mica, glass fibre, etc., gives increased mechanical and thermal strength. Alternatively, heat-conducting materials may be used such as aluminium sheet in conjunction with high-temperature epoxy adhesives permitting working temperatures as high as 350°C.

2.4.2.2 The magnet

The role of the magnet is to provide a high value of flux density B concentrated within an annular air gap just wide enough to accommodate the coil/former thickness, yet with sufficient clearance to allow freedom of movement. *Figure 2.51* shows two possible constructions having a circular centre pole piece surrounded by a front plate, completing the magnetic circuit in such a way as to establish a strong field across the gap. Increased efficiency or reduced magnet dimensions, or both, have resulted from the development of more powerful magnetic materials. Successive additions to the basic carbon steel of tungsten and cobalt were followed in the 1930s by the famous 'Alnico' series combining aluminium, nickel and cobalt with iron, as well as the independent discovery of 'ferrite' magnets. Steep rises in the price of cobalt in the 1970s led to an increased use of ferrite or ceramic magnets, despite their slightly inferior magnetic performance and the need for a particular geometry, large diameter and short length, as illustrated in *Figure 2.51(b)*.

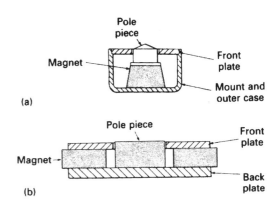

Figure 2.51 *The two basic magnet configurations: (a) conventional central 'pot' magnet; (b) flat shape used with ceramic magnets.*

If distortion is to be avoided on large-amplitude signals, it is important that the length of the coil effectively within the magnetic field is kept constant. This can be achieved either by making the coil winding (a) longer or (b) shorter than the depth of the gap by an amount which allows for the maximum peak to peak displacement expected (*Figure 2.52*). Note that the centre pole and top plate are undercut to present equal gap-facing depths and ensure symmetry in terms of flux leakage above and below the gap. Both the long-and short-coil systems are suitable for 'long-throw' drivers where large diaphragm

(a)

(b)

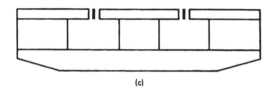

(c)

Figure 2.52 *Typical coil/gap relationships: (a) long coil; (b) short coil; (c) coil length equal to gap height.*

excursions are required at bass frequencies due, perhaps, to inadequate bass efficiency within the given cone dimensions. The short coil gives a better guarantee of linearity but is more costly because of the thicker top plate and larger magnet required. However, coil length can be reduced by about one-third for the same *Bl* values as the equivalent long-coil design. Both the long- and short-coil configurations suffer a loss in efficiency, the long coil because of power wastage in the resistance of the extra unused coil turns, the short coil because only some of the available magnetic flux is used.

Making the coil and gap lengths equal, as in *Figure 2.52(c)*, is the most efficient solution and viable in terms of acceptably low distortion in small excursion systems, such as midrange units and tweeters. The top plate thickness is made at least four times greater than the peak displacement.

2.4.2.3 The cone or diaphragm
The loudspeaker diaphragm, very often cone shaped, couples the motion of the voice coil to the air load and must be of large enough area to perform this conversion of mechanical energy to acoustic energy

with the required efficiency down to a chosen low-frequency limit. The radiated power is proportional to the volume of air moved by the pumping action of the diaphragm, i.e. the product of the diaphragm area and its peak excursion. In practice, therefore, for a given radiated power, the diaphragm displacement is inversely proportional to its area. Hence the need for a 'long-throw' design in speakers having a small cone yet operating down to low frequencies.

Displacement is also inversely proportional to frequency, each halving of frequency requiring twice the excursion amplitude. *Figure 2.53* shows the peak amplitude versus frequency required for various diameters of cone to radiate 1 W. For example, a 200-mm diameter cone would have a peak amplitude of less than 5 mm for this output level at 100 Hz, whereas a 50 mm for piston would need to move through more than 50 mm.

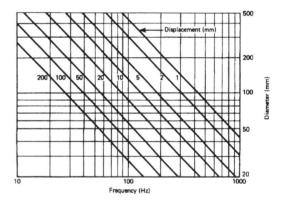

Figure 2.53 *Peak excursion versus frequency for various cone diameters to radiate 1 W.*

An ideal loudspeaker diaphragm would be infinitely light and infinitely rigid, so as to move in a piston-like manner. Then, as well as determining the acoustic power output, its dimensions will largely dictate the polar distribution. *Figure 2.54* illustrates the directivity for a rigid circular piston mounted in an infinite baffle for two different wavelength/diameter ratios. It can be seen that, at low frequencies, for which the wavelength is say more than four times the diameter, the radiation is basically omnidirectional, i.e. hemispherical if the piston is mounted in an infinite baffle. As the wavelength/diameter ratio decreases, at higher frequencies the radiation pattern narrows towards the frontal axis.

Practical loudspeakers are less sharply, or less uniformly, directional than *Figure 2.54* would suggest.

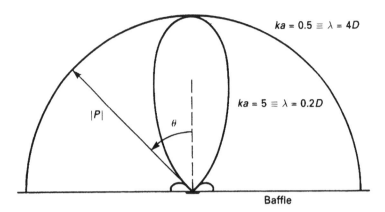

Figure 2.54 *Low- and high-frequency directivity patterns for a rigid circular piston of diameter* D *in an infinite baffle, showing the relative pressure versus off-axis angle. P, pressure.*

For one thing, the finite velocity of sound through the cone, reflections from the rim, standing waves and cone flexure tend to reduce the effective diameter at higher frequencies. For another, the finite dimensions of the baffle or cabinet progressively alter the cone loading from approximately infinite baffle to point source as the wavelength is lowered.

No single material has ideal properties for all types of loudspeaker cone. Traditionally, paper has proved itself well able to meet the principal requirements of low mass and high rigidity, and it can readily be moulded into various thicknesses and profile shapes. However, resonance break-up modes often require controlling by introducing concentric corrugations or hard resin reinforcement, etc. Metal cones can provide a higher rigidity/weight ratio and sound velocity is higher, but frictional losses are lower so that resonances tend to be sharp unless deliberately damped.

Reinforced polystyrene is of very low mass, but free-mode resonances can again arise due to low internal friction. Sandwiching the polystyrene sheet between two layers of aluminium produces an extremely stiff diaphragm with advantages at low frequencies. Bextrene combines polystyrene with neoprene and is widely recognized as a highly successful material in high-performance applications, usually coated with polyvinyl acetate to provide damping.

Cone shape is important in terms of both directivity and the suppression of resonance modes. The newer vacuum-formed polypropylene cones are easily given precise flared profiles, as perhaps determined by computer-aided design. At the same time the outer surround can be specified for optimum results. This not only acts as a cone support, but also as an energy absorbing termination to suppress unwanted standing waves whilst permitting large amplitude excursions at low frequencies. A half-roll surround of butyl rubber or moulded neoprene is commonly used, firmly bonded to the cone rim.

Most of the above descriptions apply to loudspeakers attempting to cover the full audio frequency range. However, except for low-grade applications, the difficulties of producing distortion-free radiation over such a wide spectrum (mainly with regard to gross variations in directivity and response due to cone break-up) are insuperable. The difficulties can be diminished considerably if the drive unit is asked to handle only a limited band of frequencies. This has led to the development of multiunit loudspeakers in which the lowest frequencies are handled solely by a bass driver or 'woofer' having the necessary large cone, say 200–450 mm in diameter, to move a sufficient volume of air, and a coil/suspension system capable of handling the necessary large amplitude swings. In a two-way system, the woofer will hand over at some middle 'crossover' frequency, say 2–5 kHz, to a suitably designed high-frequency unit or 'tweeter'. Here the main characteristics will be a small diaphragm, typically 19–35 mm, to produce an acceptable angular dispersion at high frequencies, extreme lightness and ability to respond to rapid transients. By further subdivision of the spectrum, one or more midrange units may be introduced to form a three- or four-way loudspeaker: or the very lowest frequencies might be allocated to a special 'subwoofer'.

Tweeter design has been refined in recent years, with the flat or conical diaphragm often replaced by a soft or hard dome for improved dispersion at the highest frequencies, assisted by contouring of the

Figure 2.55 *Dome tweeter.* (Courtesy of TDL.)

front plane (see *Figure 2.55*). Soft domes may be made of fabric or plastics, e.g. melinex bonded to a polyvinylchloride (PVC) damping layer. Hard domes are better able to provide true piston action up to supersonic frequencies and can be of very low mass. Aluminium and beryllium have proved popular and may be hard annealed or coated with gold, etc.

In pressure tweeters there is a perforated phasing plate in front of the diaphragm to give a more level response over the frequency range, or horn-loading may be used for greater efficiency.

2.4.3 Other types of driver

Moving-coil loudspeakers have been described in some detail as befits their widespread use in every application. However, many other transducer mechanisms have been developed over the years and merit brief mention here.

2.4.3.1 The electrostatic loudspeaker
The electrostatic attraction and repulsion forces which exist between electrically charged bodies according to whether they have opposite or identical polarities form the basis for the electrostatic loudspeaker. Thus if two metal plates are lined up close together and a d.c. voltage is applied across them, they will be mutually attracted. Then, if one plate is fixed whilst the other is a thin diaphragm, superimposing an a.c. voltage will set the diaphragm into to-and-fro motion.

In practice the non-linearity inherent in this unbalanced system is avoided by using a balanced push–pull structure (*Figure 2.56*). Two fixed plates or electrodes, suitably insulated and perforated to allow sound waves to pass through, are arranged on either side of the thin diaphragm. The d.c. polarizing (bias) voltage is applied equally to both fixed plates with respect to the central diaphragm, so that the 'at rest' position is exactly in the centre. The audio frequency signal is superimposed in opposite polarity to each of the outer plates, producing increased attraction of the diaphragm to one plate which is assisted by the decreased attraction to the other.

Figure 2.56 *Outline of balanced push–pull electrostatic loudspeaker.*

The diaphragm can be extremely light, since rigidity is not a requirement, and response to high frequencies and transients is very rapid. Also, since the driving force is applied over the whole diaphragm surface, unlike the localized apex drive in a moving-coil loudspeaker, break-up modes are avoided and

there is a better approximation to ideal piston motion. Note, however, that the electrostatic loudspeaker is almost invariably built without any cabinet or enclosure and therefore acts as a doublet, radiating equally from both sides to produce the familiar figure-of-eight directivity pattern. High frequencies tend to be concentrated along both the front and back axes, unless the diaphragm is subdivided into separate strips or concentric rings, each handling specific frequency bands, to broaden the dispersion.

The absence of a cabinet avoids at a stroke the various problems due to enclosure resonances but does set a limit to the low-frequency response since the forward and back waves are in antiphase and tend to cancel. For this reason most electrostatic drive units are designed solely as high-frequency components to be paired with conventional moving-coil bass drivers. Larger full-range electrostatics do feature in the high-performance market, however, notably the two models launched in 1957 and 1981 by Quad Electroacoustics.

Both the d.c. bias and the a.c. signal voltages may be of the order of several kilovolts and so it becomes important to take precautions against unduly large diaphragm excursions and possibly damaging flashover. Signal limiting and sensors which activate a cutout switch are generally included. The unusual doublet directivity with very little radiation at the sides has the incidental advantage of generating fewer room resonance modes but makes room positioning more critical.

2.4.3.2 The ribbon loudspeaker
The separate functions of a voice coil and its attached diaphragm are combined in the ribbon loudspeaker. The electrodynamic principle is again used but a transverse corrugated strip of aluminium foil acts both as conductor and radiator, being suspended in the field of a magnet having specially elongated pole pieces (*Figure 2.57*). The corrugations prevent sagging or twisting and allow the compliance to be adjusted to resonate with the mass at around 100–150 Hz. It is difficult to make ribbon loudspeakers efficient at low frequencies and so, even when they are physically quite long as in some recent designs, they are generally used as high-frequency units. Horn loading can be employed to extend the range downwards by about two octaves.

2.4.3.3 The orthodynamic loudspeaker
By an extension of the ribbon idea, it is possible to construct a flat two-dimensional diaphragm or 'current sheet' loudspeaker. The rectangular diaphragm

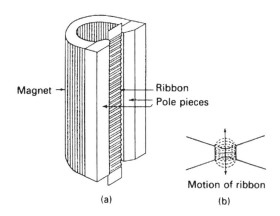

Figure 2.57 *Ribbon loudspeaker.*

is made of thin plastic with an etched pattern of copper or aluminium. This is suspended in the concentrated field of an assembly of flat magnets the polarity of which reverses with each twist of the conductor pattern to provide the correct driving force overall. This orthodynamic principle has been used in tweeters, midrange units and headphones.

2.4.3.4 The Heil air-motion transformer
This transducer, developed by Oskar Heil in the USA, has some points in common with both the ribbon and orthodynamic systems. The conductor is again etched on to a plastic diaphragm, using printed circuit technology, but is pleated in a zig-zag or concertina manner. The magnetic field is at right angles to the diaphragm, rather than coplanar, and so produces sideways motion causing the diaphragm folds to open and close, alternatively sucking in air and squeezing it out again. The transformer description refers to the enhanced matching, by a factor of 5:1, of the mechanical impedance to that of the air. The technique has so far been applied to tweeters and midrange units.

2.4.3.5 The piezo-electric loudspeaker
Various crystal materials, notably Rochelle salt, ceramics and high-polymer plastics have been found to alter their physical dimensions when subjected to an electric potential. Smoothly extended high-frequency drive units have been constructed using this principle with the material formed either into a two-layer bimorph 'bender' with a cone attached at the vibrating centre and horn loading, or a cylinder the pulsations of which closely resemble those of a perfect cylindrical source. Again the principle seems restricted to the higher frequencies but excellent transient response and high power handling are achieved.

2.4.3.6 The ionic loudspeaker

Unique in that it has no mechanical moving parts, the ionic transducer uses a radio frequency oscillator to produce an ionic discharge in the air contained inside a quartz tube. The intensity of the discharge and the consequent localized heating of the air is proportional to the radio-frequency voltage. Therefore amplitude modulation of the radio-frequency carrier with the audio frequency signals will produce air pressure variations which can be coupled from the tube to the apex of a horn and radiated as sound waves. The system has an extended high-frequency response and has been used successfully in tweeters, but needs a high-power oscillator which means that radio-frequency interference can be a problem.

2.4.4 Multiple driver systems

The near impossibility of producing distortion-free radiation over the full 20–20 000 Hz spectrum, with consistent directivity, from a single drive unit was mentioned in Section 2.4.2.3. It was also suggested that one solution is to assemble a multiple driver system in which responsibility for handling the frequency range is divided between two or more drive units each deliberately designed for a specific band of frequencies.

It then becomes necessary to introduce filters using inductors and capacitors to ensure that only the appropriate band of signal frequencies is passed to each driver: a low-pass filter will feed the woofer, a high-pass filter will feed the tweeter and, if necessary, band-pass filters will feed one or more midrange units. Practical drivers will already have a falling response beyond their intended working range but filtering is still regarded as essential to avoid distortion effects due, for example, to excessive amplitudes at low frequencies in the tweeter or cone break-up modes and intermodulation at high frequencies in the woofer. It is also important to limit the overlap region around the 'crossover' frequency where phase anomalies will occur in the dispersion pattern.

Two types of circuit configuration for these crossover filter networks are possible (*Figure 2.58*). In the more common high-level system, the various passive (i.e. unamplified) filters are connected in parallel across the loudspeaker input terminals to be fed from a single external power amplifier. In the more complex low-level system, each drive unit has its own power amplifier with an active filter at its input. This latter arrangement avoids the presentation of a complex input load to the main power amplifier

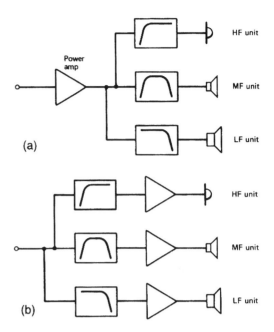

Figure 2.58 *Crossover networks: (a) high-level (passive); (b) low-level (active). HF, high frequency; LF, low frequency; MF, mid frequency.*

and allows accurate tailoring of the response to suit the given drive units in terms of their individual sensitivities, impedances, frequency characteristics and directivity.

At the crossover frequency the response from both units will usually be reduced to −3 dB. This has the effect of producing a flat response overall when the two radiated signals are added together. This is illustrated in *Figure 2.59* for three basic types of crossover network. These may be classified as first order, second order and third order, where one, two or three components are used for each drive unit

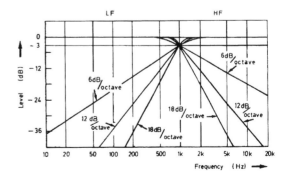

Figure 2.59 *Response curves for first-, second- and third-order crossover networks. Note that the response for all units is at −3 dB at the nominal crossover frequency.*

Figure 2.60 *Basic crossover networks giving roll-off slopes of 6, 12 and 18 dB/octave. (a) first order; (b) second order; (c) third order. HF, high frequency; LF, low frequency.*

to give increasingly steep roll-off slopes of 6, 12 and 18 dB octave, respectively. Basic circuits for these three orders of two-unit crossover networks are shown in *Figure 2.60*. The higher order filters introduce greater phase differences, are liable to 'ringing' on transients and need close tolerance components. However, they will be preferred in circumstances where the reduced overlap region is mandatory, perhaps to avoid tweeter overload and unwanted driver resonances.

The simple networks shown in *Figure 2.59* are based on the assumption that the nominal impedance (say $Z = 8\ \Omega$) is the same for all drivers and is consistent at all frequencies (i.e. resistive). In fact drive unit impedances vary quite widely with frequency (see Section 2.4.6) and the precise network component values should be calculated in conjunction with the known driver parameters, working backwards from the desired overall system response, preferably using a target function approach and computer-aided design.

Apart from the electrical considerations, all multiple driver loudspeaker designs involve physical compromises aimed at reducing the problems of interference between the waves radiated by the individual drive units at or near the crossover frequency. So far as the horizontal plane is concerned, aligning the drivers vertically one above the other (*Figure 2.61(a)*) will produce the most uniform and symmetrical response.

However, achieving uniformity of time of arrival at a typical listening position in the vertical plane can present problems. The much smaller distance between the effective source (diaphragm) and the front edge of a tweeter, compared with that in a typical woofer, has the effect of tilting the 'reference axis', on which the interunit time delay has the desired zero value. With a normal vertical front panel (*Figure 2.61(b)*), the reference axis will be tilted downwards. Where this is undesirable, the designer may resort to one of the other configurations shown in *Figure 2.61: (c)* a sloping baffle; *(d)* a stepped baffle; or *(e)* the woofer placed above the tweeter.

There are several other related problems inherent in multiple driver loudspeakers. These include interference effects in the crossover region, shifting of tonal balance at listening positions off-axis because of the different directivity patterns in the various drivers, and variable time delays altering the phase relationship between component frequencies in the sound signal. These would be reduced if the effective source position of the drivers could be located at the same point in space. *Figure 2.62* illustrates the situation in a stepped baffle system (as in *Figure 2.61(d)*) where the times of arrival have been made equal only for an on-axis listener. *Figure 2.62(b)* shows the tweeter mounted on the same axis as the woofer, as in the well-known dual-concentric arrangement employed by Tannoy. This can produce a smooth off-axis response, but there remains an interunit time delay difference on-axis. The layout shown in *Figure 2.62(c)* overcomes this problem by locating both drivers not only on the same axis but also with their acoustic centres on the same plane. This technique is used, for example, in the KEF Uni-Q system. The woofer cone in both the dual-concentric and the Uni-Q arrangements modifies the directivity factor of the tweeter to produce a very similar dispersion from both drivers in the important crossover region where radiated power is about equal.

Figure 2.61 *Methods for controlling the direction of the reference axis of a two-way loudspeaker.*

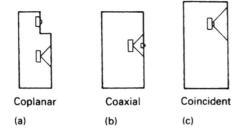

Figure 2.62 *Comparison of (a) stepped baffle; (b) dual concentric and (c) Uni-Q coincident woofer/tweeter systems.*

2.4.5 Baffles and enclosures

When a diaphragm vibrates in free air it behaves as a doublet radiator, emitting two waves rather than one, the waves from the two sides being of equal amplitude but opposite phase. At high frequencies the two waves are beamed outwards along their respective axes. However, at lower frequencies where the wavelength is greater than the diaphragm radius, the radiated waves become spherical and bend round the edges. Being in antiphase, they tend to cancel and severely reduce the low frequency response, rolling off at 6 dB/octave.

Except in low-fidelity applications, some means of enhancing low-frequency efficiency must be sought. This may be approached by preventing the back wave from reaching the listening area, or at least partly reversing its phase so that it will reinforce the forward wave. This is the role of the baffle or enclosure.

2.4.5.1 The flat baffle

Simply mounting the loudspeaker drive unit at or near the centre of a flat rigid board or baffle will keep the front and back waves apart, and extend the usable frequency range downwards by increasing the distance which the waves must travel before interfering. The lower frequency limit occurs roughly when $L = 2\lambda$, where L is the length of the shortest baffle dimension and λ is the wavelength. Taking the velocity of sound in air to be $c = f\lambda = 340 \ m/s$, this expression becomes $L = 2 \times 340/f$; e.g. for a cut-off frequency of 200 Hz, the baffle would need to measure 3.4 m across. Above the cut-off frequency the polar distribution becomes figure-of-eight in direct analogy with the large panel full-range electrostatic loudspeaker described in Section 2.4.3.1.

More manageable dimensions can be arrived at by adding sides to the baffle to form an open-backed box. Alternatively, it would be possible to mount the drive unit into a room dividing wall. This would provide a virtually infinite baffle, completely isolating the front and back waves–and supplying nearly equal sound quality in both rooms.

2.4.5.2 The closed box

The description 'infinite baffle' is sometimes applied to the closed box loudspeaker cabinet (*Figure 2.63(a)*). Certainly the wave from the back of the diaphragm is isolated from the front wave, except that internal reflections may bounce back 'through' the diaphragm or set up secondary vibrations of the box itself.

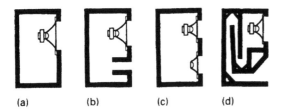

(a) (b) (c) (d)

Figure 2.63 *Loudspeaker baffles and enclosures; (a) closed box or 'infinite baffle'; (b) vented enclosure; (c) vented enclosure with auxiliary bass radiator; (d) transmission line.*

The enclosed air acts as a restoring force or spring the stiffness of which will reduce the effective compliance and raise the system bass resonance frequency. Response below resonance falls fairly rapidly at a rate of 12 dB/octave. The designer will therefore choose a drive unit with a very compliant mechanical suspension, to allow for the additional 'acoustic suspension', a high-mass long-throw cone and as large an enclosure as possible. It also becomes

necessary to introduce large amounts of absorbent material to damp dimensional resonances or standing waves, and to reinforce or brace the panels to reduce vibration. Fortunately, filling the enclosure with glass fibre or similar light absorbent material has the effect of partially changing the air operation from adiabatic to isothermal, which reduces the velocity of sound and may increase the effective volume of the box by as much as 40%.

As a result of these measures, even the best designed closed box loudspeakers tend to be of low efficiency and power-handling capacity. However, since the advent of relatively inexpensive transistor amplifiers, electrical Watts are readily available and the so-called 'infinite baffle' or 'acoustic suspension' enclosure is very widely used.

2.4.5.3 The vented enclosure

Instead of attempting to lose the backwards radiated sound energy by fitting a large baffle or a closed box to the loudspeaker, several designs actually redirect this energy so that it contributes to the total sound reaching the listener. The trick is to reverse the phase of the back wave before sending it on its way, and this will then add to the forward wave (at least at low frequencies) instead of cancelling it. This can provide greater efficiency and bass extension than the closed-box system.

In the most usual type of bass reflex enclosure (*Figure 2.63(b)*) there is a vent or port, generally let into the front panel and having an associated outlet tunnel or pipe. The dimensions of the vent are carefully calculated so that vibrations of the mass of air contained in it will resonate at the same frequency as the main cone–enclosure system. In this condition the air in the vent will move back and forth in step with the main cone at low frequencies and thus reinforce it.

Vented enclosures are relatively large, when well-extended bass is required. It may be noted that, whereas the typical bass drive unit has a much increased load impedance at resonance due to the generation of a high 'back-e.m.f.', reflex loading restricts cone movement at resonance. This produces a dip in the impedance curve at resonance with smaller peaks at either side. Further tuning (broadening) of the response can be effected by introducing acoustic resistance or air friction to the vent in the form of wool padding or other porous material.

As an alternative to the open vent, some designs employ an auxiliary bass radiator (ABR) which consists of a passive non-driven cone assembly of about the same dimensions as the main driver

(Figure 2.63(c)). This will have a greater mass than the equivalent vent air volume and can be tuned to reinforce the main driver to give more precise, smoother bass extension. ABR loudspeakers can be smaller than conventional vented types.

2.4.5.4 The transmission-line or labyrinth enclosure

A special development of the vented enclosure principle is found in the transmission-line loudspeaker *(Figure 2.63(d))*, and related designs more accurately described as labyrinth enclosures. Fairly large dimensions are needed since the back radiation from the cone is made to travel down a pipe containing absorbent material and is time delayed to such an extent that it emerges at the end of the 'line' in phase with the front wave. In theory the line should be effectively a quarter of a wavelength long at the bass resonance frequency. This would mean a 2.9-m pipe for a loudspeaker going down to 30 Hz, for example, and so transmission-line loudspeakers invariably employ a pipe which is folded back on itself once or twice. Standing waves will therefore tend to be set up at higher frequencies and careful damping and tapering of the pipe are needed to suppress these unwanted side-effects.

2.4.5.5 Horn loading

Where space is available, the back radiation, or sometimes that from the front, may be directed into a continually expanding pipe or flared horn *(Figure 2.64)*. Such horn-loaded loudspeakers, provided they are large enough, can provide greatly increased efficiency down to extremely low frequencies. Efficiency may reach 50% compared with much

Figure 2.64 *Horn loading in a corner horn enclosure.*

less than 10% in other types of enclosure, the effect being something like speaking into a megaphone. The response to transients is improved and distortion much reduced, since only small diaphragm excursions are needed. Once again the dimensions and the rate of flare are critical. The horn ideally needs to be several metres long, with a progressive opening out of the cross-sectional area. In addition, the cut-off below the critical frequency is very sharp and needs to be chosen with care. Clearly this type of loudspeaker has an application in cinemas and other large auditoria. However, a few domestic models have appeared, generally designed for corner placement when the boundaries of the room act as the final extensions of the horn. Horn loading is also applied to some tweeters, again with a view to increasing efficiency and controlling directivity.

2.4.6 The amplifier–loudspeaker interface

A loudspeaker is a low-efficiency device, the ratio of radiated acoustic power to electrical input being down around 0.5–1.5% for most domestic types, the rest of the input power being expended as heat. It must therefore be driven by a fairly high-power amplifier and all steps taken to minimize losses in cables, etc.

It is also important to achieve maximum power transfer from amplifier to loudspeaker by paying proper attention to the question of impedance matching. This is complicated by the fact that the loudspeaker presents a complex load both electrically and mechanically, as 'seen' looking into its input terminals. In practice, the load impedance quoted in amplifier specifications for correct matching to a loudspeaker, typically 8 Ω, is quite different from the actual impedance looking back into the output terminals. The latter is made very low indeed, typically 0.1 Ω, so as to be negligible compared with the voice-coil resistance and provide constant voltage operation.

The measured impedance of a loudspeaker varies widely with frequency. The nominal (quoted) impedance is usually the value measured at 400 Hz, or at the trough just above the main resonance peak. *Figure 2.65* shows a typical impedance versus frequency curve. The rated impedance (at 400 Hz) is about 20% higher than the d.c. resistance of the coil. Elsewhere the impedance is generally higher, rising to a peak at the main mechanical resonance and also increasing at a rate of about 6 dB/octave at higher frequencies due to voice-coil inductance.

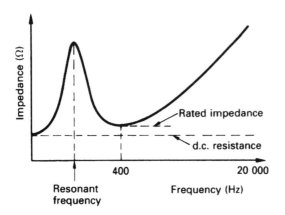

Figure 2.65 *Impedance versus frequency for a typical closed-box loudspeaker.*

Loudspeaker efficiency has already been mentioned, but the related parameter more often quoted is sensitivity. This is the sound pressure level (SPL) produced by a stated electrical voltage input, e.g. 2.83 V, which corresponds to 1 W into a nominal impedance of 8 Ω. Typical sensitivities, with SPL measured at a distance of 1 m on axis under free-field conditions, range from about 85 to 95 dB for domestic loudspeakers. The peak SPL rating prior to the onset of severe distortion, or the 'power handling capacity', will be much higher, and professional monitoring loudspeakers, for example, are typically rated at 115–120 dB.

2.4.7 Headphones

Basically a headphone earpiece may be regarded as a small-scale loudspeaker. Certainly its purpose is again to convert electrical signals into equivalent sound waves. There will be a considerable scaling down of the required acoustic level because of the close proximity of the drive unit to the ear, requiring input levels in the order of milliwatts instead of watts. In addition, the relatively small volume of approximately 30 cubic cm into which the diaphragm radiates, and resonances in this outer ear cavity, greatly influence the optimum frequency response and other design parameters.

All the main transducer types employed in loudspeakers are found in headphones, with moving-coil models again dominating and the electrostatic or orthodynamic principles applied in a few notable designs. There are even two-way designs having separate drivers to cover high and low frequencies.

Figure 2.66 shows an exploded view of a system having an electrostatic tweeter located in front of a moving-coil woofer with passive membranes which provide extra damping to smooth the main resonance peak.

Coupling to the ear can take two forms. In circum-aural headphones (*Figure 2.66*) the earpiece cushion is made large enough to enclose the ear completely and can thus provide an efficient air seal. This makes it easier to produce an extended bass response and, incidentally, up to 40 dB attenuation of ambient noise. In supra-aural headphones the cushion rests on the surface of the ear. This has led to the design of very lightweight mini-earphones and also relatively large models with porous foam cushions covering the whole ear surface.

Figure 2.66 *Exploded view of two-way moving-coil/electrostatic (electret) circumaural headphone having an array of six passive membranes.* (Courtesy of AKG.)

Another grouping exists distinguishing 'closed' and 'open' headphones. In the former, the rear surface of the diaphragm is enclosed, as in an infinite baffle loudspeaker. By contrast, open headphones are free to radiate outwards as well as towards the ear, and indeed the diaphragm itself, or its surround is transparent to sound. This type has become popular for its ability to give more realistic stereo reproduction with less of the in-head location which affects the performance of most closed headphones. However, there is very little attenuation of outside noise.

Headphone impedances fall into three groupings: low, 1–100 Ω; medium, 100–1000 Ω; and high, 10–30 kΩ. A few are designed to be driven from the same amplifier outlets as loudspeakers, but the majority are connected to line-level 'headphone' jack sockets which are fed from the main amplifier through series resistors.

Bibliography

AUDIO ENGINEERING SOCIETY, *Loudspeakers*, Vols 1 and 2, Audio Engineering Society, New York (1978, 1984)
BORWICK, J. (ed.), *Loudspeaker and Headphone Handbook*, Focal Press, Oxford (1994)
BOWERS, J., Loudspeakers. In *Radio, TV & Audio Technical reference Book* (ed. S.W. Amos), Newnes-Butterworths, London, Section 11 (1977)
COLLOMS, M., *High Performance Loudspeakers*, 4th edn, Pentech Press, London (1991)
JORDAN, E.J., *Loudspeakers*, Focal Press, London (1963)

2.5 Stereo

2.5.1 Introduction

It is the intention of this section to investigate the principles of stereophony as they relate to sound reproduction as well as to microphone design and technique. The discussion of 'stereo' is not limited to conventional two-loudspeaker reproduction, but considers it in the true sense of the Greek *stereo*, meaning 'solid', or 'three-dimensional'. Stereo techniques cannot be considered from a purely theoretical point of view, neither can the theory be ignored, the key being in a proper synthesis of theory and subjective assessment. Some techniques which have been judged subjectively to be good do not always stand up to rigorous theoretical analysis, and those which are held up as theoretically 'correct' are sometimes judged subjectively to be poorer than others. Part of the problem is that the mechanisms of directional perception are not yet entirely understood. Furthermore, most commercial stereo reproduction uses only two loudspeakers, and thus the listening situation is already a distortion of reality (since real sonic experience involves sound arriving from all around the head), perhaps leading listeners to *prefer* distorted images because of the pleasing artefacts such as 'spaciousness' which are often side-effects, just as many listeners 'prefer' sound which is distorted in other ways. Most of the stereo techniques used today combine aspects of imaging accuracy with an attempt to give the impression of spaciousness in the sound

field, and to many theorists these two are almost mutually exclusive.

In the following two sections stereo pick-up and reproduction is considered from both a theoretical and a practical point of view, recognizing that theoretical rules may have to be bent or broken for operational and subjective reasons. Since the subject is far too large even to be summarized in the short space available, extensive references are given to allow the reader greater scope for personal study.

2.5.2 Directional perception

Any study of stereo sound would be incomplete without first addressing the subject of directional perception in human hearing, since it is the aim of any stereo pick-up and reproduction technique to create the illusion of directionality in reproduced sound. A study of directional hearing will also help the reader considerably in appreciating the distinction between 'binaural', 'stereo', 'ambisonic' and 'transaural' techniques.

Directional hearing may roughly be divided into three planes: the lateral plane from left to right, the front–back or 'median' plane, and the vertical plane. In simple stereo reproduction it is the first of these that is the most important, although, as will be seen, it is not impossible to create illusions of the other two using transducers orientated in only the first plane. Front–back distinction becomes more important when considering microphone techniques which aim to offer some degree of 'surround-sound' information, and vertical distinction completes the picture if one is ever to attempt a complete reconstruction of the original soundfield, as is the case in full 'periphonic' reproduction within the Ambisonic system.

Study of directional hearing principles extends over most of the last century and, although it can be said that we now understand the mechanisms involved fairly well, it should also be said that there remains considerable debate over specific theories, especially concerning the relative importance of the different mechanisms involved at different frequencies and with different types of sound. In the following section a resumé of the principal mechanisms will be given, with specific relevance to stereo sound reproduction. Detailed studies of directional hearing principles may be found in Blauert (1983) and Moore (1989).

2.5.2.1 Lateral plane
In order to understand how the positions of sources in the lateral plane are perceived (either in front of

or behind the head) it is necessary to make a distinction between continuous repetitive waveforms (such as pure tones) and transient sounds or noises. The reason for this is that it is reasonable to talk about a 'phase difference' in degrees between two signals if they are continuous, repetitive, and have a recognizable fundamental frequency, but this concept becomes somewhat irrelevant when considering complex signals of short duration which start and stop unpredictably. In the discussion of directional hearing the existence of a phase difference between the two ears is often quoted, when what is really meant is a difference in time of arrival of the sound. When considering the mechanisms of hearing it is often convenient to use pure tones as a signal source example, but it should be remembered that one rarely listens to pure tones in real life, since real sounds are often of a more complex, noise-like nature.

One's ability to perceive directionality in sound depends almost entirely on the fact that two ears are involved, although there are more minor effects which may be perceived even if only one ear is involved. Principally, lateral distinction of direction is based on sound amplitude differences between the two ears, and on time-of-arrival differences between the two ears (which may be considered as phase differences in some cases). It is apparent that the brain relies on a combination of cues when assessing direction, and that the relative importance of the different cues changes with the frequency and nature of the source.

Amplitude difference Considering for a moment the so-called 'free-field condition', in which sound energy from a source is not reflected off any boundary surface but is free to move away from the source in all directions, there will be two potential reasons for a difference in sound pressure level (SPL) between the ears of a listener. Assuming that the sound source is at a given angle off the axis running from front to back of the head through the nose, there will be a difference in perceived level between the two ears such that the ear nearest the source will perceive a higher SPL than the distant ear. This is due firstly to the extra distance that the sound must travel between ears, and secondly to the shadowing effect of the head.

The amplitude difference which results from the extra distance travelled is very small indeed in most real circumstances, since direct sound level obeys the inverse square law of decay with distance from the source. Put in practical terms inverse square law decay results in a level drop of 6 dB for every doubling of distance away from the source. Unless the

source is very close to the head, the distance between the ears compared with the distance already travelled by the sound will be relatively small, and is unlikely to represent anything like a doubling of distance. To take an example, considering a sound source 4 m from a listener, at 90° 'off centre' (i.e. resulting in the maximum path difference between the ears), and assuming that the extra distance travelled by the sound between the ears is 20 cm, the difference in SPL which would result due to inverse square law decay will be only 0.4 dB, which may be accepted as an imperceptible difference. If the source is very close to the head, then inverse-square law decay will play a larger part in the amplitude difference between the ears.

By far the more important reason for the existence of amplitude difference is the 'shadowing' effect of the head on high-frequency sound. Acoustically, an object only begins to act as an effective barrier to sound when the wavelength of the sound is small in relation to the size of the object. Since the wavelength of audible sound ranges from around 18 m to a very few centimetres, it may be seen that at low frequencies the head is much smaller than the wavelength, and thus will not act as a barrier, but at high frequencies it will have a greater shadowing effect, resulting in a loss of high frequencies on the 'dark side' (that is the side more distant from the source). *Figure 2.67* shows the resulting amplitude difference calculated by Steinberg and Snow (1934), based on tests with pure tones made by Sivian and White (1931), and it may be seen that for frequencies above about 1000 Hz the amplitude difference between the ears becomes appreciable. This frequency–intensity response difference has variously been referred to as a 'quality' or 'timbre' difference.

Figure 2.67 *The shadowing effect of the head results in a level difference between the ears which varies with frequency. This graph shows the level difference which results between the ears when a source of pure tone is placed to the left of a listener.* (After Steinberg and Snow (1934).)

It is thus possible for the brain to base part of its estimation of direction on the difference in spectral emphasis between the ears, the ear with the greater high frequency content being that closest to the source. In a reverberant room there may be other reasons for an amplitude difference between the ears, such as the existence of standing wave patterns which result in pressure maxima and minima at points in the room. In such cases the brain becomes more confused, having to rely heavily on the transient information in the sound, using the first few milliseconds of the arrival of the direct sound upon which to base its calculation of direction.

Phase difference If one considers an imaginary sound source emitting a continuous repetitive waveform such as a sine wave, it is possible to talk in terms of a phase difference between the two ears of a listener for a given frequency and offset of the source. In such a case it should be noted that for any given position of the source the phase difference will be a function of frequency, becoming greater as the frequency increases up to a point where it becomes unclear which ear is lagging and which leading, since at the frequency where the distance d between the ears is equal to half the wavelength there will appear to be a 180° phase difference between the ears, making it difficult to tell which ear is lagging and which leading, and also at multiples of this frequency. Above the frequency where $d = 1/2\lambda$ (around 800 Hz) the phase difference cue becomes unreliable since the same phase difference may arise at a number of frequencies.

It has been shown that the phenomenon of 'phase locking' exists between the peaks of the audio waveform and the pattern of nervous discharge from the auditory fibres of the inner ear (Galombos and Davies, 1943), resulting in 'spikes' of electrical activity at positive peaks of the waveform. This provides a means for the brain to estimate the relative phase difference between the ears, since it can time the interval between these discharge spikes. Above about 700 Hz this phase-locked discharge becomes less reliable, with each fibre able to respond less often since each has a finite recovery time, although the neural discharge does not become completely asynchronous until around 4000 Hz. Between 700 and 4000 Hz the pattern of synchronous 'spikes' is maintained in the combination of outputs from many hundreds of fibres, all of which are stimulated, but not all of which 'fire' on each cycle of the waveform. Above 4000 Hz the brain can no longer rely on the timing of nervous discharge from the ear,

since the pattern becomes disorganized and somewhat random, but it is interesting to note that above this frequency a marked increase is noticed in the *amplitude* difference between the ears due to the shadowing effect of the head. There is thus a 'crossfade' area in the middle frequency range over which the brain begins to rely less on the relative timing of synchronous discharge, and more on the relative sound amplitudes between the ears. This explains the increased errors noted during subjective tests in the location of pure tones by test subjects in this middle frequency range.

It must be noted that all of the above discussion refers principally to continuous repetitive waveforms, and that there is room for debate over the relevance of 'phase' as a parameter for measurement by the brain. As previously stated, for a given position of sound source the phase difference between the ears is a function of frequency, and so it is not possible to say that a source positioned at, say, 40° off-centre will result in a given number of degrees phase difference, since the number of degrees will depend on the frequency. Similarly, if the frequency is held constant and the source is moved, the phase difference will change with position. It is clear therefore that for every position of sound source and for every frequency there will be a different phase difference, and this would require that the brain maintained an infinitely long look-up table of frequency versus phase difference versus position, which seems unlikely.

Perhaps, therefore, it is less ambiguous, and more realistic when discussing real sounds in real surroundings, to talk in terms of a *time-of-arrival* difference between the ears, since this is not a function of frequency and can be stated for any position of the source whatever its frequency. Subjective tests performed by many workers, including the author, have shown that pure tones are generally more difficult to locate than transient sounds, especially in the mid-frequency region where the phase cue becomes confusing and there is not enough amplitude difference between the ears to be helpful. Most experiments that have shown accuracy by subjects in locating pure tones have been conducted in near-free-field conditions, thus avoiding the confusion which arises out of reflections and pressure differences in the reverberant field. This is not to say that one should treat the phase cue as an irrelevance, but that its value should be regarded warily.

It should be noted that the actual response of the basilar membrane within the inner ear to a transient

click is in fact similar to a decaying sinusoidal wave-
form, and Moore (1989) suggests that it may be that
the brain is able to compare the phases of these
decaying oscillations and use this as an aid to local-
ization. He points to research by Yost *et al.* (1971)
which showed that tests using clicks which had been
highpass filtered so as to include only energy above
1500 Hz resulted in a deterioration of localization
ability, but that the same clicks low-pass filtered so as
to include only energy *up to* 1500 Hz resulted in little
change in localization ability, suggesting that local-
ization ability is largely related to the low-frequency
content of transients.

Time-of-arrival difference The ears in an aver-
age head are spaced apart by about 15 cm, result-
ing in a maximum path-length difference between
them of around 20–23 cm due to the extra distance
which must be travelled by a sound wavefront to the
more distant ear. The maximum path-length differ-
ence results when the source is at 90° to the so-called
'median plane' (running from front to back through
the nose), and at this angle, assuming that the speed
of sound is 340 m/s, the time-of-arrival (TOA) dif-
ference will be 600 μs, or 0.6 ms. At angles less than
this, the TOA difference will be proportional to the
sine of the angle θ (see *Figure 2.68*).

Although a single ear is incapable of distinguish-
ing between the arrivals of two sounds within such
a short period (Wallach *et al.*, 1949), binaural time
differences between arrivals of single clicks as small
as 10–20 μs, corresponding to angles of only a few
degrees, are perceptible in terms of a change in
apparent direction of the source. This assumes a sin-
gle point source being perceived by two separate
ears, and the situation can be verified by feeding
clicks on headphones to a listener with an adjustable
delay in the feed to one ear. The angle of perceived
offset is related directly to the number of microsec-
onds delay between the ears, being always towards
the advanced ear, up to the limit of around 0.6 ms
delay, after which a 'phasey' image or indistinct
position begins to result.

So far only the single point source has been con-
sidered, resulting in delays of up to 0.6 ms between
the ears (the so-called 'binaural delay'), but we must
also consider the situation in which more than one
source exists, since this is one of the keys to under-
standing stereo sound reproduction. In such a situa-
tion, such as that which might exist when listening
to the same sound emitted from two sources, Haas
(1951), studying the effects of echoes on the per-
ception of direction of a source, showed that the
perceived direction of the sound tended towards the

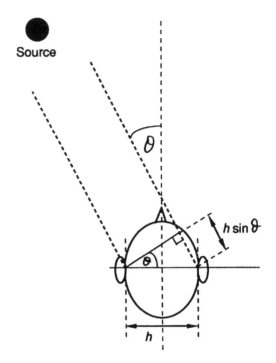

Figure 2.68 *A sound source at position X, offset from the median plane by angle θ, gives rise to a time-of-arrival difference t between the two ears of a listener, such that t = dsinθ/340.*

advanced (in time) source, and that the delays over
which the phenomenon was noticed extended up to
around 50 ms, a much greater delay than that over
which the binaural effect applies. In the case of two
sources both ears would hear both sources, and thus
the situation would not be the same as that with a sin-
gle source where one wavefront arrived at each ear at
slightly different times. Haas showed that for delays
up to about 50 ms the sounds from the two sources
would be 'fused' together by the brain, appearing as
one source with the perceived location towards that
of the first arrival. Beyond 50 ms the brain begins
to perceive the sounds as distinct, and the second
appears as an 'echo' of the first. With single clicks,
the effect breaks down more quickly than with com-
plex sounds, allowing delays of only up to around
5 ms before the 'fusing' effect disappears. Wallach
et al. (1949) have also studied the precedence effect
in sound localization, and their results bear detailed
examination since they extend the principle of the
precedence effect to a more exhaustive model of the
situation for stereo loudspeaker reproduction.

The so-called 'Haas-effect curve' (*Figure 2.69*)
shows that for the delayed source to appear equally
as loud as the undelayed source it must be a certain

Figure 2.69 *The so-called 'Haas effect' shows the combination time delay and level difference between primary and secondary sound sources for apparent equal loudness.*

number of decibels higher in amplitude to compensate for the precedence effect of the first arrival, the peak of the effect being at a delay of around 15 ms where the delayed source must be 11 dB louder than the undelayed source in order to be perceived as equally loud. We begin to see, therefore, the beginnings of a possibility for trading amplitude difference against time difference to achieve the same directional effect.

A useful recent discovery by Hafter *et al.* (1988) has shown that there is evidence to suggest that the binaural system 'locks onto' the onset of a sound (Hafter used a train of clicks), adapting its perception of direction to that suggested by the first click (and the resulting delay between the ears), largely ignoring subsequent clicks, although suitable 'triggers' may release the perception mechanism from this adaptation. In his experiments such 'release triggers' took the form of a low-intensity burst of noise, a tone burst, and a pause in the train of clicks. It is as if the brain is prepared to accept a direction suggested by the onset of a sound which can thereafter be cancelled by a suitable trigger which resets the mechanism, such triggers being related to a change in the sound. This is of considerable importance to our understanding of stereo sound.

The rôle of the pinna In addition to the primary effects noted above there are more subtle effects which also have a bearing on the perception of directionality in the lateral plane. These are principally related to the rôle of the pinna (the visible structure of the outer ear), which has recently been reconsidered. The effect of the pinna has not historically been considered very important in human sound localization, and does not feature greatly in older papers on stereo reproduction, but there is evidence to show that even though each person's ears have a different shape, the person relies considerably on information derived from reflections off different parts of it, resulting in slightly delayed components of the sound which combine additively and subtractively with the direct arrivals thus resulting in a comb-filter-like pattern of 'notches' in the perceived frequency spectrum, the position and depth of which depends on the angle of the source. Furthermore, resonances may result in the outer ear which will have a similar effect on frequency response. By learning the patterns of spectral emphasis resulting from sounds in different locations the brain gains an extra tool for localization.

In experiments where subjects' pinnae have been filled with a mouldable material, leaving simply a passage to the auditory canal, the ability to detect direction has been noticeably impaired. This will be discussed further when considering 'binaural' techniques in sound recording and reproduction.

Combining the cues The ability to detect the direction of sounds in the lateral plane therefore depends on a combination of frequency-dependent amplitude difference (due principally to the shadowing effect of the head) at mid-to high-frequencies, on frequency-dependent phase differences for continuous low frequency sounds, on TOA differences for complex transient sounds, and on the subtle spectral shaping effects of the pinna. There appears to be a crossover zone for continuous sounds between about 700 and 4000 Hz in which confusion may arise due to the breaking down of phase cues and the gradual increase in amplitude difference, above which amplitude difference becomes the most important cue. There is a need to distinguish between the 'binaural effect' in which the two ears are independently stimulated by the same wavefront (resulting in delays of between 0 and 600 µs) and the 'precedence' effect in which the two ears are both stimulated by two or more sources arriving at different times (which operates over delays up to about 50 ms depending on the nature of the source). There is some evidence to suggest that the binaural hearing mechanism adapts itself to the direction suggested by the onset of a sound, ignoring subsequent information to some extent, unless that information succeeds in retriggering the localization process. It is possibly true, therefore, that

changes in the sound have a more persuasive influence on localization than do continuous features.

2.5.2.2 Front–back or median plane

Less time will be spent on the discussion of the mechanisms for front–back and vertical perception, although they are still important for some stereophonic techniques. Initially it is hard to see what distinguishes a point source at a given number of degrees off-centre in front of the head from the same source at the same angle behind the head, since the TOA difference will be the same and so, to a large extent, will be the shadowing effect of the head. Even so, it is clear that the freedom to move the head plays an important part in localization in this plane, since, in subjective tests, people whose heads have been held perfectly still have been found to display greater difficulty in front–back distinction. The reason for this is that even a small change in the rotation angle of the head will alter the TOA difference between the ears, and for a given direction of rotation this difference will either get smaller or greater, depending on whether the source is in front of or behind the head. Considering a source at $30°$ to the right of the median plane in front of the head, we see that turning the head to the right would lessen the TOA difference, whilst for the same source at $30°$ to the right behind the head we see that turning the head to the right would *increase* the TOA difference. Thus the brain may make small movements of the head and note the change in TOA which results.

A further factor is the rôle that sight plays in this distinction, since we use our eyes much more to determine the location of a source in front, whereas we must use our ears to determine the location of a source behind. If a source cannot be seen then it must be behind (ignoring the vertical for the moment)! This factor must not be overplayed since it is still possible to determine front–back location with the eyes closed, although there is some evidence to show that blind people have better aural localization to the front than sighted people.

A final factor is the effect of the pinna and head on sounds from the rear, since the size of the pinna is such as to act as a partial barrier to very high frequency sounds from the rear, changing the spectral emphasis of a sound from the rear when compared with the same sound from the front. Furthermore, the head-related transfer function (HRTF) is slightly different for rear sounds than for frontal sounds (this being the distortion of phase and amplitude evident in the sounds reaching the two ears due to the interference effects of the head).

2.5.2.3 Vertical plane

Localization in the vertical plane is concerned partially with the effect of reflections from the ground and from the shoulders, in addition to pinna effects, since sounds at different angles of elevation will reach the ears directly and will also be reflected, reaching the ear slightly later, via indirect paths. It must be noted that there is a basic difference between the indoor and outdoor situations, since indoors there is a ceiling to consider, whereas outdoors there are likely to be no reflections from above. Furthermore, the only reliable reflection path distance to be accounted for is that from the shoulders, which for a given person remains fixed within limits of the movement of the head. For sounds from above, the difference in path length between the sound reflected from the shoulders and the floor compared with the direct path will result in cancellation and addition at different frequencies (in the same way as discussed above with reflections from the pinna, although reflections from the shoulders and floor will result in notches and peaks which are more widely spaced, and which extend to a lower frequency, due to the distances involved). It is partially in the comparison of these frequency spectra with stored templates that the brain will locate sounds in the vertical plane.

The memory of learned situations, and also the expectation that particular sounds will emanate from certain directions is also important for vertical localization, since there are not many sounds that come from below, as a general rule. The experience of standing on a mountain top whilst an aeroplane flies along a valley below is a most unusual one, since it is not the expected position of an aeroplane. Similarly, listeners played a high-volume recording of a jet aeroplane taking off over the head will tend to duck, even though there is no obvious height information in the replay system.

2.5.3 Basic directional sound reproduction

Now that the mechanisms by which direction is perceived in sound have been introduced, the principles of basic directional sound reproduction in audio systems will be discussed as a precursor to the discussion of stereo signals and microphone configurations. 'Stereophonic sound' will be considered in a broader sense than is often the case, since there is now a marked rekindling of interest in unconventional forms of directional sound reproduction such as three-channel stereo, wide-image stereo, transaural stereo and ambisonics.

2.5.3.1 Historical development

We have become used to stereo sound as a two-channel format, although a review of developments during the last century shows that two channels really only became the norm through economic and domestic necessity, and through the practical considerations of encoding directional sound easily for gramophone records and radio. A two-loudspeaker arrangement is practical in the domestic environment, is reasonably cheap to implement, and provides good phantom images for a central listening position.

Early work on directional reproduction undertaken at the Bell laboratories in the 1930s (Fletcher, 1934) involved attempts to recreate the 'sound wavefront' which would result from an infinite number of microphone/loudspeaker channels by using a smaller number of channels. In all cases, spaced pressure response (omnidirectional) microphones were used, each connected via a single amplifier to the appropriate loudspeaker in the listening room. Steinberg and Snow (1934) found that when reducing the number of channels from three to two, central sources appeared to recede towards the rear of the sound stage and that the width of the reproduced sound stage appeared to be increased. They attempted to make some calculated rather than measured deductions about the way that loudness differences between the channels affected directional perception, choosing to ignore the effects of time or phase difference between channels. Some 20 years later. Snow (1953) made comment on those early results, reconsidering the effects of the time difference in a system with a small number of channels since, as he pointed out, there was in fact a marked difference between the multiple-point-source configuration and the small-number-of-channels configuration.

It was suggested that in fact the 'ideal' multisource system recreated the original wavefront very accurately, allowing the ears to use exactly the same *binaural* perception mechanisms as used in the real-life sound field. It may be appreciated that the 'wall' of multiple loudspeakers acted as a source of Huygen's wavelets, recreating a new plane wave with its virtual source in the same relative place as the original source, thus resulting in a TOA difference between the listener's ears in the range 0–600 μs, depending on position. In the two- or three-channel system, far from this simply being a sparse approximation to the 'wavefront' system, the ears are subjected to two or three discrete arrivals of sound, the delays between which are likely to be considerably in excess of those normally experienced in binaural listening, being a number of milliseconds depending on the

original spacing of the microphones. In this case, the effect of directionality relies much more on the precedence effect described in the previous section, and also on the relative levels of the channels. Snow therefore begs us to remember the fundamental difference between 'binaural' situations, and what he calls 'stereophonic' situations (see 'Combining the Cues' in Section 2.5.2.1).

This difference was also recognized by Blumlein, whose now famous patent specification of 1931 (accepted 1933) (Blumlein, 1958) allows for the conversion of signals from a binaural format suitable for spaced pressure microphones to a format suitable for reproduction on loudspeakers, and for other formats of pick-up which result in an approximation of the original time and phase differences at the ears when reproduced on loudspeakers. This will be discussed in more detail later on, but it is of historical interest to note how many writers writing on stereo reproduction even in the early 1950s appear to have been unaware of Blumlein's most valuable work, which appears to have been ignored for some time.

A British paper presented by Clark, Dutton and Vanderlyn (of EMI) in 1957 (Clark *et al.*, 1958) revives the Blumlein theories, and shows in more rigorous mathematical detail than Blumlein's patent specification how a two-loudspeaker system may be used to create an accurate correlation between the original angle of offset of a sound source and the perceived angle of offset on reproduction by controlling only the relative signal amplitudes of the two loudspeakers (derived in this case from a pair of coincident velocity-sensitive microphones). The system also adapts Blumlein's concept of sum-and-difference processing (see Section 2.5.6.2) to increase the relative 'width' at low frequencies compared with that at high frequencies, since tests showed that increased channel level difference at low frequency in relation to that at high frequency was needed for accurate localization of phantom images. The authors discuss the three-channel spaced microphone system of the Bell laboratories, and suggest that, although it produces convincing results in many listening situations, it is uneconomical for domestic use, and that the two-channel simplification (again using spaced microphones, at about 3 m apart) has a tendency to result in the subjective 'hole-in-the-middle' effect with which many modern users of spaced microphones may be familiar (the sound appearing to come either from the left or the right but not anywhere inbetween). They concede, in a discussion, that the Blumlein method adapted by them does not take advantage of all the mechanisms of

binaural hearing, especially the precedence effect, but that they have endeavoured to take advantage of, and recreate, a few of the directional cues which exist in the real-life situation.

There is therefore a historical basis for both the spaced microphone arrangement which makes use of the precedence effect (with only moderate level differences between channels), as well as the coincident microphone technique (or any other technique which results in only level differences between channels), with some evidence to show that the spaced technique is more effective with three channels than with only two. Later, we shall see that spaced techniques have a fundamental theoretical flaw from a point-of-view of 'correct' imaging of continuous sounds, which has not always been appreciated, although such techniques may result in subjectively acceptable sounds.

Interestingly, three front channels are the norm in cinema sound reproduction, since the central channel has the effect of stabilizing the important central image for off-centre listeners, having been used ever since the Disney film *Fantasia* in 1939. Gerzon (1990) points out that people have perhaps misunderstood the intentions of Bell laboratories in the 1930s, since it is not generally realized that they were working on a system suitable for auditorium reproduction with wide-screen pictures, as opposed to a domestic system. He suggests that three-channel stereo will become more important again, but this time in the home with wide-screen TV. requiring careful optimization of the signals for the new environment.

Alternatives to three-channel stereo exist for the coverage of a wider listening area, these involving loudspeaker arrangements which are directional in such a way as to compensate off-centre listeners with a higher audio level from the more distant speaker to make up for the increased precedence effect of the closer speaker. One interesting example of this is the Canon 'Wide-Imaging Stereo' speaker system (Negishi, 1988) which makes use of shaped conical reflectors mounted above mid- and high-frequency drivers pointing upwards, in order to give the loudspeaker a polar response that favours listeners on the opposite side of the listening area, compensating for the increased precedence effect of the closer speaker.

2.5.3.2 Two-channel stereo from loudspeakers

The principles of stereo reproduction using two channels on loudspeakers (as opposed to headphones) will now be discussed in more detail, since this is the most common listening arrangement in use today.

From the above discussions it will now be clear that in most cases stereo reproduction from two loudspeakers can only hope to achieve a modest illusion of the original soundfield, since reproduction is from the front quadrant only (although see Section 2.5.4.2). It might be acceptable to state that the best illusion will be created when the sound signals present at the two ears are as similar as possible to those perceived in natural listening, even if one cannot entirely rationalize the psychoacoustic mechanisms involved. It is possible to create this illusion using either time differences between the speaker outputs or using level differences between them. It is also possible to use a combination of the two, although there is a problem with 'time-difference' stereo in that contradictions may arise between transient and continuous sounds (see Section 2.6.3). The main point to be considered with loudspeaker reproduction is that both ears receive the signals from both speakers, whereas in headphone listening each ear only receives one signal channel. The result of this is that the loudspeaker listener seated in a centre seat (see *Figure 2.70*) receives at his left ear the signal from the left speaker first, followed by that from the right speaker, and at his right ear the signal from the right speaker first, followed by that from the left speaker, the time t being the time taken for the sound to travel the extra distance from the more distant speaker.

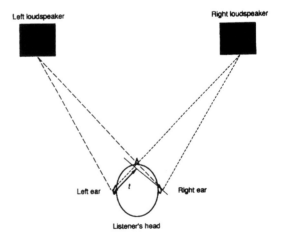

Figure 2.70 *When listening to sound on two loudspeakers, both ears receive sound from both loudspeakers, the signal from the more distant loudspeaker arriving a time* t *later than that from the nearer speaker for each ear.*

If the outputs of the two speakers differ only in level and not in phase (time) then it can be shown

(at least for low frequencies up to around 700 Hz) (Clark *et al.*, 1958) that the vector summation of the signals from the two speakers at each ear results in two signals which, for a given frequency, differ in phase angle proportional to the relative amplitudes of the two signals (the level difference between the ears being negligible at low frequency). For a given level difference between the speakers, the phase angle changes approximately linearly with frequency, which is the case when listening to a real point source. At higher frequencies the phase difference cue becomes largely meaningless (see 'Phase Difference' in Section 2.5.2.1), but the shadowing effect of the head results in level differences between the ears. If the amplitudes of the two channels are correctly controlled it is possible to produce resultant phase and amplitude differences for continuous sounds which are very close to those experienced with natural sources, thus giving the impression of virtual images anywhere between the left and right loudspeakers. This is the basis of Blumlein's 1931 stereophonic system 'invention' (Blumlein, 1958), although the mathematical analysis is quoted by Clark *et al.* (1958) and further analysed by Bauer (1961b), Bennett *et al.* (1985) and Lipshitz (1986). The result of the phasor analysis is a simple formula which can be used to determine, for any angle subtended by the loudspeakers at the listener, what the apparent angle of the virtual image will be for a given difference between left and right levels. Firstly, referring to *Figure 2.71* it can be shown that:

$$\sin \alpha = [(L - R)/(L + R)] \sin \theta_o$$

where α is the *apparent* angle of offset from the centre of the virtual image, and θ_o is the angle subtended by the speaker at the listener. Secondly, it can be shown that:

$$(L - R)/(L + R) = \tan \theta_t$$

where θ_t is the *true* angle of offset of a real source from the centre-front of a coincident pair of figure-of-eight velocity microphones. $(L - R)$ and $(L + R)$ are the differences (S) and sum (M) signals of a stereo pair, defined in Section 2.6.2 below.

This is a useful result since it shows that it is possible to use positioning techniques such as 'panpotting' which rely on the splitting of a mono signal source into two components, with adjustment of the relative proportion fed to the left and right channels without affecting their relative timing. It also makes possible the combining of the two channels into mono without cancellations due to phase difference.

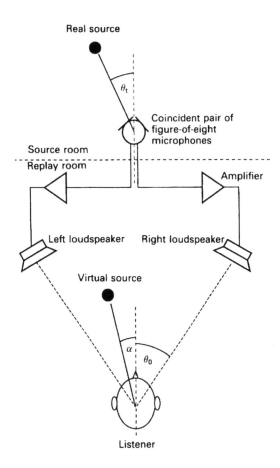

Figure 2.71 *A sound source is offset by angle θ_t from the centre-front of a pair of coincident figure-of-eight microphones, and is reproduced on loudspeakers each subtending an angle of θ_o to the listener. The virtual source is reproduced at angle α (see text).*

Experiments by the author with wideband speech signals replayed on a standard loudspeaker arrangement have shown that a level difference of approximately 18 dB between channels is necessary to give the impression that a sound comes from either 'fully left' or 'fully right' in the image (see *Figure 2.72*), and that there is a measure of disagreement between listeners as to the positions of 'half left' and 'half right', which might be expected. If a time difference also exists between the channels, then transient sounds will be 'pulled' towards the advanced speaker because of the precedence effect, the perceived position depending to some extent on the time delay. If the left speaker is advanced in time relative to the right speaker (or more correctly, the right speaker is delayed!) then the sound appears to come more from the left speaker, although this can be corrected by increasing the level to the right speaker. There is a

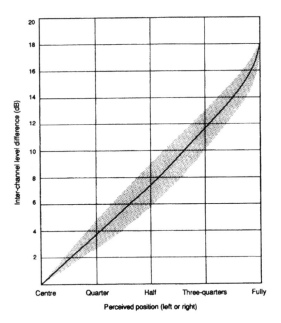

Figure 2.72 *Speech sources appear at different positions on reproduction depending on the level difference between the channels. The shaded area shows measured disagreement between listeners.*

trade-off between time and level difference, and it can be shown that if the left channel is, say, 2 ms earlier than the right, then the right must be made approximately 5 dB louder in order to compensate and bring the signal back to the centre. A time difference of between 2 and 4 ms (depending on the nature of the signal) appears to be required for a sound to appear either fully left or fully right, although this only holds true for transient signals and may be contradicted by the resulting low-frequency phase differences for continuous sounds, as argued by Lipshitz (1986). Stereo microphone techniques operate using combinations of level and time difference between the channels.

It may be shown (Clarke *et al.*, 1958) that a coincident arrangement of velocity (figure-of-eight) microphones at 90° to one another produce outputs which differ in amplitude with varying angle over the frontal quadrant by an amount which gives a very close correlation between the true angle of offset of the original source from the centre-line and the apparent angle on reproduction from loudspeakers which subtend an angle of 120° (sic) to the listening position. At lesser angles the change in apparent angle is roughly proportionate as a fraction of total loudspeaker spacing, maintaining a correctly proportioned 'sound stage'. It may also be

shown that the difference in level between channels should be smaller at high than low frequencies in order to preserve constant correlation between actual and apparent angle, and this may be achieved by using an equalizer in the difference channel (see Section 2.5.6.2) which attenuates the difference channel by a few decibels for frequencies above 700 Hz (this being the subject of a British patent by Vanderlyn (1957). Gerzon (1986) has suggested a figure between 4 and 8 dB, depending on programme material and spectral content, calling this 'spatial equalization).

The ability of a system based only on level differences between channels to reproduce the correct timing of *transient* information at the ears has been questioned, not least in the discussion following the original paper presentation by Clark *et al.*, but these questions were tackled to some extent by Vanderlyn (1979) a much later paper in which he attempted to· show how such a system can indeed result in timing differences between the neural discharges from the ears to the brain, taking into account the integrating effect of the hearing mechanism in the case of transient sounds. He quoted experimental evidence to support his hypothesis, which is convincing. If a system based only on level differences did not cope accurately with transients then one would expect transients to be poorly localized in subjective tests, and yet this is not the case, with transients being very clearly located in coincident-pair recordings.

The above theories of stereo have been termed 'summation localization theories' by Theile (1991) who dismisses such theories as not allowing for the correct natural interaural attributes of sound signals required for spatial reproduction. Theile states that the correct way to derive signals suitable for natural stereo reproduction is to generate 'head-referred' signals, i.e. signals similar to those used for binaural systems (see Section 2.5.3.3), since these contain the necessary information to reproduce a sound image in the 'simulation plane' between two loudspeakers. Subjective test results have shown that binaural signals equalized for a flat frontal frequency response produce convincing stereo from loudspeakers. There is strong disagreement between those who adhere to this theory and those who adhere more closely to traditional summation theories of stereo. Since there is not space here to cover this subject fully, the reader is referred to Rumsey (1991) for a broader discussion of the debate.

2.5.3.3 Two-channel stereo from headphones
Headphone reproduction is different to loudspeaker reproduction since, as already stated, each ear is fed

only with one channel's signal. This is therefore an example of the binaural situation and allows for the ears to be fed with signals which differ in time by up to the binaural delay (600 μs), and also differ in amplitude by amounts similar to those differences which result from the shadowing effects of the head. This suggests the need for a microphone technique which uses microphones spaced apart by the binaural distance, and baffled by an object similar to the human head, in order to produce signals with the correct differences (see Section 2.6.3.4).

Bauer (1961a) pointed out that, if stereo signals designed for reproduction on loudspeakers are fed to headphones there is a too-great level difference between the ears compared with the real-life situation, and that the correct interaural delays do not exist. This results in an unnatural stereo image which does not have the expected sense of space. Bauer therefore proposed a network which introduced a measure of delayed cross-talk between the channels to simulate the correct interaural level differences at different frequencies, as well as simulating the interaural time delays which would result from loudspeaker signals incident at 45° to the listener. He based the characteristics on research done by Weiner (1947), who produced graphs for the effects of diffraction around the human head for different angles of incidence. The characteristics of Bauer's circuit are shown in *Figure 2.73* (with Weiner's results also shown). It may be seen that Bauer chose to reduce the delay at high frequency, partially because the circuit design would have been too complicated, and partially because localization relies more on amplitude difference at high frequency anyway.

Interestingly, Bauer's example of the stereophonic versus binaural problem chooses spaced pressure microphones as the means of pick-up, showing that the output from the right microphone for signals at the left of the image will be near zero. This is only likely if the microphones are very close to the sources (as in a multi microphone balance), whereas in many spaced omnidirectional arrays there will be a considerable output from the right microphone for sounds at the left of the sound stage, thus there will also be a time delay equivalent to the path length difference between the source and the two microphones which will add to that introduced by his network. In fact what Bauer is suggesting is really a variation on Blumlein's 'shuffler' network (Blumlein, 1958) which would work best with non-spaced microphones (i.e. a directional coincident

Figure 2.73 *Bauer designed a circuit which introduced binaural delay and cross-talk into signals designed for loudspeaker reproduction so that they could be reproduced satisfactorily on headphones. The characteristics of his circuit are shown here. The upper graph shows the amount of delay introduced in the cross-feed between channels, whilst in the lower graph L_g and R_g are left and right channel gains to imitate the shadowing effects of the head. The dotted curves are Weiner's results.*

pair). Blumlein's shuffler converted the phase differences between two binaurally spaced microphones into amplitude variations to be reproduced correctly on loudspeakers, whereas Bauer was trying to insert a phase difference between two signals which differ only in level (as well as constructing a filter to stimulate diffraction effects).

Bauer also suggested the reverse process (turning binaural signals into stereo signals for loudspeakers), pointing out that crosstalk must be *removed* between binaural channels for correct loudspeaker reproduction, since the cross-feed between the channels will otherwise occur twice (once between the pair of binaurally spaced microphones, and again at the ears of the listener), resulting in poor separation and a narrow image. He suggested that this may be achieved using the subtraction of an antiphase component of each channel from the other channel signal, although he did not discuss how the time difference between the binaural channels may be removed.

Further work on a circuit for improving the stereo headphone sound image was done by Thomas (1977), quoting listening tests which showed that all of his listening panel preferred stereo signals on headphones which had been subject to the 'cross-feed with delay' processing. Such processes form the basis for 'transaural stereo', introduced below.

Work by Theile (1986) on the frequency-response equalization of headphones for optimum spatial impression has suggested that it is most important for the equalization of dummy heads and headphones to be standardized to some extent, so as not to

destroy the spectral cues which are vital for binaural localization. It is also pointed out, none the less, that the ear/brain is able to adapt to the particular spectral characteristics of a reproduction system over a period of time, after which spatial information is more readily decoded.

2.5.4 Binaural signal processing

2.5.4.1 Binaural synthesis
Recent work involving digital signal processing (Richter and Persterer, 1989) has resulted in a system which can be used for processing audio signals so as to introduce the correct 'head-related transfer function' (HRTF) for every possible angle of incidence of a monophonic sound source. In practice, the system employs a series of digital filters, delays and mixers which stimulate the effect of the pinnae on sound signals as well as simulating the delay and shadowing effects of the head. This has resulted in a system which is capable of acting as a 'binaural mixer', with the facility for positioning mono sounds at any elevation or offset angle around the listener. It is also possible to use the system as a 'room simulator' in order to test the effects of placing sources in virtual rooms of different shapes, sizes and surface characteristics, allowing a listener to place himself or herself anywhere in this virtual room and receive the correct binaural signals. This could be of considerable benefit to acoustic designers.

2.5.4.2 Transaural stereo
A recent development in the field of stereo recording and reproduction exists in the form of 'transaural stereo', a technique which has parts of its basis in the ideas of Bauer (cited above) in relation to the cancellation of cross-talk between signals designed for binaural monitoring on headphones, so as to make them suitable for loudspeaker reproduction. The basis of transaural stereo, as described by Cooper and Bauck (1989), is in the preconditioning of loudspeaker signals with a component which represents the *inverse* of the cross-talk which occurs between the two ears when listening to loudspeakers, thus allowing the ears to become independent once more, and allowing them to be supplied with the correct binaural signal relationships. He suggests the use of binaural recording techniques, or binaural synthesis, to originate the signals which are then preconditioned to create a transaural master, suitable for loudspeaker reproduction.

Experiments by Atal and Schroeder (1963, 1966) showed that it was possible to introduce a cross-talk-cancelling component into the signals fed to two loudspeakers which would compensate for the interaural cross-talk, although the system was so position critical on the part of the listener (he or she had to be within 75 mm of the 'correct' position), as well as requiring that the listening environment be anechoic, that the concept did not proceed into general acceptance. None the less, the results, for listeners in the correct environment and position, were said to be stunning, with directional information present in all dimensions.

The breakthrough described by Cooper lies in a modification of this early work, resulting in improvements in ordinary stereo reproduction derived from binaural recordings, which are less dependent on listener position, may be achieved with relatively simple electronics, and which do not require anechoic listening conditions. He also suggests that the system is capable of offering full surround-sound effects from only two loudspeakers, and that it allows for the creation of 'virtual loudspeakers' in positions other than those of the real loudspeakers. The work is too lengthy even to be summarized here, and the listener is referred to the excellent paper by Cooper and Bauck (1989). Further recent studies of similar transaural techniques may be found in Griesinger (1989), Moller (1989), and Gierlich and Genuit (1989), all of which are concerned to a greater or lesser extent with the processing of binaural signals for loudspeaker reproduction. This is an area of research which appears to promise much in the way of enhancement in stereo reproduction without necessarily demanding more complicated domestic listening arrangements, although the listening area for full spatial impression is limited to a rather narrow central position. A commercial product embodying transaural principles has recently appeared in the form of *Roland Sound Space* – a binaural mixer and transaural cross-talk canceller designed principally as an audio effects unit (Majima *et al.*, 1991).

2.5.4.3 Holophonics
A 'holophonic' system was described by Zucarelli (1983) and used on a number of projects in the early 1980s to remarkably good effect. The technique appeared to be a modified and highly optimized binaural technique (Elen, 1984), the results of which were particularly stunning on headphones, being one of the only binaural techniques capable of producing accurate headphone localization in front of the listener. Recordings were also not unconvincing when reproduced on loudspeakers. Unfortunately, a certain amount of mystery surrounded the technique, and Zucarelli's attempts at a theoretical explanation for the

success of his system (Zucarelli, 1983) were ridiculed in subsequent letters. Since unmodified binaural techniques do not work well on loudspeakers, this system appears to have sunk into the background, like most binaural techniques, despite the wide use of 'personal' cassette players with headphones.

2.5.5 Multiple-channel stereo on loudspeakers

Sound for pictures has long made use of more than two loudspeaker channels to cover a wide listening area, to 'fix' the central image where dialogue tends to reside, and to offer some additional spatial information such as rear images or 'surround sound'. Multichannel reproduction for domestic usage has had a chequered and not altogether successful history, although today there is renewed interest in it due to the launch of a consumer medium capable of multichannel sound delivery: the DVD. Quadraphonic reproduction systems appeared in the 1970s to a number of different standards, but they have not found continued favour and thus will not be discussed further here. A system which appears to have survived the test of time, although still not widely used, is the so-called 'Ambisonic' system, developed under the auspices of the NRDC in the UK during the 1970s and based mainly on work by Gerzon, Fellgett and Barton. The rights to Ambisonic technology now lie with Nimbus Records in the UK. This system is based on well-researched psychoacoustic principles and aims to reproduce the original sound field as accurately as possible in a wide variety of possible configurations. Its principal child in commercial terms is the 'Soundfield' microphone which is discussed in Section 2.6.5.

2.5.5.1 Cinema-style surround-sound

The most popular multichannel sound format for the cinema at the present time is based on four channels: left (L), centre (C), right (R) and 'surround' (S), with sound signals panned into positions around the listener during postproduction by varying the relative levels between the channels. All dialogue is normally sent to the centre channel, with music and effects using the space of the other three channels. The LCRS format for sound signals is inherent in the 'Dolby stereo' format for cinema sound tracks, and Dolby manufactures an encoder and decoder matrix which allows for the four channels to be combined into two channels for recording onto film with reasonable compatibility in the case of two-channel listening. The Dolby stereo system is covered in greater

detail elsewhere in this book, and thus will not be discussed further here. Recent digital film sound formats have extended the channel capacity of film to accommodate between five and eight channels of discretely encoded surround sound, enabling the matrix to be dispensed with and making stereo surround information available.

2.5.5.2 Ambisonics

The Ambisonic system of directional sound pick-up and reproduction is discussed here because of its relative thoroughness as a unified system, based on some key principles of psychoacoustics, and because of its relevance to the Calrec Soundfield microphone, which is somewhat unique as a commercial product. It has its theoretical basis in work by Gerzon, Barton and Fellgett, good summaries of which may be found in Gerzon (1983) and Gerzon and Barton (1984).

Ambisonics aims to offer a complete hierarchical approach to directional sound pick-up, storage or transmission and reproduction, which is equally applicable to mono, stereo, horizontal surround-sound, or full 'periphonic' reproduction including height information. Depending on the number of channels employed it is possible to represent a lesser or greater number of dimensions in the reproduced sound. A number of formats exist for signals in the Ambisonic system, and these are as follows: the A format for microphone pick-up, the B format for studio equipment and processing, the C format for transmission, and the D format for decoding and reproduction. A format known as UHJ ('Universal HJ', 'HJ' simply being the letters denoting two earlier surround sound systems), described originally by Gaskell of the BBC Research Department (Gaskell, 1979), is also used for encoding surround-sound ambisonic information into two or three channels, whilst retaining good mono and stereo compatibility for 'non-surround' listeners.

Ambisonic sound should be distinguished from quadraphonic sound, since quadraphonics explicitly requires the use of four loudspeaker channels, and cannot be adapted to the wide variety of pick-up and listening situations which may be encountered. Quadraphonics generally works by creating conventional stereo phantom images between each pair of speakers and, as Gerzon states, conventional stereo does not perform well when the listener is off-centre or when the loudspeakers subtend an angle larger than 60°. Since in quadraphonic reproduction the loud-speakers are angled at roughly 90° there is a tendency towards a hole in the middle, as well as there being the problem that conventional stereo theories

do not apply correctly for speaker pairs to the side of the listener. Ambisonics, however, correctly encodes sounds from all directions in terms of pressure and velocity components, and correctly decodes these signals to a number of loudspeakers, with psychoacoustically optimized shelf filtering above 700 Hz to correct for the shadowing effects of the head, and an amplitude matrix which determines the correct levels to be fed to each speaker for the layout chosen. Ambisonics is thus the theoretical successor to coincident stereo on two speakers, since it is the logical extension of Blumlein principles to surround sound (an area in which such principles may in fact be more appropriate than in two-channel reproduction, since the sense of 'space' which is often said to be sacrificed for the sake of imaging in two-channel Blumlein recordings can now be correctly reconstructed in the shape of reverberant signals from all angles, whilst retaining imaging).

The source of an ambisonic signal may be an ambisonic microphone such as the Calrec Soundfield, or it may be an artificially panned mono signal, split into the correct B format components (see below) and placed in a position around the listener by adjusting the ratios between the signals. Since we are principally concerned with microphones, it is intended not to cover the principles of ambisonic mixing. Good introductions to the subject of mixing may be found in Daubney (1982) and Elen (1983).

It is often the case that theoretical 'correctness' in a system does not automatically lead to widespread commercial adoption and, despite considerable coverage of ambisonic techniques, both in academic circles and in the trade journals, the system is still only used rarely in commercial recording and broadcasting. It is true that the Soundfield microphone is used widely, but this is principally because of its unusual capacity for being steered electrically so as to allow the microphone to be 'pointed' in virtually any direction without physically moving it, and set to any polar pattern between omnidirectional and figure-of-eight, simply by turning knobs on a control box. It is used in this respect as an extremely versatile stereo microphone for two channel recording and reproduction.

2.5.5.3 5.1 channel surround sound
The 5.1 channel format of L, C, R, LS, RS and subwoofer is gaining increasing prominence as the format of the future for consumer stereo listening. Stereo recording techniques for this format are being developed accordingly (McKinnie and Rumsey, 1997).

2.5.6 Basic stereo signal formats

2.5.6.1 Terminology
In the following section the nature of stereo audio signals will be discussed, together with definitions of the terms used to describe the various formats of stereo sound. Also included is a discussion of the effects of misalignment on stereo signals.

2.5.6.2 A, B, M and S signals
It is conventional in broadcasting terminology to refer to the left (L) channel of a stereo pair as the 'A signal' and the right (R) channel as the 'B signal'. In the case of some stereo microphones or systems these are called respectively the 'X signal' and the 'Y signal'. In a two-channel stereo system the A signal feeds the left loudspeaker and the B signal feeds the right loudspeaker. In colour-coding terms (for meters, cables, etc.), the A signal is coloured red and the B signal is coloured green. This may be confusing when compared with some domestic hi-fi wiring conventions which use red for the right channel, but it is the same as the convention used for 'port' and 'starboard' on ships. Furthermore, there is a German convention which uses yellow for the left channel and red for the right.

It is sometimes convenient to work with stereo signals in the so-called 'sum and difference' format, since it allows for the control of image width and ambient signal balance. The sum or main signal is denoted by 'M', and is based on the addition of L and R signals, whilst the difference or side signal is denoted by 'S', and is based on the subtraction of R from L to obtain a signal which represents the difference between the two channels (see Section 2.5.6.4). The M signal is that which would be heard by someone listening to a stereo programme in mono, and thus it is important in situations where the mono listener must be considered, such as in broadcasting. Colour-coding convention holds that M is coloured white, whilst S is coloured yellow, but it is sometimes difficult to distinguish between these two colours on certain meter types.

2.5.6.3 Derivation of stereo signals
LR, XY or AB stereo signals (they are all the same) may be derived by many means. Most simply, they may be derived from a pair of coincident directional microphones orientated at a fixed angle to each other. Alternatively, they may be derived from a pair of spaced microphones, either directional or non-directional, with an optional third microphone bridged between the left and right channels. Finally, they may be derived by the splitting of one or more

mono signals into two, by means of a 'pan-pot', which is really a dual ganged variable resistor which controls the relative proportion of the mono signal being fed to the two legs of the stereo pair, such that as the level to the left side is increased that to the right side is decreased.

MS or 'sum and difference' format signals may be derived by conversion from the AB format using a suitable matrix, or by direct pick-up in that format. A coincident pair of microphones may be used for direct pick-up, such that the S signal is derived from a sideways-facing figure-of-eight, and the M signal is derived from one of a variety of polar patterns, depending on the desired final balance. For every AB stereo pair of signals it is possible to derive an MS equivalent, since M is the sum of A and B, whilst S is the difference between A and B. Likewise, signals may be converted from MS to AB formats using the reverse process.

2.5.6.4 Conversion between signal formats
In order to convert an AB signal into MS format it is necessary to follow some simple rules. Firstly, the M signal is not usually a simple sum of A and B, as this will result in overmodulation of the M channel in the case where a maximum level signal exists on both A and B (representing a central image), therefore a correction factor is applied, ranging between -3 and -6 dB (equivalent to a division of the voltage by between $\sqrt{2}$ and 2, respectively). For example:

$$M = (A + B) - 3 \text{ dB or } M = (A + B) - 6 \text{ dB}$$

The correction factor will depend on the nature of the two signals to be combined. If identical signals exist on the A and B channels (representing 'double mono' in effect), then the level of the uncorrected sum channel (M) will be two times (6 dB) higher than the levels of either of A or B, requiring a correction of -6 dB in the M channel in order for the maximum level of the M signal to be reduced to a comparable level. In a television production where most of the sound sources are panned centrally (e.g. central dialogue) the 6 dB correction factor would be more appropriate, since there is a strong likelihood that a large number of operational situations will result in identical A and B signals close to maximum level. If the A and B signals are non-coherent (random-phase relationship), then only a 3 dB rise in the level of M will result when A and B are summed, requiring the -3 dB correction factor to be applied. This is more likely with stereo music signals. As most stereo material has a degree of coherence between the channels, the actual rise in level of M compared

with A and B is likely to be somewhere between the two limits for real programme material.

The S signal results from the subtraction of B from A, and is subject to the same correction factor. For example:

$$S = (A - B) - 3 \text{ dB or } S = (A - B) - 6 \text{ dB}$$

S can be used to reconstruct A and B when matrixed in the correct way with the M signal (see below), since $(M + S) = 2A$ and $(M - S) = 2B$. It may therefore be appreciated that it is possible at any time to convert a stereo signal from one format to the other and back again.

If either format is not provided at the output of microphones or mixing equipment it is a relatively simple matter to derive one from the other electrically. *Figure 2.74* shows two possible methods. *Figure 2.74(a)* shows the use of transformers, where A and B signals are fed into the primaries, and M is derived by wiring the two secondaries in series and in phase, while S is derived by wiring two secondaries in series and out of phase. *Figure 2.74(b)* shows the use of summing amplifiers, whereby A and B are summed inphase to derive M, and summed with a phase inversion ($-180°$) in the B leg to derive S. Both matrixes have the advantage that they will also convert back to AB from MS, in that M and S may be connected to the inputs and they will be converted back to A and B.

2.5.6.5 Effects of misalignment on stereo signals
Differences in level, frequency response and phase may arise between signals of a stereo pair,

Figure 2.74 *Stereo signals may be converted from the AB format to MS format either by using transformers wired with secondaries in phase to derive M and out of phase to derive S, as shown in (a), or by using summing amplifiers as shown in (b).*

perhaps due to losses in cables, misalignment, and performance limitations of equipment. It is important that these are kept to a minimum for stereo work, as inter-channel anomalies result in various audible side-effects. Differences will also result in poor mono compatibility. These differences and their effects are discussed below.

Frequency response and level A difference in level or frequency response between A and B channels will result in a stereo image biased towards the channel with the higher overall level or that with the better high-frequency response. Also, an A channel with excessive high-frequency response compared with that of the B channel will result in the apparent movement of sibilant sounds towards the A loudspeaker. Level and response misalignment on MS signals results in increased crosstalk between the equivalent A and B channels, such that if the S level is too low at any frequency the AB signal will become more monophonic (width narrower), and if it is too high the apparent stereo width will be increased.

Phase Interchannel phase anomalies will affect one's perception of the positioning of sound source, and it will also affect mono compatibility. Phase differences between A and B channels will result in 'comb-filtering' effects in the derived M signal due to cancellation and addition of the two signals at certain frequencies where the signals are either out of or in phase.

Cross-talk It was stated earlier that an inter-channel level difference of only 18 dB was required to give the impression of a signal being either fully left or fully right. Cross-talk between A and B signals is not therefore usually a major problem, since the performance of most audio equipment is far in excess of these requirements. Excessive crosstalk between A and B signals will result in a narrower stereo image, whilst excessive crosstalk between M and S signals will result in a stereo image increasingly biased towards one side.

References

ATAL, B.S. and SCHROEDER, M.R., Computer simulation of sound transmission in rooms. *IEEE Conv. Rec.*, Part 7, pp. 150–155 (1963)

ATAL, B.S. and SCHROEDER, M.R., Apparent sound source translator, *US Patent 3236949* (Feb. 1966)

BAUER, B.B., Stereophonic earphones and binaural loudspeakers, *Journal of the Audio Engineering Society*, **9**, 148–151 (1961a)

BAUER, B., Phasor analysis of some stereophonic phenomena, *Journal of the Acoustical Society of America*, **33**, 1536–1539 (1961b)

BENNETT, J.C., BARKER, K. and EDEKO, F.O., A new approach to the assessment of stereophonic sound system performance, *Journal of the Audio Engineering Society*, **33**, 314–321 (1985)

BLAUERT, J., *Spatial Hearing: the Psychophysics of Human Sound Localisation*, JS Allen, MIT Press, Cambridge, MA (1983)

BLUMLEIN, A.D., British Patent Specification 394325, *Journal of the Audio Engineering Society*, **6**, 91–100 (1958)

CLARK, H.A.M., DUTTON, G.F. and VANDERLYN, P.B., The stereosonic recording and reproduction system: a two channel system for domestic tape records, *Journal of the Audio Engineering Society*, **6**, 102–117 (1958)

COOPER, D.H. and BAUCK, J.L., Prospects for transaural recording, *Journal of the Audio Engineering Society*, **37**, 3–19 (1989)

DAUBNEY, C., Ambisonics – an operational insight, *Studio Sound* (Aug. 1982)

ELEN, R., Ambisonic mixing – an introduction, *Studio Sound* (Sep. 1983)

ELEN, R., Surround sound, *Studio Sound* (Feb. 1984)

FLETCHER, H., Auditory perspective, *Bell System Technical Journal*, **13** (1934)

GALOMBOS, R. and DAVIES, H., The response of single auditory fibres to acoustic stimulation, *Journal of Neurophysiology*, **6** (1943)

GASKELL, P.S., System UHJ: a hierarchy of surround sound transmission systems, *The Radio and Electronics Engineer*, **49**, 449–459 (1979)

GERZON, M.A., Ambisonics in multichannel broadcasting and video. Presented at: *74th AES Convention*, New York, Preprint No 2034 (J–1) (Oct. 1983)

GERZON, M.A., Stereo shuffling: new approach, old technique, *Studio Sound*, 122–130 (July 1986)

GERZON, M.A., Three channels, the future of stereo?, *Studio Sound* (June 1990)

GERZON, M.A. and BARTON, G.J., Ambisonic surround sound mixing for multitrack studios. Presented at: *AES Convention*, Anaheim, Los Angeles (May 1984)

GIERLICH, H.W. and GENUIT, K., Processing artificial head recordings, *Journal of the Audio Engineering Society*, **37**, 34–39 (1989)

GRIESINGER, D., Equalisation and spatial equalisation of dummy head recordings for loudspeaker reproduction, *Journal of the Audio Engineering Society*, **37**, 20–29 (1989)

HAAS, H., Uber den Einfluss eines Einfachechos und die Hörsamkeit von Sprache, *Acustica*, **2**, 49–58 (1951)

HAFTER, E.R., BUELL, T.N. and RICHARDS, V.M., Onset coding in lateralisation: its form site and function. In: *Auditory Function* (eds. Edelmann, Gall, and Cowan), Wiley, New York (1988)

LIPSHITZ, S., Stereo microphone techniques: are the purists wrong? *Journal of the Audio Engineering Society*, **34**, 717–735 (1986)

MAJIMA, Y. *et al.*, *The RSS Processor: Three-dimensional Sound Generation from Conventionally-recorded Two-channel Sources*, Roland Corporation, Tokyo (1991)

MCKINNIE, D. and RUMSEY, F., Coincident microphone techniques for 3-channel stereo reproduction, presented at 102nd AES Convention, Munich, preprint 4429 (1997)

MOLLER, H., Reproduction of artificial head recordings through loudspeakers, *Journal of the Audio Engineering Society*, **37**, 30–33 (1989)

MOORE, B.C.J., *An Introduction to the Psychology of Hearing*, 3rd edn, Academic Press, London (1989)

NEGISHI, H., The wide imaging stereo, *Proceedings of the Institute of Acoustics*, **10**, 11 (1988)

RICHTER, F. and PERSTERER, A., Design and applications of a creative audio processor. Presented at: *86th AES Convention*, Hamburg, AES Convention Preprint 2782 (U–4) (Mar. 1989)

RUMSEY, F.J., Creating new images, *Studio Sound* (June 1991)

SNOW, W., Basic principles of stereophonic sound, *Journal of the Society of Motion Picture and Television Engineers*, **61**, 567–589 (1953)

SIVIAN, L.J. and WHITE, S.D., Minimum audible sound fields, *Journal of the Acoustical Society* (July 1931)

STEINBERG, J.C. and SNOW, W.B., Auditory perspectives – physical factors, *Electrical Engineering*, **53** 12–15 (1934)

THEILE, G., On the standardisation of the frequency response of high-quality studio headphones, *Journal of the Audio Engineering Society*, **34**, 956–969 (1986)

THEILE, G., On the naturalness of two-channel stereo sound. In: *Proceedings of the AES 9th International Conference*, Detroit, 1–2 February 1991, AES, New York

THOMAS, M.V., Improving the stereo headphone image, *Journal of the Audio Engineering Society*, **25**, 474–478 (1977)

VANDERLYN, P.B., *British Patent 781 186* (14 Aug. 1957)

VANDERLYN, P., Auditory cues in stereophony, *Wireless World*, 55–60 (Sep. 1979)

WALLACH, H., NEWMAN, E.B. and ROSENZWEIG, M.R., The precedence effect in sound localisation, *American Journal of Psychology*, **62**, 315 (1949)

WEINER, F., On the diffraction of a progressive sound wave by the human head, *Journal of the Acoustical Society of America*, **19**, 143–146 (1947)

YOST, W.A., WIGHTMAN, F.L. and GREEN, D.M., Lateralisation of filtered clicks, *Journal of the Acoustical Society of America*, **50**, 1526–1531 (1971)

ZUCCARELLI, H., Holophony, *New Scientist*, 483 (10 Nov. 1983)

2.6 Microphones for stereo

2.6.1 The true stereo microphone

A stereo microphone typically comprises two or more directional capsules, and as such a stereo microphone is really at least two mono microphones in one housing. It is difficult to make a stereo microphone which operates as a spaced pair, unless the spacing is very small, for obvious physical reasons, and so stereo microphones tend to be based on the coincident pair principle. Spaced pair stereo is usually achieved using separate mono microphones, unless a 'dummy head' is used for binaural recording, in which case two small microphones may be buried in the ear canals of the head (see Section 2.6.3.4).

2.6.1.1 Coincident-pair principles

The coincident-pair stereo microphone, an example of which is shown in *Figure 2.75*, incorporates two directional capsules which may be angled over a range of settings to allow for different configurations and operational requirements. The pair may be operated in either the AB or MS modes, and a matrixing unit is sometimes supplied with microphones which are intended to operate in the MS mode in order to convert the signal to AB format for recording. The directional patterns (polar diagrams) of the two microphones need not necessarily be figure-of-eight,

Figure 2.75 *A typical coincident stereo microphone.* (Courtesy of Beyer.)

although if the microphone is used in the MS mode the S capsule must be figure-of-eight (see below).

Directional information is encoded solely in the level differences between the capsule outputs, since the two capsules are mounted physically as close as possible, and therefore there are no phase differences between the outputs except at the highest frequencies where intercapsule spacing may become appreciable in relation to the wavelength of sound.

2.6.1.2 AB pairs

AB stereo microphones (sometimes known as 'XY microphones') may be mounted vertically in relation to the sound source, such that the two capsules are angled to point symmetrically left and right of the centre of the source stage. The choice of angle depends on the polar response of the capsules used, since most stereo microphones allow for either capsule to be switched through a number of pickup patterns between figure-of-eight and omnidirectional. As shown theoretically in the papers described in Section 2.5, a coincident pair of figure-of-eight microphones at 90° gives a very accurate correlation between the actual angle of the source and the apparent position of the virtual image when reproduced on loudspeakers, but there are also operational disadvantages to the figure-of-eight patterns in many cases. We shall proceed to investigate coincident pairs which use a number of different polar patterns, bearing in mind that the theoretical arguments about correct correlation between real and phantom images can only be taken so far when there are conflicting operational considerations.

Figure 2.76 shows the polar pattern of a stereo microphone which is set to figure-of-eight pattern,

and it serves as a useful example for the discussion of a number of phenomena which are not so easy to visualize with other pairs. Firstly, it may be seen that the fully left position corresponds to the null point of the right capsule's pick-up. This is the point at which there will be maximum level difference between the two capsules, and it is a 'red herring' that the fully left position also corresponds to the maximum pickup of the left capsule, since it does not in other stereo pairs. As a sound moves across the sound stage from left to right it will result in a gradually decreasing output from the left microphone, and an increasing output from the right microphone. Since the microphones have cosine responses, the output at 45° off axis is $\sqrt{2}$ times the maximum output, or 3 dB down in level, thus the takeover between left and right microphones is smooth for music signals.

The second point to consider with this pair is that the rear quadrant of pick-up suffers a left–right reversal, since the rear lobes of each capsule point in the opposite direction. This is important when considering the use of such a microphone in situations where confusion may arise between sounds picked up on the rear and in front of the microphone, such as in television sound where the viewer can also *see* the positions of sources. The third point is that pick-up in both side quadrants results in out-of-phase signals between the channels, since a source further round than 'fully left' results in pick-up by both the negative lobe of the right capsule and the positive lobe of the left capsule. There is thus a large region around a crossed pair of figure-of-eights that results in out-of-phase information, this information often being ambient or reverberant sound. Any sound picked up in this region will suffer cancellation if the channels are summed to mono, with maximum cancellation occurring at 90° and 270°, assuming 0° as the centre-front.

The operational advantages of the figure-of-eight pair are the crisp and accurate positioning of sources, together with a natural blend of ambient sound from the rear. Some cancellation of ambience may occur, especially in mono, if there is a lot of reverberant sound picked up by the side quadrants. Disadvantages lie in the large out-of-phase region, and in the size of the rear pick-up which is not desirable in all cases and is left–right reversed (Streicher and Dooley, 1985). Stereo pairs which are made up of capsules having less rear pick-up may be preferred in cases where a 'drier' or less reverberant balance is required, and where frontal sources are to be favoured over rearward sources. In such cases the capsule responses may be changed to be nearer the

AB pair

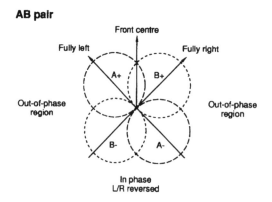

Figure 2.76 *The polar diagram of a coincident pair of figure-of-eights shows that there is a large out-of-phase region in the side quadrants, and a left–right reversal at the rear. L, left; R, right.*

cardioid pattern, and this requires an increased angle between the capsules to maintain good correlation between actual and perceived angle of sources.

The cardioid crossed pair is angled at approximately 130°, although angles of between 90° and 180° may be used to good effect depending on the 'width' of the sound stage to be covered. At an angle of 130° a centre source is 65° off-axis from each capsule, resulting in a 3 dB drop in level compared with the maximum on-axis output (the cardioid microphone response is equivalent to $0.5(1 + \cos\theta)$, where θ is the angle off-axis of the source, and thus the output at 65° is $\sqrt{2}$ times that at 0°). A departure from the theoretically correct angle is often necessary in practical situations, and it must be remembered that the listener will not necessarily be aware of the 'correct' location of each source, neither may it matter that the true and perceived positions are different. A pair of 'back-to-back' cardioids has often been used to good effect since it has a simple MS equivalent (see Section 2.6.1.3) of an omnidirectional and a figure-of-eight, and has no out-of-phase region. Although theoretically 'fully left' is at 90° off-centre, it may be appreciated that there will in fact be a satisfactory level difference between the capsules to make fully left appear at a somewhat smaller angle.

With any AB crossed pair, fully left or fully right corresponds to the null point of pick-up of the opposite channel's microphone, as already stated, and, as will be seen below, this corresponds also to the point where the amplitude of the M signal equals that of the S signal (where the sum of the channels is the same as the difference between them). It is very important to grasp the difference between the angle subtended by the two capsules and the angle between fully left and fully right. As the angle between the capsules is made larger, the angle between the null points will become smaller. Many people become confused over the effect that this will have on the width of the stereo image, and this requires some discussion.

Operationally, if one wishes to 'widen' the stereo image one will widen the angle between the microphones which is intuitively the right thing to do. This results in a narrowing of the angle between fully left and fully right, which is also correct, since sources which had been, say, half left in the original image will now be further towards the left. A narrow angle between fully left and fully right results in a very wide sound stage, since sources have only to move a small distance to result in large changes in reproduced position. This corresponds to a wide angle between the capsules. Further coincident pairs are possible using any polar pattern between figure-of-eight and omni-directional, although the closer that one gets to omni-directional the greater the required angle to achieve adequate separation between the channels. The hypercardioid pattern is often chosen for its smaller rear lobes than the figure-of-eight, allowing a more distant placement from the source for a given direct-to-reverberant ratio. Since the hypercardioid pattern lies between figure-of-eight and cardioid, the angle required between the capsules is correctly around 110°.

The theoretical foundation (see Section 2.5) should not be forgotten in that it suggests the need for an electrical 'narrowing' of the image at high frequencies in order to preserve the correct angular relationships between low and high-frequency signals, although this is rarely implemented in practice with coincident pair recording. A further consideration to do with the theoretical versus the practical is that although microphones tend to be referred to as having a particular polar pattern, this pattern is unlikely to be consistent across the frequency range and this will have an effect on the stereo image. Cardioid crossed pairs should theoretically exhibit no out-of-phase region since there should be no negative rear lobes, but in practice most cardioid capsules become more omnidirectional at low frequencies and more hypercardioid at high frequencies, and thus some out-of-phase components may be noticed in the high frequencies range whilst the width may appear too narrow at low frequencies. Attempts have been made to compensate for this in stereo microphone design, one example of which has been used by Sanken in an MS microphone (the CMS-2) (Mizoguchi and Ichikawa, 1977), by adding high and low-frequency acoustic-compensation mechanisms to maintain relatively constant directional response over the frequency range.

AB pairs in general have the possible disadvantage that central sounds are off-axis to both microphones, perhaps considerably so in the case of crossed cardioids. This may result in a central signal with a poor frequency response and possibly an unstable image if the polar response of the microphone is erratic. Whether or not this is important depends on the importance of the central image in relation to that of offset images, and will be most important in cases where the main source is central (such as in television, with dialogue). In such cases the MS technique described in the next section is likely to be more appropriate, since central sources will be on-axis to the M microphone. For music recording it

would be hard to say whether central sounds are any more important than offset sources, and thus either technique may be acceptable.

In commercial 'side-fire' AB microphones, it is common for one of the capsules to be fixed and for the other to rotate in order to set the angle between the capsules. It is sometimes difficult to see where the centre-front position is with some microphones, and this may have to be discovered through trial and error by getting an assistant to walk around the microphone making a suitable noise until equal level is obtained from both capsules. Alternatively, some stereo microphones have a small LED which moves with each capsule to identify the direction in which each is pointing.

2.6.1.3 MS pairs

Although some stereo microphones are built specifically to operate in the MS mode, it is possible to take any coincident pair capable of at least one capsule being switched to figure-of-eight, and orientate it so that it will produce suitable signals. The S component (being the difference between left and right signals) is always a sideways-facing figure-of-eight with its positive lobe facing left. The M or mono component may be any polar pattern facing to the centre-front, although the choice of M pattern depends on the desired equivalent AB pair (see below), and will be the signal which a mono listener would hear. True MS microphones usually come equipped with a control box which matrixes the MS signals to AB format if required. A control for varying S gain is often provided as a means of varying the effective acceptance angle between the equivalent AB pair (see below).

It should be remembered that MS signals may not be monitored directly: they are sum and difference components and must be converted to AB format at a convenient point in the production chain. The advantages of keeping a signal in the MS format until it needs to be converted will be discussed below, but the major advantage of pick-up in the MS format is that central signals will be on-axis to the M capsule, resulting in the best frequency response. Furthermore, it is possible to operate an MS microphone in a similar way to a mono microphone (so long as precautions are taken when moving the microphone, as detailed in Section 2.6.1.6) which may be useful in television operations where the MS microphone is replacing a mono microphone on a pole or in a boom. Hibbing (1989), amongst others, points to the reduced audible effects of variations in microphone polar pattern with frequency when using the MS pick-up technique.

To see how MS and AB pairs relate to each other, and to draw some useful conclusions about stereo width control, it is informative to consider again a coincident pair of figure-of-eight microphones. For each MS pair there is an AB equivalent. This stands to reason, since M is simply the sum of A and B signals, whilst S is simply the difference between them. The polar pattern of the AB equivalent to any MS pair may be derived by plotting the level of $(M + S)/2$ and $(M - S)/2$ for every angle around the pair ($M + S = 2A$ and $M - S = 2B$). Taking the MS pair of figure-of-eight microphones shown in *Figure 2.77*, it may be seen that the AB equivalent is simply another pair of figure-of-eights, but rotated through 45°. Thus the correct MS arrangement to give an equivalent AB signal where both 'capsules' are orientated at 45° to the centre-front (the normal arrangement) is for the M capsule to face forwards and the S capsule to face sideways.

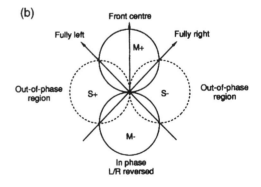

Figure 2.77 *Every AB pair has an MS equivalent. An AB pair of figure-of-eights is shown in (a) and its MS equivalent in (b). L, left; R, right.*

A number of interesting points arise from a study of the AB/MS equivalence of these two pairs, and these points apply to all equivalent pairs. Firstly, fully left or right in the resulting stereo image occurs

at the point where S = M (in this case at 45° off-centre). This is easy to explain, since the fully left point is the point at which the output from the right capsule (B) is zero. Therefore M = A + 0, and S = A − 0, both of which equal A. Secondly, at angles of incidence greater than 45° off centre in either direction the two channels become out of phase, as was seen above, and this corresponds to the region in which S is greater than M. Thirdly, in the rear quadrant where the signals are in phase again, but left–right reversed, the M signal is greater than S again. The relationship between S and M levels, therefore, is an excellent guide to the phase relationship between the equivalent AB signals. If S is lower than M, then the AB signals will be in phase. If S = M, then the source is either fully left or right, and if S is greater than M, then the AB signals will be out of phase.

To show that this applies in all cases, and not just that of the figure-of-eight pair, look at the MS pair in *Figure 2.78* together with its AB equivalent. This MS pair is made up of a forward-facing cardioid and a sideways-facing figure eight (a popular arrangement). Its AB equivalent is a crossed pair of hypercardioids, and again the extremes of the image (corresponding to the null points of the AB hypercardioids) are the points at which S equals M. Similarly, the signals go out of phase in the region where S is greater than M, and come back in phase again for a tiny angle round the back, due to the rear lobes of the resulting hypercardioids. Thus the angle of acceptance (between fully left and fully right) is really the frontal angle between the two points on the MS diagram where M equals S.

Now, consider what would happen if the gain of the S signal was raised (imagine expanding the lobes of the S figure-of-eight in *Figure 2.78*). The result of this would be that the points where S equalled M would move inwards, making the acceptance angle smaller. As explained earlier, this results in a wider stereo image, since off-centre sounds will become closer to the extremes of the image, and is equivalent to increasing the angle between the equivalent AB capsules. Conversely, if the S gain is reduced, the points at which S equals M will move further out from the centre, resulting in a narrower stereo image, equivalent to decreasing the angle between the equivalent AB capsules. This is neatly exemplified in a commercial example, the Neumann RSM 191i, pictured in *Figure 2.79*, which is an MS microphone in which the M capsule is a forward-facing short shotgun microphone with a polar pattern rather like a hypercardioid. The polar pattern of the

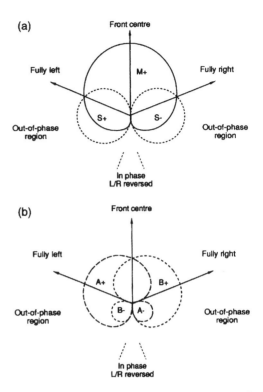

Figure 2.78 *The MS equivalent of a forward-facing cardioid and sideways figure-of-eight, as shown in (a), is a pair of hypercardioids the effective angle of which depends on S gain, as shown in (b). L, left; R, right.*

Figure 2.79 *The Neumann RSM191i is an end-fire MS microphone the control box of which allows control of S gain. (Courtesy of FWO Bauch.)*

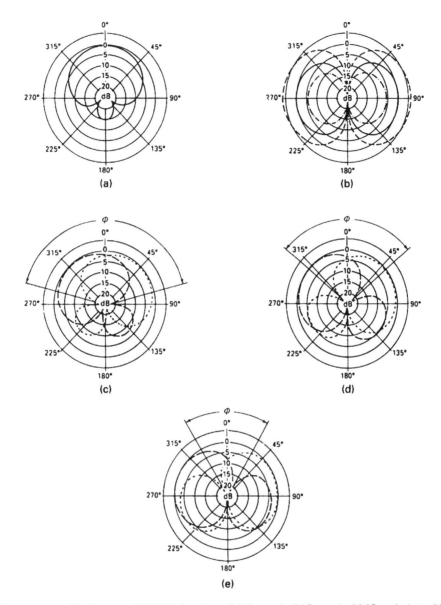

Figure 2.80 *Polar patterns of the Neumann RSM191i microphone. (a) M capsule; (b) S capsule; (c) AB equivalent with −6 dB S gain; (d) 0 dB S gain; (e) +6 dB S gain. (Courtesy of FWO Bauch.)*

equivalent AB pair is shown in *Figure 2.80* for three possible gains of the S signal with relation to M (−6, 0 and +6 dB). It will be seen that the acceptance angle (ϕ) changes from being large (narrow image) at −6 dB, to small (wide image) at +6 dB.

The other thing to be noted about the effect of changing S gain is the size of the rear lobes of the AB equivalent. It may be seen that the higher the S gain the larger the rear lobes. Therefore, not only does S gain change stereo width, it also affects rear

pick-up, and thus the ratio of direct to reverberant sound.

MS microphones, therefore, are useful as a means of deriving signals in the sum and difference format, for subsequent matrixing back to A and B at a later stage. Whilst in the MS format, the proportions of M and S may be varied in order to affect the resulting stereo width and direct-to-reverberant balance. Any AB stereo microphone may be operated in the MS configuration, simply by orientating the capsules

in the appropriate directions and switching them to an appropriate polar pattern, but certain microphones are dedicated to MS operation simply by the physical layout of the capsules. Such microphones normally include a matrix for conversion to AB format, although this conversion may be achieved in a mixing console as shown in Section 2.6.1.6. There is not space here to show all the possible MS pairs and their AB equivalents, but a comprehensive review may be found in Dooley and Streicher (1982).

2.6.1.4 'End-fire' versus 'side-fire' configurations

Physically, there are two principal ways of mounting the capsules in a coincident stereo microphone, be it MS or AB format: either in the 'end-fire' configuration where the capsules 'look out' of the end of the microphone such that the microphone may be pointed at the source (see *Figure 2.81*), or in the 'side-fire' configuration where the capsules look out of the sides of the microphone housing. It is less easy to see the direction in which the capsules are pointing in a side-fire microphone, but such a microphone makes it possible to align the capsules vertically above each other so as to be time-coincident in the horizontal plane, as well as allowing for the rotation of one capsule with relation to the other. An end-fire

Figure 2.81 *An MS microphone: (a) end-fire mode; (b) the side-fire mode.*

configuration is more suitable for the MS capsule arrangement (see *Figure 2.82*), since the S capsule may be mounted sideways behind the M capsule, and no rotation of the capsules is required, although there is a commercial example of an AB end-fire microphone for TV electronic news gathering (ENG) use which houses two fixed cardioids side-by-side in an enlarged head.

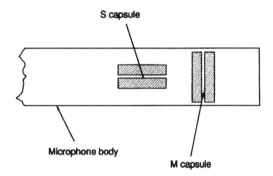

Figure 2.82 *Typical end-fire capsule configuration.*

2.6.1.5 Some commercial implementations

Although some commercial examples have already been introduced in the preceding sections, it would be worthwhile examining the approaches to stereo microphone design taken by a selection of manufacturers.

A good example of the standard side-fire microphone is the Neumann USM 69i (*Figure 2.83*), which is identical acoustically to the earlier SM 69 but incorporates switching of the polar pattern of both capsules into the microphone body, as opposed to putting this in a separate control box. The polar pattern may be switched from omnidirectional, through cardioid, to figure-of-eight in six steps. The upper capsule may be rotated so as to be at any angle between 0° and 180°. The frequency response and polar characteristics of this microphone are shown in *Figure 2.84*.

The Sanken CMS-2, pictured in *Figure 2.85* on the other hand, is claimed to be the world's smallest MS stereo condenser microphone, with an extremely low equivalent noise level of 16 dBA. It is a side-fire MS microphone, as opposed to the end-fire Neumann RSM 191, and it is made up of a fixed cardioid and figure-of-eight. Its frequency response and AB equivalent polar pattern are shown in *Figure 2.86*. In normal use the relative gains of M and S are set such that the acceptance angle equals 127°, but this may be altered by the connection of an optional MS matrix controller.

Figure 2.83 *The Neumann USM 69i is a side-fire AB microphone with variable directivity patterns.* (Courtesy of FWO Bauch.)

The AKG C34 MS microphone is a modified version of the original C34 AB microphone, in which one capsule has been fixed in the figure-of-eight mode (to make S), but where the M capsule has been turned to face out of the end of the housing and is left as a fully variable polar pattern. This is one of the few end-fire MS microphones with a variable pattern M capsule, and has proved popular in TV companies because, when mounted on a boom, the boom operator can alter the 'tightness' of the M pattern using the associated control box to suit the current scene or pick-up situation. If someone is 'off microphone' the M pattern can be made a softer cardioid to pick-up over a wider angle, and if tight coverage of a single source is required the M pattern can be made more narrowly hypercardioid.

The AMS ST250 MS microphone is based on 'Soundfield' principles, developed for the AMS/Calrec Soundfield microphone (see Section 2.6.5), and as such has a number of novel fea-

tures such as the facility for operating the microphone in either the end- or side-fire modes simply by pressing a switch which reconfigures the capsules electrically. Furthermore, if the microphone is used upside-down (usually resulting in a swapping over of left and right positions) a switch can be pressed to flip over the channels. As will be seen in the next section, this is simply a matter of phase reversing the S signal.

A similar feature is found on the control box of the Neumann RSM 191, although labelled in a confusing manner. A four-position switch on the control box is labelled −XY, XY, MS, −MS, and has the function of allowing the microphone to be used upside-down in either the XY (AB) mode or the MS mode. In the MS mode (when the control box does no matrixing), the −MS position phase reverses the S signal, with the effect of flipping over left and right in any resulting stereo image, and in the XY mode (when the box is converting the output of the microphone to XY format) the −XY position simply swaps over left and right outputs. The box also has a six-position switch to control S gain, labelled both in decibels of S relative to M (from −9 to +6dB), and in acceptance angle (fully left to fully right). The acceptance angle varies from 170° and −9 dB S gain, to 60° at +6 dB S gain.

2.6.1.6 Operational considerations with stereo pairs

The control of S gain is an important tool in determining the degree of width of a stereo sound stage, and for this reason the MS output from a microphone might be brought (unmatrixed to AB) into a mixing console, so that the engineer may have control over the width. This in itself can be a good reason for keeping a signal in MS form during the recording process, although M and S can easily be derived at any stage using a conversion matrix. If the number of channels can be afforded on the mixing console, then it would be possible to have, say, one fader for the M signal, one for the S signal, and one for a phase reversed S signal.

The diagram in *Figure 2.87* shows how it is possible to derive an AB mix with variable width from an MS microphone using three channels on a mixer. It should be noted that no MS matrices are required. M and S outputs from the microphone are fed in phase through two mixer channels and faders, and a post-fader feed of S is taken to a third channel line input, being phase reversed on this channel. The M signal is routed to both left and right mix buses (panned centrally), whilst the S signal is routed to the

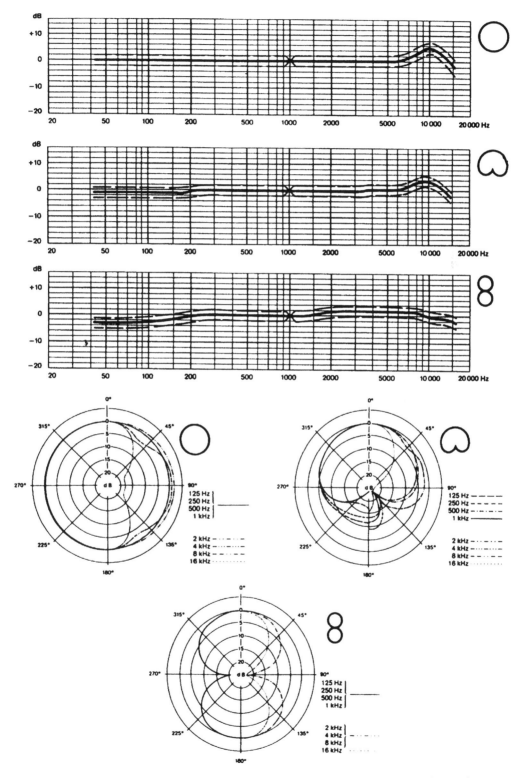

Figure 2.84 *Frequency responses and polar patterns of the USM 69i in different modes.* (Courtesy of FWO Bauch.)

Figure 2.85 *The Sanken CMS-2 is a small MS microphone with fixed directivity pattern of the MS capsules.* (Courtesy of Stirling Audio.)

left mix bus (M + S = 2A) and the −S signal (the phase-reversed version) is routed to the right mix bus (M − S = 2B). It is important that the gain of the −S channel is matched very closely with that of the S channel. A means of deriving M and S from an AB format input is to mix A and B together with the B signal phase reversed to get S, and without the phase reverse to get M. In order to convert back to A and B a similar method to the above could be employed, but this is wasteful of mixer channels and a matrix might be more convenient.

From a handling point of view, a number of points should be made. Outdoors, stereo microphones will be more susceptible to wind-noise and rumble than most mono microphones, as they incorporate velocity-sensitive capsules which always give more problems in this respect than omnidirectional microphones. Most of the interference would reside in the S channel, since this has always a figure-of-eight pattern, and thus would not be a problem to the mono listener. Similarly, physical handling of the stereo microphone, or vibration picked up through a stand, will be much more noticeable audibly than with a mono pressure microphone. Stereo microphones should not be used close to people speaking, as small movements of their heads can cause large changes in the angle of incidence, and thus considerable movement in their apparent position in the sound stage. This is especially true in cases where the S gain is high (narrow angle of AB acceptance), as very small head or microphone movements result in large swings of the sound image.

2.6.2 Near-coincident microphone configurations

It is generally admitted that the Blumlein/Clark *et al.* theory of vector summation of sounds from loudspeakers having only level differences between them only holds true for continuous low-frequency sounds up to around 700 Hz, where head-shadowing effects begin to take over (as described in Section 2.5). Indeed at the time of presentation of the Clark paper (1957) questions were asked about the ability of a system based only on level differences to cope with the important localization cues provided by short transients. The authors replied that they had decided to ignore precedence effects, because of the dependence on seating position, and that they believed that very few real sounds consisted of short pulses, many short sounds in fact being capable of being analysed into a series of spectral components (that is to say, Fourier transformation). This is interesting, since in the discussion of directional perception (see Section 2.5) it was stated that the response of the basilar membrane to transient signals resembled a decaying sinusoidal waveform, possibly allowing for phase-difference cues to be extracted. Although at the time the authors believed that their system did not exploit time difference cues, Vanderlyn (1979) later suggested a theory to explain how such a system copes with transient information.

There appears to be some basis, though, for the adoption of 'near-coincident' pairs of directional

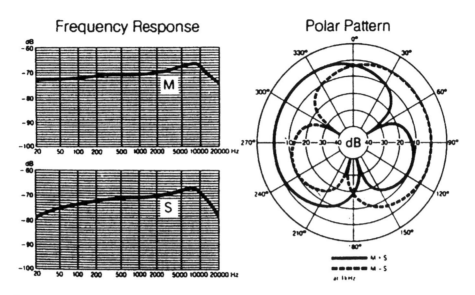

Figure 2.86 *Frequency responses and polar pattern of the Sanken CMS-2.* (Courtesy of Stirling Audio.)

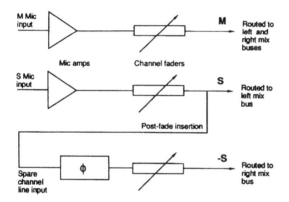

Figure 2.87 *An AB mix with variable width can be derived from an MS microphone connected to three channels on a mixer (see text).*

microphones, to introduce small additional timing differences which may help in the localization of transient sounds and increase the 'spaciousness' of a recording, whilst at the same time remaining nominally coincident at low frequencies, adequately to maintain the level differences between channels required by Blumlein *et al.* for the reconstruction of low-frequency phase differences with continuous sounds, although resulting in an upper frequency limit for this effect which is lower than that for truly coincident pairs. Subjective evaluations bear this principle out, with listeners appearing to be considerably impressed by recordings made on near-coincident pairs. One comprehensive subjective

assessment of stereo microphone arrangements, performed at the University of Iowa (Cross, 1985), consistently resulted in the near-coincident pairs scoring amongst the top few performers for their sense of 'space' and realism. Lipshitz (1986) attributes these effects to 'phasiness' at high frequencies (which some people may like, none the less), and argues that truly coincident pairs are preferable. Gerzon (1986) suggests that a very small spacing of crossed cardioid microphones (about 5 cm) actually compensates for the phase differences introduced when 'spatial equalization' is used (the technique described earlier of increasing low-frequency width relative to high-frequency width by introducing equalization into the S channel).

A number of examples of near-coincident pairs exist as 'named' arrangements, although, as Williams (1987) has shown, there is a whole family of possible near-coincident arrangements using combinations of spacing and angle. The so-called 'ORTF pair' is an arrangement of two cardioid microphones, deriving its name from the organization which first adopted it (the *Office de Radiodiffusion-Television Francaise*). The two microphones are spaced apart by 170 mm, and angled at 100°. The 'NOS' pair (*Nederlande Omroep Stichting*, the Dutch Broadcasting Company), uses cardioid microphones spaced apart by 300 mm and angled at 90°. A pair of figure-of-eight microphones pointing forwards and spaced apart by 200 mm has been called a 'Faulkner' pair, after the British recording engineer who first adopted

it. This latter pair can be found to offer good 'focusing' on a small to moderate sized central ensemble with the microphones placed further back than would normally be expected.

2.6.3 Spaced arrays

2.6.3.1 Principles

Although spaced arrays have a historical precedent for their usage, since they were the first to be documented (in the work of Clement Ader at the Paris Exhibition in 1881 (Ader, 1881)) and have been widely used, especially in the USA, since the 1930s, they are perhaps less 'correct' theoretically. Spaced arrays rely principally on the precedence effect, in that the delays which result between the channels tend to be of the order of a number of milliseconds (as opposed to the binaural delay of up to 600 μs). Wallach *et al.* discussed the relevance of the precedence effect to stereo reproduction and, as already stated in Section 2.5, it is possible to combine time and level difference to obtain a particular phantom image position, and it is known that one parameter may be traded off against the other to some extent.

With spaced arrays the level and time difference resulting from a source at a particular left–right position on the sound stage will depend on how far the source is from the microphones (see *Figure 2.88*),

with a more distant source resulting in a much smaller delay and level difference. In order to calculate the time and level differences which will result from a particular spacing it is possible to use the following two formulae:

$$\Delta t = (d_1 - d_2)/c$$

$$\Delta L = 20 \log_{10}(d_1/d_2)$$

where Δt is the time difference, and ΔL the pressure level difference which results from a source a distance d_1 and d_2, respectively, from the two microphones, and c is the speed of sound (340 m/s).

Where a source is very close to a spaced pair there may be a considerable level difference between the microphones, but this will become small once the source is more than a few metres distant. The positioning of spaced microphones in relation to a source is thus a matter of achieving a compromise between closeness (to achieve satisfactory level and time differences between channels), and distance (to achieve adequate reverberant information relative to direct sound). When the source is large and deep, such as a large orchestra, it will be difficult to place the microphones so as to suit all sources, and it may be found necessary to raise the microphones somewhat so as to reduce the difference in path length between sources at the front and rear of the orchestra.

Spaced microphone arrays do not stand up well to theoretical analysis when considering the imaging of continuous sounds, the precedence effect being related principally to impulsive or transient sounds. Because of the phase difference between signals at the two loudspeakers created by the microphone spacing, interference effects at the ears at low frequencies may in fact result in a contradiction between level and time cues at the ears (Lipshitz, 1986). It is possible in fact that the ear on the side of the 'earlier' signal may *not* experience the higher level, thus producing a confusing difference between the cues provided by impulsive sounds and those provided by continuous sounds. The lack of phase coherence in spaced-array stereo is further exemplified by phase inverting one of the channels on reproduction, an action which does not always appear to affect the image particularly, as it would with coincident stereo, showing just how uncorrelated the signals are.

It may thus be found that stereo perspective and accuracy of positioning are less realistic with spaced arrays, although many convincing recordings have resulted from their use. Lipshitz suggests that the impression of 'space' that results from the use of

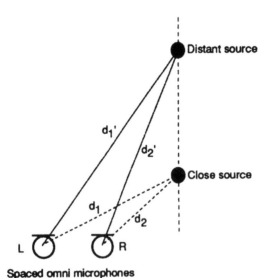

Figure 2.88 *With spaced omnidirectional microphones a source at position X results in path lengths d₁ and d₂ to each microphone, respectively, while a source in the same left–right (LR) position but at a greater distance (source Y) the path length difference is smaller, resulting in a smaller time difference than for X.*

spaced arrays is in fact simply the result of 'phasiness' and comb-filtering effects. Cooper and Bauck (1989) suggest that there is a place for both the 'spaciousness' that results from spaced techniques and the imaging which results from coincident techniques, since the highly decorrelated signals which result from spaced techniques are also a feature of concert-hall acoustics. This therefore supports to some extent the arguments for near-coincident techniques, which appear to combine the merits of both coincident and spaced microphones. Many recording engineers prefer spaced arrays because the omnidirectional microphones often used in such arrays tend to have a flatter and more extended frequency response than their directional counterparts, although it should be noted that spaced arrays do not *have* to be made up of omnidirectional microphones (see below). Mono compatibility of spaced pairs is variable, although not always as poor in practice as might be expected. Coincident pairs, from a superficial point of view, appear to promise better mono compatibility due to the lack of phase differences between the capsules, but in reverberant surroundings there is often a lot of pick-up in the out-of-phase region of such a pair, which suffers cancellation in mono.

2.6.3.2 'Decca Tree'
The so-called 'Decca Tree' is a popular arrangement of three spaced mono microphones with an omnidirectional (pressure) pick-up pattern. The name derives from the usage of this technique by the Decca Recording Company, although even that company would admit to not sticking rigidly to this arrangement in all circumstances. A similar arrangement is described by Grignon as early as 1949.

A 'tree' of three omnidirectional microphones is configured as shown in *Figure 2.89*, with the centre microphone spaced so as to be slightly forward of the two 'outriggers', although it is possible to vary

~ 75 cm

~ 138 cm

Figure 2.89 *The classic 'Decca Tree' involves three omnidirectional microphones, with the centre microphone spaced slightly forward of the outriggers.*

the spacing to some extent depending on the size of the source stage to be covered. The reason for the centre microphone and its spacing is to stabilize the central image which tends otherwise to be rather imprecise, although the existence of the centre microphone will also complicate the phase relationships between the channels, thus exacerbating the comb-filtering effects that may arise with spaced pairs. The advance in time experienced by the forward microphone will tend to 'solidify' the central image, due to the precedence effect, avoiding the 'hole in the middle' often resulting from spaced pairs. The 'outriggers' are angled outwards slightly, so that the axes of best HF response favour sources towards the edges of the stage whilst central sounds are on-axis to the central microphone.

2.6.3.3 Other spaced pairs
Spaced microphones with either omnidirectional or cardioid patterns may be used in configurations other than the Decca Tree described above, although the 'tree' has certainly proved to be the more successful arrangement in practice. The precedence effect begins to break down for delays greater than around 40 ms (depending on the nature of the source), since the brain begins to perceive the two arrivals of sound as being discrete rather than integrated, thus beginning to sound like an echo or 'phasey' effect. It is therefore reasonable to assume that spacings between microphones which exceed this delay between channels should be avoided. This maximum delay, though, corresponds to a microphone spacing of around 14 m, and such extremes have not proved to work well in practice due to the great distance of central sources from either microphone compared with the closeness of sources at the extremes, resulting in a considerable level drop for central sounds and thus a 'hole in the middle'. Furthermore, the Haas effect for longer time delays than about 10 ms begins to result in the sound being firmly in the leading speaker.

Work by Dooley and Streicher (1982) has shown that good results may be achieved using spacings of between one-third and one-half of the width of the total sound stage to be covered, although closer spacings have also been used to good effect. Bruel and Kjaer manufacture matched stereo pairs of omnimicrophones together with a bar which allows variable spacing, and suggest that the spacing used is somewhat smaller than one-third of the stage width (they suggest between 5 and 60 cm, depending on stage width), their principal rule being that the distance between the microphones should be small compared

with the distance from microphones to source. Informal experiments by the author suggest that interchannel delays of between 2 and 4 ms are adequate to give the impression that a transient sound is either fully left or right in the resulting image, depending on the nature of the signal. This agrees well with Lipshitz (1986) who also points out that for a signal to appear half-left or half-right, only about one quarter of the delay is needed compared with that needed for fully left or right, further reinforcing the likelihood of a 'hole in the middle' with spaced pairs.

Omnidirectional microphones are often used in addition to coincident pairs, as 'outriggers' to cover the extremes of an orchestra or choir, and to add some 'body' to the sound of the string section. This is hard to justify theoretically, and is beginning to move towards the realms of multimicrophone pickup, which is also hard to justify theoretically but which may be used for convenience and in order to produce a 'commercially acceptable' sound. Once more than around three microphones are used to cover a sound stage one has to consider a combination of theories, possibly suggesting conflicting sonic information between the outputs of the different microphones. In such cases the sound balance will be optimized on a mixing console and theory can to a large extent be relegated to a back seat, since what is now being created as a synthetic balance subject to the creative control of the recording engineer (see Section 2.6.4).

2.6.3.4 Binaural pairs and 'dummy head' techniques

The binaural hearing and reproduction process has already been thoroughly explained, and it is now intended to look into the means of sound pickup for binaural reproduction on headphones. Binaural recordings can have an uncanny realism when monitored on headphones, although they do not tend to work well on loudspeakers unless equalized to remove the specific pinna transforms and other 'head' effects specific to dummy-head recordings, for reasons already stated in the previous section although transaural processing, involving tailored cross-talk cancellation, may be a partial solution to this problem for listeners seated centrally between two speakers.

Binaural microphone techniques depend on the spacing of pressure microphones at the distance apart of a normal listener's ears, so as to produce two 'ear signals' which are as close as possible to a representation of the sound pressures at each ear. When these are reproduced at the same two ears on headphones, the result is a close approximation to the originally recorded soundfield, and thus a very life-like listening experience. Factors which must be considered are the effects of the pinna and ear canal on the sound to be picked-up and reproduced, as well as the similarity (or lack of it) between different people's ears. The use of artificial 'heads' and baffles to simulate the shadowing effect of the head should also be considered.

Small pressure microphones (such as those used as 'tie-clip' microphones, e.g. Sony ECM-50) may be mounted either flush with the side of a real head on a headphone-like headband, this being useful if a person wishes to make a recording using his own ear spacing, perhaps whilst walking around with a portable tape recorder. The shadowing effects of his own head would then be introduced between the microphones, although the effect of the pinna would be limited by the instrusion of the microphone and headband. Alternatively, a 'dummy head', such as the Neumann KU81i, may be used in which the microphones are mounted inside the ear canal, so that the response of the microphone is modified by the resonant and reflective properties of the outer ear, as well as by the resonant cavities of the head in some cases, and also reflections off the torso in the case of a 'head and torso simulator' such as the B&K Type 5930. Dummy heads vary in their degree of 'reality', some having imitation skin as well as the correct resonant and absorptive properties as a real head, in order that the result is as life-like as possible.

It is important to consider at which stage in the recording and reproduction process the pinna and ear canal effects will have been added – whether during recording or during reproduction. On replay, the existence of a headphone over the ear would largely suppress any pinna effects (although open-backed headphones may allow some to remain). Headphones particularly suppress the all-important conch response, the conch being the main resonant cavity of the pinna, disturbances of which have been shown to have a large effect on the perception of a sound as being 'at the ear'. To be accurate, the sound should only have been 'encoded' by the pinna and ear canal characteristics once, and thus it might be suggested that the most effective solution would be to mount microphones such that they are just at the entrance to the ear canal, in order that they are subject to pinna effects during recording, but not ear canal resonances (which are supplied on replay by the listener's own ears). For measurement purposes a slightly different approach applies, since it may be that the ear canal, pinna, and perhaps even inner

ear responses are required to affect the microphone output, since the aim is often to measure noise levels in a situation as close to that of human perception as possible, without the intention to replay that noise again on headphones. For this reason B&K makes two 'head and torso simulators', one for binaural recording and one for measurement (Type 4128). As described in Section 2.5, Theile has suggested a standard equalization characteristic for headphones to be used in high-quality listening, in an attempt to solve the difficulties outlined above.

Although not simulating all the head effects, a convincing binaural recording may be made using two microphones spaced apart by the ear distance (15–20 cm) and separated by a wooden or perspex baffle with a diameter of around 25 cm. This approximates to the shadowing and delaying effects of a real head.

2.6.3.5 Binaural microphones suitable for loudspeaker listening

As briefly described in Section 2.5.3.2, binaurally derived signals may be made more suitable for monitoring on loudspeakers by equalizing them for a flat frontal response in the free field (some sources argue for a flat response averaged over the frontal hemisphere, whilst others argue for a flat diffuse-field response). This has the effect of removing the severe coloration otherwise perceived when monitoring dummy-head recordings on loudspeakers, and also results in improved imaging without a 'hole in the middle'.

Schoeps has produced a microphone based on Theile's principles of 'head-related' stereo, the KFM 6U, which is essentially two pressure microphones mounted flush with the surface of a head-sized sphere in the ear positions. This microphone has a flat response to frontally incident sounds, but introduces time delays and head-shadowing effects for off-centre sounds which, it is claimed, produce accurate images on loudspeakers. Preliminary subjective tests conducted by the author confirm that the microphone produces a pleasing image with depth and spaciousness, although low-frequency images have been perceived to be less stable than those derived from coincident pairs. Crown has also produced a binaural microphone which produces signals suitable for loudspeaker reproduction, known as *SASS* (the stereo ambient sampling system) again producing head-related effects without the specific dummy-head equalization which results from such features as pinnae and ear canals.

2.6.4 Multimicrophone arrangements

We have so far considered the use of a small number of microphones to cover the complete sound stage, but it is also possible to make use of a large number of microphones, each covering a small area of the sound stage and intended to be sonically 'independent' of the others. In the ideal world, each microphone in such an arrangement would pick up sound only from the desired sources, but in reality there is considerable 'spill' from one to another. It is not the intention in this section to provide a full resumé of studio microphone technique, and thus discussion will be limited to an overview of the principles of multimicrophone pick-up as distinct from the more 'theoretically correct' techniques described above (sometimes known as 'purist' techniques).

In multimicrophone recording, each microphone feeds a separate channel of a mixing console, where levels are individually controlled and the microphone signal is 'panned' to a virtual position somewhere between left and right in the sound stage. The pan control takes the monophonic microphone signal and splits it two ways, controlling the proportion of the signal fed to each of the left and right mix buses. Typical pan control laws follow a curve which gives a 3 dB drop in the level sent to each channel at the centre, resulting in no perceived change in level as a source is moved from left to right due to the summation of sound intensities from left and right loudspeakers, as shown in *Figure 2.90*. The −3 dB panpot law is not correct if the stereo signal is combined electrically to mono, since the summation of two equal signal voltages would result in a 6 dB rise in level for signals panned centrally, and thus a −6 dB law is more appropriate for mixers the outputs of which will be summed to mono (e.g.

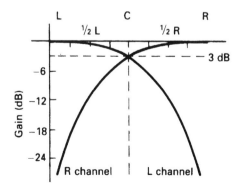

Figure 2.90 *Panpot law for sound source placement by a mixer.*

radio and TV operations) as well as stereo, although this will then result in a drop in level in the centre for stereo signals. A compromise law of −4.5 dB is often adopted by manufacturers for this reason.

Multimicrophone balances therefore rely on channel level differences, separately controlled for each microphone, to represent the positions of virtual sources in a synthesized sound stage, with relative level between channels used to adjust the 'strength' and thus perhaps the perceived distance of a source in a mix, although distance is more a function of the associated reflections and reverberant balance than of level. Artificial reverberation may be added to restore a sense of space to a multimicrophone balance. It is common in 'classical' music recording to use close microphones in addition to a coincident pair or spaced pair in order to reinforce sources which appear to be weak in the main pick-up, these close microphones being panned as closely as possible to match the true position of the source. The results of this are variable and can have the effect of 'flattening' the perspective, removing any depth which the image might have had, and thus the use of close microphones must be handled with subtlety.

The recent development of cheaper digital signal processing (DSP) has made possible the use of delay lines, sometimes as an integral feature of digital mixer channels, to adjust the relative timing of spot microphones in relation to the main pair so as to prevent this distortion of depth, and to equalize the arrival times of distant microphones in order that they do not exert a precedence 'pull' over the output of the main pair. As described in Section 2.5, it is possible to process the outputs of multiple microphones to simulate binaural delays and head-related effects in order to create the effect of sounds at any position around the head, when the result is monitored on headphones.

2.6.5 The Soundfield microphone

The AMS/Calrec Soundfield microphone (*Figure 2.91(a)*) is designed for picking up full periphonic sound in the A-format (see Section 2.5.5.2) and is coupled with a control box (*Figure 2.91(b)*) designed for converting the microphone output into both the B-format and the D-format. UHJ encoders are available for C-format encoding. The Soundfield microphone is capable of being 'steered' electrically by using the control box, either in terms of azimuth, elevation, tilt or dominance (see *Figure 2.91(c)*), and as

such it is also a particularly useful stereo microphone for two-channel work. The microphone correctly encodes directional information in all planes, including the pressure and velocity components of indirect and reverberant sounds, making it possible for these sounds to be reproduced correctly at the ears of the listener on replay.

2.6.5.1 Capsule arrangement

Four capsules with subcardioid polar patterns (between cardioid and omnidirectional, with a response equal to $2 + \cos\theta$) are mounted so as to face in the A-format directions, with electronic equalization to compensate for the intercapsule spacing, such that the output of the microphone truly represents the soundfield at a point (true coincidence is maintained up to about 10 kHz). The capsules are matched very closely and each contributes an equal amount to the B-format signal, thus resulting in cancellation between variations in inherent capsule responses.

The capsules are paired so as to produce figure-of-eight patterns, by subtracting (LF–RB), and (RF–LB), this resulting in crossed figure-of-eights at 90°, comparable with an AB pair in conventional stereo. Similarly different pairings are used to produce a vertical crossed pair of figure-of-eights, using (LF–LB) and (RB–RF).

(a)

Figure 2.91 *(a) AMS Calrec Soundfield microphone (enclosed). (Courtesy of AMS.) (b) AMS Calrec Soundfield microphone (control box). (Courtesy of AMS.) (c) Azimuth, elevation, and dominance in Soundfield microphone. (d) AMS Calrec Soundfield microphone (showing capsules). (Courtesy of AMS.)*

(b)

(c)

(d)

Figure 2.91 *(Continued)*

2.6.5.2 Derivation of B-format signals

In order to derive B-format signals from these capsule-pair signals it is a simple matter of using sum and difference technique, since it has already been shown that the sum and difference equivalent of a crossed pair of figure-of-eights is simply another pair of figure-of-eights rotated through 45° (see Section 2.6.1.3). Thus:

$$X = (LF - LB) + (RF - LB)$$

$$Y = (LF - RB) - (RF - LB)$$

$$Z = (LF - LB) + (RB - RF)$$

W, being an omnidirectional pressure component is simply derived by adding the outputs of the four capsules in phase, thus:

$$W = LF + LB + RF + RB$$

W, X, Y and Z are corrected electrically for the differences in level between them, so as to compensate for the differences between pressure and velocity components. For example, W is boosted at very low frequencies since it is derived from velocity capsules which do not have the traditionally extended bass response of omnis.

2.6.5.3 Synthesis of different polar patterns

When used as a stereo microphone, the combination of B-format signals in various proportions can be used to derive virtually any polar pattern in a coincident pair configuration, using a simple circuit as shown in *Figure 2.92*. Crossed figure-of-eights are the most obvious and simple stereo pair to synthesize, since this requires the sum-and-difference of X, and Y, whilst a pattern such as crossed cardioids requires that the omnidirectional component be used also, such that:

$$\text{Left} = W + (X/2) + (Y/2)$$

$$\text{Right} = W + (X/2) - (Y/2)$$

From the circuit it will be seen that a control also exists for adjusting the effective angle between the synthesized 'pair' of microphones, and that this works by varying the ratio between X and Y in a sine–cosine relationship.

2.6.5.4 Control of azimuth, elevation, dominance and inversion

The microphone may be controlled, without physical reorientation, so as to 'point' in virtually any direction. It may also be electrically inverted, so that it may be used upside-down.

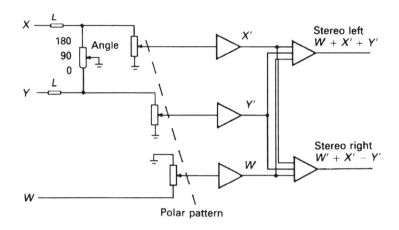

Figure 2.92 *Circuit used for controlling stereo angle and polar pattern in Soundfield microphone.* (Courtesy of Kenneth Farrar.)

Azimuth is controlled by taking X and Y components and passing them through twin-ganged sine–cosine potentiometers, as shown in *Figure 2.93*, and processing them such that two new X and Y components (X' and Y') are produced:

$$X' = X\cos\theta + Y\sin\theta$$

$$Y' = Y\cos\theta + X\sin\theta$$

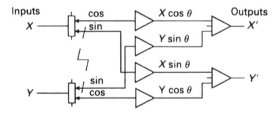

Figure 2.93 *Circuit used for azimuth control.* (Courtesy of Kenneth Farrar.)

Elevation (over a range of $\pm 45°$) is controlled by acting on X and Z to produce X' and Z', using the circuit shown in *Figure 2.94*. Firstly, a circuit produces sum and difference signals equivalent to rotations through $45°$ up and down in the vertical plane, and then proceeds to combine these rotated components in appropriate proportions corresponding to varying angles between $\pm 45°$.

Dominance is controlled by varying the polar diagram of the W component, such that it ceases to be omnidirectional, and becomes more cardioid, either favouring sounds from the front or from the rear. It has been described by the designer as a microphone 'zoom' control, and may be used to move the microphone 'closer' to a source by rejecting a greater proportion of rear pick-up, or further away by making W more cardioid in the reverse direction. This is achieved by adding or subtracting amounts of X to and from W, using the circuit shown in *Figure 2.95*.

Figure 2.94 *Circuit used for elevation control.* (Courtesy of Kenneth Farrar.)

Figure 2.95 *Circuit used for dominance control.* (Courtesy of Kenneth Farrar.)

Inversion of the microphone is made possible by providing a switch which reverses the phase of Y and Z components. W and X may remain unchanged since their directions do not change if the microphone is used upside-down.

References

ADER, C., Reprinted in *Journal of the Audio Engineering Society*, **29**, 368–372 (1981) from *Scientific American* (3 Dec. 1881)

COOPER, D.H. and BAUCK, J.L., Prospects for transaural recording, *Journal of the Audio Engineering Society*, **37**, 3–19 (1989)

CROSS, L., Performance assessment of studio microphones, *Recording Engineer and Producer* (Feb. 1985)

DOOLEY, W.L. and STREICHER, R., MS stereo: a powerful technique for working in stereo, *Journal of the Audio Engineering Society*, **30**, 707–718 (1982)

GERZON, M., Stereo shuffling: new approach, old technique, *Studio Sound*, July, 122–130 (1986)

GRIGNON, L.D., Experiment in stereophonic sound, *Journal of the Society of Motion Picture and Television Engineers* **52**, 280 (1949)

HIBBING, M., XY and MS microphone techniques in comparison. Presented at: *86th AES Convention*, Hamburg, Preprint 2811 (A–5) (1989)

LIPSHITZ, S., Stereo microphone techniques: are the purists wrong? *Journal of the Audio Engineering Society*, **34**, 717–735 (1986)

MIZOGUCHI, A. and ICHIKAWA, M., *Miniature Stereomicrophone (NHK Laboratories Note No. 213)*, NHK Research Laboratories, Tokyo (July 1977)

STREICHER, R. and DOOLEY, W., Basic stereo microphone perspectives, a review, *Journal of the Audio Engineering Society*, **33**, 548–556 (1985)

VANDERLYN, P., Auditory cues in stereophony, *Wireless World*, 55–60 (Sep. 1979)

WILLIAMS, M., Unified theory of microphone systems for stereophonic sound recording. Presented at: *82nd AES Convention*, London, preprint 2466 (H–6) (1987)

2.7 Analogue sound mixing equipment

2.7.1 Introduction

Sound mixing equipment comes in a wide variety of types and sizes, and is referred to in a sometimes confusing number of ways. There are audio mixers, sound desks, mixing consoles and control panels, used for mixing, recording, re-recording, broadcasting, production, postproduction, dubbing, editing, sound reinforcement and public address, and available in mono, stereo, on-air, self-operated, multitrack, film or theatre format.

The basic function of an analogue audio mixer is to combine two or more signals, each the electrical analogue of an acoustic waveform, into a suitable output signal. The mixer must therefore provide a means of adjusting the levels of the incoming signals and of the combined output, and also of monitoring the signals both qualitatively and quantitatively.

A simple mixer can be a small free-standing, portable or rack-mounted box; if it is mounted in a large control console housing other equipment it can be described as a mixing panel. Larger, more complex mixers are usually built as self-contained floor-standing units which may reasonably be described as mixing or mixer consoles or desks.

There are no universally accepted definitions of these various words, and the simple word mixer, with the appropriate functional adjective, will be used for all types in this section.

We will firstly examine some general considerations common to all types of mixers and describe the elements of a typical analogue audio mixer, before proceeding to detailed descriptions of some of the various specific types of mixer noted above.

2.7.2 General considerations

2.7.2.1 Microphone and line levels
Audio signal levels are measured in decibels (0 dBu = 0.775 V), and are generally categorized in two groups: microphone (low) level, and line (high) level. Although modern capacitor and electret microphones can produce very high levels, microphone level is usually taken to be around −70 dBu, compared with line level at 0 dBu (zero level). Whilst mixer inputs can be microphone or line level, mixer outputs are always at line level (unless specially attenuated, e.g. to feed into the microphone input of a second mixer).

In practice, line level can vary from less than −15 dBu on a poor telephone circuit to +24 or even +28 dBu (peak) on some mixer outputs, and microphone level can be as low as −90 dBu from a suspended ribbon microphone at a classical music concert or more than 0 dBu from a moving-coil or electrostatic microphone on a drum kit.

2.7.2.2 Dynamic range
The difference between the highest and lowest levels in an audio signal is referred to as the dynamic range of the signal. The widest dynamic range is encountered in classical music, where the difference between a full orchestral climax and a solo instrument played quietly may be greater than 75 dB; this contrasts with an experienced 'voice' who will produce a range of less than 10 dB with an absolute level accuracy of within 2 dB every time.

Although digital recording techniques can accurately reproduce dynamic ranges of more than 90 dB, current analogue broadcasting technology limits the usable dynamic range, even on the very high frequency (VHF) frequency modulated (FM) band, to about 45 dB, and there are endless debates between the purists who demand that the widest possible dynamic range be broadcast and the pragmatists who argue that ambient noise levels, particularly in motor vehicles (whose occupants represent a large proportion of the daytime radio audience), require the dynamic range to be restricted even more than the broadcast band currently requires.

It is worth noting in this context that to reproduce accurately concert-hall dynamic ranges in the home,

where the ambient noise level may be perhaps 10 dB higher than in the hall, would require listening levels approaching 120 dBA – which would in turn require a very substantial amplifier/loudspeaker system and very distant neighbours!

The dynamic range of audio signals must therefore be controlled for several reasons, including mixing together various sources at different levels and providing outputs suitable for specific listening conditions.

2.7.2.3 Balanced and unbalanced circuits
Audio signals, being very low in level (0 dBu is 775 mV) are very susceptible to interference from stray electrical and magnetic fields, e.g. lighting and power circuits, motors, etc. An unbalanced connection uses a single conductor for the signal, with the return current passing either through a coaxial wire mesh screen (as in a phono or 2-pole jack plug) or via a second unscreened wire (as in loudspeaker leads). Such a system has poor protection from external interference, and as the second conductor is usually connected to the 'ground planes' of the signal source and amplifier, a 'hum loop' will ensue if these two ground planes are also connected externally, as happens when both items are earthed via mains connections. If the hum is cured by disconnecting the loop, the two ground planes may be at different potentials which in turn causes distortion of the signal.

A balanced circuit uses two wires, one send and one return, both insulated from ground and connected to the equipment at each end by transformers (or their electronic equivalent) such that any interference (which is assumed to be picked up identically by both wires) is cancelled out (see *Figure 2.96*). This effect is enhanced by the two wires being twisted tightly together and covered by a conductive sheath of wire mesh or foil which is grounded at one end – a screened twin or pair. In addition, this arrangement isolates the different ground potentials, preventing the difference appearing as a signal, and is a major factor in eliminating ground loops (a major cause of hum) between connected items of equipment.

Microphones (except for some domestic types) are invariably 'balanced', as is most professional line-level equipment; amplifiers and other circuits within mixers are generally unbalanced, so that transformers or electronic balanced/unbalanced converters are required at the inputs and outputs of mixers.

2.7.2.4 Passive mixing
The earliest mixers were purely passive, i.e. with no electronics; the louder signals were attenuated to

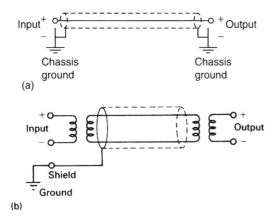

Figure 2.96 *(a) Unbalanced and (b) balanced circuits.*

match the levels of the quieter ones, and the resultant output was externally amplified as necessary to the required level for feeding the transmitter line or recording machine (see *Figure 2.97*). Such mixers worked equally well on microphone or line inputs, and fixed attenuators could be used to mix line signals with those from microphones. However, the results were not always satisfactory, as the output impedance (and therefore the output level) of such a mixer varies according to the positions of the controls, making mixing to a constant output level very difficult.

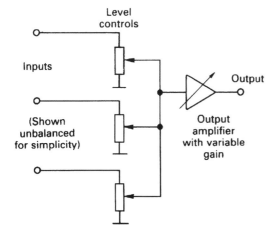

Figure 2.97 *A simple passive mixer.*

Modern mixers, therefore, have buffer amplifiers on the inputs and outputs to present constant impedances to microphones and to incoming and outgoing lines. Furthermore, the maximum usable range of a continuously variable audio level control is of

the order of 50 dB, due to the difficulty of achieving a repeatable calibration with normal production tolerances and also because any wider range with either a single-turn rotary potentiometer or a linear potentiometer of reasonable length would produce too great a change in level for the minimum movement physically possible.

This range is adequate for handling either line-level or microphone-level signals, but a mixer which is to handle both microphone and line-level signals simultaneously must provide some fixed or switched form of level adjustment before the mixing stage to bring the levels all to within the same 50 dB range.

2.7.2.5 Noise, distortion and overload margins

Noise is always present in any electrical circuit, and is a function of the amount of resistance in the circuit. Distortion occurs when an amplifier cannot accurately reproduce the signal at its input, and all amplifiers have a maximum output level above which peak distortion or 'clipping' occurs.

The difference between the maximum output level and the noise level of an electrical circuit is the dynamic range of the circuit: mixers are designed to have a 'normal operating level' at a point between these two levels. If this level is too low, the quiet parts of the signal may be lost in noise; if too high, the peaks may be clipped. The overload margin is the amount by which the signal level can exceed the normal operating level before peak distortion or 'clipping' occurs.

The specification of the operating level is thus a trade-off between a low noise level and a high overload margin: if a mixer has a maximum output level of +20 dBu and a dynamic range of 95 dB, its noise level will be −75 dBu for an operating level of 0 dBu and a 20 dB overload margin. If the operating level is raised to +4 dBu, the noise figure will improve to −79 dBu, but the overload margin will be reduced to 16 dB.

The definitions of operating level, zero level, line-up level and peak level are often a cause of confusion. Line-up level in UK broadcasting practice is zero level (0 dBu), which corresponds to PPM4. Peak level is +8 dBu, i.e. PPM 6. Using the definition above, the operating level is also +8 dBu, but this is not universally accepted, as many engineers would argue that the operating level is 0 dBu, with a peak level of +8 dBu. It is important when comparing specifications to ensure that the definitions are agreed, particularly when referring to noise level and overload margin.

The absolute maximum output level depends on the design of the output amplifier and the voltages available from the mixer power supply, and like all design criteria it is cost dependent. In broadcast systems where monitoring with PPMs and careful control of levels is the norm, a maximum output of +18 dB (providing an overload margin of 10 dB) may be considered adequate. In recording studios, with mixers fitted with VU meters feeding into analogue tape machines (said by many engineers to provide the best form of compressor/limiter available!), peak levels may exceed +20 dBu, and an output capability of at least +24 dBu is usually demanded.

However, digital recorders cannot tolerate overloading, and accurate peak metering has become an essential requirement in digitally equipped recording studios, with the extremely wide dynamic range putting a premium on low noise levels rather than high overload margins.

The noise level of a mixer depends largely on the number of channels in use; a typical high-quality stereo broadcast mixer with 16 inputs could be expected to have a noise floor of −74 dBu with all channels in use, reducing to −86 dBu if only one channel is faded up – a maximum usable dynamic range of 94 dB (86 + 8 dB).

A typical mixer designed for music recording might have 60 or more input channels, each contributing to the noise figure, so the resultant noise floor would be −66 dBu or worse. With a maximum output level of +24 dBu, a peak operating level of +4 dBu (0 VU) would provide 20 dB of output headroom and a dynamic range of 70 dB.

Overload can occur at several points within a mixer, in addition to the inputs and outputs, particularly if the mixer is operated with its gain controls outside their specified ranges. The gain structure of a mixer is a careful compromise between noise and distortion, and the settings of the various gain controls in the signal chain are part of the design specification. If a particular control is designed to operate fully open but is set say 10 dB below maximum, other controls in the chain must be adjusted to compensate if the output level is to be maintained. This means that the associated amplifiers must produce 10 dB more gain than specified, and thus either the overload margin is reduced or the noise level is increased by that amount.

Input fader settings can often differ for microphone and line inputs. Line inputs are generally from sources which have already been controlled in level, e.g. prerecorded tapes or discs, and other studios, and the faders can be set fully open for normal working, whereas microphone inputs usually require constant adjustment, both up and down, to compensate for changes in the level of the incoming music or speech; faders on microphone channels are therefore designed to be set with 10 dB 'in hand', i.e. the normal operating or 'line-up' level is obtained with the fader 10 dB below maximum.

When the incoming level increases, the fader setting is reduced to compensate; the amount by which the fader level can be reduced before the onset of distortion in the input amplifier is the input overload margin. This factor is of prime importance when mixing classical music with its very wide dynamic range, and an overload margin of at least 35 dB is generally regarded as essential, as it is unacceptable to have to adjust the gain of the input amplifier during a 'take'.

2.7.2.6 Line mixers

In its simplest form, a line mixer accepts two or more signals at line level and provides a single line-level output; level controls are provided for each input and for the mixed output. Line-level mixing is commonly used in editing suites, etc., where the sources are audio and video tape machines, effects devices and incoming lines from other studio areas. In practice, this simplest form is rarely found; the minimum requirement is for a secondary output with an independent level control for feeding a monitor loudspeaker or headphones and for some form of visual level indication such as a moving coil or bargraph meter.

Where the line mixer is used to feed a recording machine, it is convenient to be able to use the mixer's monitor output to listen to the output of the recording machine, either intermittently or continously during the recording for 'confidence' purposes or during a playback. To facilitate this, an external input may be provided, accessed by a switch in the monitor circuit; this is referred to as the 'tape monitor' or 'line/tape' switch, and often as the 'A/B' or 'B check' switch, A being the mixer output and B the output of the tape machine (*Figure 2.98*). This latter nomenclature can cause confusion when considering stereo systems, where A is used to denote the right-hand signal and B the left-hand signal.

2.7.2.7 Microphone mixers

A microphone mixer can be merely a line mixer with gain at the inputs; the necessary amount of gain, however, varies considerably, and a mixer with fixed gain which is designed for speech work at

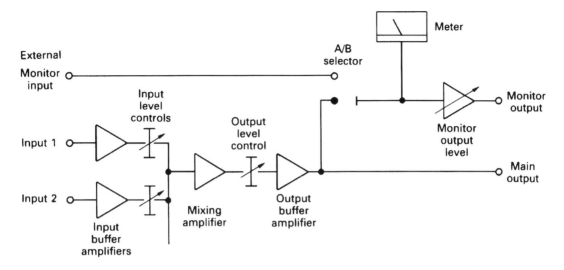

Figure 2.98 *A simple line mixer with monitor A/B switching.*

input levels of around −60 dBu will overload and distort badly if presented with music at upwards of −20 dBu. A multipurpose microphone mixer must therefore have a variable gain input amplifier before the mixing stage.

2.7.2.8 Microphone mixers

The optimum input impedance for microphones is generally stated to be approximately 1 kΩ, whereas most line circuits are designed for a bridging impedance of 10 kΩ or more. Having a single input with variable gain will therefore not produce optimum results for both microphone and line inputs; either two separate input sockets must be provided, or the input impedance must be changed as necessary.

It is often convenient to have two inputs to one mixer channel, dedicated to microphone and line, respectively, with a switch to select the one required. The microphone input can connect directly with the input amplifier, providing the correct input impedance with short leads minimizing the risk of external noise being injected, and the 'microphone/line' switch selects either the output of the microphone amplifier or the feed from the line input socket to be connected to the mixing stage.

Another design solution employs a balanced attenuator 'pad' in the line input circuit which presents a 10 kΩ impedance to the line and attenuates the signal before it is connected internally to the microphone amplifier.

These types of solution enable two sources to remain connected, effectively doubling the capacity of the mixer whilst ensuring optimum performance on both microphone and line circuits. These methods are illustrated in *Figure 2.99*.

Microphone leads are usually terminated with three-pole XLR-type connectors, whereas line connectors may be either XLRs or jack plugs, or even phono plugs if unbalanced domestic or semiprofessional equipment is being used. Some small multipurpose mixers therefore have XLR connectors for microphone inputs and jack or phono sockets for line inputs and outputs. In professional mixers it is more usual to find XLR connectors for all terminations, except in permanent studio installations where multipin connectors are often favoured for reasons of economy and ease of wiring.

2.7.2.9 Phantom power

Most mixers provide a 48 V d.c. supply connected by current-limiting resistors to the balanced microphone input for powering capacitor microphones, with an on/off switch on each channel module. Problems can arise if the microphone inputs are connected via a jackfield, as short circuiting the supply to ground (which can easily happen if a jack tip touches a socket outer) can result in the microphone transformer becoming magnetized, which can cause it to become sensitive to physical impacts and vibration (an effect known as 'microphony'). This problem can be avoided by connecting the phantom power to the microphone side of the jackfield instead of to the mixer.

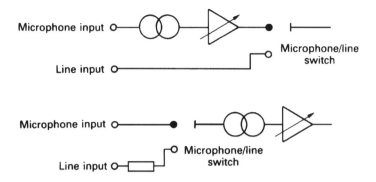

Figure 2.99 *Two approaches to microphone/line selection.*

2.7.2.10 Jackfields/patchbays

The most commonly used device for interconnecting and intercepting audio circuits is the $\frac{1}{4}$-in. (6.35 mm) jack. There are two types in use: A gauge and B gauge.

A-gauge jacks may be either two-pole (unbalanced) or three-pole (balanced line or microphone, or unbalanced stereo) and are of constant diameter, usually constructed from thin steel tubing finished in chromium plating. Their use is restricted to domestic and non-broadcast applications, e.g. electrical instruments, headphones, etc.

B-gauge jacks, which derive from a British Post Office Standard (BPO 316) are used in all professional applications, and are always used in three-pole configuration (tip, ring and sleeve), manufactured from solid brass. The tip and ring are of smaller diameter than the sleeve, and thus the two types are not interchangeable.

Standard $\frac{1}{4}$-in, jack sockets are usually mounted in horizontal rows of 20–26 jacks in a standard 19 in. (482 mm) width panel, and have a pair of 'normalling' contacts which allow a circuit to remain connected until a plug is inserted in the socket. Miniature or 'bantam' jacks of diameter 4.4 mm are also available, allowing up to 48 jacks in a single row, with identical normalling facilities.

Interconnection between jack sockets ('crosspatching') is made using a flexible three-core cable fitted with an appropriate plug at each end – a 'double-ender' or 'patch-cord', hence the term 'patch panel'.

Most mixers, except for portable types, have an associated jackfield, either integral to the mixer frame or fitted nearby in a wall-mounted rack or free-standing console, used to terminate incoming lines from remote studios, tape machines, etc., and to provide access to the mixer line inputs and outputs, insert points, etc. (see Section 2.7.3.4).

'Normalling' is used where connections are normally made but may need interrupting; full normalling requires jacks in both sockets to break the circuit, and half normalling requires only one jack (see *Figure 2.100*).

(a)

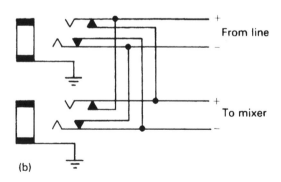

(b)

Figure 2.100 *Normalling jack sockets. (a) Half normalling; (b) full normalling.*

A 'listen' jack allows a plug to be inserted without breaking the circuit; dedicated listen jacks have no normalling contacts and are wired in parallel with the

circuit which is to be monitored. The 'source' jack of a full-normalled pair is effectively a listen jack.

For many years jackfields were regarded as unacceptable for low-level (microphone) circuits by some recording studios, and microphone inputs used XLR connectors direct into the mixer. However, the BBC has always used jackfields for connecting microphones to mixers, the only difficulties encountered being with normalling contacts becoming tarnished. If sockets and plugs are cleaned regularly, jacks are perfectly acceptable for microphone circuits, provided that normalling is not used in areas where corrosion can be a problem.

2.7.2.11 Mono and stereo

Since stereo became the norm, it is rare to find mixers that are mono only – indeed most manufactures have for many years considered a mono mixer as a custom requirement costing more than the standard stereo unit.

A stereo mixer must obviously have two outputs (left and right) but the input channels may still be mono, with panpots to adjust the signals to the required stereo position, or switches to select them to left, right or centre (both switches selected). Stereo inputs can be accommodated by using two adjacent mono channels, selecting one to each output (either by panning fully left and right or by switching), and ensuring that the gain controls are set equally and moved in exact synchronization.

Stereo input channels are effectively two identical mono channels combined in one package, with ganged controls; the left and right inputs are permanently connected to the left and right outputs of the mixer via a dual-ganged level control, with a 'balance' control to adjust the relative left/right levels if necessary to correct for any deficiencies in the input signal or the level control ganging.

Stereo mixers often comprise a mixture of mono and stereo channels, since many microphone sources (e.g. voices and solo instruments) are mono, and most prerecorded sources, e.g. tapes, discs, etc., are stereo.

2.7.2.12 Monitoring

Monitoring is carried out both visually and aurally – by meters for monitoring the level of the signal and by loudspeakers or headphones for monitoring content and quality. The minimum requirement for monitoring the signal level is for a pair of meters (or a single stereo meter) connected across the main output. All modern meters have integral buffer amplifiers which protect the mixer output from degradation.

If the mixer has an external monitor input (see Section 2.7.2.6) it is usual to connect the meters after the monitor selector switch; when the output must be monitored without interruption, e.g. in on-air mixers it is common to find dedicated meters on the main output and a second pair of meters for monitoring other sources. Additional meters may be also fitted for monitoring auxiliary and multitrack outputs, etc.

Meters are almost always mounted in the body of the mixer; monitor loudspeakers are invariably external (the only loudspeakers fitted in mixers are for special purposes such as communications (see Section 2.7.3.13) or 'signal present' checking (see Section 2.7.3.12)). A mixer's 'monitor output' is a buffered feed of the main output or of the monitor selector (where fitted) via a level control, designed to be connected to loudspeaker power amplifiers or to drive headphones directly – depending on the type of mixer.

It is common practice on larger mixers to provide separate independent outputs for loudspeakers and headphones, with individual level controls for each, and additional feeds for loudspeakers and headphones in the studio.

2.7.2.13 PPMs and VU meters

VU stands for 'volume unit'; the VU meter (*Figure 2.101*) was developed in the USA as a means of indicating the relative loudness of audio signals in radio broadcasting and recording, and was designed to bridge 600 Ω lines with no internal amplifier. The device uses a rectifier/moving coil movement and is scaled from 0 to 100% at two-thirds full scale, the top third being marked in red as +1 dB, +2 dB and +3 dB above 100%. The movement has specially defined ballistics, and a reading of 100% (0 VU) is obtained on steady-state tone at +4 dBu, i.e. 1.3 V.

Figure 2.101 *A volume-unit (VU) meter.* (Courtesy of Sifam Ltd.)

The PPM (peak programme meter) (*Figure 2.102*) was developed by the BBC to indicate near-instantaneous peak levels in an easily read format; it is scaled white on a black background with marks from 0 to 7, PPM 4 being 0 dB (775 mV) and each graduation being 4 dB (with infinity from 0 to 1). Peak level of +8 dB is thus PPM 6. In some countries the EBU scale is used; the graduations are similarly spaced but are marked in decibels from −12 to +12 dB (*Figure 2.103*). The meter movement is carefully specified and is driven by a logarithmic amplifier which gives it a rise time of 3 ms and a fall-back time of 3 s enabling fast transients to be easily monitored.

Although peak level on a VU meter reading steady-state tone (0 VU or 100% modulation) is equivalent to PPM 5, readings on programme material differ greatly, due to the different characteristics of the devices; speech peaking at PPM 6 will show only −6 to −10 VU, a difference of up to 14 dB. This difference varies according to the type of programme material, with sustained sounds such as organ music giving fairly comparable readings and percussive sounds such as piano and guitar producing the widest variations.

Although many operators maintain that experience enables them to assess levels satisfactorily using VU meters, there is no doubt that PPMs are preferable where accurate monitoring of peak levels is required, the only negative factor being the high relative cost due to the special movement and logarithmic amplifier involved.

In many types of mixer, the 'traditional' moving-coil meters with needles are still used, but since the 1970s an increasing number of linear bar-graph displays have become available, ranging from very cheap LED segment displays to elaborate multichannel devices using colour video monitors.

To achieve adequate resolution for professional monitoring (say 0.5 dB over a 30 dB range) a bar-graph display needs at least 60 segments, and current high-quality gas-plasma designs provide typically 100 segments in a display of length 5–6 in. (130–180 mm), which may be mounted either vertically or horizontally.

As such displays were not designed initially as audio meters, driver amplifiers are required to simulate either VU or PPM characteristics, and a single display can provide two or more meter characteristics, switchable at will. Some large multitrack mixers even provide a spectrum analysis facility using a number of vertical meters as level indicators for various frequencies.

Figure 2.102 *A peak programme meter (PPM)*. (Courtesy of Sifam Ltd.)

Figure 2.103 *A PPM with an EBU scale*. (Courtesy of Sifam Ltd.)

In television control rooms, to enable the mixer operator to monitor audio levels whilst watching the picture, there are various software packages which superimpose facsimile meters, either bar-graph or moving needle, on the screen, and for multichannel monitoring a dedicated visual display unit (VDU) can be used to show a colour bar-chart level display – if budgets permit!

2.7.2.14 Communications

In a fixed installation, the mixer operator usually needs to communicate with artists, presenters, etc., in the studio, and sometimes with other personnel in remote locations. 'Talkback' is a system of audio

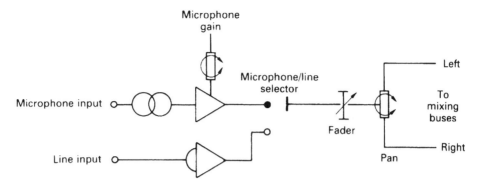

Figure 2.104 *A 'minimum' mono input channel.*

communication separate from the signal paths within the mixer; in its simplest form it allows the operator to talk to the studio via a dedicated microphone and loudspeaker. Reverse talkback enables the studio presenter to speak to the mixer operator without using the programme circuit.

Although the terms 'talkback' and 'intercom' are often used interchangeably, an intercom can often be an 'office' system which is not inherently suitable for professional use and only provides point-to-point communication, often with severe restrictions on the number of calls that can be made simultaneously. 'Talkback' implies a professional-quality system which should have clean, wide bandwidth audio, external muting facilities, call indication and the facility to call several or all destinations simultaneously (an 'omnibus' system), with no undue limitations on the number of simultaneous point-to-point calls which can be made.

Communications are of vital importance in broadcast mixers, where the operator may need to talk to other studios, outside broadcast locations, telephone callers, etc.; it has been said that a broadcast mixer is 80% communications and 20% mixing, whereas a music recording mixer is 20% communications and 80% mixing!

2.7.3 The typical mixer

As we have seen, a typical mixer may have mono and stereo inputs suitable for microphone and/or line levels, a stereo main output for feeding the destination, a separate monitor output for feeding loudspeakers or headphones, and some form of talkback.

2.7.3.1 Mono inputs
Each mono input 'channel' comprises, at minimum, a microphone amplifier with balanced input and

adjustable gain, a line input stage (either transformer or electronic balance/unbalance), a microphone/line switch, a level control (fader) and a panpot or left/right output selector switches (*Figure 2.104*). In addition it may have fine gain trim, phase change, one or more dynamic range and frequency response processing sections (compressor/limiter/expander/gate, high- and low-pass filters, simple or 'parametric' equalization), an insertion point, a direct postfader output and at least one auxiliary output with level control and pre-/post-fader switching (*Figure 2.105*).

The inclusion of such processing sections and auxiliary outputs, etc., depends on the use for which the mixer is designed (and also on the cost/benefit ratio). For example, a simple line mixer handling prerecorded material in an editing suite will not require any such additions – indeed their availability is often regarded as undesirable, as they make the mixer more complicated to operate and may tempt operators to make unnecessary changes to the sound signal.

At the other extreme, large recording and post-production mixers are usually equipped with every device which can be physically fitted in the space available, on the grounds that the time saved by having the correct processing available at the operator's fingertips (instead of having to bring in and connect up external equipment, often incurring a rental charge) more than justifies the increased cost of the mixer.

Any dynamics, equalization and filter sections must be provided with in/out switches to allow each section to be bypassed if required; switches are often also provided for the insertion point and the auxiliary outputs.

2.7.3.2 Stereo inputs
As noted in Section 2.7.2.11, a stereo channel is essentially two mono channels with ganged controls.

Figure 2.105 *A typical 'full-facility' mono channel.*

As the majority of stereo sources are line level, stereo channels are often line only, with little or no processing (as the prerecorded sources should not, in theory, require any).

A balance control is fitted in place of the panpots used on mono channels; this usually allows a swing of about 6 dB either side of centre to enable the operator to compensate for errors in recording machine line-up, etc. Where stereo channels are likely to be used for mono input signals, e.g. half-track tapes, mono land-lines, etc., 'mono L' and 'mono R' switches are fitted to connect one input to both channel outputs and disconnect the unused input to eliminate the possibility of spurious signals being received. In addition, a left/right switch may be fitted to correct errors in track selection in the original recording or the studio wiring; this swaps the left and right channels in a stereo signal.

In some mixers, selecting mono changes the function of the balance control to a panpot, enabling the mono input to be positioned in the stereo soundfield instead of being centred. A further refinement, most often used in music recording using stereo microphones, is a width control, which can narrow or widen the apparent width of the stereo signal by mixing in either sum or difference signals derived by an AB/MS matrix.

The advent of stereo sound on television coupled with the increasing use of stereo microphones has increased the pressure on manufacturers to supply 'full-facility' stereo microphone and microphone/line channels, despite the high cost. Fortunately, the concurrent advances in electronic design, including the reductions in size enabled by surface mount technology, etc., have made it possible to design a stereo channel providing almost all the facilities found in the most complex mono channel in a module of identical size and much less than twice the cost.

Indeed, some manufacturers now use identical printed circuit boards for both stereo and mono channels, 'subfitting' for mono and reduced-facility stereo channels.

Another facility resulting from the increasing demand for stereo is MS/AB conversion. The MS format is used increasingly for news and other recording where mono compatibility is vital, and tapes are often presented to the mixer operator in this format. Within the mixer, the processing must be done in the standard AB format, so it is convenient to have MS/AB conversion available in the stereo channel, both for using stereo microphones in MS format and for accepting prerecorded MS material (see *Figure 2.106*).

2.7.3.3 Processing

It is rare to find a mixer with only the basic input selection, routing and mixing facilities; most mixers have at least a basic 'equalizer' in every channel, with additional filters and compressor/limiter/expander/gate systems according to the operational requirements and price considerations.

It is perhaps worth pointing out that the word 'equalizer' was originally used to describe an amplifier with variable gain at selectable frequencies which was used to correct for deficiencies in land lines connecting studios to transmitter sites or other broadcasting areas. 'Equalizing' the circuit meant restoring its frequency response to as linear a state as possible; the modern use of the term usually implies beginning with an audio signal which has been derived from

Figure 2.106 *A typical stereo line channel showing left/right/mono switching, etc. The auxiliaries, inserts, prefader listen/level, etc., are omitted for clarity.*

high-quality microphones and tape machines with the flattest of responses, and changing it (sometimes beyond recognition!) to suit artistic considerations.

Equalizers are in fact 'tone controls'; it is regrettable (but perhaps understandable) that the BBC term 'response selection amplifier' (RSA) did not gain acceptance, as it describes the process precisely.

2.7.3.4 Insertion points

The complexity of built-in processing equipment is always limited by budgetary and size limitations, and to allow more complex external processors to be used in the signal path, most mixers have 'insertion points'. Ideally these will provide balanced outputs and inputs at zero level (0 dBu) which remain connected together unless required. This is usually done by wiring the insert point via a jackfield using normalled sockets, or by providing a prewired 'jumper' connector if the mixer is supplied without an integral jackfield.

The minimum requirement is for an insert point in every channel, usually just before the fader. Insert points in the groups and main output, again prefader, are useful for connecting compressor/limiters to control the mixer output level. More complex mixers may have several insert positions in each channel, selected by switches or internal jumper connections; postfader inserts are useful in main outputs for 'last-ditch' limiters where line overload must be avoided.

2.7.3.5 Auxiliary outputs

In addition to the main output of the mixer channel, which goes via the channel fader and panpot to the groups (where fitted) and main output, there is often a requirement for additional outputs to be sent to other destinations. e.g. effects devices such as reverberation and delay units, studio foldback systems, and telephone hybrids.

These auxiliary outputs may be derived either before or after the channel fader according to the requirement: most mixers have at least two, usually with independent level controls and pre-/postfader selection feeding auxiliary buses which connect to auxiliary output amplifiers with level controls. If stereo auxiliary buses are used, the controls on the mono channels must also include a panpot for each stereo 'auxiliary send', as both pre- and postfader auxiliary signals are derived before the channel panpot.

2.7.3.6 Direct outputs

In addition to the mixed auxiliary outputs, separate independent channel outputs may also be required, e.g. to feed another mixer in parallel, to send to a multitrack recorder, or for deriving clean feeds (see Section 2.7.3.7 below). These 'channel direct outputs' are normally derived after the channel faders and are at line level, though in some more complex mixers they may be optionally derived pre-EQ, sometimes with adjustable gain.

2.7.3.7 Clean feed/mix minus

When broadcasting telephone calls or connecting studios together for a simultaneous multiway programme, each remote source needs a feed of the programme mix without its own contribution, otherwise feedback may occur.

This 'clean feed' or 'mix minus' can be provided by using an auxiliary bus and setting appropriate auxiliary send levels on each channel except the one with the incoming contribution, which is set to zero. It is more usual on broadcast mixers to provide an additional bus dedicated to clean feed, connecting all channels except the remote source channel to this bus postfader.

If several clean feeds are required, e.g. for multiple phone-in programmes, then either the necessary number of buses must be provided, and the mixer prewired with certain channels specified for the telephone inputs, or a phase cancellation system must be used.

This system uses a single bus, to which all the channels required in the clean feed mix are selected, including the phone-in channels (as each caller may need to hear all the others as well as the studio contributors). The required clean feeds are manufactured in each channel by mixing an output from the clean feed bus in antiphase with that channel's output, thus cancelling out the signal from that channel.

2.7.3.8 Mixing buses
The input channels are mixed together by connecting them all to mixing buses (left and right), via buffer amplifiers to eliminate interaction between channels, and the resultant signals are amplified and fed to the outputs via further buffer amplifiers (and group buses and faders when fitted) and the main or master output fader.

(*Note*: the word 'bus' is an abbreviation of 'busbar' which is 'an electrical conductor usually used to make a common connection between several circuits', according to the *Collins Concise Dictionary*. ('Buss', as it is frequently misspelt, is 'an archaic or dialect word for a kiss'.) The plural may be spelt either buses or busses, according to *Collins* – but why use two letters when one will suffice?)

Mixing buses, being at low level, are as susceptible to noise and external interference as microphone cables, and in many mixers, particularly the larger types, the buses are balanced, the additional component cost being repaid in better performance.

2.7.3.9 Groups
In large mixers with many input channels it is often convenient to be able to control the overall level of several channels without affecting the others; additional mixing buses may be used to provide subgroups which themselves are combined to feed the main output. Some stereo mixers may have mono groups for reasons of cost, in which case they must be used in pairs; it is important to ascertain that the groups are stereo when comparing facilities requirements, as a 'four-group' mixer may in reality only provide two groups in stereo.

In mixers with voltage-controlled amplifier (VCA) devices in the channels, either as an integral part of the channel gain control or as part of an automation system, VCA grouping is possible. A VCA group is selected by linking together the control voltages of the VCAs in the desired channels so that any one fader controls all the linked channels simultaneously, without having a separate group fader. Without an automation system which identifies each channel, only one VCA group is possible without dedicated VCA group master faders.

2.7.3.10 Main outputs
The main outputs are simply a set of amplifiers for collecting the signals from the mixing buses and providing outputs at zero level and low impedance for feeding to tape machines or line sending amplifiers. Mixer main outputs may be mono, stereo or, occasionally, four or six tracks for film working. Most mixers except broadcast on-air types have main faders; stereo mixers generally have a single stereo main fader rather than two mono faders in order to minimize image shift when fading up or down (see *Figure 2.107*).

2.7.3.11 Meters and monitoring
As noted in Section 2.7.2.12, monitoring means watching the level of the signal using some form of visual display (usually moving coil or bar-graph meters) and listening to it via loudspeakers or headphones.

The monitoring output is used to feed loudspeakers and/or headphones via a monitor level control; there are usually also switches for dim, mono, and phase reverse, and a means of selecting at least one external input (the A/B or line/tape switch mentioned in Section 2.7.2.6). 'Dim' provides a fixed drop in monitor level of 12 dB or more, depending on the use for which it is required. It is often useful to be able to reduce the monitoring level in order to carry on a telephone conversation whilst continuing to monitor a programme or recording, and to be able to return to precisely the same level as before when the conversation is finished – not easy using a continuously variable level control. The amount of dim required will vary according to the monitoring level being used, and most mixers provide a means of adjusting the dim level, either with an internal preset or by links or resistor changes in the monitor module.

Another use for the dim facility is to provide alternative monitoring levels for checking the balance under different listening conditions; it is common practice in music mixing to monitor at very high listening levels, and it is necessary to check regularly that the mix is satisfactory at lower (domestic) levels. Some large mixers have two levels of dim,

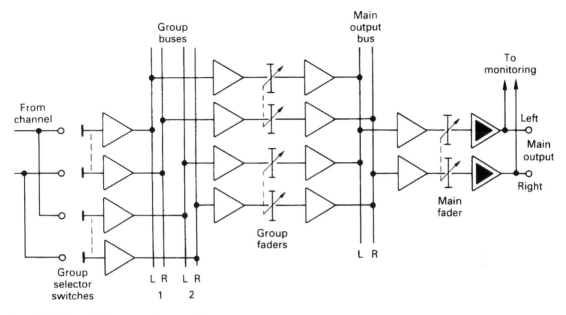

Figure 2.107 *A typical two-group stereo output.*

one of about 12 dB for mix comparison and a bigger drop of perhaps 18 dB or more for phone calls, etc.

Mono is self-explanatory, enabling the operator to check that the mix is satisfactory for listeners using mono equipment without affecting the stereo main output.

Phase reverse changes the phase in one side of the monitor output (usually the left) to enable the operator to check for phase errors in incoming programme or in microphone feeds, etc.

The selection of either the mixer output or an external input to the monitoring chain (A/B or line/tape) can include a means of monitoring other sources, including the mixer auxiliary outputs, the outputs of other studios, off-air receivers, etc. A single switch selects either the mixer main output or the output of a multi-input selector, usually a set of latching switches (see *Figure 2.108*).

It is usual to provide additional, separate, monitoring facilities for the studio. These may be a separate feed of the mixer output or of the monitor selector output via an independent level control, another independent monitor selector, or even two independent monitor chains, one for the studio loudspeaker(s) and one dedicated to studio headphones.

These studio monitor feeds obviously do not have mono, dim or phase reverse facilities, but the studio loudspeaker feed must have a *cut* facility linked to the mixer microphone channels to prevent feedback and 'howl-round' which can occur if a live micro-

phone is fed back to an adjacent loudspeaker at a high enough level.

Where artists and studio presenters need cue feeds for overdubbing, etc., it is usual to have one or more 'fold-back' systems; these are dedicated auxiliary bus feeds from the mixer which enable the operator to provide a 'personalized' mix of the required channels to selected sets of headphones or loudspeakers. Some specialized music recording and large broadcasting studios provide the studio artists with their own foldback mixers so that the mix in their headphones is truly 'personalized'!

2.7.3.12 PFL, AFL and solo

These facilities are primarily monitoring functions, but are sufficiently important (and often confusing enough) to warrant a section of their own. PFL is prefader, AFL is afterfader listen/level; the choice of listen or level depends on whether the user is more concerned with the quality and/or content of the signal or with its level.

PFL is a combined channel and monitoring function; on a mixer with PFL, each channel has a switch which can connect the channel signal path, at a point just before the fader, to the prefade bus. This bus is picked up in the monitor module and made available as an alternative signal to the mixer output. On some early mixers the PFL had to be selected by a switch on the monitor module; automatic PFL switching is now almost universal, effected by an

Figure 2.108 *A typical monitor system.*

additional bus carrying a logic voltage activated by the channel PFL switch. This triggers an electronic switch in the monitor module, selecting the output of the PFL bus to the monitor path whenever PFL is selected on a channel.

Some mixers have a small integral loudspeaker which is dedicated to PFL (or sometimes shared with reverse talkback which overrides PFL via an electronic switch). PFL always appears on this loudspeaker (unless reverse talkback is being used) and also on the mixer headphone output, but may not appear on the main monitor loudspeakers until 'PFL to LS' is selected (see *Figure 2.108*).

If the mixer is fitted with meters wired to follow the output of the monitor module, these will show the level of the prefader signal. Some mixers have a dedicated meter for PFL which is only active when PFL is selected.

In a stereo mixer, the signal in a mono channel does not become 'stereo' until after the panpot, which is usually after the fader; if the panpot is before the fader a stereo fader must be used, which adds to the cost. Stereo PFL also means additional cost, and was therefore regarded as rather a luxury for some years, with most stereo mixers having a

mono PFL bus which used a derived mono feed from any stereo channels.

PFL is used to check the level and quality of a signal before adding it to the mix by 'fading it up'. It is primarily a broadcasting function, allowing the operator to adjust the channel input gain control, and the equalizer, if necessary, so that a signal which has not been heard during the rehearsal will be at approximately the correct level and of acceptable quality when faded up into the mix 'on air'.

In recording studios there is rarely such urgency, and the operator is usually concerned with the level and quality of the signal from a particular microphone after it has been processed by the channel dynamics and equalizer systems, before committing it to multi-track tape for eventual remixing, or from a previously recorded tape, perhaps from another studio.

The 'solo' function singles out the required channel by simply muting all the other channels, leaving the selected channel to be monitored on its own. This is 'solo in place', as the channel signal remains exactly as panned; some mixers allow 'additive solo', i.e. several channels may be selected at a time, and monitored as they are mixed. As this function is dangerous if misused (it is sometimes called 'destructive

solo') a 'solo safe' switch is sometimes fitted to disable the function during recordings.

Obviously solo has no place in a broadcast mixer, but it is sometimes desirable to single out one channel from a mix without a change of level, which almost inevitably happens if prefade listen is used. AFL provides this facility; the system is identical to PFL except that the signal is derived after the channel fader instead of before. As it is in fact derived after the panpot in mono channels, AFL can be in stereo even where PFL is in mono.

2.7.3.13 Talkback

As noted in Section 2.7.2.13, talkback enables the mixer operator to communicate with the studio, and most mixers, except the smallest portable and edit suite units, have some form of talkback facility.

The minimum requirement is for an integral microphone with associated amplifier and gain control, and a 'press to talk' switch, the output of the amplifier being connected to the studio loudspeaker feed. This can be achieved by routing the talkback directly into the mixer output, either mixing in with the programme or overriding it via a relay or solid-state switch. As the studio loudspeaker is normally fed from the mixer output, it will receive the talkback whenever the switch is pressed; but so will the mixer output.

If the mixer is feeding a tape recorder, the result will be to record the talkback; this is used extensively to identify tapes and different 'takes' on a tape, and is often referred to as 'slate', a cross-over from the film industry where the name of each 'take' is written on a slate which is held in front of the camera. However, this is obviously not desirable in the middle of a take, and especially not during a live broadcast, so many mixers have some form of inhibit in the talkback circuit which prevents the system being used when the mixer is in 'record safe' or 'transmission' mode. This is usually activated by switching on a local red light.

If talkback is used when a studio microphone is 'live', i.e. faded up and being monitored on the control room loudspeaker, the conditions exist for acoustic feedback from the studio loudspeaker through the microphone and control room loudspeaker back to the talkback microphone. If the overall system gain is above the critical level, a 'howl-round' will occur, and to prevent this, the mixer must have a system for cutting the control room loudspeaker when talkback is used. Some systems dim the speaker instead of cutting it completely, allowing a two-way conversation to be carried on; this is only possible with low levels of microphone gain, and in studios set up for normal speech work a 'cut' system is usually necessary. This demands a certain discipline on the part of the users, who must wait until the previous speech has finished before replying, and particularly the producer, who must remember to release the talkback switch in order to hear what the people in the studio are saying!

Two-way conversation is of course possible when the studio personnel are wearing headphones and the studio loudspeaker is turned off; most mixers designed for studio installations have separate talkback switches for studio headphones, and often additional switches to talk to the mixer auxiliary outputs, which are used to send cue feeds to musicians for overdubbing, and to external destinations such as remote recording rooms, outside broadcast sites, etc.

During rehearsals, people in the control room can hear speech from the studio through the normal microphone/mixer/loudspeaker chain. However, if a presenter wishes to speak to the control room whilst a record is playing on-air, or if a musical director wants a private word from the rostrum, a 'reverse talk-back' system is required. This can be simply a dedicated microphone in the studio with a 'press-to-talk' switch feeding directly to a dedicated loudspeaker in the control room.

If several destinations are involved, e.g. presenter, conductor, green room, etc., the requirement leads to a full-blown talkback/intercom system which will enable the various destinations to talk to each other as well as to the control room.

In television studios it is common to require a system with at least 12 destinations, and this type of requirement is met by a separate system external to the mixer, using individual keypanels for each destination and either one or more interconnected ringwired circuits or a central matrix, either hardwired or switched, with call indication, group calls and other facilities as necessary.

For very large systems a digitally controlled central matrix is invariably used. This type requires only three connections from each keypanel (for send, return and control) instead of one for each destination, which greatly simplifies and reduces the cost of the installation, but has a central rack of electronically controlled switches instead of hard-wired tag blocks.

2.7.4 Specific mixer types

As noted in Section 2.7.1, there are many different types of, and uses for, audio mixers; the categories which follow are based on design criteria, but it must be remembered that many types are

multifunctional, and for example a mixer used in broadcasting may not be a 'broadcast mixer'.

2.7.4.1 Portable mixers

Portable mixers range from very simple two or three input mono units for electronic news gathering (ENG) work to quite sophisticated stereo mixers with 12 or more inputs for mobile music recording or audiovisual and video presentation events.

The smallest units are battery powered, sometimes with a mains input option, with rotary faders, microphone/line selection, a single meter, and a headphone monitor output. Specialist designs for radio news reporting have facilities for telephone line interfacing and an external monitor input for cue feed.

Some tape recorders have integral microphone mixers which enable high quality recordings to be made without using a separate mixer, but these are limited to one microphone channel per tape track and almost invariably have rotary faders.

For radio and television outside broadcasts or for music recording 'on location', mixers range from truly portable six- or eight-channel units up to mixers with 24 or more channels which are 'transportable', i.e. studio mixers or even consoles designed in sections for mobile use.

A portable eight-channel mixer will typically have microphone/line inputs with equalization, one or two auxiliary outputs, PFL, a stereo output with integral meters and a monitor output with tape monitor input. If it is to be used for music recording or broadcasting it will have comprehensive insert facilities, but usually a very simple talkback, adequate for 'slate' or 'talk to studio' only.

2.7.4.2 Editing mixers

Versions of standard portable microphone/line mixers are often used in edit suites, as there is a wide choice available from many manufacturers, usually from stock, and at reasonable prices. An increasingly vital requirement is an interface with the necessary protocol to enable signals from the video mixer to control the level of one or more channels in the audio mixer via VCAs in the mixer channels.

However, such mixers, though excellent value, may provide more facilities than are required and prove confusing for video editors with little audio experience. They are also usually fairly large in terms of the space available. Some manufacturers offer edit suite audio mixers (ESAMs) with an editing interface, 'line only' inputs, a limited gain range, and often no equalizers or auxiliary outputs, in a package not much larger than twice the size of the faders.

2.7.4.3 Production/on-air mixers

These are the nearest to the 'typical' mixer described in Section 2.7.3, with a mixture of mono and stereo input channels, four or more stereo groups, and very comprehensive monitoring and communications facilities. Very often such mixers will have two independent main outputs, enabling the operator to record one mix whilst broadcasting another, or to send line-up tone or test programme to the next destination whilst still recording or transmitting without interruption.

The input channels can be routed to any of the groups and to either (or both) of the main outputs directly; at least four auxiliary outputs per channel will be provided, plus at least one additional bus for clean feed. Both mono and stereo channels will require comprehensive equalizers, with compressor/limiters available for patching into groups and/or the main outputs, and often fitted as standard in the mono channels.

Monitoring will be provided for loudspeakers and headphones in control room and studio; the control room monitoring will access at least four local recording machines, the mixer main and auxiliary outputs, and usually the output of a separate selector providing access to station ring-main monitoring, off-air feeds, etc.

Communication and connection with outside sources such as outside broadcast sites, roving reporters, telephone callers, etc., will be provided, usually by 'outside source' modules fitted in the mixer penthouse. These modules interface with the outside lines, sending cue programme feed, mixed with or overridden by talkback, at the required level, and often incorporating remote controls for selecting the sources required via a centrally located switching matrix.

2.7.4.4 Self-operated/on-air mixers

These are specialized versions of production mixers, designed to be operated by a presenter who sits at the controls with a microphone. The layout usually includes a central script area, with the input channels disposed equally on either side. The number of controls is kept to the minimum, though views on this vary; some presenters like to have equalizers and special effects available, whilst others may balk at having anything more than faders accessible.

Remote start switches on the appropriate channels for turntables, compact disc players, cartridge machines, etc., are essential; these are usually wired in conjunction with microswitches in the channel faders so that a start signal can be provided either by

pressing the switch or opening the fader, as selected by a second switch.

If the machine is started on the switch, it is desirable for it to stop when the fader is closed after the item has been broadcast (and not when the fader is opened, as was the case with some early systems!).

Designing easily operated and unambiguous logic control systems is extremely important for self-operated mixers, as the operator is often under pressure with no technical assistance immediately available for operational problems. Furthermore, the operator is usually the only person monitoring the programme before it reaches the transmitter, so it is vital to provide an uninterrupted feed of the mixer output. As monitoring is invariably done mainly on headphones (as the loudspeakers are muted when a microphone fader is opened to avoid feedback), the headphone monitor output feed is usually split whenever PFL is selected, so that instead of PFL replacing programme, as it normally does on loud-speakers, selecting PFL gives the operator mono PFL in one earpiece and mono programme in the other. Any incoming talkback (from a producer or other studio, etc.) has the same effect, overriding PFL if this is already selected.

2.7.4.5 Multitrack mixers

Multitrack recording is used for most large recording sessions, including much classical music, opera, drama, etc., as well as all 'pop' music. It is also used for simultaneous recording of live broadcasts. In all applications, the principle is to record as many separate tracks as considered desirable as quickly as possible, and to produce a final mix without the time and financial pressures imposed by having musicians and singers in the studio for longer than is necessary.

Early multitrack mixers, providing only four or eight outputs or 'track sends', were designed as multigroup mixers, with the group outputs feeding the recorder. In order to obtain the best signal-to-noise ratio from the recorder, the signal levels on the tracks must be kept as high as possible. Monitoring the mixer output line outputs at these levels would therefore not give a true impression of the final result, so a separate 'monitor mixer' was required to combine the track send and recorder output signals in a more realistic way for monitoring during recording and overdubbing.

When the number of tracks doubled to 16, such mixers became very large and unwieldy to operate, and it was realized that group faders were not nec-essary, as many tracks recorded the output of only a single microphone channel. It was also more conve-nient to have the monitor controls in line with their relevant channels instead of in a separate part of the mixer; the 'in-line' multitrack mixer was the result.

Modern multitrack recording mixers are almost exclusively designed on the in-line principle; group buses become 'track sends', usually at least 24 and often 48, and every channel can be routed to any (or all) of these. A rotary track send level control may be fitted on each channel; this is effectively the group fader, but is only used as a preset to achieve the optimum recording level, particularly if several channels are being mixed onto one track send.

The return signal from each track of the recorder is fed back to the corresponding channel on the mixer, which has a monitor signal path separate from the normal or programme signal path, with level and pan controls feeding the signal to a stereo monitor bus which is picked up in the usual way by the mixer's monitor module.

The monitor path also has auxiliary send controls for sending cue mixes to artists in the studio for over-dubbing and for adding reverberation and other effects to the monitor mix as required. For remixing, the signals returning from the recorder must be routed to the mixer's main signal path; this may be done by a switch at the input of each channel, and in many mixers these switches are controlled remotely by a single master 'remix' switch.

More complex designs allow the main and monitor signal paths in each channel to be interchanged (swapped), either wholly or in sections, so that for example the equalizer section may be inserted in the monitor path. This enables the signal to be recorded unequalized whilst the operator experiments with different equalizer settings: the chosen equalization can then be switched back into the signal path for the remix.

When a multitrack recording with simultaneous stereo mix is required, as is often the case with broadcasts of large concerts, the channel 'swap' facility may be used to allow the main faders to control the stereo mix and the 'monitor' faders to become the channel multitrack send faders.

2.7.4.6 Postproduction/film-dubbing mixers

For stereo postproduction, standard multitrack mix-ers are the norm; when the final mix is for four or more channels for encoded 'surround sound' or multispeaker cinema systems, special mixers are required. These are usually based on multitrack designs but with special output routeing and monitor-ing sections; some large standard multitrack designs

incorporate four output mix buses and have front/ back panning in addition to the usual left/right on each channel.

Film dubbing is a highly specialized activity; three operators share a very large, custom-built console with separate sections for dialogue, effects and music.

2.7.4.7 PA/sound reinforcement mixers

This category ranges from basic four-input microphone/line mixers with mono outputs fitted in caravans used at village shows to the enormous rigs used at outdoor concerts. These latter provide all the facilities found in the most complex recording studio mixer, except for multitrack recording and remixing. The main outputs, instead of feeding tape machines, are connected to banks of loudspeaker amplifiers, and with the very high levels commonly used, a low noise level is vital, as is immunity to interference from lighting rigs, etc.

2.7.4.8 Theatre mixers

Modern theatre mixers are a highly specialized design, usually providing a large number of output groups which are routed to the theatre loudspeaker arrays via a mixing matrix. The continual increase in the complexity of productions has led to a high degree of automation to assist the operator with changes of level and perspective which must coincide with lighting effects and scene changes.

Minimum dimensions, especially front-to-back, are a major design criterion, as such mixers must be sited in the auditorium so that the operator can hear exactly what the audience is listening to, and every seat removed to accommodate the mixer means lost revenue.

The maximum space allowed is never more than two rows, and this includes space for the operator; channel length is therefore restricted, and external processing (dynamics, etc.) under MIDI control is widely used. As with PA/sound reinforcement mixers, a low noise level is vital; another common factor is the absence of a monitoring system (except for PFL on headphones), as the main outputs are always on loudspeakers!

2.8 Power amplifiers, control units and preamplifiers

2.8.1 Power amplifiers – historical

Figure 2.109(a) shows a British 14-W amplifier of 1938 vintage, intended for domestic high-quality

(a)

(b)

Figure 2.109 *Valve amplifiers of approximately 15 W rating: (a) 1938, (b) 1960.*

sound reproduction. At a retail price of just under £20, this yields a figure of approximately 0.7 W per £1. The Quad 15-W valve amplifier shown in *Figure 2.109(b)* cost about £22 in 1960, which also gives a figure of around 0.7 W per £1. The purchasing value of the pound had meanwhile fallen by a factor of at least 2, so that the maintenance of the 0.7 W per £1 figure evidently resulted largely from the more economical design of the later amplifier, which employed beam tetrodes with negative feedback to obtain low distortion and a low hum level.

However, the advent of transistors enabled a far more striking economy of engineering design to be effected. The Quad 405–2 transistor amplifier shown in *Figure 2.110*, which gives a total output (both channels) of 200 W, sold in 1988 for around £300,

again yielding a figure of about 0.7 W per £1. But taking into account the large financial inflation, at least seven-fold, which had occurred since 1960, it becomes evident that the true cost per watt with modern semi-conductor technology is indeed down by an impressively large factor on what could be achieved in the valve era, and this is undoubtedly a major engineering advance.

Figure 2.110 *Quad 405/2 100 W + 100 W transistor stereo amplifier, 1988.*

It should perhaps be added that the precise meaning of an amplifier rating in watts is much less straightforward than it may initially seem to be (this is considered in full detail in Section 2.8.4), but nevertheless the above comments indicate the general trends quite fairly.

2.8.2 Modern amplifiers – general features

A good modern power amplifier normally has a large amount of overall negative feedback, which gives it a very low output impedance, and great care is taken to ensure adequate freedom from crossover distortion since this might otherwise become perceptible at the lowest signal levels (Baxandall, 1991). The total remaining amplifier distortion should then be well below audibility, even for the most critical listeners, when feeding a programme to any practical loudspeaker, *provided that no overloading of any kind is*

allowed to occur (Baxandall, 1977, 1990a,b, 1991; Lipshitz and Vanderkooy, 1981; Self, 1988a,b).

Because of the above, conventional distortion measurements made on good amplifiers under non-overloading conditions tend to be of mainly academic interest (Baxandall, 1978b, 1979), and the more genuinely important aspects in assessing the performance are normally those concerned with determining the detailed features of the high-level output-producing capability under various conditions of use, and checking that any remanent crossover glitches are adequately small.

2.8.3 Clipping

As the programme input level to a power amplifier is turned up, a point is reached at which momentary overloading begins to occur now and then, and this overloading can be of either of the two distinct kinds given at (1) and (2) below. In saying this it is assumed that no slew-rate limiting occurs (Baxandall, 1978a, Cherry and Dabke, 1986), but any well-designed amplifier will satisfy this requirement with a good margin to spare. Types of overloading:

(1) The amplifier may run out of ability to produce the highest peak instantaneous output voltages required, even though it may still have something in hand with regard to supplying the peak instantaneous current demands. This is called *voltage clipping*.
(2) The amplifier may run out of ability to provide the highest peak instantaneous output currents required, even though it may still have something in hand for the peak voltages. This is called *current clipping*.

Figure 2.111 shows output voltage and current waveforms for an amplifier feeding a load consisting of R and L in series and exhibiting symmetrical current clipping. When the current is clipped, as shown by the flats on the current waveform, di/dt becomes zero, so that the voltage across L, which is $L(di/dt)$, also becomes zero. Thus, during the clipping interval, v_{out} equals v_R and is therefore constant since the current in R is constant. Hence the v_{out} waveform is as shown, and this emphasizes the point that merely looking for evidence of clipping at the top and bottom of an amplifier output voltage waveform does not necessarily tell one whether overloading is occurring.

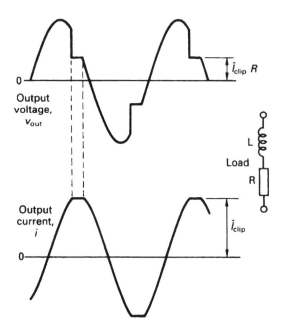

Figure 2.111 *Output voltage and current waveforms for feedback amplifier exhibiting pure current clipping, with partially reactive load as shown.*

The onset of either kind of clipping is normally quite sudden, the large amount of negative feedback keeping the distortion negligible right up to the clipping point, with the distortion increasing rapidly once this point is exceeded. Depending on circumstances, sometimes the onset of voltage clipping occurs before that of current clipping and sometimes the reverse is the case.

The peak instantaneous programme input voltage at which clipping begins is determined by a number of interrelated factors which, in addition to the design characteristics of the amplifier itself, may include the nature of the programme voltage waveform – which is dependent on the kind of music involved – and the impedance characteristics of the loudspeaker. At this point it may be remarked that many amplifier owners, usually without being aware of it, allow a good deal of clipping to occur in normal use, in the interests of obtaining the very high volume levels they desire.

Now it is possible to fit an amplifier with a visual display device which will reliably indicate the onset of clipping, no matter whether it is voltage or current clipping – the indicator circuit is operated by the negative feedback error voltage, which rapidly becomes large once either kind of clipping sets in. However, it appears to be the case that the audio industry does not much favour such clipping indicators, for an amplifier so fitted may often be

seen to be clipping even though it is judged to sound quite free from distortion. This draws attention to its output limitations, whereas competitive amplifiers without such indicators are used in happy ignorance that the limitations exist!

Many short-duration transients, such as cymbal clashes, may be allowed to exceed the clipping point by a few decibels without this being noticeable to discerning listeners, whereas with piano music a very small degree of clipping is likely to be noticed.

Clipping, however, as of an insidious nature, and in the absence of a clipping indicator it is often difficult to be quite confident that it is not causing some degradation of the sound quality. If clipping is found to occur, the proper answer, if a reduction in the volume setting is not thought to be acceptable, is simply to use a 'tougher' amplifier. Sufficient sound level for reasonable domestic listening requirements is in fact usually obtainable without recourse to clipping, even when the wide dynamic range available with digital programme sources is taken into account. My own opinion is that it ought to be regarded as quite normal to fit clipping indicators – and if this became the general practice, it would surely also be of commercial benefit to the industry as a whole.

The abrupt overload characteristics that are inherent in the straightforward use of a large amount of overall negative feedback are sometimes argued to be a disadvantage of this approach to amplifier design, it being pointed out that in amplifiers employing very little negative feedback the overload characteristics are likely to be of a relatively smooth and gradual kind, whose audible effects are often less noticeable.

Although there is an element of truth in the above argument, it should be borne in mind that an amplifier with a very large amount of overall voltage feedback, provided its current output capability is made sufficient to avoid current clipping on programme, may be made to exhibit smooth clipping behaviour by preceding it with a separate soft-clipper circuit exploiting the characteristics of diodes or transistors in an optimum manner. Then the other advantages of a high-feedback design are fully preserved.

However, such a modification will inevitably degrade the amplifier performance at signal levels considerably below the unmodified voltage-clipping point, so is better switched out when the amplifier is to be used in a civilized manner without clipping.

2.8.4 Basic high-level sine-wave notions

Much of the difficulty in trying to predict with any accuracy the true programme-handling capability of

an amplifier when feeding a specific loudspeaker arises from the fact that nearly all quantitative information about power amplifiers is at present given in terms relating to measurements made under sinewave test conditions when feeding a simple, usually resistive, load impedance. Loudspeaker impedance information is also normally sine-wave related.

The convenience of sine-wave measurements is undeniable and it is not suggested that designers and users should forgo the sine-wave approach, but nevertheless considerable benefits are to be gained by complementing sine-wave test data obtained at high signal levels with the results of tests carried out in other ways. However, before considering these other ways in detail, an attempt will be made to clarify several sine-wave-related concepts that have been known to cause confusion. Some readers will undoubtedly find this treatment unnecessarily elementary, but no apology is made for including it!

If all amplifiers had stabilized constant-voltage d.c. supplies, and if all loudspeakers had frequency-independent purely resistive impedances of standard value, then everything would be very straightforward. Such an amplifier would be designed to feed the prescribed resistive load R_L without running into current clipping, and would therefore be able to generate across R_L a low-distortion sine-wave voltage of any magnitude up to the voltage-clipping limit \hat{V}_{clip}. The output voltage waveform would then be as shown in *Figure 2.112(a)*, and the waveform of the square of this voltage would be as shown in *Figure 2.112(b)*.

The instantaneous power dissipated in R_L is V^2/R_L, so the instantaneous power waveform has the same shape as (*b*), but suitably rescaled as at (*c*). The mean or average power dissipated in R_L is half the peak value of the instantaneous power and is therefore given by:

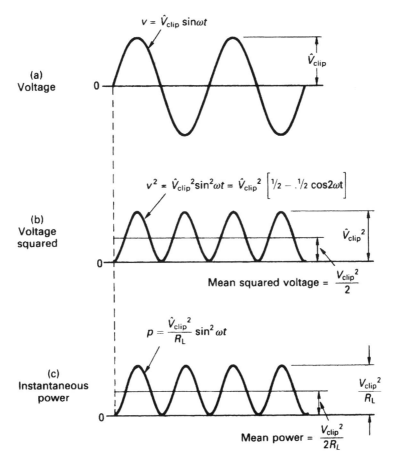

Figure 2.112 *Sine-wave output waveforms at the onset of voltage clipping, with resistive load* R_L.

$$P_{\text{mean}} = \hat{V}_{\text{clip}}^2/2R_{\text{L}} \qquad (2.13)$$

It is convenient to have an effective (from a power viewpoint) long-term voltage value V_{eff}, such that:

$$P_{\text{mean}} = V_{\text{eff}}^2/R_{\text{L}} \qquad (2.14)$$

From equations (2.13) and (2.14) it is evident that:

$$V_{\text{eff}}^2 = \hat{V}_{\text{clip}}^2/2 \qquad (2.15)$$

and, as shown in *Figure 2.112(b)*, $\hat{V}_{\text{clip}}^2/2$ is the mean squared voltage, so that equation (2.15) is equivalent to:

$$V_{\text{eff}}^2 = \text{Mean squared voltage}$$

or

$$V_{\text{eff}} = \sqrt{\text{Mean squared voltage}}$$

which is usually called the 'root-mean-square (r.m.s.) voltage', V_{rms}. Hence:

$$V_{\text{rms}} = \hat{V}_{\text{clip}}/\sqrt{2} \qquad (2.16)$$

It should be evident from the above that whereas the term 'r.m.s. voltage' is a perfectly correct one, the terms 'r.m.s. power' and 'r.m.s. watts', though they have indeed crept into widespread use in the audio world, are quite incorrect, since *mean* power is what is meant. Those of us who have stopped to think about this would do well to exert an influence against the continued perpetration of this piece of nonsense!

The output capabilities of audio amplifiers are, of course, almost universally quoted in watts, but this actually introduces quite a lot of complications when looked at closely. Under the easy conditions considered above (stabilized power supply, pure resistance load R_{L}, and sine-wave signal) the description '100-W amplifier', say, simply means that it can produce, in a resistive load of stated value, such as 8 Ω, a sine-wave output having a mean power level of at least 100 W at the onset of voltage clipping, and that it can continue to do this for a considerable period of time. If R_{L} is indeed 8 Ω, then this power level corresponds to 28.3 V r.m.s. or 40 V peak. The peak instantaneous power output (see *Figure 2.112(c)*) is 200 W.

The first complication arises when we consider not a pure resistance load, but a loudspeaker load, bearing in mind that the impedance of a normal loudspeaker varies greatly with frequency and involves reactance as well as resistance. Even under sine-wave conditions, it does not now make sense to say that a 100-W amplifier is one that can feed a genuine mean power of 100 W to the loudspeaker, for it begs the questions: 'At what frequency?' and 'On what assumptions regarding the loudspeaker impedance?'. Indeed, to take a hypothetical case, if the loudspeaker impedance was purely reactive, it would not be possible for the amplifier to feed any mean power at all to it, but only reactive volt-amperes.

Hence if amplifier output ratings are to be quoted in watts in the context of feeding loudspeakers, this has to be done on a somewhat arbitrary and artificial basis.

First of all it is necessary to define what is meant by, for example, an 8-Ω loudspeaker, and in the relevant IEC Standard (IEC 268–5; BS 5428: 1977) this is taken to mean one whose impedance does not fall below 8 Ω by more than 20% at any audio frequency, i.e. which does not fall below 6.4 Ω. In the broad region of the frequency at which the impedance is a minimum, this impedance will then look like that of a 6.4-Ω resistor in series with frequency-dependent reactance.

A 100-W amplifier intended for 8-Ω loudspeakers is then taken to be one which can produce, at any audio frequency, across a load consisting of 6.4 Ω of resistance in series with any value of reactance, a low-distortion sine-wave voltage of such magnitude as would dissipate a mean power of 100 W in a pure resistance of 8 Ω. This is equivalent to saying that the 100-W amplifier must be able to produce, across the above complex load, at any audio frequency, a sinewave voltage of at least 28.3 V r.m.s. or 40 V peak. (For a nominally 4-Ω loudspeaker (IEC minimum impedance 3.2 Ω) these figures are 20 V r.m.s. and 28.3 V peak.)

Thus the above wattage rating scheme is really a rather roundabout way of stating the peak output voltage capability of an amplifier when feeding a nominally 8-Ω loudspeaker that complies with the IEC impedance stipulation. It may be noticed that neither the 100 W mean power figure nor the 28.3 V r.m.s. figure in the previous paragraph has any direct relevance whatever when the signal is not a sine-wave one, the only directly significant figure then being the peak instantaneous output voltage capability of 40 V.

In some ways it is a pity that the present custom of describing the output capability of amplifiers in watts cannot be dropped and replaced by a peak-instantaneous-voltage scheme, so that instead

of referring to an amplifier for driving 8-Ω loud-speakers as a 100 W one, it would be called a '40 V peak (8 Ω) amplifier'.

However, even if this was done, several further complications remain:

(1) It does not follow for certain that an amplifier capable of producing 40 V peak to an 8-Ω loudspeaker under sine-wave conditions, without current or voltage clipping, will be able to do so with all types of programme waveform – current-clipping may possibly prevent this. (See Section 2.8.11.)

(2) If, as is usually the case, the power supply is not a stabilized one, then, assuming class B amplification, a higher supply voltage will be present under normal programme conditions, where the highest output levels occur only during very short transients, than under continuous sine-wave conditions. It is then likely that the peak instantaneous output voltage level obtainable on programme without clipping will considerably exceed that for sine-wave operation. This effect has given rise to the notion of a 'music power rating', higher than the continuous power rating; but expressed in this way it is still related to a sine-wave concept, since it refers to the *mean* power that can be fed to an 8-Ω resistance during a short burst of supposedly sine-wave output. In this context, 'short' normally means not longer than about 10 ms.

This notion introduces a danger of misunderstandings over the meaning of expressions such as 'peak output capability' in watts – does this refer to the mean 8-Ω-referenced output power obtainable only during a very short sine-wave burst, or to the peak instantaneous power?

(3) Quite apart from the power-supply regulation aspect just discussed, ambiguity can arise when referring to output levels that may be well below any clipping limits. What is the statement 'the amplifier output on this programme peaked up to x watts' to be taken as meaning? It could mean: (a) that the instantaneous power, referenced to 8 Ω, peaked up to x watts; or (b) that a fast-response output peak programme meter, calibrated to read mean watts in 8 Ω on a sine-wave signal, gave a peak reading of x watts; or even (c) that a quickly responding true r.m.s. voltmeter on the output gave a peak reading corresponding to x watts short-term mean power in 8 Ω. In normal practice meaning (b) is usually the one that applies.

A few words of clarification should be added in all cases where there may be genuine grounds for doubt.

However, bearing in mind that even a 3 dB change in output level, i.e. a 2:1 power change, is only just audible, it is evident that some of the above points are really of greater legal significance than true importance to the normal user.

2.8.5 Class B waveforms and power losses

Figure 2.113 shows the essential features of a normal transistor push–pull output stage, and the waveforms relating to this under ideal class B operating conditions are given in *Figure 2.114* for both a purely resistive load and a purely reactive load.

At full output level with a resistive load, peak current occurs when the collector-to-emitter voltage is

Figure 2.113 *Essential features of a push–pull output stage.*

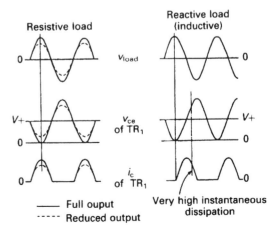

Figure 2.114 *Waveforms relating to* Figure 2.110, *with ideal class B operations.*

Table 2.7 Calculated performance figures for ideal class A and class B output stages, with 100 W ratings*

	Resistive load		Reactive load of same value	
	Class A	Class B	Class A	Class B
Mean output at full-output level	100 W	100 W	100 VA	100 VA
Mean power dissipation per transistor under no-signal conditions	100 W	Zero	100 W	Zero
Mean power dissipation per transistor at full-output level	50 W	13.7 W	100 W	63.7 W
Mean power dissipation per transistor at $(2/\pi)^2$ of full-output power, i.e. approx. 40 W	–	20.3 W	–	–
Maximum instantaneous dissipation per transistor	100 W	50 W[†]	300 W[‡]	260 W[‡]

*It is assumed that each amplifier has only two output transistors, and that these bottom perfectly at full output voltage. The output waveform is taken to be sinusoidal.
[†]For this dissipation to be reached, the output voltage swing must be at least half the full-output value. If it is more than half, this maximum dissipation figure is reached twice per half-cycle.
[‡]These values are reached only at full-output level.

ideally zero. If, however, the output level is reduced, the collector-to-emitter voltage is finite when the current reaches its peak; the result is that, for moderate reductions in output level, the mean transistor power dissipation is increased, and in fact reaches a maximum when the signal output power is about 40% of the full output level. *Table 2.7* gives some calculated figures for an ideal 100-W class B output stage under sine-wave conditions, and includes, for comparison, data on a 100-W class A output stage.

For amplifier ratings other than 100 W, the figures in the table may be regarded as percentages of the maximum mean output power. In class A, the maximum transistor power dissipation occurs when there is no signal, since the whole of the constant d.c. input power must then appear as heat within the amplifier. This amounts to 400 W for a stereo pair of such amplifiers, contributing significantly to room heating and requiring an enormous heat sink area or fan cooling! It highlights the gross extravagance, and indeed absurdity, of the class A approach at the high power levels often called for with modern low-sensitivity loudspeakers, since no quality advantage whatsoever is conferred in comparison with a well-designed class B system.

The reactive-load figures in the table of course represent an extreme condition not experienced in normal use, but nevertheless make it evident that even a partially reactive load of a given value will lead to markedly higher transistor power dissipation than will a purely resistive load of the same value. Thus, with practical loads, the mean transistor dissipation for continuous sine-wave operation is always

much less for class B than for class A, but the advantages of class B become very much more striking when programme input is considered.

With classical music, even if the occasional transient impulse fully exploits the amplifier's output capability, the mean power output (see Section 2.8.4), averaged over a period of minutes, will almost certainly not exceed about 2 W (Von Recklinghausen, 1965). Consequently, small heat sinks are quite adequate if an amplifier is to be used solely for classical music reproduction, even at very high volume levels.

With the more relentless types of pop music, on the other hand, the situation is different. Such music is usually recorded with limiters in almost continuous operation, and mean power dissipations of many watts, or even tens of watts, can occur when it is reproduced with a 100 W amplifier at the high levels often chosen. Sizable heat sinks are then necessary, but they still do not need to be as large as for continuous sine-wave operation.

2.8.6 Safe operating area curves and protective circuits

In *Figure 2.115(a)* the heavy sloping line is a plot, for a resistive load, of instantaneous load current against instantaneous load voltage – indeed nothing more nor less than an elementary Ohm's law graph for the load resistance. Referring to *Figure 2.113*, a positive-going load voltage is accompanied by a falling voltage across the top transistor, so that if

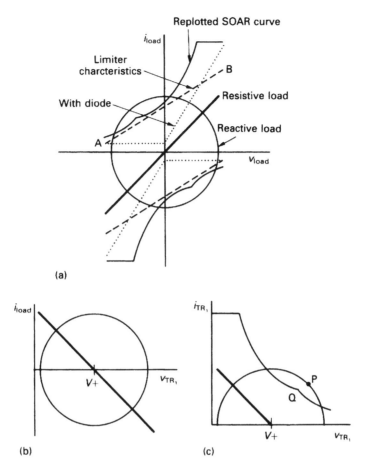

Figure 2.115 *Resistive and reactive load characteristics, plus safe operating area rating (SOAR) curves. Voltage-dependent current-limiter characteristics are also shown: ------ , circuit as in* Figure 2.116; ········, *diode added in series with R_3 (cathode earthed).*

the load current is plotted against *transistor* voltage, we obtain the line shown in *Figure 2.115(b)*. Finally, in (*c*), is shown a plot of *transistor* current against *transistor* voltage. The other half of the load current swing is handled by the lower transistor in *Figure 2.113*.

The circles in *Figure 2.115* represent the relationship between instantaneous current and instantaneous voltage, under sine-wave conditions, for a purely reactive load of the same magnitude as the 45°-slope resistive load already considered. It is evident in (*c*) that whereas with a resistive load large collector currents flow only when the transistor voltage is relatively small, with a reactive load large values of current and voltage can coexist, for example at point P. This ties up, of course, with the conclusions reached from the waveforms shown in *Figure 2.114* and the results listed in *Table 2.7*.

Transistor manufacturers provide safe operating area rating (SOAR) curves for output transistors, and an example of one of these is given in *Figure 2.115(c)*. (Usually logarithmic current and voltage scales are employed, the curves then appearing as straight-line segments.) The part of the curve to the left of the point Q represents constant power dissipation, the more rapid fall-off to the right of Q making allowance for the danger of secondary breakdown occurring in this high collector voltage region.

For short-duration excursions the operating point may be allowed to go outside the curve shown, and other curves, taking into account both the duration and recurrence frequency of such excursions, are usually provided, considerable thermal subtleties being involved (Gates and Ballard, 1974; Noble, 1978). If the SOAR curve in *Figure 2.115(c)*

Figure 2.116 *A voltage-dependent current-limiter circuit.*

is replotted on the basis of load current and voltage, it appears as the upper curve in (*a*), and the corresponding curve relating to the lower output transistor is also shown.

To avoid any danger of transistor breakdown, the operating conditions at low audio frequencies must be confined to the area between the two SOAR curves, and this can be ensured by employing a pair of voltage-dependent current-limiter circuits such as that shown in *Figure 2.116*. The resistor values are typical only. Assume initially that the instantaneous load voltage is zero, but that current is flowing in TR_1. If this current becomes sufficient to give about 0.6 V across R_1, TR_2 is turned on and acts as a clamp across the signal input, preventing any appreciable further current increase. Under these conditions R_3 has almost no influence.

However, when a considerable positive voltage is present on terminal 1, significant current flows downwards in R_3, producing a voltage drop in R_2, such that a larger voltage across R_1 is now required to bring TR_2 on. Conversely, when terminal 1 is negative, a smaller voltage across R_1 is sufficient to bring TR_2 on, limiting the output current to a smaller value than when terminal 1 is at earth. The type of limiter characteristic produced by a pair of circuits of the type shown in *Figure 2.116* is

therefore as shown in broken line in *Figure 2.115(a)*. The heights of the current ordinates at A and B may be independently determined by the choice of appropriate resistor values.

In some commercial designs a diode is included in series with R_3 (cathode earthed), producing, with appropriate resistor value changes, a limiter characteristic of the type shown dotted in *Figure 2.115(a)*. Such amplifiers, although capable of producing the rated output voltage swing across a resistive load, are liable to run into current limiting at a lower output level when feeding practical, partially reactive, loudspeaker loads, leading to a disappointing performance. An intermediate type of characteristic can, however, be produced if the diode is shunted by a resistor.

Sometimes protective circuits incorporate capacitors to make the current-limiting action less severe for short duration transients, but in other cases a capacitor may have been added merely to deal with a VHF parasitic oscillation problem.

A method for displaying the limiter characteristics of amplifiers without requiring access to the internal circuitry is described in Section 2.8.11, and Baxandall (1988) discusses how such characteristics may be related to the current and voltage demands of loudspeakers on programme.

2.8.7 Circuits based on H. C. Lin's configuration

The basic circuit configuration employed in a large proportion of transistor power amplifiers is one that may be regarded as deriving from the original quasi-complementary circuit described in 1956 by H. C. Lin of RCA (Lin, 1956). A typical example from around 1970 is shown in *Figure 2.117*, but many modifications have appeared over the years. A feature of the circuit in the form shown is that the upper or 'Darlington' pair TR_4/TR_6 has a very much more gradual turn-on characteristic (output current against v_{PQ}) than that of the lower or 'conjugate' pair TR_5/TR_7 (output current against $v_{P'Q}$). This is because TR_5 develops a gain of about 20 by the time TR_7 starts to come on, reducing the spread of the latter's turn-on characteristic by about this factor when referred to the input of TR_5. This asymmetry of behaviour gives considerable crossover distortion, which cannot be got rid of by adjustment of the bias pot P_1. The effect of this crossover distortion on the performance of the circuit shown in *Figure 2.117* as a whole is, of course, greatly reduced by the negative feedback.

Figure 2.117 *A simple Lin-type high-quality amplifier. Values shown are typical only.*

When considering the negative-feedback aspect of the circuit, it is important to appreciate that though the output stage may look like a push–pull emitter–follower type of circuit, nevertheless the local negative feedback that is inherent in a normal

voltage-driven emitter–follower does not come fully into action in the circuit shown in *Figure 2.117* until fairly high frequencies, in some cases above audio, are reached.

The above notion can best be understood by referring to the highly simplified version of the circuit shown in *Figure 2.118*, in which irrelevant details such as bias have been omitted and single transistors replace the output pairs. At low frequencies, where C has no effect, the combination of the high-output-impedance transistor TR_2 and its collector resistor R constitutes a floating signal source, of internal resistance R, connected directly between points P and Q of the output stage. Thus the signal is applied between bases and emitters, as in a normal common-emitter amplifier stage, and not between bases and earth as in an emitter–follower. The output stage therefore produces quite a high voltage gain, typically 20 or more, leading to a loop gain, for the feedback via R_{fb}, of perhaps 400 (52 dB), since the circuitry involving TR_1 and TR_2 puts further gain into the loop.

As the frequency rises, a frequency f_1 is reached, usually somewhere in the region of 1 kHz, at which the capacitor C starts to attenuate the overall-feedback loop gain, and since it also reduces the output impedance of TR_2, the drive to the output stage then ceases to be applied in the fully floating manner mentioned above, thus introducing a degree of local emitter–follower type feedback into the output stage. As the frequency continues to rise, giving less and less overall feedback, there is more and more local output-stage feedback until, by the time a frequency f_2 is reached, substantially full unity-gain emitter–follower behaviour has been established in the output stage. Frequency f_2 is in fact higher than f_1 by the common-emitter gain of the output stage, for which a figure of 20 was mentioned above, so that f_2 will usually be above the top of the audio band.

The important point to which the above discussion leads is that over the frequency interval f_1–f_2, the amount of feedback in action for reducing the distortion of the output stage is constant, the reduction in overall feedback being balanced by a corresponding increase in the amount of local output-stage feedback. In practice, unless very fast output transistors are used, a significant increase in distortion due to the output stage may occur somewhat before the frequency has risen to f_2.

Beyond f_2 the amount of local output-stage feedback ideally remains constant, but the overall feedback continues to fall off, accompanied by increased

Figure 2.118 *Highly simplified explanatory version of* Figure 2.114.

output distortion, unity loop gain being reached at a frequency f_3 of a few hundred kilohertz.

The feedback behaviour described above is illustrated in *Figure 2.119*. The broken-line characteristic may be directly measured or calculated, if desired, by breaking the connection at X in *Figure 2.118*, applying a test voltage between the bases and the point Q, and determining the resultant voltage produced between P and Q.

Some of the variations on the general theme of the circuit shown in *Figure 2.117* circuit that have appeared are described below.

(1) The use of complementary (NPN and PNP) output transistors. When this is done, the separate base-to-emitter resistors, typically 100 Ω, associated with these transistors may be replaced by a single resistor between their bases, giving faster turn-off as well as saving a resistor. Rather surprisingly, however, most designers have slavishly continued to use two resistors as in the quasi-complementary version.

(2) The addition of a diode, or a transistor connected to function as a diode ('transdiode'), in the TR_5 emitter lead in *Figure 2.117*, to give much-reduced crossover distortion when complementary output transistors are not used (Shaw, 1969; Baxandall, 1969, 1970).

(3) Replacement of the output pairs by triples, as in the Quad 303 amplifier (*Wireless World*, 74,

67, 1968). The characteristics of these triples are given in great detail together with numerous oscilloscope photographs in Amos (1977), which also gives such information for Darlington and conjugate pairs, the latter both with and without the diode modification mentioned in (2) above.

(4) Parallel connection of output transistors for increased output-current capability. Separate emitter resistors for each transistor are then necessary to ensure well equalized current sharing. However, the availability in recent years of single power transistors, or power Darlingtons (see below), with very high current ratings has made recourse to parallel connection of output devices less frequently necessary.

(5) The use of complementary Darlington transistors in place of the output pairs in *Figure 2.117*. Such transistors are really two transistors built into a single package, which also contains a resistor corresponding to R_{12}. These devices are now available in some profusion and offer a very economical solution. Tijou (1975) describes an early design of this type, with a full discussion of various circuit details. A long-tailed pair input stage is used (see (8) below).

(6) The use of a complementary output pair of metal-oxide semiconductor field-effect transistors (MOSFETs), which may be of the

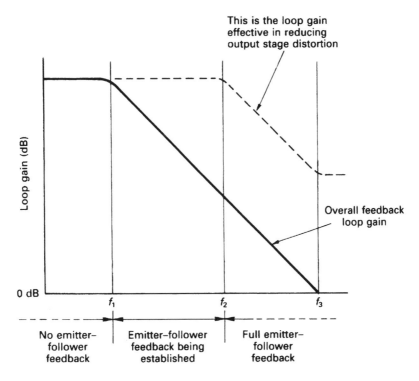

This is the loop gain
effective in reducing
output stage distortion

Loop gain (dB)

Overall feedback
loop gain

0 dB

f_1 f_2 f_3

No emitter–
follower
feedback

Emitter–follower
feedback being
established

Full emitter–
follower
feedback

Figure 2.119 *Feedback loop-gains for a circuit of the type shown in* Figures 2.117 *and* 2.118.

'vertical' DMOS variety. The basic characteristics and theory of these are clearly presented in Finnegan (1992a,b). An advantage is that there is no secondary breakdown effect, i.e. no cumulative thermal tendency to form localized hot-spots; the SOAR curves are therefore less restrictive, which beneficially influences the design of protective circuitry. Another advantage is that turn-off can be quicker, since there is no minority-carrier storage. If complementary MOSFETs are used in an otherwise conventional Lin type of circuit, with their source electrodes feeding the loudspeaker, then a quiescent current of well over 100 mA is necessary if minimum crossover distortion is to be obtained. However, this quiescent current tends to vary by a considerably smaller percentage with heat sink temperature than in bipolar designs, giving less change in crossover distortion during use. The increased quiescent current is called for because the mutual conductance of a MOSFET, at a given current, is much lower than for a bipolar transistor. An inconvenient feature of these enhancement-mode devices is that a forward gate-to-source voltage, or 'threshold' voltage, typically of about 3 V, is

necessary to start drain current flowing, and because of the relatively low mutual conductance, a gate-to-source voltage of up to 10 V may be needed to satisfy peak drain-current requirements. In the Lin type of circuit configuration, auxiliary higher-voltage power supplies are required for the drive circuitry if the full output capability of the MOSFETs is to be realized (Cordell, 1984), though this complication may be avoided if a different circuit configuration is adopted, in which the loudspeaker is fed from the MOSFET drains. MOSFETs have the advantage over bipolar transistors that they require less drive current, but a troublesome feature is that they are markedly more prone to VHF parasitic oscillation.

Although power MOSFETs were originally much more costly than bipolar transistors, prices are now fairly competitive, and with some fresh thinking about circuit design aspects, they may indeed be found to offer the best solution in some circumstances.

(7) Connection of the capacitor C_3, necessary for negative feedback stability, not as shown but instead between the collector of TR_2 and the emitter of TR_1. This much improves the

slew-rate limit of the amplifier, because the charging current for C_3 no longer has to be provided by the low-current input stage, but only by TR_2 which operates at a much larger current than TR_1. The need to watch points such as this has been known to enlightened designers of feedback amplifiers for decades, but was brought into somewhat exaggerated prominence by the coining of the terms 'transient intermodulation distortion' (TID) and 'slewing induced distortion' (SID) (Cherry and Dabke, 1986). Baxandall (1988, p. 12) gives information on the highest rates of change of programme voltage very occasionally found with digital recordings, showing that an amplifier capable of giving full output up to 20 kHz at low distortion without slew-rate limiting, i.e. having a power bandwidth of 20 kHz, is fully adequate in this respect, with a good margin to spare. This corresponds to a slew-rate limit of 0.126 V/µs per volt-peak.

(8) Replacement of TR_1 by a long-tailed pair, making the output d.c. offset voltage less dependent on component tolerances and supply voltage. Sometimes a constant-current tail circuit is employed, in the interests of the lowest possible distortion, and current-mirror or other techniques, such as are widely used in integrated circuit design, may be incorporated to make use of the outputs from both sides of the long-tailed pair. (The arrangement shown in *Figure 2.117*, involving R_3, for nominally eliminating the d.c. output offset voltage, is the most economical solution of all, but has the disadvantage that the offset voltage varies more with supply voltage than is the case when biassing networks in TR_1 base circuit are employed, as is more usual.)

(9) The addition, at the input end, of an op-amp which is included within the overall d.c. feedback loop. This reduces the output d.c. offset voltage to not more than several millivolts, without the need for any adjustment, and can be made to give some preliminary signal gain as well, allowing more negative feedback to be employed in the main section of the amplifier.

(10) The substitution of a single non-centre-tapped power supply, maybe stabilized, in place of the double-polarity supply shown in *Figure 2.117*. An electrolytic capacitor in series with the loudspeaker is then required, but this has the advantage of giving some protection of the loudspeaker against damage in the event of transistor breakdown or other faults within the

amplifier circuit. An ingenious alternative way to provide such protection is adopted in the Quad 306 amplifier, in which the tapping on the mains transformer shown in *Figure 2.117* is removed, and the power supply is 'lightly' centre tapped by what could have been just a pair of equal resistors in series, but which is actually a slightly more complex circuit involving a pair of transistors. The reservoir capacitors themselves then perform the above protective function, but do not add impedance in series with the loudspeaker at low audio frequencies because the feedback voltage is derived from the voltage appearing directly across the loudspeaker. Separate transformer windings, rectifiers and reservoir capacitors are required for each channel, but the capacitor value adopted does not need to be as large as in the normal arrangement. The rest of the 306 circuit, which employs the 'current dumping' technique (see Section 2.8.10.3), is, however, quite different from *Figure 2.117*.

(11) The incorporation of protective circuitry, such as that shown in *Figure 2.116* and discussed in Section 2.8.6.

(12) The substitution of diodes, or a thermistor network, for the 'amplified diode' or 'V_{be} multiplier' circuit shown. With the latter scheme, which is probably the most widely used, there is a good case to be made for connecting the wiper of P_1 to the lower end of R_8; then erratic contact involving either the wiper or the end connections will not result in the application of an excessive forward bias such as might lead to overheating and transistor breakdown.

(13) The addition in series with the amplifier output of L shunted by R, e.g. 1 µH and 10 Ω, plus C in series with R, e.g. 100 nF and 10 Ω, across the ouput. This reduces the effect of extreme inductive and capacitive values of load impedance on the overall negative feedback loop phase lag at frequencies in the region of unity loop gain.

2.8.8 The 'current feedback amplifier' notion

The term 'current feedback amplifier' has appeared in op-amp manufacturers' literature (Ogden, 1992) since about 1985, and also in an audio power amplifier context (Alexander, 1990). It is implied that a new basic principle is involved, but really it is an old one in disguise, as may be appreciated by referring to

Figure 2.118, the output-stage details of which are, however, irrelevant for present purposes.

With $V_{in} = 0$, disconnect R_{fb} from Q and apply a test voltage to its free end to measure the loop gain of the overall feedback loop, i.e. to determine the resultant voltage produced at Q. If it is assumed that R_e is much larger than the reciprocal of the mutual conductance of TR_1, then most of the current in R_{fb} will go to the emitter, yielding a loop gain figure which is almost independent of the value of R_e. Hence by altering R_e the closed-loop gain of the amplifier may be varied over a wide range with little effect on the loop gain or bandwidth.

Changing the values of both R_e and R_{fb} by equal factors, on the other hand, alters the loop gain and bandwidth but has virtually no effect on the closed-loop gain at low and medium frequencies.

The above are exactly the features claimed as a special virtue of the new 'current feedback' technique, but they were in fact fully appreciated by some workers even in the valve era.

In op-amps using this technique, resistors corresponding to R_e and R_{fb} are connected externally, thus allowing gain and bandwidth to be independently determined. The single transistor TR_1 mentioned above is usually replaced by a more complex circuit.

In the audio power amplifier referred to in Alexander (1990), a complete conventional op-amp is used in place of TR_1. The output of this op-amp is connected directly to its inverting input and is equivalent to the emitter of TR_1, the non-inverting input being equivalent to the base, and the supply leads to the collector.

It is felt that the choice of the term 'current feedback' in the above connection is a rather unfortunate one, since this is a long-established term employed when negative feedback is derived from the voltage drop across a small-value resistor in series with a load for the purpose of obtaining a high output impedance or, in combination with normal voltage feedback, an output impedance of some specific required value.

2.8.9 Crossover distortion – general comments

Designing for low crossover distortion is of great importance, for if this receives insufficient attention it may be found that, even though the total harmonic distortion at full-output level is well below 0.1%, nevertheless, distortion may become subtly perceptible at very low output levels, for example on decaying piano notes under quiet ambient conditions. Such distortion may manifest itself as a slight background fuzz rather than as distortion in the more obvious sense (Lipshitz and Vanderkooy, 1981; Baxandall, 1991).

The amplitude of the crossover distortion voltage waveform tends to remain nearly constant down to quite a low output level, so even though it may be a very tiny percentage of the full output level, and therefore totally inaudible during loud passages because of masking by the music, the percentage distortion rises greatly as the volume level is reduced, and it may then become audible.

Figure 2.120 shows how the total harmonic distortion due to the simple type of discontinuity there illustrated varies with level and with the value of the constant m. The distortion waveform under these conditions is a clipped sine wave, tending towards being a square wave at high levels. The information in Diggins (1990) was made use of in calculating these curves.

When overall negative feedback is applied to an amplifier for which the internal distortion mechanism is of the type shown in *Figure 2.120*, the result, for frequency-independent feedback, is simply equivalent to reducing the value of m; but since the amount of feedback in action often falls off with rising frequency, a differentiating effect is introduced and the distortion waveform then tends to take the form of alternate positive and negative pulses. When such pulses are of short duration, it is the net algebraic area of the pulse that is significant at audio frequencies.

Good conventional design along the lines described in Section 2.8.7 can produce amplifiers which are quite free from any subjective shortcomings, but techniques which offer the theoretical possibility of totally eliminating crossover distortion, or at least of reducing it to an extremely low level, without the need for preset quiescent-current adjustment, have an inherent appeal, and several approaches of this kind are discussed in Section 2.8.10.

It may be asked why modern op-amps, such as the BIFET TL071 or the bipolar NE5534 with their dual and quad equivalents, are so remarkably free from crossover distortion, despite employing class B, or class AB, operation of the output transistors. Factors involved are:

(1) the small size gives better thermal tracking, which helps to maintain optimum bias conditions;

(2) the manufacturing technique automatically achieves excellent matching of V_{be} characteristics,

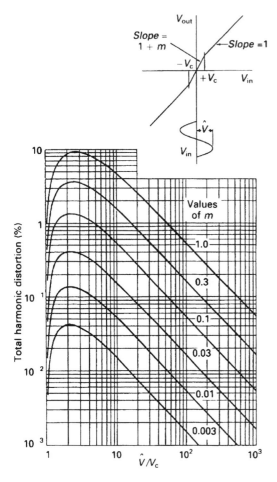

Figure 2.120 *Calculated distortion produced by the transfer chacteristic shown.*

ensuring optimum bias conditions without the need for adjustment; and

(3) the small size, with low stray capacitances, allows an exceptionally large amount of negative feedback to be employed – overall feedback at low frequencies, turning into local feedback on the output stage at higher frequencies when the stabilizing or compensating capacitor has come into action, but both kinds of feedback being fully effective in reducing the output stage distortion, as explained in Section 2.8.7.

If it were not for these favourable features, the use of an op-amp outside the main feedback loop for preliminary signal amplification, as in the Quad 405 (though not in later Quad designs), would make nonsense of using subtle techniques for minimizing crossover distortion in the main amplifier.

An important practical point is that, however good the theoretical circuit design of a class B power amplifier may be, it is inevitable that large currents with a half-wave-rectifier type waveform will be present in the output stage. If sufficient care is not taken over layout and earthing arrangements, the large distortion inherent in such waveforms may get coupled into the signal circuits, thus defeating the designer's low-distortion aims. It is not always appreciated that an inductance, or mutual inductance, of as little as 0.001 µH in the wrong place can become the dominant cause of high-frequency distortion of crossover type.

2.8.10 Eliminating or minimizing crossover distortion

2.8.10.1 The Blomley technique

A new idea in class B techniques occurred to P. Blomley of the Plessey Company in 1969 (Blomley, 1971). In a Blomley circuit both halves of the output stage have their full mutual conductance throughout the whole audio cycle. The broad principle may be exploited in a number of different ways, but *Figure 2.121* is a suitable introduction to the essence of the idea. (*Note*: I advocate the viewpoint that mutual conductance, or transconductance, should be regarded as the normal starting point in considering the behaviour of nearly all transistor circuits, and that current gain, which varies much more widely from one specimen to another, should be regarded as having relatively minor significance. The notion that a bipolar transistor is quite different from an FET or valve in being 'basically a current-operated device' is highly misleading and liable to result in trains of

Figure 2.121 *Simplified Blomley amplifier circuit.*

thought that end up with the creation of poor circuit designs.)

R_5 and R_6 are chosen so that each output transistor (which in practice may well be replaced by a pair or a triple) is biassed to have a finite collector current in the absence of any current in TR_3 or TR_4 (*Wireless World*, 74, 67, 1968; Blomley, 1971). Since TR_3 and TR_4 cannot pass reverse current, the currents in TR_5 and TR_6 can never fall below the above value. The application of an input causes TR_2 to feed a low-distortion signal current, coming from a very high source impedance, to the emitters of TR_3 and TR_4. Now whatever current is fed to these emitters must, to a very close approximation, come out at the collectors; and whichever collector the current comes out at, it produces a voltage for driving the associated always-on output transistor. Under symmetrical circuit conditions, including closely matched resistor values, the overall gain is exactly the same no matter whether the current fed to the TR_3/TR_4 emitters goes to TR_3 or to TR_4 or partly to each. Thus there is ideally a total absence of crossover distortion, and this is not dependent on the adjustment of the output stage quiescent current to any particular value.

In simple theory the forward bias on the bases of TR_3 and TR_4 shown in *Figure 2.121* is unnecessary, but with a less-than-ideal signal current source, and in the presence of transistor and stray capacitances, such forward bias is in practice desirable. Moreover, without the bias, a sine-wave input will ideally produce perfect half sine waves, with sharp corners, at the outputs of TR_3 and TR_4, and the output transistors must then be able to handle these, with their wide frequency spectrum, in a faithful manner.

However, with suitable forward bias, giving a small quiescent current in TR_3 and TR_4, all the above problems are overcome, and *Figure 2.122* shows the waveforms of the currents then produced in the two halves of the output stage, in this case employing triples. The centre-line in the photograph represents zero current, so it is evident that neither current falls right down to zero, the two currents adding together to give an output current with very low distortion.

In such an amplifier the quite large amount of local negative feedback within the triples, due to the presence of 'emitter' resistors, of say 0.5 Ω, is fully exploited in minimizing the overall distortion of the amplifier, whereas this feedback is less effective in a straightforward amplifier employing triples, because of the crossover-distortion problem. In a practical Blomley design, a large amount of overall negative feedback will be used in addition, giving a truly splendid distortion performance.

1 A/cm

0.2 ms/cm

Figure 2.122 *Observed 1 kHz current waveforms for a practical Blomley amplifier.*

The Blomley philosophy may be expressed in its most general idealized form as follows:

(1) Make an output stage which, *provided both sides remain on*, has a distortionless response to a common-mode input and zero response to a balanced input.

(2) Devise input circuitry to feed a distortionless common-mode input to the above output stage, plus a balanced input whose waveform is not highly critical but which is tailored to prevent either side of the output stage from being cut off.

(3) The terms 'common mode' and 'balanced' in the above paragraphs may be interchanged, leading to further practical circuits, e.g. with center-tapped output transformers.

The technique offers the attractive possibility of making an amplifier completely free from setting-up adjustments without any sacrifice in performance. However, to get the quiescent current to come within even quite wide limits without adjustment requires fairly high-wattage emitter resistors in the output stage and/or the use somewhere in the circuit of transistors having closely matched V_{be} values at a given current. The overall economy tends to be inferior to that achievable using the technique of Section 2.8.10.3.

2.8.10.2 The error-feedforward principle

This is an idea which occurred to H. S. Black in 1923, 4 years before he conceived the negative-feedback principle, for which a full patent was not granted until 1937. *Figure 2.123* shows an embodiment of the error-feedforward principle, which is particularly easy to understand.

For very small input levels the class B amplifier does nothing and the class A amplifier therefore simply receives an input of V_{in}, causing it to supply

Figure 2.123 *The error-feedforward principle.*

a current $V_{in}g_{mA}$ to the load. When the instantaneous input voltage V_{in} increases to the point at which the class B amplifier comes on, with mutual conductance g_{mB}, the output current from this amplifier then begins to produce a voltage drop across R which is applied, via the transformer, to oppose the input to the class A amplifier. If $g_{mB}R$ is made unity, the class A amplifier then receives no further increase in input as V_{in} continues to rise, and the required increase in load current is now supplied by the class B amplifier only. But if, in this regime, g_{mB} departs from its ideal value, this gives an input to the class A amplifier, causing it to feed forward a current component to the load which corrects for this non-ideal behaviour of the class B amplifier. It is easily shown that:

(1) if $g_{mA}R = 1$, the overall mutual conductance of the system is g_{mA}, whether or not the class B amplifier is contributing output current, and therefore independently of the value of g_{mB}; and

(2) making $g_{mB} = g_{mA}$ gives the class A amplifier the least correcting work to do.

Provided the class A amplifier has negligible distortion and the condition $g_{mA}R = 1$ is accurately

satisfied, then it is evident that the whole system must give negligible distortion.

The class A amplifier in *Figure 2.123* has to handle the full output voltage, but only a small error current. In some other error-feedforward schemes the class A amplifier has to produce only a small output voltage, inserted as a correction in series with the load, but it then has to handle the full load current, and such schemes are usually much less economical.

No negative feedback notion whatever is involved in the circuit in *Figure 2.123*, nor are any of the stability problems normally associated with negative feedback, so that it cannot be described in any other way than by using the term 'error-feedforward'.

In many practical schemes, however, such as that described in Section 2.8.10.3, negative feedback is very much involved, and though they are often regarded as examples of the error-feedforward technique (Walker, 1975; Vanderkooy and Lipshitz, 1978, 1980), there is actually no need to invoke this concept, and an explanation based purely on the notion of negative feedback that changes in amount with the instantaneous operating regime has much to commend it for clarity of thought, and it is equally sound (Baxandall, 1976; Walker and Baxandall, 1983). (See Section 2.8.10.3.)

Some schemes, which appear at first sight to be of the error-feedforward type, try to be cunning by deriving the error voltage of an amplifier, i.e. the difference between its input and a suitably scaled version of the output, and feeding this error, with appropriate sign, in series with the signal input to the amplifier, thus using this main amplifier to perform its own error-feedforward function (Davis, 1958; Cordell, 1984). Although it is readily shown that in simple theory this appears to lead to zero distortion, closer examination reveals that such schemes really involve an infinite amount of hidden negative feedback, the necessary infinite gain required for this being produced by positive feedback with unity loop gain round some path or other, leading to difficult stability problems. Such schemes are therefore not the magic panacea they may at first sight appear to be!

2.8.10.3 The Quad current-dumping technique

This technique was first used in the Quad 405 amplifier introduced in 1975, but has been retained in all subsequent Quad designs (Walker, 1976; Baxandall, 1976; Vanderkooy and Lipshitz, 1978, 1980; Walker and Baxandall, 1983).

Consider first the simple set-up shown in *Figure 2.124(a)*, ignoring the broken-line resistor for the time being. Under small-signal conditions, the dumper transistors, which constitute a crude class B amplifier with no forward bias, do not conduct at all, so that the loudspeaker is simply fed from the output of the class A amplifier via the resistor R_3, the value of which is 47 Ω in the practical design. For larger signal excursions, the voltage drop in R_3, due to the flow in it of the loudspeaker current, becomes sufficient to cause the dumper transistors to conduct during part of the cycle. Once this occurs, the voltage at Q follows that a P with little loss, most of the additional loudspeaker current being supplied by the dumpers. The overall transfer characteristic with a resistive load is therefore approximately as shown in (b), and would give gross distortion.

By applying overall negative feedback via the broken-line resistor R_1, the linearity may be greatly improved, giving a transfer characteristic as shown in (c). However, in view of the enormous amount of distortion without overall feedback, an impracticably large amount of feedback would be necessary to produce a result up to the highest standards demanded. Looking at the transfer characteristic (c), it is clear that what is wanted is to introduce a little extra overall feedback in the AB and A′B′ regimes, so as to reduce the gain in these regimes to the same value as that in the AA′ regime. This can be done by

Figure 2.124 *Diagrams for explaining the principles of the Quad current-dumping amplifier.*

the simple modification shown in (d). Now, when dumper current flows, it produces a voltage drop in the small-value resistor R_4, so that the feedback resistor R_1 receives this voltage in addition to the loudspeaker voltage itself, thus giving the wanted extra feedback. By choosing the value of R_4 correctly, the overall gain may be made independent of whether or not the dumper stage is acting. Assuming infinite forward gain in the class A amplifier, the correct value of R_4 is that which satisfies the relationship $R_2R_4 = R_1R_3$. With finite gain in the class A amplifier, a slightly different value of R_4 is required, but the overall performance is still perfect provided the class A amplifier is free from distortion and has a gain which is independent of frequency up to frequencies

much higher than the highest audio frequency. The latter requirement arises because, to give a sine-wave output, some highly non-sinusoidal waveforms have to be produced at other points in the circuit.

In a practical design, it is necessary to consider the high-frequency stability of the feedback system, and it is here that use is made of the fact that the resistors R_1, R_2, R_3 and R_4 can be replaced, more generally, by impedances Z_1, Z_2, Z_3 and Z_4. With infinite forward gain in the class A amplifier, the condition for zero distortion is now that Z_2Z_4 must equal Z_1Z_3 at all frequencies. A 120 pF capacitor is actually used for Z_2 and a 3 µH inductor for Z_4, the other two impedances being resistors as before (see *Figure 2.124(e)*). The class A amplifier has now been made into an integrator, whose gain falls off at 20 dB/decade (6 dB/octave) up to very high frequencies, thus attenuating the overall-feedback loop gain in a tame manner and conferring excellent stability on the complete amplifier.

The output impedance of the amplifier is equal in value to the impedance of L and R_3 in parallel, and rises to nearly 0.4 Ω at 20 kHz. It is independent of whether or not the dumpers are in action. In the practical design, the integrator amplifier is a very simple three-transistor circuit, of which the class A output transistor serves also as the driver transistor for one of the dumpers, the other dumper having its own separate driver.

Exactly the same circuit as that shown in *Figure 2.124(e)* may be redrawn as in *Figure 2.125* and this leads to a third way to explain the action of a current-dumping amplifier. The four impedances form a simplified Maxwell bridge – in a full Maxwell bridge series, resistance in L is balanced by shunt resistance across C. In operation, voltages appear at P and T, and may be regarded as a combination of a common-mode component appearing equally on P and T, plus a balanced component appearing between P and T. Because of the nulling action of the bridge, the balanced component produces no voltage between S (virtual earth) and Q, and therefore gives no voltage across the loudspeaker. Thus the distorted dumper waveform does not produce output distortion. The common-mode voltage component at P and T feeds a current to the virtual earth equal to I_{in} and also produces an output voltage to feed the loudspeaker, the output impedance seen by the loudspeaker being formed by L and R_3 in parallel. This is a very good example of the way in which the mere manner of drawing a circuit is liable to have a large influence on how one thinks about its operation.

The current-dumping technique is of great appeal from a production viewpoint, since it gives a consistently high-grade performance under all temperature conditions and without the need for any kind of adjustments or selection of matched transistors.

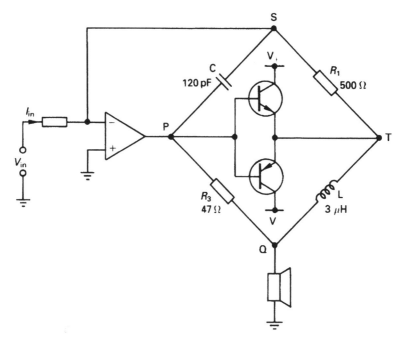

Figure 2.125 *The same circuit as in* Figure 2.121(e) *redrawn in a different way.*

2.8.10.4 The Cherry 'power amplifier improver' (Cherry, 1981)

The capacitor C_4 in *Figure 2.117* enables more over-all negative feedback to be applied with good stability than would be possible without it. It does this by feeding a current component back to the input which is proportional to the time differential coefficient of the amplifier output voltage, thus increasing the phase margin of the feedback system.

In an important fundamental paper, Cherry (1978) showed that by employing a more elaborate active circuit arrangement, involving differentiation, in the feedback path of an audio amplifier, the amount of stable feedback that can be employed may be very greatly increased. A loop gain of 88 dB at 20 kHz is reported, even though the f_T of the output transistors was only 2 MHz.

One way to exploit this principle is to make a separate circuit unit, called by him an 'improver', for connection externally to an existing commercial amplifier. This enables the distortion of a typical amplifier of the type shown in *Figure 2.117*, at audible frequencies, to be reduced by 25–30 dB, converting a perhaps indifferent performance into a superbly good one.

2.8.10.5 The McLoughlin technique

An amplifier employing a class B output stage running at zero quiescent current, with no associated class A amplifier, would normally be regarded as quite unsuitable for high-quality applications. However, McLoughlin (1990) has shown that very good results can in fact be obtained this way, at least in a low-power experimental version using fast output transistors, provided appropriate modifications are made to the basic Lin circuit shown in *Figure 2.118*. These modifications are:

(1) connect the right-hand lead of the capacitor C to the output point Q, so that this capacitor no longer prevents the collector of TR_2 making very rapid traversals of the output stage 'dead zone';

(2) replace the resistors in TR_2 collector lead by a constant-current circuit; and

(3) apply more overall feedback than usual, making up the lost gain, if necessary, by a preamplifier.

2.8.11 Current and voltage displays

A technique which I developed in 1984 (Baxandall, 1988) enables the instantaneous current and voltage

Figure 2.126 *Set-up used for displaying current-limiter characteristics on a cathode-ray oscilloscope.*

output capabilities of power amplifiers to be clearly displayed on an oscilloscope and compared with those for loudspeakers. This test method is now specified in an IEC Standard (268–3: 1988).

The principle is shown in *Figure 2.126.* The amplifier under test is fed with an input having the top waveform, consisting of a low-frequency sine wave at 20 Hz, or some frequency in that region, plus alternate positive and negative 50 µs pulses at a much higher recurrence frequency such as 500 Hz.

The reactance of the 40 µF capacitor at 20 Hz is about 200 Ω, so that the amplifier gives only quite a small 20 Hz output current component. The 20 Hz input is made large enough, however, to give the full output voltage swing, mild voltage clipping being allowed to occur.

For the 50 µs pulses, on the other hand, the effective load impedance is very low, and large pulse currents are turned on by the amplifier. The pulse input voltage is made large enough to drive the amplifier output stage into alternate positive and negative current-limit values as determined by the protective circuitry.

By switching the circuit to a different arrangement, lines representing 4 Ω and 8 Ω resistive loads may be displayed, composite photographs then being obtained by making separate exposures. By adopting standard scaling for the *V* and *I* axes, the size of the display gives an immediate and vivid impression of the potency of the amplifier.

Figure 2.127 shows examples of displays obtained for four amplifiers. The display in (*d*) shows that the technique has value as a diagnostic tool for disclosing faulty design – in this case the positive-side current-limiter circuit bursts into parasitic oscillation.

Full details, with many more photographs, including information on the current demands of loudspeakers under both sine-wave and programme conditions are given in Baxandall (1988). A digital technique based on this work is described in Fairwood and Reed (1991).

2.8.12 The Carver amplifier

The American Carver M–400 stereo power amplifier is a highly innovative commercial design which gives a sine-wave mean output power of 200 W (57 V peak inst.) per channel to 8-Ω loudspeakers, despite an overall size of less than 18 cm (7 in) cube and a weight of only about 4.5 kg (10 1b). The two main novel features are described below.

(a)

(b)

(c)

(d)

Figure 2.127 *Examples of displays obtained with the set-up shown in Figure 2.126. (a) Quad 303. (b) Quad 405/1. (c) Quad 405/2: this has a time-dependent current limiter; the dim traces were obtained with 50 µs pulses at 100 Hz PRF and the bright traces with 50 µs pulses at 2 kHz PRF. (d) Amplifier in which parasitic oscillation of one limiter circuit occurs. The intended protection characteristic is, moreover, too severe and would give current clipping when feeding most loudspeaker loads.*

(1) In normal practice the mains transformer primary is connected to the mains all the time the amplifier is on, although it really only does its job during the short times for which the rectifier diodes are charging the reservoir capacitors. In the Carver amplifier a triac in series with the transformer primary, controlled via an optocoupler by circuitry on the secondary side, switches the transformer on only when it is needed for capacitor charging. This allows a remarkably small mains transformer to develop the required output power without the peak flux density having to exceed normal values. The triac circuitry is also used to regulate the d.c. output voltage, preventing the unwanted voltage rise on light loads given by oridinary power supply circuits.

(2) Three power transistors are connected in series in each half of the class B output stage, their inter-connection leads being taken via diodes to lower-voltage supply lines. For small output voltage swings, only the pair of output transistors whose collectors go via diodes to the lowest-voltage supply lines operate, but for larger swings one or both pairs of 'outer' transistors come into action, the associated diodes ceasing to conduct. This scheme much reduces the total power dissipation and hence the required heat sink area.

2.8.13 Class D amplifiers

An approach which offers the possibility, in ideal simple theory, of 100% power efficiency is that of the pulse-width-modulated (PWM), or class D, amplifier.

Referring to *Figure 2.113*, suppose that TR_1 and TR_2 are switched hard on, or bottomed, alternately, in square-wave fashion, at some frequency well above the top of the audio band. If this switching is on a 50:50 basis, then the mean voltage fed to the loudspeaker will be zero. By modulating the switching ratio linearly at audio frequency over a maximum range of 100/0 to 0/100, the mean output voltage may be made to vary from $V+$ to $V-$ with theoretically zero distortion.

Since there is ideally no voltage across the transistors when on, and no current through them when the voltage is large, no transistor power dissipation ideally occurs. Provided lossless filtering is added to ensure that only the wanted audio component reaches the loudspeaker, the efficiency of the system, in simple theory, is 100%.

In practice there are many formidable problems to be solved before a satisfactory design, capable of meeting the highest standards of sound quality, can be produced on this basis. An excellent account of these problems and their solution is given by Attwood (1983). Commercial amplifiers, largely based on this work, are now being made in the USA.

An interesting and clear exposition of some of the problems to be faced, as seen from an earlier viewpoint, is given by Birt (1963).

2.8.14 High-output-impedance power amplifiers

It is usually taken for granted that a good audio power amplifier should have a low output impedance, in order to give good loudspeaker damping. However, it is by no means impossible to design a passive loudspeaker to produce first-rate sound quality, with good transient response, when driven by an amplifier with a very high output impedance, and this technique has the advantage that changes in coil resistance due to heating, and the impedance of connecting leads, then have no effect on the response. The non-linearity distortion may also be reduced.

There are various possible approaches to the design of such high-output-impedance amplifiers (Mills and Hawksford, 1989a,b), but one very satisfactory solution, which I adopted in a design for use by KEF Electronics Ltd in a research project, is based on taking an existing commercial d.c. coupled low-output-impedance amplifier as the starting point, as shown in *Figure 2.128*.

Additional negative feedback is derived by inserting a low-value resistor in series with the load, and to obtain enough of this current feedback an op-amp preamplifier is incorporated.

With the loudspeaker disconnected, and in the absence of the broken-line components, it is found that the op-amp offset voltage and bias current are liable to cause an unacceptably large voltage offset at the main amplifier output. This problem is solved by the broken-line addition, which does not affect the performance at signal frequencies.

The inductor L very greatly eases the problem of ensuring good Nyquist stability margins in this additional feedback loop, for any value of load impedance from zero to infinity and for any load phase angle. Its presence is quite harmless so far as output impedance is concerned, but it is liable to reduce the maximum load voltage obtainable at high frequencies, and it is this consideration that determines the maximum inductance value that should be used.

Figure 2.128 *Scheme for obtaining high-output-impedance drive for a loudspeaker.*

The design outlined above gives an output impedance of at least 250 Ω at all audio frequencies, with extremely low distortion. Ordinary wire-wound resistors of low value often generate appreciable harmonic distortion, due both to end-connection non-linearity and to magnetic B–H curve effects in ferrous end caps and/or in the wire itself. Special 0.1-Ω resistors were, therefore, made. Careful layout is required, it being especially important to avoid even a small fraction of a picofarad of stray capacitance from the output of the main amplifier back to the input of the op-amp.

2.8.15 Control units and preamplifiers

A stereo control unit may either be a separate entity, or it may be integrated with the power amplifier and/or radio receiver. Furthermore, in some modern designs, the actual controls are on a hand-held remote-control unit which communicates with the circuitry in the parent unit via a pulsed infrared data-transmission link. When this is done the parent unit may have no manual controls at all, so that it is then better called a 'preamplifier' rather than a 'control unit'.

The use of soft-touch push buttons with CMOS electronic switching plus LED or other visual displays, not to mention remote control, leads to far more complex circuit diagrams than in the past – and these are often very tiresome to interpret because insufficient thought has been devoted to evolving the clearest possible presentation!

Apart from the above aspects, however, the basic nature of a normal control unit has not changed much during the past decade or two (Amos, 1977), although the advent of compact discs, video recorders, NICAM TV stereo and domestic digital sound recording has considerably increased the number of signal sources that may have to be accommodated by the switching scheme.

Figure 2.129 shows a simplified block schematic of a very versatile control unit, in which the push-buttons and their labelling have been arranged in the easiest possible manner to assimilate and use. The scheme permits:

(1) listening to any one of the seven sources; and
(2) entirely independently of (1), selecting any one of the seven sources for recording; this is then fed to both recorders, so that two recordings,

Figure 2.129 *Block schematic diagram of a versatile control unit.*

say ordinary cassette and DAT, may be made at once if desired.

If, however, 'Record from tape 1' is pressed, for making a transcription onto recorder 2 of a tape being played on recorder 1, the switching is arranged to prevent the recorder 1 replay output being fed round into its own input, thus avoiding the possibility of an unstable feedback loop being set up. Similarly, when 'Record from tape 2' is pressed, the switching prevents the output of recorder 2 being fed round into its input.

The facility for listening, for example, to a radio programme while a tape transcription is under way, or listening to a compact disc while a radio programme is being recorded for a friend, is much valued by some users. Many control units do not provide such a facility.

2.8.15.1 Volume control and signal levels

The simplest volume-control arrangement is that shown in *Figure 2.129*. If such a control unit is intended to be usable with associated items of all makes, without the need for preset adjustments, then the fixed gains A_1 and A_2, before and after the volume control, must be very carefully chosen.

The product $A_1 \times A_2$ must be large enough to cope with low-output sources and insensitive power amplifiers. But it is desirable that with the volume control at a low or zero setting, the noise level at the output of a high-sensitivity power amplifier should be below the maximum obtainable output by at least 90 dB. This sets a limit to the maximum value of A_2 that can be adopted. (If the power amplifier sensitivity is 200 mV, i.e. a 200 mV r.m.s sine-wave input will give full output, then a noise level 90 dB

below this is 6.3 μV r.m.s. Using normal bipolar op-amps, the maximum permissible value of A_2 will then be about ×10.) The value of A_1 must then be chosen to suit low-output sources when the control unit is feeding a low-sensitivity power amplifier.

The problem with a versatile design evolved on the above basis is that the value of A_1 then required severely restricts the maximum input level that can be accepted by the A_1 amplifier without clipping – in other words, the 'headroom' is rather small. This may lead to distortion on programme peaks with high-output signal sources.

However, if suitable preset gain adjustments are provided for A_1 and/or A_2, such a control unit may be set up to give an excellent low-distortion performance with any associated equipment, and the preset adjustments, when correctly carried out, will also ensure that the normally used settings of the volume control knob are neither confined too closely towards the anticlockwise end, nor will the user ever feel a desire to turn the knob beyond the fully clockwise limit.

In practice, even if preset adjustments of the above kind are provided, not all owners will take the trouble to utilize them correctly, and the adoption of a negative-feedback volume control scheme is highly advantageous in ensuring very low distortion and excellent noise performance under all conditions of use or misuse.

The simplest arrangement is that shown in *Figure 2.130*. With the slider at or near end 2, the output noise level becomes extremely low – at least 110 dB below full output level for a power amplifier of 500-mV sensitivity. Signal input levels up to a few volts r.m.s. can then be accepted. With the slider at end 1, a gain sufficient to cope with the lowest-output sources and the least-sensitive power amplifiers is provided. All danger of clipping within the control unit is thus completely avoided.

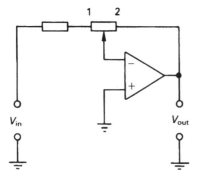

Figure 2.130 *Simple negative-feedback volume-control circuit.*

Stud type gain-control potentiometers, or electronically switched resistor networks, are nowadays fairly widely used, ensuring very accurate tracking between left and right channels and enabling a desired law relating gain to knob setting to be obtained whatever the associated circuit arrangement may be.

Gain control circuits of several kinds exploiting negative-feedback have been used for over forty years, and these, together with some new contributions, are described in detail in Baxandall (1980a,b). It is shown that an excellent gain-control law can be obtained over a wide range using an ordinary linear potentiometer, and that by employing a circuit slightly different from the one shown in *Figure 2.130*, the gain at all settings may be made independent of the absolute resistance of this linear potentiometer and dependent only on its tapping ratio. This enables very good stereo tracking to be obtained using cheap linear-law ganged carbon potentiometers, either rotary or slider, the two sections of which often differ considerably in absolute resistance but only slightly in law. A very satisfactory circuit of this type is shown in *Figure 2.131*.

Figure 2.131 *More subtle negative-feedback volume-control circuit.*

The appropriate input sensitivities which the designer of a control unit should aim to provide need to be carefully considered. The trend in recent years has been to abandon making signal connections between units by means of DIN leads at various, sometimes fairly low, signal levels, and to standardize on so-called 'line-level' interconnections via phonoplug cables. This has simplified the design of control units and has also been beneficial in

mitigating the hum, noise and interference problems that were sometimes experienced in practice.

The highest-level signal source is usually compact disc, and a fully recorded disc gives a peak instantaneous output voltage from a normal player of nearly 3 V. Tests I have made using a very fast-response peak programme meter (Baxandall, 1988) show that most compact discs in fact produce peak levels within 3 dB of this maximum level, and that some, not necessarily of music that is naturally loud, go right up to it. Thus a sensitivity of 1 V r.m.s. for the compact disc input to a control unit should be comfortably adequate, though a sensitivity of 300 mV is more typical of commercial practice.

Interesting information obtained from a manufacturer is that equipment which in early production had a compact disc input sensitivity of 600 mV resulted in complaints from owners that the volume obtainable with the knob fully clockwise was not high enough, and later production was modified to provide 300 mV sensitivity. This, most regrettably I feel, bears out the contention made in Section 2.8.3 that many owners habitually allow a good deal of clipping to occur.

2.8.15.2 Pick-up equalization

Because digital recording is very much superior to analogue recording, both technically and musically (when compared on a fair basis), interest in analogue pick-up equalization has waned in recent years. A six-page treatment of the principles and correct design procedures for RIAA (Record Industries Association of America) equalization networks, including noise and headroom aspects, is given in Amos (1977).

More recently, an excellent paper by Lipshitz (1979), with 36 references, has been published, which gives an exhaustive and rigorously accurate analysis of RIAA equalization circuits of the widely used non-inverting negative-feedback type. The IEC amendment to the original RIAA replay characteristic is taken into account; this introduces an extra 7950 μs low-frequency roll-off time constant corresponding to a corner frequency of 20 Hz. The analysis also allows for finite active-device gain and for the several minor response errors caused by the presence of the usual small-value resistor in series with the earthy end of the R–C feedback network.

The above, which needs prolonged study to assimilate in full detail, caused me (Baxandall, 1981) to put forward an alternative equalization circuit which is very much easier to analyse and design but which nevertheless gives equally accurate results. Favourable comments on this circuit by Lipshitz were also published. The practical circuit is shown in *Figure 2.132*.

With the switch in the position shown, the gain at 1000 Hz is ×10, allowing plenty of headroom to cope with pick-ups of unusually high sensitivity playing records with very high peak recorded velocities. Moving the switch wiper downwards increases the gain in 10 dB steps while fully preserving the high accuracy of the IEC-amended RIAA equalization characteristic.

Figure 2.132 *An unusual RIAA equalization circuit.*

Baxandall (1981) also gives details of a high-precision 'inverse RIAA' circuit for use in checking the accuracy of the RIAA equalization in control units. Frequency response curves obtained with this circuit in cascade with the circuit shown in *Figure 2.132* are also presented, together with perfect-looking square-wave responses. The effects of introducing 5% errors in component values are also shown.

2.8.15.3 Moving-coil pick-ups

Low-impedance moving-coil pick-ups came into fairly widespread use after the treatment in Amos (1977) was written, necessitating the use of either step-up transformers or low-noise preamplifier circuits. Contrary to widely held beliefs, such transformers, of immaculate performance, can be very small and of quite low cost, and are, indeed, particularly easy to design. Using ordinary Mumetal laminations with '2 × square' stack and maximum dimension about 2 cm, two such transformers can be contained in a single deep-drawn cylindrical Mumetal screening can with lid. The coil axes should be at 90° to the can axis, with the cores slightly spaced from each other (1 mm is plenty) and from the can surfaces. The transformers should be towards the closed end of the can, away from the lid. Provided that care is taken to avoid even small-area loops in the external wiring, no hum problems will be experienced.

The circuit shown in *Figure 2.133*, designed in 1977 for Lentek Audio Ltd, can be claimed to be the simplest possible moving-coil pick-up preamplifier capable of giving a first-class performance.

The battery current consumption per stereo channel (0.25 mA) is only that of a single transistor, since the transistors are in series, but the two equivalent voltage-noise resistances act in parallel giving a reduced effective value. By using driver-type transistors, the r_{bb}, noise contributions (Baxandall, 1968a, b) are minimized, and the total equivalent noise resistance for the whole circuit does not exceed 150 Ω, of which 100 Ω is due to the feedback resistor in the emitter circuit. This noise is well below the level of the record noise.

Because of the accurate equality of the mean currents in the two transistors, even-harmonic distortion is reliably balanced out to a high degree, keeping the total distortion, of a benign smooth-curvature kind, well below 0.05% under all normal conditions of use. This distortion level is well below perceptibility (Baxandall, 1991), and is very much lower than the distortion levels present on analogue discs, so that elaboration of the circuit to yield even lower distortion levels, though it could easily have been done, did not seem to be rationally justified.

The ceramic capacitors across input and output suppress VHF interference from radio stations, light switches, etc., and the earthing arrangements for the

Figure 2.133 *Preamplifier for moving-coil pick-ups.*

two channels are such that hum-loops are avoided. Such considerations are of great importance in the design of good audio equipment (Ott, 1976).

2.8.15.4 Anti-hum inputs

If, as may sometimes happen, an item of equipment and the control unit or preamplifier it feeds are both independently connected to third-pin mains earths, a hum loop will be set up if the signal connection between them is made in the normal coaxial-cable manner. However, by arranging the input circuit as shown in *Figure 2.134*, this problem is overcome since there is then nominally zero response to common-mode hum input. The ceramic capacitor suppresses switch-click interference, etc. This scheme is used on two of the inputs in the Quad 66 preamplifier.

Figure 2.134 *'Anti-hum' input circuit.*

2.8.15.5 Tone controls

Tone controls may be used either because of technical shortcomings introduced within the programme chain, or simply because a modified tonal balance is felt to be more pleasing. The widespread tendency in recent years has been to use the description 'equalizer' to cover all such circuits, whereas logically the term implies the correction of a non-flat frequency response; hence its use in relation to circuits whose function is purely artistic really seems rather inappropriate.

Normal bass and treble tone controls A passive bass and treble tone control without switches was described by Voigt in 1940, though without curves. The circuit, with carefully measured responses, is given in Amos (1977). Subsequently, circuits advantageously employing negative feedback have been widely used, most of these being based on a design which I published in 1952 (Baxandall, 1952).

The exact shape of the response curves obtained depends on the circuit design details; two limiting

cases are shown, in somewhat idealized form, in *Figure 2.135*. For adjustment of the general bass and treble musical balance, the type of characteristic shown in (*a*) is usually preferred, but for dealing with background hiss, high-frequency response peaks, or tracing distortion on gramophone records, the adoption of a control with the type of high-frequency characteristic shown in (*b*) enables it to take over the function of a low-pass filter tolerably well in economical equipment not fitted with a separate filter.

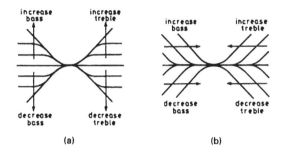

Figure 2.135 *Two basic types of tone-control characteristic.*

My original published design is shown in *Figure 2.136*, although all impedance values have been reduced by a factor of 10 to make them more typical of the semiconductor era. The cheapest versions employ just a single transistor, but lower distortion, more-symmetrical curves and lower output impedance may be obtained by using a transistor pair or an op-amp. In practice it is usually beneficial to add blocking capacitors to prevent the flow of base current, etc. in the control potentiometers, otherwise scraping noises are liable to occur when the knobs are rotated.

The idea of using a centre-tapped linear potentiometer, of adequately low value, to apply variable feed-in or feed-back via a capacitor, thus varying the effective lift or cut corner frequency, did not occur to me until a year or two after the circuit with shelving-type high-frequency characteristics had been evolved. It was pointed out in the original article that by adding a switch in the centre-tap lead, both types of characteristic could be made conveniently available.

The tone-control circuit configuration now most widely used, both in domestic equipment and in mixer desks, is either that the one shown in *Figure 2.136*, with the treble-control centre-tap omitted, or the alternative configuration shown in *Figure 2.137*. This latter version, which I did not publish, is actually the arrangement first used, the

change to the circuit shown in *Figure 2.136*, even with no centre-tap, being made largely on the grounds that it saved a capacitor.

(a)

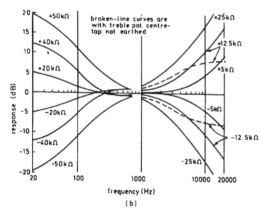

(b)

Figure 2.136 *Baxandall tone contol (Baxandall, 1952). To suit solid-state usage, the impedances have all been reduced by a factor of 10 compared with those in the original valve version. A resistor, e.g. 2.2 kΩ in series with the 1000 pF capacitor, may be necessary for high frequency stability when using op-amps.*

Figure 2.137 *Earlier form of Baxandall tone control.*

The bass-control characteristics given by all these circuits approximate to the type shown in *Figure 2.135(b)*, although the actual curve shapes at intermediate lift and cut settings are more complex, as shown in *Figure 2.136*. Such characteristics have been retained in most contemporary designs. However, bass characteristics of exactly the shape shown in *Figure 2.135(a)* may readily be obtained if preferred, and a bass-only tone-control circuit doing this is given by Amos (1977: Figure 14.26). This could be combined with a treble control in the same circuit, but now that gain has become so cheap, there is a good case to be made for employing separate circuits for bass and treble. This avoids interaction between controls and tends to give better noise performance because large impedance ratios may be avoided. Figure 14.27 in Amos (1977) gives a suitable complementary shelving-type treble circuit.

Accelerated slope tone controls In tone controls of the type described above, simple first-order networks are employed, with the result that, in the case of treble characteristics, the transition from level response at low frequencies to a lifting or falling response at high frequencies is rather gradual, some change in response occurring even at midband frequencies such as 800 Hz. Hence adjustment of the treble control affects the response at these latter frequencies to some extent, and a similar effect occurs on altering the bass control. Although in practice this is not normally found to be objectionable, a recent American contribution (Bohn, 1992) shows how, at the expense of 8 op-amps per stereo channel, this midband gain variation with control settings may be avoided.

The essence of the scheme is that the feed-in and feed-back, mentioned above, take place via low-Q *second-order* C–R circuitry, giving a sharper response transition and an enhanced rate of lift or cut. The much increased internal phase shifts that would normally accompany the use of second-order circuits are prevented from having a highly adverse influence on the shapes of the response curve, owing to partial nulling effects caused by inappropriate phasings, by incorporating additional time-constants to give compensatory phase shifts. A tone control giving virtually zero variation of midband response with knob settings is claimed.

'Tilt' controls A negative feedback tone control of an unusual kind was published by Ambler in 1970. In this ingenious 'tone balance' control, the circuit and measured curves for which are presented in Figure 14.28 of Amos (1977), a single knob varies the relative levels of bass and treble. A slightly more elaborate version, which I evolved in 1979, is shown in *Figure 2.138*, and its response curves extend, with nearly constant slope, over a wider frequency range

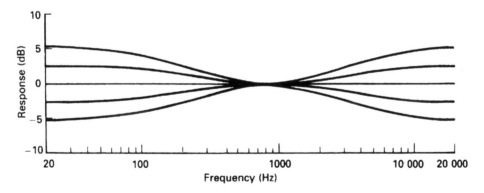

Figure 2.138 *'Tilt' control.*

than for the original design, which used only two capacitors. This type of control has been found very appropriate when pure musical balance is involved, rather than technical shortcomings of a more severe nature.

2.8.15.6 Filters
With inferior quality sources a low-pass filter is sometimes advantageous. Third-order Butterworth response is usually appropriate, and the design procedure for Sallen and Key active filters of this type, using a single op-amp, is presented in Figure 14.21 of Amos (1977) in the easiest possible form to use.

The design procedure for third-order Butterworth high-pass filters is also given.

2.8.15.7 Parametric equalizers
So-called 'parametric equalizers' are frequently provided in mixer desks to allow the response over one or more relatively narrow frequency bands to be raised or lowered. They may be either separate circuits, or incorporated as a modification to the circuitry of bass and treble tone controls to permit variation of the midband response independently of bass and treble. A similar facility may be provided in domestic control units.

(a)

(b)

Figure 2.139 *Parametric equalizer circuits. In (b), a resonance frequency of 1.00 kHz may be obtained with* $R = 10\ k\Omega$, $C_1 = 45.0\ nF$, $C_2 = 5.63\ nF$. *The theoretical Q value is then 1.414, giving a one-octave bandwidth and maximum peak and dip magnitudes of 19.1 dB.*

To obtain symmetrical 'bell-shaped' lift and cut curves, the circuit used should involve the notion of adjustable feed-in or feed-back via a constant-Q tuned circuit. The simplest version is shown in *Figure 2.139(a)*. The L–C tuned circuit may, however, be replaced by an active circuit, as in (b).

Because this very simple selective circuit, unlike the L–C one, is phase inverting, its output is fed to the non-inverting input of the op-amp.

The response of the selective circuit itself is:

$$\frac{V_2}{V_1} = -\frac{2QpT}{1 + (pT/Q) + p^2 T^2} \qquad (2.17)$$

where $T = 1/\omega_o = R\sqrt{C_1 C_2}$, $Q = 1/2\sqrt{C_1/C_2}$, and $p = j\omega$. (Comparably simple non-phase-inverting circuits can be made, with C_1 equal to C_2 and positive feedback, but they are less desirable because the sensitivity to component tolerances is much greater.)

The response of the complete circuit shown in *Figure 2.139(b)*, for lift settings of the control, i.e. with the slider to the left of the centre-tap, is:

$$\frac{V_{out}}{V_{in}} = -\frac{1 + \left[\dfrac{1}{Q} + 4kQ\right] pT + p^2 T^2}{1 + pT/Q + p^2 T^2} \qquad (2.18)$$

The on-tune gain is $-(1 + 4kQ^2)$, where k is as shown on the circuit diagram. The sharpness of the peak is constant for all settings and is determined by the Q value of the denominator.

With the slider moved to the right, the response is given by the reciprocal of equation (2.18), and the constant sharpness of the dip is now determined by the Q value of the numerator.

2.8.15.8 Graphic equalizers

A graphic equalizer, sometimes incorporated in control units, may be regarded as the most versatile of all tone controls, though practical versions vary greatly in their effectiveness.

The best graphic equalizers are of the one-third-octave constant-Q type, and when optimally designed produce a frequency-response curve almost free from ripples which conforms well with the 'graph' depicted by the positions of the vertical slider-type potentiometers (Bohn, 1986).

A series of separate parametric equalizer circuits of the type shown in *Figure 2.139* connected in cascade is one way to make a good graphic equalizer, and one-octave spacing and bandwidth is a reasonable compromise to adopt. The caption to *Figure 2.139* gives values for this condition.

However, the almost universal practice is to associate several, sometimes all, controls with a single main op-amp. (To minimize ripples in the response curves at some control settings, it is preferable not to have the selective circuits for adjacent frequency bands connected to the same main op-amp). When this is done broadly in accordance with *Figure 2.139*, the tuned-circuit-stimulating circuits must be replaced by ones of the non-phase-inverting kind, so that their outputs may be added by feeding them to the inverting input (virtual earth) of op-amp 1. This non-phase-inverting requirement can of course be satisfied, if preferred, by retaining the circuitry associated with op-amp 3 in *Figure 2.139(b)* and adding a phase-inverting stage to it.

The most refined technique for producing active tuned circuits and filters, which is capable when necessary of giving very high Q values with low sensitivity to component tolerances and active device parameters, is based on the two-integrator-loop principle, to which the rather curious term 'state variable' is nowadays often applied. This technique was originated in Britain in the early 1950s by two of my colleagues (Girling and Good, 1970a,b), who designed many two-integrator-loop filters of high precision for use in radio astronomy and other applications. These used valve op-amps.

2.8.15.9 Artificial reverberation

An outstanding development in recent years is the ability, using digital techniques, to produce natural-sounding artificial reverberation, free from the unpleasant colorations often associated with springs and plates. The addition of artificial reverberation, if done with discretion, can sometimes greatly increase the musical enjoyment to be obtained from classical recordings, which not infrequently lack the kind of ambient support one experiences and enjoys in a good concert hall.

Since it is now possible to provide such a facility at a fairly low cost, which need not exceed about £100, it seems likely that it will become available to an increasing extent in future control units. Deciding on the most appropriate characteristics and simplified controls to offer provides much food for thought and experiment, guided, it is to be hoped, by a sensitive musical ear.

References

ALEXANDER, M., A current-feedback audio power amplifier. Presented at: *88th Convention at Montreux*, AES Preprint 2902 (March 1990)

AMBLER, R., Tone-balance control, *Wireless World*, **76**, 124–126 (1970)

AMOS, S.W., *Radio, TV and Audio Technical Reference Book*, Newnes-Butterworth, London (1977). (See Chap. 14 by P.J. Baxandall)

ATTWOOD, B.E., Design parameters important for the optimization of very-high-fidelity PMW (class D) audio amplifiers, *JAES*, **31**, 842–853 (1983)

BAXANDALL, P.J., Negative-feedback tone control, *Wireless World*, **58**, 402–405 (1952)

BAXANDALL, P.J., Noise in transistor circuits, Part 1, *Wireless World*, **74**, 388–392 (1968a)

BAXANDALL, P.J., Noise in transistor circuits, Part 2, *Wireless World*, **74**, 454–459 (1968b). Note: in Fig. 9, middle diagram, the current generator labelled $\sqrt{2qI_c B/g_m}$ should be $\sqrt{2qI_c B/r_{b,e} g_m}$

BAXANDALL, P.J., Letter to Editor, *Wireless World*, **75**, 416–417 (1969). (Shaw (1969) requires a power diode, whereas this scheme uses a small diode or transdiode. Six transfer-characteristic photographs are presented)

BAXANDALL, P.J., Letter to Editor, *Wireless World*, **76**, 386 (1970). (This corrects a small error and replies to correspondence raised by Baxandall (1969))

BAXANDALL, P.J., Current dumping audio amplifier, *Wireless World*, **82**, 60–61 (1976) (letter). (This comments on the background to the evolution by Quad of their current dumping technique and puts forward the alternative feedback explanation of *Figure 2.124*)

BAXANDALL, P.J., Audible amplifier distortion is not a mystery. *Wireless World*, **83**, 63–66 (1977)

BAXANDALL, P.J., Audio power amplifier design, Part 1, *Wireless World*, **84**, 53–57 (1978a)

BAXANDALL, P.J., Audio power amplifier design, Part 5, *Wireless World*, **84**, 53–56 (1978b). (This article and Baxandall (1979) deal with the theory of non-linearity distortion in negative-feedback amplifiers)

BAXANDALL, P.J., Audio power amplifier design, Part 6, *Wireless World*, **85**, 69–73 (1979). (This presents an 'inverse distortion' theorem, as used in Cherry and Dabke (1986))

BAXANDALL, P.J., Audio gain controls, Part 1, *Wireless World*, **86**, 57–62 (1980a)

BAXANDALL, P.J., Audio gain controls, Part 2, *Wireless World*, **86**, 79–83 (1980b)

BAXANDALL, P.J., Comments on 'On RIAA equalization networks', *JAES*, **29**, 47–53 (letter) (1981)

BAXANDALL, P.J., A technique for displaying the current and voltage output capability of amplifiers and relating this to the demands of loudspeakers, *JAES*, **36**, 3–17 (1988)

BAXANDALL, P.J., Letter to Editor, *Electronics World and Wireless World*, **96**, 1044–1045 (1990a). (This makes further comments relating to the topic of Baxandall (1990b))

BAXANDALL, P.J., Letter to Editor, *Electronics World and Wireless World*, **96**, 520–522 (1990b). (This is a criticism of subjective amplifier assessments made by a well-known reviewer)

BAXANDALL, P.J., Letter to Editor, *Hi-Fi News*, **36**, 8–9 (1991). (This describes a subjective crossover-distortion test and a test in which distortion was added to the sound of a real piano. *Note*: paragraph (a) should end with 'when presented by itself')

BIRT, D.R., Modulated pulse AF amplifiers, *Wireless World*, **69**, 76–83 (1963)

BLOMLEY, P., New approach to class B amplifier design, *Wireless World*, **76**, 57–61, 127–131 (1971)

BOHN, D.A., Constant-*Q* graphic equalizers, *JAES*, **34**, 611–626 (1986)

BOHN, D.A., Accelerated slope tone control equalizers, *JAES*, **40**, 1018–1024 (1992)

CHERRY, E.M., A new result in negative-feedback theory and its application to audio power amplifiers, *International Journal of Circuit Theory*, **6**, 265–288 (1978)

CHERRY, E.M., A power amplifier 'improver', *JAES*, **29**, 140–147 (1981)

CHERRY, E.M. and DABKE, K.P., Transient intermodulation distortion – Part 2: soft nonlinearity, *JAES*, **34**, 19–35 (1986). (There is a summary of Part 1 and 35 references)

CORDELL, R.R., A MOSFET power amplifier with error correction, *JAES*, **32**, 2–17 (1984). (There are 20 references)

DAVIS, J.C.H., Total differential feedback, *Electronic and Radio Engineer*, **35**, 40–44 (1958)

DIGGINS, J.E., Clipped-sine Fourier analysis, Letter to Editor, *Electronics World and Wireless World*, **96**, 1046 (1990). (See errata in **97**, 151 (1991))

FAIRWOOD, R.C. and REED, M.J., Digital measurement and comparison of current-voltage output capability of amplifiers and demands of loudspeakers, *JAES*, **39**, 344–354 (1991)

FINNEGAN, T., Going linear with power MOSFETs, Part 1, *Wireless World*, **97**, 635–640 (1992a)

FINNEGAN, T., Going linear with power MOSFETs, Part 2, *Wireless World*, **97**, 778–782 (1992b)

GATES, T.W. and BALLARD, M.F., Safe operating area for power transistors, *Mullard Technical Communications*, **13**, 42–65 (1974)

GIRLING, F.E.J. and GOOD, E.F., Active filters: Part 7, *Wireless World*, **76**, 76–80 (1970a). (This and the following reference are parts of a 16-part series extending from August 1969 to December 1970)

GIRLING, F.E.J. and GOOD, E.F., Active filters: Part 8, *Wireless World*, **76**, 134–139 (1970b)

IEC, *IEC 268-5 Loudspeakers*. (Equivalent to BS 5428: 1977)

IEC, *IEC 268-3 Amplifiers* (1988)

LIN, H.C., Quasi-complementary transistor amplifier, *Electronics*, **29**, 173–175 (1956)

LIPSHITZ, S.P., On RIAA equalization networks, *JAES*, **27**, 458–481 (1979)

LIPSHITZ, S.P. and VANDERKOOY, J., The great debate: subjective evaluation, *JAES*, **29**, 482–491 (1981). (This is a very well-presented plea for much greater care in the carrying out of subjective assessments, and the discarding of unfounded beliefs. There are 49 references)

McLOUGHLIN, M., Reducing crossover distortion, *Electronics World and Wireless World*, **96**, 879–882 (1990)

MILLS, P.G.L. and HAWKSFORD, M.O.J., Distortion reduction in moving-coil loudspeaker systems using current-drive technology, *JAES*, **37**, 129–148 (1989a). (There are 41 references)

MILLS, P.G.L. and HAWKSFORD, M.O.J., Transconductance power amplifier systems for current-driven loudspeakers, *JAES*, **37**, 809–822 (1989b). (There are 28 references)

NOBLE, P.G., The safe operation of power transistors, *Mullard Technical Communications*, **14**, 346–375 (1978)

OGDEN, F., 'The current alternative to operational amplifiers, *Electronics World and Wireless World*, **97**, 643–644 (1992)

OTT, H.W., *Noise Reduction Techniques in Electronic Systems*, Wiley, New York (1976). This is a commendably lucid treatment of the fundamental principles of interference production and suppression, by a member of the staff of BTL

SELF, D.R.G., Science v. subjectivism in audio engineering, *Electronics and Wireless World*, **94**, 692–696 (1988a). (This

is an eloquent attack on contemporary irrational subjectivist attitudes. There are 19 references)

SELF, D.R.G., Self assured, *Hi-Fi News*, **33**, 28–29 (1988b). (This is a shorter version of Self (1988a). There are 14 references)

SHAW, I.M., Quasi-complementary output stage modification, *Wireless World*, **75**, 265–266 (1969). (Distortion curves are given)

TIJOU, J.A., High quality 25 W audio amplifier, *Mullard Technical Communications*, **13**, 280–288 (1975)

VANDERKOOY, J. and LIPSHITZ, S.P., Current dumping – does it really work?, *Wireless World*, **84**, 38–40, 83–85 (1978)

VANDERKOOY, J. and LIPSHITZ, S.P., Feedforward error correction in power amplifiers, *JAES*, **28**, 2–16 (1980). (There are 45 references)

VOIGT, P.G.A.H., Getting the best from records, *Wireless World*, **46**, 210–213 (1940)

VON RECKLINGHAUSEN, D.R., Class B amplifier dissipation, instantaneous and steady state, *IEEE Transactions on Audio*, **AU13**, 83–87 (1965)

WALKER, P.J., Current dumping audio amplifier, *Wireless World*, **81**, 560–562 (1975). (See also Letters to Editor, **82**, 54–55 (1976))

WALKER, P.J. and BAXANDALL, P.J., Letters to Editor, *Wireless World*, **89**, 49–50 (1983). (These two letters discuss criticisms of the Quad 405 amplifier by M. McLoughlin *et al.*)

2.9 Limiters and compressors

The operator or sound engineer in a studio controls the level of different sound sources to produce a well-balanced programme. That is, the listener at home when listening to a recording or a broadcast programme does not want to have to continually adjust their listening level during a piece of music or between items. From the threshold of hearing to the threshold of pain the dynamic range is about 140 dB. Even with digital recording techniques this dynamic range is too great if noise on low level signals and distortion on peaks are to be avoided. This is shown by the graph in *Figure 2.140* of the equal-loudness curves for the human ear. The ear is least sensitive at low and high frequencies and most sensitive to the middle speech band. This can result in 10–20 dB extra level being required if bass instruments are to sound as loud as a vocal line in a mix.

The operator always needs to be aware of sudden programme peaks that can cause overload distortion in a sound desk or saturation on tape. In a radio studio, for example, a guest speaker can change from being quiet and timid to excitable and shouting in just a few moments. It may not be possible to predict this sudden explosion of passion and it is inevitable

Figure 2.140 *Equal-loudness curves.*

that, unless the signal level of the output of the microphone is reduced quickly, distortion will result. However, if the programme level is too low it may be difficult to appreciate a performance fully because noise may be audible throughout. Each item within the programme needs to be loud enough to be heard, yet not too loud. Thus one important function of the sound-desk operator is to control the dynamic range.

There is a range of studio equipment that can change the dynamic range of a programme item far quicker than a human hand can react. These devices are collectively known as *dynamics*. The gain of a dynamic device is dependent on either the level of the programme passing through it, or on some other audio signal that is used to provide a control signal.

The most common dynamic device found is the compressor which reduces the dynamic range of a programme. Another dynamic device, called an 'expander', increases the dynamic range. The function and audible effect of these two devices are very different but their design is very similar. There are other less common dynamic devices which insert filters into a programme chain. These devices include dynamic noise filters and de-essers.

2.9.1 Compressors

A compressor is a device that reduces the dynamic range of an audio signal with no perceptible distortion when operated correctly. A compressor can be thought of as an automatic fader that is capable of controlling the level of an audio signal very

much faster and more predictably than is possible by manual methods. Compressors are installed in many sound desks and they also exist as separate 'out-board' pieces of studio equipment. As an audio processing tool, compressors can be protective or creative.

(1) *Protective* The dynamic range of an audio signal can be reduced to prevent overload of sound desks or tape saturation, so reducing the risk of distortion. Compressors are used in studios to reduce the dynamic range of unpredictable signals like the human voice or some live musical instruments. Compressors that prevent any programme peaks above a maximum level are known as *limiters*. They are used in studios and are fitted at the input of transmitters and some digital equipment.

(2) *Creative* In a studio, compressors can be used creatively to reduce the dynamic range of audio signals so as to produce a more evenly controlled or differently balanced sound (i.e. altering the natural dynamics of an instrument or voice). Most important of all, compressors can increase the perceived loudness of a signal without increasing the peak programme level.

Compressors can also be used to produce special effects such as 'voicing over' a music item in a programme. This is an effect where a compressor is used to reduce the level of a piece of music whenever a presenter speaks or a vocalist sings.

2.9.2 Compressor characteristics

Low-level signals entering a compressor pass through to the output unaffected. As the input level increases, a point called the *threshold* is reached. When the input level exceeds the threshold the gain of the compressor is reduced and so the output level is altered relative to the input level. This is shown by the two graphs in *Figure 2.141* which represent the static or steady-state characteristics of a compressor.

The threshold of most compressors can be varied over a wide range from as low as −30 dBm up to about +12 dBm. The rate at which the output level of a compressor increases when the input level exceeds the threshold can usually be varied and is known as the *compression ratio*. This is the ratio of increase in input level (above the threshold) to increase in output level (above the threshold) (in decibels):

$$\text{Compression ratio} = \frac{\text{Increase in input level}}{\text{Increase in output level}}$$

(a)

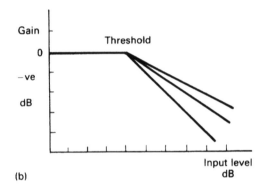

(b)

Figure 2.141 *(a) Input/output characteristic of a compressor. (b) Input gain characteristic of a compressor.*

With a compression ratio of, say, 2:1 an increase in the input level of 2 dB *above the threshold* will produce a corresponding increase in the output of only 1 dB. Compression ratio is measured against a logarithmic scale and so a 2:1 compressor is also known as a *square-law compressor*. The compression ratio is usually variable either continuously or in switched steps up to as high as 50:1 on some equipment. A compression ratio of 20:1 or greater is known as *limiting*. The function of a limiter is to prevent any programme peaks above the threshold.

Compressors exist with characteristics as shown in *Figure 2.141*, but they are not always liked by operators as the onset of compression is sudden and can be audible. This type of compressor is said to have a 'hard knee'. Many compressors have a less stylized characteristic as shown in *Figure 2.142*. The onset of compression is less sudden and the compression ratio increases as the level of the input signal increases. This is known as a 'soft knee' compressor. The compression ratio is usually quoted for a point where the output level has been reduced

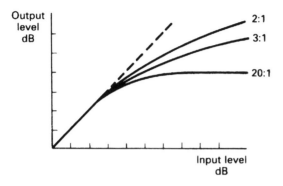

Figure 2.142 *Soft-knee characteristic of a compressor.*

by 4 or 6 dB. The exact point at which the ratio is calculated is not that critical, as the quality of the output signal is more dependent on the amount of gain reduction rather than the individual settings of the threshold and ratio.

Reducing the compression ratio or increasing the threshold level will both reduce the amount of compression. For any piece of programme material it is possible to choose many combinations of threshold and ratio that will produce a similar quality of sound output with similar amounts of level reduction. The choice of threshold and ratio setting depend on the dynamic range of the programme required after compression. If the compressor is used in its protective role a compression ratio of 20:1 or greater will be required. In this mode the compressor is operating as a limiter and the programme will be prevented from exceeding the threshold level. If a lower compression ratio is used, some dynamic range will be preserved above the threshold which will have to be lower to produce the same amount of compression. In this mode the compressor can be used more creatively.

2.9.3 Compression meters

Most compressors have a meter to indicate the amount of gain reduction that has been introduced by the compressor. This meter may be a conventional moving coil meter or a column of LEDs. This shows the 'amount of compression' or the 'decibels of compression'. The instantaneous value of gain reduction is often of little interest except to warn if an excessive amount of compression is being used. The movement of the gain reduction meter will indicate how rapidly the level of the programme is being changed. For example, if a large amount of gain reduction is being shown, yet the meter indication is moving slowly, the compressor is effectively acting as an attenuator.

2.9.4 Compressor systems

There are two paths for the audio signal: a programme path and a control path or 'side-chain' (*Figure 2.143*). The audio signal at the input to the compressor passes through a variable gain amplifier to the output. The gain of this amplifier depends on the control signal from the side-chain and its operational control adjustments. The input to the side-chain can be derived from either the input to the compressor or from its output.

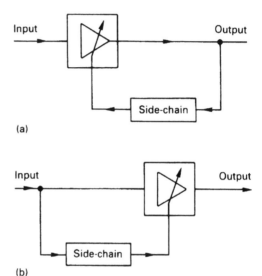

Figure 2.143 *(a) Feed-backwards compressor. (b) Feed-forward compressor.*

2.9.4.1 Feed-backwards compressor

The most apparent drawback with a feed-backwards compressor, as shown in (*Figure 2.143(a)*) is that a sudden transient above the threshold level will have to pass through to the output before the side-chain can detect the peak and reduce the gain of the amplifier. The side-chain is always too late to react to transients and a real programme signal is full of transients.

The advantage of the design is that it is a 'closed loop servo system' and so the characteristics of the compressor can be tightly controlled. The feed-backwards design is used in fast acting peak level limiters where the main requirement is that little if any programme will exceed the threshold level.

2.9.4.2 Feed-forward compressor

It may appear that the feed-forward compressor (*Figure 2.143(b)*) overcomes some of the problems

of transient overshoot that are found in the feed-backwards design. However, a transient signal reaches the input to the amplifier at the same time as the signal reaches the side-chain. The propagation delay of the variable amplifier is in the order of nanoseconds, and so is insignificant. The result is that transients still reach the output before they can be controlled.

In practice, there is no real advantage to the design of either the feed-forward or the feed-backwards compressor, and there are roughly equal numbers of both designs in production. Some compressors make use of both designs, with a feed-forward compressor for creative effects and a feed-backwards peak limiter (*Figure 2.144*).

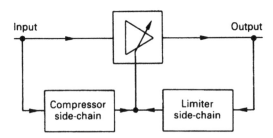

Figure 2.144 *A compressor/limiter with two side-chains.*

2.9.5 Gain controls

The action of a compressor reduces the level of programme material. To achieve a specific effect a large amount of gain reduction may be required, which may result in the output of the compressor being undesirably low in level. It is therefore common to find compressors which have a variable gain amplifier on output and a level control known as *gain make-up*. Another approach is to keep the output gain fixed and amplify the input signal so that more of the programme material is above the threshold. The input gain control therefore directly adjusts the amount of compression.

Some compressors do away with the threshold control altogether and instead use input and output gain controls. One advantage of this arrangement of operational controls is that setting up the device is very straightforward. The input level can be adjusted quickly and creatively by ear to give the amount of compression desired. The output gain control adjusts the peak level of the programme.

The input and output gain controls are combined on some compressors to act as a threshold control. As the input level is increased a dual or 'ganged'

control also reduces the output level. This arrangement is shown in *Figure 2.145*. If the input signal is amplified by, say, 10 dB, more of the programme will be above the threshold and will be compressed. If the output level is attenuated by the same amount, 10 dB, it will appear to the operator that the ganged control is adjusting the threshold level.

Figure 2.145 *Gain controls used instead of threshold.*

2.9.6 Variable gain amplifier

The simplest variable gain amplifier is a single field-effect transistor (FET) acting as a potential divider (*Figure 2.146*). Unfortunately, the characteristics of an FET are far from linear (*Figure 2.147*). This simple design can therefore really only be used in feed-backwards compressors where it is in a closed-loop feed-back system. The design is very common and has been used to great effect in some very successful compressors. The maximum amount of attenuation produced by an FET is about 20 dB, which can be used to great effect in compressor design.

Figure 2.146 *Field-effect-transistor (FET) variable gain amplifier.*

FETs have been used in feed-forward designs with a modified control circuit to take into account the

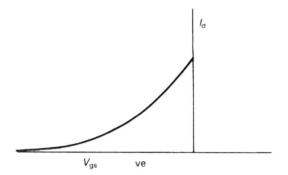

Figure 2.147 *Field-effect-transistor characteristic.*

non-linear characteristics. *Figure 2.148* shows two FETs acting as variable gain amplifiers. The main programme path passes through the second variable gain amplifier (FET2) with a side-chain taken from its input. This side-chain is arranged to form a feedback loop around the first amplifier (FET1). The two

FETs need to be the same type and must have very closely matched characteristics. The non-linearity of the FET in the programme path is compensated for by the FET in the side-chain.

There have been many designs of transistor-based variable gain amplifiers. One of the oldest designs uses a modified 'long-tailed pair' in a configuration known as a *variable transconductance amplifier (Figure 2.149)*. The gain is given by:

$$\text{Gain} = \frac{-g_m RL}{2} \quad (2.19)$$

where g_m is the transconductance of a transistor (in mA/V). For most practical purposes:

$$g_m = 35 \times \text{Collector current} \quad (2.20)$$

The result is that if the current passing through the long-tailed pair is changed, the gain of the amplifier is also changed. The collector current is altered by controlling the current generator transistor (Tr3 in *Figure 2.149*).

Figure 2.148 *A modified control system.*

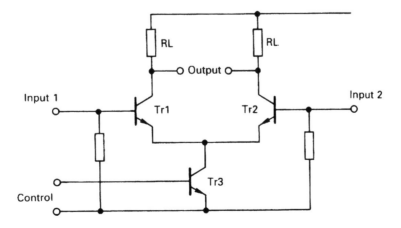

Figure 2.149 *A variable transconductance amplifier.*

The problem with this design is that the input signal must be kept small or distortion will result. A line-level input signal has to be attenuated and the output then amplified. The result of this action is that the signal/noise ratio will suffer.

The long-tailed pair circuit is also known as a 'two-quadrant multiplier'. That is, the audio input signal can swing both positive and negative, yet the control signal is always positive. An improved design uses a four-quadrant multiplier in the negative-feedback circuit of an operational amplifier. This is shown in *Figure 2.150*. The operational amplifier is used to change a control signal that rises with input signal level into a controlled, reduced output level:

$$\text{Output of the multiplier} = kV_xV_y \qquad (2.21)$$

Figure 2.150 *A quadrant multiplier as a variable gain amplifier.*

where k is a constant, and V_x and V_y are the input signals to the multiplier.

Treating the operational amplifier as a virtual earth:

$$\frac{V_{in}}{R_2} = \frac{-kV_xV_y}{R_1} \qquad (2.22)$$

$$V_{out} = \left(\frac{-R_1}{kR_2}\right)\left(\frac{V_{in}}{V_{cont}}\right) \qquad (2.23)$$

That is, the output signal is proportional to the input signal divided by a control voltage. The control voltage needs to have a positive bias for unity gain through the amplifier. Also care must be taken to ensure that the control voltage is not too low, as the high gain produced could cause distortion.

Some compressors use a purpose-built voltage-controlled amplifier (VCA) that is available from many manufacturers. A VCA is basically a three-terminal device with an input, an output and a control input (*Figure 2.151*). The advantage of these VCAs is that they are relatively cheap, the noise level is very low, the dynamic range is wide and the control voltage will give an accurately known amount of attenuation.

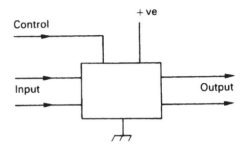

Figure 2.151 *A single-chip voltage-controlled amplifier.*

The design of the compressor is therefore very much simplified.

Most purpose-built VCAs are known as 'log-antilog' devices. That is, the input audio signal is passed through a logarithmic amplifier, a control signal is added and the resulting signal is passed through an antilog circuit. This is shown in *Figure 2.152*. Addition or subtraction of a number from a log is the same as multiplication, and multiplication in audio terms is the same as changing the level of a signal. If the control signal is also passed through a log amplifier we have a VCA that will produce a given amount of attenuation (in decibels) for a linear change in control signal. A typical value is 3 mV per decibel of attenuation.

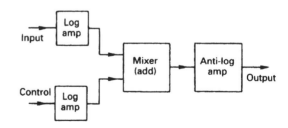

Figure 2.152 *A log-antilog voltage-controlled amplifier.*

2.9.7 Side-chains

The function of the side-chain circuit is to convert the input (or output) audio signal into a control signal that reflects the level of this audio signal. A

side-chain circuit typically consists of the following elements:

(1) a full-wave rectifier,
(2) a threshold circuit (comparator),
(3) dynamic control (attack and recovery), and
(4) ratio control (gain).

The order of these items changes from one design to another, but the design requirements are usually the same. *Figure 2.153* shows typical side-chain arrangements.

2.9.7.1 Full-wave rectifier

A full-wave precision rectifier (*Figure 2.154*) is required because audio signals are far from symmetrical and a half-wave rectifier would miss many transients. A full-wave rectifier can be used to give a signal that is proportional to either the mean or the peak level of the audio signal. It is the peak level signal which is an essential requirement for a peak limiter.

In some compressors a root-mean-square to d.c. converter is used instead of the precision rectifier. The r.m.s. level can give a better indication of the effective loudness of programme. This is especially important if the compressor is to be used for creative or artistic purposes. Single-chip r.m.s./d.c. converters are readily available and are widely used in compressor designs.

2.9.7.2 Threshold comparator

The threshold circuit compares the output of the rectifier with a reference voltage. A simple comparator circuit formed by an additive mixer and a rectifier is shown in *Figure 2.155*. The control signal level must be greater than the threshold level if a signal is to pass through the rectifier. Within some compressors

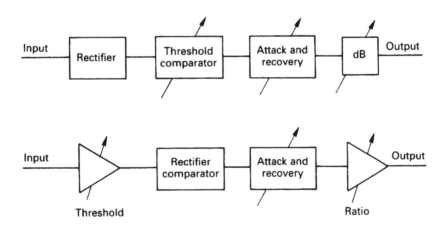

Figure 2.153 *Typical side-chain arrangements.*

Figure 2.154 *A full-wave precision rectifier.*

Figure 2.155 *A threshold comparator.*

the comparator function is performed by the full-wave rectifier, which is designed to only pass signals above a set level.

2.9.7.3 Ratio

The compression ratio controls the gain of the side-chain circuit. If no control signal feeds through to the VCA the compression ratio will be 1:1. If the full control signal feeds the VCA the device will act as a limiter. If only half the amplitude of the control signal reaches the VCA the compression ratio will be 2:1, and so on.

The gain of the side-chain circuit is calibrated to avoid over-reduction of the compressor's output level. If the gain is too great, overlimiting would result because the level of the output signal is increasingly reduced as the input level increases. This is shown in *Figure 2.156* and would give a very strange and probably undesirable effect.

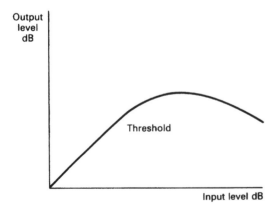

Figure 2.156 *Overlimiting.*

2.9.7.4 Dynamic control

The dynamic control section affects the way in which the compressor reacts to changes in input programme

level. When a transient which exceeds the threshold enters a compressor, it takes a finite time for the gain to be reduced. This is known as the compressor *attack* (or *attack time*). The rate at which the compressor gain is returned to normal when the input signal falls below the threshold is called the *recovery* (or *recovery time*).

The dynamic controls have a major effect on the quality of the sound produced by the compressor and are the most difficult to set up operationally. The adjustment of the attack and recovery times are therefore described here in detail.

2.9.8 Attack

A compressor or limiter will take a small but finite time to react to a signal that suddenly rises to exceed the threshold level. The time taken for the device to react and reduce its gain is known as the 'attack time' and on some compressors this is adjustable.

Figure 2.157 shows the output of a limiter where an input signal has suddenly exceeded the threshold level. On a typical compressor, attack times can be adjusted between about 100 µs (fast) and 25 ms (slow). During the time it takes for the gain of the limiter to be reduced, the output level exceeds the threshold level. This could cause an overload of a sound desk, etc., and distortion of the audio signal may result. This can be a particular problem with programme material that has many short and high peaking transients, e.g. a piano or harpsichord. With a slow attack time virtually all programme peaks will exceed the threshold level.

It may appear that a very fast attack time is required to control these transient overshoots. However, if a compressor has a fast or near-zero attack time, transients that peak above the threshold level will be altered by the sudden change in compressor gain. The way the human ear recognizes the character of a sound is very dependent on the shape of transients. Very fast attack time, shorter than about 100–150 µs, can over-reduce transients and this effect is audible as a form of distortion.

The audible effect of adjusting attack time, apart from the distortion that can arise if it is too fast, is fairly subtle. It is possible to hear a slight tonal quality change with some material due to the frequency content of transients that passes through the compressor uncontrolled. Many compressors have either a fixed attack time of about 1 ms or the attack can only be switched between a few values, none of which will cause distortion.

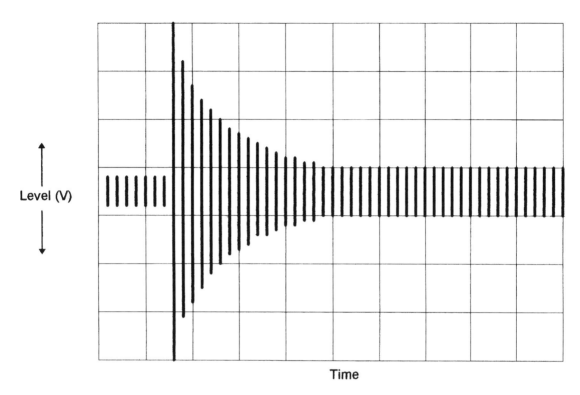

Figure 2.157 *The output of a limiter showing attack.*

There is no single agreed method for the measurement of attack time. Some manufacturers specify the time for a given overshoot to reach its final level or 80% of the final level. One method is to measure the time for the gain to be reduced by 4 dB from a 12-dB overshoot. This allows easy measurement using an oscilloscope as the 4-dB point is halfway between the 12-dB overshoot and the final signal level (*Figure 2.157*). Different measurement techniques can give different values for the same device but, as the attack time can be varied up to 100:1, a general order of magnitude is all that is necessary for an operator who will adjust the control 'by ear'.

A slow attack time which avoids unpleasant distortion effects will allow sudden programme peaks to pass through uncontrolled. Some equipment, such as digital coders and MF transmitters, cannot tolerate any programme peaks above a maximum level. Overshoot-free compressors have therefore been developed that do not produce the undesirable side-effects of a fast attack time, yet control all programme peaks.

The input signal to the compressor is delayed for a short time (typically 200 µs). The input to the side-chain is taken before the delay line, so that

the gain of the compressor can be smoothly reduced ahead of the signal peak (*Figure 2.158*). When the delayed signal peak enters the compressor's VCA the gain will have already been reduced. The result is a distortion-free output with no programme overshoots of the compressor threshold.

Figure 2.158 *A delay line compressor.*

The delay line compressor can be thought of as a perfect protection device as no programme peaks will appear above the threshold and the rate of gain reduction does not cause distortion. However, there are several disadvantages to this compressor design. The delay line is expensive and is usually constructed from coils and capacitors to simulate a long length

of cable. Charge-coupled devices have been used to produce the delay, but these are noisy and degrade the performance of the compressor. In addition, the very success of the limiter to catch all peaks above the threshold tends to change the tonal quality of the output signal. In studios, some operators complain that the compressor sounds 'dull' or lacking in high frequency when compared with other compressors that allow transients to pass through. Delay line compressors are therefore mostly used as protection limiters rather than for creative uses.

2.9.9 Recovery

The recovery time (or release time) is the time taken for a device to return to normal when the input signal falls below the threshold (*Figure 2.159*). As with attack time there is no agreed system for the measurement of recovery time. Again, some manufacturers quote the time for the gain to recover completely and others quote 80% recovery. A simple method using an oscilloscope is to measure the time for signal to reach within 4 dB of its final value from 12 dB of gain reduction. The 4-dB point is halfway between the initial level and final level when measured using an oscilloscope.

Accurate calibration of the front panel controls is not that important, as an operator will adjust the recovery by ear to give the desired result. Some compressors do no more than label a control with 'fast' and 'slow' at the extremities.

The action of compressor recovery changes audio signals. This change is often audible and can be creative or unpleasant depending on control settings. The recovery time can usually be adjusted over a very wide range from, say, 25 ms (fast or short) up to 3.5 s (slow or long). The choice of a recovery time depends very much on the type of programme being compressed. A short recovery time will allow the gain of a compressor to return to normal, during breaths in speech for example. This can have the very useful effect of increasing the subjective loudness of a signal, especially speech, without an increase in peak programme level.

Although the compressor reduces high-level signals, the ear is aware of low-level background noises rising and falling. This rapid rise and fall can draw attention to noises that might have been otherwise acceptable, such as studio rumble, ventilation noise,

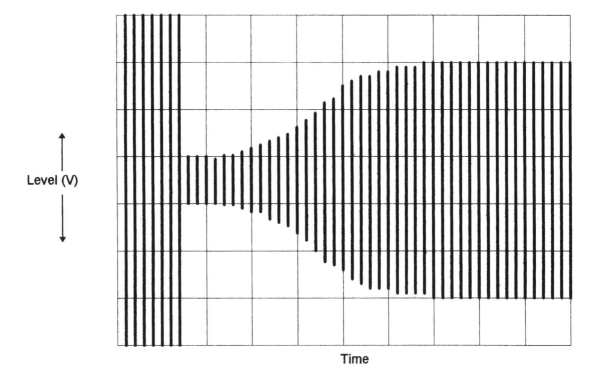

Figure 2.159 *The output of a limite- showing recovery.*

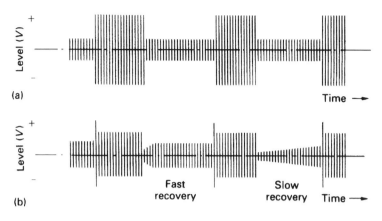

(a)

Time ⟶

(b) Fast recovery Slow recovery Time ⟶

Figure 2.160 *Input (a) and output (b) signals showing modulation.*

tape hiss, or print-through (*Figure 2.160*). This modulation effect is known as 'pumping'.

A long recovery time will prevent pumping, but will also reduce the subjective loudness. If the recovery is long the gain reduction is very nearly static so that the compressor is acting as an attenuator. If this is the case the programme level could be reduced with an attenuator and the compressor be put to better use elsewhere. A very undesirable feature of a long recovery time is that a short duration high peak can 'punch a hole' in a programme. The short-duration peak which caused gain reduction can be long past, yet the level of the programme is still reduced and only slowly returns to normal.

There is no single correct recovery time, but with speech it may be possible to use a fairly fast recovery (e.g. 200 ms) to give the greatest increase in subjective loudness. Music and drama will require different recovery times, depending on the reason for using the compressor in the first place. An operator therefore tends to find that recovery time has to be adjusted continually as the nature of the audio signal changes.

2.9.10 Auto-recovery

As we have seen, recovery time has to be adjusted according to programme content. This can cause problems when choosing an optimum recovery time for a fixed limiter. Some compressors have an auto-recovery facility which allows programme-controlled recovery. Short duration programme peaks tend to be high-level transients and a fast recovery time of between 20 and 100 ms is well suited for such signals. This would produce an increase in perceived loudness and is ideal for speech and some 'pop'

music. A long duration piece of high level programme would benefit from a slower recovery from around 500 ms up to 1 s in order to minimize pumping. This is the function of auto-recovery which gives a two-stage recovery: a fast stage, followed by a much slower one. Some compressors even have three stages of recovery time with the longest recovery of around 10 s. Auto-recovery can be a good compromise for most types of programme, and operationally can be a good starting point when setting up equipment.

2.9.11 Control of attack and recovery time

The control of both attack and recovery times is usually based on the charging and discharging of a capacitor (*Figure 2.161*). During the attack period the capacitor is charged through the series resistor. When the programme level falls the capacitor is discharged through the parallel resistor. The voltage across the capacitor falls exponentially with time, which produces a logarithmic change to the

Figure 2.161 *Control of attack and recovery time.*

VCA, but as the human ear's perception of loudness is proportional to logarithmic changes in level, this sounds like a smooth change in programme level.

Auto-recovery circuits can be created with a double capacitor discharge circuit (*Figure 2.162*). The exact values of the capacitors and discharge resistors are decided following subjective assessments made at listening tests. A general compromise is for about two-thirds of the voltage to be developed across the top capacitor (C1) which has a discharge time constant of around 30 ms. The lower capacitor (C2) has a discharge time constant of around 700 ms. The result is that the recovery is initially fast but then slows down. This helps to give an increase in perceived loudness, reduces pumping, and helps to avoid too much interference with long reverberation tails.

Figure 2.163 *A constant-current recovery.*

Figure 2.162 *An auto-recovery circuit.*

Figure 2.164 *Auto-recovery for single-chip voltage-controlled amplifiers.*

2.9.12 Side-chain inputs

The description of recovery circuits has so far been based on VCAs that have a linear relationship between the control voltage and the amount of output level reduction. Most single-chip VCAs have a logarithmic relationship between the control voltage and level reduction, i.e. changes in control voltage give a linear change in attenuation (in decibels). These VCAs use a different recovery circuit. The attack time is still determined by charging a capacitor through a resistor, but the recovery time is controlled by discharging the capacitor with a constant-current source (*Figure 2.163*). This produces the same audible effect as the conventional resistor discharge circuit for a linear controlled VCA. Some compressors then have an auto-recovery circuit that is based around a single resistor discharge circuit, similar to the conventional recovery circuit. The logarithmic relationship between control voltage and attenuation produces a recovery that starts fast and then slows down (*Figure 2.164*).

Compressors often allow access to the side-chain circuit that will allow a different audio signal from the one passing through the main programme chain to control the gain of the device. This is known as the 'side-chain' or 'voice-over' input, as one programme source is said to 'voice-over' another. For example, the level of a piece of music can be reduced whenever a presenter speaks.

It is not difficult to use a compressor as a voice-over device. The compressor is inserted in the sound-desk channel of the audio signal that is to be controlled, e.g. a source of music. The voice-over input to the compressor is taken from a spare feed of the controlling signal which could be a microphone channel. This is shown in *Figure 2.165*. Whenever the presenter speaks, the level of the music will be reduced. The success of the effect is very dependent on the recovery time of the compressor, which determines how quickly the music returns to its normal level when the speech stops.

Figure 2.165 *A voice-over system.*

One problem with the voice-over process is that the amount of gain reduction applied to the music is dependent on the level of the speech. If the presenter speaks quietly, the level of the music will hardly be reduced and it may be difficult to hear what is said. If the presenter shouts it will be much easier to hear the voice because the music will be greatly reduced in level by the voice-over process. The technique can be greatly enhanced by compressing the side-chain signal into the compressor (*Figure 2.166*). The level of the controlling signal is kept more constant by the limiter on the side-chain input. The result is that the music is reduced in level by a near-constant amount, regardless of the loudness of the speech.

Figure 2.166 *A compressor in the side-chain of a voice-over system.*

It must be said that the voice-over effect produced by a compressor is a little 'mechanical' and, if not set up carefully, the level of the music can jump up every time the presenter takes a breath. Many operators prefer to control the level of the music manually, as it is generally possible to predict what is going to happen and, with experience, to produce a smoother result.

The side-chain access to a compressor can also be used to apply equalization to the controlling signal.

One example of the use of this is the improvement of the quality of a sibilant voice where there is too much high frequency. If the voice is filtered to remove the sibilance the low level high frequencies will also be removed and the quality of the voice can be affected. With high frequency boost in the side-chain of a compressor it is possible to reduce the level of the voice only when high level, high frequency sibilance is present. The compressor is then acting as a simple 'de-esser'. The attack and recovery times need to be adjusted so the non-sibilant programme is not affected too much by the compressor. As the unwanted signal is high frequency a fast attack will be needed to catch the beginning of the sibilance. The recovery time will also probably need to be short to avoid reduction in the level of the rest of the programme when the sibilance has passed.

2.9.13 Stereo

When using compression on stereo programme material it is necessary to use two devices, one for the left channel and one for the right, and link (or gang) the operation of the two units. Stereo ganging or linking is the process whereby the gain reduction of two dynamic devices are kept equal at all times. If the operation of the two compressors is not linked unpleasant movements in the stereo image may result due to the unequal and changing gain in each channel.

To recognize the need for linked compressors, imagine a stereo drama that is being controlled by a pair of unlinked compressors set to limit. Suppose there are three actors positioned fully left, centre stage, and fully right, respectively. If the left actor speaks alone the level will be controlled by the left limiter only, and if the right actor speaks alone the level will be controlled by the right limiter only. When the central actor speaks, however, the level will be controlled by both limiters.

If the central character is speaking and, say, the left actor loudly interrupts, the left limiter only will react to the shout. A reduction in the level of the left channel only is equivalent to panning the central actor to the right whenever the left character is shouting. The result of using unganged limiters is movement of central images within the stereo sound stage. With some programme material this can be subtle, often being no more than slight 'flicks' of the stereo image.

Although linking will keep the gain reduction of the two units equal, most compressors require all

control settings to be the same. Stereo linking usually takes place after the attack and recovery circuits in the side-chain. If the recovery and attack times are different in the two units, programme peaks on the left of the sound stage will cause the two compressors to react differently from programme peaks on the right. The two compressors react together, but different effects will be produced by the compressors, depending on where a loud sound is within the stereo sound stage.

2.9.14 Expanders

An expander is a device that increases the dynamic range of an audio signal. As audio processing tools expanders can be remedial or creative.

(1) *Remedial* Expanders can help to reduce the background noise of an audio signal by reducing unwanted low-level signals. Low-level noise could be tape noise or disc surface noise, but more often expanders are used to remove spill (i.e. pick-up from other unwanted sources) in a microphone's output.

(2) *Creative* Expanders can be used to change a sound by removing parts of a transient or by cutting natural decay. This effect can be unpleasant if used with some types of programme material, but can greatly enhance others.

The internal structure of an expander (*Figure 2.167*) is very similar to that of a compressor. A variable-gain amplifier is controlled by a side-chain circuit. The side-chain contains a full-wave rectifier, a threshold circuit, attack, recovery and ratio controls. The side-chain is arranged so that if the programme level is *below* the threshold the output level will be reduced. As with compressors the expander acts to reduce the level of the programme.

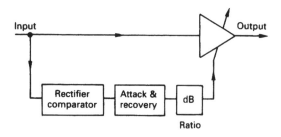

Figure 2.167 *An expander.*

The static input/output characteristic of an expander is shown in *Figure 2.168*. High-level signals, above a threshold level, pass through the expander unaffected. Below the threshold level the output signals are reduced in level relative to the input. The amount of change in output level relative to input level is known as the *expansion ratio*:

$$\text{Expansion ratio} = \frac{\text{Change in input level in dB below the threshold}}{\text{Change in output level in dB below the threshold.}}$$

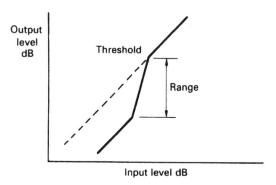

Figure 2.168 *An expander input/output characteristic.*

With an expansion ratio of 1:2, a 1-dB change in input level below the threshold will cause a 2-dB change in output level. This has the effect of making low-level signals quieter and increases the dynamic range. Usually both the threshold and the expansion ratio are adjustable. It is worth noting that, although expansion ratios should be quoted as 1:2 or 1:3, it is common practice to refer to these ratios as 2:1 and 3:1.

Very high expansion ratios will cause the output of the expander to be greatly reduced once the input level falls below the threshold. A high expansion ratio of 1:20 or greater is known as *gating* and this can produce an audible effect similar to cutting the output of the expander. Some expanders also refer to lower expansion ratios of around 1:10 as a *soft gate*. Gating can produce undesirable audible effects such as clicks or the loss of transients. These effects can be reduced by preventing the audio signal from being attenuated excessively. The action of a gate is therefore restricted to a limited amount of attenuation, rather than simply cutting the output. The amount the low-level signals are attenuated by is known as the *range* of the gate; the range is usually adjustable. A variable range facility makes an expander/gate a very flexible and useful device.

Expanders and gates are usually set up by ear and are simple to use if operated in a logical manner. If the programme level is high, low-level background noise will not be audible, because high-level signals conceal the low-level noise. This effect is known as 'noise masking'. If noise masking can conceal noise or spill when the programme level is high, an expander or gate can be used to reduce the noise during gaps and natural pauses in the programme. One of the main difficulties experienced when setting up expanders or gates occurs when the programme does not successfully mask the noise or spill. The noise is audible even when the programme level is high and the variations in gain of the expander are heard as changes in the background level, which can be unpleasant and draw attention to the noise.

If the range of the expander is too great, a click may be produced due to the loss of the original edge of the transients. These unpleasant side-effects can be reduced if the range is decreased, but this will also reduce the amount of noise reduction. A lower expansion ratio is less audible in operation than a gate, and with care it is possible to gain modest amounts of noise reduction even in difficult cases. It is probably not necessary to attempt to remove all spill and noise totally. Even small reductions can be useful.

Like compressors, expander/gates are dynamic devices and take time to react to changes in programme level. The attack time is the time taken for the expander to react when programme level exceeds the threshold. Attack time can usually be adjusted from about 25 µs to around 50 ms, and may be switched or continuously variable. The recovery time is the time taken for the expander to react when the programme level falls below the threshold. This can usually be continuously varied from about 25 ms up to 5 s. Both attack and recovery times have a large effect on the successful operation of the equipment.

Expanders/gates are more impressive devices when they are used as creative tools rather than for remedial purposes. They can be used with great success to change the sound of percussive instruments. A slow attack time can be used to remove a transient from a signal which dramatically changes the sound of an instrument. For most purposes it is usual to use a gate with a fast attack time so that either the transient is preserved or a new transient is created. A fast attack will allow the expander/gate to open without significant loss of the initial transient. If the attack is too slow the first syllables of speech could be lost and a good drum kit can be made to sound like a set of cardboard boxes.

The recovery time determines how much of the natural decay of an instrument will be heard. If the recovery is fast a gate can close as soon as the programme level drops below the threshold. With a percussive sound this can be just after the initial transient and a resonant drum becomes a click. To make expanders and gates easier to use, some devices set the recovery threshold to, typically, 6–10 dB lower than the attack threshold. This allows more of the natural decay of an instrument to be kept before recovery starts and, subjectively, this can improve the performance of the expander. Another technique is to have a *gate-hold* control which is used to 'hold' the gate open for a selected time after the signal falls below the threshold level. This gives an operator full control over how much of the natural envelope of a sound will be heard before the expander/gate recovers.

2.9.15 Key inputs

At times it can be difficult to adjust the threshold of a expander/gate so that the spill from other instruments does not also trigger the device to open, especially if the spill is of near-equal level to the desired sound. In the case of a drum kit, even with good microphone positioning the sound of the snare drum can be easily picked up by the kick drum microphone. It is possible to reduce the level of spill by filtering the signal before the input to the expander/gate. The only problem with equalization is that there may be a conflict between the desire to remove background spill and to creatively improve the sound of the desired instrument.

Most expanders/gates allow external access to the side-chain through a key input. It is possible using this input to control a gate with a different signal to the one passing through the device. One of the uses of the key input is to insert equalization into the side-chain circuit of the gate. For example, with a kick drum it is possible to boost the level of the low fundamental frequencies and cut the higher spill frequencies. The resulting sound may be unpleasant but is only used to open the gate. The gate allows an optimum quality drum sound through (unaffected by the filtering) but only low frequency kick drum fundamentals will cause the gate to open. Equalization is often fitted into the side-chain circuits of more sophisticated expanders/gates. When using this filtering technique care needs to be taken not to remove the natural attack of the drum from the side-chain, otherwise the expander/gate will open late. Removing too much of the high frequencies has the same effect as slowing down the attack time.

Another use of a key input is to trigger a continuous sound so that it is only heard when a note is played or a drum is hit. For example, a low-frequency tone can be mixed into the sound of a kick drum. If the tone is gated by the drum it will only be heard during the impact of the drum, and this can enhance an otherwise weak sound.

2.9.16 Compressor expanders

There is a wide range of dynamic devices available that combine a single variable gain amplifier with several side-chains to give compression and expansion within one piece of equipment. This is shown in *Figure 2.169* which shows three separate side-chain circuits. With this type of equipment it is possible to:

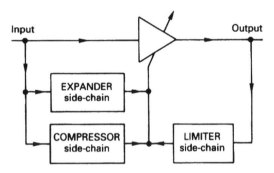

Figure 2.169 *A compressor/expander system.*

(1) use the limiter to catch programme peaks;
(2) use the compressor creatively at a threshold below the limiter to make the programme sound louder; and
(3) use the expander (or gate) to remove low-level noise during programme pauses.

2.9.17 De-essers and dynamic noise filters

Dynamic devices that are at present, less widely available but operate on the same basic principles as compressors and expanders are de-essers and dynamic noise filters. Both these devices combine the action of a variable gain amplifier with equalization or filtering. A de-esser is like a compressor which only reduces the level of high-level, high-frequency signals. A dynamic noise filter is like an expander that only reduces the level of programme within a selected frequency range and below a threshold level.

The simple de-esser described earlier (Section 2.9.12), consisting of a compressor with equalization in the side-chain circuit, can be useful when controlling sibilant speech, but has problems with a full-frequency range programme. If the equalized high-frequency signals exceed the threshold the rest of the programme material is reduced in level. The full deesser only reduces a selected frequency range of signals, leaving the rest of the audio spectrum unaffected. This device operates as if the variable gain amplifier has been replaced by a variable gain filter.

Variable gain filters are a little difficult to design, so some de-essers combine a traditional variable gain amplifier with static filtering (*Figure 2.170*). Programme signals pass through the static filter, which is bypassed by the variable gain amplifier. The side-chain circuit contains a filter with a complementary response to that in the main programme path. If the filtered side-chain signal exceeds the threshold level the gain of the amplifier is reduced and the filtered signal dominates on the output. The system shown in *Figure 2.170* is a little simplistic, and feedback is usually incorporated to give a flat frequency response when the de-esser is not acting. The advantage of a de-esser design of this type is that it can become a switched facility on a flexible compressor and need not necessarily form an extra expense as a stand-alone unit.

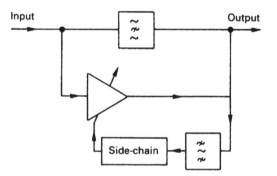

Figure 2.170 *A de-esser (also a dynamic noise filter).*

Dynamic noise filters (DNFs) are used to filter out unwanted low-level noises when programme level is low. Unlike static filters, DNFs automatically remove themselves from the circuit when the programme level is high and the noise is masked by the high-level programme. As a remedial tool DNFs are powerful and are very useful when removing surface noise and hiss from old archived recordings. DNFs are also a useful tool when cleaning up actuality

recordings that have a high level of background effects such as traffic noise. The disadvantage of using DNFs is that they are generally fiddly to set up and need constant adjustment as the nature of the programme changes.

DNFs are very similar in design to de-essers except that the side-chain circuit is designed to reduce the level of low-level signals and that the equalization is usually in the form of high- or low-frequency filters.

2.9.18 Digital compressors

Dynamic devices are now available that carry out the processing of the audio signal in the digital domain. They are available as part of digital sound desks and as stand-alone units. These devices usually have conventional analogue inputs and outputs as well as digital ones to make them as flexible as possible. The use of digital processing offers a wide range of operational controls. The attack and recovery times, for example, can often be adjusted from zero upwards over a wide range of values. As zero attack and recovery times sound unpleasant with even small amounts of compression this facility seems to be available simply because it can be done. Extra facilities that are found on some digital compressors include variable delay, which is a simple digital effect, that can be inserted in either the programme path or the side-chain.

The internal block design of digital compressors is very similar to that of analogue devices. The compressor's variable gain amplifier is formed by a binary multiplier. The side-chain has to perform the same task as an analogue side-chain and derive the control signal for the multiplier. Digital processing has to simulate the functions of rectification, the theshold comparator and the dynamic control circuits. The difficult part of the design is attack- and recovery-time control. The rates of change have to be calculated over many samples and log–antilog converters are required to give the characteristics of a capacitor charging or discharging.

As with all forms of digital processing the digital compressor does not exist as an identifiable set of components. The usual design approach is to use digital processor chips and construct the desired effects in software. At present the amount of processing required by dynamics is very high. Digital sound desks, or out-board digital processors, that offer compression or expansion as an effect are unable to offer all other facilities at the same time.

Bibliography

BASKIND, D. and RUBENS, H., *Techniques for the Realisation and Application of Voltage Controlled Amplifiers and Attenuators*, Audio Engineering Society Presentation, New York (1978)

BEVILLE, M., Compressors and limiters: their uses and abuses, *Studio Sound*, **19**, 28–32 (1977)

DUNCAN, B., VCAS investigated, *Studio Sound and Broadcast Engineering*, **31**(7) 58–62 (1989)

GLEAVE, M.M. and MANSON, W.I., The development of sound-programme limiters in the BBC, *BBC Engineering*, **107**, 9–19 (1977)

HARTLY-JONES, M., *A Practical Introduction to Electronic Circuits*, 2nd edn, Cambridge University Press, Cambridge (1985)

3 Recording and reproduction

Andy Wilson
Studer UK Ltd
3.1 Analogue tape recording 3/2

Kenneth Gundry
Principal Staff Engineer, Dolby Laboratories Inc.
3.2 Noise reduction systems 3/25

Jan Arts
Sony Service Centre Europe
3.3 Compact disc 3/44

Jan Maes
Head of Training and Technical Support Engineer, Sony Service Centre, Europe
3.4 Digital audio tape 3/64

Douglas Ford
Skywalker Sound, a division of Lucas Digital Ltd
3.5 Audio editing 3/86

Peter Skirrow
Designer and Proprietor, Lindos Developments
3.6 Audio measurements and test equipment 3/94

3.1 Analogue tape recording

3.1.1 Aims of the recording process

Magnetic recordings are used as a storage medium. There are two essential requirements in audio recording: firstly, that a recording made on any tape machine can be replayed on any other tape machine; and secondly, that the best performance possible is obtained from the recording system. In this section we will look at how the recording to tape process is carried out, and get an understanding of the requirements of the record/replay chain. This involves looking at the basic principles and properties of magnetic materials.

Figure 3.1 shows the basic elements of a tape machine. The magnetic tape travels from the supply reel to the take-up reel at a constant speed generated by the capstan motor. On the way it passes in front of the erase head which demagnetizes the tape to clear it of any previous recording. The audio input signal is converted from an electrical waveform to a magnetic waveform which is then stored on the tape via the record head, and this stored information induces a replica of the original voltage signal which can be monitored via the replay head, magnified by the replay amplifier and sent to the output of the tape machine.

3.1.2 Some magnetic principles

In magnetic materials, the smallest volume of material which can act as a magnet, i.e. with a north and south pole, is called a magnetic domain (Hammond,

1978). These are infinitesimally small groups of molecules. If these domains are randomly polarized then the overall magnetism of the material is zero.

If a magnetizing force is applied to the material, the domains will tend to turn in the direction of the field. If the force is weak not all the domains will be aligned with the field; but if a very strong field is applied all the domains will align and the material is then said to have reached magnetic saturation. On removing the magnetizing force, with some materials the domains may eventually resume their random orientations. This material is no good for magnetic recording, and is called magnetically soft. If most of the domains remain in position after the force has been removed, the material is then a permanent magnet, unless an opposing magnetizing force is applied. This is a magnetically hard material. The agent that creates magnetic flux in a magnetic circuit is the magnetomotive force (m.m.f.). In an air-core coil, this is the total force acting in a circuit; it is proportional to both the current flowing through the coil and the number of turns:

$$\text{m.m.f.} = NI$$

where N is the number of coil turns, and I is the current flowing (amperes).

The unit of m.m.f. is therefore amperes (A) or, more usually, ampere-turns (At).

The intensity or field strength is the total force acting per unit length l of the magnetic circuit. This is called the magnetizing force, H:

$$H = \text{m.m.f.}/l = NI/l \ (\text{At/m})$$

It is this magnetizing force created in a recording head which imposes the signal magnetically onto a

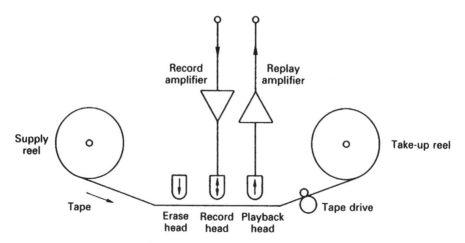

Figure 3.1 *The basic elements of a recording system.*

tape. It gives rise to a magnetic flux ϕ, which has a flux density, i.e. the amount of flux per unit area B measured in tesla (webers/m^2). The magnetomotive force is the driving force and together with the flux density, it has an analogy with electrical circuits in which:

$$I = V/R$$

So: $\phi = $ m.m.f.$/S$

where S is the reluctance of the circuit.

The relationship between the magnetizing force H and the resulting flux density B is shown in *Figure 3.2*, and is of vital importance to the recording process (Wilson, 1985). If we start at the origin of this graph there is no magnetizing force and therefore no resulting flux density B. As H is increased, the magnetic domains start to align with the increasing field and there is some resulting flux density. If H is increased still further it reaches point 1 where the flux density reaches a maximum, and further increases in H do not result in any further increase in B. All the domains are aligned with the magnetic field and the material is saturated. If H is now reduced, some domains flip back to random positions, while others remain aligned even when $H = 0$ (point 2). There is residual B which is known as the 'remanent flux density' $+B_r$. It is just this retentive effect which makes magnetic recordings possible, and the material is said to possess remanence.

To restore the flux density back to zero, i.e. random polarization of the domains, the magnetic field H must be applied in the opposite direction. The value of B then eventually falls to zero, which requires a field of $-H_c$ (point 3). This is known as the *coercive force*. Increasing $-H$ above $-H_c$ starts to align the domains in the opposite direction, so B becomes negative and, again, eventually the material reaches saturation (point 4). The same thing happens if the polarity of H is reversed again towards zero and increased back in the original direction. The loop (1–6) is known as the BH loop or hysteresis loop.

Looking again at the BH loop we see that the flux density changes in a non-linear way with respect to the magnetizing force. So if the magnetizing force is proportional to the electrical audio signal, the resulting flux density will be distorted. Imagine a relatively slowly changing sine wave signal which is applied as a sine wave of magnetizing force H to the tape. A large number of tape elements are moving very quickly into and out of the narrow record field so that the instantaneous value of the signal field does not change significantly whilst the element is under its influence. Therefore, one particular element experiences a magnetizing force which increases from zero outside the field to a maximum value proportional to the instantaneous amplitude of the input signal. This rapidly drops to zero again as the element moves out of the field. *Figure 3.3* shows this effect on all the tape elements affected by one cycle of the signal field. This shows that the magnetic flux

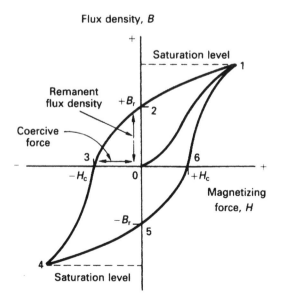

Figure 3.2 *The flux density/magnetizing force (BH) loop (or hysteresis loop).*

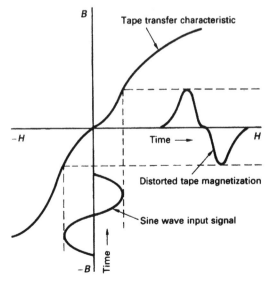

Figure 3.3 *A tape transfer characteristic.*

remaining on the tape is not proportional to the input signal level.

We know from *Figure 3.2* that a magnetizing force H_m produces a remanent flux density B_r. For values of H below H_m, smaller loops are produced which result in different values of B_r. A complete curve for positive and negative values of H is known as the tape transfer characteristic.

3.1.3 Physical construction of magnetic tape

The tape itself is the heart of the recording medium. There are several standard tape widths (*Table 3.1*). *Figure 3.4* shows the various layers which make up a typical tape. The tape consists of a base film with a coating of magnetic powder and lubricants bonded onto one side. The backing is usually a conductive material to prevent the build-up of static electric charges in high-speed spooling. The rough texture of the backing helps hold the tape pack in one plane when it is wound onto a spool. For the base, cellulose acetate was once used, but its strength was poor and thus any edge damage could result in a tape breakage. Polyvinyl chloride (PVC) has also been used as a base material, but rather than breaking it stretches and whereas a break can be spliced back together, a stretched tape is useless. A commonly used material is Mylar which has the right degree of strength and elasticity. The thickness of the base can be from about 25 to 40 µm, and it is resistant to fire, and fungal and bacterial attack.

Table 3.1 Standard tape widths

Tape width		
(in.)	*(mm)*	*Example of use*
5/32	3.8	Cassettes
1/4	6.25	Mono full track, 1 or 2 track, stereo
1/2	12.5	Stereo mastering, 4 track
1	25	4-track mastering, 8 track
2	50	16 track, 24 track

The coating, usually 10–15 µm thick, is commonly made from ferric oxide (Fe_2O_3). It has acicular or needle-shaped particles which are roughly 0.6–1 µm in length and less than one-tenth that in diameter. Long, thin magnets retain their magnetism better than short, fat magnets, and therefore the acicular particles retain their magnetic pattern better than spherical particles under the influence of stray fields, mechanical

shock or thermal cycling. The way in which the magnetic powder is prepared, its impurities, either deliberately or accidentally introduced, and the physical size and shape of the particles, have a lot to do with the audio performance of the finished tape.

The magnetic powder is mixed with a binder. This is a type of glue which not only bonds the magnetic material to the base, but also holds the particles apart from each other when wet, so that they remain evenly dispersed throughout the coating. Once dry, the binder must not adhere to adjacent layers even when the finished tape is tightly wound and stored in conditions which expose it to thermal cycling. The characteristics of the tape should remain stable for many years.

In the manufacturing process, the wet mixture of binder and magnetic particles is precisely coated onto large rolls of film. For audio tape, where information is recorded longitudinally, more signal can be recovered from the tape if the particles are aligned with their major axis parallel to the recorded tracks, so the rolls are then passed over an intense unidirectional field to orientate the needle-shaped particles uniformly. The coated film is then passed through drying ovens and eventually between sets of rollers to polish the surface and ensure a uniform thickness of coating. The large rolls of finished tape are then slit into the required width, depending on their application, and wound onto reels.

3.1.4 Magnetic tape heads

The elementary features of a single tape head are shown in *Figure 3.4*. The construction of the tape head is similar to a transformer with a single winding. It is built on a ring-shaped magnetic core with two gaps in the magnetic circuit, one filled with a non-magnetic material that is in contact with the tape, and the other at the rear of the head. The core is formed from a special laminated metal alloy which has low remanence and coercivity and is therefore magnetically soft; which is in contrast to the oxides used in the tape which are magnetically hard, with high remanence which retains as much of the signal as possible when the recording field is removed, and high coercivity to prevent erasure from stray magnetic fields (*Figure 3.5*). The purpose of the core is to provide a low reluctance path for the magnetic flux set up by the signal current when recording, or induced from the tape in replay. Because the flux varies at the signal frequency, it must be capable of changing quickly with the core material retaining hardly any residual magnetism.

Figure 3.4 *The elements of a single tape head with a magnetic tape.*

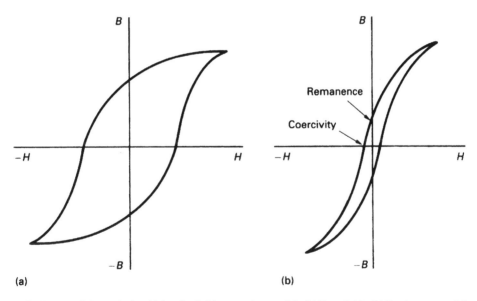

Figure 3.5 *Typical hysteresis loops for hard (a) and soft (b) magnetic materials. (a) Tape Oxide. (b) Head core material.*

3.1.4.1 Record and erase heads

The tape coating bridges the front head gap so that the flux which would otherwise be forced to flow through the comparatively high reluctance of the gap finds an easier path through the much lower reluctance of the oxide of the tape coating. The magnetizing force H leaves a remanent flux in the tape according to the BH loop for the particular magnetic material. This tape flux is expressed in terms of the width of the track in nanowebers per metre (nWb/m).

The magnetic reluctance of the core varies with different levels of flux, i.e. it is non-linear, so in the record head, a rear gap is built in to introduce a high reluctance in series with the low reluctance of the core material. This high reluctance swamps the core reluctance and hence its variations.

We have already seen that m.m.f. $= \phi S$

where m.m.f. $= NI$, ϕ is the resultant flux, and S is the reluctance. The total reluctance S_{tot} is given by:

$$S_{\text{rg}} + S_{\text{c}} + (S_{\text{fg}}||S_{\text{t}})$$

where S_{rg} is the rear gap reluctance, S_{c} is the core reluctance, and $S_{\text{fg}}||S_{\text{t}}$ is the front gap in parallel with the tape coating. Therefore:

$$NI = \phi S_{\text{tot}}$$

S_{rg} is the largest component and so linearizes the relationship between the flux in the magnetic circuit and the current driving it.

In *Figure 3.6* we can use the analogy of bar magnets within the tape coating to show the result in magnetic terms of applying a sine-wave signal to a record head (Arnold, 1979). The magnet length depends not only on the frequency of the wave, but also on the speed of the tape. If the current in the coil varies at a frequency of f hertz and the tape moves at v metres per second, then one cycle will be recorded on the tape as a variation of magnetic intensity over the distance λ metres:

$$\lambda = v/f \text{ metres}$$

λ is known as the recorded wavelength. As an example, if the tape is moving at 38 cm/s (15 in./s)

and the recorded frequency is 1 kHz, then $\lambda = 0.38/1000 = 0.38$ mm.

3.1.5 Erasure

Erasure is the restoration of the magnetic domains to their random polarities after alignment in the recording process. As we shall see later, the replay head responds to the rate of change of flux on the tape, so the erase process must make $d\phi/dt = 0$.

3.1.5.1 d.c. Erase
One way to do this is to apply a unidirectional field, either using a permanent magnet or electromagnet so that the tape is saturated in one direction and therefore gives a constant flux at the replay head. The problem with this method is that if the tape has an irregular surface contour (and all tape does to some extent), then the main body of the coating will vary in spacing from the replay head, and so the replay head only sees the high value of the saturation remanence through the varying head to tape spacing (i.e. a varying high reluctance air-gap), and so a large $d\phi/dt$ is produced. The result is that, although all the previous information has been erased, the process

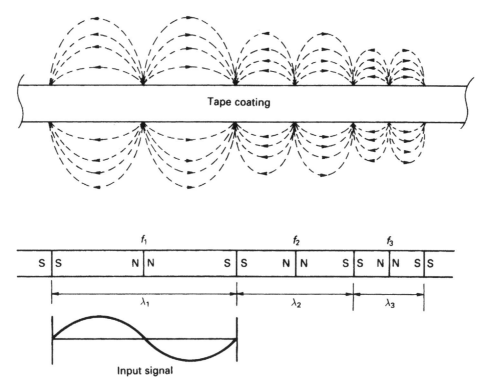

Figure 3.6 *Magnetic result of applying a sine-wave signal to a record head.*

still gives rise to noise in the form of this coating generated $d\phi/dt$.

3.1.5.2 a.c. Erase

A better method of erasure is to use an a.c. field. In this, the tape coating is subjected to several cycles of an alternating magnetic field which is large enough for saturation generated across an erase head gap. (*Figure 3.7(a)*). Each tape element is brought into the high frequency saturating a.c. field, which decays slowly away as the element moves away from the head. In effect, the domains of the tape are subjected to cycles of magnetization which can be described by the ever decreasing BH loops (*Figure 3.7(b)*). Ultimately, the loop is so small that the values of H become insufficient to reverse the magnetization of the domains and they cease to become affected. This occurs in a random manner. So that the tape crossing the head gap experiences several cycles of saturation, a comparatively large erase head gap is required, which means that the head needs no rear gap because the high reluctance is now provided by the front gap.

If we assume that any given tape element needs to undergo, say, 15 erase cycles for complete erasure, then we can work out the required gap length:

$$l_{gap} = nv/f_{erase}$$

where n is the number of erase cycles, v is the tape speed (m/s), and f_{erase} is the erase frequency. There-fore, at 7.5 in./s (19 cm/s), with an erase frequency of 150 kHz, the required gap length is $l_{gap} = 15 \times 190 \times 10^{-3}/(150 \times 10^3$ m$) \simeq 20$ μm.

As the BH loop is highly non-linear at low field levels, any asymmetry in the erase waveform, i.e. even harmonic distortion, will be rectified by the tape characteristic to leave a unidirectional magnetization which, as we saw with d.c. erasure, gives rise to noise. The erase waveform should therefore have low distortion of even harmonics.

3.1.6 Record process

The aim of the record process is to overcome the non-linearity of the tape transfer characteristic and to put enough level into the tape to fully use the available dynamic range (between tape overload and residual tape noise).

3.1.6.1 Linearity improvement with bias

d.c. Bias One method which enables the use of the linear part of the tape transfer characteristic is the use of a d.c. bias, i.e. a d.c. producing a constant value of $+H$ or $-H$ to the signal. Used in conjunction with d.c. erase, the head receives a tape already saturated in one direction. However, this only uses the positive or negative half of the characteristic (*Figure 3.8*).

Used in war-time wire recorders, d.c. biasing has some disadvantages. A residual unidirectional field is still present on the tape causing noise for the same reason as d.c. erase produces noise. In addition, the bias is only correct over part of the coating depth so

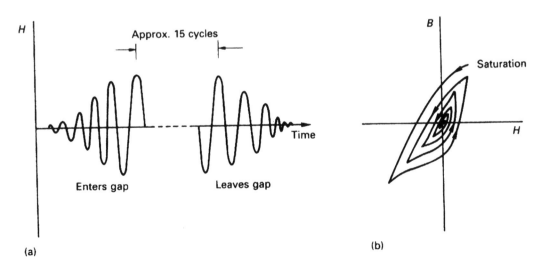

(a) (b)

Figure 3.7 *Effects of a tape element passing the erase head. (a) The flux experienced by the tape element moving across the erase gap. (b) Diminishing hysteresis loops as the tape element leaves the erase-head gap.*

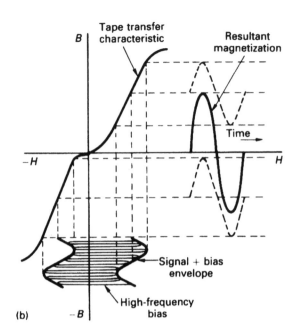

Figure 3.8 *Action of (a) a.c. bias and (b) d.c. bias.*

that distortion, although reduced, is still present. The bias required is critically dependent on the magnetic tape properties.

a.c. Bias This is a method which has been used successfully where a sine wave bias waveform is linearly added to the signal current feeding the record head so that the bias amplitude varies as shown in *Figure 3.8(b)*. The frequency of the bias signal is approximately 4–6 times the highest recorded audio frequency, to prevent intermodulation between the bias and harmonics of the signal frequencies. In this way the modulation products are placed outside the audio range. As an example, if we have a bias frequency f_b of 25 kHz and an audio frequency f_a of 15 kHz, then the intermodulation products are given by $f_b + f_a = 40$ kHz (which is no problem as it is out of the audio band) and $f_b - f_a = 10$ kHz, right in the middle of the audio band.

3.1.6.2 Record head with a.c. bias
The bias waveform is usually tapped off from the erase oscillator and applied to the record head in addition to the audio signal (*Figure 3.9*) (Harris, 1982). The bias current is shown as typically 10 times the signal current in the head. Since the head impedance (mainly ωL) is also much greater at bias frequencies, then the bias voltage at the head is quite large. A tuned filter called a 'bias trap' is incorporated in the signal path to prevent this large

Figure 3.9 *Addition of a.c. bias at the record head.*

signal from feeding backwards into the audio drive amplifier.

The flux encountered by an element of tape passing the record head is similar to that at the erase head, but the mean of the high-frequency signal is displaced from zero by the audio signal (*Figure 3.10*). The diminishing hysteresis loops are asymmetrical, and so as an element leaves the gap the tape is left with a remanence corresponding to the audio signal.

It can be seen from *Figure 3.8(b)* that the very non-linear portion of the transfer characteristic as it passes through the origin is not used, and also that there is more magnetization and therefore greater signal level onto the tape because both the positive and negative sections of the tape transfer characteristic contribute. Obviously the amplitude of the bias is critical to obtain correct operation in the linear part of the characteristic, and so the bias must be set up for each particular type of tape.

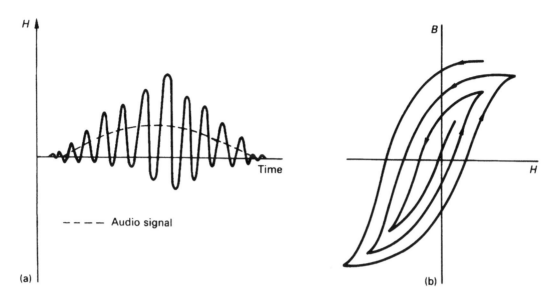

Figure 3.10 *a.c. Bias with an added signal. (a) The effect of an audio signal (exaggerated). (b) Diminishing hysteresis loops as the tape element leaves the record-head gap through the recording zone.*

Figure 3.11 *The field distribution around the record gap.*

With regard to the tape, the bias creates a field within the record-head gap which then spreads out into the tape coating. The field in the coating is strongest near the gap and reduces to, typically, half amplitude in the coating nearest the base. If the flux spreads further from the gap it then has to pass through the high reluctance of the non-magnetic base, and so the bias field rapidly decays to zero (*Figure 3.11*). The bias is adjusted until the field it produces is just sufficient to saturate the coating nearest the base film. All the tape elements entering the record field are thus taken through saturation loops opposite the record gap. Recording takes place on the trailing edge of the bias field as the field rapidly decays to zero. This area of the decaying bias field is known as the recording zone.

At short recorded wavelengths, where the time which an element spends in the recording zone is a significant fraction of the periodic time of the signal, the remanence will be an 'average' of the signal excursion which occurs whilst the element is in the recording zone. This is one of the losses that will be looked at later. *Figure 3.12* shows the effect of increasing bias, from zero to the optimum value, on the efficiency of the record process, the percentage distortion of the output signal and the signal/noise ratio (the difference between the signal and the noise curves). From these curves we can see that for optimum performance the bias is adjusted for a peak signal, then increased further until there is a 1 dB loss on a 1 kHz input frequency, or a 3 dB loss using a 10 kHz frequency (which gives a more accurate setting). As with many things, the bias setting is a compromise. The lowest noise figures do not necessarily correspond to the lowest distortion or maximum output levels.

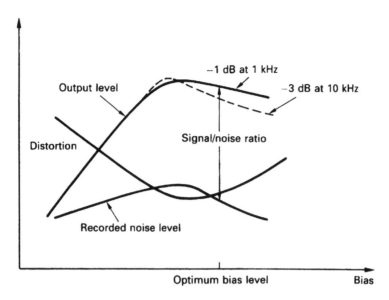

Figure 3.12 *Optimum bias setting.*

As the recorded signal current is increased, the remanent flux increases until the onset of saturation. At this point, distortion begins to rise rapidly with any further increase in record current. The maximum output level (MOL) of a tape is the magnetization in nanowebers per metre that corresponds to a 3% third harmonic distortion of a 1 kHz tone. The percentage harmonic distortion produced by a tape is almost proportional to the output, so if 1% distortion is generated with a recorded flux that gives an output of 1 V, then it will be approximately 2% when the output is 2 V. The third harmonics are measured because the symmetrical nature of the recording process when using a.c. bias produces few even harmonics.

Since the background noise is about as low as possible with the chosen coating materials, the maximum flux level used to record the peaks of the program signal will determine the signal/noise ratio. The MOL of a 1 kHz tone varies with bias current; a low bias level restricts the mid-band MOL, although at high frequencies, the MOL is slightly improved. As the bias is increased, the 1 kHz MOL reaches a plateau while the MOL of high signal frequencies decreases. The MOL performance of a tape determines the relative distortion levels of tape at lower signal levels. Some recording engineers drive the tapes hard at high peak levels for optimum signal/noise ratio, while others use lower levels to get lower distortion and then use a noise reduction system to get the noise levels down.

3.1.6.3 An ideal recording

If we consider a recording amplifier being fed with a constant-voltage frequency sweep over the audio range it would be ideal if this was recorded on tape as a constant r.m.s. flux level at all frequencies. The tape path would then reach saturation flux at the same input voltage for all frequencies as the signal input was increased. But there is no such thing as an ideal recorder, as we shall see.

3.1.7 The replay head

The replay head has a very low reluctance (the core) between its pole tips and this produces a magnetic short circuit to the flux from the magnetization pattern on the tape, so the flux flows around the core. Changes in the flux give rise to an e.m.f. (*e*) generated in the coils according to the basic principle of generators:

$$e = N \mathrm{d}\phi / \mathrm{d}t$$

where N is the number of windings, $\mathrm{d}\phi/\mathrm{d}t$ is the rate of change of flux in the core.

If we now take our ideal recording with constant-amplitude sine-wave flux at all frequencies, the head output voltage should rise by 6 dB/octave, since the rate of change of the flux doubles with each octave. Unfortunately, in the high frequency region, a combination of losses will mean that the head core will not see all of the flux recorded on the tape and will produce less than the ideal 6 dB/octave.

3.1.8 Replay losses

3.1.8.1 Gap length loss

This is an important effect of the replay head gap at short recorded wavelengths λ, where $\lambda = v/f$ and v is the tape speed. For a given tape speed, there will come a frequency where $\lambda = \lambda r = l$ (head-gap length). At this point the replay head will see no net change in the flux at its pole tips and therefore will produce zero output. This point is known as the extinction frequency and defines the upper limit of the usable frequency response for a given head-gap length. The replay gap width must therefore be less than the recorded wavelength of the maximum frequency to be replayed. The ratio l/λ has a considerable effect on the output level.

$$\text{Output level} = 20 \ \log_{10} \sin[180° (l/\lambda)](\text{dB})$$

So at $l/\lambda = 1$

$$\text{Output level} = 20 \ \log_{10} \sin 180° = -\infty(\text{dB})$$

at $l/\lambda = 0.5$

$$\text{Output level} = 20 \ \log_{10} \sin 90° = 0 \ \text{dB}$$

at $l/\lambda = 0.1$

$$\text{Output level} = 20 \ \log_{10} \sin 18° = -10.2 \ \text{dB}$$

It can be seen from *Figure 3.13* that the head output rises at 6 dB/octave at first. As the gap length begins to become significant with rising frequency,

the curve turns over with maxima at $l/\lambda = 0.5$, 1.5, 2.5, etc. In practice, the output above the first zero point is unusable.

The effective gap length l of a tape machine can be determined by reproducing a test tape which has constant magnetization with frequency and observing the first zero in the output at f_0. So

$$l = (v/f_0) \times 10^6 \ \mu\text{m}$$

where v is the tape speed (in m/s).

3.1.8.2 Coating thickness loss

At short recorded wavelengths, the remanent magnetism stored deep in the oxide of the tape coating away from the surface is unable to contribute to the flux seen by the replay head. At long and medium wavelengths, a half-wavelength stored on the tape is much longer than the thickness of the coating, therefore, when the low reluctance of the replay head appears across the stored magnetization, all of the flux from the north to south poles in the coating, flows via the low reluctance of the head which it sees as nearly a magnetic short circuit.

If we look at the situation that occurs at very short wavelengths (*Figure 3.14*) we see that the flux path to the head pole-tips from the magnetization at the back of the coating away from the surface is of comparable magnetic length to the path via the base. So as the recorded wavelength is reduced, the external flux from the tape as seen by the replay head reduces and it is mainly the surface magnetization of the coating that contributes to the core flux. If we

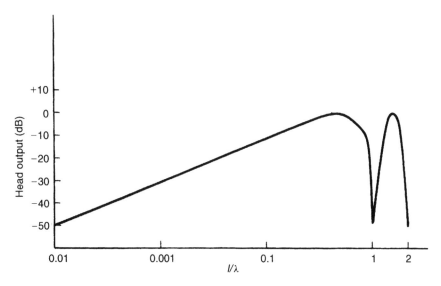

Figure 3.13 *Theoretical response of a replay head, including gap loss.*

Figure 3.14 *Flux path at the replay head at short wavelengths.*

now take our 'ideal' recording of constant amplitude, sine wave flux from low to high frequencies, and it is replayed with a perfect head (i.e. the head has no other losses), the core flux will show the loss due to the coating thickness. The effect becomes significant at $\lambda = 2\pi c$, where c is the coating thickness (*Figure 3.15*).

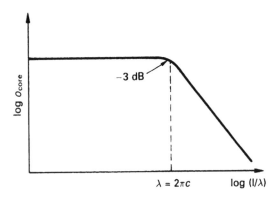

Figure 3.15 *Tape-coating thickness (c) loss as monitored by a perfect replay head.*

Because it is wavelength dependent and the frequency represented by a wavelength depends on the tape speed, then if the speed is halved, the onset of coating loss occurs at half the frequency for a given coating thickness.

3.1.8.3 Pole-piece loss
At long recorded wavelengths the replay head pole pieces are not long enough to cause a magnetic short circuit across the recorded flux pattern. As we have seen, at shorter wavelengths the low remanence of the replay head causes the flux on the tape to flow round the core of the head where it is sensed by the coils, but at longer wavelengths a high reluctance air-gap is introduced between the magnetized tape and the pole pieces of the replay head.

As the recorded wavelengths get longer, i.e. the signal frequency gets lower, then there is a drastic fall off in the output from the replay head. Together with gap length loss at high frequencies this shows that there is an upper and lower frequency limit set by the replay gap dimensions, which is approximately 10 octaves in bandwidth.

3.1.8.4 Spacing loss
If the replay head loses contact with the tape by a distance d, due to dirt on the head, a rough tape surface, or a worn head profile, then the loss of head output is proportional to d/λ. The effect of the head clog is therefore more severe at high frequencies.

3.1.9 Record losses

We have seen before in the description of the action of the decaying bias field in the recording zone that a remanence proportional to the instantaneous value of the signal current is left on each tape element as it moves out of the recording zone. However, as the signal frequency increases, the instantaneous value of the signal current changes appreciably in the time that the given element spends in the recording zone. This causes an effect due to erasure by the bias called self-demagnetization where the signal current leaves a reduced remanence on the tape at the higher signal frequencies.

3.1.10 Recording characteristics

Knowing that the ideal replay head when presented with an ideal recording on tape of a constant flux level at all frequencies will produce a head output rising at 6 dB/octave, what we want to produce from the electrical output of the tape machine is a flat response with frequency, and to be able to take any tape recorded on any machine and replay it on any other machine. For optimum recorded performance, the recorded response is equalized to decrease with increasing frequency in a standardized way. Therefore the replay chain must have a response that rises with increasing frequency in a standard complementary way, so that an overall flat frequency response is obtained.

The standard characteristic responses are commonly identified by the initials of the organization that wrote the standard. In the USA, this has mainly been the National Association of Broadcasters (NAB), and in Europe mainly the International Electrotechnical Commission (IEC). Most countries

Table 3.2 Standard equalizations in IEC, NAB and AES Standards

Tape speed	IEC1	IEC2 and NAB
9.5 cm/s; 3.75 in./s	–	50 Hz; 1800 Hz 3150 μs; 90 μs (NAB only)
19 cm/s; 7.5 in./s	0 Hz; 2240 Hz (∞, 70 μs)	50 Hz; 3150 Hz (3150 μs, 50 μs)
38 cm/s; 15 in./s	0 Hz; 4500 Hz (∞, 35 μs)	50 Hz; 3150 Hz (3150 μs; 50 μs)
76 cm/s; 30 in./s	[(IEC1 not used)] [0 Hz; 4500 Hz] [(∞, 35 μs)]	AES 1971 0 Hz, 9000 Hz (∞, 17.5 μs)

submit their proposed standards to the IEC for approval (MRL 1985). The Consultative Committee Internationale Radiotelephone (CCIR) Rec. 408–3 1974 is the same as IEC1 except that the CCIR publication only covers speeds of 19 and 38 cm/s. The Deutsche Industrie Normen (DIN) standard is more or less the same as IEC1. The IEC redesignated NAB as 'IEC2'. For all new 30 in./s (76 cm/s) tapes, the equalization follows the Audio Engineering Society (AES) proposal which is now also designated 'IEC2'.

Standards do change over the years and identification of the equalization by organization name can be ambiguous. Therefore the equalization can be quoted in the time constants τ or their inverse corresponding transition frequencies: $f_t = 1/(2\pi\tau)$ (see *Table 3.2*).

These time constants are a convenient way of describing the required flux characteristics by analogy to simple CR timing circuits where $\tau = 1/(2\pi RC)$. From this, we can see that standards are specified for each tape speed; for example, IEC1 has transition frequencies at 2240 Hz at 7.5 in./s and 4500 Hz at 15 in./s. The standard specifies that the recording process shall give a recorded flux which if measured at the tape surface by a 'short circuit flux meter' gives the flux/frequency plot (*Figure 3.16*) corresponding to the constant-input-signal level at all frequencies. The maximum flux must be recorded at low frequencies where playback head voltages are low and the signal/noise ratio is low. The loss at high frequencies due to thickness loss sets a limit to the linearly recorded flux. From a recording made to the characteristic of *Figure 3.16*, the replay characteristic is shown in *Figure 3.17(a)*. The ideal head response and the replay amplifier response required to equalize this to an overall flat response is shown in *Figure 3.17(b)*. The replay

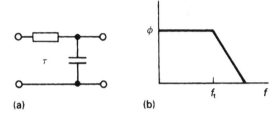

Figure 3.16 *The equivalent circuit (a) of a standard record flux equalization (b).*

equalization is made up from a fixed equalizer giving a falling gain/frequency response up to the transition frequency, the constant gain above it.

In a 'real' replay head response there are spacing, gap and core losses which are a function of the replay head type and age. Therefore the total replay equalization is made up of the fixed standard plus a variable high-frequency equalizer to compensate for the losses (*Figure 3.17(c)*). Similarly, in the record head, the self demagnetization, spacing and core losses must mean that a constant current in the record head over the audio frequency range does not result in a constant r.m.s. flux being recorded. There is an increasing reduction in the recorded flux as the frequency is increased. Advantage can be taken of the fact that, in general in the audio band, the high-frequency signals are lower in amplitude, i.e. the signal consists of lower amplitude harmonics. In addition, the harmonic distortion produced by high-frequency signals lies outside the audio band so we need only consider the in-band intermodulation (IM) distortion produced. These factors enable a certain amount of high-frequency boost or pre-emphasis to be applied in the recording amplifier which helps to offset the losses, and so give the correct recording characteristic (*Figure 3.18*).

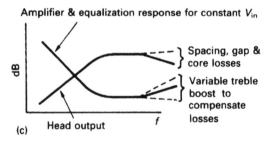

Figure 3.17 *Replay amplifier equalization. (a) Equivalent circuit of a standard replay response. (b) Replay head and amplifier response for an ideal head. (c) Real replay head and amplifier responses.*

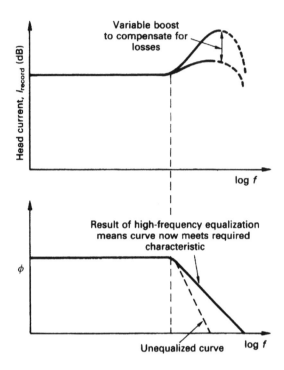

Figure 3.18 *Use of a high-frequency boost to produce the recording characteristic.*

The NAB standards, besides using different time constants, also apply a bass boost below a 50 Hz turnover point. The response of the replay amplifier therefore does not have to rise below 50 Hz as must happen in the IEC1 replay amplifier. The risk of record overload distortion at bass frequencies is low since most bass notes are rich in harmonics and contain a relatively small part of their energy at the fundamental frequency.

3.1.11 The tape machine

The tape transport should be designed in such a way that the tape travels from the supply reel via the tape guides, past the soundheads and onto the take-up reel at constant speeds which should be kept within tight tolerances. There should be sufficient tension that the tape to head contact is adequate and constant. This

ensures minimum head wear and the best signal, and the tape should be moved in a precisely controlled way without snatching or throwing loops and so producing a good 'pack' in the take-up reel.

Some of the major requirements which need to be satisfied in the tape machine design are as follows (in no particular order of importance).

(1) The idler and guide rollers must be positioned in such a way that the tape wrap around the soundheads is optimal.

(2) The rollers must be positioned so that the tape can easily be threaded.

(3) All machined parts which come into contact with the tape must be exactly perpendicular to the tape path. This requires an absolutely flat base, i.e. a stable and precision machined tape deck chassis. The precision with which a professional tape deck has to be built is of the order of a few micrometres.

(4) The tape should be moved across the rollers and heads at the correct height, i.e. without tilting movement and without damaging the edges of the tape.

(5) No fixed guidance elements should be located near the soundheads and particularly not in

front of the capstan drive, otherwise longitudinal oscillations can occur (see later description of mechanical performance).

(6) Any tape tension measuring devices (tension sensors) must function accurately from play speed to maximum spooling speed. This requires a robust design with low inertia.

(7) All rollers must run true with a tolerance of a few micrometres. In addition, they must be absolutely cylindrical so that the tape does not drift off the ideal path.

Additional requirements are long life of the bearings and wearing parts, and minimal maintenance (the machine should be designed so that servicing and adjustment can be performed easily and quickly). The machine should also be easy to clean; this last point may seem insignificant but it can have major consequences.

3.1.11.1 Capstan drive

The capstan motor is the heart of the tape drive system. It is the component that governs the speed of the tape past the soundheads. The tape is driven by being pinched between the capstan shaft and the rubber pressure roller. At one time capstans were commonly synchronous motors which were locked to the mains frequency reference. Different rotational speeds were accomplished by selecting the number of poles switched into the circuit. Speed accuracy depended on the accuracy of the mains frequency. Machines with these motors often had a belt drive from the main motor flywheel to a substantial capstan flywheel. This allowed the capstan motor to be mounted further away from the headblock and so reduce hum induction and the transmission of flutter from the main drive motor.

The successor to the synchronous motor was the asynchronous, tacho controlled motor with its own frequency reference (*Figure 3.19*). In a typical system, a frequency is derived via the tacho heads from the capstan rotation and is compared with a stable oscillator. any error being used to modify the drive to the capstan motor. This produces very good speed stability which is not affected by mains frequency or voltage. 'Varispeed' may be implemented either by applying a variable voltage v_{ext} or a frequency f_{ext} from an external oscillator to the capstan drive. Typically a variation of about ±7 semitones is available from varispeed units.

Some machines now use brushless d.c. motors (*Figure 3.20*). The servo system of these may

Figure 3.19 *Asynchronous capstan servo control.*

Figure 3.20 *A brushless d.c. motor with microprocessor control. PWM, pulse width modulation.*

incorporate a microprocessor speed controller, a power amplifier, the motor and a tacho speed sensor. The motor should have characteristics of fast acceleration (0–76 cm/s in about 100 ms), high efficiency, smooth running with minimal torque pulses and low inertia. Brushless motors require quite a lot of electronic control circuits. In *Figure 3.20* the inverter stage comprises Hall elements which continuously detect the current position of the rotor. These elements feed back to give logic control that determines which drivers are to be switched on. A single bit determines the direction of rotation. The inverter output stage and d.c. motor together constitute a normal d.c. motor and also have the same control characteristics. The tacho heads pick up information concerning the speed and direction of rotation in the form of variable capacitance. This information is transmitted to the servo control circuits. The advantages of brushless motors are long life, near silent operation, no brush sparking to cause interference, and no brush friction.

3.1.11.2 Spooling motors
The spooling motors are usually heavy-duty, high-torque d.c. motors. Again, modern machines make use of tacho sensors to allow fine control over the speed and direction of the motors. Together with tape tension sensors, servo control enables proper tape tension to be maintained before and after the capstan, even under extreme spooling conditions. Some machines only have one spooling speed in each direction, whilst others have a choice of speeds or are fully variable. These are advantageous since spooling a tape which is full of spliced edits at high speeds

could cause destruction of a master tape if an edit were to fail.

It has already been mentioned that all guides and rollers need to be perfectly concentric and vertical; the spooling motor spindles are no exception, as any misalignment can cause the tape to ride up and down while the spindle is rotating, causing an uneven wind. The tape should wind onto the centre of the spool without touching either flange. Bad spooling can be caused by air becoming trapped when the tape winds onto the previous layer on the spool. Matt-backed tapes are better in this respect than shiny-backed tapes.

In wind mode, the tape tensions should not be excessive or the tape may stretch, particularly on the inside diameters. This may be a problem that occurs with small cine spools. If the spools are made of plastic then high wind tensions can result in the spools becoming cracked.

Normal spooling should not be performed with the tape in contact with the heads as this severely increases the head wear. The tape machine may offer different wind modes which allow the tape to be in or out of contact with all the heads or just the replay head during spooling, for monitoring purposes. In this case, the high frequencies in the replay amplifier are usually attenuated so that the high-speed output does not cause destruction of the loudspeakers.

It is important to maintain the brakes that act on the spooling motors. Braking should be even and the tape should stop gradually without throwing loops or snatching the tape. This is especially important on machines which rely on mechanical brakes to bring the tape to a halt. Later machines use some form

of tension-dependent d.c. braking to bring the tape to a gradual halt, then apply the mechanical brakes to park the tape and prevent spillage. The test of good braking characteristics is to see what happens at full wind speed when the mains power is suddenly removed. A good machine will automatically apply the brakes and come to a controlled halt. Poor machines may deliver the tape on the floor.

Tape spools should always be tightly clamped to the spooling platform. Unclamped spools can ride up and down causing bad packing, and in some cases loose spools can upset the tape tensioning servo mechanism causing the tape tension sensors to judder.

3.1.12 Causes of poor mechanical performance

3.1.12.1 Tape speed
As the capstan motor is the speed controlling element in the tape path, the pinch roller pressure must be maintained within specification. A high take-up tension will drag the tape from between the capstan and pinch roller so that the capstan cannot control the tape speed. Similarly, a high back tension will also prevent the capstan from doing its job. If the capstan is a synchronous motor then a mains frequency drift will cause tape speed variations. Non-synchronous capstans with servo systems for speed control can also have failures which cause speed errors.

3.1.12.2 Wow and flutter
Variations in the tape motion can give rise to flutter which is caused by variations in the tape speed that have a frequency greater than 10 Hz. Wow is caused by tape speed variations between 0.1 and 10 Hz and drift which means variations below 0.1 Hz of a non-regular nature, extending down to a steady d.c. error speed change.

The wow and flutter can be caused by eccentricity in any rotating component, worn bearings, damage or poor-quality tape, incorrect adjustment of tape tensions, varying friction in the tape path, or winding variations including sticky edited joints. The larger bearings such as those in the spooling motors which rotate slowly can contribute to wow, as can sticking brake bands. The smaller, faster bearings can add to the flutter. Incorrect back tension can also indirectly contribute to flutter by causing the tape to vary its pull on the capstan bearing; the bearing then rocks backwards or up and down. It is this movement that generates flutter. The wow and flutter speed will usually indicate the rotational speed of the part

causing the main component of the frequency variation. There are different measurement standards for wow and flutter. It can be quoted as either an r.m.s. or peak-to-peak value over a specific band of frequencies; usually the value is weighted. Commonly, a peak weighted value according to DIN 45507 is used.

As the tape passes across the head it can resonate at the record head if the length of the tape between the left hand roller bearing and the capstan shaft is not sufficiently damped. The resonant frequency depends on the length of the tape, the tape type and the applied back tension. With some types of tape this resonance can cause the tape to vibrate against the record head and cause scrape flutter which modulates the audio recording by adding noise. The effect can be greater if the tape guides are too narrow for the tape or the tape is slightly too wide, or badly slit.

Some manufacturers use scrape flutter rollers on their head blocks. These consist of rotational bearings with extremely low friction which lightly touch the tape between the record and replay head (sometimes situated between the erase and record head). The effect is to damp out the resonance in the tape, leading to cleaner recordings. The scape roller must be maintained properly; if it becomes worn then it can produce flutter which is proportional to its rotational speed.

3.1.12.3 Tape tension
Early professional machines controlled forward and back tension purely by reel motor energization, so that considerable variation in the tape tension could be encountered at different parts of the reel. This caused wow, especially towards the end of the reel. There are a number of solutions to this problem. The mechanical braking on the supply motor could be altered to reduce the back tension produced by the motor at small diameters while it revolves without any energization. In this way the tension can be maintained within about ±15%.

A better system is to use electronic servo control involving motorized back tension and applying feedback through various circuits to the motor. This feedback system can be mechanical or it can be electronic using photoelectric devices or inductive or capacitive proximity circuits. Part of the feedback elements can be tension-control levers which are connected to potentiometers that give a voltage proportional to the instantaneous tension value. This value is then fed back to the tension-control circuits. Constant forward and back tensioning allow different spool sizes to be used. Other machines have a switch to change tensions depending on the spool size used;

smaller spools requiring lower average torque. A start-up boost is sometimes applied to the take-up spool for a couple of seconds after the transport is energized. This gives a high torque value to accelerate the spools so that play speed can be reached more quickly. After the boost, the tension is restored to its normal running value.

A different approach to tape tensioning is to use a dual capstan assembly in which the tape drive capstan is rotating slightly faster than the back tension capstan, allowing constant tape tension across the heads. Adjustment of dual capstan machines is very critical for the system to work well.

3.1.13 Headblocks and soundheads

Nearly all professional recorders incorporate three separate heads, the first for erase, the second for record, and the third for replay. The erase head may be either full track or split track allowing recording on all or only selected tracks. Split track heads vary in the amount of unerased tape between the tracks and so should not be used to erase full track recordings, otherwise the original recording may break through on replay.

Professional machines may include a sync playback facility which allows individual tracks of the record head to become replay tracks either through the normal replay amplifiers or through separate sync amplifiers. The playback quality from the record head is not to the same standard as the proper replay head. The record head gap is wider than the replay gap so the sync playback response is usually low-pass filtered, otherwise the cross-talk between channels can become quite severe. Some manufacturers allow this low-pass filtering to be switched out, enabling the full audio range on the sync channel, albeit with worse cross-talk performance. The sync facility is used for monitoring material previously recorded on certain tracks, so that new recordings can be dropped in without the time delay that would be caused by monitoring from the replay head. The sync head is also used for 'track bouncing' of two or more tracks onto one new track. This enables the original tracks to be free for more recordings. This is useful if there are only a limited number of tracks available on the machine, although there is obviously quality loss each time this is done.

Soundheads have many different configurations depending on tape widths and track formats (*Figure 3.21*). They may also be constructed from different materials, such as Mumetal, ferrite and cobalt. Ferrite heads have lower distortion at high

Figure 3.21 *A selection of head track formats.*

levels than Mumetal and can last longer. They are also available with closer tolerance gaps than Mumetal so that they can be used for recordings that require an even level of magnetization across the tape width, such as test tapes.

The development of new magnetic tapes imposes more and more stringent demands on the soundheads, and so head manufacturers are always looking for new alloys to give the required characteristics of high permeability, adequate saturation induction, low remagnetization losses, high mechanical strength, as well as resistance to corrosion and abrasion.

One such material is the so-called 'glass metal' which is an amorphous alloy of cobalt with boron or silicon (Hanggi, 1988) in which the atomic arrangement has no order, i.e. it has no crystal lattice like conventional metallic materials. This alloy is almost free of magnetostriction and so has only a very low magnetic sensitivity to mechanical stress. Its advantages include significantly lower head wear and, because of greater dimensional accuracy due to spark erosion of the core in the manufacturing process, the gap depth can be kept smaller which improves the efficiency of the soundhead. The hardness of amorphous alloys mean they have a long life-span and recalibration made necessary by the change in gap depth can be performed at longer intervals. The relatively high resistance and the material thickness minimizes the eddy current losses so that the recording currents can be decreased to approximately 60% of those required with conventional heads, while the reproduce level is about 3 dB higher at 1 kHz and 6 dB higher at 20 kHz. These higher reproduce levels mean an improved signal/noise ratio.

It can be seen from *Figure 3.21* that there are two different head configurations for recording two tracks of audio onto tape. In the late 1950s, when stereo sound reproduction began to appear, standards were written to specify the track formats for recording the required two channels (Dorner, 1986). The NAB 1965 standard for magnetic tape makes first mention of this where it states that the recorded tracks for two-track monophonic or stereophonic recording will be 0.082 in. (2.1 mm) in width with a centre spacing of 0.156 in. (from a reference tape edge). This results in an unrecorded width of 0.071 in. (1.88 mm).

The standard mentions two-track monophonic recording; with the tracks being separated by 1.88 mm, the cross-talk would not be a problem. (A figure of not less than 60 dB from 200 Hz to 10 kHz was specified.) These track dimensions compare well with those mentioned in DIN 45500: Part 1: 1971, where the two-track dimension is specified with 2 mm for each track and giving a track separation of 1.85 mm whilst also specifying the same 60 dB crosstalk figure. In both cases, the idea was that separate monophonic recordings should be possible as well as stereo. This meant an erase head design

was required to permit individual or simultaneous track erasure. With IEC Recommendation 94: 1986 for both two-track monophonic and stereophonic formats, only the guard band is specified, which should measure at least 0.03 in. (0.75 mm), with a small note saying a guard track of 0.08 in. (2 mm) is sometimes used. No reference to cross-talk is given.

The DIN specification (1955) for recorded flux level at 15 in./s was given as 32 millimaxwell per millimetre (mM/mm) which added up to 200 mM on a 0.246 in. (6.25 mm) wide tape. Actually the flux totalled 189 mM because the recorded area on the tape is only 0.232 in. (5.9 mm) wide.

When German broadcasting stations began making their recordings in stereo, their stations were still transmitting in mono and it was found to be a problem that the stereo recordings when being played back on a full-track reproducer did not produce the same signal level as that available from a 200 mM full-track recording. The two 2-mm tracks with their flux levels of 64 mM each added up to an r.m.s. value of $\sqrt{(64^2 + 64^2)} \approx 90.5$ mM instead of the 189 mM of the full-track tape, i.e. the output fell short by 6.4 dB. To compensate for this it was decided to make the tracks wider, resulting in less separation. The worse cross-talk performance could be tolerated in stereo recordings. The result was a guard band of 0.03 in. (0.75 mm) as mentioned in IEC94. The same figure for the guard band is specified in DIN 45511 with an individual track width of 0.102 in. (2.6 mm).

These wide tracks now contained 83.2 mM/track thus totalling 117.7 mM r.m.s. This is still 4.1 dB short of the 189 mM of the full-track tape. To compensate for this, the oxide formulations were improved until 4.1 dB more magnetic flux could be recorded without any increase in distortion. Raising the 32 mM/mm by 4.1 dB gives 51.4 mM/mm or 514 nWb/m. The two 2.6-mm-wide tracks when combined now yield 189 mM and mono-stereo level compatibility is achieved.

So why does the stereo head have this 'butterfly'-like construction? With only 0.75 mm separation of the two core sections, the only way to accommodate the windings is to spread the two cores apart in a V-shape.

3.1.14 Audio electronics

3.1.14.1 Record path
The record amplifier chain converts a voltage at the line input into a current in the record head. The flux produced on the tape is approximately proportional to

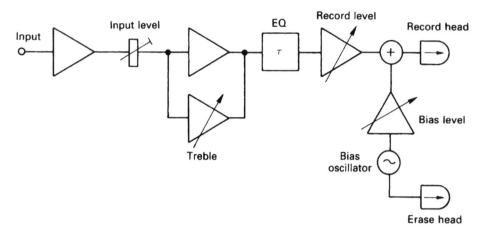

Figure 3.22 *A simplified record chain. EQ, equalizer.*

the current flowing through the head. As the recorded frequency increases then the inductance of the head increases and so, in order for the amplifier to produce a constant current through the head at all frequencies, the record amplifier must have an output impedance which is much greater than the record head at the highest audio frequency.

After the external input levels are matched to the internal operating levels of the tape machine with the input level controls (*Figure 3.22*), the signal then goes through a treble control for loss compensation, and then through a standard equalization network (which is usually selectable depending on the equalization required), to give the correct recording characteristic. The recording level control sets the required drive signal level which is then mixed with the bias signal and passed through to the record head.

The bias may be obtained from the erase oscillator. Its waveform must have as low an even harmonic distortion as possible to reduce modulation noise and background hiss on the tape. Odd harmonic distortion is not important in this application since all that is required from bias and erase is for the area of the waveform above zero current to equal the area below it. It would be possible to use a perfect squarewave which has no even harmonics for bias and erase, although this is not really practical. Often, the erase and bias circuits have adjustable symmetry controls to minimize the even harmonics.

The bias current path should always present a high impedance to audio frequencies so that any high-frequency signals from the record amplifiers do not cause cross-talk on other channels. Since the bias frequency is approximately 5–10 times the highest audio frequency, this can usually be done

with a low-value series capacitor in the bias feed (McKenzie, 1980). What is more difficult is to keep the bias frequency from blocking the output circuits of the record drive amplifier because the bias current through the record head is always greater than the audio current. A tunable bias trap located in the output circuit of the record amplifier is adjusted for the lowest high-frequency bias voltage possible on the output of the record amplifier, since serious breakthrough at this point can be passed back and cause severe distortion. The erase head has usually a much lower inductance than either the record or replay heads and considerable erase current passes through it. The head is usually designed to be part of the tuned circuit of the erase oscillator. Therefore, if the erase head is replaced, the circuit will need retuning as the new head may have a slightly different inductance, which alters the erase and, therefore, maybe the bias frequency. It is not usually necessary to replace an erase head for magnetic performance reasons. It is only necessary if the head is so worn that flats have been scored across the front of the head causing shedding of the oxide coating of the tape.

When the tape machine is switched into record the bias oscillator should build up to full output within a fraction of a second. Similarly, when the machine is dropped out of record the bias oscillator should ramp down slowly to prevent d.c. magnetization of the record head. When the tape machine is being used for playback there should be no net recording audio current passing through the record head. The machine may employ a mute relay for the record head which shorts the record head drive current to earth when the head is not switched to record.

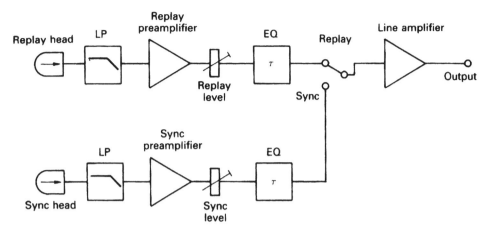

Figure 3.23 *A simplified relay/sync chain. EQ, equalizer; LP, low-pass filter.*

3.1.14.2 Replay/sync path
Figure 3.23 shows a simplified replay/sync path. If we consider the replay head, this chain has to provide considerable amplification of the output voltage of the head to drive fully the line output stage. The heads can be thought of as modified current generators in series with their own inductance. The output voltage varies in proportion to frequency within the audio range. At low frequencies the head gives a low output voltage, but from a low source impedance. High frequencies give a higher output but from a higher source impedance. From the head the signal goes through a low-pass filter which attenuates any remaining bias and erase frequencies still present in the signal. The signal is then amplified, level set and equalized. The equalization characteristics have an inherent 6 dB/octave bass lift which must roll off at the extreme bass end to prevent transistor flicker noise from becoming a problem. The first-order slope terminates at one of the characteristic transition frequencies, so that the overall response is flat. The signal then passes through the drive amplifier to bring it up to studio operating levels and provide a balanced, low impedance output.

3.1.15 Alignment

3.1.15.1 Mechanical considerations
Before starting on audio calibration of the tape machine it is assumed that tension adjustments have been carried out according to the manufacturer's instructions. Once set, these should only need checking every so often during routine maintenance. The tape path should be regularly demagnetized using an appropriate tool. Using a small hand demagnetizer

that has a small tip designed for cassette decks on a 2 in. multitrack machine will result in leaving more residual magnetism on the deck components than it removes because its magnetic field is not strong enough to bring the domains through their saturation loops. The demagnetizer should always be switched on away from the tape deck, then slowly brought towards the heads and taken to each part of the deck in turn. It should be kept on until it has been drawn away from the deck where it can be safely switched off without causing any residual magnetization.

It is very important to keep the tape path clean at all times. It is always a good idea to have some cotton buds and isopropyl alcohol to hand near the machines in a studio. The tape path should be cleaned before each recording session. Not only will oxide build-up on the heads cause severe high-frequency loss, but also, if the oxide is allowed to accumulate on rotating guides then they can act as rotating magnets causing partial demagnetization of the tape, resulting in modulation at the rotational speed of the roller.

3.1.15.2 Head alignment
The soundheads can usually be adjusted in four ways (*Figure 3.24*). Wrap is where the head is rotated so that the tape skims across the profile of the head symmetrically on either side of the head gap. The height of the head is very important to get the maximum output from the tape and minimize cross-talk between tracks. The zenith is the amount of tilt into or away from the tape. This results in uneven signal output across the head and uneven head wear which can prove expensive to put right, i.e. relapping, which involves an abrasive process to take away head

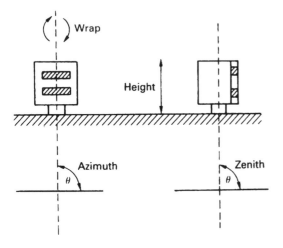

Figure 3.24 *Mechanical adjustments of a sound head.*

material and so restore the correct head profile. This may not be possible if the head is so badly worn that relapping destroys the head gap and so renders the head useless. In this case, the only option is to fit a new head.

Azimuth adjustment is the tilt parallel to the tape path. It is this adjustment that lines up the head gap so that it is perpendicular to the audio track and ensuring that signal phasing is accurate. Excluding the azimuth which is aligned using a test tape, all these adjustments are normally carried out with the aid of precision machined gauges with the headblock and gauge sitting on a very flat surface such as a glass block. There are set screws for each of the adjustments.

3.1.15.3 Calibration of the audio path
First the reproduce and sync paths are aligned with the aid of a reference tape containing calibrated recordings. Subsequently, the recording channel is adjusted using the previously calibrated reproduce channel as a measuring instrument. In this way, the reference tape with its magnetic reference fluxivity serves as a base for all subsequent adjustments.

Manufacturers' operating manuals often use different terms for different levels through the chain. Line level is the level that appears on the output of the correctly aligned replay amplifier when a tape with reference fluxivity is reproduced. When this line level is fed to the input of a correctly adjusted record amplifier it produces the reference fluxivity on the tape. The operating level has different definitions depending on whether NAB or CCIR characteristics are used. The NAB operating level is correct when

the line level is set for +4 dBu (0 dBu = 0.775 V r.m.s.) and the Vu meter reads 0 Vu. The CCIR operating level is when the line level is +6 dBu resulting in a meter reading of +6 Vu. Peak level is usually taken as +6 dB relative to the operating level.

3.1.15.4 Replay and sync
Magnetic reference tapes are available with different reference fluxivities. They are usually magnetized full track and come with different test sequences. Some of these reference tapes feature a blank segment after the frequency response section, a so-called 'record pad', so that the record adjustments can be done. This is a dangerous practice and should not be attempted because of the risk of erasing the reference section of the tape (these tapes are expensive). It is best to use virgin tape of the type to be used in the session for record alignment.

3.1.15.5 Level
The level tone section of the tape is usually at a frequency of 1 kHz or 700 Hz. It has been recorded at an accurate, well-defined level. This is taken as the 0 dB reference level. If this is not the reference fluxivity required, then the following example shows how the tape can still easily be used.

Assume that we need to align a machine to NAB characteristic at 320 nWb/m for 0 Vu = +4 dBm. The reference fluxivity on the tape is 200 nWb/m which will not give the required line level at the output of the machine and therefore a quick conversion formula must be applied. If any tape deck is to be calibrated with a different reference level, i.e. one that is generally higher than the one available, then the reference fluxivity difference is calculated as:

$$20 \log_{10} \text{(Desired reference fluxivity/Reference fluxivity on tape)} = \text{Difference (dB)}$$

So, in this example,

$$\text{Difference (dB)} = 20 \log_{10}(320/200) \approx 4 \text{ dB}$$

So with NAB operating levels, when the reference tape is replayed, the output level is adjusted so that the meter reads:

$$\text{(Operating level – Difference)} = (+4 \text{ dBu} - 4 \text{ dB}) = 0 \text{ dBu} = -4 \text{ Vu}$$

3.1.15.6 Azimuth
The azimuth test section of the reference tape is used to adjust the vertical position of the soundheads: The test recording usually consists of two segments, the first segment of around 6.3 kHz allows the course

azimuth to be adjusted. If a dual-beam oscilloscope is connected to two output channels of the tape machine the head azimuth can be adjusted so that the output levels reach a peak voltage with both sine waves covering each other. If the head is considerably out of alignment, then a second, smaller voltage peak will occur, so care must be taken to find the correct peak. A fine azimuth section at about 12 kHz is then used to minimize the phase difference between the channels.

On multitrack heads, the phasing problems are more noticeable as the number of tracks is increased. The azimuth of 2-in. tape is best done using the outside tracks. The inner ones can then be checked against the outer ones. It may be that phase jitter becomes apparent with this test. Serious jitter is produced when tape-guidance problems exist, such as worn guides or if the heads have zenith problems. The tape should ride evenly around the pressure roller and should not ride up and down. Much greater precision is required in the manufacture of multitrack heads if they are not to show serious phase errors between the outside tracks. Any slight movement of tape up, down or skew has considerably more effect on phase than that caused by similar motion on narrower tape widths. Severe phasing sounds can be caused by mixing together tracks which have similar information but are widely spaced. Therefore phase-sensitive sounds should be placed as close together as possible on the multitrack recording.

Record-head azimuth can be aligned in the same way using the replay test tape, but this time switching the channels to sync, or it can be done later on during the record alignment, using the replay head to monitor the tape output.

3.1.15.7 Frequency response
This tape section contains the frequency segments. The equalization of the reproduce or sync channel is adjusted to give a flat frequency response when this section is replayed. Sometimes the level of this section is 10 or 20 dB below that of the reference tone section to prevent high-frequency crushing of the tape.

It is important to know that if a full-track reference tape is used for multichannel machines with different head configurations a measuring error occurs at bass frequencies. This effect depends on the actual reproduce head dimensions and is called the fringing effect. It occurs because of extra bass signal that is induced in each replay head core by parts of the recording on either side of the track. Some reference-tape manufacturers include a list of compensation factors which provide correction for fringing effects. Often it is better to leave the replay bass adjustment until after the record side has been aligned. It is possible to obtain multitrack test tapes which have individually recorded tracks which do not produce bass boost problems. However, the magnetization of the different tracks of the test tape may not be consistent at high frequencies because of the difficulty in aligning different vertical gaps in the manufacture of the record head. Alternatively, full-track, bass-compensated tapes are available where the levels at bass frequencies have been adjusted to compensate for the fringing effect. This type of tape is said to have compensation for multitrack recordings.

3.1.15.8 Handling of test tapes
It may be obvious that test tapes should be handled carefully. They should not be exposed to magnetic fields (do not store next to motors, speakers, headphones, etc.), and stored at room temperature with no exposure to direct sunlight or humidity. It has already been said that the tape transport must be kept clean and demagnetized. In addition, the audio channels should be switched to 'safe' so that there is no chance of recording over reference tracks. Unnecessary fast-wind operations or rewind in edit mode (with the tape against the heads) should also be avoided.

3.1.15.9 Record alignment
With a line level at 1 kHz applied to the input of the recording channels, and with the machine switched to record, the record level is adjusted to give unity gain through the record/replay chain. If the azimuth alignment has not already been carried out using the sync path in the replay chain, then the course azimuth alignment can be done by applying 6.3 kHz to the record head whilst monitoring the replay channels with a dual-beam oscilloscope and adjusting the record-head azimuth for maximum peak voltage.

3.1.15.10 Bias
The bias is one of the most critical audio adjustments on the machine. Its setting is a compromise and needs to be set to allow the particles on the tape to accept the audio part of the magnetic field with the minimum distortion that is compatible with a good response at the high-frequency end of the audio spectrum.

A 10 kHz signal at 10–20 dB below line level is fed to the machine. This lower level is to again ensure that there are no compression effects on the tape. Whilst metering the output of the replay amplifier, the bias level for each channel is increased

slowly until a peak voltage is measured. The bias level is then further increased until the output level starts to fall again. The amount of overbias required for a particular brand of tape is usually that recommended by the manufacturer of the tape to achieve the best performance. Typical figures for overbias at 10 kHz are 1.5 dB at 30 in/s, 3.5 dB at 15 in/s, and 6 dB at 7.5 in/s (*Figure 3.25*).

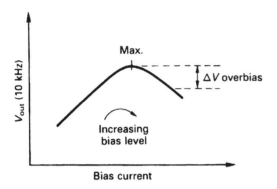

Figure 3.25 *Overbias.*

We can see from this that the more bias that is applied to the tape, the lower the output level at the high-frequency end. Note that the bias adjustment should not be used as a quick form of record equalization and that the bias should always be correctly set before any record equalization is carried out. At this point it is worth checking and readjusting the record levels again at 1 kHz line level, as the bias will have affected the output level.

3.1.15.11 Frequency response

If frequencies between 10 kHz and 15 kHz are applied at levels of 10–20 dB below line level, the high-frequency equalization can be adjusted to give an optimally flat response. The fine azimuth can be carried out now with a frequency in the range 12 kHz–16 kHz. It is done after the bias because varying the bias affects the apparent azimuth of the record head. The mechanical and effective magnetic gap are not in the same position, the distance depending on the magnitude of the bias current.

At this point the replay bass response can be adjusted (if it was not done previously) using a full-track reference tape with multitrack compensation by recording a frequency in the range 60–100 Hz and adjusting the replay bass for a flat response. It is not always possible to get an absolutely flat response; often there are small peaks and troughs throughout the frequency range. It is usually considered better

to get the channels to look as identical as possible rather than go for the flattest channel response.

3.1.15.12 Erase

The erase head and associated circuitry do not often need to be adjusted, but the erase efficiency of the head may be checked by recording a high level of 1-kHz tone onto the tape and erasing it with no input signal to the record head. Measuring the output of the replay amplifier of each channel with a 1 kHz notch filter, the erase field should be such that the original recording is reduced in level by, typically, greater than 75 dB. If the channel with the worst erase efficiency is selected the erase head itself may be turned slightly until the best erase efficiency results on this channel. Once again, the worst channel is selected, and the erase current adjusted to give the required erase depth. Some erase heads do not provide a sufficient field for complete erasure, and increasing the erase current or changing the frequency will not necessarily improve results since the head may be saturating.

3.1.16 Cassettes

The cassette tape is a bidirectional system of recording offering twin-track stereo recordings. The side-one pair of tracks are recorded in one direction and the side-two pair in the other (*Figure 3.26*). The cassette contains tape which is 3.8 mm (0.15 in.) in width and which travels at a speed of 4.75 cm/s ($1\frac{7}{8}$ in/s). Each magnetic audio track is 0.6 mm wide.

Figure 3.26 *Audio cassette tape.*

Originally, the oxide material was ferric oxide, but to meet the demand for extended high-frequency response at the low speeds found in cassette recorders. Du Pont developed a new material, chromium oxide (CrO_2), in the late 1960s. In the mid-1970s, TDK introduced a cobalt doped ferric oxide particle, Super Avilyn, that matched the high-frequency response of

CrO_2 but had a higher overall MOL. Pure metal particle (non-oxide iron) tape was introduced in 1978 by the 3M Company to provide better still high-frequency response at higher recording levels.

3.1.16.1 Tape duplication

There are several steps required to produce a pre-recorded cassette (Siegenthaler, 1984). The original recording is made with a multitrack reel-to-reel machine at 15 or 30 in/s. This is then mixed down to a master recorder at 15 or 30 in/s (this is also the master tape used for disc manufacture). The tape is then copied to a 1 in, master recorder running at 3.75 in/s which has a track configuration analogous to the cassette tape. It is then loaded into a reproduce loop bin storage chamber. The end of each master tape is joined to a short length of cue tape which has a 6 Hz signal on it. This is then spliced to the beginning of the tape and so completes the loop.

When this tape is played at high speed, special slave recorders loaded with virgin tape in a conventional reel-to-reel way copy the tape to cassettes at a speed of $64 \times 1\frac{7}{8} = 120$ in/s (64:1), or more than 3 m/s. Since the running master is a continuous loop, each duplicated tape pancake contains a series of copy recordings of the master, separated by the 6 Hz cue tone. Each pre-recorded cassette tape pancake then goes on to a tailoring and winding machine where it is packed into empty cassette shells.

In order to save copying time, the copy speed should be as high as possible. But the frequency range of the programme material to be transferred also changes with higher copying speeds so the slave recorders have to deal with frequencies 64 times greater than that of an ordinary audio recorder, i.e. the frequency band of 20 Hz to 20 kHz shifts to 1.28 kHz to 1.28 MHz and these frequencies are largely in the radiofrequency medium waveband. This poses problems for the recording heads and electronics.

The master recorder needs to record at as low a speed as possible since to playback a tape, made at say 15 in/s, at the copying speeds given above would mean having a reproducer replaying at 24.38 m/s which is a very difficult task. Therefore a speed of 3.75 in/s is used. This just shifts the problems to the other side as the master recorder must feature extremely low wow and flutter and outstanding stability in the head region. Problems caused by mechanical deformities in the heads, vibration, flutter caused by tape transport guides, etc., build up between the recording and the soundheads and so the absolute error increases with lower tape speeds

because the recording density increases proportionally. The master recorder must therefore be a precision machine.

References

ARNOLD, A.P., *Principles of Magnetic Tape Recording (BBC Engineering Training Information Sheet 45T)*, BBC, Evesham (1979)

DORNER, J., Stereo or two track recording. *Swiss Sound*, **16**, 8–10 (July 1986)

HAMMOND, P., *Electromagnetism for Engineers*, Pergamon Press, Oxford (1978)

HANGGI, R., Frozen chaos. *Swiss Sound*. **23**, 5–7 (July 1988)

HARRIS, P.J., *Magnetic Recording (BBC Engineering Training Information Sheet 21P)*, BBC, Evesham (1982)

MCKENZIE, A., Tape Machines. In: *Sound Recording Practice* (ed. J. Borwick), 2nd edn. Oxford University Press. Oxford (1980)

MRL. *Magnetic Reference Laboratory Product Summary 1985–08*, MRL, Mountain View, CA, USA (1985)

SIEGENTHALER, M., Precision. *Swiss Sound*. 2/84, 3–4 (June 1984)

WILSON, F.A., *Audio*, Bernard Babani, London (1985)

3.2 Noise reduction systems

3.2.1 General principles of noise reduction

3.2.1.1 Companders

Generally we want the output signal from an audio transmission or recording medium to resemble the input closely. However, the medium often impairs the programme material by adding noise. When the reproduced volume of the loudest signal is set as desired, this noise may be audible and at times obtrusive; the quietest input signal may be below the noise level of the medium, and may therefore be lost in the noise. It may be impractical or uneconomical to improve the basic signal/noise ratio of the medium, and so electronic noise reduction (NR) may be used to improve the subjective signal/noise ratio.

Analogue NR systems can be divided into single-ended and complementary (*Figure 3.27*). Single-ended ones receive the programme material already polluted by noise, and attempt to remove the noise without impairing the wanted signal. Such systems include filters, both fixed and programme controlled, and noise gates. Since some wanted signals are noise like and some unwanted noises resemble wanted signals, the separation of noise and programme is in general impossible, although in some cases intelligent 'steering' of the equipment can do a useful job.

(a)

(b)

Figure 3.27 *Noise-reduction systems: (a) single ended; (b) complementary.*

The subject of this section is complementary systems. These require that the wanted signal be available prior to pollution by the noisy medium (tape-recorder, land line, film soundtrack, satellite link, etc.). The noise reduction involves two stages of processing: at the input of the noisy medium, encoding in some manner to render the signal less susceptible to damage, and after passage through the noisy medium, decoding to undo the previous processing and thus recover the original material. The simplest form of complementary NR is fixed pre- and deemphasis, but the term 'noise reduction' is usually reserved for systems in which the processing includes adaptation to the signal, that is, time-variant characteristics.

A pair of such complementary processors is known as a compander (compressor-expander). Companders have been known since the 1930s (e.g. Cushman, 1933), but apart from a few experimental systems (e.g. Fletcher, 1941) it was only with the widespread introduction of high-quality magnetic tape recording in the late 1940s and early 1950s that the principle began to be considered seriously for purposes other than telephone communication.

In its most elementary form, a compander can be considered as an electronic volume control at the input of the medium, turning up the level when the input signals are quiet, together with an inverse volume control at the output turning down the level for quiet signals. It is obvious that this variation in the reproducing gain must be accompanied by a rise and fall in the noise introduced by the medium. (Any noise present in the source is not affected; since overall the gain and loss of the compressor and expander

cancel; only noise and distortion introduced between the compressor and expander will be affected.)

When applied to telephone lines, the variation in the background noise is a small price to pay for the improvement in intelligibility; quiet voices no longer get drowned in the noise. However, for music reproduction the modulation of the noise by the signal can be very disturbing because varying noise draws attention to itself in a way that a fixed noise level, even if high, does not.

All NR systems yield varying, programme-modulated noise levels. The most difficult problem in designing an NR system for high-quality audio is making this noise modulation inaudible.

3.2.1.2 Noise masking

NR systems are intended to create the illusion that the signal/noise ratio of a medium is greater than it really is. This is only possible with an understanding of what we can hear and what we cannot. The solid line in *Figure 3.28* (Ehmer, 1959) shows the sound pressure level corresponding to the typical human hearing acuity, i.e. the threshold of hearing. In the absence of loud sounds, sound pressures below this curve are inaudible, while those above are audible. The transition from inaudible to audible as the level of a sound is raised, crossing the threshold, is abrupt, occurring over very few decibels.

The dashed curves in *Figure 3.28* show how the acuity is modified by the presence of moderately loud (65 dBA) sounds at a number of frequencies. At frequencies close to a loud sound, the acuity curve rises dramatically, so other sounds in that region of the spectrum must be much louder before they can be heard at all. Sounds that would otherwise have been audible but are rendered inaudible by the presence of a louder sound are said to have been masked. At frequencies far from the loud sound hearing acuity is not affected by its presence; in these areas there is no masking. The extent and degree of masking obviously depend on the frequencies and levels of the loud sounds: every instant of music will give a different acuity curve. Quiet sounds may provide little or no masking of other quiet sounds.

With a simple wide-based compander as described above, in which the expander merely changes gain at all frequencies equally, it is difficult if not impossible to eliminate audible noise modulation. At the moment of a bass drum stroke, for example, the decoder gain rises and thus noise at all frequencies rises. The low frequencies in the noise are masked, so the change there will not be heard. High-frequency noise is not masked, so its change may be

Figure 3.28 *Sine-wave masking at 65 dBA acoustic sound pressure level (0 dBA = 20 μPa).* (Courtesy of Ehmer (1959).)

audible. The designer of an NR system must have an understanding of masking if the inevitable variations in noise level are to be inaudible.

3.2.1.3 Attack behaviour

Consider first the compressor of an elementary compander (*Figure 3.29(a)*). The variable-gain amplifier receives the input audio and a control signal which may be derived from the output as shown or from the input. If an input signal suddenly rises in level (the envelope is shown in *Figure 3.29(b)*). the control signal rises more slowly, reaching a new value over a period known as the *attack time* (*Figure 3.29(c)*). During this time the gain falls from its previous value to a new one, so the output is initially the new input signal amplified by the *previous* gain (*Figure 3.29(d)*).

The initial excess output is known as an *overshoot*. If the attack time is substantially longer than a cycle of the input signal, the magnitude of the overshoot is equal to the degree of gain reduction and its duration is equal to the attack time. Overshoots at low levels are innocuous, because they do not reach the overload of the medium, but at high levels they are obviously undesirable because they cause transient overload. A superficially attractive way of reducing them is to use an attack time which is short compared with the period of the highest input signal, substantially an instantaneous attack.

In general, very fast attacks are unsatisfactory. Consider the expander of the elementary compander (*Figure 3.29(a)*). The output of the variable gain amplifier is the product of its two inputs. If the

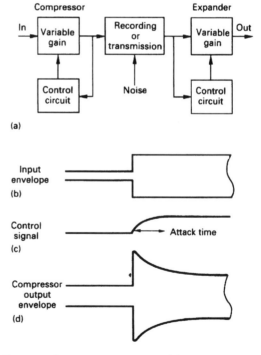

Figure 3.29 *Elementary compander (a) and the dynamic response of its compressor (b–d).*

control signal were a constant (d.c.), the output would merely be a scaled version of the audio input, with exactly the same spectrum. However, in practice the control signal is varying and has a spectrum of its own: thus the non-linear process of multiplication yields a different spectrum.

The system can be considered as an amplitude modulator, in which the audio input is the carrier and the control signal the modulating waveform. The process of modulation therefore introduces side-bands; for each frequency f_a in the audio and each frequency f_c in the control signal there is added a pair of sidebands at $f_a + f_c$ and $f_a - f_c$. The magnitude of these side-bands depends on the degree of modulation, i.e. the amount of gain change occurring. These sidebands are known as *modulation distortion*. If the control signal moves slowly, it contains only low frequencies, so each pair of side-bands is close in frequency to the corresponding 'carrier', within the masked region, and the distortion will, therefore, be inaudible. However, if the control signal can move rapidly (short attack time), the side-bands will be remote from the 'carrier' and may not be masked (depending of course on the masking properties of the audio signal at that instant). Since a fast attack usually occurs briefly, the audible result is a brief burst of wideband energy not present in the source, which the ear interprets as a click.

Note that modulation distortion is not usually audible at instants when overshoots would have been a problem unless a fast attack had been applied, because those are the instants when the signal masks noise and distortion over a wide band. It is more likely to be audible when the signal does *not* provide much masking, as for example a tone without extended harmonics but with a periodic smooth rise and fall in amplitude: after compression (or expansion) each rise is accompanied by a click. A good example might be a held piano note, in which the amplitude changes result from beats between the multiple strings.

Since the compressor and expander of an NR system will have inverse characteristics, one might hope that the modulation distortion generated by the compressor would be cancelled by that of the expander. Indeed, if a compressor is connected directly to an expander, the distortions often do cancel more or less, but since an NR system must work with the imperfections of a real medium, including bandwidth limitations and phase shifts, in practice the cancellation is frequently not enough to make the clicks inaudible.

Some NR systems take no precautions against high-level overshoots, accepting that the medium will be driven into transient overload and that occasionally this may lead to audible distortion, while others employ various techniques, lumped under the term 'overshoot suppression', to prevent transient overload.

3.2.1.4 Recovery behaviour

Figure 3.28 illustrates masking in the presence of a loud signal, when all NR systems inevitably lose some of their noise reduction. If the loud signal stops suddenly, full acuity is not restored instantaneously; it takes at least 100 ms for the ear to recover. Thus an NR system does not have to restore full noise reduction immediately; the circuitry may take some time, provided noise reduction has been restored before the ear has recovered. If, however, the ear recovers first, the sudden cessation of a loud sound may be followed by a brief audible burst of noise; this is called a *noise tail*. In general, if the recovery of the system is substantially complete within 50–100 ms, no noise tails will be perceived. However, in many designs the recovery time of the processor (the fall time of the control signal) determines the ripple on the control signal, and hence harmonic and intermodulation distortions, particularly at low frequencies. Some systems achieve low distortion only at the expense of noise tails.

3.2.1.5 Philosophies of noise reduction

There are two basic approaches in designing NR systems.

In the first approach the aim is to lower the noise so that at all frequencies it is always below the current (masking) acuity curve; if noise from the noise-reduced medium is always below the threshold of hearing, noise modulation clearly cannot be audible. This is the more ambitious of the approaches, since the object is to eliminate audible noise altogether and at all times. Whether this is achieved depends partly on the initial, unreduced noise level of the medium, as reproduced to the listener.

Consider a hypothetical system which yields little or no audible noise modulation on a particular item of music when used with a medium with a signal/noise ratio of 60 dB and with a particular setting to the playback volume control. The reduced but modulated noise level will be rising just to the acuity curve but rarely crossing it. However, if the medium has instead a signal/noise ratio of 54 dB, or if the volume is raised by 6 dB, the same system on the same music will exceed the acuity curve by 6 dB, yielding gross and disturbing modulation. Note that a degradation in the signal/noise ratio of 6 dB corresponds to four passes through the '60-dB' medium, such as a fourth-generation copy.

It is apparent that a system of this type must inevitably suffer from audible noise modulation if the reduced noise level in the absence of signal is audible, i.e. already above the no-signal acuity curve.

The second approach is to provide an unchanging degree of NR, and hence an unchanging noise level, in all unmasked areas of the spectrum, and to confine the gain changes (and thus noise changes) to the parts of the spectrum where there are wanted signals to provide masking. This is the more conservative method, aiming for a fixed and predetermined amount of noise reduction irrespective of the signal/noise ratio before NR, and permits the design of a system free from audible noise modulation even under unfavourable conditions such as many passages through the medium. Systems following this method may more easily be considered as variable pre- and de-emphasis, where the degree of emphasis is fixed except where high-level signals result in noise masking; these are of course the frequencies where overload of the medium might occur unless the emphasis is reduced.

3.2.1.6 Compression and expansion characteristics

Practical companders can be divided into *constant-slope* systems and those with *bilinear* characteristics. Other characteristics are possible but seem to have no virtues in analogue NR systems, although some bit-saving techniques in digital audio use compression and expansion with more complicated transfer functions.

Constant-slope devices These have the same compression or expansion ratio at all input levels (*Figure 3.30*). For example, a 2:1 constant-slope compressor turns a 2-dB input change into a 1-dB output change whether the input change was from −62 to −60 or from −10 to −8 dB (with respect to some arbitrary reference). They have the

property that absolute levels do not need to be matched in the compressor and expander; if the level entering the expander is different from that leaving the compressor, the whole output level will be in error by that difference multiplied by the expansion ratio, but all *relative* dynamics within the programme will be reproduced correctly. It is clearly a virtue not to have to know during reproduction anything about the absolute levels during the compression process, which may have occurred at a different time and/or place. However, this independence from level matching has a number of implications.

(1) The dynamic characteristics must also be independent of level, so that the compressor can make no distinction between signals that are likely to overload the medium and those that are not. Since high-level overshoots give transient distortion in the medium, the attack behaviour of a constant-slope device must be chosen for high levels, and accepted for all other conditions. In practice loud transient signals are more likely to mask modulation distortion than quieter ones, so masking may be inadequate in relatively quiet passages, leading to a vaguely distorted sound.

(2) Compression continues, at least in theory, down to no signal. For very quiet signals, the input to the expander is necessarily different from the output of the compressor: it has had the noise of the medium added to it. Hence under no or very low-level signal conditions, the gain introduced by the compressor will be determined by the input signal but the loss in the expander will be less, set by the noise of the medium. This failure in complementarity will occur only in very quiet passages, giving an overall compression to material with very wide dynamic range. Note that if the input signal (or the compressor itself) has a high enough noise level that after compression it exceeds the noise of the medium, the effect will not occur.

(3) For any given input spectrum, the compressor and expander must adopt the same shapes of frequency response (complementary ones), irrespective of the absolute signal level. Thus if the level of the input signal with the given spectrum changes, the compressor and expander must change gain at *all* frequencies equally, and the (reduced) output noise must change equally at all frequencies.

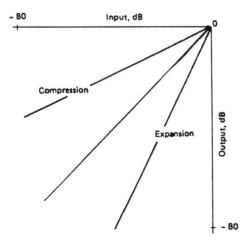

Figure 3.30 *Constant-slope compression and expansion.*

For example, consider a simple 2:1 constant-slope compander handling a long-held note at say middle C (about 261 Hz). For an input level of −20 dB the compressor will deliver an output of −10 dB (*Figure 3.30*). If the expander receives this at −10 dB, it will deliver its output at −20 dB, equal to the compressor's input. However, if there is a gain of 2 dB through the medium, the decoder will receive −8 and its output will instead be −16 dB. Everything will be reproduced correctly, but 4 dB higher than it entered the compressor.

Now consider that the signal entering the expander rises from −10 to −8 dB not because of a gain error in the medium but because the input to the compressor actually rose from −20 to −16 dB. The expander treats the change as before (it has no way of knowing why its input is at −8 rather than −10 dB), so the gain at all frequencies rises, increasing the noise at all frequencies. In the areas where the signal does not provide masking, we will perceive the increase as noise modulation.

Note that this mechanism of noise modulation is a property of a constant-slope compander, and applies not only to a wide-band device, in which the whole audio spectrum is treated together, but also, contrary perhaps to one's intuition, to constant-slope multiband devices.

Bilinear devices The commonest other characteristic is the bilinear curve, in which the device has a constant gain at low levels, a different constant gain at high levels, and an intermediate range within which gain changes occur (*Figure 3.31*). Since the compression and expansion slopes vary with signal

level, it is necessary to ensure that the decoder receives the correct absolute level. The procedure of gain setting to achieve this is known as *calibration*. In general, calibration involves the use of predetermined reference levels in the processors and in the medium (e.g. a defined fluxivity on magnetic tape or frequency deviation in frequency-modulated systems), gains being adjusted so that the processor and medium references coincide (see Section 3.2.2.5). The inconvenience of having to calibrate the system is offset by the elimination of some of the problems described above.

(1) The dynamic behaviour can be different for high and low levels. In particular, the response can be deliberately sluggish whenever overshoots can be passed without overloading the medium, thereby removing any chance of audible modulation distortion. The compressor becomes linear at high levels; signal changes within the linear range do not result in gain changes, so there are no overshoots. Sudden transitions from low to high levels, which give masking over a wide range of frequencies, can be accommodated by a short-term fast attack.

(2) The system becomes linear at low levels, so the addition between the processors of noise or other spurious signals below the threshold does not lead to mistracking (see Section 3.2.1.7).

(3) Below predetermined thresholds, the decoder has a fixed loss, so the amount of NR at each frequency is known and can be made independent of signal components at other frequencies. This makes possible an NR system subjectively free of noise modulation.

3.2.1.7 Tracking and robustness

Practical noisy media have imperfections other than noise; they usually have bandwidth restrictions (at both high and low frequencies), non-linear phase characteristics (i.e. frequency-dependent time delay), non-flat amplitude responses, and absolute gain errors. As a result the control signals generated in an NR decoder may not be identical with those in the encoder (or in the case of constant-slope devices, may not differ merely by the effect of a fixed level difference), and the decoder may not undo perfectly what the encoder did. The discrepancies between the original and reproduced signals are known as tracking errors. There are two factors to be considered: mechanisms leading to mistracking, and how audible that mistracking will be.

Mechanisms of mistracking All NR systems employ compression in some form or other. A

Figure 3.31 *Bilinear compression and expansion.*

fundamental parameter of a processor is its compression or expansion ratio, which may be a constant or varying over the range of input levels. During the decoding, level errors in the medium are magnified by the expansion ratio: to avoid excessive tracking errors from this mechanism. NR systems rarely use expansion ratios of more than 2 or 2.5. A system in which the processors can only change gain by a modest and predetermined amount is obviously much less likely to give significant mistracking than one whose gain can change by tens of decibels.

If compression continues down to very low levels, where noise and/or spurious out-of-band signals may dominate, or continues up to high levels, where the medium may be non-linear (particularly at high frequencies), mistracking is probable. A system with a linear (no gain changing) compression characteristic at very low and very high levels is immune to these mechanisms.

The type of rectifier used to generate control signals also affects the probability of mistracking. Non-linear phase response results in changes to the shapes of waveforms, and hence to different peak levels. Thus peak level detection leads to mistracking; this is another reason to avoid very fast attack, which inherently responds to the peaks of the waveform.

True root mean square (r.m.s.) detection would respond only to the amplitudes of the spectral content and would ignore the phase; however, it requires integration over time from minus infinity to plus infinity! Practical smoothing over a few milliseconds removes its virtue at low frequencies, where much of the dispersion occurs. In addition it can be shown that errors in the amplitude response alter the resultant control signal more with r.m.s. detection than with simple averaging. Thus averaging the absolute magnitude leads to least mistracking.

Response errors in the medium more often occur at the extremes of the audio band, so systems which measure the level at the extremes are more likely to suffer from mistracking. Spurious signals in the source at or beyond the top of the audio band which actuate the compressor but are lost in the medium and never get to the expander can lead to gross mistracking. Spurious signals that get into the expander but which were not present in the compressor can also cause mistracking. Such signals may include TV horizontal synchronizing pulses, recording bias, and numerous artefacts from digital sources. Excessive energy in the programme material above 15 kHz, for example from a synthesizer or from very close microphones, can also be troublesome.

Robustness A given degree of mistracking may or may not be audibly disturbing. The key factor is the extent to which mistracking in one part of the spectrum leads to gain variations in other parts.

For example, consider a medium with a loss of 2 dB at 10 kHz passing a musical item which contains only lower frequencies most of the time, but has occasional instants when 10 kHz is the dominant signal, say a struck cymbal. With a wideband NR system, there will be no mistracking most of the time, but at those critical moments when the 10 kHz is dominant, the whole output from the decoder will suddenly fall by 2 dB multiplied by the expansion ratio, probably 4 dB. Subjectively the whole band will be modulated by the cymbal; such mistracking is audibly obvious and disturbing. At the other extreme, a system in which changes of gain at 10 kHz do not affect other frequencies at all will still mistrack, but the effect will merely be that the cymbal will be quieter than it was in the source. The rest of the music will be unaffected, and the mistracking will probably pass unnoticed.

The property of tolerating spurious signals or imperfections in the medium is called *robustness*. Other things being equal, the more independently a system treats different parts of the spectrum the more robust it is. Robustness can to some extent be judged by listening to the compressed signal. If the gain is obviously pumping up and down, and high-frequency instruments alter the level of low-frequency ones or vice versa, the complete NR system is not robust. If the compressed signal has limited dynamic range but no obvious artefacts of compression, the system is probably robust.

A system whose compressor yields a 'listenable' sound, albeit with a reduced dynamic range, has further commercial attractions. It permits 'single inventory', i.e. the issuance of sound recordings such as audio cassettes and film release prints in encoded form only. The encoded material can be played satisfactorily without decoding equipment.

In summary, independent treatment of different parts of the spectrum leads directly to:

(1) freedom from audible noise modulation;
(2) robustness; and
(3) 'compatibility', i.e. acceptable reproduction without decoding.

3.2.2 Practical noise reduction systems

In what follows, the operation at the input of the noisy medium may be called any of the following:

'compression', 'encoding', or 'record-mode processing'. Similarly, the terms 'expansion', 'decoding' and 'playmode processing' are used interchangeably.

3.2.2.1 Common features of the Dolby noise reduction systems

Since the mid-1960s, one name has been almost synonymous with noise reduction: Dolby Laboratories. The Dolby systems have certain features in common, so those features will be discussed first.

An important principle underlying these features is 'least treatment'. The input signal should be treated as gently as possible, with dynamic modifications performed only in those areas of the spectrum where NR is necessary and then as slowly as possible (consistent with psychoacoustic requirements).

Bilinear compression characteristic First used in Dolby A-type noise reduction, the virtues of this characteristic are discussed in Section 3.2.1.6.

Dual-path configuration Since noise constitutes a very small fraction of the total signal coming out of the noisy medium, it seems illogical to process high-level signals which are well above the noise (40 dB or more). The dual-path configuration used in all the Dolby systems has a main path via which the whole signal is passed without dynamic modification, and a subsidiary path, known as the 'further path' or 'differential network', which passes only low-level signals. As shown in *Figure 3.32*, the output of the further path is added to or subtracted from the main path to encode or decode. Note that the (identical) blocks G_1 and G_2 are not mere control voltage generators, but convey low-level signals to be added to or subtracted from the main path. In the presence of high-level signals, the further path does not add or subtract significantly because it contains a compressor or limiter which prevents its output from rising above a small fraction of the maximum possible main

path signal (*Figure 3.33(c)*). Below a predetermined threshold, the further path has constant gain.

This configuration inherently gives a bilinear compression characteristic. The encoder gain at low input levels is the sum of the gains of the main and further paths (*Figure 3.33(a)*), while at high levels it is the gain of the main path alone. At increasing intermediate levels above the threshold of the further path's compressor, the output of the further path progressively becomes smaller relative to that of the main path, so the overall encoder gain falls smoothly to that of the main path alone. Similarly, the decoder gain rises from a minimum at low levels (a maximum loss) to that of the main path (*Figure 3.33(b)*). By deriving the input of the further path as shown in *Figure 3.32*, the encoding and decoding operations are complementary.

The design of the compressor in the further path is such that the compression or expansion ratio of the whole processor, after summation of main and further paths, rises from 1 (no compression) at low

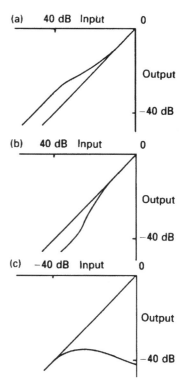

Figure 3.33 *Input-output characteristics of dual-path system: (a) compression; (b) expansion; (c) differential component from the further path. The compression characteristic is formed by adding the differential component to the input signal. The expansion characteristic is formed by subtracting the differential component shown in* Figure 3.32.

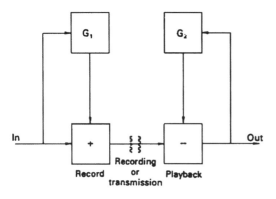

Figure 3.32 *Layout of dual-path noise-reduction systems.*

levels to a maximum of 2 or 2.5, depending on the particular Dolby system, and then falls back to 1 (again no compression) at high levels.

Non-linear smoothing and overshoot suppression The rectifiers of Dolby systems measure the level of the signal in the further path averaged over a few tens of milliseconds (the exact times vary from system to system). They use two-pole smoothing to achieve low ripple and hence low distortion without the need to compromise recovery time. Considering the further path as a closed-loop, output-controlled servo, the Q of this smoothing is chosen to give a sensible degree of damping to the step response. The response time is sufficiently slow that modulation distortion is insignificant. This slow attack is satisfactory while the resultant overshoots, lasting tens of milliseconds, remain well within the range that the medium can handle. Only when overshoots may overload the medium is the dynamic response modified.

To understand this modification, we need to consider the effect of various different sorts of transition. Consider the encoding curve shown in *Figure 3.33(a)*. It is assumed that the noisy medium can accept signals at least up to 0 dB on this graph. Note that in this example the maximum gain change available is 10 dB, so overshoots can never exceed 10 dB, and those will occur only when the signal changes from below threshold (full encoder gain of 10 dB) to a high level where the final gain is that of the main path alone (0 dB, unity gain). We see that an input transition from −50 to −40 dB involves no gain change and hence no overshoot; one from −10 to 0 also involves no gain change and no overshoot; one from −30 to −20 dB demands a few decibels of gain reduction, but the resultant overshoot is well within the capabilities of the medium. It is only large transitions that finish with input levels above −10 dB which can yield an output overshoot which might exceed 0 dB and hence overload the medium. Under these conditions we need overshoot suppression.

The circuitry detects the conditions under which overshoots might be troublesome, and under those conditions only, adapts the dynamic response of the further path to restrict its contribution to the encoder output. In most Dolby systems, this restriction consists of the combination of a progressively more rapid attack time, down to values in the region of 1 ms (where modulation distortion is in general still not a problem) and under still more extreme transients, a non-linear limiting (clipping) of the further path output. Note that when this limiting occurs, the signal

from the further path is forced to be small compared with that in the main path, so the non-linearity of the signal from the further path is greatly diluted by the (linear) main path. With the overshoot suppressed, the signal can now be conveyed through the medium without overload, so that the decoder will receive what the encoder delivered, and accurate expansion can occur.

In the more recent systems, Dolby SR (spectral recording) and S-type NR, the restriction consists of 'pumping up' the control signal to just that level which contains the overshoot to an acceptable amount. The result is the same as in the earlier method as far as the overshoots are concerned, but this technique allows the processor to retain its slow (i.e. gentle) averaging for a larger proportion of the time.

Antisaturation Many noisy media have reduced headroom at the extremes of the audio spectrum. It is desirable that NR systems be tolerant of this imperfection. From C-type NR onwards, Dolby systems have employed circuit techniques to improve their tolerance.

At low levels, the output of a dual-path encoder is primarily that of the further path, the gain of which is usually substantially greater than that of the main path. At high levels, the output is primarily that of the main path, the further path having been heavily attenuated. Thus frequency-dependent attenuation in the main path alone has the effect of reducing the maximum output at the attenuated frequencies, without significantly reducing the low-level gain and hence noise reduction (*Figure 3.34*). This technique is known as 'antisaturation', since the frequency-dependent attenuation can be chosen to match the saturation curves of the medium (e.g. cassette tape), so reducing the probability of non-linearity.

Figure 3.34 *Placement of antisaturation networks in the main paths of dual-path processors.*

Spectral skewing The ear is much less sensitive at low and very high frequencies, so less (or no) NR is required in these regions. Therefore little or no processing is needed, making it possible to trade the NR effect for reduced sensitivity in the processor to signals at the extremes.

By preceding the encoder by a network which attenuates these extremes, and following the decoder by an inverse network to restore the overall response, the encoder's control signal puts less weight on signal components at the extremes. Thus the decoder has a better chance of generating an accurate replica, even if the response of the medium is uneven at these frequencies (*Figure 3.35*). This method of modifying the input spectrum before processing is known as 'spectral skewing'. Like antisaturation, it was first used in Dolby C-type NR.

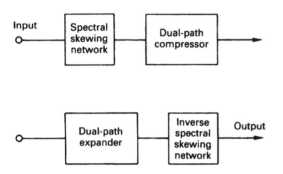

Figure 3.35 *Placement of spectral skewing networks.*

3.2.2.2 The individual Dolby noise reduction systems
There are, at the time of writing, five distinct Dolby systems, two professional and three primarily for home use. They differ mainly in their approach to achieving the frequency dependence vital to the elimination of audible noise modulation. They are presented here in the order of their introduction.

A-type noise reduction The original Dolby noise reduction system (later called 'A-type' when it became necessary to distinguish it from other Dolby systems) was introduced in 1966. This was the first NR system the performance of which satisfied the critical listeners of the audio industry. Previous systems from other companies (and many since) suffered from noise modulation and transient distortion. It is a professional system offering 10 dB of NR, a modest degree by later standards. However, it is worth pointing out that improving the signal/noise ratio of magnetic tape recording by this amount by using wider tracks would require a 10-fold increase.

A-type NR achieves adequate frequency dependence by splitting the audio spectrum into four bands (*Figure 3.36*).

Band 1: low-pass at 80 Hz
Band 2: band-pass 80 Hz to 3 kHz
Band 3: high-pass at 3 kHz
Band 4: high-pass at 9 kHz

Bands 1, 3 and 4 use two-pole Butterworth filters, while band 2 has such a response that the summation of all the bands yields a smooth response. The linear limiter in each band uses programme-dependent attack times and is followed by overshoot suppression (non-linear limiting) as described above.

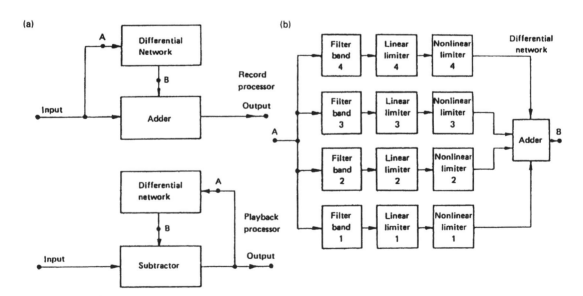

Figure 3.36 *(a) Block diagram of Dolby A-type noise reduction. (b) The differential network is the same for encoding and decoding.*

The primary application of this system was originally in professional audio recording. However, because the system provides reduction of hum and rumble as well as high-frequency hiss, it has also found uses in improving landlines and other long-distance links, has been extensively used on the audio tracks of professional video recorders, and has become a vital part of film production, including use on optically recorded release prints ('Dolby Stereo'). A special impetus came from the spread of 16-track and, later, 24-track recording, starting in the early 1970s. At the time of writing, about 200 000 channels of A-type processing are in use round the world.

B-type noise reduction Except for an inferior signal/noise ratio, home tape recording (open reel) in the late 1960s had a performance comparable in most respects with the best audio media available in the home. Dolby A-type NR was far too expensive for home products. However, at the low speeds used (typically 9.5 cm/s) the primary noise consisted of high-frequency hiss, prompting investigation into simpler circuits reducing only high-frequency noise.

Tests showed that a 10-dB fixed band such as band 3 of A-type NR did not provide enough noise reduction. However, if the band was extended downwards to start at, say, 500 Hz, the result was audible noise modulation. The solution was to use a new technique, termed 'sliding band' (Berkovitz and Gundry, 1973). In a dual-path processor the further path of which contains one high-frequency band only, the compressor or linear limiter of A type is replaced by a variable high-pass filter (*Figure 3.37*). When the input signal is below threshold, the filter slides below the frequency of the fixed band-determining filter (1.5 kHz high-pass in B-type NR), so the response of the further path is set by the fixed filter. The summation of the flat main path with the high-pass further path gives a response shelf 10 dB high, as shown by the left-most curve in *Figure 3.38*.

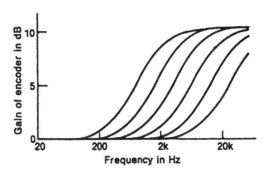

Figure 3.38 *Sliding-band responses of Dolby B-type noise reduction.*

When there are high-level middle-frequency signals, the control signal causes the filter to slide upwards, reducing the gain for middle frequencies to avoid overload of the medium, but retaining full boost and therefore noise reduction at higher frequencies. Under the extreme case of high-level high-frequency signals, the filter slides up beyond the audio band, effectively turning off all NR: however, the high-frequency signals are then sufficient to mask the high-frequency hiss of the medium.

Note that the family of responses shown in *Figure 3.38* can be considered as a variable pre-emphasis of fixed magnitude with moving break points. The system retains the programme-adapting attack time and the overshoot suppression of A-type NR.

At about the same time (1969) as the development of this system, experiments showed that the Philips compact cassette, introduced some years earlier primarily as a speech recorder, had the capability to offer wide frequency response and satisfactory speed stability. However, these developments were not worthwhile because of the poor signal/noise ratio.

One of the side benefits of this sliding-band processing is that middle-frequency material does not modulate higher frequencies, so that the varying degree of high-frequency boost does not lead to the audible artefacts normally associated with compressors. The major record companies performed experiments and decided that B-type encoded material was sufficiently 'compatible' that they could issue prerecorded tapes in encoded form only, thereby avoiding the need for double inventory.

The marriage of B-type NR, precision cassette mechanisms and improved tape formulations led to the mass marketing of home tape recorders for the first time (market penetration of open-reel recorders had been limited). Dolby Laboratories entered into licensing agreements with all the major

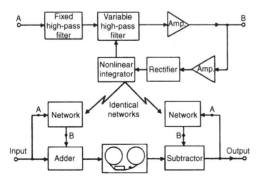

Figure 3.37 *Block diagram of Dolby B-type noise reduction.*

manufacturers of home audio equipment, and since then some hundreds of millions of cassette recorders and players with B-type NR have been made and many thousands of millions of B-type encoded prerecorded cassette tapes sold. The large-scale production made economical the development of integrated circuits for B-type NR; these are now available from about 15 manufacturers. More recent applications for B-type NR include the stereo linear tracks of VHS video cassettes and (modified) the rear channel of Dolby Surround.

C-type noise reduction By, 1980, the development of the cassette recorder had progressed to the point where the signal/noise ratio, even improved by B-type NR, was again perceived as the weakest parameter, at least at the high end of the market. A single sliding band, like that of B-type NR, is inadequate to provide substantially more than 10 dB of NR; problems include excessive compression ratio, unsatisfactory overshoot performance and noise modulation. These can only be solved by making a more complex system.

C-type NR (Dolby, 1983) therefore employs two 10-dB sliding bands stages in tandem, each similar to B-type but extending two octaves lower, with staggered thresholds so that the area of maximum compression ratio in one corresponds to an area of low ratio in the other. Each stage employs the same techniques of adaptive attack and overshoot suppression as A- and B-type NR. It can be shown that

staggered tandem stages yield worst-case overshoots which are little greater than those of the higher level stage alone. When middle-frequency high-level signals are present, the two stages provide a steeper filter action, and therefore maintain NR closer to the middle frequencies than would a single sliding filter.

With increased expectation from the cassette, it was desirable to make the new system more tolerant of the headroom and bandwidth limitations of the medium, so high-frequency antisaturation and spectral skewing were incorporated. The result is about 19 dB of noise reduction, together with a marked improvement in high-frequency headroom, and hence much more accurate reproduction of material rich in high frequencies. A block diagram of a C-type encoder containing these features is shown in *Figure 3.39*.

C-type NR has been built into a large number of home recorders, but it has not become a standard feature of mass-marketed playback-only cassette machines, since it appeals chiefly to the more critical consumer who makes his or her own recordings. Furthermore, with the increased amount of NR goes a corresponding extra degree of compression, and the record industry has not considered C-type material sufficiently 'compatible' to permit single inventory. Thus prerecorded tapes have continued to employ B-type NR (but see further discussion under 'Dolby S-type NR' below). The system has

Figure 3.39 *Block diagram of Dolby C-type noise reduction. N_1 and N_2 are similar to the differential network of B-type NR.*

also found a niche in the audio tracks of professional video cassette formats. Betacam and MII.

The Dolby spectral recording system The 1980s saw the introduction of professional digital recorders and of the compact disc, with potential signal/noise ratios in the region of 90 dB. Professional analogue recording with industry-standard A-type NR offers a range of about 80 dB on a direct stereo recording, and a little less on a multitrack mix-down. However, analogue recording has operational and economic advantages over digital, and in some cases performance advantages as well. Therefore, in 1986 Dolby Laboratories introduced a new analogue process, called 'spectral recording' (Dolby, 1987). Dolby SR in conjunction with a good analogue recorder offers a dynamic range of well over 90 dB with no audible noise modulation or modulation distortion, and with headroom which (like digital) is sufficiently independent of frequency.

Spectral recording depends upon a new circuit technique known as 'action substitution'. When a high-frequency sliding band (e.g. B-type NR) responds to a high-level signal, it retains full NR above the frequency of the signal, but loses most of its NR below that frequency. When a high-frequency fixed band (e.g. band 3 of A-type NR) responds to an input signal, it loses NR uniformly through its operating band. Action substitution combines a fixed and a sliding band in such a way that their better properties are retained without their less desirable ones; in *Figure 3.40* compressor 1 is a fixed band and compressor 2 a sliding one (or vice versa) with the appropriate interconnection, the combination then forming the further path of a dual-path processor. A stage employing action substitution permits a closer approach to the ideal of providing fixed boost at all frequencies within the defined range covered by the stage, except in the immediate frequency range of high-level signals.

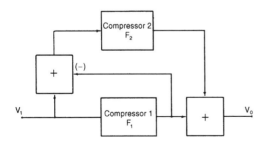

Figure 3.40 *Block diagram of action substitution. Compressor 1 has a fixed band and compressor 2 has a sliding band (or vice versa).*

Like C-type NR, Dolby SR uses multiple tandem stages, but in this case three at high frequencies and two at low, with a crossover at 800 Hz. Each stage contains a fixed and a sliding band with action substitution (*Figure 3.41*). The system gives 25 dB of weighted NR and, since each stage only has to provide about 8 dB, the overshoot behaviour (approximately that of the high-level stage) is excellent. The control-signal generation, which constitutes most of the circuitry, includes modulation control (marked modulation control circuits 1–9 in *Figure 3.41*), a means of generating additional information about the level and spectrum of the input signal so as to modify the main control signals before they alter the boost or cut applied to that input. The system incorporates anti-saturation at both high and low frequencies, flattening the overload characteristic of professional recording media (including elimination of the ill-effects of the 3180 µs time-constant used in some playback standards). Mistracking resulting from response irregularities is reduced by spectral skewing at both ends of the audible range. A detailed description of spectral recording is beyond the scope of this book but is given in Dolby (1987).

Since its introduction, SR has been enthusiastically taken up by the recording and film industries, and at the time of writing over 120 000 processors are in use. Comparative listening tests between analogue recording with SR and digital recording often show a slight preference for the SR. Two major manufacturers, Studer and Otari, have incorporated the system within multitrack recorders. It is widely used during film production and is the basis of Dolby Stereo SR films, where it is applied to the optical soundtracks of 35-mm film release prints to give a performance similar to that of 70-mm magnetic prints at about one-tenth of the cost.

Dolby S-type noise reduction The success of Dolby spectral recording naturally led to thoughts of a home system. Tests of spectral recording round a cassette recorder were very impressive, but it was clear that a system employing five stages (10 bands) could not be manufactured at a low enough price, whatever the quantities. As with the development from A to B type, certain simplifications are possible because the noise spectrum and level of cassette tape are well defined. A major difference from 1969, when B type was first developed, is that it was possible in 1990 to use a custom integrated circuit from the start. The first two or three years of B-type production used discrete components, but a system as complex as the envisaged one could not have

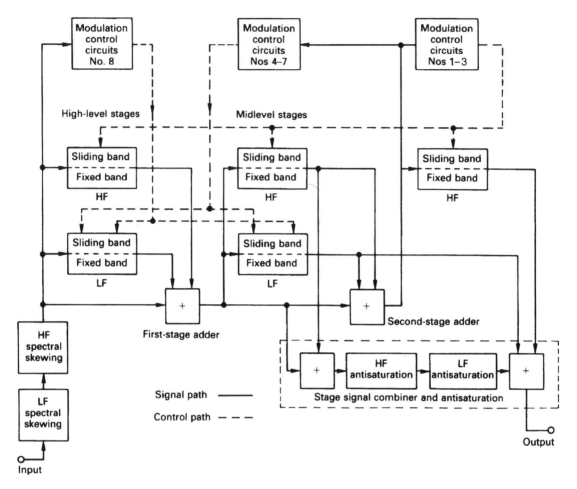

Figure 3.41 *Block diagram of an encoder for Dolby spectral recording, showing the signal paths and modulation control. HF, high frequency; LF, low frequency.*

been built economically or reliably using off-the-shelf parts.

S-type, as developed from spectral recording, employs two high-frequency stages, each with an action substitution pair of bands providing 12 dB of NR above 400 Hz, plus a single low-frequency band providing 10 dB of NR below 200 Hz (*Figure 3.42*) (Cossette, 1990). It employs all the previously described circuit techniques and gives up to 24 dB of NR on cassette tape. The resultant dynamic range, in the region of 85 dB, is close to that obtainable from compact disc and digital audio tape (DAT).

An important result of the improved frequency agility provided by multiple stages of action substitution is that, despite the large amount of NR, the output of the encoder does not suffer from the standard artefacts of compressors, but sounds steady despite the reduced dynamic range. At the time of writing,

record companies are evaluating S-type for 'compatibility', which would enable them to continue selling their recordings in the form of the economical cassette rather than changing to some new format costing them and the public more. Prerecorded S-type cassettes offer very high quality reproduction to those who invest in an S-type machine, excellent reproduction to those who play the tapes on their existing B-type machines, and compressed dynamics to those who play the tapes on a cheap pocket portable or in the car (where the compression is frequently desirable). In addition, S-type has been adopted for the 'narrow gauge' multitrack formats (1-in. 24-track, etc.).

3.2.2.3 The dbx systems

The dbx systems, type I, II and 321, are all wide-band constant-slope companders, the first two with

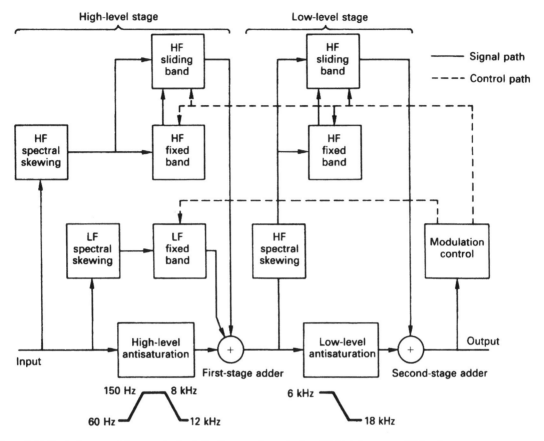

Figure 3.42 *Block diagram of the encoder for Dolby S-type noise reduction. HF, high frequency; LF, low frequency.*

a 2:1 compression ratio and the third with a 3:1 ratio. *Figure 3.43* shows the general configuration of all the systems. The signal path contains fixed high-frequency pre-emphasis in the compressor and deemphasis in the expander; these provide some NR at high frequencies in the presence of high-level low and middle frequencies, at the cost of exaggerating low-frequency noise in the presence of high-frequency signals.

All the dbx systems employ control signal generation (level sensing) in which single-pole smoothing is performed in the logarithmic domain and following a multiplication by 2, giving an average of the logarithm of the square of the magnitude of the waveform. This is equivalent to measuring the short-term r.m.s. value of the signal, and reduces the probability of mistracking due to high-frequency dispersion; however, since the averaging time is only a few milliseconds, the effects of low-frequency phase shifts remain. No provision is made to reduce overshoots. The level sensing is preceded by weighting networks

so that the gain of the compressor is forced down at high frequencies, offsetting the effect of the signal pre-emphasis and reducing the probability of overloading the medium at high frequencies.

The voltage-controlled amplifiers are arranged to operate in opposite senses; that is, in the encoder increasing control voltage decreases the gain, whereas in the decoder it increases the gain.

dbx I In the encoder of dbx I (Blackmer, 1972), the voltage-controlled amplifier is preceded by a fixed pre-emphasis, a 12-dB shelf rising from 400 to 1600 Hz. To counter the resultant rising output level when the encoder is fed with a rising frequency sweep, the level sensor is preceded by a (different) preemphasis or control weighting, which rises at 6 dB/octave above 1.6 kHz (see *Figure 3.43*).

dbx II dbx II, the system intended for home recording, is very similar to type I, except that the control weighting shown in *Figure 3.43* is supplemented by a band-pass filter to reduce response to signals outside the range 30 Hz to 10 kHz. This

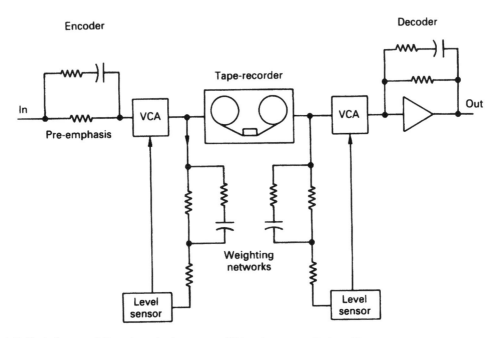

Figure 3.43 *Block diagram of dbx noise-reduction systems. VCA, voltage-controlled amplifier.*

is intended to reduce mistracking due to irregular frequency response in the medium. The method (unlike spectral skewing, which is in the signal path) has the potential disadvantage that *isolated* signals at these extremes are filtered out of the signal to be measured and are therefore not given their full weight. Thus under rather rare conditions, the compressor gain is not turned down appropriately and the medium can be driven into saturation.

An alternative version of the dbx II expander, intended for reproducing dbx-encoded long-playing (LP) records, contains an additional input high-pass filter at 30 Hz to reduce mistracking caused by rumble from the record and turntable.

dbx 321 This is a professional system designed for use only on analogue satellite links, and has been used to distribute network radio programming in the USA (Talbot, 1981). Its chief distinction from dbx I is that it uses a (constant) compression ratio of 3:1. This unusually high ratio is permissible in this application because the frequency-modulated transmission system yields a very predictable overall gain and frequency response, so that even when magnified three times the tracking errors are within an acceptable tolerance.

3.2.2.4 Other systems

Over the years, designers do not seem to have grasped the drawbacks of constant-slope and/or wide-band devices. The past three decades have seen a multiplicity of companders, both professional and for home use, that have failed in the marketplace. They include Burwen Noise Eliminator (Burwen, 1971), Telefunken HighCom, Toshiba Adres, Sanyo Super-D, Nakamichi HighCom II, EMT NoisEx (Aldous, 1965) and CBS CX and CX-TV. The systems described below are three surviving examples.

telcom c4 Telefunken (now ANT) developed telcom c4 (for a four-band compander) about 10 years after the introduction of A-type noise reduction, with the aim of offering more than 10 dB of NR (Wermuth, 1976). To reduce the magnification of errors in the medium it uses a constant slope of 3:2, achieved by a reinvention of a technique from Bell Laboratories (Cushman *et al.*, 1933). A string of identical voltage-controlled amplifiers (VCAs) connected in tandem with common control signals has its output servoed to a constant level (*Figure 3.44*). If an output is taken from between any two of the VCAs, the resulting compression ratio is the total number of VCAs divided by the number downstream of the pick-off point. The virtue of this approach is that the VCAs do not need any particular relationship between gain and control voltage, provided they are identical.

The telcom c4 decoder (*Figure 3.45(c)*) uses four expanders with their outputs summed together,

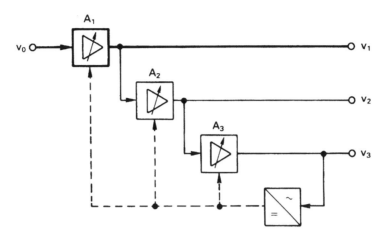

Figure 3.44 *The Cushman compressor used in telcom c4. A_1 to A_3, identical voltage-controlled amplifiers; v_3, servoed to a constant peak level; v_1, compressed with a ratio 3:2; v_2 compressed with a ratio 3:1.*

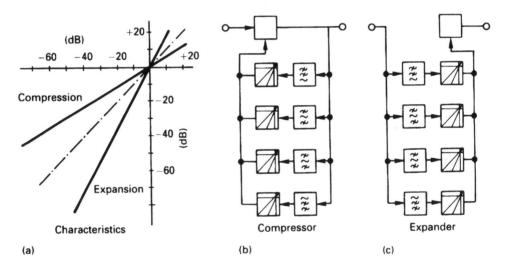

Figure 3.45 *The ANT telcom c4 noise-reduction system.*

operating in four bands; the bands are defined by singlepole band-pass filters as follows.

Band 1: 35–215 Hz
Band 2: 215–1450 Hz
Band 3: 1450–4800 Hz
Band 4: 4800–16 000 Hz

The corresponding encoder is derived by connecting this decoder in the feedback of an amplifier (*Figure 3.45(b)*).

The response times of each band are set by single capacitors, and are scaled to the frequency range covered by that band. In each case the attack time is chosen to be short compared with a quarter-period

of the highest frequency entering that band, in effect giving an instantaneous peak detection and inviting large amounts of modulation distortion and sensitivity to dispersion. The recovery time is chosen to keep ripple on the control voltage and hence harmonic distortion of constant sine waves down to an acceptable level. For the lowest frequency band this results in a recovery time much longer than that of human hearing; for example, it takes 1.4 s to recover 10 dB of gain.

The single-pole band splitting means that the synthesized responses formed by the combinations of the bands can never have filter slopes steeper than 6 dB/octave, and of course the actual degree of NR at any frequency is changing continually. In the

presence of middle-frequency high-level signals, the shallow filter slopes can only provide very modest noise reduction at high and low frequencies.

The system has enjoyed some commercial success in Germany, Switzerland and Denmark, primarily in the broadcast industries.

VHS Hi-Fi It is not widely realized that the frequency-modulation subcarriers laid down with the video on VHS hi-fi require NR to achieve a tolerable signal/noise ratio. With high-frequency pre- and deemphasis (56 μs) but no NR, the signal/noise ratio is only of the order of 50 dB, similar to that of the conventional linear tracks.

VHS hi-fi uses a conventional 2:1 constant-slope compander, with pre-emphasis in the signal path and frequency weighting in the control path (*Figure 3.46*). In fact it closely resembles dbx, except that the control signal is generated by peak rectification and smoothing in the linear rather than the logarithmic domain; the attack time is about 5 ms and the release 70 ms. With this form of control signal, it is rather more convenient to construct a 2:1 expander than a compressor, so the compressor is produced by connecting the expander in the feedback of a high-gain amplifier.

Figure 3.46 *VHS hi-fi noise-reduction decoder. E_1, deemphasis, $240 + 56$ μs; E_2, de-emphasis, 56 μs; W, weighting, $240 + 24$ μs. Det., peak rectification and smoothing; G_2, buffer amplifier; VCA, voltage-controlled amplifier.*

The audio is recorded as frequency-modulated tracks using heads on the rotating drum. As a result the signal is recovered in segments corresponding to each picture field, and it is almost impossible to ensure that the segments fit together perfectly when the temperature or humidity change, or a tape is played on a different machine from that which

recorded it. Hence the reproduced signal is liable to be accompanied by a buzz at TV field rate which is in turn modulated by the varying NR of the expander, rising to a maximum in the presence of isolated high-frequency signals.

8-mm Video cassette The 8-mm video cassette format has provision for three types of audio tracks, conventional linear, frequency-modulated and digital. They all use a constant-slope compander very similar to that used in VHS hi-fi, with changes of pre-emphasis to suit the different noise spectra of the various media. (The digital track uses, in addition, non-linear quantizing, i.e. digital companding.)

3.2.2.5 Calibration of noise-reduction systems

As discussed in Section 3.2.1.6, Dolby systems need calibrating. To make this more convenient, professional Dolby NR processors include a generator to produce a calibration signal (Dolby tone) at a level which has a predefined relationship to the compression and expansion characteristics. They also contain a calibration display which indicates when the appropriate calibration signal is present at that same level.

Calibration involves sending the Dolby tone from the encoder and adjusting the equipment so that the tone reads correctly on the calibration display in the decoder. The A-type Dolby tone is an 850 Hz sine wave frequency modulated upwards in pitch by 10% for 30 ms every 750 ms; the characteristic warble announces that this source has been encoded with A-type NR, and permits adjustment of levels so that this calibration signal reads 'on the mark' of the calibration displays. Since the warble is not accompanied by a change in level, meter readings are not disturbed by the modulation.

Tones similar to A-type Dolby tone are defined for B-, C- and S-type NR, although the generators for these are rarely incorporated in home equipment; in all cases they consist of 400 Hz, modulated upwards by 10% in a periodic manner.

For audio and video cassettes, there is a defined fluxivity on the tape corresponding to the Dolby tone: on the Philips compact cassette it is 200 nWb/m ANSI, and on VHS it is 120 nWb/m DIN. The manufacturers of home recorders adjust their products in the factory so that standard test-tapes play at the defined reference level in the Dolby NR circuits. This standardization permits all prerecorded tapes to be played on home machines without further adjustment.

Dolby spectral recording uses a calibration signal called 'Dolby noise', which consists of pink noise

interrupted periodically by brief 'nicks'. The level is again defined in terms of the compression characteristics of spectral recording. The interruptions are sufficiently short (about 2 ms) that the level as displayed on a calibration display is steady.

When this Dolby noise enters a decoder set up to receive it, the calibration display shows its level, as with Dolby tone, permitting the appropriate adjustment. In addition, 4-s stretches of the reproduced Dolby noise appear at the output of the decoder alternating with 4-s stretches of reference (locally generated) pink noise. Any discrepancy in level and/or high-frequency response can be identified easily by ear; this feature is called 'auto-compare'.

It is good practice to lay down a few seconds of Dolby tone/noise at the beginning of every recording using A-type NR or SR, or to send tone/noise just prior to using a land line or other link.

ANT uses a similar approach in telcom c4, with a calibration tone which alternates between 550 and 650 Hz at a period of roughly 1 s.

3.2.3 Digital bit-rate reduction

A properly designed digital system using linear pulse-code modulation behaves as a linear system with a signal/noise ratio of approximately $6n$ dB, where n is the number of (binary) bits per sample. There are many potential applications where the bit-rate required to give high-quality audio (at least 600 kbit/s) cannot be accommodated, and where bit-saving techniques are necessary. The details of such techniques are beyond the scope of this chapter, but the principles are the same as those of analogue NR.

The simplest bit-saving techniques employ non-linear coding, in which the available quantizing levels are close together (small error) for small absolute input voltages, and progressively further apart (increasing error) as the input rises; this is standard practice in digital telephone communication. More sophisticated systems operate on blocks of a millisecond or so, scaling the quantizing range to accommodate the largest sample in each block.

The bits generated and the resultant quantizing error are in both cases similar to what would have been generated by a fixed quantizing interval and a wide-band analogue compressor. Indeed, early attempts at bit-saving used analogue companding in conjunction with 8- or 10-bit analogue-to-digital converters to achieve wide dynamic ranges (but poor instantaneous signal/noise ratios). All these systems inherently give rise to noise modulation, which is

likely to be obtrusive if the signal/noise ratio in the presence of full-scale signals falls below about 60 dB (equivalent to 10 bit/sample, discounting scaling bits).

Recently, much more advanced coding schemes have been developed. They are similar to advanced analogue NR processes in that they divide the audio spectrum into a large number of narrow bands and allocate bits to provide in each band only that signal/noise ratio sufficient to render noise inaudible by masking (*Figure 3.28*).

There are two fundamental approaches to the band-splitting: sub-band coding (Stoll and Dehery, 1990), which employs digital techniques to synthesize the required narrow filters directly; and transform coding (Davidson *et al.*, 1990), which employs methods derived from Fourier transforms to generate spectrum information. Both are capable of excellent results, although the latter requires less digital processing. These methods permit very high quality at 128 kbit/s per channel or less, a saving of a factor of 5 or more.

References

ALDOUS, D., NoisEx recording system, *International Broadcast Engineering*, **10**, 510 (1965)

BERKOVITZ, R, and GUNDRY, K.J.; Dolby B-type noise reduction, *Audio Magazine* (Sep./Oct. 1973)

BLACKMER, D.E., A wide dynamic range noise reduction system, *db Magazine*, **6**, 54–56 (1972)

BURWEN, R.S., Design of a noise-eliminator system, *Journal of the Audio ngineering Society*, **19**, 906–911 (1971)

COSSETTE, S.G., A new analog recording process for use with consumer recording formats. Presented at *Audio Engineering Society 89th Convention*, Preprint 3004 (Sep. 1990)

CUSHMAN, R.W. *et al.*, Volume control system, *US Patent 2 035 263* (26 May 1933)

DAVIDSON, G., FIELDER, L. and ANTILL, M., Low-complexity transform coder for satellite link applications. In: *Audio Engineering Society 89th Convention*. Preprint 2966 (Sep. 1990)

DOLBY, R.M., An audio noise reduction system. *Journal of the Audio Engineering Society*, **15**, 383–388 (1967)

DOLBY, R.M., A 20 dB audio noise reduction system for consumer applications, *Journal of the Audio Engineering Society*, **31**, 98–113 (1983)

DOLBY, R.M., The spectral recording process, *Journal of the Audio Engineering Society*, **35**, 99–118 (1987)

EHMER, R.H., Masking of tones vs. noise bands. *Journal of the Acoustical Society of America*, **31**, 1253–1256 (1959)

FLETCHER, H., The stereophonic sound-film system. *Journal of Society of Motion Picture Engineers*, **37**, 331–352 (1941)

STOLL, G. and DEHERY, Y.F., High-quality audio bitrate reduction system family for different applications. In: *Proceedings of the IEEE Conference on Communications*, p. 937, Atlanta, GA (Apr. 1990)

TALBOT, D.B., A satellite communications broadcast quality amplitude compander, *Journal of the Audio Engineering Society*, **29**, 690–698 (1981)

WERMUTH, J., Compander increases dynamic range, *db Magazine*, **10**, 25 (1976)

3.3 Compact disc

3.3.1 Introduction

'CD' is the abbreviation for the 'compact disc digital audio system', announced in 1980 as the result of a cooperative venture between Sony and Philips. It is the CD that has introduced most people to digital audio reproduction. *Table 3.3* shows the main CD parameters.

Table 3.3 CD parameters

Disc diameter	120 mm
Rotation speed	568–228 r.p.m. (at 1.4 m/s)
	486–196 r.p.m. (at 1.2 m/s)
Playing time (maximum)	74 min
No. of tracks	20 625
Track spacing	1.6 µm
Lead-in diameter	46 mm
Lead-out diameter	116 mm
Total track length	5300 m
Linear velocity	1.2 or 1.4 m/s

The CD contains a spiral-shaped track of successive depressions, called 'pits', in a reflective layer. The encoded digital audio information is stored in the length of pits and the gaps between them. The disc rotates at a constant linear velocity (CLV) of 1.2 or 1.4 m/s. A laser beam emitted by a solid-state GaAlAs laser (wavelength about 780 nm), focused on the disc is reflected by the information on the disc, is detected by photodiodes and will be processed further electronically.

The readout is contactless; electromechanical servo systems focus the laser spot on the disc and follow the tracks within the specified accuracy. This CD system enables recording and reproduction of a 16-bit two-channel stereo signal of maximum 74 min on a 12 cm diameter single-sided disc. This remarkable characteristic is obtained by the high-density recording EFM modulation method and the highly efficient CIRC error correction method (see later).

3.3.2 Recording and readout system on a CD

The data on a CD are recorded on the master by using a laser beam photographically to produce pits in the disc surface, in a clockwise spiral track starting at the centre of the disc. Length of the pits and the distance between them form the recorded information, as shown in *Figure 3.47*. In fact, on the user disc, the pits are actually bumps. These can be identified by focusing a laser beam onto the disc surface: if there is no bump on the surface, most of the light that falls on the surface (which is highly reflective) will return in the same direction. If there is a bump present however, the light will be scattered and only a small portion will return in the original direction. The disc has a 1-mm thick protective transparent layer over the signal layer, i.e. the pits. More importantly, the spot size of the laser beam is about 1 mm in diameter at the surface of the disc, but is as small as 1.7 µm across at the signal layer. This means that a dust particle or a scratch on the disc surface is literally

Figure 3.47 *Pits on a CD.*

out of focus to the sensing mechanism. *Figure 3.48* illustrates this. Obviously, control of focus must be extremely accurate.

Figure 3.48 *Illustration of the way in which a dust particle on a disc surface is out of focus.*

3.3.2.1 Signal parameters
Before being recorded, the digital audio signal (which is a 16-bit signal) must be extended with several additional items of data. These include:

(1) error correction data;
(2) control data (time, titles, lyrics, graphics and information about the recording format or emphasis);
(3) synchronization signals, used to detect beginning of each data block; and
(4) merging bits (added between each data symbol to reduce the d.c. component of the output signal).

3.3.2.2 Audio signal
The audio signal normally consists of two channels of audio, quantized with a 16-bit linear quantization system at a sampling frequency of 44.1 kHz. During recording, first order pre-emphasis may be applied. Pre-emphasis standards agreed for the CD format are 50 and 15 µs (or 3183 and 10 610 Hz). Consequently, the player must in this case apply a similar de-emphasis to the decoded signal to obtain a flat frequency response. A specific control code recorded along with the audio signal on the compact disc is used to inform the player whether pre-emphasis is used, and so the player switches in the corresponding deemphasis circuit.

Alternatively, audio information on the CD may comprise four music channels instead of two; this is also identified by a control code to allow automatic switching of players equipped with a four-channel playback facility. Although, on launching CD, there were no immediate plans for four-channel discs or players, the possibility for later distribution was provided in the standard.

3.3.2.3 Additional information on the CD
Before the start of the music programme, a 'lead-in' signal is recorded on the CD. When a CD is inserted, most players immediately read this lead-in signal which contains a 'table of contents'. This table contains information on the contents of the disc, such as the starting point of each selection, the number of selections, and the duration of each selection. This information can be displayed on the player's control panel, and/or used during programme-search operation. At the end of the programme, a lead-out signal is similarly recorded which informs the player that playback is complete. Furthermore, music start flags between selections inform the player that a new selection follows.

Selections recorded on the disc can be numbered from 1 to 99. In each track, up to 99 indexes can be given, which may separate specific sections of a selection. Playing time is also encoded on the disc in minutes, seconds and 1/75 of a second; before each selection, this time is counted down. There is further space available to encode other information, such as titles, performer names, lyrics and even graphical information.

3.3.2.4 CD production
The CD production process involves seven main stages (*Figure 3.49*).

(1) A glass plate is polished for optimum smoothness.
(2) A photo-resist coating is applied to its surface. The roughness of the glass surface and the thickness of the coating determine the depth of the pits on the compact disc.
(3) The photo-resistive coating is then exposed to a laser beam, the intensity of which is modulated with digitized audio information.
(4) The photo-resist layer is developed and the pits of information are revealed.
(5) The surface is silvered to protect the pits.
(6) The surface is plated with nickel to make a metal master.
(7) The metal master is then used to make mother plates. These mother plates are in turn used to make further metal masters, or stampers.

Figure 3.49 *Stages in the 'cutting' of a CD.*

The stamping process, although named after the analogous stage in LP record production is, in fact, an injection moulding, compression moulding or polymerization process, producing plastic discs. The signal surface of each disc is then coated with a reflective material (vaporized aluminium) to enable optical read-out, and further protected with a transparent plastic layer which also supports the disc label.

3.3.3 CD encoding

3.3.3.1 Basic requirements
After the sampling and quantization circuitry in the analogue-to-digital converter a 16-bit sample signal is obtained. These signals are in the so-called NRZ format (non-return to zero); they may have a high d.c. content and are not self-clocking, as the 'run length' (the distance between transients in the signal) is unlimited. Therefore, they are not suited for direct recording and have to be converted into another code which should meet the following special requirements.

Correct read-out at high information densities
The modulation transfer function or frequency characteristic of the CD read-out system can be modelled as a low-pass filter with linear phase. The cut-off frequency (ideal optics) is simply given by:

$$fc = \frac{2NA}{\lambda V}$$

where NA is the numerical aperture of the objective lens, λ is the wavelength, and V the linear scanning velocity. Non-ideal optics, e.g. due to defocusing or disc skewing with respect to the optical axis, will degrade the phase and amplitude characteristic. A modulation system should satisfy the requirement of 'low sensitivity to tolerances' of the optical light path even at high information densities.

Clock content The bit clock is one of the most important signals in the CD system. It is used for the synchronization of the digital data and the motor control. The bit clock must be regenerated from the read-out signal itself by detecting the pit edges. The signal must, therefore, have a sufficient number of transitions.

Low-frequency contents As the signal of the CD is covered by a 1-mm thick transparent layer, it is protected against dirt and scratches. The dirt and scratches on the surface of the disc change the envelope of the read-out signal, giving rise to low-frequency noise. These low-frequency disturbances can be filtered out provided that the signal contains no low-frequency components of its own. Another reason for imposing the low-frequency constraint is the use of servo systems in the CD. Low-frequency components cause interference in the servo systems, making them unstable.

Error propagation The error propagation of the modulation system must be as small as possible and should match the CIRC error correction system (see later) that treats 8 data bits as a unit. Error propagation should consequently be limited to the eight consecutive data bits forming a symbol. In this way, error correction can work in an optimal way.
The above requirements have led to the adoption of the so-called 'eight-to-fourteen modulation' method for the CD. This code converts eight consecutive data bits into 14 'channel' bits which are then sent to the cutting laser.

3.3.3.2 Principles of error correction
The error-correction scheme is an essential part of the CD system. With analogue techniques the effect of errors such as dropouts is that the sound is merely skipped, and the listener may not hear it, unless he or she is very attentive. However, with digital audio, the dropouts are heard in the form of irritating pulse noise.

Roughly categorized, there are two types of error. The first type are random errors, caused in disc production by inaccurate photo-resist coating, cutting, microscopic bubbles, etc. The second type are the burst errors resulting from scratches, dust and finger-prints on the disc surface; when these errors occur some bits or even bytes may be missing. This

means dropouts are produced and this leads to critical read-out errors in the digital data.

Errors must be detected by some error-detection mechanism. After detection, an error-correction circuit attempts to recover the correct data. If this fails, an error-concealment mechanism will cover up the faulty information so, at least, there is no audible disturbance. These three basic functions – detection, correction and concealment – are illustrated in *Figure 3.50*.

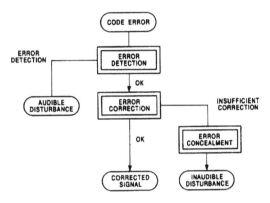

Figure 3.50 *The three basic functions of error compensation: detection, correction and concealment.*

Error detection A very simple way to determine whether a detected data word contains an error is to add one extra bit to it before transmission. The extra bit is given the value 0 or 1, depending upon the number of 1s in the word itself. Two systems are possible: odd parity and even parity checking.

(1) *Odd parity*: where the total number of 1s is odd. For example,

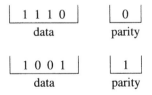

(2) *Even parity*: where the total number of 1 s is even. For example,

1 1 1 0	1
data	parity

1 0 0 1	0
data	parity

The detected word must also have the required number of 1s. If this is not the case, there has been a transmission error. This rather elementary system has two main disadvantages:

(1) even if an error is detected, there is no way of knowing which bit was faulty; and
(2) if two bits of the same word are faulty, the errors compensate each other and no errors are detected.

The most efficient error-detection coding system used in digital audio is cyclic redundancy check code (CRC), which relies on the fact that a bit stream of n bits can be considered an algebraic polynomial in a variable x with n terms. For example, the word 10011011 may be written as:

$$M(x) = 1x^7 + 0x^6 + 0x^5 + 1x^4 + 1x^3 + 0x^2 + 1x^1 + 1x^0$$
$$= x^7 + x^4 + x^3 + x + 1$$

Now to compute the cyclic check on $M(x)$, another polynomial $G(x)$ is chosen. Then, in the CRC encoder, $M(x)$ and $G(x)$ are divided:

$$M(x)/G(x) = Q(x) + R(x)$$

where $Q(x)$ is the quotient of the division, and $R(x)$ is the remainder.

Then (*Figure 3.51*), a new message $U(x)$ is generated as follows:

$$U(x) = M(x) + R(x)$$

so that $U(x)$ can always be divided by $G(x)$ to produce a quotient with no remainder.

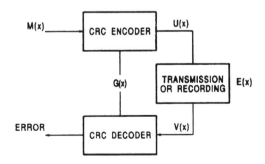

Figure 3.51 *The CRC checking principle.*

It is this message $U(x)$ that is recorded or transmitted. If, on playback or at the receiving end, an error $E(x)$ occurs, the message $V(x)$ is detected instead of

$U(x)$, where:

$$V(x) = U(x) + E(x)$$

In the CRC decoder, $V(x)$ is divided by $G(x)$, and the resultant remainder $E(x)$ shows there has been an error.

Example
The message is:

$$M(x) = x^9 + x^5 + x^2 + 1$$

The check polynomial is:

$$G(x) = x^5 + x^4 + x^2 + 1$$

Now, before dividing by $G(x)$, we multiply $M(x)$ by x^5; or, in other words, we shift $M(x)$ five places to the left, in preparation of the five check bits that will be added to the message:

$$x^5 M(x) = x^{14} + x^{10} + x^7 + x^5$$

Then the division is made:

$x^5 M(x)/G(x)$

$$= \frac{(x^9 + x^8 + x^7 + x^3 + x^2 + x + 1)}{Quotient} + \frac{(x+1)}{Remainder}$$

So that:

$$U(x) = x^5 M(x) + (x+1)$$
$$= x^{14} + x^{10} + x^7 + x^5 + x + 1$$

which can be divided by $G(x)$ to leave no remainder. In fact, the original data are unmodified (only shifted), and the check bits follow at the end.

CRC checking is very effective in detection of transmission error. If the number of CRC bits is n, detection probability is $1-2^{-n}$. If n is 16, the detection probability is $1-2^{-16} = 99.9985\%$. This means that the CRC features almost perfect detection capability. Only if $E(x)$, by coincidence, is exactly dividable by $G(x)$, will no error be detected. This obviously occurs only rarely and, knowing the characteristics of the storage medium, polynomial $G(x)$ can be chosen such that the possibility is further minimized.

Although CRC error-checking seems rather complex, the divisions can be done relatively simply using modulo-2 arithmetic. In practical systems LSIs are used which perform the CRC operations reliably and fast.

Error correction In order to ensure later correction of binary data, the input signal must be encoded

by some encoding scheme. The data sequence is divided into message blocks, then, each message block is transformed into a longer one, by adding additional information, in a process called 'redundancy'. The ratio (Data + Redundant data)/Data is known as the 'code rate'. Basically, correction (and detection) of code errors can be achieved by adding to the data bits an additional number of redundant check bits. This redundant information has a certain connection with the actual data, so that, when errors occur, the data can be reconstructed again. The redundant information is known as the 'error-correction code'. As a simple example, all data could be transmitted twice, which would give a redundancy of 100%. By comparing both versions, or by CRC, errors could easily be detected, and if some word were erroneous, its counterpart could be used to give the correct data.

The development of strong and efficient error-correction codes has been one of the key research points of digital audio technology. A lot of experience has been used from computer technology, where the correction of code errors is equally important, and where a lot of research has been spent in order to optimize correction capabilities of error-correction codes. However, the design of strong codes is a very complex matter, which requires thorough study and use of higher mathematics: algebraic structure has been the basis of the most important codes.

Combinational (horizontal/vertical) parity checking If, for example, we consider a binary word or message consisting of 12 bits, these bits could be arranged in a 3×4 matrix as shown in *Figure 3.52*. Then to each row and column one more bit can be added to make parity for that row or column even (or odd). Then, in the lower right-hand corner, a final bit can be added that will give the last column an even parity as well; because of this the last row will also have even parity.

Figure 3.52 *Combinational parity checking.*

If this entire array is transmitted in sequence (row by row, or column by column), and if during

transmission an error occurs to one bit, parity check on one row and on one column will fail; the error will be found at the intersection and, consequently, it can be corrected. The entire array of 20 bits, of which 12 are data bits, form a code word. There are $20 - 12 = 8$ redundant bits. All error-correction codes are more or less based on this idea, although better codes, i.e. codes using fewer redundant bits, can be constructed.

Crossword code Correction of errors to single bits is just one aspect of error-correction codes. In digital recording errors very often come in bursts, with complete clusters of faulty bits. It will be obvious that, in view of the many possible combinations, the correction of such bursts or errors is very complicated and demands powerful error-correcting codes. One such code, developed by Sony for use in its PCM-1600 series, is the crossword code. This uses a matrix scheme similar to the previous example, but it carries out its operations with whole words instead of individual bits, with the resultant advantage that large block code constructions can be easily realized

so that burst error correction is very good. Basically, the code allows detection and correction of all types of error with a high probability of occurrence, and that only errors with a low probability of occurrence may pass undetected. A simple illustration of a crossword code is given in *Figure 3.53*, where the decimal numbers represent binary values or words.

Error concealment Next comes a technique which prevents the uncorrected code errors from affecting the quality of the reproduced sound. This is known as concealment, and there are four typical methods.

(1) *Muting*: the erroneous word is muted, i.e. set to zero (*Figure 3.54(a)*). This is a rather rough concealment method, and consequently is rarely used.

(2) *Previous word holding*: the value of the word before the erroneous word is used instead (*Figure 3.54(b)*) so that there is usually no audible difference. However, especially at high frequencies where sampling frequency is only 2 or

Figure 3.53 *Illustration of a crossword code.*

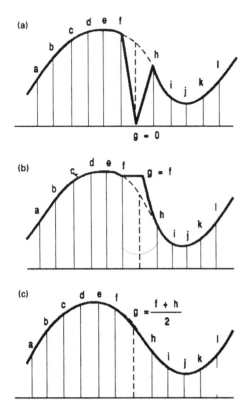

Figure 3.54 *Three types of error concealment: (a) muting; (b) previous word holding; (c) linear interpolation.*

3 times the signal frequency, this method may give unsatisfactory results.

(3) *Linear interpolation* (also called *averaging*): the erroneous word is replaced by the average value of the preceding and succeeding words

(*Figure 3.54(c)*). This method gives much better results than the two previous methods.

(4) *Higher order polynomial interpolation*: gives an estimation of the correct value of the erroneous word by taking into account a number of preceding and following words. Although much more complicated than previous methods, it may be worth using in very critical applications.

Interleaving In view of the high recording density used to record digital audio, dropouts could destroy many words simultaneously, as a burst error. Error correction that could cope with such a situation would be prohibitively complicated and, if it failed, concealment would not be possible as methods like interpolation demand that only one sample of a series is wrong. For this reason, adjacent words from the analogue-to-digital converter are not written next to each other, but at locations a sufficient distance apart to make sure that they cannot suffer from the same dropout. In effect, this method converts possible long burst errors into a series of random errors. *Figure 3.55* illustrates this in a simplified example. Words are arranged in interleave blocks, which must be at least as long as the maximum burst error.

3.3.3.3 CIRC encoding

Figure 3.56 illustrates the encoding process and shows the various information to be recorded. As there are usually two audio channels with 16-bit coding, sampled at 44.1 kHz, the bit rate, after combining both channels is: $44.1 \times 16 \times 2 = 1.4112 \times 10^6$ bit/s. To cope with both random and burst errors during CD playback, the cross interleave Reed–Solomon error-correction code (CIRC) was

Figure 3.55 *Illustration of how interleaving of data effectively changes burst errors into random errors.*

Figure 3.56 *The CD encoding process.*

developed. CIRC is a very powerful combination of several error detection and correction techniques such as crossword code and interleaving (described above).

The CIRC principle is as follows (see *Figure 3.57*).

(1) The audio signal is sampled (digitized) at the analogue-to-digital converter and these 16-bit samples are split up into two 8-bit words called 'symbols'.

(2) Six 16-bit samples from each channel, i.e. 248-bit symbols, are applied to the CIRC encoder, and stored in the random access memory (RAM).

(3) The first operation in the CIRC encoder is called 'scrambling'. The scrambling operation consists of a two-symbol delay for the even samples and a mixing up of the connections to the C2 encoder.

(4) The 24 scrambled symbols are then applied to the C2 encoder which generates four 8-bit parity symbols called Q words. The C2 encoder inserts the Q words between the 24 incoming symbols, so that at the output of the C2 encoder 28 symbols result.

(5) Between the C2 and the C1 decoder there are 28 8-bit delay lines with unequal delays. Due to the different delays, the sequence of the symbols is changed completely, according to a determined pattern.

(6) The C1 encoder generates further four 8-bit parity symbols known as 'P words', resulting in a total of 32 8-bit symbols.

(7) After the C1 encoder, the even words are subjected to a one-symbol delay, and all P and Q control words are inverted. The resultant sequenced 32 8-bit symbols is called a 'frame' and is a CIRC-encoded signal that is applied to the EFM modulator.

The bit rate at the output of the CIRC encoder is:
$1.4112 \times (32/24) = 1.8816 \times 10^6$ bit/s.

3.3.3.4 The control word

One 8-bit control word is added to every 32-symbol block of data from the encoder. The CD standard defines eight additional channels of information or subcodes that can be added to the music information; these subcodes are called P, Q, R, S, T, U, V and W. At the time of writing, only the P and Q subcodes are commonly used.

(1) The P subcode is a simple music track separator flag that is normally 0 (during music and in the lead-in track) but is 1 at the start of each selection. It can be used for simple search systems. In the lead-out track, it switches between 0 and 1 in a 2-Hz rhythm to indicate the end of the disc.

(2) The Q subcode is used for more sophisticated control purposes; it contains data such as track number and time.

The other subcodes carry information relating to possible enhancements, such as text and graphics, but are not discussed here.

Each subcode word is 98 bits long and, as each bit of the control word corresponds to each subcode (i.e. P, Q, R, S, T, U, V and W), a total of 98 complete data blocks or frames must be read from the disc to read each subcode word (*Figure 3.58*).

After addition of the control word, the new data becomes: $1.8816 \times (33/32) = 1.9404 \times 10^6$ bit/s.

3.3.3.5 The Q subcode

Figure 3.59(a) illustrates the structure of the 98-bit Q subcode word. The R, S, T, U, V and W subcode words are similar. The first two bits (S0 and S1) are

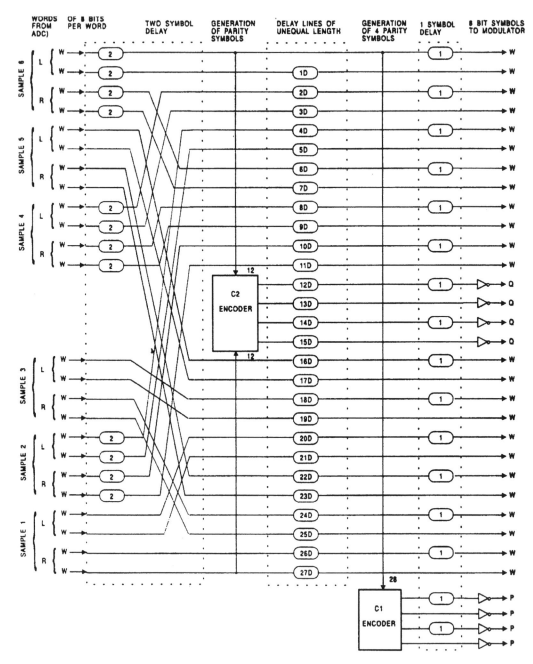

Figure 3.57 *The CIRC encoder.*

synchronizing bits. They are necessary to allow the decoder to distinguish the control word in a block from the audio information, and always contain the same data.

The next four bits are control bits, indicating the number of channels and pre-emphasis used, as follows.

0000 Two audio channels/no pre-emphasis.
1000 Four audio channels/no pre-emphasis.
0001 Two audio channels/with pre-emphasis.
1001 Four audio channels/with pre-emphasis.

Four address bits indicate the mode of the subsequent data to follow. For the Q subcode, three modes are defined.

Figure 3.58 *Each of the eight subcode bits are present in every frame.*

At the end of the subcode word a 16-bit CRC error-correction code, calculated on control, address and data information, is inserted. The CRC uses the polynomial:

$$P(X) = X^{16} + X^{12} + X^5 + 1.$$

The three modes of data in Q subcode words are used to carry various information.

Mode 1 (address = 0001) This is the most important mode, and the only one which is of use during normal playback. At least 9 out of 20 consecutive subcode words must carry data in mode-1 format. Two different situations are possible, depending whether the subcode is in the lead-in track or not. When in the lead-in track, the data are in the format illustrated in *Figure 3.59(b)*. The 72-bit section comprises nine 8-bit parts:

(1) **TNO** – containing information relating to track number: two digits in BCD form (i.e. 2 × 4 bits). Is 00 during lead-in.
(2) **POINT/PMIN/PSEC/PFRAME** – containing information relating to the table of contents

(TOC). They are repeated three times. POINT indicates the successive track numbers, while PMIN, PSEC and PFRAME indicate the starting time of that track. Furthermore, if POINT = A0, PMIN gives the physical track number of the first piece of music (PSEC and PFRAME are zero); if POINT = A1, PMIN indicates the last track on the disc, and if POINT = A2, the starting point of the lead-out track is given in PMIN, PSEC and PFRAME.

(3) **ZERO** – eight bits, all 0.

In music and lead-out tracks, data are in the format illustrated in *Figure 3.59*. The 72-bit section now comprises the following.

(1) **TNO** – current track number: two digits in BCD form (01 to 99).
(2) **POINT** – index number within a track: two digits in BCD form (01 to 99). If POINT = 00 it indicates a pause in between tracks.
(3) **MIN/SEC/FRAME** – indicates running time within a track: each part, of digits in BCD form. There are 75 frames in a second (00 to 74).

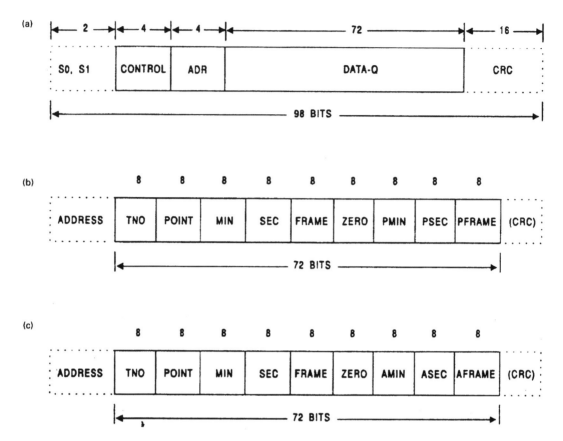

Figure 3.59 *Data format of the Q subcode. (a) Overall format. (b) Mode-1 data format in the lead-in track. (c) Mode-1 data format in music and lead-out tracks.*

Time is counted down during a pause, with a value zero at the end of the pause. During lead-in and lead-out tracks, the time increases.

(4) **AMIN/ASEC/AFRAME** – indicates the running time of the disc in same format as above. At the start of the programme area, it is set to zero.

(5) **ZERO** – eight bits, all 0.

Figure 3.60 shows a timing diagram of P subcode and Q subcode status during complete reading of a disc containing four selections (of which selections three and four fade out and in consecutively without an actual pause).

Mode 2 (address = 0010) If mode-2 data are present, at least 1 out of 100 successive subcode words must contain it. It is of importance only to the manufacturer of the disc, containing the disc catalogue number. The 98-bit, Q subcode word in mode 2 is shown in *Figure 3.61*. Structure is similar to that of mode 1, with the following differences:

(1) **N1 to N3** – catalogue number of the disc expressed in 13 digits of BCD, according to the UPC/EAN standard for bar coding. The catalogue number is constant for any one disc. If no catalogue number is present, N1 to N13 are all zero, or mode-2 subcode words may not even appear.

(2) **ZERO** – 12 bits, all zero.

Mode 3 (address = 0111) Like mode-2 data, if mode 3 is present, at' least 1 out of 100 successive subcode words will contain it. Mode 3 is used to assign each selection with a unique number, according to the 12-character International Standard Recording Code (ISRC), defined in DIN-31-621.

If no ISRC number is assigned, mode-3 subcode words are not present. During lead-in and lead-out tracks, mode-3 subcode words are not used, and the ISRC number must only change immediately after the track number (TNO) has been changed.

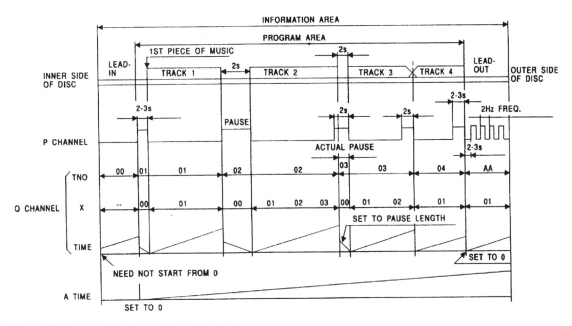

Figure 3.60 *Diagram of the timing of P and Q subcodes.*

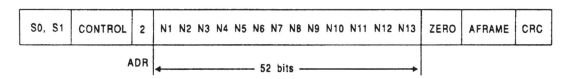

| S0, S1 | CONTROL | 2 | N1 N2 N3 N4 N5 N6 N7 N8 N9 N10 N11 N12 N13 | ZERO | AFRAME | CRC |

ADR ←———————— 52 bits ————————→

Figure 3.61 *Q subcode format with mode-2 data.*

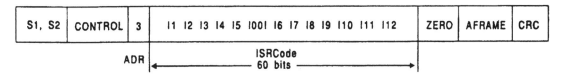

| S1, S2 | CONTROL | 3 | I1 I2 I3 I4 I5 I00I I6 I7 I8 I9 I10 I11 I12 | ZERO | AFRAME | CRC |

ADR ←———— ISRCode ————→
←———— 60 bits ————→

Figure 3.62 *Q subcode format with mode-3 data.*

A 98-bit, Q subcode word in mode 3 is shown in *Figure 3.62*. Structure is similar to that of mode 1, with the following differences:

(1) **I1 to I12** – the 12 characters of the selection's ISRC number. Characters I1 and I2 give the code corresponding to country. Characters I3 to I5 give a code for the owner. Characters I6 and I7 give the year of recording. Characters I8 to I12 give the recording's serial number. Characters I1 to I5 are coded in a 6-bit format while characters I6 to I12 are 4-bit BCD numbers.
(2) **00** – two bits, both zero.
(3) **ZERO** – four bits, all zero.

3.3.3.6 EFM encoding

EFM, or eight-to-fourteen modulation, is a technique which converts each 8-bit symbol into a 14-bit symbol, with the purpose of aiding the recording and playback procedure by reducing required bandwidth, reducing the signal's d.c. content, and adding extra synchronization information. A timing diagram of signals in this stage of CD encoding is given in *Figure 3.63*.

The procedure is to use 14-bit codewords to represent all possible combinations of the 8-bit code. An 8-bit code represents 256 (2^8) possible combinations, while a 14-bit code represents 16 384 (2^{14}) different combinations. Out of the 16 384 14-bit codewords,

Figure 3.63 *Diagram of the timing of EFM encoding and merging bits.*

only 256 are selected, having combinations which aid processing of the signal. For instance, by choosing codewords which give low numbers of individual bit inversions (i.e. 1 to 0, or 0 to 1) between consecutive bits, the bandwidth is reduced. Similarly, by choosing codewords with only limited numbers of consecutive bits with the same logic level, overall d.c. content is reduced. A read-only-memory (ROM)-basedlook-up table can then be used to assign all 256 combinations of the 8-bit code to the 256 chosen combinations within the 14-bit code.

In addition to EFM modulation, three extra bits, known as 'merging bits' are added to each 14-bit symbol, with the purpose of further lowering d.c. content of the signal. Exact values of the merging bits depend on the adjacent symbols. Finally, the data bits are changed from NRZ into NRZI (non-return to zero inverted) format, by converting each positive-going pulse of the NRZ signal into a single transition. The resultant signal has a minimum length of 3T (three clock periods), and a maximum of 11T (11 clock periods), as shown in *Figure 3.64*.

The bit rate is now: $1.9404 \times (17/8) = 4.12335 \times 10^6$ bit/s.

3.3.3.7 The sync word

To the signal, comprising 33 symbols of 17 bits (i.e. a total of 561 bits), a sync word and its three

$T_{min} = 3T$, where $T = \dfrac{1}{4.3218}\ \mu s = 231.39\ ns$

d_{min} (at V = 1.2 m/s) = 3T X $V_{1.2}$ = 0.833 μm

(at V = 1.4 m/s) = 3T X $V_{1.4}$ = 0.972 μm

d max (at V = 1.2 m/s) = 11 T X $V_{1.2}$ = 3.05 μm

(at V = 1.4 m/s) = 11 T X $V_{1.4}$ = 3.56 μm

Figure 3.64 *The effects of adding three merging bits to the EFM encoded signal. T, clock period.*

merging bits are added, giving 588 bits in total. Sync words have two main functions: (1) they indicate the start of each frame; and (2) sync word frequency is used to control the player's motor speed. The 588-bit long signal block is known as an 'information frame'.

The final bit rate, recorded on the CD becomes: $4.12335 \times (588/561) = 4.3218 \times 10^6$ bit/s.

The frame frequency F_{frame} is: $4.3218 \times 10^6/588 = 7350$ Hz.

As subcodes are in blocks of 98 frames, the subcode frequency F_{sc} is: $7350/98 = 75$ Hz.

Playing time is calculated by counting blocks of subcode (75 blocks = 1 s). A 60-min CD thus contains: $60 \times 60 \times 7350 = 26\,460\,000$ frames. As each frame comprises $33 \times 8 = 264$ bits of information, a 1-h CD actually contains 6 985 440 000 bits of information, or 873 180 000 byte! Of this, the subcode area contains some 25.8 kbyte (200 Mbit). This gigantic data storage capacity of the CD medium is also used for more general purposes like CD-ROM (compact disc read-only memory), which is derived directly from the audio CD.

3.3.4 Opto-electronics and the optical block

3.3.4.1 The injection laser diode

The basic operation of any laser consists of pumping the atoms into a stimulated state, from which electrons can escape and fall to the lower energy. In a solid-state laser, the input energy populates some of the usually unpopulated bands. When a photon of energy equal to the band gap of the crystal passes through it stimulates other excited photons to fall in step with it. Thus a small priming signal will emerge with other coherent photons.

The laser can be operated in a continuous wave (CW) oscillator mode if the ends of the laser path are optically flat, parallel and reflective, so forming an optical resonant cavity. A spontaneously produced photon will rattle back and forth between the ends, acquiring companions due to stimulated emission. This stimulated emission can be like an avalanche, completely draining the high-energy states in a rapid burst of energy. On the other hand, certain types of lasers can be produced which will operate in an equilibrium condition, giving up photons at just the input energy pumping rate. The injection laser diode (ILD) is such a device.

The material in an injection laser diode is heavily doped so that under forward bias the region near the junction has a very high concentration of holes

and electrons. This produces the inversion condition with a large population of electrons in a high-energy band and a large population of holes in the low-energy band. In this state, the stimulated emission of photons can overcome photon absorption and a net light flux will result. In operation, the forward current must reach some threshold beyond which the laser operates with a single 'thread' of light, and the output is relatively stable but low. As the current is increased, light output increases rapidly.

The ILD characteristic is highly temperature sensitive. A small current variation or a modest temperature change can cause the output to rise so rapidly that it destroys the device. A photodiode, monitoring the light output, is commonly used in a feedback loop to overcome this problem. Like the light-emitting diode (LED), the ILD must be driven from a current source, rather than a voltage source, to prevent thermal runaway. With a voltage source, as the device junction begins to warm up the forward voltage drop decreases, which tends to increase the current, in turn decreasing the forward voltage drop, and so on until the current tends towards infinity and the device is destroyed. In addition, the ILD has a tendency to deteriorate with operation. Deterioration is greatly accelerated when operating the ILD outside of its optimum limits. A typical ILD characteristic is shown in *Figure 3.65*.

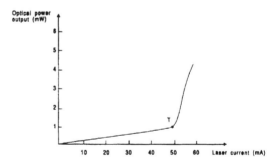

Figure 3.65 *The characteristic of a typical injection laser diode.*

3.3.4.2 The optical pick-up

A schematic illustration of an optical pick-up of the three-beam type used in CD players is shown in *Figure 3.66*. This drawing shows the different optical elements composing a pick-up and indicates the laser beam route through the unit. The laser beam is generated by the laser diode. It passes through the diffraction grating, generating two secondary beams called 'side-beams', which are used by a tracking servo circuit to maintain correct tracking of the disc.

Figure 3.66 *A three-beam optical pick-up.*

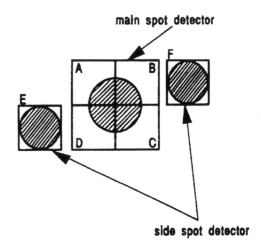

Figure 3.67 *The relative positions of the six photodetectors of an optical pick-up.*

The beam enters a half mirror prism which is a non-polarizing prism that reflects half of the incident light and lets pass the other half. The light beam, still divergent at this stage, is converged into a parallel beam by the collimation lens. The laser beam is then focused by a simple lens onto the pit surface of the disc. The simple lens is part of a servo-controlled mechanism known as a 'two-axis device'. The beam is reflected by the mirrored surface of the disc and converged by the two-axis device lens into a parallel beam. After passing through the collimation lens the laser beam is converged onto the half mirror prism where half of the light is reflected towards the detector part of the pick-up device. The resulting light beam intensity at the detector part is $\frac{1}{2} \times \frac{1}{2} = \frac{1}{4}$ of the original beam.

In the detector part the cylindrical lens focuses the laser beam in only one place. As a result, the cross-section of the beam after passing through the lens is elliptical with the degree of ellipticity varying according to the distance from the lens. By detecting the degree of ellipticity an indication of whether or not the beam is focused on the disc surface can be obtained. Six photodetectors enable the read-out of the pit information from the disc. *Figure 3.67* shows the arrangement of the four main spot detectors (A, B, C and D) and two side spot detectors (E and F).

Optical pick-ups contain an actuator for objective lens position control. The CD player, due to

the absence of any physical contact between the disc and the pick-up device, has to contain auto-focus and autotracking functions. These functions are performed by the focus and tracking servo circuits via the two-axis device, enabling a movement of the objective lens in two axes: vertically for focus correction, and horizontally for track following. The principle of operation is that of a moving coil in a magnetic field. Two coils, the focus coil and the tracking coil, are suspended between magnets (*Figure 3.68*) creating two magnetic fields. A current through either coil, due to the magnetic field, will cause the coil to be subjected to a force, moving the coil in the corresponding direction.

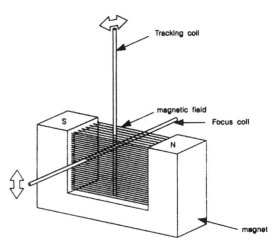

Figure 3.68 *The operating principle of a two-axis device.*

3.3.5 Servo circuits in a CD player

A CD player typically uses four distinct servo circuits (*Figure 3.69*).

(1) Focus servo circuit: this servo circuit controls vertical movement of the two-axis device and guarantees that the focal point of the laser beam is precisely on the mirror surface of the CD.
(2) Tracking servo circuit: this circuit controls the horizontal movement of the two-axis device and forces the laser beam to follow the tracks on the CD.
(3) Sled servo circuit: this circuit drives the sled motor which moves the optical block across the CD.
(4) Disc motor servo circuit: this circuit controls the speed of the disc motor, guaranteeing that the optical pick-up follows the CD track at a constant linear velocity.

The optical pick-up is the source of the feedback signals for all four servo circuits.

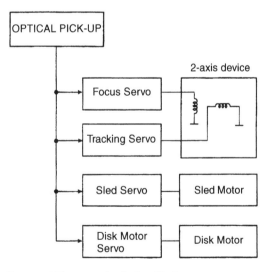

Figure 3.69 *The servo circuits in a CD player.*

3.3.5.1 The focus servo circuit
The reflected laser beam is directed to the main-spot detector (*Figure 3.70(a)*), an array of four photodiodes (A, B, C and D). When the focus is correct the beam falls equally on the four diodes, and the focus error signal $(A + C) - (B + D)$ is zero. On the other hand, when the beam is out of focus (*Figures 3.70(b)* and *3.70(c)*), an error signal is generated, because the beam passes through a cylindrical lens, which makes the beam elliptical in shape. The resultant focus error

signal from the main-spot detector $(A + C) - (B + D)$ is, therefore, not zero.

When a disc is first loaded in the player the distance between the two-axis device and the disc is too large: the focus error signal is zero, making the focus servo circuit inactive. Therefore, a focus search circuit is used which, after the disc has been loaded, moves the two-axis device slowly closer to the disc. Outputs of the four photodiodes are combined in a different way $(A + B + C + D)$ to form a radiofrequency signal, which represents the data bits read from the disc. When the radiofrequency signal exceeds a threshold level, the focus servo is enabled and now controls the two-axis device for a zero focus error signal.

3.3.5.2 The tracking servo circuit
Figure 3.71 shows the three possible tracking situations as the optical pick-up follows the disc track. In *Figure 3.71(a)* and *3.71(b)* the main-spot detector is not correctly tracked, and so one or other side of the spot detectors gives a large output signal as the pit is traversed. In *Figure 3.71(c)*, on the other hand, the main-spot detector is correctly tracked, and both side-spot detectors give small output signals.

Side-spot detectors consist of two photodiodes (E and F) and generate a tracking error signal: TE = E − F. The tracking servo acts in such a way that the tracking error signal is as small as possible, i.e. the mainspot detector is exactly on the pits of the track. Tracking error, focus error, and resultant focus signals are shown, derived from the optical pick-up's photodiode detectors, in *Figure 3.72*.

3.3.5.3 The sled motor servo circuit
The two-axis device allows horizontal movement over a limited number of tracks, giving a measure of fine tracking control. Another servo circuit, called the sled servo circuit, is used to move the complete optical unit across the disc for coarse tracking control. It uses the same tracking error signal as the tracking servo of the two-axis device. However, the output of the tracking servo circuit is linearly related to the tracking error signal, whereas the output of the sled servo circuit has a built-in hysteresis: only when the tracking error signal exceeds a fixed threshold level does the sled servo drive the sled motor. Tracking error and sled motor drive signals are shown in *Figure 3.73*.

3.3.5.4 The disc motor servo circuit
Each frame of information on the disc starts with a sync word. One of the functions of the sync words is

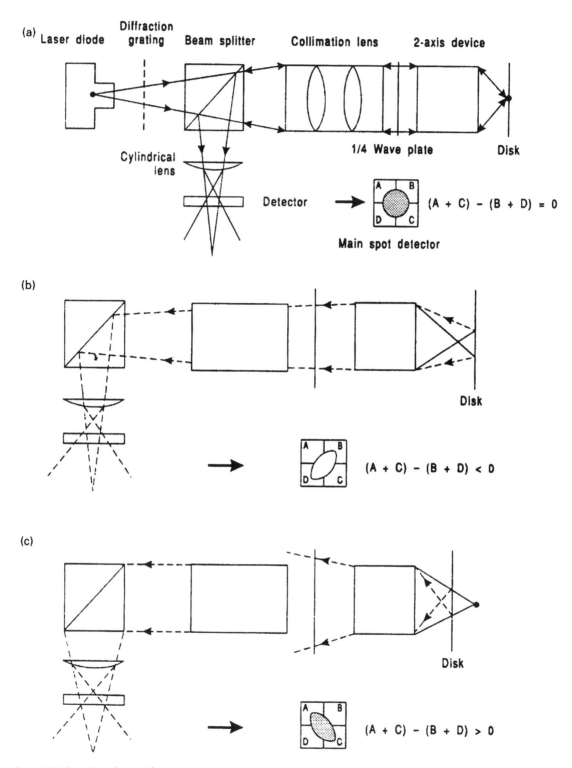

Figure 3.70 *Detection of correct focus.*

Figure 3.71 *Three possible tracking scenarios.*

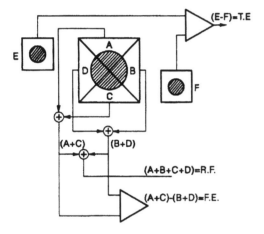

Figure 3.72 *The various error signals obtained from the photo-detectors of the optical pick-up.*

to control the disc motor. The sync word frequency is compared against a fixed frequency (derived from a crystal oscillator) in a phase comparator and the motor is driven according to any frequency or phase difference. As the length of the tracks increases linearly from the inner (lead-in) track to the outer (lead-out) track, the number of frames per track increases in the same respect. This means that the frequency of the sync words also increases, which causes the motor speed to decrease, resulting in a constant linear velocity.

3.3.6 Signal processing

The block diagram in *Figure 3.74* represents the basic circuit blocks of a CD player. The signal that is read from the compact disc and contains the data information is the radiofrequency signal, and consists of the sum of signals $(A + B + C + D)$ from the main-spot detector. At this stage, the signal is a weak current signal and requires current-to-voltage conversion and amplification.

The radiofrequency signal from the radiofrequency amplifier needs to be converted into a square wave signal. This signal contains a certain amount of irregularities due to disc tolerances, and a low-frequency component due to the changing pit length and distance. A wave-shaper circuit (*Figure 3.75*) is used to convert the radiofrequency signal into a square wave signal. The d.c. offset component is removed with a coupling capacitor and the signal is applied to a comparator. The output signal of the comparator is integrated in the feedback loop through a low-pass filter to obtain an equal occurrence probability for high and low levels. The output d.c. voltage is applied to the inverting input of the comparator. As a result, the slice level of the shaper will depend on the signal contents in order to extract a correct square wave.

Figure 3.76 shows the relationship between the radiofrequency signal and the resulting square wave. After wave-shaping the radiofrequency signal is applied to a phase-locked loop (PLL) circuit in order to detect the clock information from the signal

Figure 3.73 *Tracking error and sled motor drive signals within CD player.*

Figure 3.74 *Signal decoding within a CD player. RF, radio frequency; D/A, digital to analogue; PLL VCO, phase-locked loop voltage controlled oscillator.*

Figure 3.75 *Wave-shaper circuit.*

which, in turn, synchronizes the data. In addition, the radiofrequency signal is applied to an EFM demodulator and associated error stages in order to obtain a demodulated signal. In more recent CD players all these functions are performed by a single integrated circuit, the CXD-1135 Digital Signal Processor (*Figure 3.77*). The digital signal processor performs frame sync detection, fourteen-to-eight demodulation to a parallel 8-bit data output, subcode Q detection and generation by an internal counter and timing generator of the synchronization signals.

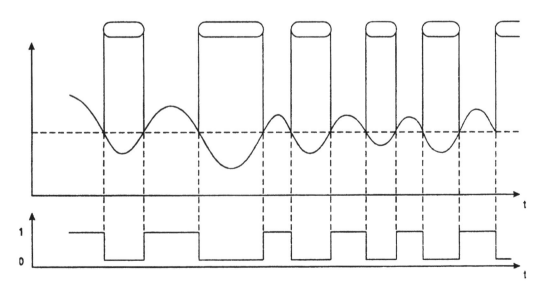

Figure 3.76 *Diagram of the timing of the radio frequency signal before and after wave-shaping.*

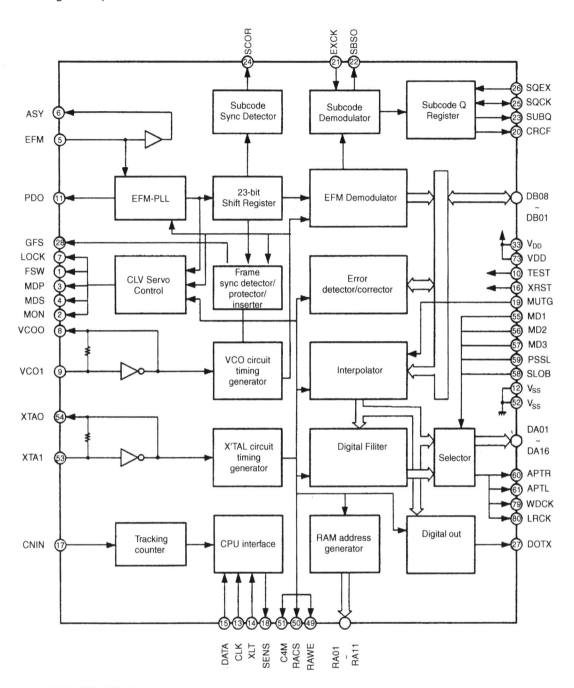

Figure 3.77 *The CXD-1135 digital signal processor.*

CIRC decoding is performed on the data stored in the RAM. The data are checked, corrected if necessary, and deinterleaved during read-out. If uncorrectable errors are found, a pointer for this data word is stored in memory and the circuit corrects the data by interpolation.

The digital signal processor outputs left and right digital audio data in 16-bit format. The serial or parallel output (selectable) can directly be applied to the digital-to-analogue converters.

Additional functions performed by the digital signal processor include the following.

(1) Constant linear velocity (CLV) servo: servo circuit to control the rotation speed of the disc motor.
(2) Central processing unit (CPU) interface: serial interface for communication with an external microcontroller.
(3) Digital filter: 20-kHz cut-off filter for aperture effect compensation. Pass band ripple ±0.07 dB.
(4) Digital output: digital audio interface according to the Electronic Industries Association of Japan format.

Bibliography

CLIFFORD, M., *The Complete Compact Disc Player*, Prentice-Hall, Englewood Cliffs, NJ (1987)
LENK, J.D., *Complete Guide to Compact Disc (CD) Player Troubleshooting and Repair*, Prentice Hall, Englewood Cliffs, NJ (1986)
POHLMANN, K.C., *Principles of Digital Audio*, Macmillan, New York (1985)
SONY SERVICE CENTRE, *Digital Audio and Compact Disc Technology*, 2nd edn, Newnes, Oxford (1992)

3.4 Digital audio tape

3.4.1 Introduction

3.4.1.1 History
Digital audio tape (DAT) is not a new product. The history of DAT starts nearly at the same time as compact disc (CD) players. It seemed logical that if a CD disc was to replace the analogue audio disc, there was also a need for a digital tape to replace the analogue audio cassette. But the origin of the DAT system lies also in the professional audio world, where the use of pulse code modulation processors putting the PCM information on VCR systems (U-matic, Betamax, etc.) meant a boost to the quality of recording and transmitting audio in the professional world, but were too expensive and cumbersome to use for the general public. Around June 1983, two possibilities were agreed for further development, the S-DAT format (stationary head) and the R-DAT format (rotary head). Of these two formats, only the R-DAT format was developed into a commercial product. The DTC–1000 was Sony's first commercialized DAT recorder. Several professional models followed, and DAT became widely used in the professional audio recording business. Opposition from the recording companies, which feared that the commercialization of an almost perfect recording medium, allowing

multigeneration copies to be made without significant loss in quality, would seriously hurt their business, had stopped DAT from being sold in the consumer market. In 1989, however, a copy-protect system, the serial copy management system (SCMS) was agreed upon, and since then Sony has introduced various types of DAT recorders and players (deck, portable and in-car). All of these models feature the copy management system which enables only one first-generation digital copy and prohibits any further digital copies. As of that moment DAT was considered a consumer item, and not just a professional item.

3.4.1.2 Basic principles
The aim when developing DAT was to achieve a digital audio cassette minimal in size, superior to the analogue audio cassette, and with a quality level comparable to the CD format. Further requirements were ease of use, including features not possible with the analogue cassette, and economy in cassette and running costs.

Basically, one can describe a DAT system as a mixture of analogue-to-digital conversion and data processing on the electronics side and video recorder technology on the mechanical side. People with a background in audio will be more familiar with the electronics side. It is clear that, like CD technology, any digital audio processing system will need appropriate encoding, decoding and error processing. The data format, however, needs to be adapted to the recording medium.

As for the mechanical side, which is more familiar to video people, helical scan azimuth systems were developed for video use in order to store a maximum of information on a rather small tape. For very high recording densities a system using a stationary head was hardly applicable. The big difference between a standard video recorder and a DAT system using helical scan is that the recorded signal is analogue in the case of video, and digital in the case of DAT.

It is also important to notice that a part of the digital information can be recorded without changing the audio contents.

3.4.2 Helical scan azimuth systems

3.4.2.1 Helical scan
In order to achieve high recording densities, the helical scan system was developed. This system uses two (or possibly more) heads which are placed opposite each other in a rotating cylindrical drum. These heads rotate at a speed which is determined by the parameters of the system. The tape is guided over

the drum, which is composed of an upper half and a lower half (upper drum and lower drum). Between the two halves there is a small gap enabling the heads to make contact with the tape which is running over the drum. Even if the drum is composed of two halves, it is usually seen and referred to as one unit. The drum assembly is not placed horizontally but at an angle. As a result of this, the signal is recorded on the tape in a discontinuous way.

Figure 3.78 shows how the tape is guided over the drum: it is obvious that in order to guide a tape over a tilted cylinder there is a need for accurate tape guiding. For this reason several roller guides are used as well as inclined guides. The inclined guides

Figure 3.78 *Layout of the helical scan system.*

are needed to give the tape an inclination suitable to move over the cylindrical drum without causing any discontinuity or tape damage. The figure also shows that the tape is guided only partly over the drum. The angle that the tape runs over the drum (seen from above) is called the 'wrap angle'.

Figure 3.79 clarifies the above-mentioned items. The lower part shows the tilted drum in front view with the tape running over it. It can be seen clearly how the heads will move over the tape. The resulting track pattern is seen on the upper part of the drawing. The pattern on the tape is effectively as shown with each track at a certain angle. The previous and following tracks are of course at the same angle.

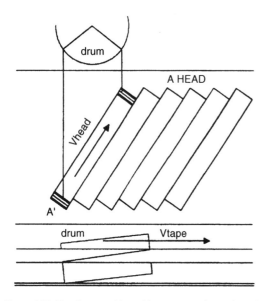

Figure 3.79 *The drum position with respect to the track on the tape.*

3.4.2.2 Azimuth system

When looking at the track pattern shown in *Figure 3.80*, one possible problem becomes clear: tracks which are written so closely to each other may cause a cross-talk problem. Therefore in professional video recorders a gap is left between the tracks. In order to maximize tape usage, however, all consumer video recorders and also DAT leave no gap at all, but avoid cross-talk by means of the 'slanted azimuth' recording system. This means that each head has a different azimuth angle. The result is that the information on one track will be written with a certain azimuth angle and the next (as well as previous) track will be written with a different azimuth angle. *Figure 3.80* shows the two azimuth angles used in the DAT format, i.e. $\pm 20°$. It also

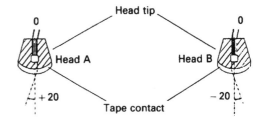

Figure 3.80 *The head azimuth system.*

becomes clear that when reading out the information it is necessary to have the correct head over the correct track, otherwise there is mistracking and corrective action must be taken.

The theory behind this system is as follows. If a playback head reads out signals with a wrong or incorrect azimuth angle the high frequencies will be degraded. This also happens with an analogue audio cassette with a stationary head, as will be well known by audio engineers. But in this case different azimuth angles are wanted, because the frequencies recorded are very high, so if a head reads part of the adjacent track with an azimuth angle which is incorrect (in the DAT case there is thus a total azimuth difference of $40°$) it will have a very low cross-talk content. Of course, as noted before, it is crucial that each head reads the tracks with the same azimuth angle as its own angle. If a head is positioned over a track with the opposite azimuth angle the rf level will be very low, and can therefore not be used.

3.4.3 Basic mechanical specifications

3.4.3.1 VTR and DAT comparisons

As stated before, the DAT mechanical side is derived from VTR and video 8 mechanisms and, as a consequence, the track format is similar. It is therefore useful to make a comparison between several different existing systems. *Figure 3.81* shows the differences

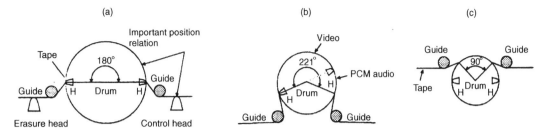

Figure 3.81 *Comparison between systems with different wrap angles: (a) Betamax and VHS; (b) 8-mm VTR; (c) DAT.*

in wrap angle between Betamax, VHS, 8 mm VTR and DAT. Note also that in Beta/VHS systems erase as well as control heads are included, which is not the case for a DAT system. In the DAT system there is no need for a control head, and erase is performed by so-called 'overwrite', so a separate erase head is not needed.

One very important note needs to be made here: the wrap angle shown for the DAT format is 90°. This is usually the case in the normal home deck player, but in DAT systems using small size drum assemblies such as walkman DAT and car DAT the wrap angle can be 180°. The wrap angle can theoretically be any angle, provided that the data format on the tape is respected. This particular (90°) wrap angle is possible as the signal on the tape is time-compressed. The wrap angle of 180° or more in video players is needed since in these cases the signal is analogue and

uncompressed. Other benefits of the reduced wrap angle include lower tape tension, less tape damage and increased head life.

Figure 3.82 gives a comparison between the track formats on tape in the different systems. Each system uses different track and azimuth angles, but the main idea remains the same. Beta and VHS systems have control tracks and separate audio tracks, and for this reason the non-rotating erase and control heads need to be included. The optional tracks defined in the DAT format are not used.

A comparison of the basic mechanical specifications is given in *Table 3.4*. It is clear that the size of the DAT format is very small compared to, for example, Betamax or VHS. It should be noted that a manufacturer can use different specifications as long as the format on the tape is respected. The head cylinder diameter, for example, is no strict specification.

Figure 3.82 *Comparison between different track layout systems: (a) Betamax BII; (b) VHS; (c) 8-mm VTR; (d) DAT.*

Table 3.4 Specifications for DAT and VTR

	DAT	8-mm Video	VHS	Betamax (BII)
Tape width (mm)	3.81	8	12.65	12.65
Tape running speed (mm/s)	8.15	14.345	33.35	20.00
Head cylinder diameter	30	40	62	74
Relative speed (m/s)	3.133	3.8	5.8	6.993
Track pitch (μm)	13.591	20.5	58	29.2
Effective track width (mm)	2.613	5.85	10.07	10.2
Head azimuth	±20°	±10°	±6°	±7°
Cassette dimensions (mm)	73 × 54 × 10.5	95 × 62.5 × 15	188 × 104 × 25	159 × 96 × 25
Cassette volume ratio	1	2.2	11.8	9.0

Note: compact audio cassette tape width 3.81 mm, dimensions 102 × 63 × 12 mm.

It is perfectly possible to use other head diameters (the head cylinder diameter used in the small walkman DAT and car DAT are actually 15 mm). One of the most important specs is the relative speed; this is the combination of tape speed and head rotation; this will determine (along with drum size and wrap angle) how the data will be written on and read out from the track.

Figure 3.83 shows the exact data track format. It has been noted before that there are several mechanical possibilities such as different wrap angles, different drum diameters, different head rotation speeds, etc. All these mechanical specifications can be chosen freely as long as the track format as shown below is respected. This will ensure correct recording and playback as well as compatibility. The R-DAT format specifies a track width of 13.6 μm, but the head width is about 1.5 times this value. Therefore, the recording is called overwrite recording, where one head records over the track recorded by the previous head. This means that as much tape as possible is used; rotating head recorders without this overwrite record facility must leave a guard band between each track on the tape. Therefore systems using overwrite recording used techniques are called 'guard-bandless'.

(A) Tape width : 3.81mm
(B) PCM center : 1.905mm
(C) Optional track I width : 0.5mm
(L) Track length : 23.501mm
(P) Track pitch : 13.591μm
(R) Optional track II width : 0.5mm
(W) Effective recording width : 2.613mm
(a) Head gap azimuth angle : . 20°
(q) Track angle : 6° 22' 59.5'

Figure 3.83 *The DAT track format.*

3.4.3.2 R-DAT cassette tape

The R-DAT cassette is totally different from a normal analogue audio cassette. *Figure 3.84* shows the sizes of the most used cassette systems. The R-DAT cassette is of the closed, flangeless type. *Figure 3.85* shows the outside of the normal cassette. The main outside parts are the upper shell, the slider shell which will slide backwards in order to open the cassette bottom and the cassette lid which will pivot upwards to open the front of the cassette where the tape runs. The slider and lid can only be opened when inserted correctly in the player and there are locks to prevent opening them by accident. The reels (take-up reel and supply reel) are also locked when the cassette is closed.

In the bottom section of the cassette there are several holes which were specified originally for information such as tape type, track pitch, etc., but as the same information is also included in the data stream which is recorded, the use of these holes is entirely at the manufacturer's discretion (see *Figure 3.86* and *Table 3.5*). Note that identification holes 1–4 can only be set once. Opening a hole means that it remains open permanently. Hole 5, however, can be opened and closed by means of a slider located on the back of the cassette.

An important feature of the DAT cassette is the tape end section (*Figure 3.87*). Each DAT cassette

has two built-in prisms: these can be used for tape-end detection or tape-beginning detection. However, it is again up to the manufacturer whether this item is used or not.

Each tape must contain a transparent leader and trailer part (60 mm long). Only during this leader and trailer part can light pass through the tape. A light-emitting diode (LED) is used to emit light into the prisms. This light is guided back and, depending on tape transparency, it will or will not reach the detector. It is therefore easy to determine whether the tape is ended and if so the system can determine on which side it has ended. It is also possible to detect if the tape is broken. Another possibility is to have one LED in the middle of the cassette and which emits on both sides. In this way the prisms are not used. Such a system is used in walkman DAT players to economize on space and parts.

A point worth mentioning is the fact that there is a certain overlap between take-up space and supply space. *Figure 3.88* shows that there is a common area between the supply and take-up diameter. This is one way to minimize the format volume whilst maximizing tape length. As the tape will be either on the take-up or on the supply side there is no physical problem.

Figure 3.84 *Cassette formats.*

Figure 3.85 *A DAT cassette.*

Figure 3.86 *Positions of the discriminating holes in a DAT cassette.*

3.4.3.3 Mechanical assembly requirements

There are many possible variations on the mechanical aspects of DAT assembly. As mentioned already there is a variety of items which can be defined by each manufacturer; this of course leads to different mechanical approaches. There are, however, basic ideas which one will find in any DAT player. The mechanical side will look similar to the one shown in *Figure 3.89*. The drawing shows the tape in the threaded position, which is the correct working position. When the cassette is inserted this is not yet the case and, therefore, the tape needs to be taken out of its shell and guided into the correct position. Whenever a DAT cassette is inserted into the player the following needs to happen: the bottom part of the cassette (the slider) and the front lid have to open. It is then possible to insert some guides of the player into the cassette in order to engage the tape and guide it to the correct position. The act of guiding the tape

Table 3.5 Settings of the discriminating holes in DAT cassettes

Hole 1	Hole 2	Hole 3	
0	0	0	Metal tape or equivalent/tape thickness 13 μm
0	1	0	Metal tape or equivalent/thin tape
0	0	1	Track pitch 1.5 × /tape thickness 13 μm
0	1	1	Track pitch 1.5 × /thin tape

Hole 4	
1	Prerecorded music tape
0	Non-prerecorded tape (regular tape for recording)

Hole 5	
1	Recording not possible
0	Recording possible

Figure 3.87 *Tape-end-detection system in a DAT cassette.*

to its position is called 'threading'. Unthreading is of course necessary whenever ejecting the cassette.

The diagram also shows the tape in the threaded position (solid line) as well as unthreaded position (dotted line). When not yet threaded but with the cassette inserted, roller guides 1 and 7 and the drawer roller guide will be positioned at their outermost position, i.e. inside the DAT cassette in the gap between the reel and the tape running in front of the cassette (dotted line). The guides will be moved

Figure 3.88 *Tape overlap in a DAT cassette.*

Figure 3.89 *Tape path: threaded (———) and unthreaded (- - - - -).*

mechanically to their inner position, near to the drum and capstan and will take the tape and position it correctly. In this example there is only one inclined (or slanted) guide, but the number of these depends on the particular design; there may be more.

Note the position of the identification holes (see above). Needless to say, at these positions one may find detectors. There is also the possibility that there will be a detector which is not located at a hole position, but in between two holes. This is usually for cassette-in detection.

Figure 3.90 shows the positions of motors. The drum motor is needed to rotate the drum; normally the speed is 2000 r.p.m. In long-play modes, however, 1000 is also possible. The capstan motor moves the tape and ensures the correct relative speed. The capstan speed is therefore regulated by

Figure 3.90 *The positions of the motors.*

feedback from the data processing side. The supply and take-up motors are needed for normal operation, but also for fast forward and reverse operations. Of course it is possible to have only one motor taking care of take-up and supply through correct mechanics, as only one motor is used at a time. Where only one motor is used this is referred to as the 'reel motor'. Apart from the motors shown in *Figure 3.90*, there is also a need for one or more loading and cassette-handling motors.

There will usually be some tension-regulating mechanism to give the tape the correct tension. The assembly shown in *Figure 3.91* shows a possible tension-regulating assembly. In this case the assembly will also push the pinch roller against the capstan. This is needed in order to squeeze the tape between the capstan and the pinch roller and to enable correct tape movement. The controlled motor which will take care of guide movements, etc. is not shown in this figure.

Figure 3.91 *The regulator-arm assembly.*

3.4.4 DAT data format

The DAT data format was specially developed in order to ensure high recording densities, low error contents and a variety of other features. As noted previously, the data format has been influenced by the mechanical design. Due to the helical scan azimuth system the data must be divided into separate tracks.

Part of the information recorded on one track may be re-recorded without changing the real data contents, i.e. the digitized music, and this has an influence on the data format. Furthermore, the data themselves are divided into audio data (PCM) and subcode data. All these items are discussed below.

3.4.4.1 Main DAT specifications

Some of the data given in the specifications table (*Table 3.6*) have already been discussed in the mechanical section. *Table 3.6* gives the original specifications. The DAT specifications include several options. Apart from the standard mode, these options are included to enable more possibilities such as four tracks, long play and prerecorded tapes. The standard mode is two-channel, 48-kHz sampling frequency, 16-bit linear quantization. It is clear that whenever a 12-bit non-linear quantization is used, the frequency range will be reduced to approximately 14.5 kHz instead of the full 20 kHz. The sampling rate of 32 kHz enables the long-play mode as well as four-track recording.

The system allows high recording densities and a relatively high subcode capacity. In reality only part of this capacity is actually used. The transmission rate is 2.46 Mbit/s, but the data are compressed by a factor of 3 and processed at 7.5 Mbit/s; this enables the signal to be recorded continuously. The modulation system uses an 8–10 bit conversion. This is to overcome low-frequency problems of coupling transformers in the record/playback head. In addition, proper clocking can be assured.

The correction system will be explained later, but this is very similar to the data processing used in CD formats. The tracking system has been 'borrowed' from video 8 (see below). Basically, we can say that a reference signal is recorded on the tape in order to check whether the correct head follows the correct track at the correct moment. Based on this information the capstan speed can be adjusted.

Some explanation is required concerning prerecorded tapes. As shown in *Figure 3.92* there is a normal track and a wide track possibility depending on the dubbing technology: remember that, starting from one master tape, a large number of copies are made. With the most used dubbing technique (contact copy) there is some loss of signal, partly because the tape used in this case is not a metal powder tape but an oxide tape.

In order to have the same error rate and radiofrequency input the wide-track option has been defined. This is possible with the recording heads, but several specifications need to be adapted in this case and

Table 3.6 The main DAT specifications

Item	DAT (REC/PB mode) Standard	Option 1	Option 2	Option 3	Pre-recorded tape (PB only) Normal track	Wide track
Channel number (CH)	2	2	2	4	2	2
Sampling frequency (kHz)	48	32	32	32	44.1	
Quantization bit number (bit)	16 (Linear)	16 (Linear)	12 (Non-linear)	12 (Non-linear)	16 (Linear)	16 (Linear)
Linear recording density (KB/l)	61.0		61.0		61.0	61.1
Surface recording density (MB/l^2)	114		114		114	76
Transmission rate (MB/S)	2.46	2.46	1.23	2.46	2.46	
Subcode capacity (KB/S)	273.1	273.1	136.5	273.1	273.1	
Modulation system	8–10 conversion					
Correction system	Double Reed–Solomon code					
Tracking system	Area sharing ATF					
Cassette size (mm)	73 × 54 × 10.5					
Recording time (min)	120	120	240	120	120	80
Tape width (mm)	3.81					
Tape type	Metal powder					Oxide tape
Tape thickness (μm)	13±1					
Tape speed (mm/s)	8.15	8.15	4.075	8.15	8.15	12.225
Track pitch (μm)	13.591				13.591	20.41
Track angle	6°22′59.5″					6°23′29.4″
Standard drum specifications	Ø30 90° Wrap					
Drum rotations (rpm)	2000		1000	2000	2000	
Relative speed (m/s)	3.133		1.567	3.133	3.133	3.129
Head azimuth angle (°)	±20°					

Figure 3.92 *Track format. ATF, automatic track finding; IBG, interband gap.*

the recording time is decreased (maximum 80 min). These specification changes for wide-track operation are needed so that the heads follow the tracks correctly with a maximum radiofrequency input.

3.4.4.2 Track format

Each head runs over the tape with a specified angle; during this tracing period several different signals need to be recorded. (See *Figure 3.93*.)

Figure 3.93 *Basic track organization.*

Margin area As in any data-transmission system, retrieval of clock information, and locking of phase-locked-loop (PLL) circuitry is important. The margin area is used to achieve PLL pull-in and to smooth the transition from one track to the next one. PLL pull-in means that the clock of the DAT circuitry which has to handle the incoming data will be matched (phase-locked) to the clock component of the incoming signal. In this way variations of the incoming data clock, or momentary loss of incoming clock signals (which happens during transitions from one track to another) will not cause the system to lose control over correct timing data.

Subcode area 1 and 2 In these areas the specific subcodes are written. These subcodes could contain much more information than is actually used. At present it is mainly used for identification code information: this means that several identity codes can be included; these codes can be used to identify track numbers, to make track jumps with extreme precision, and to stop or to skip a sequence. These identification codes can be rewritten without changing the music (PCM) contents; that is why these subcode areas must be separated from the main music area.

IBG areas Interband gaps (IBGs) are separator areas, but they are also used like margin areas for PLL purposes. These IBG areas are needed as buffers between re-recordable subcode areas, tracking (ATF) areas and main areas to maintain the distinction between these areas. In the case of re-recording the subcodes, avoid overlapped recording or partial erasure.

ATF areas The automatic track finding (ATF) areas are represented as one area, but they contain different frequencies; in them the information is a sequence of various analogue frequencies. The ATF will enable the system to check if the heads are correctly positioned over the data track. Basically the system works as follows: whenever triggered by one of the specific frequencies in the ATF zone the system will start looking for other specific frequencies which are at that moment on the left and right adjacent tracks. By making a level comparison the system can evaluate the head's position. Equal levels mean that the head is centred correctly on the current track and unequal levels mean that the head is off centre; in this case it is also possible to find out to which side it is off-centred. Based on this information the tape speed can be adjusted and the correct head location ensured. It has been mentioned that the special azimuth system is used to prevent cross-talk between adjacent tracks, but here it is the small part of residual cross-talk that is used for the ATF process.

Main area In this area the actual audio signal is recorded and is, of course, the most important area. Each track contains a well defined and specified number of blocks, each block containing some sync, identity coding address and PCM data (digitized music) which has been encoded for error correction.

Some more details about the track format are shown in *Figure 3.94*. Note that the subcode area also contains a PLL pull-in area and a post-amble. This allows for the re-recording of this part of the track.

3.4.4.3 PCM block format

The main data area is the centre area of a track. This area contains 128 blocks. Each block contains 288 bits. The first 8 bits are sync bits. The following 16 bits contain main identification coding; 8 bits of parity are then inserted which are based on the main identification bits. In this way some error detection is possible in the identification area. The largest part of the main data area is the digital music area: 32 symbols of 8 bits each are included in each area.

Main identification codes Sync, parity and audio data are fairly self-evident but the main identification area is somewhat complex. In order to have a full block of main identification coding, eight consecutive blocks have to be read. A group of eight consecutive blocks, each having 16 bits of main identification (ID) coding, can give the full information as shown in *Figure 3.95*. The first block contains the first 16 bits: format ID, ID1, frame address and block address. The second block contains eight unused bits

1	2	3	4	5	6	7	8	9 (PCM)	10	11	12	13	14	15	16

		Frequency	° Angle (deg)	Number of blocks	Time (µs)
1	MARGIN	1/2 fch	5.051	11	420.9
2	PLL (SUB)	1/2 fch	0.918	2	76.5
3	SUB-1		3.673	8	306.1
4	POST AMBLE	1/2 fch	0.459	1	38.3
5	IBG	1/6 fch	1.378	3	114.8
6	ATF		2.296	5	191.3
7	IBG	1/6 fch	1.378	3	114.8
8	PLL (PCM)	1/2 fch	0.918	2	76.5
9	PCM		58.776	128	4898.0
10	IBG	1/6 fch	1.378	3	114.8
11	ATF		2.296	5	191.3
12	IBG	1/6 fch	1.378	3	114.8
13	PLL (SUB)	1/2 fch	0.918	2	76.5
14	SUB-2		3.673	8	306.1
15	POST AMBLE	1/2 fch	0.459	1	38.3
16	MARGIN	1/2 fch	5.051	11	420.9
	Total		90	196	7500

Recording density 61.0 KB/I ° Values for 30s, 90° wrap angle, 2000 rpm cylinder

fch 9.408 MHz

Figure 3.94 *Track format details. fch is the basic channel frequency. All the frequencies used in each track have a basic mathematical relationship to fch, in order to maintain clock consistency.*

and a block address. The third block contains ID2, ID3, frame address and block address. In this way each consecutive block contains part of the information. Note that part of the data is noted as 'not used'. The bits recorded here are random, and they are disregarded by the actual DAT players.

The identification codes indicate several items: use, emphasis, sampling rate, etc. (see *Figure 3.95*). Note also that some of these items can be repeated by the identification holes in the DAT cassette.

The frame address requires some explanation. Up to now only tracks have been mentioned; a frame is

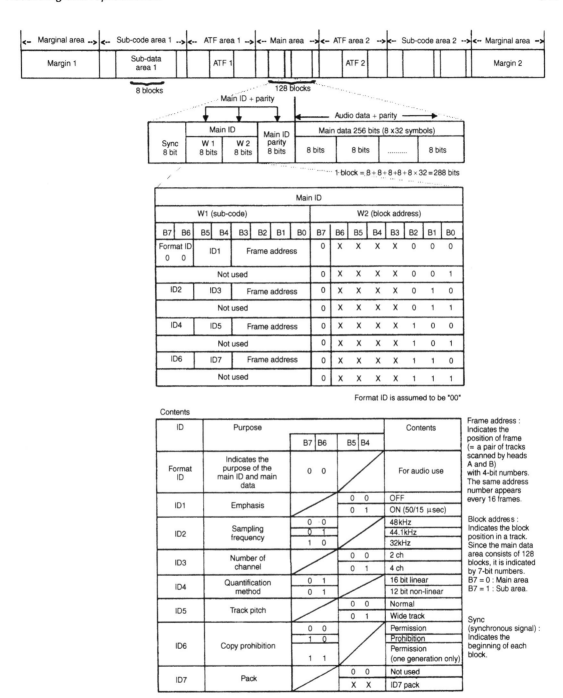

Figure 3.95 *The main area data contents.*

a combination of two adjacent tracks. One of the error-avoiding methods is to interleave the audio data, as in CD technology. Since the interleave in DAT is performed over two tracks it is normal that we consider two tracks which are related via the interleave as one unit, called a *frame*. The frame address, however, has four bits. This means that there are 16 possible frame addresses, which will be given sequentially. The block address gives the address of the particular block within the track.

The first bit (b7) designates whether this is sub-code data or main data. The next 7 bits give the block address. Seven bits give a possibility of 128 addresses, which is just enough as each track has exactly 128 blocks.

3.4.4.4 Subcode format

The subcode information is written in the subcode areas 1 and 2, but the format is very similar to the information in the main area. One subcode block contains 288 bits. The first 8 bits are synchronizing bits, the next 16 bits are identification codes; 8 parity bits based on the subidentification bits are then inserted for error detection. There are 256 bits for subcode data (or 32 blocks of 8 bits). Each subcode area contains 8 subcode blocks. The subcode areas 1 and 2 are identical.

Subidentification codes Like the main identification codes, the subidentification codes require some further explanation. For full subidentification coding two consecutive blocks are needed. One subcode area contains eight blocks. Consequently there are in each subcode area four full subidentification combinations.

In subidentification the first 4 bits contain control identification to designate the presence of a table of contents (TOC) but also to indicate skip, start and priority. These identification codes are the ones that can be used to control programmed start, stop and skip functions; remember also that this is important for the re-recordability of the subarea, especially for the identification codes. The data identification will usually be a fixed code as this designates the use.

Programme number zones contain programme number information and lead-in/out information. Pack identification tells how many packs are used. One pack is a byte (8 bits) in the subdata area. Bit 7 of the second byte will also be set to '1'. In the main data area the same bit will always be set to '0'. In this way the distinction between the subarea and the main area can be made. The block address zone takes 4 bits, but only one bit is really needed.

Subdata area The subdata area contains several packs, each pack containing 8 bits. For complete information see *Figure 3.96*. Note, however, that part of the pack area is not used. Note also that there is a link between the subidentification area and the pack area as the subidentification area says whether there is a TOC or not. If there is a TOC, it is given in the pack area. The subidentification area tells how many packs there are and the subdata area gives the packs. Similar to the coding in CD technology, the

pack area can give programme time, absolute time and running time, all accurate to a frame. The last pack gives a parity code based upon the previous pack contents for error detection.

3.4.5 Error encoding and correction

It is obvious that a powerful error strategy is needed. Two main strategies are used, similar to the CD signal processing. The first strategy is to add information which is calculated from the original digital audio data; the calculated data are inserted in the data stream and if some data are lost or incorrect, the calculated data will enable the microprocessor to recover the lost data. This calculation is performed with a certain polynominal. This method is called a 'double-encoded Reed–Solomon code'. The Reed–Solomon codes produce C1 (32, 28) and C2 (32, 26) parity symbols.

The second strategy is called 'interleaving'. The data stream is already made up of small blocks, but normally this data stream is in a continuous, logical sequence. By mixing the data blocks or in other words, by putting them in a non-logical but known sequence, burst errors, etc., can be recovered more easily. There are two types of interleave: the C1 and C2 parity symbols are interleaved within blocks and within one entire track, and the PCM data themselves are interleaved between two tracks (one frame) (*Figures 3.97* and *3.98*).

One track contains 128 blocks consisting of 4096 (32 × 128) symbols. Of these, 1184 symbols (512 symbols, C1 parity and 672 symbols, C2 parity) are used for error correction, leaving 2912 data symbols (24 × 104). In fact, C1 encoding adds four symbols of parity to every 28 PCM data symbols, C2 adds six symbols of parity to every 26 PCM data symbols. The above figures give the data allocation for the C1/C2 interleave and PCM data interleave. It is obvious that this Reed–Solomon processing and interleave processing requires powerful microprocessors, but it also requires a certain amount of memory space. As one frame contains 64 kbit (128 × 32) symbols × 8 bit × 2 tracks = 65 536 bits the minimum memory space is therefore 128 kbit. *Figure 3.99* shows how this memory space is used during recording and replay.

3.4.6 Automatic track finding processing

The automatic track finding (ATF) areas have already been noted and briefly explained. It is, however,

Figure 3.96 *Subcode area contents. TOC, table of contents; ATF, automatic track finding.*

necessary to give a more complete idea of this part of the track format. Each track contains at top and bottom an ATF area. These contain a sequence of frequencies, depending on whether the track has an odd or even frame address (remember that one frame has two tracks).

f1:	fch/72 (pilot)	130.36 kHz
f2:	fch/18 (sync1)	522.67 kHz
f3:	fch/12 (sync2)	784.00 kHz
f4:	fch/6 (erase)	156.8 kHz
f1–f4:	duty cycle 50%	

In the above, fch is the basic channel frequency (see legend to *Figure 3.94*).

Figure 3.100 shows the frequencies in the ATF areas and gives an indication of their lengths. It can be seen that the pattern will be different, depending on whether the track is odd or even numbered. When following the head movement over the track, note that the head will see first either sync1 (f2) or sync2 (f3) in the ATF1 and ATF2 areas. Upon detection of one of these sync signals the system will check the pilot signal (f_1) on the two adjacent tracks (right

Figure 3.97 C1/C2 code insertion.

and left) sequentially; after that the pilot signal on the current track is also traced and in this way the processor can make a level comparison.

Figure 3.101 gives a more detailed example. Suppose the A head is on the middle track.

First the head detects the f_2 sync. This triggers the comparison sequence. First the f_1 signal on the right side is detected and secondly the f_1 signal on the left side. Hence the f_1 signal on the middle track itself will be detected. If the two adjacent f_1 levels are equal, the head is centred on the track. If the right-hand f_1 signal is higher than the left-hand f_1 the tape is running too slowly as the head is already partly on the right track and therefore the tape speed

has to increase. If the f_1 coming from the left side is higher than the f_1 coming from the right side the tape is running too fast and the head is off-centred. The system can also define which track (odd or even) it is scanning currently by seeing if the sync is f_2 or f_3.

The azimuth system has been explained above. This system is included to prevent intertrack cross-talk. The ATF system seems contradictory with the azimuth system. That is one of the reasons for the relatively low f_1–f_4 frequencies. Azimuth systems avoid high-frequency cross-talk, but even with a large azimuth angle ($\pm 20°$) lower frequencies still create cross-talk. The f_1–f_4 frequencies are low

Figure 3.98 *Data interleave format.*

Figure 3.99 *RAM organization. As interleaving is completed in two tracks, one drum rotation of 30 ms (minimum) is required before a playback sound is obtained after reading the radiofrequency signal.*

frequencies, but they are also the only analogue signals on the DAT track. They do not contain any data.

Note that there are already existing systems (like DTX-10) which do not use the tracking system. These are called 'non-tracking systems' (NT systems). The principle is that even if a head does not stay correctly on the track it will read part of the information on that track anyway. If the head

goes over the same track several times it is possible to reconstruct the full information. For this purpose these systems can use a drum rotating at, for example, 4000 r.p.m. instead of 2000 r.p.m. In this case the PCM information will be read twice and, even if this should be partly unprocessable, the combination of the two readouts can provide a correct audio output. This system of course requires more memory space, but that is no problem nowadays.

Figure 3.100 *Automatic-track-finding (AFT) frequencies. IBG, interband gap.*

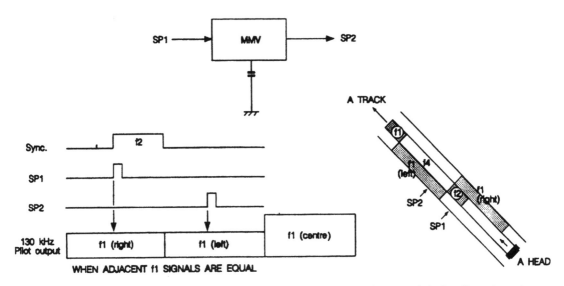

Figure 3.101 *Automatic-track-finding (ATF) switching system. When adjacent f₁ signals are equal, the head is on the track.*

3.4.7 Basic block diagram

Until now the DAT system has been looked at in small sections. With the basic knowledge thus gained one can easily grasp the operation of a normal DAT recorder/player. *Figure 3.102* gives an impression of the integration scale that was used in the first-generation players (DTC-1000) and the much higher integration density in the current generation (DTC-55). The large blocks represent one IC in the latest versions, whereas the small blocks were separate ICs in the previous generation of DAT machines. We can distinguish two main parts in the block diagram, one audio-related part and one mechanical-related part; of course, both are linked as there is an indication between mechanics and data. The main central processing unit (CPU) is also embedded in the mechanical processor. Communication between the main blocks is in the form of standard computer-like signals. The audio-related block is composed of an input analogue-to-digital converter/output digital-to-analogue converter, a PCM processor integrated circuit, RAM, a radiofrequency amplifier, an ATF IC and the link to the heads in the drum assembly. The mechanical-related block comprises several motors: drum motor, capstan motor, reel motor(s) for the real record/playback actions, but other motors are included for cassette and tape handling (cassette eject, tape threading/unthreading, etc.). Apart from these motors there is mainly one processor,

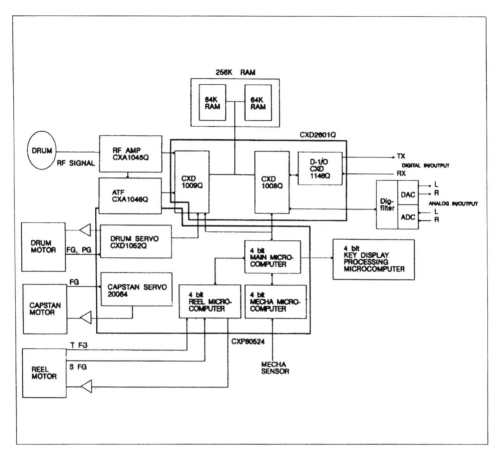

Figure 3.102 *Block diagram of a DAT system.*

which includes also the main CPU. Inputs to this processor are from sensors which monitor motors, cassette position, cassette compartment condition, dew detection, etc., as well as key inputs from the front panel.

3.4.7.1 Audio-related block diagram

In the DAT format it is of course possible to have digital audio input/output. Therefore apart from the analogue input/output there is also digital input/output, which is connected directly to the PCM processor as this has the necessary circuitry for such signals.

Consider an analogue line input; this is analogue-to-digital converted and filtered digitally, then the digitized audio input is fed into the PCM processor. This performs proper modulation, C1/C2 encoding and interleaving and carries out all the necessary processing to put the proper track format on the tape. When the correct track format

is obtained the data are transferred to the drum assembly through the radiofrequency amplifier. During audio processing, the RAM area is continuously used. Note that in the diagram the RAM is a 256 K RAM instead of the minimum 128 K. This is needed for improved error correction. Likewise, digital data retrieved from the tape during playback are demodulated, error decoded and de-interleaved in the PCM processor, which of course also uses the RAM area during this processing. Digital output audio is then fed to the digital filter, digital-to-analogue converter, de-emphasized if necessary, and sent to the line output.

The digital input and output are connected directly to the PCM processor. The transmission format is the S/PDIF format which is generally used for coax/infrared transmission of digital audio. This format was jointly defined by Sony and Philips and is currently becoming a world-wide standard format. The PCM processor is also linked to the

system controller (included in the mechanical IC). Part of the digital audio output is connected to a meter section for display purposes. Based on the information retrieved from the PCM processor the mechanical part is operated through the system controller. (See *Figure 3.103.*)

3.4.7.2 Mechanical related diagram

Part of the mechanical assembly has been explained previously. *Figure 3.104* gives some specific details on servo processing, the most important motors such as drum motor and capstan motor being driven through servo circuits. The drum assembly has built-in frequency generator (FG) and phase generator (PG) feedbacks. These give a frequency/phase feedback related to the speed. The combination of FG and PG can detect that a head is at the beginning of the track area.

The mechanical processor calculates an error signal based on the FG and PG input signals together with internal reference and a reference derived from the digital signal read-out by the heads and processed in the PCM processor. This error signal will be low-passed as it is a pulse width modulated (PWM) signal and this is the driving signal for the drum motor. The AGC circuit is needed to control gains with respect to the ATF section. The capstan is driven likewise but, as noted before, capstan driving is also performed partly through the ATF section. This can also be seen in the diagram.

The detailed block diagram of a home deck player (*Figure 3.105*) is an example of a possible mechanical assembly. There are sensors for cassette table in/out, cassette in, thread limit, tape end (take up and supply side) and reel motion. In this case there is only one motor for take-up and supply motion. Therefore, a reel solenoid is included to lock and unlock the appropriate mechanism. The drum motor and capstan motor have been mentioned earlier. There is of

Figure 3.103 *Block diagram of a DAT system.*

Figure 3.104 *Block diagram of servo processing. FG, frequency generator; PG, phase generator; SWP, switching pulse.*

course also a cassette compartment motor, controlled by the sensors.

The part not mentioned so far is the control motor and rotary encoder. As mentioned before, the tape is taken out of the cassette and wrapped around the drum. The pinch roller has to squeeze the tape against the capstan. This implies several mechanical movements performed by guides, and the reverse movements also when stopping and unloading. In order to achieve these mechanical movements a control motor

is needed which will drive the mechanical parts described earlier. There is also a need to know the mechanical position of these parts. Therefore a rotary encoder is used. This is a mechanical control unit which monitors the position of the mechanical parts and sends feedback signals according to the detected situation. Depending on the feedback from the rotary encoder the system knows the status of the mechanical assembly at any given moment. Based on this information, mechanical control can be carried out.

Figure 3.105 *Block diagram of the mechanical aspect of a home DAT player.*

3.5 Audio editing

3.5.1 Introduction

It is likely that only moments elapsed between listening to the first recording of a performance and a request from the producer 'to remove a cough from between two words'. Straight away the requirement for editing was born! The longevity of analogue tape as a recording and editing medium must be due, in part, to the ease and elegance which it lends to the editing process. Using the most rudimentary of tools, analogue cut-and-splice editing is ideally suited to the requirements of the majority of speech-based editing tasks. Initially, digital, nondestructive editing was only feasible for the most lavish of projects due to its use of sophisticated and expensive hardware, derived from video recording technology. Although the costs involved in performing edits in the digital domain have fallen, it is the

flexibility of these edits which far exceed anything achievable by physical cuts that has won over the analogue die-hards to digital.

It is interesting to note the cross-fertilization between the audio, video and computer industries. Early digital audio recording and editing relied on video-based hardware. More recently, digital-audio-tape recorders are using the same *azimuth recording* techniques as *D2 video tape-recorders* (VTRs). Today's manufacturers of tapeless recording and editing systems crave high capacity, fast access, low-cost storage just as much as the computer industry.

By far the greatest volume of edits created is in the production of material for broadcast. The cost of editing technology meant that for several decades performers and politicians were saved from their own lack of vocal dexterity by the humble razor blade. Today the performance of personal computers has risen dramatically, and their price has fallen to below the cost of the cheapest professional reel-to-reel tape machine. The physical destruction of the

recording medium to modify a performance is on the wane.

This section describes the evolution of techniques and equipment for editing recorded sound.

3.5.2 Analogue disk editing

The ability to edit recordings predates the popularity of magnetic tape recording. In the 1930s the BBC was using direct cut lacquer disks as a means of recording and repeating programmes. These 13-in. disks, rotating at 78 r.p.m. gave a usable playing time of only 5 min. In order to reproduce programmes of longer durations, operators had to be adept at playing consecutive disks to achieve a seamless join. A turntable designed to ease this task had a parallel tracking arm, with a vernier scale for precise groove location. The audio output was fed through a fader, which contained a switch to start the turntable motor as the fader was opened. To perform an edit it was necessary to note the vernier reading for the last piece of audio prior to the edit point on one turntable, and set the vernier to the start of the first piece of audio to be joined on a second. As the first turntable approached the edit point, the two faders would be manipulated, simultaneously starting and cross-fading from the old to the new material.

To accommodate edits that might lie within a recording which resided on one disk, it was common practice to make two simultaneous recordings of the original performance. The BBC TD/7 reproducer, of around 1935 vintage, which included twin turntables and a mixer with remote groove indication and pick-up lowering, provided within one unit all the elements of the most sophisticated of today's hard disk digital editors. These elements are: random access to source material, non-destructive edits, accurate sample location and variable cross-fades.

3.5.3 Analogue tape editing

The editing of recordings made on analogue tape is a process which is destructive to the recording. This risk has to be balanced against the operational requirements of the programme maker which may preclude the use of a duplicate tape due to pressure of time and the reduction in quality which occurs whenever a copy is made in the analogue domain. Most analogue editing takes place on mono or stereo material held on either $\frac{1}{4}$-in. or $\frac{1}{2}$-in. tape. However, the same techniques are applicable across the entire range of formats.

The editing of analogue, $\frac{1}{4}$-in. tape can be performed on almost any tape machine, although one which readily allows the spool brakes to be released whilst still applying tension through the motors is preferred. A splicing block, a soft marking pencil, a razor blade and some suitable adhesive tape are the only additional tools required.

The tape is first played normally until it reaches within a second or so of the proposed edit point. The transport is then stopped and, with the brakes released, the feed and take-up spools are rotated manually, drawing the tape back and forth across the replay head to locate the edit precisely. This process is referred to as *reel-rocking* or *scrubbing*.

In speech editing it is easiest to locate the gaps between words or percussive syllables such as Ps, Bs or Ds. In music, drum-beats are an easy and reliable reference. The lack of strong rhythm tracks in classical music has defeated many an operator who would consider himself or herself accomplished in the art of editing pop music!

Having located the first edit point, a vertical mark is placed on the back of the tape at the point where it rests on the replay head. The tape is then placed in a splicing block, which is a jig designed to hold it firmly and allow an accurate cut to be made across its width at the mark. The unwanted portion of tape is now pulled through by hand, or wound onto another spool if long distances occur between wanted 'takes'. The next edit point is again located by reel-rocking, then marked and cut in the block. The two wanted portions are next butted together and joined using adhesive tape. The adhesive is specially designed to provide a strong mechanical joint, whilst at the same time permitting the edit to be peeled apart if necessary. The tape can now be laced back on the machine and the resulting edit auditioned. (See *Figure 3.106.*)

There are few possible refinements to this method, although they do exist in the form of marking pens attached to the tape deck and mechanically operated scissors which spring forth at the press of a button. Most operators prefer the simpler approach and can make edits at a remarkable speed.

It is worth examining the physical nature of the cut produced by a splicing block. At first one might assume that the cut would be vertical; however; such a cut is rarely used since it sounds unnatural and can cause mechanical problems such as fouling on guides around the tape deck or causing the tape to lift off the heads at the edit point.

In the early days of mono, when a recording was stored across the entire width of the tape, an angled

Figure 3.106 *A Studer headblock and scissors. This machine is fitted with an on-board tape marker and scissors (located just behind the pinch wheel). The more traditional tools are also close at hand.*

cut of 45° was adopted. At the common professional tape speeds of 19 and 38 cm/s this resulted in a cross-fade at the edit point of 33 and 16.5 ms, respectively. Such a cross-fade was considered to be acoustically more pleasing than the transition produced by a vertical cut. (See *Figure 3.107.*)

A mono splice with 45 cut

A stereo splice with 60 cut

Figure 3.107 *The effects of an angled cut on the audio transition at an edit: 45° and 60° splices.*

The situation arises with stereo recordings, where the left and right channels each occupy a little under half the tape width, that a 45° cut will still produce a cross-fade. However, it will be slightly delayed on the left channel with respect to the right. This will be most noticeable if the edit takes place on a sibilant

sound and has the effect of causing such a sound to 'jump' into its correct stereo position across the edit. The effect is more noticeable at slower tape speeds and is reduced by performing cuts at an angle of 60° to the horizontal. At 38 cm/s such a cut results in a cross-fade duration of 10 ms and a left/right offset of only 5 ms.

If an edit proves artistically unsatisfactory, it is possible to unpick it and try again at different points. A practical limit to this is enforced either by the physical proximity of splices or by the loss of a particularly vital piece of the recording amongst the debris on the floor!

Whilst it is possible to splice pieces of multitrack tape together, the problem of the edit occurring during different portions of the material is obviously exacerbated across the width of a piece of 2-in. tape and must therefore be performed with caution.

A notable example of multitrack splicing is the 1970s song 'I'm not in love' by the British group 10cc. In the making of this record sections of multitrack tape were spliced to form continuous loops. Each track contained a lush, multivocal note. The backing of the song was achieved by feeding the outputs of the tape machine through a mixing console and playing the tune on the faders of the desk, each fader producing a separate note.

3.5.4 Electronic digital tape editing

The advent of digital audio recording saw a new approach to editing, modelled on the techniques used to edit pictures recorded on videotape. The large signal bandwidth produced by transforming an analogue signal into a stream of digital samples led to the use, early on, of helical-scan videotape machines as the recording device. A PCM adaptor took the analogue audio signal, and converted it into a pseudo-video waveform, called a *composite digital signal*, which was suitable for recording by a conventional VTR. In the interests of the international exchange of material it was decided that such recordings would be based on the American NTSC video standards with a 60-Hz frame rate. The VTRs used are industrial-grade U-Matics and the format is commonly known as Sony 1610 or 1630 format. It is one of that rare breed, a true industry standard, and its prime use is in compact-disc mastering.

The use of VTRs for recording implies that the same techniques used for editing video could be applied to digital audio. Whilst this is basically true there is a little more to the matter!

1630-type editor

Figure 3.108 *An edit system for composite digital recordings: a 1630-type editor.*

The editing of video tapes is a copying process. The required portions from a collection of source tapes are transferred, in the required order, onto an initially 'blank' master tape. The timing and synchronization requirements of professional VTRs mean that SMPTE *time-code*, recorded on a separate analogue track, is used to define edit points and position machines during editing. (See *Figure 3.108*.)

A VTR-based digital audio editor has requirements over and above those of normal video edit controllers. For example, although accurate machine synchronization is already catered for in a video editor, there is the restriction that VTRs will only enter or exit record mode at video frame boundaries, approximately every 33 ms in the NTSC standard. This is not a suitably fine tolerance for audio where edits should, ideally, be achievable to sample accuracy. In any case, it is not possible to simply join two different sets of samples together, as any discontinuity at the join will produce an audible click. With regard to edit point location, the tape transport of a helical scan VTR can be jogged (incremented one frame at a time) to identify a picture edit. Jogging a tape containing a composite digital audio signal will not produce bursts of audio yet reel-rocking is an accepted way of identifying edit points.

The importance of a cross-fade in the join between two pieces of edited audio has already been mentioned. In analogue tape the duration of this is limited by the effect of the offset between edit timing on the two stereo channels. Many editors feel constrained by this and crave the flexibility of longer cross-fades without the penalty of left/right offsets. (See *Figure 3.109*.)

Figure 3.109 *The Sony DAE-3000 digital audio editor.*

Happily the above limitations and requirements are satisfied by a common mechanism. Within the edit controller there is an area of random-access memory (RAM) which fulfils a number of functions during the editing process. Whilst playing audio from tape the RAM is constantly loaded and reloaded. When a button is pressed to identify an approximate edit point, audio continues being written into memory for a period equal to half the total RAM capacity. If for example there are 12 s of RAM then having marked an edit point there will be 6 s of audio before and after this mark in RAM. The operator is provided with a wheel which controls the replay of audio from RAM and this gives a tactile and aural simulation of reel-rocking. The edit controller keeps

track of sample timing as the position of the edit is adjusted. A further area of RAM is needed to hold the audio relating to the edit point of the second portion of audio which is to be joined to the first. With both halves of the edit available in memory it is easy to trim and audition the join many times in quick succession, since there is no need for the VTRs to be prerolled and synchronized. As well as being able to define timing of the edit it is also possible to adjust the cross-fade duration asymmetrically (typically between 1 and 999 ms) and control level and stereo offsets. (See *Figure 3.110*.)

Figure 3.110 *A close-up of the DAE-3000 display, showing the cross-fade parameters.*

During an edit pass the machines are synchronized and the record VTR enters 'record' ahead of the edit, at a point defined by the video frame rate. From this point, across the edit, to the next suitable video frame boundary, audio will be sourced from RAM. The audio will then continue to be transferred from the play VTR up to the next edit point.

Early examples of this sort of editor were introduced at a time when RAM was extremely expensive. In order to reduce costs, all audio used for rehearsing edits was in mono and at reduced bandwidth. More recent systems have no such limitation.

Although the system described above provides both technical and creative benefits over physical tape editing, its detractors felt that the need to copy sections of audio from one tape to another slowed down an editing session. They also pointed out that if a new section were to be inserted part way through an assembly, then all edits downstream of that point would have to be remade. Thus analogue cut and splice editing continued to thrive and the digital research and development departments continued to burn the midnight oil to produce a system which recorded digitally, but was capable of being cut for editing.

3.5.5 Physical digital tape editing

The *DASH* and *Pro-Digi* open-reel formats were devised to replicate typical analogue-tape operations,

particularly with respect to editing. Both systems are available in two track form and will support physical editing by making use of powerful error correction and concealment within the tape machine. The formats have a number of digital tracks which carry the audio signals and synchronizing information as well as restricted quality analogue tracks for use in edit point location.

Having sampled audio into a digital signal, odd and even samples are recorded on separate tracks which are offset relative to one another. At a splice point there will inevitably be a disturbance in the signal being read off the tape. Due to the odd/even sample offset this disturbance produces a period of good odd-only or even-only samples. The audio is reconstructed by interpolating the available good samples. Unfortunately, such a technique, which relies on *concealing* the error, can produce the noticeable side-effect of a reduction in audio bandwidth and *aliasing* of high frequencies.

Twin-DASH is a more robust format (*Figure 3.111*) which allows for true error *correction* at edit points. It provides two-channel recording on $\frac{1}{4}$-in. tape, running at 38 cm/s and uses four digital tracks per audio channel. Again odd and even samples are allocated to individual tracks and in addition the odd/even interleave is inverted between adjacent tracks. At

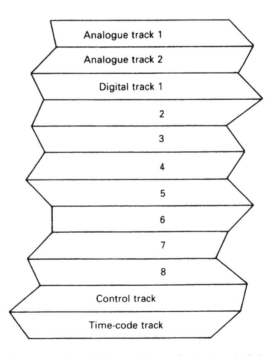

Figure 3.111 *Twin-DASH track format. Each channel of the stereo audio signal occupies four of the digital tracks.*

a splice point there are always good odd and even samples available, thus no interpolation need take place.

Edit points are located using the conventional analogue method of reel-rocking whilst monitoring the low-quality analogue tracks recorded alongside the digital information. For cutting and joining the tape a special jig which is similar to, but more sophisticated than, a conventional analogue splicing block is recommended. Unlike analogue practice, the cut is vertical, and a physical gap of about 0.5 mm, but no greater than 1 mm must be left between portions of tape.

When compared with analogue editing there is a need for increased cleanliness when handling the tape. The ability to 'unpick' an edit and try again, whilst not impossible, is reduced. These facts, coupled with the high capital cost of Twin-DASH machines, continue to restrict the penetration of the format as a replacement for analogue editing.

Electronic copy-editing of DASH tapes is also possible. Some machines incorporate an edit panel, which uses time-code for edit point location and is capable of controlling two machines, with RAM for edit point rehearsal and trimming. This arrangement does little to bring digital editing in line with analogue costs.

3.5.6 Jump editing

A relatively new format, aimed at producing digital recordings which may be edited on a single machine utilizes a technique known as 'jump editing'.

Kudelski, manufacturers of the renowned Nagra tape-recorders, have outlined a technique in which no attempt is made to correct errors in the area of an edit. Instead an electronic pointer on an auxiliary track of the tape signals a forthcoming edit. This causes the linear tape speed to increase, loading advanced audio samples into a buffer, which continues to output audio at true speed. During this period a physical splice takes place and another pointer indicates the join between the presplice and postsplice audio. Given a sufficiently large buffer, certain types of edit, involving the elimination of, say, a cough or stutter would not require removal of any tape at all.

As it stands, the format places some restrictions on the proximity of edits. It is intended for use on a helical scan tape transport which, in turn, puts mechanical complexities on any physical joining of tape that may be involved.

3.5.7 Editing in the DAT format

The acceptance of digital audio tape (DAT) into professional circles was probably eased by the incursion, some 10 years earlier, of the consumer *F1* machines which recorded a *PCM*, pseudo-video signal and were widely used as a field acquisition device.

Determined manufacturers have shoe-horned the additional features such as off-tape confidence monitoring and SMPTE/EBU time-code necessary for professional acceptance into this tiny format. A machine with these features is capable of electronic copy-editing when coupled with suitable controlling hardware.

DAT recording benefits from the technique of azimuth recording. This makes it much simpler to produce an electronic edit on a DAT tape than on a video tape using conventional helical scan recording methods which necessitate a system of 'flying' erase heads on the spinning head drum. The DAT head drum has four heads located at 90° intervals. These may be used in a read-after-write mode for off-tape confidence check during recording, or write-after-read during editing. The signal path in write-after-read takes the off-tape audio through a cross-fade circuit where it can be combined with new incoming audio and routed back to the record heads. Simple editing with limited cross-fades (around 10 ms) can be achieved by linking two machines to a synchronizer and control interface. Timecode, resolved to sub-frame accuracy is used to identify edit points.

Unfortunately, without the provision of large amounts of RAM it is not possible to provide the reel-rocking feature conventionally used for locating edit points. Edits have to be marked 'on the fly', i.e. identified whilst the audio is playing. This method is surprisingly accurate when working on rhythmic musical material. Edit points can be previewed and trimmed, although each preview will require both player and recorder to be rolled in sync which is not as fast as a RAM-based method.

Some DAT machines can be interfaced to the helical scan video editing system described earlier.

3.5.8 Hard disk editing

As already stated, the introduction of digital audio saw the need to record what was considered at the time to be a large-bandwidth signal. When viewed purely as digital data the volume is still huge. A swift calculation shows that for a sampling rate of 44.1 kHz, with 16-bit linear quantization, roughly 7 Mbyte of storage is required to record 1 min of

mono audio. Fortunately, the computer industry was also hungry for large-volume storage with fast access time, and thus in the mid-1980s recording and editing of audio on Winchester and, more recently, optical disks became a reality.

When recording audio onto a hard disk it is worth remembering that, although the system may have been set to record continuously for, say, 5 min, the resulting data may well occupy non-contiguous blocks across the surface of the disk. This means that in order simply to replay the recording, the heads will be continually seeking data blocks, possibly located at extreme regions of the disk surface. Even if the blocks were located adjacent to one another the data would still appear in bursts. To produce a constant flow of audio, RAM buffers are placed between the disk and the system digital-to-analogue converters (DACs). Provided these buffers have a capacity which exceeds the worst-case disk access time all will be well.

In order to produce an edit it is necessary to instruct the system to collect and output data in a different order to that in which it was recorded. *Figure 3.112* shows an example of a simple edit.

This is all very well if wanted and unwanted portions of audio sit neatly within block boundaries. To achieve editing resolution to sample accuracy it will be necessary partly to empty one buffer and then switch to another. Again the importance of cross-fades to editing means that the buffers are fed not through a switch, but through a variable cross-fade device.

To be readily acceptable to audio editors the hard disk system should be capable of reel-rocking emulation when identifying edit points. Providing this whilst avoiding the use of large areas of RAM requires the system to control disk access in a way that ensures the buffers are kept topped up as commands to play audio at non-linear speeds, both forwards and backwards, are produced by the operator. The tasks of reel-rocking emulation and complex cross-fades are achieved using high speed,

high performance digital signal processor (DSP) technology. (See *Figure 3.113*.)

Some manufacturers provide an on-screen display of the audio waveform as the method for locating edit points instead of reel-rocking. Such systems require edits to be trimmed whilst listening to audio playback and watching a cursor cross the screen.

Hard disk editors have all the advantages of the electronic, copy-editing machines already described in that edits are non-destructive and can be auditioned and adjusted minutely in a manner impossible with physical splicing. They also overcome the two main disadvantages of tape-based electronic editing. Firstly, since no copying takes place to perform an edit, it is possible to proceed from one edit to the next without the need to re-record the intervening section. This makes it possible to assemble a programme in a period of time which is less than its linear duration. Secondly, if it is decided to extend a programme, material can be inserted at any point without the need to remake subsequent edits.

The data which make up the edit decisions themselves, timing, level and cross-fade information can be saved as files on a floppy disk and alternate versions of the same material readily constructed and compared, earning the equipment the soubriquet the 'What if?' machine!

The use of DSP technology allows the hard disk system to provide another 'impossible' feature, i.e. time compression or expansion *without* pitch change. This powerful tool can be used to match the tempo of two dissimilar recordings of a piece of music. It can also be employed to great effect when making commercials, allowing an advertiser to squeeze yet one more price into his precious 30 s of airtime!

Hard disk editing systems fall broadly into two categories; multitrack systems for use in motion-picture and music production and stereo-capable units for music editing.

Having already seen that electronic editing surpasses the physical method by virtue of its

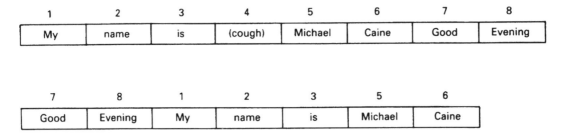

Figure 3.112 *Editing by means of rearrangement of block replay order.*

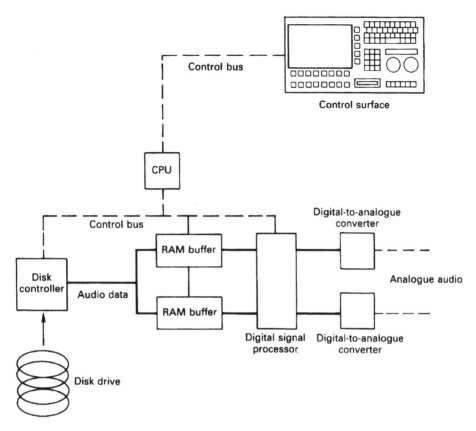

Figure 3.113 *A conceptual view of a hard disk editor (replay only). CPU, central processing unit; RAM, random-access memory.*

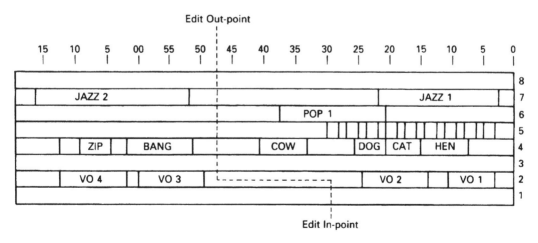

Figure 3.114 *A Z-shaped edit.*

ability to produce non-symmetrical cross-fades with independent overlaps, some multitrack systems go a step further by permitting what might be thought of as a 'Z-shaped edit' (*Figure 3.114*). Operators may specify an edit which starts at one time address on

the 'multitrack tape' but which finishes at another, whilst retaining the relative synchronization of the elements either side of the join.

Figure 3.114 shows a view as presented to the operator of a multitrack hard disk editor. The named

blocks represent segments of audio that have been placed relative to one another on a virtual section of multitrack tape. An edit can be specified on one track (in this case, track 2) which will cause all audio downstream of the edit point to 'ripple', whilst maintaining the sync relationship between segments.

Whilst the challenge to manufacturers of producing a viable method of recording and recovering high-quality audio on hardware intended for use in computers cannot be ignored, many systems stand or fall by the way they display and control the varying parameters of an edit for the operator. The permutations of operator interface seem almost endless; touch screen, track-ball, screen with multifunction 'soft' keys, stylus, liquid-crystal-display (LCD) screen, windows and mouse.

The popularity of the Macintosh computer and the intuitive uniformity it lends to software developed to run on the machine has meant that a 'desktop audio' revolution has occurred in much the same way as for desktop publishing. Despite the apparent acceptance of this popular interface, many manufacturers believe that a machine based on their own proprietary hardware and software is better suited to the rigours of the professional recording environment and offers greater scope for product development by not being reliant on third parties.

At the start of the 1990s there were around 60 different hard disk editing systems commercially available. This figure is significantly higher than the total number of analogue tape machine manufacturers at the peak of popularity of that format.

As already noted it is the operator interface, rather than the technical performance which tends to separate the systems. Many manufacturers produce a variety of application-specific software for such varied tasks as feature film dialogue replacement, and automatic track-laying based on picture-defined edit lists.

Bibliography

BORWICK, J. (ed.), *Sound Recording Practice*, Oxford University Press, Oxford (1977)

RUMSEY, F., *The Audio Workstation Handbook*, Focal Press, Oxford (1996)

WATKINSON, J., *The Art of Digital Audio*, Focal Press, Oxford (1994)

3.6 Audio measurements and test equipment

3.6.1 Introduction

Audio measurements are made for many purposes. Sound level measurements to assess background noise, and measurements made on studios to assess reverberation times are better described as 'acoustic measurements' and will not be covered here. Other measurements fall into two classes: those made operationally, such as programme level measurement, and those made on equipment. Measurement is vital to every stage of equipment design, specification, marketing, and maintenance. It provides the designer with figures to improve on, the user with specifications to ensure compatibility and adequate performance, and the maintenance engineer with minimum performance standards to maintain.

Most measurements made on equipment are concerned with quantifying audio quality, or fidelity, and they generally employ artificial test signals rather than real music and speech which are too changeable to be of much use. Strictly, perhaps, they should be referred to as tests. The newcomer to the field of audio testing may be forgiven for wondering why it seems so complicated, with so many odd terms like 'wow & flutter', 'frequency response', 'headroom', etc. Why not just put a music signal into the equipment under test, subtract input from output in a differential amplifier, and specify the maximum instantaneous error in the signal waveform as a percentage? This method, revived from time to time in the popular audio press as a critical test of power amplifiers, is of very limited use. Accurate waveform preservation will certainly guarantee fidelity of reproduction, but gross changes of waveform do not necessarily imply audible impairment, and in the practical world of tape machines and lines we need to know which distortions matter and which can be ignored.

It is this link with psychoacoustics that makes audio testing such a complicated and fascinating discipline, and it must always be remembered that the purpose of testing is the minimization of subjective sound impairment, not the maximization of figures. The reader who glances at current magazines and sales literature will find there a confused world of specmanship coupled with outright hostility to measurement techniques. On the one hand, advertisers claim ultra-low levels of distortion which are quite irrelevant, while, on the other, reviewers

claim to hear differences between amplifiers that cannot be measured, favouring 'audition' and rejecting normal measurement techniques. Measurements can show degradations before they become audible, and this is important in broadcast and recording studio environments where a signal may have to pass through dozens of pieces of equipment: if each were only just good enough to pass a listening test then the final output could sound very bad indeed. The author has never heard any impairment of sound that could not be accounted for by the basic measurements to be described, and in the world of professional recording and broadcasting this view is widely upheld.

Even among professionals, however, we have a long way to go to achieve standard test methods worldwide, and the rapid pace of development in the digital era now upon us seems to have changed the rules. As in so many fields, Britain once set standards for the world to follow, and the BBC (British Broadcasting Corporation) and the IBA (Independent Broadcasting Authority) in particular were pioneers in the field of audio measurement for many decades, doing much of the work on which present international standards are based. Both organizations had very similar codes of practice up until the 1990s, when they underwent drastic change, becoming more answerable to business pressures and less a focus for standardization. The IBA in particular has been disbanded and its activities distributed among several other organizations (ITC, ITV, NTL). This is unfortunate because the IBA produced a code of practice for commercial radio and television which was thorough and realistic. Specifications are only useful if they can be compared, and all too often comparison is made impossible even for the expert, because different measurement standards have been adhered to or because something has be left out. A good core of generally applicable measuring techniques does exist, though, and it is to be hoped that they will not be forgotten in the rush for new technology.

3.6.2 Measurement techniques

The ear and brain analyse sounds largely in terms of their frequency components, with little regard for their phase relationship (except between channels in stereo, phase difference being a major directional cue). This explains why detailed waveform preservation is not important. Of the other forms of signal modification that occur, some are more tolerable than others, and some are of particular significance in

certain pieces of equipment, so they are measured and quoted individually.

The main parameters of importance in testing are: level, frequency response, noise, cross-talk, distortion, phase difference, and wow & flutter. Changes in level affect loudness, of course, while changes in *frequency response* have the same effect as tone controls or graphic equalizers, increased relative levels of higher-frequency components producing a 'brighter' sound, increased levels of low frequencies, more 'bass'. Most equipment adds *noise* to some degree. It is the natural result of random electronic processes in components, and generally sounds like a hiss, but because of the psychoacoustic phenomenon called 'masking' it is most prominent in quiet passages and so is best measured in the absence of signal. *Distortion*, on the other hand, refers to the addition of extra frequency components through the effect of a non-linear transfer function on the signal itself. Strictly speaking it should be referred to as 'waveform distortion' or 'non-linear distortion'. Some distortion products, especially low-order harmonically related ones, are masked much more effectively than others, and so different techniques have been devised to distinguish total harmonic distortion (THD), intermodulation distortion (IM) and quantizing distortion (QD). *Cross-talk* refers to the unwanted leakage of signals between channels in a stereo signal, leading to loss of stereo effect, or between different signal paths in a studio or mixer, while *wow & flutter* is peculiar to mechanical tape recorders and gramophones where small speed variations cause audible changes in pitch. While the term 'wow' originally referred to slow variations (0.1 to 1.0 Hz) and 'flutter' to rapid ones (4 to 300 Hz) they are now generally combined as a single weighted measurement and referred to as *wow & flutter*. For all these parameters, standards exist which lay down preferred methods of measurement. The most useful standards for current reference are those produced by the IEC (International Electrotechnical Commission) but what they contain is the best of the British (BS), German (DIN), French (CCIF) International (CCIR/CCITT/ITU) and American (ASA, ANSI and NAB) standards which have generally been adopted by the other countries. Increasingly, the separate national standards are being identified as literal copies of IEC standards, published locally only for convenience. The EBU (European Broadcasting Union) also publishes important standards. A list of standards relating to measuring techniques is provided at the end of this section, and the essential methods involved are described below.

3.6.2.1 Signal level

The measurement and specification of levels should be simple, and yet it is an area of appalling confusion. Signal levels are commonly measured in dBu, meaning decibels relative to a reference level of 0.775 Vrms. This odd figure represents the voltage needed across a 600 Ω load to dissipate 1 mW of power, and originated historically because most audio was once carried by matched 600 Ω transmission lines. Modern studio practice dispenses with matched lines in favour of low-impedance outputs (10–40 Ω) and high impedance inputs (10–20 kΩ) and 0 dBu or +4 dBu are commonly adopted as nominal levels for professional equipment, with maximum levels on signal peaks reaching +20 or even +26 dBu. Consumer equipment tends to operate at lower levels, but here chaos reigns with levels of −6 dBu, −7.5 dBu, −10 dBu, and −20 dBu all in common use, and no agreement on what these levels mean in terms of maximum level. A German DIN standard actually specified current rather than voltage, using high-impedance sources and low-impedance inputs, as a means of providing for both low-level microphones and high-level line signals with a single input socket, but it failed to meet the low noise requirements of modern systems and is now completely obsolete. In the 1980s it looked as though Japanese manufacturers were adopting a nominal level for cassette decks of −10 dBu (245 mV), but when CD players first appeared they were specified as having 2.0 Vrms maximum output. Despite the apparent difference (which spawned many bogus interface adapters), surveys of discs have shown that they tend to be recorded with 18 dB of headroom, so that it would be more appropriate to quote a nominal level of 251 mV or near enough −10 dBu! More recent CD players specify 2.5 or 3.0 V output, the latter corresponding to −6 dBu nominal level. Many televisions, however, adopt an audio input level of 0.5 V nominal or −3.8 dBu, and video recorders are adopting numerous levels, such as −7.5 dBu on Hi8 video machines and −6 dBu on many video cassette recorders! In the author's opinion the adoption of a nominal level of −6 dBu (387 mV) for all home audio systems would put an end to this mess, solving many interconnection problems and also allowing direct compatibility with half the output from a centre-grounded balanced professional output. Many items of professional equipment have adequate gain adjustment to accommodate −6 dBu input signals from consumer equipment.

Every input to or output from a piece of equipment is designed to operate up to a maximum voltage limit without serious distortion, but for reasons of design this limit varies. There are two ways of setting signal levels when interconnecting equipment. The first gives priority to keeping the meter on each piece of equipment 'out of the red', to avoid clipping, and so arranges for all equipment in a chain to overload simultaneously. Under this scheme gain settings are adjusted so that the maximum level from an analogue tape (around +10 dB) drives a digital recorder to its maximum level (around +18 dB). A compact disc is broadcast at a low enough level for the loudest peaks (+18 dBu) to fall within the +9 dB headroom of an FM transmitter. It is totally unsatisfactory because it ignores all notions of subjective loudness and forces programme material originated on systems with a lot of headroom to be transmitted at ridiculously low levels to avoid clipping, yet it is rapidly becoming the accepted method in most people's minds, and level meters are increasingly being calibrated with 0 dB at the top and no other markings to suggest any alternative mode of use. What happens in practice, though, is a very different matter, with compressors and limiters being used to restore loudness, arbitrarily and often to the highest level possible, this being the perceived way to capture audiences!

The second method, much to be preferred in the author's opinion, adopts the concept of a *'nominal level'* (0 dBnom), to which all equipment in a signal chain is aligned. Different pieces of equipment may represent nominal level differently (commonly as 0 dBu or +4 dBu), but their gain settings are such that they all function at their nominal level together, despite different amounts of *headroom*. Line-up, using a test tone, is then to nominal level at all points. In principle, with this scheme, programmes from any source can be broadcast or copied without regard to metering or gain adjustment (just as is the case with video signals) and they will maintain the levels set by the recording engineer based on artistic considerations. Of course, this means that a CD or other digital recording is likely to overload the broadcast channel on maximum levels, but this is a problem to be tackled by the careful incorporation of compressors or limiters into the chain, the purpose of which is to match *headroom*. Programme level meters should ideally measure subjective loudness, not peak level, with peak level meters serving the separate function of ensuring that the signal path is not driven into overload.

The BBC and Independent (formerly IBA) broadcasters have always tended to adopt the second scheme, with 0 dBu as nominal level (PPM4), but they only made it work because of the special

properties of analogue tape machines (which ensured that most programme material was soft-limited to around +10 dBnom on peaks) and PPMs (Peak Programme Meters) which were deliberately designed not to register the fast transients that tape machines would at some stage compress acceptably. Even then, tables were issued to recording engineers listing 'target levels' for different types of programme; classical, pop and speech, to correct for the anomalies between perceived loudness and PPM reading. A *'line-up level'* of PPM6 (+8 dBu) has sometimes been adopted, further confusing matters.

In an attempt to clarify matters for programme exchange the CCITT (now disbanded) defined the term dBm0s (a most cumbersome term that has not helped its adoption!) to mean 9 dB below maximum allowable level at any point in a system. This definition in terms of maximum level is most unfortunate, because it conforms to the 'out of the red' method, though it was probably never intended to apply to systems with anything other than the 9 dB headroom that became the EBU (European Broadcasting Union) standard. The term dBnom is self-explanatory and much more suitable for common use. Nominal, of course, need not mean average or typical, but it is a mark common to all systems and all points in a signal path signifying a reasonable intermediate loudness against which the recording engineer can make artistic judgements, as opposed to just 'keeping out of the red'!

3.6.2.2 Programme-level meters

Programme-level meters are an integral part of most mixers and tape machines. Steady-state levels should measure the same on any a.c. voltmeter, but music and speech will produce different readings on different meters unless these have the same 'dynamics' (also referred to as 'ballistics' when referring to mechanical meters). Dynamics depend on the type of rectifier (mean, peak, r.m.s.) as well as the speed of response which is characterized in terms of *rise-time, integration time, delay time, overshoot*, and *decay time*. Of these, *integration time* is perhaps the most important, and is usually defined as the length of a tone burst needed to achieve a reading of 80% of the steady-state reading on the same tone, or −2 dB. *Rise-time* is a term best reserved for simple R-C time-constants, and is not the same as integration time because a meter response is often determined by multiple time-constants and delays. It should also be noted that integration time in a peak follower circuit is considerably longer than the charging time-constant of the circuits involved,

because charging only occurs during a small part of the cycle.

The well-known VU (volume unit) meter has dynamics as specified in IEC 60268-17. With an integration time of about 165 ms, and a similar decay time, it largely ignores peaks and tends towards a better indication of loudness than a PPM, but is rather 'bouncy' and difficult to read. The PPM (peak programme meter), on the other hand, has an integration time of 10 ms (IEC type 2a, or BBC type) and a decay time-constant of 1.0 s (specified as 24 dB fall in 2.8 s) making for more relaxed reading, but being basically peak-reading it is more useful for avoiding overload than as a loudness indicator. Its integration time of 10 ms is actually much too slow to show true peaks, and was chosen to reduce the influence of brief peaks that could advantageously be allowed to overload analogue tape without significant audible effect. Programme-level meters require steady-state calibration on a 1 kHz sine-wave, followed by a series of tone-burst tests to determine peak readings for several different burst durations, usually at 5 kHz, as well as repetitive tone-burst tests. Details of these tests, which are usually adequate when verifying performance, are summarized in *Table 3.7*, but the relevant standards also describe tests for polarity-reversal error and overload which are necessary in their design.

Digital devices usually incorporate a much faster peak-reading meter which is good for avoiding the severe distortion of digital clipping, but of little value as a loudness indicator. In the author's opinion, both the VU meter and PPM are relics of a bygone age, compromised by the limitations of past technology. Recent attempts to produce a more useful meter have tended to aim for a loudness indication using a bargraph display with a trailing peak hold superimposed. One such attempt, backed by two major UK television companies, involved listening-panel tests to determine the optimum integration characteristic for subjective loudness, but unfortunately the criterion set for loudness seems to have been 'annoyance', the object being to find a way of controlling the level of television 'commercials' to minimize viewer complaints. This resulted in a surprisingly long integration time of around 5 s, and it is probably not suitable for general programme balancing. Loudness is also recognized as being dependent on the 'fullness' of sound, several simultaneous frequencies sounding louder than a single tone of the same r.m.s. value, and attempts have been made to sum the output of a filter bank, although quasi-peak measurement would seem to produce much the

Table 3.7 Target levels for testing meter dynamics with tone-bursts

Meter type	Level	Burst duration (ms)									
		0.5^1	1	1.5	2	5	10	20	50	100	200
IEC286-10 PPM											
-Type I	+8 dB	–	–	–	–	-2 ± 1	-1 ± 0.5	–	–	–	–
-Type IIa	+8 dB	$1\frac{3}{4}$	–	$3\frac{3}{4}$	–	5	$5\frac{1}{2}$	–	–	6	–
-Type IIb	+8 dB	-9 ± 2.0	–	-1 ± 1.0	–	$+4 \pm 0.75$	$+6 \pm 0.5$	–	–	+8	–
CCIR468-4 Quasi-peak											
target value	0 dB	–	−7.4	–	−3.5	0.0	+1.6	+2.3	+3.4	+4.7	+6.1
(tolerance)		–	±2.0	–	±1.5	±1.3	±1.3	±1.3	±1.3	±1.2	±1.2

Additional repetitive burst tests for CCIR468 quasi-peak (use 5 ms bursts of 5 kHz):

2 bursts per second −6.4 dB ±0.9
10 bursts per second −2.3 dB ±0.6
100 bursts per second −0.25 dB ±0.25

Notes:
1. Tone bursts start on a zero-crossing and contain an integral number of full periods. Bursts of 5 kHz tone at +8 dB should be used, except for 0.5 ms bursts where 10 kHz should be used.
2. Target levels and tolerances for PPM type IIa are the same as for type IIb but the former are shown in terms of scale numbers (PPM 0 to 7).
3. Strictly, for CCIR468 the burst amplitude applied should be such that a steady signal would give a reading of 80% (−1.9 dB) of full scale, but here all levels have been adjusted to a scale reference of +8 dB for easy comparison and applicability to a PPM type scale range. Tolerances have been very slightly altered to avoid different plus and minus values. Actual input levels should be low enough (−20 to −60 dBu) to avoid clipping.

same result because it registers the maximum sum of component peaks.

How are levels to be set and metered in the new age of digital broadcasting? Hopefully, digital radio and television systems will standardize on more than 9 dB of headroom, otherwise much of the benefit of digital audio will be thrown away, yet it is hard to see how they can work optimally without changes to current practice. Video recordings in particular, are currently 'finished' in post-production ready for broadcast on existing systems, with maximum levels that never exceed +8 dBnom; but surely these cannot do justice to digital broadcast systems. Either the tapes must be produced and transmitted in two versions (not practicable) or working methods must change so that all programmes are exchanged with signal peaks preserved as far as possible until the last stage where peaks should be limited only at the interface to low-headroom systems such as existing analogue transmitters. The much-respected PPM, optimized as it was for analogue tape recording, indicates neither the loudness nor the peak reliably, and a new standard is badly needed for a programme meter that reads both subjective loudness (using frequency and time weightings) and fast-peak levels. Just as dBm (power) levels are no longer appropriate in the studio, dBu can no longer be applied to digital systems, and so numeric values need to be defined for nominal level on all the new digital standards, 16-bit to 24-bit, NICAM, Musicam, MPEG, and other compressed systems. In some cases this has already been done, but even transfer between digital formats may not be free from problems because each can have a different amount of headroom and some even have varying headroom that reduces with frequency because of pre-emphasis (NICAM uses the J17 curve and some CDs also have pre-emphasis). It is to be hoped that the 'nominal level' concept will gain favour as it offers the most sensible way of coping with these issues.

3.6.2.3 Frequency response

The measurement of frequency response involves plotting a graph or listing a number of values. A sine-wave generator with constant output over the audio frequency range is used as the test signal, and the output level plotted against frequency. Manual measurements are best carried out at IEC 1/3-octave standard frequencies which follow the series 20, 25, 31.5, 40, 50, 63, 80, 100, 125, 160, 200 Hz and so on repeating. Automatic frequency response plotting may be carried out using chart recorders, but these days it is easier to use microprocessor-controlled

sweep generators with screen display and printout facilities. Swept response testing can give misleading results if the sweep is too fast, so if rapid sweeping is carried out the results should be checked against a slower test. There are several reasons for this. On rapidly falling 'brick wall' filters there may be a limit to how fast the measuring rectifier output can decay. It must also be remembered that a rapid sweep is in fact a frequency modulated (FM) signal and so contains side-bands which can distort results. Finally an effect is often observed at low frequencies which results from phase shift that causes the outputs from different input frequencies to arrive time-displaced, interfering to produce a peak that is not really there in steady state measurements. Notwithstanding this it is possible to carry out a valid 1 s sweep on most devices such as analogue tape recorders, though 5 s is to be preferred.

Frequency response is often tested at several levels, especially on analogue tape machines or where companders or noise reduction systems are involved. Open-reel tape machines give the most valid results if tested at 10 dB below nominal level to avoid high-frequency saturation. Cassette tape needs to be checked at −10 dBnom or even −20 dBnom for the same reason.

The correct specification of frequency response requires a graph or a listing of points. If abbreviated to a bandwidth specification it must include response limits, e.g. 20 Hz to 20 kHz +2/−3 dB. Specifications without such limits are quite meaningless – but surprisingly common.

3.6.2.4 Noise

Early measurements of noise in audio systems failed to correlate with subjective annoyance value because the ear has its own frequency response curve which peaks at 3–4 kHz, and depends on level, as determined by Fletcher and Munsen (Moir, 1976), and later by Robinson and Dadson (see Section 1.4). Improved correlation came when the 'A' weighting curve was tried. Originally devised as an approximation to the low-level hearing-response curves, for use in acoustic noise measurements, it is in fact a very poor representation of them, lacking the peak in sensitivity of around 9 dB at 4 kHz that is so important to our perception of noise and distortion. As this deficiency became apparent two somewhat harsher weighting curves came into use. The German DIN standard has long been abandoned, but collaboration between Germany and the UK resulted in the international CCIR Recommendation 468 which is now universally adopted, and commonly

referred to as CCIR468-4 (though the CCIR and CCITT were recently disbanded leaving their parent body International Telecommunication Union (ITU) as one of the issuing bodies. *Figure 3.115* shows this weighting curve, along with the 'A' weighting curve, and a further bandpass curve specified by CCIR468 for so-called 'unweighted' measurement (unweighted measurements must still be band limited or inaudible components will produce a meaningless result). Developments in the 1960s confirmed the validity of CCIR468 when Dolby™ B noise reduction became popular on cassette recorders, the measured 10-dB improvement over non-Dolby material corresponding well with subjective impressions, while other measurements failed to show much difference. It should be noted that the 468 weighting curve was developed by subjecting listening panels to noise of varying types in the presence of signals.

CCIR 468-4 Noise Weighting Curves

CCIR 486-4 Weighted Response

Hz	dB	Tolerance
31.5	-29.9	±2
63	-23.9	±1.4
100	-19.8	±1.0
200	-13.8	±0.85
400	-7.8	±0.7
800	-1.9	±0.35
1k	0	±0.2
2k	+5.6	±0.5
3.15k	+9.0	±0.5
4k	+10.5	±0.5
5k	+11.7	±0.5
6.3k	+12.2	0
7.1k	+12.0	±0.2
8k	+11.4	±0.4
9k	+10.1	±0.6
10k	+8.1	±0.8
12.5k	0	±1.2
14k	-5.3	±1.4
16k	-11.7	±1.6
20k	-22.2	±2.0
31.5k	-42.7	+2.8 -∞

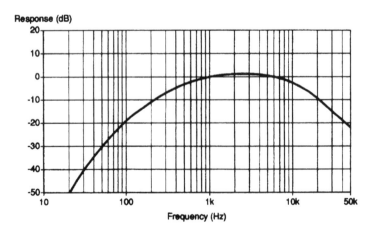

A Weighting Curve (IEC179)

Hz	dB	Tolerance
20	-50.5	±2.0
40	-34.6	±1.0
80	-22.5	±1.0
100	-19.1	±0.5
200	-10.9	±0.5
400	-4.8	±0.5
800	-0.8	±0.5
1k	0	0
1.6k	+1.0	±0.5
2.5k	+1.3	±0.5
4k	+1.0	±0.5
8k	-1.1	±0.5
10k	-2.5	±0.5
12.5k	-4.3	±1.0
20k	-9.3	±1.0
50k	-22.0	±1.0

Figure 3.115 *Noise weighting curves.*

It therefore reflects low-level 'intrusiveness', which is not quite the same thing as loudness; hence the departure from the hearing sensitivity curves.

CCIR468 also specifies the use of a special quasi-peak rectifier in place of the usual root-mean-square (r.m.s.) measurement, the aim of which was to take proper account of impulsive noise bursts experienced on telephone lines by responding to them with carefully defined attack and decay times. Quasi-peak measurement is less vital when measuring the uniform random noise of most audio equipment, but it does give readings around 5 dB higher than are obtained with r.m.s. measurement, a fact which must be allowed for if it is not used. For many years the BBC used the PPM for CCIR weighted noise measurement, and this does give results close to the true quasi-peak values provided there are no short bursts or clicks. The tone-burst parameters of the CCIR quasi-peak rectifier are shown in *Table 3.7*.

Specifications of noise performance should always be relative to nominal level. It is lamentable that most specifications for noise on CD players mention no noise weighting, and are relative to peak level, without mention of the 18 dB of headroom normally employed on CD to avoid harsh clipping. A true CCIR-weighted figure relative to nominal level allows much fairer comparison with other devices, as will be shown later, and this is another good reason why the nominal level concept makes sense, since headroom is always increased at the expense of signal-to-noise ratio, which is arguably more important than dynamic range (a confusing term best avoided).

It is useful to note that a noise measurement to CCIR468 quasi-peak will generally read 11 dB higher than an audio band limited flat r.m.s. measurement. This is true on white noise, and also on most tape and amplifier noise. A method of noise measurement referred to as CCIR ARM-2k was devised by Dolby Labs which gained some popularity for measurements on domestic equipment, and this employed an average reading meter together with a CCIR weighting filter shifted in level to give unity gain at 2 kHz. The aim was to combine cheapness with lower readings (preferred by marketing departments) which were more in line with unweighted measurements. Measurements made in this way are generally also 11 dB lower than true CCIR figures, and to avoid confusion it is better to stick to the true standard.

It is worth noting here that the 'A' weighting curve, as well as being inappropriate for equipment measurement, is also quite unsuitable for most environmental noise measurements. Originally recommended only for low-level noise measurements, it is now in general use at higher levels. Even at the 60-phon level it reduces low frequencies by some 15 dB too much at 100 Hz, underestimating the significance of most road-traffic and industrial noise!

3.6.2.5 Cross-talk or channel separation

Cross-talk is measured by driving one channel of a device with sinusoidal signals at varying frequencies, and usually nominal level, and measuring the relative level of the breakthrough at the output of the other channel. Cross-talk below −20 dB is usually acceptable, provided it is undistorted, making this a relatively easy measurement. Cross-talk between different sources on a mixing console or within a studio is a different matter however, and figures below −100 dB are sometimes demanded. Measurement is then made using narrow band filters to distinguish the cross-talk from noise.

3.6.2.6 Harmonic distortion

Harmonic distortion results when an audio signal is passed through any device with a non-linear transfer function, and is so called because testing with a single sinusoidal test tone will produce measurable additional tones or harmonics at multiples of the test frequency. The distortion produced by transfer functions which are symmetrical about the x-axis is referred to as 'odd-order', a reference to the mathematical representation of the function, and it contains only odd harmonics, that is, they are at multiples of 3, 5, 7, etc., times the fundamental. Even-order distortion contains only even harmonics. Analogue tape recorders produce almost exclusively third harmonic distortion as a result of their gentle and symmetrical saturation curve. Total harmonic distortion is measured by driving the device under test with a pure sinusoidal test tone and removing the fundamental component at the output with a notch filter before measuring what remains. Traditionally an r.m.s. figure was quoted, but quasi-peak weighted measurement is now to be preferred as explained later.

Over the years there has been much debate over the usefulness of total harmonic distortion (THD) measurement as a measure of quality, and yet it remains a key measurement, partly because it is relatively easy to carry out and universally understood, partly because a look at the distortion residue with an oscilloscope is often instructive in identifying its cause, but also because in practice it is adequate for most applications.

The main area of debate has been power amplifiers, and it is true that these need to be treated with caution, because the kink in the transfer function caused by 'crossover distortion' represents a very high-order function and gives rise to high-frequency harmonics which sound harsh because they are not masked by the original signal and tend to fall at the high end of the frequency spectrum where the ear is most sensitive. This problem was well described long ago (Shorter, 1950) in a paper which suggested that good correlation with subjective degradation could be obtained by weighting each harmonic according to its number squared, divided by four. Thus the second harmonic was multiplied by 1, the third by 2.25, the fourth by 4, the tenth by 25 and so on. The IEC have recently incorporated weighted THD measurement, using the CCIR468 noise weighting curve and quasi-peak rectifier, into IEC60268 and this seems highly sensible because the curve is already available in many instruments and any attempt to emphasize high harmonics can only improve subjective correlation. The fact that it would also produce higher distortion figures should not be allowed to get in the way, and it is to be hoped that its use will be accepted. It is interesting to speculate that crossover distortion tends to measure low but sound bad because it produces a 'spike' which only lasts a short part of each cycle and so has a low r.m.s. value. The ear responds quickly to clicks, and seems to respond to their instantaneous value rather than their spectral analysis, so the use of CCIR weighting and quasi-peak detection should produce results much more in line with subjective assessment.

Total harmonic distortion has traditionally been quoted in per cent, but the trend is towards a figure in dB, relative to the level of the fundamental output. The frequency and level at which the measurement is made must always be quoted, together with the bandwidth of the measuring device. It is convenient to limit measurement bandwidth with the CCIR 22-22 kHz bandpass filter. Measurements of THD at 100 Hz and 1 kHz are generally thought sufficient to show up most system defects, although swept measurements are also made. Measurement of THD above 6.3 kHz is pointless in most systems because then all harmonics fall outside the band making a null result inevitable: a 6 kHz square-wave sounds the same as a 6 kHz sine-wave!

3.6.2.7 Intermodulation distortion

The main consequence of a non-linear transfer function is not the production of harmonics but the production of intermodulation products in the presence of two or more tones. This is because music and speech produce a great many such products, adding roughness to the sound whereas harmonics are generally already present so that small extra amounts may go unnoticed.

If a non-linear system is driven with two sinusoidal tones it produces intermodulation products which may be at sum and difference frequencies derived from the two tones and their harmonics. Many papers have been written on the subject of intermodulation and its measurement (Scott, 1945; Hilliard, 1946; Wigan, 1961) and they generally compute possible terms mathematically by making use of the fact that any transfer curve can be represented as a power series. Odd-order terms in this series, corresponding with symmetrical curves, are shown to produce one set of products, whilst even order terms produce a different set.

An early test for intermodulation, still in use and referred to as the 'SMPTE' method (SMPTE PH22.51: 1961), was devised by the Society of Motion Picture and Television Engineers, originally for measurements on optical film soundtracks. It involves driving the system under test with two tones, usually 50 Hz and 7 kHz with amplitudes in the ratio 4:1. After filtering out the low tone the high tone is demodulated rather like an AM radio carrier and the result passed through a low-pass (500 Hz) filter and its r.m.s value measured. A feature of the method is that it also shows up modulation noise on tape machines, but its weakness is that it really only measures static non-linearities of the type that can be quantified by THD measurement.

Dynamic non-linearity results from the fact that the transfer function of many devices changes at high rates of change of voltage. This may be because a circuit 'slews', i.e. fails to follow, or because of loss of loop gain at high frequencies, or because of non-linear capacitance in semiconductors. This is something that only high-frequency intermodulation tests can reveal.

One such test, the Difference Frequency Method, use two equal amplitude high-frequency sine waves, often swept over the 5–20 kHz region with a constant difference of 1 kHz or 70 Hz. A low-pass filter picks out the resulting difference products, but it is not easy to get a result for both even- and odd-order products with this method, and it is not widely used.

Perhaps the most promising method is called the 'total difference frequency' method (Thiele, 1975).

This employs two tones at (usually) 8 kHz and 11.95 kHz, (800 Hz and 1195 Hz can also be used). Because of the choice of frequencies a selection of both odd- and even-order intermodulation products appears around 4 kHz, in two slightly separated groups which beat in a way that avoids the possibility of cancellation. A steep 4 kHz low-pass filter is used to separate out these components for r.m.s. measurement. Because of this, and the just sufficiently demanding choice of frequencies, good subjective correlation has been claimed (Small, 1984) and the method is sensitive and fairly easy to implement.

3.6.2.8 Quantizing distortion and noise

Quantizing noise and quantizing distortion are difficult to define separately. The inevitable errors that result from digitizing a signal into discrete levels produce an objectionable granular noise in the presence of signal, which can also be interpreted as a form of high-order non-linearity. In most practical systems though, dither is added to the original signal to eliminate this cause of distortion while introducing a largely signal-independent and less intrusive noise. Quantizing distortion can still remain over and above this noise, as the result of extra errors introduced by unequal step size in analogue-to-digital (A–D) or digitial-to-analogue (D–A) converters and it can also result from other irregularities arising, for example, in a near-instantaneous companding system such as NICAM where varying step size is employed. The simplest method of measurement involves a sinusoidal test signal at 40 Hz and a 400 Hz high-pass filter so that all harmonics below the tenth can be rejected. High-order products are thus distinguished from low-order harmonics arising from other parts of the signal chain. The test is best carried out at full signal level and then 20–60 dB below nominal level in order to emphasize low-level effects which tend to be more intrusive. A more complicated method employs a 2 kHz sine wave and a comb filter to remove the fundamental and all harmonics from the measured residue, leaving only the non-harmonically related components that result from quantizing errors. The method claims the advantage that it can be used on lines employing the J17 pre-emphasis curve that could cause overload problems with a 40 Hz test signal. Quantizing noise and distortion figures are rarely quoted at present, and it can be argued that normal noise and distortion measurements encompass these digital artifacts though this is only true provided weighted quasi-peak THD plus noise is measured

at several levels. Like crossover distortion, quantizing distortion can be 'spiky', sounding a lot worse than it measures unless CCIR-weighted quasi-peak measurement is used.

3.6.2.9 Wow & flutter, rumble, frequency, phase difference

Wow & flutter is usually measured by recording a 3.15 kHz tone and replaying it into a fluttermeter which contains a frequency demodulator. The demodulated output contains mostly low-frequency components corresponding to speed irregularities, and these are weighted according to a curve defined in IEC386: 1971 as shown in *Figure 3.116*. This emphasizes variations around 4 Hz to which the ear is found to be most sensitive, and a special quasi-peak rectifier is then used for measurement (not the same as that used for noise!).

Rumble is a measurement peculiar to gramophones which tend to produce high levels of very low frequencies due to turntable vibration and bearing irregularities. Measurement is made using two weighting curves as defined in BS 4852: 1972 or IEC98: 1964 (BS 4852 uses a preferred slow meter response) also shown in *Figure 3.116*.

Frequency, and phase difference between channels can be estimated by viewing waveforms on a dual-channel oscilloscope, but a more modern approach is to use digital techniques to time between zero-crossings of a sinusoidal test signal using two comparators and a microprocessor. Frequency is calculated by measuring the time between zero crossings on one channel over several cycles, and then dividing the time into the number of cycles, a method sometimes referred to as 'reciprocal counting'. A feature of this method is the speed with which low frequencies can be accurately measured, as only one cycle is needed. High frequencies are usually measured over several cycles to obtain higher accuracy, but phase measurement does place heavy demands on processor speed, requiring the relative position of crossings on the two channels to be measured to an accuracy of around a microsecond or so for one degree accuracy. Averaging techniques can be devised to improve accuracy though.

3.6.2.10 Group delay, and transient intermodulation distortion

Variations in phase with frequency between the input and output of a device constitute a frequency-dependant time delay, sometimes referred to as 'time smearing' or 'group delay'. Group delay is of importance on telephone lines where it can amount to

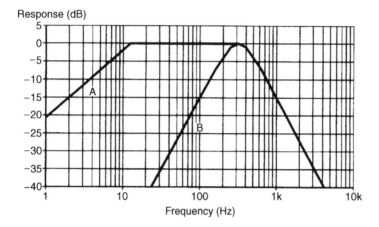

Rumble Weighting Curves (IEC98)

IEC98 Rumble Curve A

Hz	dB	Tolerance
2	−15	±1.5
10	−0.5	±0.5
20	0	±0.5
315	0	0
400	−1	±1
630	−7	±1
1k	−15	±1
3.15k	−35	±1

IEC98 Rumble Curve B

Hz	dB	Tolerance
31.5	−35	±1
100	−15	±1
160	−7	±1
250	−1	±1
315	0	0
400	−1	±1
630	−7	±1
1k	−15	±1
3.15k	−35	±1

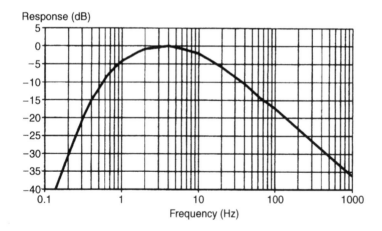

Wow & Flutter Weighting Curves (IEC 386)

IEC386 Wow & Flutter Weighting

Hz	dB	Tolerance
0.2	−30.6	+10-4
0.315	−19.7	±4
0.4	−15.0	±4
0.63	−8.4	±2
0.8	−6.0	±2
1	−4.2	±2
2	−0.9	±2
4	0	0
6.3	−0.9	±0.2
10	−2.1	±2
20	−5.9	±2
40	−10.4	±2
63	−14.2	±4
100	−17.3	±4
200	−23.0	±4

Figure 3.116 *Flutter and rumble weighting curves.*

tens of milliseconds, but its relevance to most audio devices is questionable. Measurement techniques can be found by reference to telecommunications literature.

Transient intermodulation distortion (TIM) measurement was much discussed in the 1970s (Otala, 1977; Jung *et al.*, 1979), but has gone out of fashion. What was claimed by some to be a new form of distortion was in fact only an aspect of intermodulation, as discussed above, and some of the tests devised involved signals with rise times hundreds of times faster than are ever found in recorded music, making them unrealistic. The concept of using a single specially devised waveform to design valid tests for IM is interesting though and both sawtooths and asymmetrical rounded waveforms have been investigated (Antoniazzi *et al.*, 1981; Hirata 1981).

3.6.3 Test equipment

In the 1970s test equipment tended to exist as a large number of separate oscillators, millivoltmeters, noisemeters, distortionmeters and so on, each in a box with a meter and variety of knobs and switches. The Ferrograph test set (*Figure 3.117*) was perhaps the first attempt at an all-in-one test set, but when, in 1977, I designed and introduced the Lindos LA1 and coined the term 'Audio Analyser' I found considerable resistance to change, and even dropped the term for a while because so many people confused it with the spectrum analyser. Today, to my great delight, there are half a dozen units on the market calling themselves 'audio analysers', and individual instruments have all but disappeared.

Figure 3.117 *The Ferrograph test set.*

The audio analyser offers obvious advantages in cost saving and compactness through the use of a single display for many measurements, but it also offers ease of use without the need for repeated connection and disconnection of equipment. The concept really comes into its own, though, with the use of microprocessor control, soft-touch keys, and full graphic display. The Lindos LA100 (*Figure 3.118*), launched in 1985, was the first unit to provide full measurement facilities together with graphic display of frequency response in a portable unit using a liquid crystal display (LCD). A slightly different approach, using a standard desktop PC and menu driven operation is represented by the Audio Precision System One

(*Figure 3.119*). This manufacturer also specializes in 'dual domain' testing via a standard serial digital connection.

Figure 3.119 *The Audio Precision System 1.*

3.6.4 Automated testing

In addition to manual measurement of many parameters, these units provide for automated testing. This may refer to the plotting of a parameter such as distortion against frequency or level, or to completely automatic testing of all the parameters of a system. In a development or production environment, the user will often prefer to write a specific test program, and in some cases this can be interactive, instructing an operator to make certain adjustments as tests proceed. In the vast area of quality control, however, a more rigid pass/fail approach is often desirable, and problems of synchronizing the source and measuring set arise when lines, satellite links and tape machines are to be tested. This led to the development of sequence testing around 1984.

3.6.5 Sequence testing and frequency shift keying synchronization

Sequence testing refers to the generation of a set sequence of carefully timed test signals to

Figure 3.118 *The LA100 test set.*

be measured automatically. An early Australian test set made by AWA, no longer in existence, adopted the use of FSK (frequency shift keying) at the start of its sequence for synchronization and identification, and this idea was taken up by the EBU in its recommendation R27-1982, later to become the CCITT O.33 standard for automatic lines testing.

Lindos has refined the technique for general use in the LA100 Audio Analyser, through the concept of 'segmented sequence testing'. A key feature of this system is the use of 'segments', or standard building blocks from which sequences are assembled. Each segment consists of a short sequence of tones or a sweep preceded by an FSK 'header' code lasting around 100 ms. The header acts as the synchronizing trigger for a particular measurement as well as identifying the segment code, so that the measuring set can configure itself and make the right measurement at the right time automatically. Segments can be sent in any combination over any audio path or recorded and played back from tape and any LA100 will still make the correct measurements without any programming. The use of short segments brings many advantages over a single long sequence. Speed errors on tape replay, which could put a long sequence out of synchronism, are well tolerated as each segment is retimed from its own FSK header. Perhaps most important, though, is the way in which segmented sequences combine flexibility with standardization. A relatively small number of segments can be put together to form an almost unlimited number of test sequences, and yet every one is recognized automatically. Special needs are catered for by programmable segments which carry parameters in FSK.

3.6.6 Typical applications and results

The LA100 can store and print out sequence results on any dot matrix printer, and a typical test of a cassette deck, completed in under one minute, is illustrated in *Figure 3.120*. This also shows the use of tolerance limits with automatic pass/fail indication. As already suggested, only a unified approach to testing, with standard tests and methods of presentation, can fulfil the need for rapid verification and comparison between equipment, and to illustrate what is possible a summary of sequence test results made on seven standard items of equipment is presented in *Table 3.8*. To those who like to refer

to 'CD quality', 'FM Quality' etc. this may contain some surprises, since for once all parameters have been measured in a way that permits comparison. In particular, the common claim that 16-bit digital audio permits 96 dB of dynamic range is shown to be nonsense when conventional measuring standards are adopted (for a theoretical justification of this see the notes under *Table 3.8*). Surprisingly, for many, the VHS home video deck is shown as having 10 dB less noise than CD, and 8-bit Hi8 pulse code modulation sound also comes out remarkably well. To the author, these results are entirely in line with experience: analogue tape is a little too noisy, especially for multitrack use, which is why noise-reduction systems like Dolby A and Dolby SR exist. Frequency-modulated (FM) broadcasts also suffer noise problems. Compact cassette can be very good except for phase stability problems, and compact disc is excellent. There is a little more to the story, however. Compact cassette at its best may be excellent, but it is rarely at its best because of tape skew, azimuth drift, and dropouts on worn tape – reliability aspects that also have to be considered in appraising any medium. It also suffers a high level of noise modulation on high-frequency tones that builds unacceptably on copying. Systems involving companding also need a little caution because they require tests for compander tracking and transient effects which are not in widespread use. VHS and SVHS hi-fi sound have a particular problem caused by brief chopping of the frequency modulated (FM) carrier at the end of each helical scan. This results in a barely audible loudness-modulated buzz which worsens when tapes recorded on one machine are played on another, or when they are copied. Hi8 PCM sound suffers from audible whistles appearing as artifacts on certain sounds, and from a tendency to overload briefly on sibilant sounds because of slowness of the companding system used. All these effects can be measured quite easily, but they demonstrate the need to be on the lookout for special problems, and the difficulty of devising tests that will show up everything. Loudspeakers, still the weakest link in the chain for most people, suffer from limited low-frequency headroom and also from coloration, an effect caused by delayed echoes and resonances for which special tests have been devised, but that is a whole other area of measurement in which the special technique of time domain spectrometry (TDS) is commonly used to measure delayed responses to an impulse or other specially devised test signal.

```
LINDOS AUDIO SEQUENCE TEST        EXAMPLE PRINTOUT (Chrome Tape)
SOURCE 0518A                      LA102 0518  V5.5  14 JUN 1991
SEGMENTS +TPCINWZ   TPCIN         TOLERANCE 12 IBA TAPE/RESTRICTED FAILED

FREQUENCY RESPONSE (P) -20dB
    30Hz    +1.27    +0.64    +6
    40Hz    +2.12    +1.59    +3.5-2.5<3
    50Hz    +1.20    +0.70    +3.5-2.5<3
    63Hz    +2.09    +1.64    +3.5-2.5<3
   100Hz    +0.78    +0.46    +3.5-2.5<3
   125Hz    +0.59    +0.28    +3.5-2.5<3
   250Hz    +0.23    +0.11    +-1.5<2
   500Hz    +0.21    -0.12    +-1.5<2
    1kHz     0.00     0.00    +-1.5<2
    2kHz    +0.29    +0.07    +-1.5<2
    4kHz    +0.31    -0.02    +-1.5<2
  6.3kHz    +0.27    +0.11    +-1.5<2
    8kHz    +0.33    +0.04    +-1.5<2
   10kHz     0.00    -0.14    +-1.5<2
 12.5kHz    -0.67    -0.82    +3.5-2.5<3
   14kHz    -0.67    -1.50    +3.5-2.5<3
   15kHz    -1.14    -1.83    +3.5-2.5<3
   16kHz    -1.24    -1.94
   18kHz    -1.64    -2.75
   20kHz    -5.45    -5.53
TEST LEVEL OUT (T) 0dB
           -0.78    -0.96dBu +-1.5
NOISE CCIR 468-3 Q-Pk (rel) (N)
PK WTD     -42.7    -39.0    -36
PK UNW     -46.7    -42.8    -40
MN WTD     -42.9    -39.5    -36
CROSSTALK (rel) (C) 0dB
    40Hz   -36.6    -36.7    -17
   100Hz   -44.2    -44.7    -25
   315Hz   -43.7    -43.2    -36
    1kHz   -42.6    -42.5    -36
  6.3kHz   -32.4*   -32.8*   -36
   10kHz   -32.4    -29.6*   -32
DISTORTION (I) +8dB
100 +8     -16.4*   -17.0*   -28
1k  +8     -21.2*   -23.1*   -28
100-10     -42.0    -40.1    -30
1k -10     -42.1    -40.6    -30
WOW & FLUTTER IEC386 Q-Pk (W) 0dB
PK WTD    0.111%            .15
MN WTD    .0777%            .15
PHASE       +8d            +-90
SPEED      0.0%            +-.5
PHASE (mean) (Z) 0dB
    40Hz     0d            +-180
   100Hz     0d            +-90
    1kHz    +1d            +-90
  6.3kHz   +13d            +-90
   10kHz   +16d            +-90
   15kHz   +24d            +-180
```

Figure 3.120 *The Lindos audio sequence test. Typical results for a cassette deck.*

Table 3.8 Comparison of measurements made on typical good equipment in 1991

Path	Bandwidth at −3 dB (kHz)	Flatness 40–12.5 kHz (dB)	Noise CCIR468 wtd (dB)	Headroom (dB)	Distortion THD 1 kHz, −10 dB (dB)	Phase/Jitter (°)	Wow and flutter (%)
16-bit compact disc	20–20	±0.3	−60	+18	−60	0 ± 0	0.00
Professional analogue tape, 38 cm/s	20–22	±0.5	−50	+10	−60	20 ± 10	0.03
Compact cassette, chrome, Dolby B	30–16	±2.0	−53	+8	−60	90 ± 30	0.08
FM broadcast, off air, with NICAM link	20–15	±1.0	−50	+9	−50	0 ± 0	0.00
VHS hi-fi, FM sound carrier system	20–20	±1.0	−71	+15	−50	1 ± 0	0.00
Hi-8 video, 8 bit digital, companded	20–15	±1.0	−65	+12	−50	4 ± 0	0.00

Notes:
1. All figures relate to the total path, input to output, as measured using the Lindos LA100 Audio Analyser.
2. The theoretical noise level on CD can be deduced assuming a Gaussian dither signal of r.m.s. value $\frac{1}{2}q$ (where $q = 1$ quantizing step). The maximum signal is ±32767q or ±32767.2q r.m.s. Dynamic range is therefore 93 dB measured r.m.s. Gaussian noise reads +11 dB CCIR weighted quasi-peak, and allowing 18 dB headroom (typical) gives a CCIR weighted noise level of −64 dB. Measurement on a dithered test CD gave −61 dB. In practice additional digital dither is usually added at the mixer and in final level adjustment, and imperfect analogue-to-digital and digital-to-analogue converters add noise, hence −60 dB.
3. FM broadcast measured on BBC Radio 1, continuity to Woodbridge, Suffolk, off air, using LA100 sequence test.

References

ANTONIAZZI, C., *et al.*, Measuring transient intermodulation in audio amplifiers, *Wireless World*, 45–47, 53 (May 1981)

HILLIARD, J.K., Intermodulation testing, *Electronics*, **19**, 123–127 (July 1946)

HIRATA, Y., Quantifying amplifier sound, *Wireless World*, (Oct. 1981)

JUNG, W.G., STEVENS and TODD, An overview of SID and TIM, *Audio* (June 1979)

MOIR, J., Acoustic noise units, *Wireless World* (Dec. 1976)

OTALA, M., Nonlinear distortion in audio amplifiers, *Wireless World*, 41–43 (Jan. 1977)

SCOTT, H.H., Audible audio distortion, *Electronics*, **18**, 126–131 (Jan. 1945)

SHORTER, D.E.L., The influence of high-order products in nonlinear distortion, *Electronic Engineering*, 152 (April 1950)

SMALL, R.H., Total difference frequency distortion, *AES 75th Convention Preprint* (March 1984)

THIELE, A.N., Measurement of nonlinear distortion in a band-limited system, *IREE Convention Digest* (Aug. 1975)

WIGAN, E.R., New distortion criterion, *Electronic Technology*, **38**, 128–137, 163–174 (April and May 1961)

Standards relating to audio measurement

(Preferred source is listed first)

Level measurement

IEC60268–10, second edition, Sound System Equipment, Part 10: Programme Level Meters (1991)

IEC60268–17 Part 17: Volume unit (VU) meters

IEC60268–18 Part 18: Digital level meters Parts 1 to 18

IEC60268 Parts 1 to 18 have been recently updated and cover many related topics.

BS 6840 Part 10 (alternative source for above).

BS 4297: 1968, The Characteristics and Performance of a Peak Programme Meter (1968) (Earlier version now obsolete).

ITU-R Rept 292 (was CCIR292-2) *Use of IEC type 1 PPM*

ITU-R Rept 820 (was CCIR820) *Use of the VU meter*

ANSI C16.5–1954 (ASA C16.5) original VU Meter Specification.

Noise measurement

REC ITU-R BS468–4 Measurement of Audio-frequency Noise Voltage Level in Sound Broadcasting etc. (formerly CCIR REC468–4) (1990) (Earlier versions had slightly different tolerances but were essentially the same)

CCITT 0.41, Specification for a Psophometer for Use on Telephone Type Circuits, Fascile IV.4, (applicable to telephone noise only). (Also equivalent to CCITT 0.41)

IEC60651, Precision Sound Level Meter (A weighting) (1993), (applicable to environmental noise measurement)

Rumble measurement

IEC98-1964

BS 4852, Part 1, 1972, roughly equivalent to IEC98

Wow & flutter measurement

IEC 386, Method of Measurement of Speed Fluctuations in Sound Recording and Reproducing Equipment, January 1971

BS 4847: 1972, Equivalent to IEC386

CCIR 409-3, Measurement of Wow and Flutter in Recording Equipment and in Sound Reproduction. Earlier equivalent

Intermodulation distortion measurement

SMPTE PH22.51–1961, Intermodulation Tests for 16mm Variable-Density Photographic Sound Prints, Journal of the SMPTE, **70**, 536 (1961)

Frequency intermodulation measurement

DIN 45 411, Frequency Intermodulation of Records – Measuring Methods (May 1976).

Automatic test equipment

CCITT 0.33, Specification for an Automatic Equipment for Rapidly Measuring Stereophonic Pairs and Monophonic Sound-Programme Circuits, Links and Connections, CCITT Fascile IV.4 (July 1985)

4 Digital equipment

Terry Clarke
Technical Director, mc² Audio Ltd
4.1 Artificial reverberation devices 4/2

David Mellor
Media Consultant
4.2 Digital mixing and filtering 4/8

Francis Rumsey
Senior Lecturer, University of Surrey
4.3 MIDI 4/17

4.1 Artificial reverberation devices

4.1.1 Reverberation

Reverberation consists of multiple sound reflections which re-reflection causes to build up into a dense statistical clutter wherein individual reflections can no longer be identified by the listener. The size and shape of the acoustic environment determines the spacing of the early reflections and the time taken for these to build up into a statistical clutter. The acoustic properties of the room boundaries, the room contents and the air within the room affect the frequency content of the reverberated sound and also its decay time. Much of what we hear is reflected sound, so control over reverberation is an essential part of sound engineering.

4.1.1.1 The identifiable components of reverberation

First we have a finite delay or predelay between the initiation of any sound and the first reflection from that sound reaching the listener. Also, in a conventional concert situation, the direct sound will arrive at the listener slightly before the first reverberant reflections. Secondly, this reflection will be followed by other discrete early reflections, the spacing and intensity of which will depend on the physical properties of the room and the position of the listener relative to the sound source. However, these early reflections themselves will be surrounded by a halo of reverberation due to diffusion occurring at reflective surfaces. As we shall see later, these early reflections will need to be considered separately to the main body of reverberation.

The properties of the room will also determine the way in which the frequency content of these reflections will be modified so we have to find a way of simulating this effect. Finally, the process of reflection and re-reflection will cause these early reflections to build up into a dense statistical clutter of decaying amplitude, the frequency content of which will be further modified by the reflective characteristics of the wall materials and the air absorption within the building. So we have to devise a way of implementing a statistical clutter generator. The actual decay time is a function of the reflective properties of the walls, the contents of the room and the room dimensions. *Figure 4.1* illustrates the way in which reverberation develops in response to a short impulse of sound. In order to make our hypothetical reverberator capable of simulating a variety of natural acoustic environments, not only must

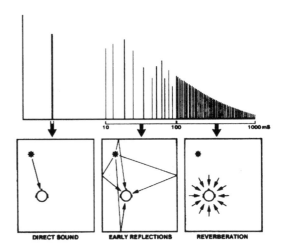

Figure 4.1 *Build up of reverberation from a sound impulse.*

these parameters be simulated, they must also be variable.

4.1.1.2 Early reverberation devices and their development

The most common early reverberators were spring or plate devices which worked on electromechanical principles rather than purely electronic ones. Both relied on a transducer feeding mechanical energy into a structure that could propagate and reflect sound; in the case of the spring system this was a helical spring or springs under tension and, in the case of a plate, a thin sheet of metal suspended in a frame. The vibrations in the spring or plate were then converted back to an electrical signal by another transducer and subsequently mixed with the dry signal in the desired proportions. Stereo effects could be obtained by using several springs and routing their outputs to one or other side of the mix and, in the case of the plate, using two pick-up transducers mounted on different parts of the plate surface.

Although the operating principle is very similar in both cases, the plate gave generally better results due to its lower coloration and was almost universally adopted by sound studios for a number of years. The plate sound is, however, not an accurate simulation of true room reverberation as it is characterized by a very rapid build up of reflections which attain a high density very early on. Also some plates tended to add more coloration to the sound than others and the effect is rather percussive. However, the achieved result was very usable and some control over the decay time could be exercised by damping the plate. What the plate could not do though was to simulate a

specific acoustic environment as there was no control over the build up or spacing of the early reflections, the reverberation density or the way in which high frequencies decay relative to the low frequencies other than the changes that could be made by altering the plate tension and mechanical damping. The only variable parameters were relative level, decay time and overall equalization. Furthermore, they were extremely susceptible to mechanical vibration and were bulky.

4.1.2 Artificial reverberation – the basic theory

4.1.2.1 Schroeder's pioneer work

Schroeder's AES paper published in 1961 and entitled 'Colorless artificial reverberation' was the starting point. Here he discussed the behaviour of rooms in response to transient sounds and came up with a set of conditions which an artificial reverberator would need to satisfy before it would sound natural. These were:

(1) the frequency response must be flat when measured with narrow bands of noise, with a bandwidth corresponding to that of the transients in the sound to be reverberated;

(2) the normal modes of the reverberator must overlap to cover the entire audio frequency range;

(3) the reverberation times of the individual modes must be essentially equal so that different frequency components within the sound will decay at equal rates;

(4) a short time after the excitation, the echo density must be high enough so that individual echoes cannot be resolved by ear;

(5) the echo response must be free from periodicity which would manifest itself as flutter; and

(6) the amplitude/frequency response must not exhibit any apparent periodicities, because otherwise unnatural colorations would be generated.

Schroeder then goes on to introduce the *all-pass filter* and the *comb filter*, which were covered again in his 1962 paper on the subject. In 1962, Schroeder published another AES paper entitled 'Natural sounding artificial reverberation' to follow up his previous work. This paper started out by underlining the shortcomings of the existing systems which were mainly electromechanical or relied on magnetic recording in the form of tapes or discs.

These suffer from two serious defects.

(1) Their amplitude responses are not flat leading to unnatural coloration.

(2) The number of echoes per second in response to a single input pulse is too low compared with the manifestations in a real room. This leads to the characteristic 'fluttering' which is particularly evident on percussive sounds.

Schroeder then went on to describe the all-pass reverberator which would exhibit a flat frequency/amplitude response and so solve the problem of coloration. This was based on the all-pass filter, a block diagram of which is shown in *Figure 4.2*.

x = in
y = out
* = multiply
Σ = sum
□ = memory access
▶ = signal flow
t = time delay
G = gain

*Σ = 4
mem acc = 6
FREQUENCY RESPONSE IS FLAT

Figure 4.2 *The all-pass filter.*

The concept involved connecting several all-pass filters in series in order to create the required reverb density and, as each filter section has a flat frequency response, the overall system should preserve this characteristic. He suggested making the delay of each filter section about one-third of that of the previous section and with five filters in series, the calculated echo density for his values of delay was 810 per second. He originally postulated that something in the order of 1000 echoes per second was sufficient to create a natural sounding reverberation. It is now accepted that between 1000 and 3000 reflections per second are required for a truly convincing simulation.

Exploration of the possibility of connecting several comb filters in parallel to simulate the behaviour of rooms that do not have a perfectly flat amplitude/frequency response led to the block

diagram of a comb filter as shown in *Figure 4.3*. Those who have experimented with digital delay devices will recognize this as the same effect obtained by setting up a delay time of several milliseconds and then feeding some of the output of the delay line back to the input. If it is desired to make the reverberation time a function of frequency for greater realism, it is a simple matter to insert filter components in the feedback path. However, the bank of comb filters alone does not provide a great enough echo density, and so a further section consisting of several cascaded all-pass filters must be inserted in series with the comb filter, as shown in *Figure 4.4*. The paper concludes with a description of how a matrix system could be devised for simulating a multichannel output to be fed to speakers placed around the room to give an enhanced effect.

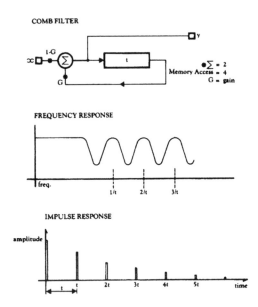

Figure 4.3 *The comb filter.*

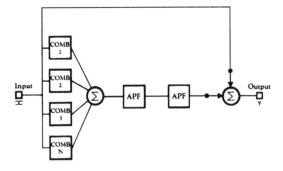

Figure 4.4 *Block diagram of a reverberator using both all-pass and comb filters.*

Some years later, in 1970, Schroeder suggested using a separate reverberator to simulate the first few reflections and this was followed up by Moorer.

4.1.2.2 Moorer's work

'About this reverberation business' (Moorer, 1979) took up from where Schroeder had left off and, with the help of the more advanced computer technology that had developed in the intervening years, Moorer set out to narrow the gap between natural and synthetic reverberation. The first step was to examine the problem areas in Schroeder's original designs, and these were as follows.

(1) With a reverberator built up only from a number of all-pass filters in series, the decay did not start with a dense sound and die out exponentially. In fact, the higher the order, the longer the echoes took to build up to a pleasing level. This lag in reverberation could be as much as several hundreds of milliseconds.
(2) Even slight changes in delay time between the successive filters could cause the smoothness of the overall decay to vary enormously.
(3) The tail of the decay exhibited a metallic ringing.

These parameters defied precise mathematical analysis and the most satisfactory combination of values had to be arrived at empirically.

In his paper, Moorer examined the properties of the Boston Symphony Hall, widely held to be a particularly good example of concert hall acoustics, and published the following observations. 'The reverberation time of over 1.7 s suggests that the sound has travelled over 600 m before dying away and, given typical atmospheric conditions, the 4-kHz signals will be attenuated some 60 dB more than 1-kHz signals.' This shows up the importance of simulating the correct frequency characteristics for the late decay. Furthermore, to simulate a good listening environment, the reverberation must be smooth and dense with no apparent resonances. In addition, to simulate the early reflections accurately, these would have to include a halo of reverberation to simulate the effect of diffusion at the reflecting surfaces.

Moorer then moved on to consider the simulation of these early reflections using a finite impulse response (FIR) filter (*Figure 4.5*). This is in effect a tapped delay line where each of the taps has separate gains. The spacing of the taps must be irregular so as to avoid any obvious periodicity in the reflection pattern but the optimum delays have to be

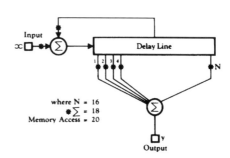

Figure 4.5 *The finite impulse response (FIR) filter.*

chosen empirically; attempts to calculate mathematical spacings based on prime numbers or to simulate the behaviour of actual rooms did not prove to be satisfactory. These taps may then be summed and treated with an all-pass/comb reverberator to build up the required density.

Taking all the above factors into account, the block diagram of our reverberator might look something like that shown in *Figure 4.6*. Using modern computer technology, these components can be simulated in real time, but the computer needs to be capable of very fast arithmetic processing if it is to generate a convincing, dense reverberation with an acceptable bandwidth.

4.1.2.3 Tone control
In addition to the three basic but important elements, we also need to introduce frequency-selective filters

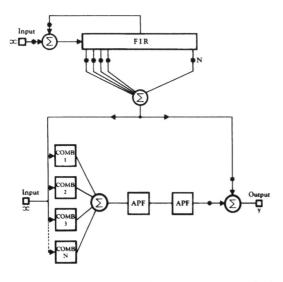

Figure 4.6 *A reverberator with finite impulse response (FIR) and comb and all-pass (APF) filters.*

into the processing path. The basic low-pass filter and the full low-frequency/high frequency (LF/HF) control are shown in *Figure 4.7*. The full LF/HF control configuration is very flexible as it gives control over both HF and LF cut and boost and both the HF and LF turnover points are variable. Again this imposes a demand on processing time which is two multiplier/accumulator operations and three memory accesses per sample period for a simple low-pass filter; seven multiplier/accumulator operations and 10 memory accesses are needed for the full LF/HF control filter.

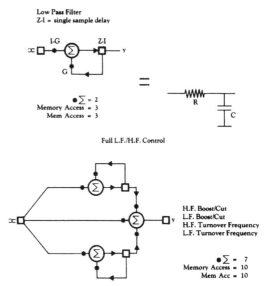

Figure 4.7 *Low-pass and low frequency/high frequency (LF/HF) control filters.*

4.1.2.4 Summary
There are three main elements or building blocks which can be used in combination to simulate reverberation: the all-pass filter, the comb filter, and the

finite impulse response filter. In addition, some way of modifying the frequency response of feedback loops within the system must be implemented. The characteristics of each of these building blocks significantly affects the subjective quality of the result and these parameters cannot be optimized purely by mathematical means. Some empirical adjustments must be made. The implementation of a high quality reverberation simulator makes heavy demands on processing speed.

4.1.2.5 The complete basic reverberator
Figure 4.8 illustrates a complete reverberator combining all the elements so far discussed. Note that some taps from the FIR are fed only to the comb filter bank and that later taps are only summed into the all-pass-filter (APF) section. Also note that the feedback around the FIR is routed via an APF and a low-pass filter to add diffusion and to tailor the frequency response of the decay. By duplicating the APF section and summing different taps from the FIR and outputs from different comb filter taps within the bank, two different outputs can be synthesized. These two outputs can be used in stereo applications and the results from a carefully designed system can be most effective, yielding a greatly enhanced sense of depth.

Figure 4.8 *A complete digital reverberator. APF, all-pass filter; FIR, finite impulse response filter; HF, high frequency; LF, low frequency.*

In terms of computing power, the mathematical implications of this system, using 12 combs, are that we need to perform 140 multiplier/accumulator operations and 200 memory accesses per sample for a stereo unit.

4.1.3 A modern approach

In order to obtain a satisfactory audio bandwidth, the sampling rate could reasonably be expected to be at least 48 kHz and it transpires that 16-bit minimum (18- to 20-bit preferred) sampling is needed to give the required resolution for serious professional work. A floating point system is not adequate for the task as lack of accuracy in the ensuing arithmetic processing leads to a noticeably 'digital sound'. Furthermore, the internal architecture should be at least 32-bit to prevent digital overload or truncation. Before pursuing the subject of hardware any further, it is interesting to examine the algorithms required to construct an effective reverberation program with particular regard to their demands on processor speed.

4.1.3.1 Digital signal processing
If an attempt was made to design a processor-based system that would compute 3000 + individual reflections every second and assuming a 48 kHz sampling rate, it would soon be realized that it is well beyond the capability of even the fastest microprocessors currently available. However, by using a combination of the building blocks previously discussed, much more efficient use can be made of the available computing speed. The reader will probably have gathered by this point that the building blocks do not exist as discrete components, but exist only in software, the outcome of which is that different block diagrams can be set up for different reverb effects without any changes in hardware being necessary. Even so, a dedicated 'number crunching' device will be required. There are several dedicated Digital Signal Processing (DSP) 'chips' (integrated circuits) on the market which can be configured to carry out all the necessary mathematics described above. Most of them can be run in multiples, either parallel or serial, to increase their word length or processing power. These are pure 'number crunchers', and a separate general-purpose microprocessor or controller will be required to interface with the front panel controls and supply the relevant data and instructions to the DSP section. *Figure 4.9* shows a very simplified block diagram of a DSP/microprocessor configuration that would be found in a typical digital reverberator.

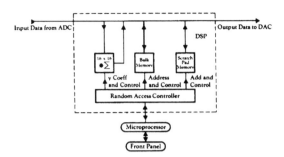

Figure 4.9 *Block diagram of a digital signal processor.*

Some of the digital reverberation devices on the market use their own custom 'chips' which include all the processing requirements of the unit, as well as the A/D and D/A convertors, in one or two large integrated circuits, thus reducing the cost considerably. In most cases, performance or versatility has to be compromised.

4.1.3.2 Summary
It is not practical to calculate the required number of reflections independently to simulate natural reverberation; consequently, the building block approach is adopted. The building blocks are implemented entirely in software. Even using the building-block approach, careful design is necessary to devise a processor system that is both fast enough and accurate enough to synthesize high-quality reverberation. A dedicated DSP 'number cruncher' will be needed to achieve this.

4.1.4 Control parameters

There are so many variables within a modern digital reverberation system that if they were all made available to the user it is unlikely that he or she would even get a recognizable reverberation sound out of it at all, let alone the one wanted. On the other hand, the old spring and plate devices were far too inflexible, so some middle ground has to be sought where the user can quickly and accurately set up a wide variety of natural and unnatural reverberation simulations.

By looking back over the explanation of the mechanism of reverberation we can gain clues as to what parameters we need to be able to adjust and which are best left to the software writers. We have already seen that the early reflections have a profound effect on the character of the reverberation, and so we need to have a choice of reflection patterns in order to

convincingly simulate rooms, halls, plates and so on, and of course gated and non-linear treatments will also require their own specific patterns. Also, adding a user definable amount of predelay means that the time lag between the direct and reverberant sound can be duplicated. By varying the spacing between these reflections, we can alter the impression of room size and it is also an advantage to make their amplitude variable. However, this is not simply a matter of turning the level of the early reflections up or down en masse, as this sounds unnatural. What is required is a system that varies the initial level of these reflections, but allows them always to decay to the same level at the point where the statistical clutter becomes the most significant part of the sound.

In a concert hall, it is surprising what effect the input from our eyes has on perceived sound, and tests indicate that the initial reflections need to be of greater amplitude in situations where sound is the only stimulus. To put it simply, in order to create the effect of a concert hall on record, it may well be necessary to set the early reflections at a higher level than their natural counterparts to create the same subjective effect. Furthermore, the intensity of these early reflections will vary depending on whereabouts in the room the listener is situated.

Reverberation decay time is a fairly obvious parameter, but some control also needs to be exercised over the frequency content if a really convincing simulation is to be achieved. In the case of the reverberator shown in *Figure 4.10*, the designers have made it possible for the user to vary the high and low frequency decay times relative to the midband decay time. Furthermore, they have built in a single 'room size' control which varies all the necessary parameters automatically when the simulated environment needs to be changed in size. Something in the order of 60 parameters are changed by this one control saving the user a lot of unnecessary work. This ability to simulate the acoustic properties of various sized rooms rather than merely being given control over the reverberation decay time is of paramount importance not only in music and drama productions but also in film work where what the

Figure 4.10 *A typical digital reverberation unit.*

audience sees must be reinforced by what they hear and vice versa.

4.1.5 Applications

Most of the applications of reverberation are similar, whether in a live sound application or in a recording studio. In general, reverberation is added for the benefit of the end-listener, but there are exceptions. For example, it has been shown that a number of musicians find it easier to pitch sounds correctly in some rooms than others, and current thinking suggests that the early reflections within a room are used subconsciously to help pitching. By feeding a pattern of early reflections generated by a digital reverberator to the stage monitoring system, the situation is often much improved. A parallel situation occurs in the recording studio where a singer may perform better when a little reverb is added to his voice in the headphones, even though this may not be recorded onto tape.

As intimated earlier, much of what we normally hear is reflected sound, even in a fairly dead room, though there may be no apparent presence of reverberation in the accepted sense. These reflections, however, give the original sound much of its spatial identity and a dryly recorded or close-miked sound can be brought alive by the addition of a short reverberation treatment. In any event, it is usual to pan the two outputs from the reverberation unit hard left and right to create the best stereo effect. The reason that most reverberators have only a single input is that, in real life, reflections come from all sides, regardless of the position of the sound source. Of course the reverberation returns may be panned differently if the aim is to create a specific effect.

At decay times of up to 0.5 s or so, the addition of reverberation creates the effect of a room without the listener being aware that any processing is taking place. This is a useful treatment for voice in a broadcast situation and makes the programme more listenable without reducing intelligibility. Likewise, in music production close-miked drums or drum machines can be given a sense of substance without being obviously affected.

Longer decay times up to 1.5 s are commonly used to thicken vocal or drum sounds and here the treatment takes on the guise as an obvious, though possibly natural sounding, effect. When enhancing drums with reverberation it is common practice to avoid putting reverberation on the bass drum as this tends to clutter the mix with unwanted low-frequency sound. The same is true of any bass instrument

so if it is required to add a significant amount of reverberation to a bass guitar say, then it might be advisable to equalize the effect-send line in order to attenuate the level of low-frequency signal being fed into the reverberator.

Longer reverberation times are flattering to choral vocals, flute and various synthesized sounds, and it is up to the producer to decide what level of treatment is artistically viable for a given set of circumstances. Also increasing the relative level of the early reflections can be particularly effective in lending presence to a vocal line.

Because the reverberant drum sound was first established when plate reverberators were standard studio equipment, a plate setting is often thought to be the best treatment for drums. This is purely a matter of tradition of course as there are no absolutes in the world of art, but the dense, percussive nature of the plate setting is nevertheless very flattering to drums.

Gated effects can be applied to all the drums of a kit as the abrupt cut-off prevents the sound from getting cluttered, and this is a very popular production trick. Originally this sound was brought to the public attention by the musician Phil Collins and, though it is no longer new and is used extensively, it has established itself as a standard effect and looks set to stay with us for some time.

Reference

MOORER, About this reverberation business, *Computer Music Journal*, **3** (1979)

4.2 Digital mixing and filtering

4.2.1 Introduction

The move from analogue audio to digital recording which gathered momentum during the 1980s will probably be completed, apart perhaps for music studios, during the 1990s and will be seen as being wholly beneficial. Digital recording places enormous demands on mechanical systems, disk drives, tape, tape transport mechanisms and heads, and on systems necessary to keep the data flowing at a rate which is sufficiently even to be within the handling capabilities of the digital processing circuitry. In professional equipment, audible errors and drop-outs are unacceptable and it has been shown that the major manufacturers are quite capable of producing equipment which, when properly maintained, can reach

and surpass the standard formerly set by analogue recording equipment. Although digital recording was a major achievement, digital sound processing is an even more demanding problem, if it is to be capable of every technique and trick that is available in the analogue domain. Where formerly engineers were content to work around the limitations of analogue mixing consoles and processors, because there was no alternative, digital equivalents must provide the characteristic benefits that digital operation can offer *and* perform all the existing analogue functions. This is a tall order, and digital mixing consoles on the scale of large analogue consoles as found in a multitrack recording studio are only now becoming commonplace.

Digital mixing facilities can also be found in smaller scale equipment. Lower-cost digital mixing consoles are available and can be a practical proposition for even the home studio. Digital mixing also takes place in equipment such as hard disk editors and samplers. While it is unlikely that the practising sound engineer will ever find a workplace application for his or her knowledge of how digital mixing works, as might be the case for an equivalent degree of knowledge of analogue consoles, which are frequently modified by their owners to suit the special needs of each studio, knowledge for its own sake will assist in the selection of digital processing equipment, in its operation, and in giving feedback to the manufacturer on how a product can be improved. Unfortunately, existing literature on the subject is densely populated with mathematical concepts and equations which may often be above the level that a working sound engineer can be expected to have attained, sound engineering being a complex discipline in its own right and seldom attempted by mathematicians! In this section, bearing in mind that this is not meant to be an instructional text on how to design a digital console, all mathematics has been eliminated so that the concepts may be explained and understood more easily.

4.2.2 The need for digital mixing

The need for digital mixing is motivated more than anything else by the fact that analogue consoles are becoming increasingly large. Of course, this is not without good reason. In all fields of sound engineering activity the consumers of the technology, the producers and musicians for instance in the fields of music recording and concert PA, are becoming more demanding. These increased demands lead to a requirement for more control over the quality of the sound and more flexibility in routing. In an analogue console this inevitably means that there have to be more knobs and switches and the control surface must be larger. Furthermore, in order to maintain an acceptable sound quality, the controls must be physically close to the circuitry. The process of mixing two or more analogue signals together is very susceptible to cross-talk and one manufacturer has taken pains to place an important component only millimetres from its associated control surface switch to combat this. In addition, if the distance between controls and circuitry is increased, then the risk of interference entering the cabling increases in proportion. One solution which has been attempted is to mount the circuitry remotely and have it controlled by non-audio signals from the console. This is not entirely satisfactory as the sound quality may be slightly, but significantly, compromised by circuit components such as voltage-controlled amplifiers and attenuators.

The key advantage of digital mixing is that the circuitry can be mounted remotely from the console with no loss of quality. This frees the console designer considerably, even more so when it is realized that the console is now released from the straitjacket of one control/one function. A digital mixing console need not be an array of thousands of knobs and switches. In theory the controls could be reduced to a data entry wheel or slider, a button, and a display to indicate the values of the various parameters. There would be no need for this to be any larger than an A4 sheet of paper. Perhaps it would not be convenient to operate, but at least it may be realized that the opportunity now exists for the designers of digital mixing consoles to offer any operator interface that can be imagined. Perhaps the thousand-knob interface may prove to be the most popular, or maybe an assignable console with eight or so channels worth of controls, but mapping onto any number of real channels, would be preferable.

Digital mixing consoles would have other advantages, including the promise of increased sound quality over analogue systems. As will be shown later, the microphone amplifiers and analogue-to-digital convertors may be placed remotely from the console (*Figure 4.11*). A fibre-optic link could replace the conventional bulky multicore cable in a PA system, for example. Also in connection with microphones, it has long been realized that the conventional method of multimiking an orchestra suffers from the time delay between the sectional microphones and the microphones placed for overall pick-up. While patching digital delay lines into every channel in use in an

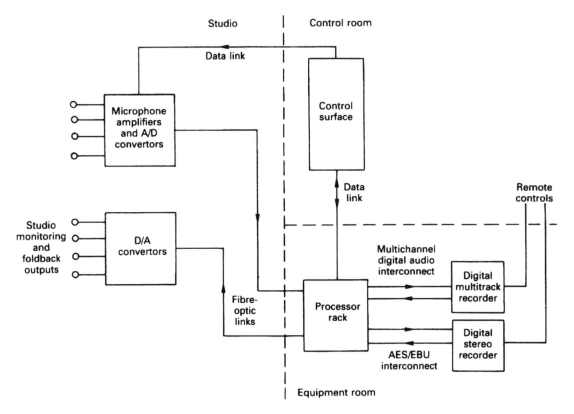

Figure 4.11 *The digital mixing console in a studio setting. Since there is no signal degradation in digital links, the microphone amplifiers, analogue-to-digital convertors, console and processor rack can be in their optimum positions.*

analogue console is virtually unheard of, it becomes exceedingly practical to have a control on each channel of a digital console labelled simply 'delay'. One speculates whether this could be set automatically using a test signal.

It must always be remembered that, besides providing good sound quality, and being straightforward to operate, a mixing console is a business tool. Like any plant or equipment, it is only earning money while it is in use. For instance, in a music recording studio, a considerable investment in time is made in setting the controls of the console during recording or mixing. To retain these settings, clients often pay for a 'lock out' so that they can finish work one day and come back the next knowing that the settings will be unchanged. A comparable problem exists at concerts where there are several acts. It is not uncommon for two consoles to be set up so that, while one is in use, the other can be set for the next band. There are of course a number of systems which will memorize the settings of the console's controls, but these have to be reset by hand, under the guidance of a visual display, which still takes considerable time.

Programmability is almost inherent in a digital console, the processor must 'know' the positions of all the knobs and switches simply to mix. And since this information is known it can be memorized either internally or recorded onto a transportable medium such as a floppy disk or memory cards. Resetting the console can be done automatically and almost instantaneously. The net result is that the console can be used for more hours in the day and, therefore, even if its initial cost is greater than an analogue console, it has the potential to generate income more efficiently.

4.2.3 Digital console architecture

Before examining the internal arrangements of digital mixing consoles it is necessary to understand the ways in which digital data can be transmitted. There are basically two choices: serial and parallel. When an analogue signal is sampled and converted into digital form the result is a sequence of 16-bit digital words, with present-day technology. There

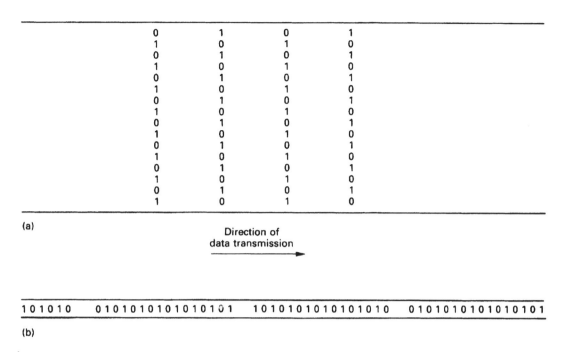

(a)

Direction of
data transmission
————————▶

(b)

Figure 4.12 *(a) Parallel data transmission requiring 16 wires and an earth reference for 16-bit digital audio. (b) Serial data transmission requiring just one pair of wires and screen for 16-bit stereo digital audio (AES/EBU standard).*

are then 44 100 or 48 000 16-bit words every second, according to the sampling rate used. (These two figures represent the most common rates.) For the sake of understanding the concepts of serial and parallel data transmission it would be reasonable to say that the form of the original data is as shown in *Figure 4.12(a)*. It will be evident that to send this data from one place to another in this form will require at least 16 wires plus an earth reference. This is parallel data transmission, which is not a problem within one piece of equipment, but for transmission over long distances the impracticality is self-evident. The preferred method of transmission of digital audio is specified in the AES/EBU standard which uses one pair of wires to carry two channels of audio data. To fit data 16 bits wide (actually 32 bits if we consider the other half of the stereo pair) onto one electrical circuit, each 16-bit word is turned 'sideways' and transmitted bit by bit, as in *Figure 4.12(b)*. This is serial data transmission. It will be obvious that, while serial transmission is more convenient, parallel transmission is faster, which is why digital mixing consoles use parallel data buses.

In *Figure 4.13*, the analogue-to-digital convertors are remotely mounted. The gain of the microphone amplifier will need to be adjusted, as it is in an analogue console, and it is possible to do this via

a data link from the digital console's control surface processor. Gain control may be linked to the channel fader to provide extra dynamic range without the operator's conscious intervention. As mentioned earlier, the resulting digital signals may be multiplexed onto a fibre-optic link for serial transmission to the console, rather than being linked via a multicore cable. Fibre-optic links have the advantage of being immune to electrical interference. There will undoubtedly also be digital sound sources, recorders, compact-disc players, etc., which can be linked directly via AES/EBU interconnects. All these inputs will enter an input interface of some kind the function of which is to separate multiplexed inputs and to place them, together with the inputs from the AES/EBU sources, onto the digital bus. In contrast with analogue consoles which have many buses, one for each signal, there need only be one bus carrying many digital signals. There may be, for instance, 128 signals all multiplexed onto the same set of wires and travelling through the system as parallel data streams. All it takes is (very) precise timing.

The function of the mixing processor is to combine data from the control surface with the audio data so as to make possible the functions that are required of a mixing console. Digital mixing is a matter of multiplying the numbers representing samples by

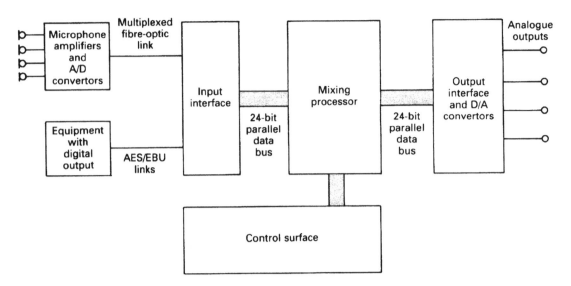

Figure 4.13 *Simplified block diagram of a digital mixing console.*

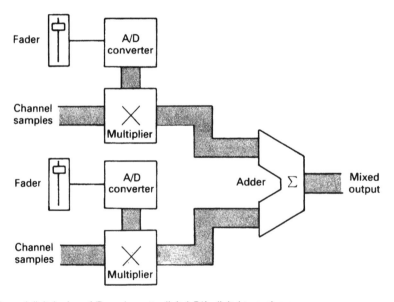

Figure 4.14 *Two-channel digital mixer. A/D, analogue-to-digital; D/A, digital-to-analogue.*

coefficients generated from data provided by the faders, and of adding the resulting numbers together. It will be appreciated that when 16-bit numbers are multiplied and added, the end result will very likely be a number which is greater than the highest possible 16-bit value. For this reason, the processor must work to a much higher number of bits, typically 32 giving a dynamic range of 192 dB. For reasons of headroom, the digital buses will operate to at least 24 bits with a dynamic range of 144 dB.

To repeat the above information in another way, hopefully to assist the reader's understanding, *Figure 4.14* shows a much less ambitious two-channel digital mixing circuit, but in more detail. The gain coefficients are generated from faders via analogue-to-digital convertors. The digital samples are multiplied by these coefficients and summed. The result is the mixed digital audio output. As a result of all this mixing and processing we have a digital data stream up to 32 bits wide. To convert this back

to 16 bits, care needs to be taken to avoid distortion at low levels. This is explained in Section 4.2.5.

4.2.4 Digital filtering

4.2.4.1 The advantages of digital filtering

The principal advantage of digital filtering is the absence of the disadvantages of analogue filtering. Simple analogue filters are easy to design and construct, and they will most probably perform as expected. This would apply to filters with slopes up to 18 dB/octave. With steeper slopes, component values become increasingly critical. The design response is unlikely to be achieved with accuracy and will drift with changes in temperature and with time as the components age.

Although analogue filters are usually constructed to achieve a desired frequency response, the phase response and impulse response will also have a bearing on the sound quality. A change in frequency response necessarily leads to a change in phase response, but in a perfect filter the phase shift will be proportional to frequency. Such a filter would be described as *phase linear*. Unfortunately, many filters suffer from group delay error, meaning that phase shift is no longer proportional to frequency, and they require a group delay equalization stage which may be every bit as complex as the filter itself. In a perfect filter, every frequency will be delayed by the same amount of time. If there is a group delay error, different frequencies will be delayed by different amounts, the consequences of which are obviously undesirable. With digital filtering, it is relatively easy to design a filter which is phase linear, and since numerical coefficients do not age or drift, the response of the filter will be maintained indefinitely.

Digital filters are not described or understood easily, and not described fully without the application of complex mathematics, but a lot can be achieved by turning one's attention from the frequency response of a filter, which is the main item of interest in the analogue domain, to the *time* response, and a good feeling for how a digital filter works can then be attained. First we need to look at the basic components of a digital signal processing system.

4.2.4.2 Building blocks of digital signal processing systems

The building blocks of analogue mixing consoles are resistors, capacitors, integrated circuits and transistors, with the odd sprinkling of inductors and transformers. With the integrated circuit operational amplifier, usually known as the 'op-amp.', analogue circuits can be very simple, with few circuit elements necessary to implement a filtering or mixing stage. The components of an analogue mixer are physical things, they can be picked up and looked at. However, in a digital mixing system, the building blocks are intangible theoretical constructs which have been conjured into existence in a universe consisting of digital signal processing (DSP) chips, digital buses and random-access memory (RAM). The conjurer is the person who programs these elemental devices into the system in a shape and form which will carry out the intended function.

Digital signal processing algorithms can be made up from three basic processes: summing, multiplication and delay (*Figure 4.15*). Summing corresponds to simple signal mixing and is the addition of two or more digital words. The multiplication operation involves a *coefficient* where, in analogue terms, a series of coefficients corresponds roughly to a set of component values. The most common delay is one sampling period. A delay of one sample period is here denoted by T. Elsewhere it may be notated as z^{-1}, since this is more meaningful mathematically.

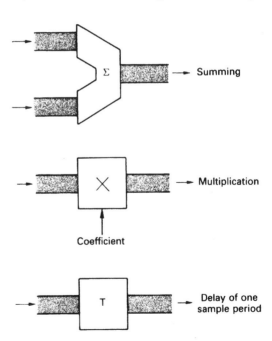

Figure 4.15 *The basic components of a digital-signal-processing system.*

4.2.4.3 Simple filter conceptualization

Using the simple building blocks we can devise a simple filter (*Figure 4.16*). In this filter each sample

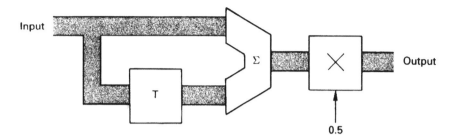

Figure 4.16 *Simple digital filter.*

is added to the previous sample and the sum divided by 2. In other words, an averaging process is carried out which will tend to reduce the amount of variation in the output signal. We have a low-pass filter.

4.2.4.4 Frequency response and impulse response

Consider the circuit shown in *Figure 4.17(a)*. It is of course a simple integrator, or low-pass filter if one prefers. Its frequency response is flat in the low frequency region, and falling at a rate of 6 dB/octave at higher frequencies. Looking at the same filter in a

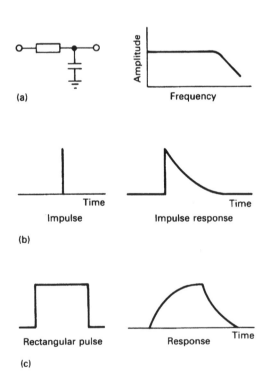

Figure 4.17 *(a) Frequency response of an RC network. (b) The response to an infinitesimally short impulse. (c) The response to a rectangular pulse input.*

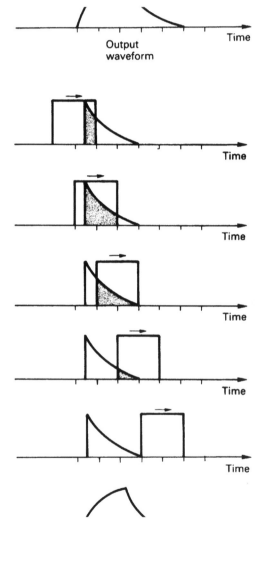

Figure 4.18 *Convolution of the impulse response of an RC network with a rectangular pulse input signal.*

different way we can examine its *impulse response*. Instead of drawing a graph of amplitude against frequency we draw a graph of amplitude against time (*Figures 4.17(b)* and *4.17(c)*).

These responses will be familiar, but the interesting point is that the response to a rectangular pulse input, or any input, can be obtained from the impulse response by a process that mathematicians know as convolution. One does not have to go into the heavy theory of convolution here, but just consider a graphical example, step by step. Instead of sending an electrical signal through a network, consider sending the rectangular pulse 'through' the impulse response. Suspend judgement for a while until we see how this works out. As the pulse moves across the impulse response we shall measure the area where they coincide and plot that (*Figure 4.18*).

Amazingly, the result is exactly the same as that shown in *Figure 4.17(c)*, and is just what we would expect. We can thank mathematicians for working

this out for us, and now examine how this can be achieved digitally. We know now that if an input signal is passed 'through' the impulse response (or *convolved* with the impulse response) of a filter as in *Figure 4.18*, the output will behave the same way in time as would an analogue signal through a real network, and therefore have the same frequency response.

Finite impulse response filters *Figure 4.19(a)* shows a finite impulse response (FIR) low-pass filter with coefficients describing the impulse response of the filter, and therefore its frequency and phase response. The signal is fed into the filter sample by sample, convolved with the impulse response, summed and sent to the output. *Figure 4.19(b)* shows the impulse response from which the coefficients have been calculated. Mathematically, an impulse consists of three things; an infinite time period before the impulse, the impulse itself and an infinite time afterwards. Because of this, the impulse response

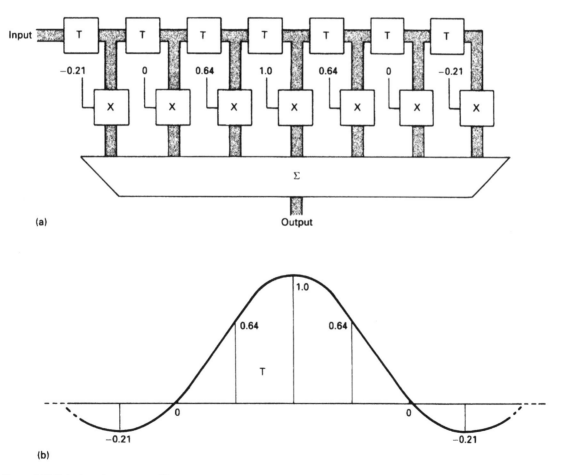

Figure 4.19 *Finite impulse response filter.*

appears to start before the impulse itself actually occurs and therefore the impulse response is symmetrical. In a real system, a delay is introduced so that the system is *causal* – it only responds after the impulse has occurred. This would seem to make good engineering sense.

Although the example given shows a seven-point FIR, in practice the number would be increased to as many as 96. With few points, there will be an unwanted ripple in the filter's response, both in and out of band. With many points, this ripple is reduced as the impulse response of the filter, which in theory extends to plus and minus infinity, can be more accurately represented. With any number of points,

however, the impulse response will be cut short and this results in Gibb's phenomenon which causes a peak just before the cut-off frequency and ripples in the filter's response. Fortunately, this effect can be reduced by 'massaging' the impulse response with the aid of *window function*. There are a number of window functions which have been developed for various purposes, not all of them for audio, and names which will be encountered include: the Hanning window, the Hamming window, the Bartlett window and family of Kaiser windows. The window function simply multiplies every coefficient by a given value and optimizes the impulse response to fit into the number of points available.

Figure 4.20 *Simple infinite impulse response filters and equivalent analogue circuits.*

Infinite impulse response filters Although the FIR filter is relatively easy to understand, it is hardware intensive and often impractical to implement for audio purposes. The infinite impulse response (IIR) filter is difficult to understand but easier to make. The mathematics necessary for an understanding of IIR filters are too complex in this context, but two simple examples of IIR filters are given. The essential difference between the two classes of filter is that the FIR filter works sequentially on the samples which are input and is, therefore, intrinsically stable. The IIR filter has an element of feedback which, just like analogue feedback, can introduce instability. This will be examined further later.

Figure 4.20(a) shows an IIR filter and its equivalent analogue network, which is a simple integrator and, by varying the coefficients, low pass filter responses can be obtained. *Figure 4.20(b)* shows a network using similar elements configured as a differentiator, and once again by varying the coefficients a range of high-pass responses are available.

The equalization section of an analogue mixing console controls the frequency response of various filters by varying component values. In a switched equalizer this could be resistor and capacitor values. If the equalizer has sweep controls then the only values that are changed are resistances. In a digital equalizer the equivalents of the component values are the coefficients that are used in both the FIR and IIR filter classes. Varying the equalizer is a matter of changing the coefficients, which can be stored in a read-only memory (ROM).

4.2.5 Source of errors

In digital mixing and filtering, errors can occur which will either compromise sound quality or result in instability. When an analogue signal is quantized, the least-significant bit will contain an error with up to half the value of that bit which is signal related and, therefore, can be classed as distortion. To overcome this a noise signal is added, known as *dither*, which randomizes the last bit and replaces the distortion with noise. As a by-product, dithering allows signals which would otherwise have fallen below the level of the least significant bit to be captured. If during some digital signal processing operation this dither is lost, for example when a signal is attenuated, then quantization distortion will result, as though the dither had never been there. The solution is to perform digital arithmetic to a much greater degree of precision than the samples themselves, for example using 32-bit arithmetic for 16-bit samples, and ensuring that when a sample is truncated then dither, which may be digitally generated, is added correctly.

In FIR filters the error caused by using only a portion of the complete impulse response of a filter can be minimized, as we have seen, by using an appropriate window function. In IIR filters, however, the problem is that multiplication operations are carried out and the results are carried back to an earlier point and then operated upon again. The problem is not so much that the coefficients themselves are not calculated to sufficient precision, as the inaccuracies can be known and adjustments made. Inaccuracies in the digital audio data are more problematical because the word length increases on each successive multiplication as the signal goes round and round. Sometimes it can happen that, instead of decaying gracefully, a signal may build up and then decay again and become locked in an endless sequence known as a *limit cycle*. This can be prevented by using a digital dither signal appropriately.

4.2.6 Summary

The analogue mixing console was not developed overnight and neither were analogue filters that meet the needs of the working sound engineer. The same will apply to digital mixing consoles and to digital equalization and filtering. Digital technology gives us the power to do many things. The role of the sound engineer in future developments is to work out what the practical requirements are, and to pass that information on to companies engaged in research and manufacture in the digital field.

Bibliography

POHLMANN, K.C., *Principles of Digital Audio*, Howard W. Sams, Indianapolis (1985)
WATKINSON, J., *The Art of Digital Audio*, 2nd edn, Focal Press, Oxford (1994)

4.3 MIDI

4.3.1 Introduction

MIDI, the musical instrument digital interface, is a standard for the communication of control information between electronic musical instruments. Its success in this area has resulted in the adoption of MIDI in other equipment such as effects units, mixer automation systems and synchronizers.

Predominantly, it is used for the transference of performance and control information around a music system, but secondary functions of the interface involve the transference of timing information, and the transference of miscellaneous data relating to instrument voice parameters, digital audio samples, and set-up information.

The standard defines the electromechanical interface and the communications protocol to be used. In other words, it defines the type of connector to be used, the type of wire, the voltages/currents involved, the hardware interface, and the 'language' to be spoken by all the instruments. This has resulted in the unification of different manufacturer's systems, such that the various different communication standards which previously existed have largely disappeared in favour of MIDI. It means that virtually any electronic musical instrument can be incorporated into a system with others without difficulty.

The MIDI 1.0 standard has been formulated by joint agreement between manufacturers, and ratified by the International MIDI Association (IMA), the MIDI Manufacturers Association (MMA), and the JMSC (the Japanese MIDI Association). It has been subject to certain modifications over the past few years, to include functions for which a need has arisen. It should be borne in mind that the large volume of MIDI-controlled equipment in use throughout the world makes it difficult for manufacturers to upgrade software of older equipment to work with revisions to the standard, and therefore some incompatibilities may arise in cases where recent equipment is used with older equipment. Theoretically, incompatibilities should not arise, as very few of the original aspects of the standard have been changed, but incompatibilities are often due to older software not catering for the possibility of the arrival of data that did not exist in the standard at the time of writing.

Complementary to this problem is the one of adherence to the letter of the standard, as not all devices that claim to use MIDI do so either to the fullest degree, or to the letter of the standard. Some aspects of the standard exist as guidelines rather than rules, and it is in these areas that some incompatibilities may arise.

4.3.2 Hardware interface and specification

In this section the nature of the MIDI interface will be discussed, both in hardware terms, and in terms of its function as a serial communications interface between networked devices.

4.3.2.1 A serial interface

A serial interface is one in which digital data are transferred sequentially, i.e. one bit after another. It is thus different from a parallel interface in which a number of bits may be transferred concurrently, using a number of parallel links. Effectively, a serial link between two devices requires a single communications channel, whereas a parallel link requires many, thus the serial link is useful in cases where economy in wiring is paramount, or where only a single channel exists (over a telephone link, for example).

The adoption of a serial standard for MIDI was dictated largely by economic and practical considerations, as it was intended that the interface should be installed on relatively cheap items of equipment, and that it should be available to as wide a range of users as possible. A parallel system might have been more professionally satisfactory and allowed a higher communications speed, but would have involved a considerable manufacturing cost overhead per MIDI device, as well as parallel cabling between devices which is more expensive and bulky than serial interconnection. The simplicity and ease of installation of a MIDI system has been largely responsible for its rapid proliferation as an international standard.

4.3.2.2 Asynchronous communication

MIDI uses what is known as 'asynchronous serial communication'. In asynchronous communication data are transmitted without an accompanying clock signal, and the sending device expects the receiver to lock onto the incoming data whenever they arrive. This makes it very easy to interlink two devices, as there is only really a need for a single wire to carry the data (plus an earth return).

In an asynchronous interface the clocks of the transmitter and the receiver must be running at exactly the same rate, otherwise data may be lost due to timing errors. This close tolerance in clock rates is fundamental to the satisfactory operation of any asynchronous interface, as it is vital that the receiver can rely on a particular bit of a received word occurring at a particular time. The tolerance in clock frequency allowed for MIDI is ±1%. Also, data may be transmitted at any time, and there may be long gaps in between messages. Thus there is a need for a means of signalling the start and end of a data word. This is achieved using start and stop bits.

A start bit precedes every byte transmitted, and a stop bit follows every byte. On recognition of the leading edge of the start bit, the receiver adjusts the phase of its clock such that it thereafter accepts

the bits of the following word in their correct time slots, and the stop bit terminates the word (see *Figure 4.21*).

Figure 4.21 *The format of a MIDI message byte. MSB, most significant bit.*

Every byte transmitted over a MIDI interface is preceded and succeeded in this way, even if the message is longer than 1 byte in total. A device known as a UART (universal asynchronous receiver/transmitter) automatically handles the timing of the serial interface and converts data to and from the parallel form required internally by most equipment and the serial form described above.

It will be appreciated, therefore, that in order to transmit an eight bit word the UART actually transmits ten bits, and that the MIDI bus idles in the 1 state, i.e. with no current flowing. It is important to remember this when attempting to calculate the length of time required to send a particular string of messages.

The MIDI standard specifies a unidirectional serial interface running at 31.25 kbaud. The baud rate is the maximum number of bits per second that can be transferred over the interface. Data are transmitted unidirectionally; that is it has only one direction (from the transmitter to the receiver). Standard MIDI messages typically consist of 1, 2 or 3 byte, although there are longer messages for special purposes.

4.3.2.3 The hardware interface

The hardware interface used by MIDI is shown in *Figure 4.22*. Most equipment using MIDI has three interface connectors: IN, OUT, and THRU. The OUT connector carries data which the device itself has generated. It originates from the device's UART. The IN connector receives data from other devices and relays it to the device's UART. Data messages received will be acted upon by the instrument concerned if it is capable of handling the particular message, and if it is destined for that instrument (see Section 4.3.3.2). Not all MIDI equipment can handle

Figure 4.22 *The MIDI hardware interface.*

every type of MIDI message. The THRU connector is a direct relay of the data that is present at the IN. As can be seen from the hardware interface diagram (*Figure 4.22*), it is simply a buffered feed of the input data, and it has not been processed in any way. A few cheaper devices do not have THRU connectors, but it is possible to buy 'MIDI THRU boxes' which provide a number of THRUs from one input. Occasionally, devices without a THRU socket allow the OUT socket to be switched between OUT and THRU. The THRU socket can be used to 'daisy-chain' MIDI devices together, so that transmitted information from one controller can be sent to a number of receivers without the need for multiple outputs from the controller.

The optoisolator between the MIDI IN (that is the receiving socket) and the device's microprocessor system ensures that there is no direct electrical link between devices, and helps to reduce the effects of any problems that might occur if one instrument in a system were to develop an electrical fault. In the MIDI specification, the optoisolator is defined as having a rise-time of no more than 2 μs. A current loop is created between a MIDI OUT or THRU and a MIDI IN, when connected with the appropriate cable, and data is represented by the turning on and off of this current by the sending device – a binary zero is represented by a current flowing through the current loop, whereas a one is represented by a lack of current. This principle is shown in *Figure 4.23*.

A small amount of rise time distortion can arise in the interface's optoisolator, and also in long cables due to increased capacitance and susceptibility to interference, thus care should be taken not to connect too many instruments in series, otherwise corruption of data may occur, resulting in false messages being

Figure 4.23 *A current loop is formed between the MIDI OUT of the transmitter and the MIDI IN of the receiver.*

received at the end of the daisy-chain. The delay in data passed through the optoisolator contributes little to any audible delays perceived in the musical outputs of some instruments in a large system. The bulk of any perceived delay is due to other factors, which are covered later.

4.3.2.4 Connectors and cables

The connectors used for MIDI interfaces are like the five-pin DIN plugs used in some hi-fi systems, and it is possible to use these (depending on the way that they are wired) although better quality connectors are to be preferred. The specification also allows for the use of professional XLR connectors, although these are rarely encountered in practice. Only three of the pins of a five-pin DIN plug are actually used in most equipment (the three innermost pins). The cable should be shielded twisted-pair with the shield connected to pin 2 at both ends, although within the receiver itself the MIDI IN does not have pin 2 connected to earth. This is to avoid earth loops, but makes it possible to use a cable either way round. It is recommended that no more than 15 m of cable is used for a single cable run in a simple MIDI system, although deterioration of the signal is gradual and depends on the conditions, cable, and equipment in use. Longer distances may be accommodated with the use of buffer or 'booster' boxes.

In the cable, pin 5 at one end should be connected to pin 5 at the other, and likewise pin 4 to pin 4, and pin 2 to pin 2. Unless any hi-fi DIN cables to be used follow this convention they will not work. Professional microphone cable terminated in DIN connectors may be used as a higher quality solution, because domestic cables will not always be a shielded twisted-pair and thus are more susceptible to external interference, as well as radiating more themselves which could interfere with adjacent audio signals. It is recommended that the correct cable is

used in professional installations where MIDI cables are installed in the same trunking as audio cables, to avoid any problems with cross-talk.

4.3.3 Simple system principles

4.3.3.1 Basic interconnection

In the simplest MIDI system, one instrument could be connected to another as shown in *Figure 4.24*. Here, instrument 1 would send information relating to actions performed on its own controls (notes pressed, pedals pressed, etc.) to instrument 2 which may imitate these actions as far as it is able. This type of arrangement can be used for 'doubling-up' sounds, or 'layering', such that a composite sound can be made up from two synthesizers' outputs. It should be pointed out that the audio outputs of the two instruments must be mixed together for this effect.

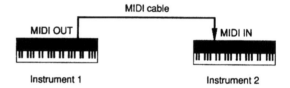

Figure 4.24 *A simple MIDI daisy-chain interconnection.*

Larger MIDI systems may be built up by further daisy-chaining of instruments, such that instruments further down the chain all receive information generated by the first (see *Figure 4.25*). The limit to the number of instruments depends on whether they are all to receive on different channels (in which case it is 16), or whether they are all receiving in omni mode (see below) in which case the limit is that at which the data becomes impossibly corrupted by successive THRU to IN links. Distribution matrices may be used in large systems to avoid long daisy-chains.

Figure 4.25 *More devices may be added in series on a daisy chain.*

4.3.3.2 MIDI channels

Looking at the daisy-chained system in *Figure 4.25*, it will be appreciated that instruments 2, 3 and 4 all receive the information sent out by instrument 1. Instrument 1, in this case, could be considered as the

controlling device, or master keyboard. The problem is how to define which instrument(s) in the chain will respond to which data.

MIDI messages are made up of a number of sections or bytes. Each byte of the message has a specific purpose, and one of these is to define the channel that the message refers to. In this way, a controlling device can make data device-specific. This is most important in large systems which use a sequencer as a master controller, where a large amount of information will be present on the MIDI data bus, not all of which is intended for every instrument. If a device is set in software to receive on a specific channel it will act only on information which is 'tagged' with its own channel number. Everything else it will usually ignore. There are 16 MIDI channels. Instruments can usually be set to transmit and receive on any specific channel, or to receive on all channels (omni on mode).

4.3.3.3 Message format

There are two basic types of MIDI message byte: the status byte, and the data byte. Status bytes always begin with a binary 1 to distinguish them from data bytes which always begin with a 0. As shown in *Figure 4.26* the first half of the status byte denotes the message type and the second half denotes the channel number. It should be appreciated that there are really only seven bits per byte to represent real information, as the most significant bit of each byte is reserved to denote the type (status or data). Seven bits allow $2^7 = 128$ possible messages.

Figure 4.26 *MIDI message data format.*

The first byte in a MIDI message is normally a status byte, which contains information about the channel number to which the message applies. It can be seen that 4 bits of the status byte are set aside for the representation of the channel number, which allows for 24 (or 16) possible channels. The status byte is a label which denotes which receiver the message is intended for, and it also denotes which type of message is to follow (e.g. a NOTE ON message). It will also be seen that there are three bits to denote the message type, because the first bit must always be a one. This theoretically allows for

eight message types, but there are some special cases in the form of system messages (see below).

Standard MIDI messages can be up to 3 bytes long, but not all messages require 3 bytes, and there are some fairly common exceptions to the rule which are described below. *Table 4.1* shows the standard MIDI messages by describing the common titles of the messages followed by the values that represent them in hexadecimal. The channel number is represented by the letter *n*, being any hexadecimal value from &0 to &F. Since space in this chapter is limited, the messages shown in the table will not be discussed in detail. For a detailed explanation of MIDI messages the reader is referred to the book by Rumsey (1994).

4.3.3.4 Channel and system messages compared

Two primary classes of message exist: those which relate to specific channels and those which relate to the system as a whole. One should bear in mind that it is possible for an instrument to be receiving in omni mode, in which case it would ignore the channel label and attempt to respond to anything that it received.

Channel messages start with status bytes in the range &8n to &En (they start at hexadecimal eight because the most significant bit (MSB) must be one for a status byte). System messages all begin with &F, and do not contain a channel number. Instead the least significant nybble of the system status byte (that which normally would contain the channel number) is used for further identification of the system message, such that there is room for 16 possible system messages running from &F0 to &FF.

System messages are themselves split into three groups: system common, system exclusive, and system real-time. The common messages may apply to any device on the bus, depending only on the device's ability to handle the message. The exclusive messages apply to whichever manufacturer's devices are specified in the message (see below), and the real-time messages are all time-critical. The status byte &F1 is reserved for MIDI time-code use (see below).

4.3.3.5 Running status

When a large amount of information is transmitted over a single MIDI bus, delays naturally arise due to the serial nature of transmission wherein data such as the concurrent notes of a chord must be sent one after the other. It will be advantageous, therefore, to reduce the amount of data transmitted as much as possible, in order to keep the delay as short as

Table 4.1 MIDI data messages

Message	Status	Data 1	Data 2
Channel-specific messages			
NOTE OFF	&8n	Note number	Velocity
NOTE ON	&9n	Note number	Velocity
POLYPHONIC AFTERTOUCH	&An	Note number	Pressure
CONTROL CHANGE	&Bn	Controller number	Data
14-bit controllers MSByte			
(Examples)	&Bn	01 (modulation wheel)	Data
		02 (breath controller)	Data
		04 (foot controller)	Data
		05 (portamento time)	Data
		06 (data entry slider)	Data
		07 (main volume)	Data
14-bit controllers LSByte			
	&Bn	21 (mod. wheel), etc.	Data
7-bit controllers/switches			
(Examples)	&Bn	40 (sustain pedal)	}00–3F (off)
		41 (portamento)	}40–7F (on)
		42 (sostenuto pedal)	}40–7F (on)
		43 (soft pedal)	}40–7F (on)
		60 (data increment)	7F
		61 (data decrement)	7F
		62 (non-regulated parameter control)	LSByte
		63 (non-regulated parameter control)	MSByte
		64 (regulated parameter control)	LSByte
		65 (regulated parameter control)	MSByte
		79 (reset all controllers)	7F
Channel modes			
	&Bn	7A (local)	00 off/7F on
		7B (all notes off)	00
		7C (omni off)	00
		7D (omni on)	00
		7E (mono)	
		(No. of channels)	00–0A
		7F (poly)	00
PROGRAM CHANGE	&Cn	Program No.	–
CHANNEL AFTERTOUCH	&Dn	Pressure	–
PITCH WHEEL	&En	LSByte	MSByte
System messages			
System exclusive			
SYSTEM EXCLUSIVE START	&F0	Manufacturer ID	Data $+++\cdots\cdots$
END OF SYS. EXCL.	&F7	–	
System common			
SONG POINTER	&F2	LSbyte	MSbyte
SONG SELECT	&F3	Song number	–
TUNE REQUEST	&F6	–	
MIDI time-code			
QUARTER-FRAME	&F1	Data	–
System real-time			
TIMING CLOCK	&F8		
START	&FA		
CONTINUE	&FB		
STOP	&FC		
ACTIVE SENSING	&FE		
RESET	&FF		

possible and to avoid overloading the devices on the bus with unnecessary data.

Running status is an accepted method of reducing the amount of data transmitted, and one which all MIDI software should understand. It involves the assumption that once a status byte has been asserted by a controller there is no need to reiterate this status for each subsequent message of that status, so long as the status has not changed in between. Thus a string of 'note on' messages could be sent with the 'note on' status only sent at the start of the series of note data:

[NOTE ON − DATA − VELOCITY][DATA

− VELOCITY][DATA − VELOCITY] . . .

It will be appreciated that for a long string of note data this could reduce the amount of data sent by nearly one-third. But as each 'note on' is almost always followed quickly by a 'note off' for the same number, this method would clearly break down, as the status would be changing from 'note on' to 'note off' very regularly, thus eliminating most of the advantage gained by running status. This is the reason for the adoption of 'note on', 'velocity zero' as equivalent to a 'note off' message, because it avoids a change of status during running status, allowing a string of what appears to be 'note on' messages, but which is, in fact, both 'note on' and 'note off'.

Running status is not used at all times for a string of same-status messages, and will often only be called upon by an instrument's software when the rate of data exceeds a certain point. Indeed, an examination of the data from a typical synthesizer indicates that running status is not used during a large amount of ordinary playing.

4.3.3.6 Channel modes

Although grouped with the controllers, under the same status, the channel mode messages differ somewhat in that they set the mode of operation of the instrument receiving on that particular channel.

Local on/off is used to make or break the link between an instrument's own keyboard and its own sound generators. Effectively there is a switch in between the output of the keyboard and the control input to the sound generators which allows the instrument to play its own sound generators in normal operation when closed (*Figure 4.27*).

If the switch is opened, the link is broken and the output from the keyboard feeds the MIDI OUT while the sound generators are controlled from the MIDI IN. In this mode the instrument acts as two

Figure 4.27 *Data routing in 'local off' mode.*

separate devices: a master keyboard without any sound, and a sound generator without a keyboard. This configuration can be useful when the instrument in use is the master keyboard for a large sequencer system, where it may not always be desired that everything played on the master keyboard results in sound from the instrument itself.

All notes off performs the function of turning off all the notes that might be sounding on the specified channel. It can be useful for ensuring that an instrument is silenced completely when problems occur in a MIDI system.

Omni off ensures that the instrument will only act on data tagged with its own channel number(s), as set by the instrument's controls.

Omni on sets the instrument to receive on all of the MIDI channels. In other words, the instrument will ignore the channel number in the status byte and will attempt to act on any data that may arrive, whatever its channel. Devices should power up in this mode.

Mono mode sets the instrument such that it will only reproduce one note at a time, as opposed to poly (phonic) in which a number of notes' may be sounded together. The mono mode comes into its own with so-called 'multi-timbral instruments', which are capable of generating more than one voice (independent mono sound) at a time; in other words, that not only can they produce, say. 16 notes at once, but each note could have a different sound quality (each acting as a different monophonic instrument). The data byte that accompanies the mono mode message specifies how many voices are to be assigned to adjacent MIDI channels.

For example, if the data byte is set to 4, then four voices will be assigned to adjacent MIDI channels, starting from the basic channel which is the one

on which the instrument has been set to receive in normal operation. Exceptionally, if the data byte is set to 0, all 16 voices (if they exist) are assigned each to one of the 16 MIDI channels. In this way, a single multitimbral instrument (i.e. one capable of more than one simultaneous voice) can act as 16 monophonic instruments, although on cheaper systems all these voices may be combined to one audio output.

In *Poly* mode the instrument will sound as many notes as it is able on one MIDI channel at the same time. Instruments differ as to the action taken when the number of simultaneous notes is exceeded; some will release the first note played in favour of the new note, whereas others will refuse to play the new note. The more intelligent of them may look to see if the same note already exists in the notes currently sounding, and only accept a new note if it is not already sounding. Even more intelligently, some devices may release the quietest note (that with the lowest velocity value) to make way for a later arrival.

It is becoming common to run a device in poly mode on more than one receive channel, provided that the software can handle the reception of multiple polyphonic channels. A multi-timbral expander unit may well have this facility, commonly referred to as 'multi' mode.

4.3.3.7 System exclusive

A 'system exclusive' message is one which is unique to a particular manufacturer, and often a particular instrument. The only thing that is defined about such messages is how they are to start and finish, with the exception of the use of system exclusive messages for universal information such as MIDI time-code and sample dump. System exclusive messages generated by a device will naturally be produced at the MIDI OUT, and not at the THRU, so a deliberate connection must be made between the transmitting device and the receiving device. Occasionally it is necessary to make a return link from the OUT of the receiver to the IN of the transmitter so that two-way communication is possible, and so that the receiver can control the flow of data to some extent by telling the transmitter when it is ready to receive and when it has received correctly.

The message begins with status &F0, and this should always be followed with the relevant manufacturer identification, which is a number defining which manufacturer's message is to follow. Data of virtually any sort can follow the identification, and it does not have to conform to the MIDI data format of STATUS–DATA–DATA, so that it can be used for

a variety of miscellaneous purposes which have not been defined in the MIDI standard, and the message can have virtually any length that the manufacturer requires. Exceptions are data bytes which look like other MIDI status bytes, as they will naturally be interpreted as such by any receiver, which might terminate reception of the system exclusive message. The message should be terminated with &F7, although this is not always observed in which case the receiving device should 'time-out' after a given period, or terminate the system exclusive message on receipt of the next status byte. It is recommended that some form of error checking is employed for long system exclusive data dumps, and many systems employ means of detecting whether the data has been received accurately, asking for re-tries of sections of the message in the event of failure, via a return link to the transmitter.

Examples of applications for such messages can be seen in the form of sample data dumps (from a sampler to a computer and back again for editing purposes), and voice data dumps (from a synthesizer to a computer for storage of user-programmed voices). It is sometimes possible for instruments of like kind from the same manufacturer to accept each other's programmed voice data, such that a 'black-box' instrument without many controls of its own (often known as an 'expander') may be programmed remotely.

The three highest numbered manufacturer identification codes have been set aside to denote special modes. These are the universal non-commercial messages (ID: &7D), the universal non-real-time messages (ID: &7E) (among other things these handle sample dumps to a specific format), and the universal real-time messages (ID: &7F) which handle some aspects of MIDI time-code.

4.3.3.8 Active sensing

Active sensing messages are single status bytes sent roughly three times per second by a controlling device when there is no other activity on the bus. It acts as a means of reassuring the receiving devices that the controller has not disappeared. Not all devices transmit active sensing information, and a receiver's software should be able to detect the presence or lack of it. If a receiver has come to expect active sensing bytes then it will generally act by turning off all notes if these bytes disappear for any reason. This can be a useful function in the case when a MIDI cable has been pulled out during a transmission, in which case notes will not be left sounding for very long. If a receiver has not seen

active sensing bytes since last turned on, it should assume that they are not being used.

4.3.3.9 Priority
A standard priority should exist in the handling of message types which ensures that the most urgent are handled first. The order is:

> *Reset*
> *System exclusive*
> *System real-time*
> *System common and channel messages.*

4.3.4 Synchronization

A major function of MIDI is the transference of timing data from place to place within a system. This takes the place of the various older standards for synchronization on drum machines and sequencers, which used separate 'sync' connections carrying clock signals at a number of rates, usually measured in pulses-per-quarter-note (PPQN).

It perhaps goes without saying that not all devices in a MIDI system will be interested in timing information; it depends on the function fulfilled by each device. Those whose job it is to control the replay and recording of MIDI information played elsewhere in the system will need some reference to determine the rate at which that information is replayed, to ensure that music results at the intended speed. Such data recording devices are usually called sequencers. Drum machines should also take notice of timing information, but that is because a drum machine usually contains a sequencer which stores the patterns of rhythm which a player may lay down. On the other hand, a normal synthesizer, effects unit or sampler is not concerned with timing information, because it has no functions which would be affected by a timing clock. Such devices do not normally store rhythm patterns, although there are some keyboards with onboard sequencers which ought to handle timing data.

As MIDI has become more involved with the professional recording studio, means have arisen by which MIDI timing information may be locked to a prerecorded time-code which may reside on a multitrack tape or video tape. Using such methods, the MIDI system and the recording or postproduction system may be fully integrated. Recently, the need for timing information relating directly to real time as opposed to musical time has resulted in the development of an extension to the MIDI standard called 'MIDI time-code' (MTC).

4.3.4.1 System real-time
A group of system messages called the 'system real-time messages' control the execution of timed sequences in a MIDI system, and these are often used in conjunction with the song pointer (which is really a system common message) to control the movement around a recorded passage. The system real-time messages are:

&F8	Timing clock
&FA	START
&FB	CONTINUE
&FC	STOP

The MIDI standard allows for the transmission of a MIDI clock byte, which is a single status byte (&F8) to be issued by the controlling device six times per MIDI beat. A MIDI beat is equivalent to a musical semiquaver (or sixteenth note), so the increment of time represented by a MIDI clock byte is related to the duration of a particular musical value, not directly to a unit of real time. At any one musical tempo, a MIDI beat could be said to represent a fixed amount of time, but this would necessarily change if the tempo changed. The status byte &F8 is somewhat unique in the MIDI standard, as it may temporarily interrupt the current status, the status reverting to the previous status automatically after &F8 has been handled by a receiver.

START is used by a controller to signal the start of execution of a prerecorded song to any synchronized device. It should result in the playback of the sequence from the very beginning. *STOP* is used to halt the execution of a song which is running on a synchronized instrument. *CONTINUE* is used to restart the execution from the point at which the sequence was stopped, not necessarily from the beginning. For example, a sequencer may be controlling a number of keyboards, but it may also be linked to a drum machine which is playing back its own stored sequence internally. The two need to be locked together, and thus the sequencer could send the drum machine a start message at the beginning of the song, followed by &F8s at the correct intervals thereafter to keep the timing between the two devices constantly related. If the sequencer is stopped it should send *STOP* to the drum machine, whereafter *CONTINUE* would carry on from the stopped position, and *START* would restart at the beginning.

4.3.4.2 Song-position pointers
Song-position pointers (SPPs) are used in more complex systems in which one device needs to tell another where it is in a song. A good sequencer or

synchronizer should be able to transmit song pointers to other synchronizable devices when a new location is required or detected. For example, one might 'fast-forward' through a song and start again 20 bars later, in which case the other timed devices in the system would have to know where to restart. An SPP would be sent followed by start and then regular clocks.

The SPP represents the position in a stored song in terms of number of MIDI beats from the start of the song. Thus again it relates directly to musical time rather than real time: it does not necessarily represent the number of seconds into the song. It uses two data bytes, and the MSB of data bytes must be zero, so the resulting 14 bits can specify up to 16 384 (= 2^{14}) MIDI beats. SPP is a system common message (not a real-time message). It is often used in conjunction with &F3 (song select), which is used to define which of a collection of stored songs (in a drum machine, say) is referred to.

SPPs are fine for directing the movements of an entirely musical system, in which every action is related to a particular beat or subdivision of a beat, but not so fine when actions must occur at a particular point in real time. If, for example, one was using a MIDI system to dub music and effects to a picture in which an effect was intended to be triggered at a particular visual occurrence, that effect would have to maintain its position in time no matter what happened to the music. If the effect was to be triggered by a sequencer at a particular number of beats from the beginning of the song, this point could change in real time if the tempo of the music was altered slightly to fit a particular visual scene. Clearly some means of real-time synchronization is required either instead of, or as well as the clock and song pointer arrangement, such that certain events in a MIDI-controlled system may be triggered at specific times in hours, minutes and seconds.

4.3.4.3 Real-time synchronization and MIDI time-code

There are a number or ways of achieving real-time synchronization in a MIDI system, so as to solve the problem introduced above. One centres around the generation by a synchronizer of MIDI time-code (MTC), which is an extension to the original MIDI standard, and another centres around the incorporation of a time-code reader within a system's controlling sequencer. The two approaches need not be mutually exclusive, since the sequencer in the latter case may also output MTC to controlled devices, if this is thought necessary.

MIDI time-code has two specific functions. Firstly, to provide a means for distributing conventional SMPTE/EBU time-code around a MIDI network in a data format which is compatible with the MIDI protocol, and, secondly, to provide a protocol for transmitting so-called 'set up' messages which may be downloaded from a controlling computer to receivers in order to program them with cue points at which certain events are to take place. The intention is that receivers will then read incoming MTC as the program proceeds (probably derived by converting SMPTE/EBU code from a tape recorder track), executing preprogrammed events at certain times.

In order to transmit time-code information over MIDI, it has to be turned into a format which is compatible with other MIDI data (i.e. a STATUS byte followed by relevant DATA bytes). Essentially, there are two types of MTC synchronizing message: one which updates a receiver regularly with running time-code, and another which transmits one-time updates of time-code value for situations such as exist during the high-speed spooling of tape machines, where regular updating of each single frame would involve too great a rate of transmitted data. The former is known as the *quarter-frame message* and makes use of an unused status byte (&F1), whilst the latter is known as a *full-frame message* and operates under the system exclusive banner, as part of a special subset of messages known as the *universal real-time messages*.

The quarter-frame format transmits one SMPTE/EBU frame value in eight discrete MIDI messages. Each message of the group of eight represents in turn the frames most significant nybble (MSN), frames least significant nybble (LSN) seconds MSN/LSN, minutes MSN/LSN, and hours MSN/LSN.

Each quarter-frame message takes the format of:

[&F1] [DATA]

The data byte begins with 0 (as always), and the next seven bits of the data word are made up of a 3-bit code defining whether the message represents hours: minutes:seconds:frames, MSN or LSN, followed by the bit representing the binary value of that nybble (*Figure 4.28*). The three data bits defining the nature of the message start at 000 for the frame's LSN, rising sequentially to 111 for the hour's MSN. In order to reassemble to correct time-code value from the eight quarter-frame messages, the LSN and MSN of hours, minutes, seconds and frames are each combined within the receiver to form 8-bit words as follows:

Frames : rrr qqqqq

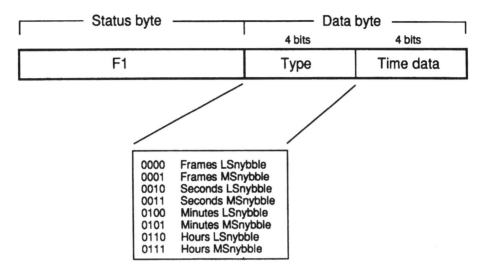

Figure 4.28 *Structure of MIDI time-code quarter-frame messages.*

where rrr is reserved for future use and qqqqq represents the frames value from 0–29,

 Seconds : rr qqqqqq

where rr is reserved for future use and qqqqqq represents the seconds value from 0–59,

 Minutes : rr qqqqqq

as for seconds, and

 Hours : r qq ppppp

where r is undefined, qq represents the time-code type (see below), and ppppp is the hours value from 0–23.

The time-code frame rate is denoted as follows in the hours count:

 00 = 24 fps
 00 = 25 fps
 10 = 30 fps drop-frame
 11 = 30 fps non-drop-frame

Unassigned bits should be set to 0.

At a frame rate of 30 frames per second, quarter-frame messages are sent over MIDI at a rate of 120 messages per second. As eight messages are needed fully to represent a frame, it can be appreciated that $30 \times 8 = 240$ messages really ought to be transmitted per second if the receiving device is to be updated every frame, but this would involve what has been considered to be too great an overhead in transmitted data, therefore the receiving device is updated every

two frames. If MTC is transmitted continuously over the bus it will take up approximately 7.5% of the available data bandwidth.

Quarter-frame messages may be transmitted in forward or reverse order, to emulate time-code running either forwards or backwards, with the frames LSN message transmitted on the frame boundary of the time-code frame that it represents. It may be appreciated that the receiver must in fact maintain a two frame offset between displayed time-code and received time-code since the frame value has taken two frames to transmit completely.

Further MTC message types exist for transmitting full-frames, user bits and set-up messages but space does not permit their coverage here.

4.3.5 System configurations

A number of the systems aspects of MIDI control will be introduced in an overview form below. For a detailed discussion of the implementation of MIDI in each system device the reader is referred to Rumsey (1994).

4.3.5.1 Master keyboard and expanders

All MIDI systems must be controlled by a source of MIDI data, which will feed the inputs to some or all of the devices in the system. This source will be referred to as the 'controller', whatever it happens to be in physical terms. Most commonly, in a recording or composition environment, the controller will be a desk-top computer, running one or more of a number

of possible commercial MIDI software packages. In a performance environment it is possible that the controller might be a musical keyboard, but many performers also make use of presequenced tracks for backing.

Even if the overall system controller is a computer, there will nearly always be the need for a master keyboard, since the computer must originally derive its data from a MIDI source. It is possible that the master keyboard could simply be one of the sound-producing keyboards within the greater system, in which case it will also receive data from the computer as well as providing the computer with a source of musical and control information. This keyboard should be a comprehensive piano-type kyeboard with all the facilities required to control other instruments, since this will be the most often-used source of MIDI information for the rest of the system. A number of considerations arise from this suggestion. Firstly, the master keyboard must be able to send all the different controller messages that are commonly required, such as velocity, after-touch, pitch-bend (continuous variation of the musical pitch), modulation wheel (a controller for continuously varying the depth of modulation of a note), pedals, program changes and perhaps messages from other analogue sliders or wheels. Secondly, the master keyboard should have a full-range keyboard to make maximum use of the receivers' sound generators. Thirdly, one must remember that the master keyboard's own audio output may sound when its keyboard is played, even though it is transmitting to another device, so either its output level would have to be cut, or a means of implementing the 'local off' function would be required; and finally, the other instruments in the system might not need to have keyboards of their own, as they could be 'played' by MIDI control from the master.

The concept of the expander, i.e. a device which can generate sounds under MIDI control but which is without a keyboard of its own, has proved popular in recording systems where a large number of sounds are required, and where the controller is a sequencer. Expanders often contain all the MIDI and voice generation circuitry in a fraction of the space taken up by the keyboard-equipped equivalent.

4.3.5.2 The role of the sequencer

The sequencer is a device which stores MIDI information taken from one or more MIDI inputs, with the ability to send it out again at some later point in time from one or more MIDI outputs. In between these two operations a certain amount of editing and manipulation of the data may take place, and this can be performed in a number of ways: either musically on a visual 'score', numerically in a table of MIDI data values, or by replacing certain notes with others using individual note input from a keyboard. The sequencer may have some automatic manipulation functions, such as the ability to 'time-align' or 'quantize' the rhythm of information being transmitted. Importantly, many sequencers have the ability to 'channelize' retransmitted data whilst recording, so that the master keyboard input can be made to control different devices by selection of different tracks on the sequencer. The retransmit channel need not be the same as the master keyboard's original transmit channel.

Other automatic or selectable functions may include the ability to filter out certain information that is not required, either before or after storage, the ability to lock the data output to a time reference such as MIDI time-code or MIDI song pointers, and the ability to perform automatic 'drop-ins' for recording specific sections of music as a particular time passes. This is by no means a comprehensive list, as once the MIDI data is in a computer virtually any function can be executed depending on the features of the software which is run. Features of commercial sequencer software change regularly, and thus any attempt at a full description would serve little useful purpose.

4.3.5.3 Delays in MIDI systems

Data delays may arise in large MIDI systems principally due to the fact that the interface is serial. A six-note chord in which all the notes are played concurrently must be transmitted one note after the other. A 3-byte message takes nearly 1 ms to transmit, thus a six-note chord would take 6 ms (although running status might reduce this by a useful degree). One can see that in an example of a system with many devices, each being sent six-note chords from a sequencer over the same interface, plus the addition of real-time messages and control changes, careful organization of data transmission order is required. The more data a sequencer is expected to transmit over one MIDI bus, the more difficult it will become to output messages for every channel at the correct time.

Further delays may arise due to data backing up in the input buffers of receiving devices, and not being processed quickly enough by the receiver. Also, some devices have longer internal processing delays than others, there being a number or MIDI keyboards on the market which exhibit many milliseconds delay between receiving a message and acting on it.

4.3.5.4 Routing matrices

Two types of routing matrix should be considered: the intelligent and the non-intelligent. The intelligent matrix allows one or a number of MIDI INs (or a high-speed link from the controlling computer) to be routed to a number of THRUs or OUTs with some intermediate processing. This could involve 'channelizing' the data, as well as perhaps instituting some degree of modification or filtering of the MIDI data, similar to that found in some sequencers (transposing, or filtering out after-touch, for example).

The unintelligent device would simply take in data from its inputs and route it to specific outputs. It is common for an input to be routable to many outputs, so that one controller could be fed to a number of receivers. Even a device such as this can be very useful in a system as it helps to reduce the replugging of cables that is required if one wishes to assign a different device as the master keyboard or temporarily alter the system configuration. It also eliminates the long daisy-chains which would otherwise be necessary.

4.3.5.5 Practical system examples

To conclude, two examples of practical MIDI systems will be illustrated, one simple and one quite complex. The system shown in *Figure 4.29* incorporates a master keyboard, a sequencer running on a desk-top computer, a single multi-timbral expander, and a drum machine. A basic set-up such as this would allow for a certain amount of experimentation with voicing for simple backing tracks, especially if the master keyboard itself were to be a good synthesizer.

If we assume that the expander is a multi-timbral device then it would be possible to obtain from it a number of different polyphonic voices, each responding on a different MIDI channel. The master keyboard's voice generator would act as a further expander. The drum machine could be used for rhythm tracks, storing performance data either internally (many drum machines contain their own sequencers) or using the main computer. Replay of the drum sequencer would be locked to replay of the music sequencer by simple timing messages, as discussed above.

The outputs from each sound generator would be fed into a basic mixer, and greater mixing flexibility could be achieved if the expander and drum machine provided separate outputs for each voice, as opposed to simple mono or stereo outputs, since each sound could be equalized, level-controlled, and panned by the operator. It can be seen that the means of data distribution is a simple daisy-chain, which is quite acceptable in this case, since the number of devices is small. In this case, each device receives all the data generated by the master keyboard, and each device will be set to receive on a separate MIDI channel. The expander may be set to receive on a number of channels, depending on voice allocation.

Figure 4.30 *A complex MIDI system for synchronized recording of tracks with mix-down automation. TC, time-code.*

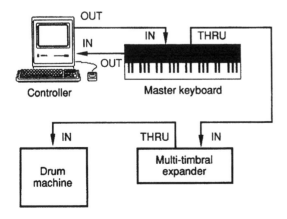

Figure 4.29 *A simple sequenced MIDI system.*

Figure 4.30 shows a complex recording system, synchronized to time-code from a tape-recorder, and incorporating mixer automation and audio routing. In this system, the same computer which handles music sequencing also handles mixer automation data, which are transmitted from the mixer to the computer on a single MIDI channel. A comprehensive MIDI implementation is assumed for this mixer (fader levels, mutes, routing, etc.), but the system would accommodate a lower level of MIDI implementation on the mixer (e.g. mutes only) equally well.

Routing of data gets a little complicated here, and there are a number of options, but in normal operation the mixer would simply be another destination for data from the computer, taking its feed from the MIDI matrix, and receiving on a unique MIDI channel. In order to record the mix data onto the sequencer it would be normal to use the mixer's own faders and other controls, and to route the MIDI OUT of the mixer to the input of the computer, recording on a unique track of the sequencer, whilst the sequencer replayed any prerecorded tracks. Possibly the recording of mix data would be done after all the music tracks had been laid, in which case the mix data could simply be merged with the timing data from the synchronizer in place of the master keyboard data, and the mix data could be recorded while a sync replay of the prerecorded tracks was taking place. It is now common to use a multiport MIDI interface connected to the controlling computer using a high-speed interface (not MIDI). Such interfaces have a number of independent MIDI ports, each of which can be fed with data for a specific device or collection of devices. Software setup applications are available which allow the computer to keep track of device and port assignment. The use of such interfaces effectively increases the number of available MIDI channels by the number of ports ×16.

Reference

RUMSEY, F.J., *MIDI Systems and Control*, (Second edition), Focal Press, Oxford (1994)

5

Studios and their facilities

Malcolm Nelson
Head of Operational Training, Centre for
Broadcast Skills Training, BBC Wood Norton

Peter Harris
Formerly Technical and Managing Director of
CTS Studios

Larry Goodson
Formerly Head of Sound, BBC Television

John Howell
Senior Sound Supervisor, BBC Television

Peter Jackson
Technical Strategy Adviser, Merlin
Communications International Ltd

Geoff Atkins
Consultant Engineer, the Black Mountain
Mobile. (Formerly BBC)

Yasmin Hashmi
Independent Digital Audio Consultant, SYPHA

Tim Robinson
Consulting Engineer in Building Services

5.1 Radio studio facilities

5.1.1 Introduction

Radio makes pictures in the mind, using sound to stimulate the listener's imagination. Whatever the type of programme, whether it is a compelling radio play, a dynamic Top 40 countdown, an intriguing documentary feature or an informative drivetime show, we derive a great deal of information about where the action takes place, about the character of an interviewee or about the personality of the presenter from the subtleties of the broadcast sound.

Radio studios come in all shapes and sizes and the equipment in them must be flexible enough for the programme maker to have the utmost creativity in building those sound pictures. It must impose no limitations of quality or scope and should help to convey the message in the most effective way.

Radio is also a very accessible medium, not only from the listener's point of view, but also when it comes to setting up a studio facility. Certainly, the world's major broadcasting organizations have invested many hundreds of thousands of pounds building prestige music or drama studios, capable of the most complex productions. However, for a campus station or a hospital radio studio the simplest arrangements will often suffice and they can be installed at modest cost. So long as the facilities enable the right sort of programmes to be made, without hindrance or drawback, then the studio will be doing its job. Unlike video and film where the technology is complex and often fairly costly, sound equipment, even of the digital type, is usually comparatively cheap. Indeed, much professional audio gear is compatible with domestic hi-fi equipment which can often be adapted for use in a broadcast environment.

General-purpose TV and film studios usually have to be large, expensive places capable of creating the illusion of anything from a fireside chat to the sinking of the *Titanic*. To achieve the same range of illusions in radio would require only a relatively modest space, although a good range of pre-recorded sound effects would be a great help! Since the early days of the medium, radio broadcasters have built general purpose and also specialist studio areas geared to the making of just one type of programme. With the current proliferation of radio stations not only small community stations but also whole networks may be dedicated to a single style of programming.

When choosing equipment and when deciding what features to install in a particular area, it helps to focus on the style of programmes which will be made there. If everything works well together, the studio will be a transparent part of the process of conveying ideas from the mind of the programme master to the mind of the listener.

5.1.2 General-purpose studios

This broad category probably covers most on-air and preparation studios for radio, including those used for news broadcasts, interviews, presentation announcements and music shows – in fact any type of programme where live speech is an essential part, but in which its principal purpose may be to provide a link between recorded or live items originating elsewhere.

Much of what follows is also applicable to facilities which can now be provided in a presenter's own living room, linked to a radio station proper by high-quality ISDN lines.

5.1.2.1 Layout

The studio may consist of just one acoustically treated room where the programme presenter speaks, operates the technical equipment and interviews guests. Alternatively, there may be two linked rooms, one of which is designated as a control room containing the majority of the technical equipment. From here an operator or engineer can coordinate and control the programme's technical aspects, working closely with the production team. In this case, although the studio itself may still contain a small mixer and other technical equipment, its principal purpose is to provide an acoustically controlled area where speech may be picked up on one or more microphones. A sound lobby between the studio and the control room is vital to provide a degree of acoustic separation and to prevent a potential howl-round ('feedback') when someone walks from one area to the other while a microphone is live.

An adjacent area for editing or preparation of programme inserts is useful in a news environment, or it may perhaps be used for the production of commercials. If the style of programming demands it, a phone-in area where incoming calls are intercepted and vetted before being put on-air may also be necessary. In any small studio complex an adaptable arrangement of technical areas allows for maximum flexibility and utilization (see *Figure 5.1*).

The orientation of the control desks, particularly in relation to the observation windows is also something which needs careful thought. If positioned at right angles to the window this may help the presenter or

Figure 5.1 *Layout of a small studio complex.*

operator concentrate on what they hear rather than on what they can see and, incidentally, may also provide a better acoustic environment for the loud-speakers. Whatever the geographical arrangement, it is always vitally important that there are good sightlines between each of the key areas. Presenters need encouragement by subtle visual feedback from their colleagues and in a fast-moving programme or network, communication between those involved may have to rely on simple hand-signals. Neither of these will happen unless everyone can see each other easily. However, large parallel areas of glass should be avoided if possible, to help prevent flutter echoes and standing waves in the studio or control room.

5.1.2.2 Control of background noise

Background noise is clearly a major consideration for any broadcast area, coupled with the associated problem of acoustics, or how sound behaves in the studio itself. Although these topics are dealt with elsewhere in this book, it is worth appraising the particular requirements of a general-purpose radio studio.

A self-operated studio where generally a close-microphone technique is used is probably the most tolerant of extraneous noise. This is especially true for a music station where the presenter very often speaks against some sort of music 'bed', which effec-tively masks any unwanted sound. In this type of studio, of course, extraneous noise may also come from the operation of equipment which is positioned within arm's length of the presenter or from the handling of scripts or running orders. Only careful choice of microphone type and position can mini-mize this – supplemented, perhaps, by a little train-ing! Acoustically, though, a relatively poor separa-tion from the outside world may still give perfectly acceptable results in a presenter-operated studio.

Where a more distant microphone technique is likely, where more than one microphone may be on air at the same time, or where there is no benefit from any background masking, unexplained noise from the outside world can be very distracting for the listener. This is especially true if the programme is being broadcast in stereophonic sound, where the disturbing noise may be heard loud and clear and also be located within the stereo image. For this more critical requirement, it will probably be necessary to achieve a background noise level of no more than 30 dBA.

Isolation from the outside world is partly achieved by preventing paths through which airborne sound can reach the room. The inevitable effect of this is to necessitate some form of ventilation (if not full

air conditioning), otherwise the presenter and guests may not survive! However, the process of moving or removing air causes noise and the required capacity of such a system is determined in turn by the likely amount of heat generated by people and equipment in the studio. In planning an installation, therefore, it is worth being mindful of the total heat dissipation of the equipment and considering ways of minimizing this load by, perhaps, siting some of it remotely in a separate apparatus room.

5.1.2.3 Studio acoustics

In a studio intended principally for speech, acoustic treatment is usually designed to control the reverberation time and to prevent undesirable standing waves by achieving maximum diffusion. While specialized absorbers can be fitted to the walls and perhaps the ceiling, the shape and size of the studio, the positioning of observation windows, the nature of the floor coverings, the type of furnishings and the placing of equipment will all have a bearing on the way in which sound behaves in the room.

The aim of controlling the acoustics is that the sound heard by the listener is as natural as possible, whether there's just one person speaking or a discussion. In essence, radio of this nature has always tried to create the impression that the presenter and contributors are in a similar acoustic environment to the listener's own, perhaps a domestic sitting room. Today, though, when listening to the radio is often an adjunct to some other activity, it is much more likely that the audience will be in much less controlled surroundings, perhaps in a car or in the kitchen. The requirement for this type of programming is simply that speech from the studio must be as clear as possible, complemented by the acoustics and not in any way impeded by them. Some reverberation is necessary in order to make the speech sound natural, but too much would confuse and give an unreal impression of spaciousness. A reverberation time of around 0.4–0.5 s is usually effective for a speech studio.

Control of acoustics is a complex science and, as mentioned above, everything in the studio will contribute to the sound picked up by the microphone. Studio tables can give problems by reflecting sound into the microphone, giving rise to undesirable coloration. Attempts to reduce this by covering the table in, for example, an absorbent cloth will inevitably be unsuccessful as such a cloth provides no absorption in the frequencies where the coloration occurs. Instead, the table should be constructed with a perforated mesh surface, covered with cloth for cosmetic reasons if desired, to reduce the proportion of reflected sound reaching the microphone. However, where a microphone is positioned above a mixer and other technical equipment, as in a self-operated studio, such a technique may not be practicable and only the use of a directional microphone can minimize the problem.

5.1.2.4 The control desk

Unless it is intended simply as an acoustically treated room where a single microphone allows a newsreader or contributor to provide an insert to another studio or a network area, the technical heart of a radio studio will be its control desk or mixing console. As with any mixer, this is used to bring together the various sources of audio material, whether live or pre-recorded, so that an operator or presenter can balance their relative levels to give the artistic effect that's needed. It also provides the means of exercising overall control of signal level to avoid distortion at subsequent points in the broadcast chain.

When selecting a particular model of control desk – no doubt weighing up the facilities available from different manufacturers against the likely cost – it is also important to think about the ease of operation and the robustness of the gear. In radio, or in television for that matter, equipment may be subjected to much harder use than in the recording industry, because broadcasting is often a round-the-clock operation. Also, in the stressful atmosphere of live broadcasting, equipment may not receive much tender, loving care!

Preparation and setting up often has to be done in a very short time, so the ergonomics of the console must allow for quick and intuitive operation, especially as working in live radio means there is usually only one chance to get it right. Certainly for self-op work, where the programme presenter is carrying out all the necessary technical operation, a comfortable, logical layout with no unnecessary controls is absolutely vital (*Figure 5.2*). The engineering requirements of such a mixer, in terms of its amplification, frequency response, noise performance, headroom, distortion levels, input and output impedances, and so on, will be similar in radio to those of any other professional audio area. These parameters are described in Section 2.7 of this book. However, a desk used for radio broadcasting is required to do a rather specialized job and, in addition to the general facilities available on any audio mixer, it will incorporate a number of distinctive features.

Channels and groups The number of channels and groups on the desk is the first obvious consideration and, except in the case of an assignable console,

Figure 5.2 *Plan view of a small presenter-operated studio.*

this generally has a major bearing on the physical size of the equipment. Even in the smallest studio at least two microphone channels will be needed, so that main and spare microphones can be rigged. Coupled with a need for sufficient record and replay sources to cope with a variety of programme material means that the smallest size of mixer needed in a speech studio is around eight channels, while between 16 and 24 is much more typical. This also allows for some redundancy in the case of a channel fault so that a programme can be kept on the air by replugging a source to another channel.

There may be a requirement for a number of mono channels, for microphones and obvious mono sources such as telephone calls, complemented by several stereo input channels for Digital Audio Tape (DAT), compact disc (CD), MiniDisc (MD) and so on. Channels intended mainly for use with microphones need to provide phantom power and a full range of amplification, say 70–80 dB of gain, while channels for line-level sources need only 20 dB or so gain.

Being able to control all microphone channels, outside sources or replay machines in separate groups also helps make life easier for the presenter or operator, meaning that for most purposes at least two stereo groups are required. For complex news or network continuity operations a desk with four groups will be much more useful.

The layout of those channel and group faders is a very important consideration in selecting or designing a console for radio. Many presenters prefer a central script space with the more important faders

positioned on either side, while some operators prefer to have what they see as the most important tools of the trade, the faders, right in front of them. Although well rehearsed, the argument about the direction of fader travel (i.e. whether faders should move away from or towards the operator to increase level) is one which has no definitive answer, as people seem to adapt very quickly to whichever type they are using. It is only when having to move from one method of operation to the other that confusion seems to occur. No matter which is chosen, standardization within a radio station or studio complex is obviously a good idea!

Console systems where the operator can assign, say, a single set of channel controls to any one of a number of channels or groups at the touch of a button undoubtedly have a use in music or drama studios. However, assignable desks probably have no real advantage in the fast-moving world of a live programme containing a mix of speech and prerecorded sources.

Prefade listen A prefade listen (PFL) facility on each channel or group, operated by a fader backstop switch as well as a separate push-button is vitally necessary for radio work. This allows the operator or presenter to listen to a source without fading it up on air, in order to check its incoming level, make sure it is the right one or perhaps even liaise with a distant contributor prior to conducting an interview. Without this feature, many radio programmes, especially live broadcasts, would quickly fall apart. After-fader listen (AFL), however, is generally not as useful in a live programme; loudspeaker monitoring of each of the desk's group outputs is probably better.

Jackfield A comprehensive jackfield or patchbay allows the interconnection of a variety of audio sources and destinations, using plugs and cords. The individual jack sockets will be mainly associated with the balanced audio circuits carrying the inputs and outputs of the control desk. Where programmes generally follow an established and well-known format it may be helpful to arrange to carry normalling through the jackfield so that no plugging is necessary for, say, microphone 1 to appear on the input of channel 1, microphone 2 on channel 2, a CD machine on channel 3, and so on. However, being able to trace faults, set up alternative paths in the event of such a fault or to provide an unusual facility for a particularly complex programme is an essential means of flexibility provided by the jackfield.

Although they can occupy a substantial amount of space, the best type of connector for this purpose is undoubtedly the traditional $\frac{1}{4}$-in. B-gauge type

Figure 5.3 *Jack plug and socket. (a) A balanced audio pair is connected to the tip and ring; the screen/earth is connected to the sleeve. (b) The wiring to the inners and outers allows a normalled circuit through the jack socket. (c) Inserting the plug breaks the normalled connection and substitutes the circuit from the plug.* (Courtesy of Talbot-Smith (1990).)

316 jack plug and socket (*Figure 5.3*). Not only do the connections available on the socket allow the normalling mentioned above, but the system has been proved to be robust and reliable for handling even the small voltages associated with microphone level signals.

Many commercially-available desks also feature a separate jackfield for access to signal paths, sometimes electrically unbalanced, which are within the desk itself. Such an insert jackfield is enormously useful to operators, enabling them to add external devices such as compressor/limiters to specific channels, groups or even the main output. In the interests of saving precious space in the physical structure of the desk, such jackfields are often made up of rows of miniature bantam jack sockets.

Equalization It is useful to have some general-purpose equalization available on each channel of the desk, although it is sometimes thought that for self-op work by unskilled operators this is a dangerous distraction. Although not intended to provide the sophisticated creativity of a graphic or parametric equalizer on a music recording console, the ability to remedy minor problems of quality and improve the characteristics of a voice is the main purpose of such equalization, although not as a substitute for getting

the microphone in the right place first! It may also enable the operator to make a poor quality telephone call more intelligible or minimize objectionable noise on a long-distance circuit. Above all, equalization must be easy to use and quick to set up.

High- and low-frequency shelf filters with at least 15 dB of lift and cut, mid-frequency presence boost or cut together with high- and low-frequency roll-off will provide a useful range of effects. A means of deliberately switching this equalization out of circuit on each channel is essential to prevent accidental processing when none is intended.

Auxiliary outputs At least two auxiliary outputs from each channel, with a master control of level on the overall output, are also vital to the versatility of a typical radio studio desk. They provide the means of setting up artificial reverberation effects, deriving foldback feeds on headphones to presenters or guests, or even selective recording of some of the sources passing through the desk. Clean feed (mix-minus) outputs to outside contributors can also be derived in this way.

Loudspeakers Being able to monitor the broadcast output of the studio by listening on loudspeakers is obviously important, not only for hearing the programme but also for assessing sound balance and audio quality. As with all the other desk facilities, maximum flexibility is desirable and so it is useful to be able to switch the same pair of speakers (assuming stereo working) to monitor such things as the main studio output, the transmitted output of the station, the received signal from the transmitter, the end of the previous programme on the network, individual group outputs of the desk, prefade listen signals and auxiliary outputs.

In order to monitor a stereo programme correctly, the loudspeakers themselves need to be properly aligned. This involves, for example, making sure that a sound intended to be heard centrally in the stereo image does in fact appear so to the operator or presenter. The desk should, therefore, provide a balance control for the speakers, as well as a means of deliberately selecting a mono source of programme material (*Figure 5.4*). Used together these will enable the operator to adjust the relative speaker volumes to give a central image. This will ensure that stereo programme material is heard in the studio precisely as it is intended to be heard by the listener. Controls allowing a mono programme to be heard on just one loudspeaker (conventionally the left of the pair) are also useful, as is a way of deliberately reversing the phase of one speaker to check for a suspected phase reversal in the programme chain.

The positioning of loudspeakers in the control room must be carefully planned at the initial design stage of the room. While the speakers need to be in a convenient operational position in relation to the mixing desk and any observation windows, it is important that they deliver a stable, reliable and consistent stereo image. There must be no adverse effects on the frequency content of the sound as a result of standing waves or undesirable acoustic reflections from equipment, windows or other hard surfaces. It is also important that the correct stereo balance can be heard, not just in one crucial spot, but over a reasonable range of sitting positions, allowing everyone involved to appreciate the way the programme sounds. It will also avoid the programme balance and quality appearing to change radically when the operator moves slightly out of the normal

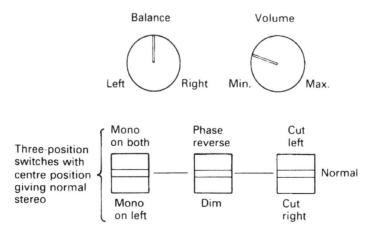

Figure 5.4 *Typical desk controls for stereo loudspeaker monitoring.*

listening position, perhaps to adjust a control at one end of the desk.

It may take considerable trial and error to achieve the best combination of loudspeaker and control-desk position in a studio, but it is a very worthwhile exercise. Without the confidence that what is heard in the studio is a faithful reproduction of the programme signal, it is impossible to judge what the listener might be hearing.

Most radio sound desks will also arrange to mute the audio feeds to loudspeakers whenever a microphone in the same room is faded up, thus avoiding a potentially disastrous howl-round. This may be achieved by using backstop switches on the microphone faders to operate a muting relay.

Headphones Separate monitoring feeds to headphones are also helpful, both in the studio and in the control room, for checking incoming feeds and for cueing up the next item to be played. Headphones for the presenter, especially in a self-operated studio will be a necessity. Not only do most radio professionals need to hear their own voice as they speak, but while microphones are faded up in the studio the presenter must use headphones rather than loudspeakers to hear telephone contributors or incoming news reports as well as the effects of voice-over mixes. However good they are, headphones cannot be used satisfactorily for judging audio quality – that is what the speakers are for. Some control of their volume is indispensable, as too quiet may mean a missed cue while too loud risks a howl-round, not to mention hearing damage!

Metering To make an objective measurement of the audio signal level some form of metering device

is needed. Information from this allows the studio operator to control programme signal levels within the permitted dynamic range, avoiding problems of noise and distortion in the studio as well as in subsequent stages of the equipment chain between studio and transmitter. There are two main types of audio meter, each available as either mechanical pointers or as illuminated light-emitting diodes (LED) or plasma bar graphs.

The volume-unit (VU) meter is a simple and fairly cheap voltmeter, easily connected across the desk output. The pointer is often claimed to give a rough indication of subjective loudness, but its rapid movements and complicated scale can make it difficult or tiring to read. Because it does not read peak signals accurately it can introduce a tendency to occasional overload and distortion, even with a skilled operator.

The peak programme meter (PPM) has a needle with a fast rise time and slow decay time, allowing it to register even short duration peak signals quite accurately and proving much kinder on the eyes (*Figure 5.5*). Although more complex and therefore more expensive than the VU meter, the PPM is very useful for the broadcaster.

If the studio is designed for working in stereo, meters must be available to monitor signal levels in the left and right channels as well as any derived mono output. An indication of the difference between the two halves of the stereo signal can also provide useful information about width and phase, especially when used in conjunction with loudspeaker monitoring.

Outside sources The control desk of a general purpose studio must also have facilities for handling

Figure 5.5 *Peak programme meters for stereo.*

a variety of outside sources. These may include telephone calls from the public network as well as contributions by line or radio link from outside broadcast venues, radio cars or fixed installations in police headquarters, motoring organizations or a local weather centre. While the incoming audio signal can generally be treated like any other input to the desk, the problem of how the distant end hears what is going on in the studio must not be ignored.

So that they can hear verbal cues or take part in a two-way discussion, the remote contributor must be able to hear some sort of feed of the studio's output. If this feed consists of the complete programme signal, perhaps derived from an off-air receiver, this also gives them the confidence of knowing that they are on-air when they hear themselves faded up. If an off-air feed is not practical for any reason, then the desk must be able to derive a return feed of cue programme, or studio output, to each outside source. While the major international broadcasters may sometimes rent a separate circuit from the national telecommunications company for this purpose, more usually public phone lines are used for such return signals. A circuit is established by dialling up on a normal telephone in the studio, which is then switched to receive an output from the desk.

Long-distance circuits, especially those via satellite channels, may include an appreciable delay in the signal path. In such cases a delayed return feed of the complete programme signal, including their own contribution may be very confusing for the distant end. Usually a mixer designed for broadcasting will be able to provide a clean feed (mix-minus) signal for each outside source, either generated by a separate matrix within the desk or derived by the operator from auxiliary channel outputs. A clean feed signal consists of the mixed output of the desk, minus that particular source, so that they hear everything except for their own contribution – hence the term 'mix-minus'.

Although audio signals from outside sources are essentially the same as any other, telephone calls from the public network do require special arrangements. A number of telephone balance units are available commercially and they provide an essential interface between the telephone line and the mixing desk. Such units provide the important functions of:

(1) holding the call when it is diverted from a normal telephone handset to the mixing desk;
(2) terminating the incoming telephone line with the correct impedance, not only for greatest electrical efficiency but also to meet the requirements of the telephone company;
(3) electrically isolating the telephone network from the studio equipment;
(4) separating the send and receive signals from the two-wire telephone circuit for connection to the desk;
(5) automatic gain control of send and receive signals; and
(6) providing a voice-over circuit to reduce the level of the telephone call when the presenter speaks, thus minimizing unwanted coloration.

Some broadcasting organizations also insist on using an audio delay, sometimes called a 'profanity delay', of several seconds in the transmitted output of any studio using live telephone calls. The purpose of this is to allow the presenter time to react to any unacceptable language from a caller by cutting the phone call before it is broadcast.

If the desk is capable of independently recording phone calls or other outside sources while a tape or CD is being broadcast then this will greatly increase programme flexibility. This is particularly effective in, for example, a sports report where several correspondents may be providing results over the telephone while other events are being covered.

Line-up oscillator A line-up oscillator built into the desk provides an essential means of sending a steady reference signal on the desk's main outputs to some distant technical area. This not only allows a check on signal levels between source and destination, but is also a way of identifying a stereo circuit. To do this the oscillator may deliver only a single frequency tone of accurate level, or a known sequence of tones which uniquely identify left and right stereo channels as well as providing a check on their level and relative phase. The same oscillator may be used to send line-up tones to local recording devices for alignment purposes or for adding a reference identification to recordings of programmes. On most desks designed for radio, the loudspeaker monitoring is flexible enough to allow rehearsal of a programme to continue while line-up tone is being sent on the studio output.

Communications facilities To make everything run smoothly, there has to be a good flow of information between everyone involved in putting a programme on the air. This means that the control desk must incorporate some essential communications facilities. A quick and fool-proof talkback system is necessary in any radio studio, allowing an operator and presenter to speak to each other or to

an adjacent studio or to any outside sources. Sometimes it will be possible to speak on the actual main output of the desk itself, which can be a useful way of identifying an outgoing circuit from the studio when contributing to a programme in another broadcast centre. A dedicated talkback-to-tape facility can also be helpful in providing an audio identification on a local recording.

An intercom network, quite separate from the desk, may link the studio to a central technical area, to the newsroom and to any phone-in or preparation areas. All these systems are useful but must be used with care if they are not to be an irritation to the receiving end. It must also be remembered that talkbacks and intercoms are fairly public ways of passing a message – carelessness can lead to embarrassment! If security or privacy is important, and speed is not essential, a telephone is better.

Computer-based, visual communications systems are a useful adjunct to talk back and are nowhere near as intrusive. They can be used to display the name and details of a telephone caller or, perhaps coupled with a networked information system, they can show a running order for the programme and its commercial breaks, details of music items, newsflashes or even an updated copy of the presenter's script.

Perhaps the simplest form of visual communication, apart from the wave of a hand, is a cue light in the studio. Often coloured green and operated by a footswitch or a key adjacent to the channel fader, this can be a useful and easily assimilated signal to the presenter. By either a steady illumination or a series of flicks it can indicate that the studio microphone is live, or about to be faded up or whatever other message has been arranged previously. It is best if the circuitry for such lights, especially if they are mains powered, is operated by a relay which is distant from the desk in an effort to avoid any problems of induced clicks on the audio output.

Transmission safeguards A number of transmission safeguards will need to be incorporated in any desk intended for radio broadcasting, in an effort to make its operation as foolproof as possible in the stressful moments of live programmes. These may include:

(1) a means of disabling the line-up oscillator which is normally used to put a reference signal on the desk's output;
(2) facilities to route talkback signals only to studio headphones in the presence of live microphones and not on to loudspeakers or the desk output; and

(3) Conversion of any solo facility which works by cutting other channels at their input to afterfader listen, affecting only the loudspeaker monitoring.

5.1.2.5 Recording and playback facilities

Not all the material for a radio programme is likely to be generated live in front of the studio microphones – even the simplest news bulletin or discussion has an opening jingle or signature tune. Consequently, high-quality equipment will be needed to record and/or play back speech and music inserts, as well as programme trails, jingles, commercials, network ident announcements and indeed anything which it is more usual or more convenient to prerecord.

Recording and playback formats In the early days of radio the functions of recording and playback were often carried out in a central recording room, using vinyl discs or magnetic tape machines, with signals routed to the appropriate studio by a central control room. However, the modern general-purpose radio studio must incorporate a wide variety of professional recording and playback formats within its own technical facilities. These are likely to include the following.

(1) *Vinyl disc players*. With the rapid advance of other recorded music formats most stations now feel that this is a facility which need not be provided in an on-air studio, especially as in unskilled hands they are difficult to cue-up accurately and start cleanly. However, as there is still a significant amount of material available only on vinyl, at the very least there needs to be some sort of disc to tape copying facility in a back-up or preparation studio.
(2) *Quarter-inch, open-reel, tape recorders*. These are usually equipped with facilities for headphone monitoring and editing as well as remote starting by means of switches built into the desk faders. If used for both recording and replay, at least three machines will probably be needed. Although the $\frac{1}{4}$-in. format is being rapidly superseded by digital audio tape or computer-based recording and editing systems, it still remains the workhorse of some small studios.
(3) *CD players*. Apart from their obvious advantages of high audio quality and comparative robustness the CD player is the broadcaster's godsend. Despite an occasional problem with the technology, the benefits to a presenter or operator are immense. Even domestic CD players can cue-up and play from virtually any point

on a disc, quickly, reliably and accurately. They can also display the elapsed time or the time remaining on a track. Broadcast versions of the CD player may have back-cueing facilities and a means of storing several cue-points in memory. Some models are designed to accept a CD which has been pre-loaded into a plastic cartridge. This not only allows more convenient handling but it also protects the disc itself from undesirable fingermarks and damage. CD recorders are now relatively cheap, as are the discs themselves. As a result some stations now insist on all pre-recorded material being presented in this format.

(4) *Cartridge recorders solidus players.* Although this originally referred only to NAB cartridges containing an endless loop of $\frac{1}{4}$-in. tape in a plastic box, there are now several excellent digital alternatives. These can record short segments of audio on floppy discs, computerised hard discs or even magneto-optical discs. As well as the traditional advantage of instant starting at the press of a button (essential for jingles, stings, commercials and signature tunes), the newer devices offer quicker cueing and sophisticated information displays showing titles and cue words. The Sony MiniDisc format, originally intended for the domestic market, is now a very effective example of this kind of machine.

(5) *Compact cassette recorders.* With many good quality portable recorders available for location interviews this remains a useful and cheap format to have available in the studio. However, because of the inherent difficulties in locating items and cueing-up a cassette tape it is rarely a good idea to broadcast material directly from this medium – far better to copy it to another format, which may be necessary anyway if the material requires editing before transmission.

(6) *DAT recorders.* Digital audio tape (DAT) recorders not only have the same digital audio quality advantages as CDs, but they now also provide facilities for precise cueing and instantaneous playback. Although editing a DAT tape involves copying selected material on to a second tape, rather like editing video, sophisticated control systems can make this a relatively painless process for the complication of programme tapes in a preparation studio. This is, however, too slow for a fast turnaround of news or feature items. Hard disk and PC-based systems are much quicker and more flexible for editing items of short duration, although the time

taken to load audio material from DAT into the editing system may be a drawback.

As well as offering long recording times, one major advantage of using the DAT format in conjunction with computer based editing systems is that audio can remain in the digital domain all the way from recording to transmission. Ideally, professional DAT machines in the studio should be capable of operating at a sampling frequency of either 44.1 or 48 kHz so that they can interface correctly with other digital systems.

DAT was originally developed for the consumer market, but has been taken up enthusiastically by professional users. It remains the format of choice for acquisition in the field.

(7) *Digital mass storage.* For programme items of relatively short duration, networked hard disk systems are now available. These may form part of an integrated system for the whole radio station, using a central fileserver for the storage of digital audio, accessible from several computer terminals throughout the building. The individual workstations will probably contain analogue-to-digital and digital-to-analogue conversion processors as well as digital bit-rate reduction to save on storage capacity. In this way it is possible to compile and edit inserts to a programme in a production office or back-up studio, transmitting them later by means of a dedicated playback console in the studio. Meanwhile, the running order and selection of items can be under the control of the programme's editor at yet another terminal in the newsroom (*Figure 5.6*).

It is a short step from such a system to a highly automated radio station where whole sequences of music, jingles or commercials can be played back either manually by the presenter or automatically in response to some external cue. A computer-driven system can also be used to control such things as transmitter switching for regional opt-outs, as well as supervising scheduling and logging of music items and even sending out invoices to advertisers!

Whatever the range of recording and playback devices selected for a particular studio, it is important that they are properly and professionally installed so that there are no problems of level, earthing or impedance incompatibility between them. Nothing is more frustrating to the programme maker than to find that certain combinations of devices just will not work.

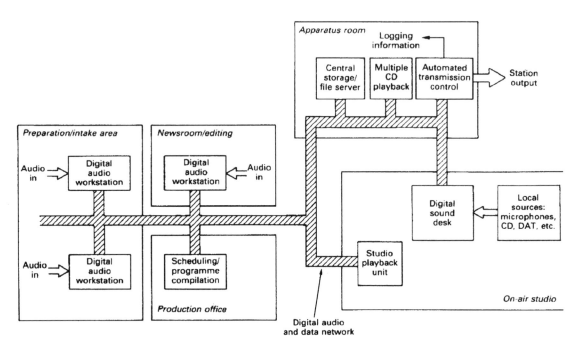

Figure 5.6 *A semi-automated radio station complex.*

Audio switching matrix An audio switching matrix which routes individual machine outputs to specific desk channels can be very useful operationally, avoiding the need for plugging. It can also be constructed to provide a mix of similar sources (e.g. all CD players) to just one channel of the desk. In the absence of a dedicated switcher a jackfield incorporating several parallel strips can be almost as effective, although slower to set up.

Another related facility which the matrix may provide is that of being able to record from one device to another independently of the mixing desk. For example, in this way a CD can be copied to another format while the studio is on-air without interfering with the normal operation of the main console. Suitable aural and visual monitoring to be able to line up and check the copying process is also helpful. In whatever way this facility is achieved, of course, it must be sufficiently foolproof to help avoid unintentional broadcast of what is being copied!

Positioning of recording and replay devices The positioning of recording and replay devices in relation to the mixing desk is also something which can help or hinder the programme maker. This is especially so in a self-operated studio, for example, where the presenter must start a CD, while at the same time speaking into the microphone. Anything

intended to be used in a live programme should be positioned within easy arm's reach, so that the presenter does not have to move off the axis of the microphone to operate them. At the same time, it is important to be aware of the likelihood of these machines generating undesirable mechanical noise, either continuously or on start-up. This might be picked up by the microphone even if they are positioned on its dead side. It is useful to bear this in mind when choosing a microphone – highly directional types may be a desirable option.

5.1.2.6 Transmission facilities

In addition to all the specialized equipment mentioned so far, a radio studio needs a few other basic facilities to make it satisfactory for live broadcasts:

Visual signalling Visual signalling, such as red or amber lights to indicate that the studio is on-air or that microphones are live. Not only are these a useful indication to the personnel working on the programme, but with luck they will act as a deterrent to undesirable interruptions from visitors!

Clocks Accurate, easy to read clocks in the sightline of everyone working on the programme. It is obviously helpful for time-checks if such clocks are crystal-locked or linked in some way to standard time, perhaps by receiving a reference

signal transmitted from a national telecommunications authority. Whether they are analogue, with sweep hands, or digital, or some combination of both is open to choice. Perhaps the simplest, yet most important, requirement is for every clock in the studio complex to be synchronized so that they all read the same time.

Network switching In a relatively small radio station, a network switching matrix may be needed in the studio so that the operator can access outgoing circuits to the appropriate transmitting stations or hand over control to another studio. As mentioned previously, however, this may be under the control of a scheduling computer which also gives some degree of studio automation. In larger studio centres, network switching may be controlled from a central technical area, with simply a display in the studio of where its output is routed. A direct-to-network facility may allow a prerecorded programme to be broadcast independently from a tape machine while preparation of the next programme carries on in the studio.

5.1.3 Specialist studios

The range of output of the larger national broadcasting organizations means that they often have a requirement for prestigious studio areas, employing large-scale resources to handle the most complex productions. They are equipped very comprehensively, often with the latest technology, and are able to provide every facility for the origination of high-quality music or radio drama. In size they may vary from something approaching a concert hall capable of accommodating a large orchestra, to a relatively modest studio area for multitrack overdubs. In virtually every case, there will be a control room or mixing area separate from the acoustic environment in which the players or musicians perform.

5.1.3.1 Rock music
Similar in almost every respect to commercial recording studios, these are designed to provide an appropriate venue for artists who are accustomed to working in the music industry. Apart from major public concerts, it is unusual for such music to be broadcast live and so the same facilities for multitracking, overdubbing and mixing, either in the analogue or digital domain are needed as in any recording studio.

As with other types of radio studio, robustness, reliability and ease of operation are paramount because time is never on the broadcaster's side. What may take several weeks to lay down and

mix in a commercial studio will probably have to be completed in a few hours when working for radio – four or five numbers in a single working day is typical. That is not to say that the end-product for radio is necessarily inferior, but simply that it is intended to be used in a different way. Good results must be achieved as quickly as possible, without taking up valuable studio time for experimentation which could be done before the session. The studio itself must look good and sound good, providing a comfortable working space which suits current industry practice.

Acoustics Traditionally, the acoustics of a pop-music studio were made as dead as possible, so as to achieve maximum separation between individual instruments using screens and close-microphone techniques. Currently, much more flexible working arrangements are provided, with both live and dead areas complemented by isolation booths for vocals or acoustic instruments. The best effect from a drum kit, for example, is generally a large, open sound, best suited to a combination of close and distant microphones in the live end of the studio. The sound of electric bass and rhythm guitars is normally picked up from a combination of direct injection and microphones facing their loudspeaker cabinets. They are best placed in the dead area, which itself may be screened off to increase separation from the drum microphones even further. Acoustic guitars, on the other hand, will need to be played in an isolation booth to achieve a satisfactory pick-up against the high noise level in the studio. Although multitrack techniques would allow each instrument to be laid down separately, thus achieving maximum possible separation, this would lead to a very sterile and unmusical end-product. It is far better to allow the musicians to perform as an ensemble, adding extra layers later if required.

Acoustic isolation from the outside world must also be effective, although there may be more of a problem of sound leaking out than unwanted sound finding its way in. There must also be good sightlines from the control room to anywhere in the studio – either directly or, very much second best, by means of a closed-circuit TV system. Good communications, coupled with flexible foldback arrangements for the performers are vital to ensure the smooth running of a session.

The acoustics of the control room are particularly critical because, in essence, it is here that the final sound is really created. Lack of coloration and accurate stereo imaging are vital. Most rock music includes significant amounts of low-frequency

energy and, as a result, standing waves can be a real problem in the control room unless steps are taken to avoid them. This can be done by means of diffusers and low-frequency traps, paying particular attention to the absorption of first reflections.

Control room Ideally, the control room itself must be sufficiently spacious to accommodate everyone involved when it comes to listening to a playback or mix-down, while at the same time providing the best possible monitoring for the balance engineer and the producer. A separate apparatus room for multitrack tape machines, desk power supplies, digital processing racks or any other noisy or heat generating machinery will improve the working environment immensely. There must also be sufficient space in the mixing area for instrumentalists to rig keyboards, samplers and the like. This suggests a need for sufficient direct injection boxes, supported by comprehensive tie-line and patching facilities within the control room.

The ability to distribute computer and MIDI control data signals around the whole studio area is an essential requirement for current music-making techniques. For an all-digital area, a comprehensive AES/EBU digital signal distribution system must also be installed, with analogue-to-digital convertors at appropriate points for analogue signal inputs and digital-to-analogue convertors for outputs and monitoring. Some desk automation systems may also require a distribution of timecode signals.

As with any commercial music studio the balance, perspective and stereo image of the finished product are largely created in the control room. Consequently, the mixing desk must be extensively equipped for the job, with sufficient channels and groups to handle the most complex close-microphone techniques or intricate mix-downs. Although individual desk channels may contain sophisticated equalization and dynamics processing, there must also be enough auxiliary sends to be able to drive a comprehensive array of outboard effects devices. These may include programmable equalizers, compressor/limiters, noisegates, samplers and a selection of audio delay and reverberation devices.

In a broadcast environment, some means of storing information about the desk set-up at the end of a session can be an absolute godsend. This may take the form of a computerized notebook from which the operator resets the controls manually at a later date or, in the case of assignable consoles, a complete memory of the entire desk configuration which can be automatically recalled on demand. Either of these will help save precious time when there may be a fast turn around from one session to another to maximize the use of the studio.

Listening levels The creative process of recording and mixing rock music usually demands high listening levels in the control room as well as in the studio. Sound pressure levels may frequently exceed 100 dBA and this raises important concerns about the potential for hearing impairment amongst performers and technical staff. Damage to human hearing is a cumulative effect which depends not only on the absolute sound level but also on the duration of the exposure. Depending on the current legislation, it might be necessary to consider some precautionary measures where the daily personal exposure is as low even as 85 dBA. It may therefore be appropriate to consider installing a warning-light system to indicate dangerously high sound pressure levels in the studio or control room. In the extreme, such equipment can be used to disconnect the mains supply to the studio if a preset noise threshold is exceeded.

5.1.3.2 Serious music

Here too, the broadcaster must provide an environment in which the musicians can perform comfortably and which is sympathetic to their music. It is likely to be similar to a small concert hall in size and appearance – perhaps even with a raised stage and facilities for an audience. After all, most music of this nature was originally written for public performance.

Acoustics A studio for serious music must have fairly lively acoustics with good diffusion. A reverberation time in the region of 1.2–2.0 s is typical – as flat as possible over the range of 50 Hz to 10 kHz. Some studios have hinged wall panels, with different acoustic absorption on each side. These can be used to give some limited control of reverberation time, adapting it to suit the style and scale of the music. For a serious music studio, the criteria for acoustic isolation from the outside world will need to be quite exacting, perhaps with a permissible background noise as low as 25 dBA.

It is usually essential for the various sections of the orchestra to see and to be seen by the conductor. Raised wooden rostra are often provided for this purpose, but they can give rise to a number of practical problems. Inevitably, they will produce an irritatingly wide range of noises and creaks once they are covered by an orchestra and its instruments. Rostra also introduce a large area of reflective surface which may have an adverse effect on the acoustics by producing standing waves or flutter echoes. Any permanent installation will need to incorporate sufficient

microphone points, sockets for orchestra lights and perhaps even headphone facilities.

Microphone techniques The aim of broadcasting serious music is to create for the listener at home the opportunity of sitting in a good seat in the concert hall. The balance and perspective of the separate sections of the orchestra, as well as their positioning, are controlled largely by the conductor. Consequently, studio microphone techniques are often based on coincident or spaced pair methods to capture the internally balanced sound. This will require some safe and convenient means of slinging the microphones above the orchestra, choir or chamber ensemble. Nevertheless, multi-microphone techniques coupled with multitrack recording are by no means ruled out, especially where soloists are featured, and so the mixing desk must be able to support just as many microphone channels as a rock music studio. Hybrid techniques, where coincident pairs are supplemented by individual spot microphones are quite commonplace, relying on the skill of the sound engineer to create a satisfactory balance.

Given the essentially purist approach to recording classical music, a wide range of effects devices is unlikely to be needed and there will be much less reliance on equalization, compression or limiting. However, digital reverberation, if used with great care, can complement the natural acoustics. Subtle use of audio delay can also help, by maintaining the correct perspective for spot microphones.

Live broadcasts Live broadcasts from such a studio may not be unusual, requiring many of the transmission safeguards and facilities specified for general-purpose studios. Good communications, by talkback, cue lights and telephone, between the sound mixer and the conductor will be essential to coordinate a transmission or facilitate the smooth running of a recording session. Other transmission facilities such as clocks and red lights will also be needed.

5.1.3.3 Radio drama

Although comparatively rare, it is here that the full scope and inventiveness of the medium of radio can be realized. Voices, music and sound effects are woven together in the theatre of the mind, with a flexibility and range of settings which television simply cannot match. A radio play may be transmitted live, recorded as a continuous take or, as is more usual, recorded scene by scene with subsequent editing and post-production. A typical layout of a radio drama studio is shown in *Figure 5.7*.

Acoustics and perspective Manipulation of acoustics and perspective is probably the most

fundamental tool of the trade, helping to create an enormous variety of scenes and locations in the listener's imagination. One of the basic requirements of a studio designed for radio drama is a flexible range of acoustic areas.

(1) *A dead room.* Highly absorbent acoustic wedges line the walls, ceiling and floor, with a wire mesh grid for the actors and crew to walk on. The virtual absence of any sound-wave reflections (a reverberation time of around 0.1s is typical) allows a simulated outdoor acoustic to be created, especially when supplemented with appropriate background sound effects. Although a very realistic effect can be achieved in such a room, it can be a very unpleasant and claustrophobic environment to work in. A better approach, particularly if no dead room is available, may be to create a 'tent' of acoustic screens in a moderately dead area in the main body of the studio itself.

(2) *A live area.* This may form the main body of the studio, but with scope for adjusting the reverberation time of part of it, say between 0.5 and 0.8 s. This can be achieved by lifting carpets. pulling back thick curtains or moving hinged wall panels which have an absorbing side as well as a reflective side.

(3) *Effects areas.* Many drama studios contain purpose-built kitchens and telephone booths, much in demand in modern drama. Sound-effects stair-cases with a variety of surface coverings on the treads are another major asset, if space permits.

(4) *Narrator's booth.* A self-contained and acoustically isolated room where announcements or narration can be read is very worthwhile. Like a speech studio it should have a reverberation time of around 0.4 s. This facility enables recording to continue while the main studio is reset or while the cast move to different positions.

A great deal of space is often needed in a drama studio, especially when working in stereo. To achieve the right movements across the sound stage as well as convincing approaches to the microphones, the actors need sufficient working space around them. In some cases it is useful to mark out the floor around a stereo microphone to help the cast make, and be able to repeat, the correct movements. The only way to organize a complex production, with many scenes, may be to build a series of acoustic

Figure 5.7 *Layout of a typical radio drama studio.*

'sets' in which the action takes place, each with their own microphone positions.

Live sound effects A well-equipped drama studio also needs to have the means of generating a variety of live sound effects. These might include working effects doors with a selection of bells, locks, bolts and handles, complemented by walking areas surfaced with wood, gravel and concrete. These will be supplemented by a host of other objects, from crockery to bead curtains, telephones to buckets of water, which are used to create a whole library of realistic and bizarre sound effects in front of the microphone. The provision of effects using water, of course, must be considered carefully in terms of electrical safety.

Acoustic screens These can be used either to provide some degree of acoustic separation between two scenes or to modify the local reverberation time of a particular area of the studio. They can also

helpfully define the major acting areas in the studio for the benefit of the cast. A typical screen will consist of a rigid aluminium frame, about 2 m high by 1 m wide, with high-density absorbing material on the dead side and a hard plastic-faced coating on the live side. Such screens can be positioned to achieve a variety of effects, such as a localized dead area as mentioned above, or a small room, a railway carriage, or the interior of a motor car. Some screens incorporate window sections to help with sight-lines and they generally have wheels for ease of movement. When not in use, screens should be stacked away so that they do not have an unwanted effect on the acoustics of the studio.

Studio floor The technical facilities on the studio floor will be conventional enough, but flexibility is again paramount. There must be sufficient microphone and talkback lines to every part of the studio. Other parts of the building in which the studio is

situated can also be equipped with tie-lines if they might prove useful for recording in realistic surroundings. Equally important are foldback circuits, allowing prerecorded scenes or sound effects to be played into the studio by means of loudspeakers and picked up on the microphones. This technique helps to blend the recorded effects into the live action, since the background sounds acquire the same acoustic as the actor's voices. It can also provide a more life-like atmosphere for the cast to respond to.

Communications As in any radio studio good communications are obviously vital, helping to coordinate the different elements of the production. There must be easy visual contact between the various acting areas and the control room, by closed-circuit TV if necessary. Infrared or radio talkback systems, which allow complete freedom of movement within the studio, are often employed. If staffing arrangements permit, a member of the technical crew can act as a floor manager, relaying the producer's instructions to the cast. Flexible cue-light signalling facilities are essential.

Control room In the control room, many of the requirements for recording rock music are equally valid for radio drama. Control desks must be able to accommodate a large number of sources, perhaps with a multitrack capability, supported by comprehensive arrangements for subgrouping, auxiliary outputs and signal processing. Assignability and facilities for storing and recalling desk settings from memory will be an enormous help for particularly demanding productions and for speed of operation. Not only do they allow for a quick set-up at the start of a session, but they also enable the instant resetting of the desk configuration for a particular scene within the production.

The creative use of sound effects helps paint the scenery for radio drama. Although live effects in the studio are often the most appropriate, there must also be an adequate range of playback devices in the control room for recorded sounds. As well as being able to construct and then play back background atmospheres, the operator must also have the facilities for firing off instantaneous effects on cue.

Whole libraries of recorded effects can be purchased on CD, so there must be enough CD players to run several effects simultaneously when compiling a complex montage of sounds – creating anything from a naval battle to a car crash. Digital samplers, recording on hard disk or optical disk, are particularly useful for cueing short segments of audio very precisely. Complemented by instant-start DAT recorders and MiniDisc recorders/players they allow

dedicated compilations of effects to be assembled and retained, if necessary, for another studio session or for a series of productions.

Recording of the studio output may be on $\frac{1}{4}$-in. tape, DAT, one of a variety of analogue or digital multi-track formats or direct to a digital hard-disk system. An audio switching matrix, with its own memory capability, will help considerably in setting up the right combinations of recording and playback machines and then interfacing them with the control desk. However, the complexity of managing and mixing all of these record and replay sources may usefully be delegated to a second technical operator who has a separate audio mixer with its own monitoring and machine remote controls. This enables a premix of all recorded sounds to be presented to the main control desk, allowing the member of technical staff there to concentrate on mixing the whole programme and to manage what goes on in the studio.

For the utmost creativity, a large range of audio processing devices will also be needed, similar to those found in a rock music studio. If used sensitively, digital reverberation devices can be useful in supplementing the natural acoustics created in the studio and can also help suggest unreal or unnatural locations if required by the storyline. Dedicated devices for creating telephone effects are often a great help in modern drama, enabling either end of the conversation to be heard, as well as generating the right kind of ringing tones, and so on.

Reference

TALBOT-SMITH, M., *Broadcast Sound Technology*, pp. 128–142. (Second edition), Focal Press, Oxford (1995).

5.2 Facilities required in recording studios

5.2.1 Recording studio

The term 'recording studio' generally refers to an establishment principally dedicated to music recording. It may serve the record, advertising, broadcasting and film industries. It can encompass anything from a one-man business to a multistudio complex. Many of its facilities are similar to those found in establishments dedicated to radio, television or film applications. What makes it different, is the variety of work which the average studio can be called upon to undertake. The demands of the hard-rock record producer may be vastly different to those of a film-score composer. Applications for recorded

music continue to grow: CD-ROMs, DVD, computer games, all with their special requirements. The studio has to make profits, to finance future re-equipment and reward its investors. Like any other business, it needs to have a predominantly positive cash flow to survive.

5.2.1.1 Mobile recording

In addition to permanent recording studios, much work is undertaken by mobile recording units. These range from relatively simple facilities, transported to the remote venue and set up in a suitable room in the building, to complex installations in customized vehicles, offering all the facilities and comfort to be found in a permanent control room, albeit in rather less space.

With a mobile recording vehicle, thought must be given to mundane things such as ease of parking, and coping with a variety of mains-supply conditions. Adherence with local authority regulations is necessary, especially if an audience is present at the recording. This may complicate the running of cables from the vehicle to the performance area. A driver with a heavy goods vehicle licence will be required for large vehicles. It may consist of separate tractor and trailer sections. If the unit is operating in conjunction with a television scanner, time-code and picture links will be required.

Good communications are a must. A CCTV link and comprehensive talkback are the basics. Radio communication, via hand-held units, can be an asset during rigging, or when problems arise during performance.

Radio microphones may be required for performers, when recording a live pop concert. Easy identification, by coloured markers, will reduce the chances of missing an opening, because the artist has picked up the wrong microphone.

5.2.1.2 People

However comprehensive and costly the facilities, the studio operational and technical staff hold the key to the success or failure of the business. For balance engineers, a thorough knowledge of the facilities and a 'good pair of ears' are essential. Musical ability is an obvious asset, although some respected recording engineers are unable to read music or follow a lead sheet. Unflappability and a reassuring 'bedside manner' are worth their weight in gold, especially if clients or artists are nervous or insecure.

Maintenance engineers need to be able to 'get the show on the road' rapidly, in the event of a breakdown during a session. The instincts of a master detective may be required when attempting to diagnose a fault from the assumptions of non-technical operational staff.

5.2.2 Console

The console, mixer or desk, is at the heart of the studio installation; it must be suitable for the type of work the studio expects to undertake (*Figure 5.8*). It may be totally analogue in its functions, totally digital, or a hybrid. Its function is to generate technically and artistically satisfactory outputs, for recording, in whatever format is chosen. If television and film scoring is anticipated, a console with an additional four or six group mix-down section would be appropriate for the task. Some individual console features and functions are considered below, insofar as they are affected by the studio application.

Figure 5.8 *The new 60-channel Neve VRP console with recall and flying faders installed into Studio One, CTS Studios, London.*

5.2.2.1 Console layout

If a standard manufacturer's console is purchased, the layout may be immovably fixed, or may allow some degree of modification to suit the user's requirements. With a custom console the field is wide open. Depending on the variety of work likely to be undertaken, specialized features may be incorporated for several applications. The use of mock-ups can be helpful in deciding on layout. Human arms can only reach so far! When single-manning of a control room is envisaged, comprehensive remote controls should be provided, at or in the console, for equipment sited elsewhere in the room.

If large music lead sheets are in use, then a suitably wide producer's area will be appreciated – and also

save the engineer from having to unearth channels from beneath the producer's paperwork.

A rigid mechanical construction is essential. Modules should be easily removable and replaceable, and interchangeable – if the console frame becomes distorted, this may not be possible. Connectors should continue to mate fully and accurately, throughout the console's life.

A suitably placed area for synthesizers, etc., with appropriate services, may be required within the control room. A matching customized position is frequently constructed behind the console, within the audio and video monitoring areas. Appropriate power, signal and MIDI feeds should be built-in.

5.2.2.2 Input stages
The microphone input stages of the console should have adequate headroom, to cope with the differing output levels of anything from a ribbon microphone, slung high above a classical orchestra, to condenser (electrostatic) microphones at close quarters to an energetic brass section. An input stage should also be able to contend with a direct-injection box, fed from an electronically amplified instrument. These requirements call for a low-noise design, with high overload capabilities.

Gain adjustment may be stepped or continuous. Whichever it is, accuracy of calibration is a must. A continuously variable control should use a precision potentiometer; it may be coupled to a detent wafer, to produce repeatable settings. Arguments continue over whether transformerless or transformer inputs should be used. The pros and cons of the particular design being considered should be evaluated before making a choice. The provision of 48 V phantom power, for condenser microphones, may be less complex with transformer inputs.

Line inputs should also have a sensible gain adjustment range, and sufficient headroom to be able to cope with the maximum span of input levels likely to be experienced, bearing in mind that these can include off-tape signals, for remix or overdubbing purposes, and direct outputs from electronic instruments.

5.2.2.3 Equalization and filtering
The many possible types and combinations of equalizers and filters are covered in detail elsewhere. Sufficient to say that recording engineers will have their own ideas as to what constitutes the ideal package. Extensive listening tests with a sample module, in one's own surroundings, should be undertaken before reaching a decision. Most of the major manufacturers

will be pleased to arrange the loan of a demonstration module, or complete channel strip. Attention should be paid to the ergonomics of the design. Will the operator need especially slender fingers, to operate the overcrowded collection of tiny knobs? Is the layout logical? Can one read the legends without a magnifying glass?

Again, accurate repeatable settings are a must. Variable controls should be as silent in operation as possible, and remain so for the useful life of the console. Obviously, operating a signal-carrying switched control may introduce an audible step into the programme material, but hopefully this type of adjustment will not be carried out during a 'take'.

5.2.2.4 Dynamics
The choice of dynamic range control circuitry is considerable. Personal preferences and technical performance will determine one's choice. What combination of limiting, compression and expansion is wanted? How much control over attack and recovery times? Are noise gates required on every channel? What type of indication of operation is required? Is the facility to insert an equalizer into the sidechain, for frequency-dependent dynamics control, required? Is look-ahead circuitry required, for zero overshoot? Is stereo-linking wanted? Will a sidechain keying facility be required?

A demonstration module will again aid the choice. It should be experimented with, at length, on as many different types of programme material as possible, with all the potential users present.

5.2.2.5 Foldback
The foldback, or cue, facilities should be as flexible as possible, to cope with musicians' and vocalists' preferences. Overdubbing artists frequently request a full stereo orchestral mix on headphones, with their voice, with reverberation added, pulled forward. Each section of a band may request a different foldback mix. The engineer should be able to monitor the foldback outputs, on speakers or headphones. When an assistant engineer is present, he/she may be delegated the task of foldback monitoring, on headphones.

5.2.2.6 Headphones
In a large studio, large numbers of headphones may be required, e.g. when an orchestra is overdubbing an existing rhythm track. Robust reasonable-quality headphones are necessary to cope with the rigours of studio use and abuse. Single earpiece units, with a balancing pad, allow the musician to hear his own

performance, as well as the guide track. The units should be comfortable to wear for long periods, and be fitted with long leads. Medium impedance inserts make it practical to bridge many sets of headphones across a power driver amplifier. Care should be taken not to damage the hearing of headphone wearers through misoperation of the system.

In a large set-up, linked distribution boxes, using XLR connectors, should be used. There may be several different foldback mixes being fed to the studio; colour coding of leads and boxes can reduce confusion when changes have to be made during a hectic session.

With smaller combinations, individual mini foldback mixers allow performers to produce their own optimum mix between elements arriving down, say, four foldback feeds.

Headphones may be used in the control room for checking purposes during a session. An engineer not familiar with a particular studio and its monitoring may carry his own reference headphones to assess the sound balance. Assistant engineers may use headphones to check certain outputs, while the engineer is listening to the main monitors.

5.2.2.7 Solo
Most consoles provide several varieties of solo, including stereo 'in-place', with reverberation, in addition to normal prefader listen (PFL) or afterfader listen (AFL) facilities. These are all tools to help the engineer achieve the best results. They can be particularly useful when a problem occurs during a live recording session. Additive solo will allow the balance between several microphones (say those covering a trombone section) to be assessed.

5.2.2.8 Metering
The choice of console metering is particularly important in the case of digital recording. Very fast attack, peak reading, metering is necessary, preferably with a peak hold facility. High-resolution bargraph devices, based on fluorescent or light-emitting-diode (LED) technology, are widely used. The display will typically cover a range of 60 dB. It may be calibrated in dB, to match the metering of the digital recorders, or in percentage terms. The driving electronics can also be arranged to provide a volume-unit (VU) type scaling and ballistics, for use with analogue recorders.

Displays are available which will operate and display two different modes simultaneously; this is advantageous when both digital and analogue recorders are in use together.

Mechanical VU meters or peak programme meters (PPMs) may be the only options available on some standard consoles. A careful watch on the input overload indicators of digital recorders used with such consoles is required, in view of the different ballistics of different types of metering on dynamic signals. Auxiliary outputs, such as foldback and reverberation sends, may use smaller, less accurate mechanical metering.

Whatever the system, the engineer needs to know how the console meters relate to the recorders being driven from the console outputs.

5.2.2.9 Panning
When stereo recording is the sole end product of the studio, standard two-channel pan-pots on the routing modules are all that is required. If film or video scoring for multi-channel release is undertaken, three- or four-channel panpots may be required, usually as separate units which can be patched into the desired location.

5.2.2.10 Patching
The choice between standard- and bantam-sized patch bay components is often determined by the vastness of the patching facilities in an analogue console. With inboard patching, bantam jacks are the obvious choice. Sufficient space between rows should be left to allow fingers in! Labelling should be legible.

Patching components need regular cleaning to maintain perfect contact. Normalling contacts should be of the highest quality. Staff must be brainwashed to not remove patch cords by pulling on the leads.

5.2.2.11 Talkback
Recording studio talkback requirements will depend on the type of operation. With one studio and one control room, a simple system may suffice. With multiple studios, mixing and machine rooms, a complex system will be needed. The balance engineer should be able to communicate with every location he may be working in conjunction with.

The producer, at the console, will probably need slightly less talkback access. Routing the producer's talkback output through the engineer's controls will give the engineer overriding priority on common destinations.

The ability to receive reverse talkback should be present at all times. This may mean arranging for it to appear on an alternative set of monitors,

when its normal speakers are in use for some other function.

5.2.2.12 Eavesdrop/slate
The engineer should be able to listen to the output of the conductor's talkback microphone, at will (eavesdrop). Conductors are busy people and appreciate not always having to operate a key, to talk to the engineer. This output should also be routeable to 'talk-to-tape' (slate), fed to the recorder. This will allow the conductor's count-in to be recorded, which can be invaluable during subsequent overdubbing sessions.

The use of eavesdropping feeds, from the engineer's talkback microphone to assistants in machine rooms, can allow a degree of anticipation of the engineer's needs (say for a particular playback, or a fresh 'take').

5.2.2.13 Automation
The choice of console will be influenced by a current or future need for automation. If it is likely that automation will be needed at some later date it is sensible to acquire an automation-ready console, simplifying the eventual installation. A number of excellent systems are available, providing control over functions such as fader levels, muting, insertions, linking and grouping, plus machine control.

The automation computer will remember and recall large numbers of mixes, allow updating, modification and merging of those mixes. The system will have many sophisticated features to aid the process. Regular use of a complex system is the only way to make its operation second nature to the engineer.

A choice has to be made between servo-driven faders or voltage-controlled amplifiers (VCAs) for level control. The servo fader has the advantage of having no additional circuitry in the signal path, plus giving an instantaneous indication of its setting and allowing instinctive updating. Some VCAs may require additional operations to update a mix.

Album mixes can be retained on large-capacity floppy discs or other alternative storage systems which can travel with the master tape to other studios with similar facilities. A digitally controlled console (hybrid or all-digital) lends itself particularly well to automation.

An all-digital console can achieve an immediate total reset of all controls, including routing, grouping and monitoring. It becomes possible to change the console between preset 'snapshots' within a session at the push of a button. Mixing sessions can be picked up precisely where they left off, without the time-consuming matching operation of a 'recall' type console.

5.2.3 Outboard equipment

Whatever facilities exist within a console, they are unlikely to provide 100% of the studio's requirements. The console's facilities therefore need to be augmentable, by inserting outboard equipment (or 'black boxes') into the signal path. Outboard equipment may be mounted in standard 19-in. racks in the control room, in portable trolleys with umbilical cable links, in spare space within the console frame or in specially built enclosures sited within reach of the console.

Appropriate line-level interfacing is required. Insertion points should be able to handle balanced or unbalanced inputs and outputs. The outboard devices should always have power earthing, to avoid leakage currents flowing through signal grounds. With problematical interconnections of outboard equipment, it may be necessary to earth signal cable screens at one end only. Outboard equipment often contains ventilation fans, which should be near-silent in operation. Sufficient space should be allowed around the unit for the fans to do their work.

A studio is unlikely to own all the possible devices, and will hire in special units requested by its clients, hopefully remembering to pass the charges on to them. Attention should be paid to verifying that hired-in units adopt the same connector wiring practice as the studio.

5.2.3.1 Reverberation devices
The recording studio may need to provide more choices in these devices than its counterpart in other fields. Fashion can influence the choice. Few studios retain a natural 'echo chamber'. Artificial devices are now so convincing and flexible that their use is practically universal. Large 'mechanical' plate reverberation units still find favour in some studios, but their physical size and need for regular time-consuming adjustment weigh against them.

Compact solid-state devices dominate the reverberation scene. Programs can be selected, generally from a hand-held controller, to simulate many different types of auditoria, with control over all parameters within that program, including predelay for increasing the apparent size of the auditorium. Some products even have programs to simulate named auditoria. Most are configured as stereo devices. As with all computer-based technology, manufacturers

release improved software, sometimes with additional features, within the life of a product.

5.2.3.2 Additional equalization

Specialized equalizers, in addition to those within the console, may be required. Devices such as third-octave graphic equalizers and tuneable steep notch filters for removing unwanted narrow-band noise. An artist or producer may insist on using a particular equalizer which they believe is ideal for the artist's voice. An outboard equalizer may be cascaded with an internal equalizer to achieve a special effect. The danger of excessive amounts of equalization should always be borne in mind. It is possible to finish up with an end-product which will not transfer to another media system in its existing form. For example, in analogue disk cutting, excessive low frequencies would considerably limit available playing time. Excessive high frequencies could produce accelerations which most arm/cartridge/stylus combinations would be unable to track.

5.2.3.3 Additional dynamics

Many specialized dynamic range controlling devices exist, from prized valved compressors of yesteryear, onwards. They will often be used to obtain a special effect, rather than as a straightforward dynamic range controlling device.

5.2.3.4 Harmonizers

As their name suggests, these devices were originally designed to produce a pitch-shifted output, to 'harmonize' with an input signal – usually a vocal, to 'thicken it up'. Harmonizers aim to achieve pitch-shifting without noticeable glitches. They are used for producing a large number of special effects on programme material, using sampling, looping and delay circuitry, usually with two-channel operation, stereo linked or unlinked.

5.2.3.5 Stereo enhancers

Various 'black boxes' are claimed to improve the effectiveness of the stereo signal. They may utilize a mixture of equalization, frequency-dependent cross-feeds between channels, phase manipulation and psychoacoustics to achieve their aim. From the user's point of view, a lack of mono compatibility may prove to be an embarrassment in certain applications. Use with care!

5.2.4 Microphones

Studios tend to have a greater selection of microphones than might be the case in other fields.

Individual engineers' preferences frequently dictate the choice. Particular microphones may be in vogue – early valved condenser (electrostatic) microphones, with their cumbersome power units, are much sought after. Manufacturers have produced new versions of their old devices to take advantage of this market. Certain artists may demand a particular microphone, because they believe it particularly suits their voice. The television requirement for unobtrusive microphones does not apply in the recording studio. Solid-state, phantom powered condenser microphones have the lion's share of the studio market.

Pick-up patterns may be fixed or variable. With directional microphones, the off-axis sensitivity should be sensibly constant throughout the frequency range. The device's overload capabilities become especially important in situations such as close-miking a brass section at full blast. Many condenser microphones incorporate a switchable pad, for high-level use.

Moving-coil or ribbon microphones tend to be used for specialist applications. An engineer may find that he or she achieves a perfect bass drum sound with a moving-coil microphone inside the drum frame, with an old army greatcoat stuffed in behind it.

Stereo microphones are popular for overall orchestral pick-up purposes, or may be used only on a string section. Some stereo devices allow MS (middle and side outputs) or AB (left and right outputs) operation.

Microphone cables should be of a low-noise type, such as those using conductive plastic screening. Using a variety of different coloured cables can facilitate tracing a particular cable, for rerouting or replacement, in a large set-up.

5.2.4.1 Microphone checks

A studio should check its microphones frequently for damage or loss of performance. Phase checking is particularly important. Whenever a new unit is acquired, it should be confirmed that it is in phase with the other microphones in the studio. When microphone cables are made or repaired, they should similarly be checked for phase. Commercial phase checkers exist, or comparisons with a standard microphone may be made, by placing the devices in close proximity and comparing their outputs on speech, via two phase-checked channels mixed to mono.

5.2.4.2 Microphone care

Microphones are valuable items, and should be treated as such. Recording studios are notorious

for their rate of damage and loss. Ideally, at the conclusion of a session, microphone heads should be removed and returned to their protective boxes in a secure place. This may not always be possible, but is highly recommended. It has the added advantage of reducing the opportunity for theft.

An inventory of the studio's microphone stock, with serial numbers recorded, should be kept and frequently checked. Should a unit go missing, details should be passed to the manufacturer or his local agent, enabling detection should it be subsequently repaired by them. Trade associations should also be notified, to allow details to be circulated within the industry.

5.2.4.3 Microphone stands

Stands should provide a degree of isolation from the studio floor, especially when the floor is not rock solid. Additional isolation may be achieved by the use of a resilient mount for the microphone head. Most manufacturers produce such mounting devices for their microphones. The maximum usable height or extension from a particular stand is governed by the weight of the microphone. The combination must be stable – the expensive outcome of ignoring stability could be a damaged microphone, a damaged instrument or a damaged musician! To prevent accidents when large fixed booms are in use their heavy counterweights should be made very obvious by attaching an eye-catching (not literally) warning device.

5.2.5 Recorders – general

While hard-disk recording is increasingly used in studios, particularly for synthesizer work and stereo editing, the resulting programme material, in whatever track configuration, eventually needs to be archived to tape. The majority of recording sessions are still recorded directly to tape. Many recorder functions can be remote controlled, either using simple extensions of internal controls or using sophisticated RS 232/422 systems. When a project is particularly important, such as a live concert when retakes are impossible, 'cutting double' (running two recorders in parallel) is recommended to avoid problems due to recorder or tape malfunction.

5.2.5.1 Digital recorders

For digital multitrack recording, the DASH (Digital Audio Stationary Head) format is currently the preferred option. The maximum number of tracks available in DASH format is 48, on $\frac{1}{2}$-in. tape.

Compatibility exists with 24-track DASH tapes. In PRO-DIGI. 32 tracks on 1-in. tape is the maximum. The tape cost in digital multitrack recording, track for track, can be less than analogue costs. Tape storage requirements are also reduced with the narrower tapes used.

Multiple machines may be linked via time-code synchronizers to increase the number of tracks available for large productions. The recording studio may opt for one or other standard and rely on hiring the alternative type when requested by its clients.

The studio choice for stereo digital recorders is split between the $\frac{1}{4}$-in. tape stationary-head formats and the various cassette-based rotary-head machines, such as U-Matic, digital audio tape (DAT) and F1. The stationary-head machines offer the possibility of cut-splice editing, although studios generally prefer non-destructive editing systems. Digital tape must be kept squeaky clean – any form of dirt or grease (including fingerprints) will raise error rates and impair performance. Cleaning digital tape heads must be strictly in accordance with the manufacturer's instruction. Incorrect procedures can easily damage digital heads beyond repair.

Retrofitting of improved analogue-to-digital and digital-to-analogue converters, not necessarily from the original manufacturer, can improve the audio performance of older digital recorders. Alignment of digital recorders is not required on a session-to-session basis. Adjustments are best carried out by personnel who have completed a technical course provided by the machine manufacturer.

When a digital console is in use the recorders may use a different sampling rate to the console. Sampling rate converters are then necessary, between console and recorder. One-to-one digital-to-digital copies will be identical to the source tape, assuming low error rates. Such identical copies are referred to as 'clones'. Being of master quality, clones can be provided to pressing or duplicating plants at home or overseas, effectively providing original stereo masters to each one.

5.2.5.2 Analogue recorders

Analogue multitracks provide up to 24 tracks on 2-in. tape. Masters are recorded at tape speeds of 30 or 15 in./s. The predicted demise of analogue recording, because of the superior quality of digital recording, has been averted by developments such as Dolby HX Pro circuitry, Dolby SR noise reduction and the introduction of higher-flux tapes. Cut-splice editing of analogue tape may be carried out, in any width from $\frac{1}{4}$ to 2 in., using a precision editing block.

Caution should be exercised in handling heavy 2-in. tape reels, especially the 14-in. diameter variety. Apart from possible injuries from dropping the tapes, damage to the tape itself will result if the reel becomes bent.

When compact cassette copies are produced for clients, producers and artists to listen to in their own surroundings it is as well to remember that the output of the studio may be judged by these tiny tapes. Good-quality, well-maintained, cassette recorders are a must in any studio.

When the studio is involved in picture-related projects it may be asked to provide feeds to clients' equipment, such as Nagra recorders. A plentiful and diverse supply of signal leads and 'sex-changers', plus a knowledge of the interfacing requirements of such devices, saves an embarrassing delay while leads are made and tried.

5.2.5.3 Analogue alignment

Before any session, analogue recorders should be aligned. On a number of top-of-the-range recorders, full electronic auto-alignment is built in. The heads and tape path must be cleaned before alignment. For original recordings the replay chain should first be calibrated using an appropriate test tape. Any noise-reduction system should be switched out of operation during alignment.

Replay azimuth setting is the starting point. The engineer should peak the channel outputs on machine or desk meters. A phase meter should be used to achieve fine azimuth setting. On a multitrack recorder upper and lower bank channels can be averaged by mixing to left and right groups on the console, with the phase meter across these groups.

Replay levels and responses follow, levels having previously been determined according to the characteristics of the tape in use and system interfacing levels adopted. On full-width recorded test tapes, a phenomenon known as 'fringing' occurs at lower frequencies. This results in a false rise in response because of pick-up of flux from outside the head track width. In this case it is normal to make the final replay low-frequency setting at the end of the record section alignment for a flat overall low-frequency response.

In the recording studio, master tapes are frequently received from other studios. The originating studio should have recorded a series of test tones at the commencement of the reel and marked the tape box with details of the tones. In this situation, the replay chain should be calibrated using the supplied test tones.

The record chain is aligned using an audio oscillator. On a partially recorded master a blank section of the tape, or a blank reel of the same batch, should be used. Record-head azimuth should be set using similar techniques to those recommended for replay alignment in respect of phase correlation. To check that a false azimuth has not been selected, sweep down the frequency range; the phase relationship should remain constant. Patience may be required to achieve the best setting. If the tape is in good condition, the tape path is clean, the heads in good condition and the tape transport functioning correctly the fluctuation of output should be within 1 dB at 15 kHz. The phase meter should remain in its positive sector at all times.

Bias should be set at the tape manufacturer's recommended frequency. The peak of output, corresponding to peak bias, is sharper at higher frequencies. A typical optimum setting could be a 3 dB overbiasing at 10 kHz.

The middle frequency (usually 1 kHz) record level setting is next aligned for unity gain through the recorder. The record frequency response can then be tackled. The replay low-frequency response can then be set, as mentioned previously.

The final check should be to erase the test tones on the section used for record alignment, while monitoring replay outputs for any lack of erasure.

Any noise-reduction system can now be switched back into circuit. There may also be appropriate tones on received master tapes for setting the correct operating level of the noise reduction units.

5.2.5.4 Noise reduction

Recording studios tend to receive tapes from many sources for continuation of recording, overdubbing or mixing. Any of the current analogue noise-reduction systems, Dolby A, Dolby SR, Telcom or DBX, may be encountered. Fortunately, most systems are produced on a card which will operate in Dolby racking frames. Caution should be exercised in providing adequate ventilation for the frames, particularly as some later cards produce higher temperatures than the frames were designed for. The studio may not own a full range of cards but rely on hiring in for specific sessions.

5.2.6 Tape

The technical department should monitor the quality of tape supplies. Problems due to coating and slitting errors are better detected before the tape

has a valuable master recording on it. Tape samples from different manufacturers should be evaluated from time to time, especially when a new product becomes available. All analogue tapes are a compromise between factors such as basic noise, modulation noise, distortion and print-through. The technical engineer should help his operational colleagues to make the right choice. Other important factors are availability (especially in an emergency situation), quality of back-up and, last of all, price.

5.2.6.1 Tape storage
A studio will normally store master tapes for work in progress. It may also be asked to undertake continuing storage, after the conclusion of recording projects. A secure room should be set aside for storage, with a strictly supervised booking-in and booking-out system. Master tapes are valuable items! Storage systems should make the most of the available space. They may take the form of fixed steel racking or consist of sets of racks running on tracks, with only one gap at a time accessible within the system, allowing greatly increased storage density in a given area. Whatever system is selected it must be strong and stable enough to cope with the considerable weight involved. The maximum permitted floor loading must not be exceeded.

5.2.6.2 Tape labelling
Each tape used in a studio must be properly labelled to allow rapid identification of its contents. Below the studio's own details, spaces should be designated for information such as the project name/artists involved, the commissioning company, relevant dates, a job number (to tie-up with invoicing), the recording format (including noise reduction details), calibration tones and studio staff involved. The remainder of the label should be for continuity purposes. Details of each 'take', including timing and its position within the tape, and any comments as to performance. Details of track allocation, and studio layout/microphones used, will aid subsequent session set-ups or overdub/mixing sessions.

5.2.6.3 Tape categorization
Apart from original tapes recorded on live sessions many further tapes may be generated in the process of producing an end-product. These may include safety copies, various mix-downs, production masters, copies/clones for distribution to pressing plants/duplicators. With such a proliferation of 'masters' and copies, confusion can occur when a request is received for a particular tape. This may result in an inferior copy being used when a superior one is available. The favoured solution is the APRS (Association of Professional Recording Services) tape labelling system, which enjoys almost world-wide acceptance.

5.2.6.4 Tape insurance
The intrinsic value of a multitrack session master is considerable. A prudent studio will protect itself against accidental tape damage by a specially tailored insurance scheme. Clients' tapes should be returned to them, whenever possible, at the conclusion of the recording project. Tape insurance may not cover long-term storage.

5.2.7 Audio monitoring

The monitoring requirements are tied to the type of work to be undertaken. If the studio is aimed solely at the rock and pop record market, stereo monitoring will suffice. The choice of monitors for studios serving the record industry may, however, be governed by additional factors, such as which monitor is the flavour of the month! Monitors should be mounted in a manner that decouples them from the room structure. Their driving amplifiers, which may be sited in an adjacent room, need to be linked via multistrand heavy-gauge cable, away from low-level signal leads. Low-frequency units may be driven by separate high-powered units. Cross-over networks may be active or passive.

Monitors with a low failure rate should be chosen. Having to change a drive unit in the middle of a session is best avoided. Spooling analogue tape with the tape held against the heads, for location purposes, may damage high-frequency units by subjecting them to excessive energy from the elevated pitch of the main programme content.

5.2.7.1 Film-scoring monitoring
If film scoring and mixing is to be performed, three or five front monitors may be specified, with additional surround-sound speakers to the rear, to cope with six-channel sound-track release requirements. In film and television scoring the ability to play back a music section in a rough mix with the section's dialogue and effects guide tracks is frequently required.

When a film sound-track is destined to be released in Dolby SVA (Stereo Variable Area) optical format, a special monitor matrix can be inserted into the monitoring feeds. This simulates the effect of the

format's circuitry, particularly with respect to inter-channel cross-talk, and allows optimum results from the system.

5.2.7.2 Monitor equalization
To achieve the required response from the system and accurate stereo location, equalization may be applied prior to the power amplifiers. This may take the form of continuous boost or cut equalization, or be by full, half- or third-octave band control. Calibration is effected by the use of an analyser which measures the relative sound amplitudes, produced by bands of white or pink noise, in the console monitoring position. Monitor calibration should be regularly checked. High-level monitoring can noticeably modify speaker response.

The control room should be acoustically designed such that the absolute minimum of monitor equalization is used; sizeable amounts of equalization will adversely affect the phase response of the monitoring system. Significant room reflections are necessary for a realistic monitoring situation; careful design will ensure that they are sufficiently dispersed so as not to confuse the monitoring field across the listening area. The use of a 'live-end/dead-end' design of control room may produce fewer problems with room reflections, but many recording engineers prefer a 'traditional' room.

5.2.7.3 Hearing damage
Attention should be paid to recommendations concerning monitoring levels. In the UK, the Health and Safety Executive produces mandatory regulations covering all aspects of hearing protection. Guidance on the effect of these regulations on recording studios is available from the Association of Professional Recording Studios.

Permanent damage to the ears can result from continuous high level monitoring. Staff working in control rooms should have regular precautionary hearing checks. These can also highlight any normal loss of hearing range which could affect their work.

5.2.7.4 Near-field monitors
Many producers of music for records and television wish to hear playbacks on small speakers, to assess the sound in consumers' living rooms. Considering the variation between different makes of domestic equipment and the varying conditions under which the material will be heard, this can only be an imperfect guide.

Provision should be made, in the console, to select between main speaker monitoring and one or two other small speaker systems, plus headphone monitoring, preferably with independent level controls.

5.2.8 Mains power

Each studio should have separate power feeds from the building's electrical intake for technical equipment, non-technical equipment, and lighting, to minimize mains spikes. Interference may still be generated by other consumers on the same supply. Whilst allocating different mains phases for each use might seem attractive, the danger of severe shock, from being able to touch different phases while carrying out maintenance work, must not be ignored. Another factor to be considered is balancing the phase loadings. This is particularly important if the use of an emergency generator is possible.

5.2.8.1 Emergency generators
Few studios can afford to be incapacitated by significant power failures. An emergency generator, driven by diesel power, is the alternative. It should be rated for the maximum power requirement of the studio(s); otherwise manual load shedding may be required if, say, the building is heated by electricity, or uses a great deal of lighting.

Generator autostart on mains supply failure is preferable. This involves the installation of heavy duty contactors to effect the changeover. The resultant glitch, between mains failure and the generator reaching operating speed, may result in technical equipment, particularly that using computers, needing to be reset or reprogrammed. The use of an uninterruptible power supply (UPS) should prevent this unwanted disruption if the UPS has sufficient internal power to bridge the gap and is glitchless in its operation – not all are!

The generator should be speed stabilized to maintain the correct power frequency. Metering of supply voltage and frequency should be provided at a convenient point within the studio.

5.2.8.2 Isolation transformers
Mains power is required in the studio for driving electronically amplified instruments, synthesizers, etc. As their condition is generally outside the control of the studio management, mains isolation transformers should be provided. This will prevent dangerous electric shock in a situation where a musician whose instrument has a faulty or wrongly wired mains plug touches an earthed microphone. An alternative to isolation transformers is a properly installed

and thoroughly tested earth leakage circuit-breaker system.

5.2.8.3 Power wiring
Frequently in a recording studio new equipment has to be installed or added. This may require additional power feeds. Care should be taken not to exceed the temperature rating within power ducting by cramming in more and more cables.

5.2.8.4 Earthing policy
All equipment provided with frame earthing points must be connected to mains earth. The only exception is double-insulated equipment, specifically designed to be operated power earth free. It may be necessary to connect signal cable screening to the chassis of such equipment.

A studio's signal earthing must be run separately from its power earthing to avoid unwanted noise injection. The provision of a technical earth, grounded to a well buried copper mat or rods, is sometimes adopted but must be proof against hazards such as the ground drying out in a drought.

5.2.9 Timing aids

In the world of music for film, television and advertising, timing is all important. Cue points have to be met to match picture cuts or action changes. There are various aids that can be used.

5.2.9.1 Click tracks
Click tracks were originally developed by the film industry as an interlocked metronome, synchronized to the film section being projected, for music scoring. They were produced by an editor punching holes in opaque film stock at appropriate intervals for the tempo required, in the area of the optical soundtrack. As each hole was scanned, a click was produced at the audio output of the optical sound system. This was fed to the conductor, and sometimes to the rhythm section, on headphones. The click-track tempo was expressed in frames and part-frames, assuming a 24 frames/s projection rate. Thus an 8-frame click would correspond to a tempo of 180 beats/min.

Electronic devices are now generally used, unless a click track has to follow a varying tempo guide track which had been used during the filming of a musical scene. Then a 'laid' track is prepared by the music editor, matched to the guide track, to enable the conductor to replace the guide track with music matching the screen action.

Electronic click generators should be operated in synchronism with the picture source/recording system. Some versions allow the first beat of a bar to be emphasized.

5.2.9.2 Timing devices
Stop clocks with digital displays will normally be a standard feature within consoles and conductors' rostrums. They will generally be crystal referenced. Local and/or remote control should be possible. Optional linking of control between a producer's and a conductor's stop clock can be advantageous when the conductor is too involved to start his own clock. Optional blanking of the tenths of seconds display, while the clock is running is desirable. A present count-down to zero is useful in music scoring. Starting the count from a remote contact closure can be used to link the stop clock to time-code controlled equipment.

Some conductors have a preference for sweep-hand timing clocks rather than digital displays. These are less amenable to remote control.

Timing films have been produced which count in minutes and seconds when projected at 24 frames/s. They may also have a calibrated wipe (a line travelling across the screen) to give the conductor a cue for commencing his down beat. In use, they are projected in synchronism with the cutting copy of the film section being scored and superimposed on the picture.

5.2.9.3 Video timing aids
By the use of picture insertion techniques, time-code can be displayed for timing purposes during scoring. Suppression of the hours and frames digits may make it more readable. More sophisticated devices can provide an electronic equivalent of the timing film, locked to whatever frame rate is in use. Click-track generation can also be synchronized to the timing information.

5.2.9.4 Studio clock
The studio should be equipped with an accurate time-of-day clock, in prominent view of both studio and control room occupants. Extremely accurate radio clocks are now available at reasonable prices. Session musicians' agreements are tied to specific-length sessions. If everyone can see the same clock, life may be a little more straightforward!

5.2.9.5 Time-code
The recording studio may have to deal with time-codes of all types. Typical uses are locking a video playback machine to an audio multitrack, or locking

two or more audio machines together. Audio-to-video or audio-to-audio interlocking systems should be capable of handling 24, 25 and 30 drop-frame rates. When copies of timecoded audio tapes are made the code on the copy must relate to the original. Timecode should be reshaped or regenerated from the original input.

Before the start of a fresh session using time-code locking, the blank tape should be 'prestriped' with the appropriate time-code. Depending on the level of the code, and the cross-talk performance of an analogue recorder, it may be necessary to leave a blank guard track between the code track and the nearest music track.

When working to picture, with synthesizers, a time-code-to-MIDI interface will be required. It is then possible for a music section to be completed within the hard disk storage of the synthesizers before committing it to tape.

Automation systems rely on time-code for their operation. They may require a special wide-band replay amplifier, able to read code during high-speed tape spooling or shuttling.

5.2.10 Film projection

The use of direct film projection in music scoring studios has decreased as the use of video-based picture sources increased. Where projection is available there are four possible combinations of basic format. The film gauge may be 16 or 35 mm; the film speed may be 24 or 25 frames/s (for material filmed solely for PAL or SECAM television use). With cinema film the various image aspect ratios should be catered for although the use of 70 mm projection in studios is virtually unknown.

Two projectors may be run in synchronism, when one carries the 'cutting copy' picture and the other carries a timing film for on-screen superimposition. The projected light level needs to be high enough to be viewed against the ambient light level in the studio.

There may also be a requirement for 'sound heads' (magnetic film reproducers) to run in synchronism with the picture, to provide guide tracks during recording or the subsequent playback of a music section. Film recorders may also be required to produce the sprocketed music masters, either live on a session or during mix-down from time-coded multitrack tape, for later manual editing and inclusion in the film dubbing process.

Synchronism between sprocketed film transports was originally achieved by the use of mains level three-phase control equipment and individual interlock motors on each transport, with all rotors and stators linked to common buses. Current techniques use signal level buses to carry the interlocking information, generally biphase signals, with electronic power circuitry to drive individual transport motors.

The ability to control the film motion from the engineer's position of the console may be advantageous. The provision of some basic film editing equipment is desirable, to enable last-minute changes to be carried out during scoring.

5.2.10.1 Video projection
As an alternative to direct film projection video projectors may be used. The screen luminosity achievable may require careful attention to screen siting and ambient lighting levels. Remote control of routine adjustments will generally be necessary, in view of the inaccessible positions in which video projectors tend to be mounted.

5.2.11 Digital editing

Many systems exist for digital editing, ranging from audio versions of simple video editors to sophisticated systems, utilizing large amounts of hard disk storage and including comprehensive digital mixing facilities. In a commercial operation the time saving brought about by down-loading the programme material on to hard disk, for editing, can be a significant factor in the cost of editing a complex production. Some systems allow simultaneous editing of one production while another is being down-loaded. The ability to locate any point in the programme material virtually instantly has both economic and artistic advantages. In the simpler tape-based systems the waiting time while tape winds, shuttles and finally locates the desired edit point, can be tedious.

The hard-disk-based system provides additional flexibility, such as the ability to instantly change the running order of items in an album. Cross-fades of any desired length can be effected. 'Impossible' edits can be produced, using visual waveform assistance. Digital noise reduction can be applied to existing material, with special software for click and other unwanted noise removal – comprehensive restoration of archive material for reissue becomes possible.

Real-time sampling rate conversion during down-loading can be undertaken. Edit decision lists can be generated. Digital editing is not confined to two-channel operation, although the bulk of work is in this format.

5.2.11.1 CD preparation
The principal use of digital editing is in the preparation of programme material for transfer to compact disc (CD) or other media. PQ coding can be integrated with the editing process to add track data and index marks configured to the standard CD specification (usually referred to as the 'red book').

A separate operation allows verification of the PQ coding on the final digital stereo master tape together with a report on error rates within the tape. A copy of this printed-out report should accompany the tape to the CD plant. Apart from the usefulness of the information it contains, it may nip possible disputes ('it was fine when it left us...') in the bud.

There is a choice of recording media for pre-mastering. The U-Matic format is losing ground to Exabyte, CD-R and Sony PCM-9000. Caution is required before adopting any particular format, as it could have inherent limitations. In any case, it is essential to liaise with the glass mastering plant to ensure compatibility. Any 20- or 24-bit material should be converted to 16-bit resolution before delivery to the plant.

The digital versatile disc (DVD) format presents a fresh set of technical requirements to studios servicing it, whether for audio or audio/video applications.

5.2.11.2 Digital editing for picture
Sound-tracks destined for film or television use, synchronized to picture action, can be produced on a digital editing system which has the facility to lock to time-code from the picture source. Background sound effects (continuous ambience tracks) can be generated by selecting a short section of the desired effect and having the system repeat it for as long as required, making seamless joints between the ends of the section. Time or pitch changes can be applied to achieve accurate matching between on-screen action and the accompanying sound-tracks.

5.2.12 Disk cutting

The days of the vinyl disk are numbered. Few recording studios now offer disk cutting facilities. Those that do are mainly concerned with producing white label disks for use by disk jockeys. Cellulose acetate, rather than copper, blanks are preferred for this application because of superior high-level performance.

5.2.12.1 Cutting room
The room will normally include a small console giving control over gain, equalization, dynamics and noise reduction. Programmable equalizers, which automatically change to previously selected settings for each band of an album, may be found in some cutting rooms. This is the last point in the recording chain at which changes can be made. Remote control for lathe functions and tape machines will be present.

Very fast and accurate peak-reading metering, with a wide dynamic range, is necessary. The cutting system must be calibrated against standard test discs. Controls are configured to allow repeatable settings when recuts of the same material is required. A logging system for control settings needs to be operated.

Rather than relying on the logging system the provision of an additional recorder, on the output of the console, allows a tape copy to be made of the equalized and otherwise modified signal. This tape can then be recut if necessary, without further changes being required. Alignment tones on the tape are essential here, as in previous operations.

A first-class monitoring system is naturally required.

5.2.13 Noise isolation

Recording studios, particularly the small private type, are often housed within existing buildings not purpose built for studio use. Effective isolation of recording areas from one another, and from other activities inside and outside the building, is required. This is a field where the services of a qualified architect/acoustician is highly desirable.

A problematical situation can arise in a multistudio complex during the rebuilding of one area while the remainder remain operational. Apart from scheduling the noise-producing work outside normal recording hours an extension of the recording-light system to a very visible/audible alarm in the constructional area can be effective – if the recording engineer(s) remember to use the system!

5.2.13.1 Studio acoustics
Successful acoustic design is a specialized field. It embodies specifying the necessary absorbent/reflective/dispersive surfaces dictated by the expected usage and combining them with attractive decor, with all the necessary services integrated within the scheme. All of this must fulfil the requirements of the appropriate building and fire-safety regulations. A control room presents a different set of demands to a studio area with its requirement for accurate monitoring over a wide area. Strict supervision during

the construction phase is required to ensure that the design detail is strictly adhered to.

The acoustic requirements of a recording studio can differ from establishments in other fields in that the type of work undertaken may vary dramatically from session to session. Therefore, some ability to vary the apparent room acoustic is desirable. This may take the form of designed-in variable peripheral panels, reflective on one side and absorbent on the reverse. These may be manually operated or motor driven. This approach may require considerable fine-tuning when initially put into service. The general experience is that much less use is made of this variability once the studio is in everyday operation.

The alternative is to design for the longest reverberation time deemed necessary and to modify the apparent acoustic as required for the work in hand by the use of a mixture of portable acoustic screens, removable carpets and close proximity microphone placement.

5.2.13.2 Acoustic screens

Screens should embody sufficient mass for them to be effective isolators at middle to low frequencies and should have absorbent surfaces effective at middle and high frequencies. A stable design within a sturdy frame with quality castors is essential. Full-height screens may include double-glazed windows for visual contact between musicians. Visual contact with orchestral conductors is sometimes overlooked when laying out a studio for orchestral recording, using screens or isolation rooms. Closed-circuit TV can be useful in difficult situations. Visual contact between conductor/musicians and recording engineer is also desirable.

5.2.13.3 Isolation rooms

As their name implies these are areas associated with a studio but with a degree of sound isolation from the main studio floor. Their purpose is to allow better individual control over particular instrument(s) or vocal(s), or to prevent sound leakage into the main studio microphones in order to facilitate later replacement of the isolated track.

Isolation rooms should preferably have visual contact with the studio or control room. They should have an isolated air-conditioning feed and a full range of inputs and outputs available. Closed-circuit or programme video monitors may be required. There is nothing worse for an artist than to be marooned in a badly sighted or equipped isolation room, feeling out of touch with the session.

5.2.14 Ancillary services

Although not strictly the province of the audio engineer there tends to be a considerable degree of overlap between areas of responsibility in recording studios. It is therefore worth considering the requirements for some of the services required within such an establishment.

Whilst recording studio operation is primarily a blend of creativity and electronic engineering, with a dash of acoustics, no studio can exist in isolation. It has to be housed in a suitable shell and be able to look after the creature comforts of those working within that shell.

5.2.14.1 Furnishings and decor

Colour schemes should be easy to live with, throughout an extended working day. Quality furniture is a worthwhile investment, especially musicians' and engineers' chairs – people spend a considerable part of their working lives in a sitting position. Attention should be paid to comfort, including back support. Avoiding plastic seat coverings will minimize perspiration problems. The absence of creaking sounds from musicians' chairs is a welcome bonus. Music stands should be stable and rigid, not cheap folding devices that collapse at the critical moment.

Special seating is necessary for certain instrumentalists such as double bass and cello players; provision of a wooden surface linked to their seating to receive the spike of their instruments will prevent damage to the studio floor. Keyboard players require standard adjustable-height piano stools to achieve the optimum playing position.

Engineers' chairs should be high-quality gas-lift types mounted on a stable five-castor base to enable them to move rapidly along a lengthy console. A hard floor surface should be provided behind the console for the castors to run on. Easy chairs should be provided in the control room for clients not occupying positions at the console.

Rostra for certain orchestral sections or choirs should be portable and stackable. Safety rails must always be fitted at the rear of the highest section to prevent injuries and expensive subsequent lawsuits.

A conductor's rostrum should provide such facilities as adequate score space, room for seating, stop clock, headphone facilities and talkback communication with the control room isolation booths and foldback circuits. A telephone link for private discussions with the producer is sometimes requested.

5.2.14.2 Pianos

The only instruments which many studios own are pianos. Invariably this will be a larger model grand piano from one of the top makers. Studio use generally entails moving the instrument frequently to fit in with different session requirements. To withstand this the piano should have an additional sturdy mounting frame linking the legs and be equipped with heavy duty castors.

Frequent tuning is required in studio use. Many factors can cause tuning variations over remarkably short periods, including the manner in which the instrument is played. A stand-by tuner may be required on important sessions.

5.2.14.3 Lighting

Lighting, adjustable by switching or dimming, can be set according to the activity taking place in the control room. Maintenance activities need good general illumination, whereas a late-night mixing session may require soft lighting, centred on the console area. Rock musicians may be content with moody lighting in the studio; orchestral players will wish to read their music without developing eye-strain.

5.2.14.4 Air conditioning

In control room and studio areas a constant temperature in the 'comfort zone' is desirable throughout the seasons. Apart from its beneficial effect on the occupants, many instruments lose their tuning as temperatures change. Humidity control is also desirable. Very low plant noise, with freedom from draughts, is essential. Rapid back-up is essential if failure occurs.

5.2.14.5 Effects library

Requests for sound effects crop up from time to time in recording studios. The effects may be required for film/video postproduction work or for atmospheric use on a pop record. The acquisition of a good CD-based library containing a wide selection of effects is to be recommended. Using a professional CD player, any effect can be located rapidly by using the indexing facility. If a longer section than exists on the disc is required, seamless 'looping' can be achieved electronically.

5.2.14.6 Catering

Staff and clients often spend long hours in stressful activity. Provision should be made for the supply of refreshments on whatever scale is appropriate. A studio in the middle of a busy metropolis may have many catering establishments on its doorstep, lessening the need for in-house catering. A plentiful supply of hot and cold drinks is a minimum requirement. A quiet refrigerator in the control room can work wonders for the studio image.

5.2.14.7 Recording media suppliers

Apart from the technical performance and relative cost of the various recording media, factors such as short notice availability and technical back-up can be just as important. Most tape manufacturers' products have problems from time to time. The ability to sort the problems out satisfactorily, and come up with rapid replacements, is most important.

5.2.14.8 Equipment technical back-up

Manufacturers should be able to provide rapid and real help if their console, recorder or outboard device develops a major fault which cannot be cured in-house. This important factor needs to be taken into consideration when making purchasing decisions.

5.2.14.9 Fire safety

Recommended practice should be followed with regard to fire safety. Particular attention should be paid to keeping escape routes clear. Studios are notorious for parking equipment and instruments in corridors. Management and staff should pay more than lip-service to fire safety. Emergency lighting systems should cut-in automatically on failure of the mains supply and provide adequate illumination to enable all occupants of the building to reach safety. Regular fire drills and testing of alarms, appliances and emergency lighting must be carried out at regular intervals, and logged.

5.2.14.10 Telephones

Studio clients may or may not wish to be disturbed during a session by incoming telephone calls. A flashing light to indicate ringing is preferable to an audible calling system in a control room. The client should be provided with the opportunity of taking the call in reception, rather than add to the unwanted sounds reaching the engineer's ears.

5.2.15 Maintenance

As in any technology-based business, good maintenance is paramount. Unnecessary egg on faces is to be avoided.

5.2.15.1 Maintenance engineers

Engineers should have a good general theory grounding upon entry. Completion of specialized studies in

studio engineering, preferably a 'sandwich' course, is a bonus. Early familiarization with the particular equipment in the studio is essential. Access to manufacturers' courses should be provided by the management.

Maintenance cover should be provided for all recording sessions, either by suitably trained staff being on the premises or being rapidly available via a pager. The possible consequences of a breakdown should be assessed when deciding the degree of cover to be provided.

5.2.15.2 Technical department
The technical department should contain the necessary tools of the trade for electronic and mechanical servicing. A well-equipped workshop, able to cope with the majority of on-site repairs and the construction of specialist small items of studio equipment, should be provided. It should preferably be sited on the same floor as the technical areas it serves. Electronic and mechanical areas should be physically separated – iron filings in multitrack recorder innards are to be avoided at all cost!

5.2.15.3 Test equipment
A high-quality combined audio test set providing measurement of response, amplitude, distortion and phase relationships can be a better investment than separate signal generators, millivoltmeters, distortion analysers and phase meters. Wow and flutter analysis is also available within some combined test sets. Digital multifunction test meters are reasonably priced and accurate. They are also more resistant to being dropped than are their mechanical-movement counterparts.

A very high-speed dual-beam oscilloscope is required for digital audio equipment servicing. Manufacturer's test leads and probes should be used, whenever possible, as their particular characteristics may be important for accuracy.

For monitor-speaker alignment a suitable analyser is required. Its microphone should be of high accuracy and regularly checked.

Specialist test gear/jigs/tools dedicated to particular equipment, may be necessary. The manufacturer concerned should acquaint the purchaser with any such requirement.

5.2.15.4 Alignment tapes and films
Test tapes should be carefully stored away from heat and significant magnetic fields, and carefully handled when in use to avoid mechanical damage. They

should be from a reputable manufacturer, and be of guaranteed accuracy. Calibration should be checked at sensible intervals. When a new alignment tape is acquired compatibility with the tape it is replacing should be checked. Should there be noticeable differences in levels or azimuth setting further investigation is called for.

5.2.15.5 Other facilities
A comprehensive technical library, including books on general audio topics and techniques, manuals for all equipment and suppliers catalogues and data sheets, should be assembled. Items should be filed such that rapid retrieval is possible. A small drawing board, for producing scaled drawings of items to be manufactured in-house, or by an outside contractor, is worthwhile.

5.2.15.6 Spares
A stock of essential spares should be held. Manufacturers' recommended holdings should be carefully considered in making the decision as to the level of spares to be purchased. The aim is always to keep the session going – somehow.

5.2.15.7 Preventive maintenance
Despite the popular maxim of 'if it ain't broke, don't fix it', regular preventive maintenance should be undertaken. Full testing of every function on every channel of a studio console can save reputation-damaging problems during an important session. Similarly, every piece of control room and studio equipment should be checked conscientiously.

5.2.15.8 Cleaning aids
Studio equipment needs cleaning if it is to continue functioning correctly. Heads and drive-path components on recorders are especially vulnerable to dirt, grease, and oxide shedding. Follow the manufacturer's cleaning recommendations before using the machine. The heads on digital recorders need particular care. They may be irrevocably damaged by incorrect cleaning techniques; the direction of the cleaning movement across the head is critical. The most widely used cleaning agent is isopropyl alcohol, coupled with lint-free pads or wipes (beware – alcohol is highly flammable). Only small quantities should be carried into technical areas. Bulk storage must be in locked fire-proof cupboards.

Consoles suffer from all manner of debris and contaminates in their 'works'. A ban on smoking

in the control room is desirable from all points of view, including the console's. Keeping drinks away from the console is also desirable, but often impractical where the client calls the tune. Spillage into faders and modules should be attended to at once. The treatment may be as simple as washing a fader track with distilled water, or it may require extensive stripping and careful cleaning of a complex module.

Bibliography

Audio Media, St Ives, Cambridgeshire (monthly)
BORWICK, J. (ed.), *Sound Recording Practice*, 4th edn, Oxford University Press, Oxford
Pro Sound News, P.S.N. Publications Inc., New York (monthly; European edition available from Link House Publications)
Recommended Practice for the Interchanges of CD Premasters, APRS, Reading, Berkshire
Sound Pro, SOS Publications group, Cambridge (monthly)
Studio Sound and Broadcast Engineering, Link House Publications (monthly), London

5.3 Facilities required in television sound control rooms

5.3.1 Introduction

Television production sound control rooms have much in common with other sound control and mixing facilities, but differ in a number of important ways. The differences are a reflection of the way in which the business of making television programmes is conducted, the dominance of the pictures in the final product being indicated by the priorities given to the sound and picture elements during the production process.

The most obvious difference at first sight is the presence of picture monitors, placed in a prime position and, therefore, having a significant effect on the layout of the remainder of the equipment in the room and on its acoustics (*Figure 5.9*). The number of picture monitors in all control rooms shows a tendency to rise and in the sound control room this can cause particular difficulties for

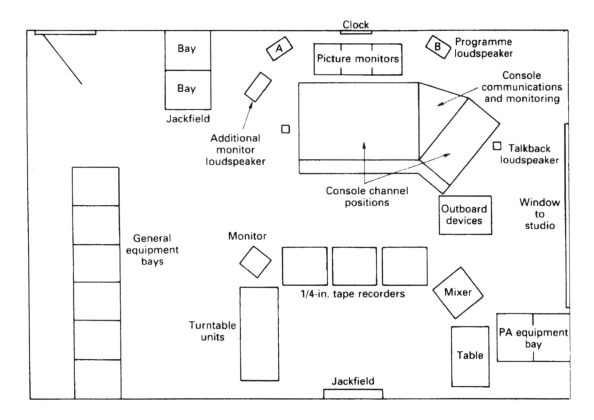

Figure 5.9 *Typical layout of a television sound-control room.*

the placement of programme monitoring loudspeakers. The second important difference is in the quantity of communications equipment and the location it occupies in the room. The television production process proceeds in several locations simultaneously, and good communications are vital to its success.

Television studios are expensive to build and equip and are therefore rarely specialized. A general-purpose facility can expect to work on a high proportion of available days. Some are, however, purpose-built, typically in the news and current-affairs areas. The increased prominence of sports and news programmes has led to the conversion of some facilities into dedicated linking and production areas capable of handling multiple programme origination, while the utilization of the studio floor has tended to shift to simple linking using virtual backgrounds behind presenters and to the accommodation of support services. In most other studios the scale of equipment in the sound control room will reflect the size of the studio floor and the general standard of equipping of the studio.

5.3.1.1 Acoustics
Control-room acoustics generally fall into two distinct groups: those intended to support the characteristics of the sound to be produced, and those intended to be completely neutral in nature. Many music recording studios have a control-room acoustic controlled by diffusers and carefully angled reflecting surfaces rather than absorbers. This can produce a light and relatively live (i.e. reverberant) room well suited to the monitoring of music, but which colours the sound of speech so that the original acoustic surroundings of the speaker are masked. The technique also requires precise positioning of large pieces of equipment in the room in a way completely at odds with television operations.

In the alternative and much more common approach to control room acoustics the sound of the room is controlled by the use of absorbers, designed to reduce the naturally occurring resonances of the room so as to produce a neutral uncoloured sound. For most rooms a reverberation time of 0.3 s or a little less is the aim, since this will match most listeners' experience of their normal living room environment. The reverberation characteristic will generally show a tendency to rise at low frequencies and fall gently above middle frequencies, exactly as in a living room with soft furnishings.

The very great majority of television sound control rooms fall into the latter category. The reasons are two-fold: first because the quantity and positioning of equipment in the room makes any other acoustic design very difficult, and second because the wide range of programme material encountered in the typical television sound-control room makes tailored design impossible. The costs of effective acoustic treatment are under constant pressure and become increasingly difficult to support when contrasted with the low-cost alternative of location and drive-in working.

5.3.1.2 Interior design
In an acoustically specified environment the interior design is inextricably linked to control of the acoustics; at the very least the final surface finishes must be chosen knowing their acoustic performance. Thus wall finishes, drapes and carpet are part of the overall acoustic design, as are any large reflecting surfaces. This difficult feature of control room design has begun to ease as miniaturization and computer-aided control of equipment has led to reduced physical bulk of both mixing consoles and equipment bays. Though many more pieces of ancillary equipment, both permanent and temporarily imported, can be expected than in the past most are physically smaller than those of previous generations. Where reflecting surfaces of any significant size are unavoidable, it is normal to arrange that reflected sound waves do not arrive at the listening position as early discrete reflections, but as diffuse waves well separated from the original in time.

Pieces of equipment with large exterior panels must be regarded with suspicion on two counts: both as reflectors and resonators. In the latter case, purpose-made damping pads are usually applied. Acoustic considerations apart, the control-room interior finishes must be durable and provide a pleasant working environment. Control of temperature is normally by way of control of air conditioning, which should preferably be of the constant volume type so that air movement and the noise it causes are constant. Lighting too should be under the control of the operator who will need a general or high lighting condition for non-production activities, and a low light condition in which only essential working surfaces are lit and reflections in picture monitor screens are eliminated. More recently it has become necessary to consider the effect of lighting on LCD displays and on dual-intensity LED indicators.

5.3.1.3 Fixed equipment
The capital cost of building and equipping a television studio makes it imperative that the facility is

used to the fullest extent possible. This in turn means that the great majority of studios must be general purpose, and able to service every style of programme. Whilst this may seem wasteful when considering a simple interview taking place in a studio capable of accommodating a major light-entertainment spectacular, it must be seen in the context of building a second, smaller, studio for the interview, perhaps leaving the larger studio standing idle.

The 'fixed equipment' described here is not simply that which is literally fastened down, but also that which you would expect to find in almost any operational studio. Thus in addition to the console and loudspeakers, equipment such as tape-recorders, compact disc (CD) players and outboard devices are included, although they may actually be moveable.

5.3.2 The console

As a general, though not infallible, rule the capacity of the mixing console will increase with the size of the studio. A small studio of perhaps 100 m^2 or less will be used for speech-based programmes of limited scope and will typically be equipped with a console of 8–12 channels feeding two or four groups. A 1000 m^2 studio which can house large-scale musical and light-entertainment productions will need a console with a minimum of 50 channels and may well have twice that number. The alternative, preferred by some owners, is to have a fixed console of moderate size supplemented when necessary with an additional mixer to handle the extra sources from, for instance, an orchestra. However, finding a suitable position for the extra mixer which the operator can reach whilst still being in a correct monitoring position can be very difficult; the alternative of a second operator represents an extra cost and may also be operationally unsatisfactory.

Consoles designed for broadcast work differ in some details from recording studio desks. Television sound in the UK is virtually all stereo, so consoles must accommodate stereo signals on many or even all channels, usually by means of one fader generated control signal being simultaneously applied to two selected signal paths. Because of the need to produce a complete output which will be routed onward to a videotape recorder or transmission chain, it is normal to have one or more main programme bus-bars onto which all submixed outputs are finally routed and which are hardwired to the principal monitoring selector. Thus more than one audio version of the programme may be simultaneously produced for different purposes though one will normally be the

main and thus most closely monitored. Monitoring, both visual and auditory, is normally also provided for the large variety of other sources which may form part of the final programme, although having originated outside the studio, and for sources which may at some point be required as cue material for contributors to the programme.

Auxiliary output chains are often required in greater number on a broadcast desk than elsewhere; in addition to providing the means of feeding outboard devices such as reverberation systems, they are also called on to send clean feeds (mixed minus feeds) to outside contributors or telephone callers and to send a wide variety of cue programme feeds to presenters, musicians, scoreboard operators and so on. Some of these auxiliaries may be dedicated to a particular purpose, for instance public address. The reason lies in some additional switching enabled by fader backstop microswitches which can be used to establish and break circuits in certain situations.

A further difference between recording and broadcasting desks lies in the choice of meters for monitoring, the recording industry by and large preferring volume-unit (VU) metering, whereas broadcasters need the accurate indication of peak values and so prefer peak programme meters which may be of the mechanical or inserted-in-video form.

Larger broadcast consoles frequently include the ability to send signals to multitrack recorders in addition to the main programme chain. Many television programmes go through an extensive postproduction process at which the final mix is created; the multitrack recording allows the production of a sound track not compromised by decisions made at the time of the studio production.

The need for extensive communications both within the studio and with the outside world will be clearly seen on the broadcast console. Talkback facilities for production direction will be supplemented by completely separate sound and lighting talkbacks, plus intercoms for virtually every group of staff engaged on the programme.

5.3.3 Loudspeakers

The most critical test of a loudspeaker is to listen to a sound with which the individual is completely familiar. Nothing fits this description better for most people than the human voice; a loudspeaker to be used for general programme monitoring should therefore pass this stringent test. Generally this will favour loudspeakers with a neutral, uncoloured sound which may at first listening seem unexciting. Loudspeakers

which make some types of programme, for instance music, seem more attractive should be avoided, since they cannot be relied upon to tell the truth when listening for faults, a very important consideration in the broadcasting environment. A further consideration when choosing loudspeakers should be consistency, allowing matching between areas, since parts of programmes will often be made in different studios or locations. If the acoustics of the control rooms and the loudspeakers within them are carefully matched, then close matching of sound sources will be much easier.

The main monitoring loudspeakers should be positioned according to the equilateral triangle rule for stereo monitoring, that is with the principal listening position being at one of the points of an equilateral triangle of sides of 2–3 m and the loudspeakers at the other two. If this placing means that they are partially or completely behind the console, they should be raised on substantial stands so that at least all of the cone area of the loudspeaker units is above the desk, and leaning forward if necessary so as to bring the axis of the tweeters to the operators' head height.

In addition to the main programme monitoring loudspeakers, there will usually be another pair fed from a matrix selector to allow for the monitoring of other sound sources independently of the programme. These should normally be of the same type as the main pair of loudspeakers if space and other considerations allow.

5.3.4 Tape and disc operator's position

Television programmes often include quite complex sequences of material played from prerecorded tapes and discs. It is common practice to concentrate all this activity in an area of the sound-control room where tape-recorders, turntables, CD players and cartridge machines, etc., can be organized so as to be used by one operator. The outputs of all of these devices are combined by a small mixing console under the control of the operator and offered as one or more premixed feeds to the main console. The submixer offers the same general facilities as the larger console but on a reduced scale, so that, for instance, feeds of individual sources can be given to foldback when needed or reverberation added to one item and not the remainder of the output.

The same operator is usually also responsible for making any recordings required in the course of production or for the postproduction activity.

5.3.4.1 Tape recorders
Quarter-inch standard tape recorders are usually twin track rather than stereo in television. This means that the recorded tracks are narrower than on stereo machines, thus allowing a wider guard band in the centre of the tape. Although the signal/noise ratio is slightly poorer as a result, the extra physical separation on the tape gives greater signal separation between the recorded tracks which may, therefore, carry unrelated signals. Longitudinal recording of SMPTE or EBU time-code may be made in the guard band area down the centre of the tape. Recordings made in this way are frequently used in postproduction programme dubbing when a time-code synchronizer may be employed to allow the tape to be run once more in perfect synchronism with the picture, though this function has largely been taken by digital recorders usually of the R-DAT format.

5.3.4.2 Turntable units
Turntable units in sound-control rooms are of a type developed to allow accurate cueing of recordings and rapid starts when needed. Facilities incorporated in the design include fast start, reverse rotation for setting up, fader start option and prefade listen. A means is also sometimes provided whereby the precise position of the pick-up on the disc may be noted so that rehearsed settings may be rapidly repeated. Between two and four disc players are usually provided, though the increased use of compact discs, samplers and mini-disc players will cause a consequent reduction in numbers of conventional turntables.

5.3.4.3 Compact disc
Compact disc (CD) players having the same sort of facilities as standard turntables have also been developed for professional use. The CD, however, has additional features which can be very valuable for broadcast use. In particular, track numbers and running times recorded in the subcode make setting up particularly simple, whilst the use of an output buffer in the player can make the start truly instantaneous. The clocking out of the digital signal by reference to a very stable oscillator completely eliminates wow and results in extremely accurate control of both pitch and running times, while electrical or optical digital outputs permit the export of complete audio and subcode information to other digital devices. Some players will give track time as either elapsed time or remaining time, and some have memories for sequences of cues. So-called 'CD juke-boxes' are manufactured with the ability to store many CDs and

select and play tracks from them according to a pre-set sequence; with the addition of a purpose-designed front-end and specially written software these are now offering controlled replay of long sequences of cues, cue listing and editing on a visual display unit (VDU) and complete library management of, for instance, a sound effects library.

5.3.4.4 Cartridge machines

Once the automatic choice for playing sound effects and music cues into programmes, cartridge machines are now largely superseded by other systems, although many will still be found in use due to the extensive libraries of material available. Their relatively poor stereo performance caused by intertrack phase errors, limited signal/noise ratio and tape wear problems give them limited life expectancy now that digital equivalents and new formats such as mini-disc offer superior performance at low cost. It should be noted, however, that these latter systems gain their large storage capacity at the cost of heavy bit rate reduction whose impairments and artefacts become readily apparent as copy generations increase.

5.3.4.5 Digital audio tape

Digital audio tape (DAT) recorders of both the rotating head (R-DAT) and stationary head (S-DAT) types have been produced and have found a place in the broadcast sound control room. R-DAT is now a standard format offering 2 hours of stereo recording with time-code on professional machines. The R-DAT cassette is very small as is its drive mechanism, and the development of very large scale integration (VLSI) electronics has permitted the manufacture of extremely compact recorders, very popular in field recording. Studio machines are usually larger and have more facilities such as a choice of sampling frequencies and output formats, and sophisticated cueing systems. Both share the advantage of the high storage density of the small cassette. Although intended originally as a consumer format, R-DAT is now supported as a professional standard by a number of manufacturers who include in their range machines capable of remote control by standard protocols, time-code synchronization and perfect editing. The addition of four-head machines with flying erase has made possible write-before-read (confidence replay), read-modify-write (single machine dubbing) and insert recording in the manner of video recorders. It is even possible to time-code stripe a tape using subcode recording after the audio

has been recorded. These are very powerful facilities especially in postproduction. R-DAT has another advantage in postproduction use – its very low linear tape speed (8 mm/s) and tightly servo-controlled transport make for very rapid spooling and short lock-up times. For this reason they are very popular as *'snoop'* recorders both on location and in the studio.

Stationary head machines using a cassette of similar proportions to the compact cassette have been produced, though mainly for domestic use. However, since prerecorded music will be issued on this format, the broadcast sound control room will have need of at least replay facilities. This format is unlikely to gain acceptance as a professional broadcast standard for a number of reasons, amongst them its use of a bit-rate reduction system of uncertain multigeneration performance and the lack of fully specified professional machines.

5.3.4.6 Multitrack

Multitrack recorders of both the analogue and digital varieties first found their way into television sound control rooms for use on music programmes, for exactly the same reasons that the record industry used them. More recently, the growth in postproduction activity has seen an increase in their use, as sound operators have found it necessary to retain all possible flexibility to the very end of the production process. The degree to which they can be used successfully in a television environment will be governed to a degree by the ability of the console to produce the necessary feeds, which will usually be required to be from before the fader in order to allow full flexibility of control later. Many broadcast consoles cannot meet this requirement without very convoluted routeing and plugging, and later consoles are often specified as a hybrid of recording and broadcasting types so as to incorporate this facility. The physical bulk of both machine and tape has led to the abandonment of full broadcast machines in favour of semi-professional digital multitrack cassette recorders in many situations.

5.3.4.7 Samplers

Samplers have been popular in the music industry for a number of years, and have become valued in the broadcast field as their capabilities and memory sizes have increased. Some, which can be provided with additional storage in the form of optical or Winchester disk drives, can hold for instant access a considerable amount of audio. Multiple channels of output and excellent digital quality make them

very powerful tools in both production and post-production. Since most were produced first for the music business, MIDI is the usual control standard, although MIDI-to-time-code interfaces are available either from the manufacturer or third-party suppliers. In addition to storing audio in random access memory (RAM) and disk stores, samplers will also manipulate it in a variety of ways. Pitch change, lengthening and shortening, editing and looping are all easily performed.

Like the majority of equipment first provided for the domestic market, these devices normally have unbalanced inputs and outputs. A number of manufacturers make professional interfaces designed to integrate such equipment into professional systems.

5.3.5 Outboard equipment

From small beginnings when perhaps a graphic equalizer would have been balanced on the end of the script ledge, this has grown to become a major part of sound control room equipment. Outboard equipment is usually taken to mean any equipment not directly built into the console or bays, but for which additional accommodation is provided. The accommodation is usually in the form of either 19-in. racking or trolleys and in broadcast studios is placed within easy reach of the seated operator. Equipment in this category will include some or all of the items described below.

5.3.5.1 Reverberation systems
Early digital systems were large, expensive and tended to have individual and characteristic sounds, and were therefore chosen when their characteristics enhanced the programme material. Current devices have much more highly developed software, and competition has brought down prices. Older reverberation systems based on electromechanical principles have been consigned to the museum.

Digital reverberation systems now offer a wide range of options when categorized by their setting-up routine. Some are set by reference to the more obvious parameters (reverberation time, early reflection spacing, low- and high-frequency content), whilst others will accept room dimensions as the starting point and allow variations from there. Some intended mainly for music use have many preset programs with helpful names such as 'concert hall' and 'recital room', or general-purpose programs such as 'ambience'. The majority will allow the storage in battery-backed memory of presets of the users' own devising or variations of factory-set programs.

5.3.5.2 Digital effects systems
Although often cased for portable use and brought or hired in as necessary, it is common to find at least one of these versatile devices on the control-room outboard rack. A typical system will provide limited reverberation programs, pitch change (for harmonizing and flanging), automatic double tracking (ADT) and straightforward delay, amongst other programs. However, only one such effect can generally be set up at a time, unless combinations are allowed for in the programing capabilities, as they are for instance with ADT which is a combination of delay and pitch change.

User programing is normally accomplished by the use of front panel keys and a liquid crystal (LCD) or plasma display. Memories are provided for the storage of combinations of effects parameters, with some systems having battery back-up.

5.3.5.3 Equalizers
Mixing consoles will provide equalizers in every full facility channel, normally offering control of low- and high-frequency shelving and two or three bands of peaking or notching. These are now usually of the parametric type. Operators frequently have favourite outboard equalizers whose qualities they know and which fall into two groups; graphic and parametric.

Graphic equalizers These have vertical sliding controls arranged in a horizontal row on the front panel so as to indicate by their position the curve selected. Each slider represents a centre frequency in a series normally at either octave or one-third octave intervals, the latter allowing a fairly fine degree of control. The controls normally have a range of about 15 dB either side of zero gain, and can therefore give a misleading idea of the severity of the effect created if, for example, adjacent controls are set to opposite extremes of their travel, thus apparently indicating a setting beyond the capability of the circuitry. Graphic equalizers are extensively used where the object is to create a curve which is the inverse of the response indicated on an analyser the centre frequencies of which match those of the equalizer, as for instance when setting room or loudspeaker equalization.

Parametric equalizers In these, continuously variable rotary controls set not only the degree of effect but also the centre frequency of operation. Thus the centre frequency may be swept through the band. It is normal for the complete frequency range of the equalizer to be covered by several overlapping bands. Other controls vary the degree of peak or notch and its Q, or sharpness. Parametrics are particularly valuable when attempting to eliminate an

unwanted sound with a distinct pitch or tone; a notch is selected at a frequency near to the offending noise and then swept along the band until the two coincide. Conversely, some musical instruments can be given much greater presence by carefully tuning a peak in their output whilst listening to the effect.

Although they are professional devices produced to high standards, these equalizers often have a characteristic sound of their own, perhaps due to the way in which the phase relationships within the signal change as the settings are varied.

5.3.5.4 Dynamics

Dynamics processors comprise limiters, compressors, expanders and gates. They may be selected for use for one of two reasons: to control the excursions of the signal within desired limits; or to modify the sound in some chosen way.

Taking first the control of signal parameters, those under consideration are absolute peak level, dynamic range and noise floor. Broadcasters need to control peak levels for several reasons, some legal and some technical. Under the terms of their licences they must not overmodulate amplitude-modulated (am) transmitters, because to do so generates high levels of spurious *side-bands* which will interfere with adjacent channels, quite apart from the distortion of the audio signal. Frequency modulation (fm) of transmitters is similarly constrained by the need to keep side-bands within limits, though the distortion problem is not so great. Digital systems are sensitive to excessive levels, but are often protected by extra headroom to obviate the necessity for limiting in either the analogue or digital domains.

Control or modification of the dynamic range of a signal is usually done at the production stage, and is often applied to one component of the mix in order to give it improved presence within the balance. All the forms of control described below may be used and some consoles are equipped to offer these facilities on every channel. As a general rule, the most used of the four dynamic units at the acquisition point is the limiter, because it gives the confidence that over-modulation distortion can be avoided, and so allows higher general levels of recording or transmission. The other forms of dynamic control are more often employed where signal excursions are precisely known, and hence in postproduction.

Limiters The prime function of a limiter is to set an absolute ceiling on level, above which the output signal will not rise, although the input may. Below its limiting level the limiter should be transparent to the signal. In theory, therefore, the input–output transfer characteristic should be 1:1 and linear from noise floor to limiting level and then infinity beyond that point. The 'knee', or point at which the characteristic changes should preferably be sharply defined, or at least capable of being so when required. In reality, such characteristics are rarely achieved. The sharp knee is impossible in analogue devices and the infinity ratio may be undesirable for reasons of signal quality. However, slopes of 100:1 are common, and onset of gain reduction at about 6 dB below peak can be realized.

Two further signal parameters are important: attack and release times. Once again, in theory, the attack time should be infinitesimally small for accurate control of signal peaks; however, very short attack times (1 μs is a practical possibility) produce audible distortion on some waveforms and in any case remove transients too brief to be of importance in analogue systems. An attack time of 1 ms sounds better and is completely effective. Distortion apart, very fast attack times can be a cause of signal 'pumping' – rapid gain reductions in sympathy with programme content caused in this case by the limiter acting on high-frequency transients. Release or recovery time is the time taken for the signal path to be returned to zero gain following limiter action, and is normally an operator selectable parameter.

Many theories have been advanced to explain how the choice should be made, though in truth the choice will depend on the mixing technique of the operator. The choice of settings when the limiter is to be left unattended, for instance as a protection device at a transmitter, may well follow different rules. In the television sound control room, when the effect of the device is immediately monitored and the input is under constant control, the choice of parameters will be those of the individual operator. It may be, for instance, that very fast attack times are effective in controlling levels but make the signal sound lifeless, so a longer one allowing leading transients to 'escape' is chosen. Equally, a short recovery time of 100 ms will often exaggerate a 'pumping' effect as the gain recovers before the input signal has settled to a new level, whereas a longer time would appear smoother; however, a skilled operator using 100 ms recovery and intelligent fader control will avoid this problem where a fast recovery is preferred for other reasons.

Compressors These are used to control the dynamic range of a signal, i.e. to reduce the range of the excursions between the low and high points of

the signal. The compressor is essentially an ampli-
fier which is non-linear in the gain sense, and the
input-output transfer characteristic of which above
a threshold point is a simple arithmetic ratio greater
than 1. This 'compression ratio' is operator selectable
to one of several preset values. Thus a compressor
with a ratio of 3:1 will divide by 3 the decibel values
of all signal excursions above its threshold. A wide
range of ratios is usually offered; from 1.5:1 to 20:1,
stepped or continuously variable, is common. On a
compressor for control room use all parameters may
be selected by the operator. Threshold may be a dis-
crete control or may be set by operation of input- and
output-level controls whilst both attack and recovery
are given wide range controls. Since the intention is
to reduce the dynamic range but not necessarily the
peak level of the signal, gain make-up is provided to
restore the signal to its former level.

Compressors must be used with great care, espe-
cially in large open studios such as are common
in television. The action of compressing dynamics
brings up to a higher level the noise which lies below
the wanted signal: included in that noise is unwanted
microphone pick-up of such sounds as the movement
of people, air conditioning and the reverberation of
the studio.

Expanders These are like compressors with
their functions inverted; signals below a threshold are
reduced in level due to the effect of the input–output
transfer characteristic which has an arithmetic ratio
of less than 1. The effect therefore is to increase
the dynamic range of signals below the threshold. In
general, expanders are used to separate signals from
unwanted background noise, and are therefore dif-
ficult to set effectively unless the dynamics of the
signal can be predicted accurately. For this reason
they find their principal application in postproduc-
tion operations where the prerecorded programme
material has known levels and dynamic variations,
although with care they do find some use in live
work. Controls similar to those on a compressor are
usually provided.

Gates Gates (or noise gates) can be used in a
similar way to expanders in that they reduce the
signal level below a threshold. However, in this
case the level is reduced in one step by the action
of a soft switch which brings into circuit a preset
amount of attenuation. In addition to controls similar
to those found on expanders, one more is usually
provided, called 'hold'. This sets a delay time after
the signal has fallen below the threshold and before
the gate closes, and is used to prevent the gate closing
between, for instance, the notes of a piano and thus

cutting off the natural decay, but allows the gate
to close after a brief delay once the piano passage
is over.

Expanders and noise gates, used separately or
together, are sometimes referred to as 'single-ended
noise reduction' systems and in this role may include
frequency sensitive components and multiple bands.

All these dynamics controlling devices have a gain
control element at their heart. For this reason it is
common to find combinations of two or more types
in one unit, all controlling the same signal path. The
various circuits produce control voltages according
to their intended action and these are then resolved
to one control signal which is applied to the gain
control element.

5.3.5.5 Delay lines

Although a great many digital audio devices which
use storage as a fundamental part of their process-
ing also offer simple delay as one of their operating
modes, they can only be used in this way when not
required for more comprehensive processing. Con-
sequently, it is common to have available one or
more units providing delay alone, usually adjustable
in small increments to cope with the many needs
which will arise. Delay systems discussed here are
RAM based and have limited (though no longer
small) memories. Hard-disk based equipment capa-
ble of many minutes or even hours of storage are
not included. Delay lines offer increments of a few
milli-seconds and total delays of several hundreds of
milli-seconds, although for special uses, such as pre-
venting the broadcast of profanities on public-access
programmes, much longer delays can be obtained.

Some delay lines offer several simultaneous out-
puts (multitap delays), whilst others contain two
quite separate circuits in one chassis. It is impor-
tant to note the difference between two independent
delays and one true stereo delay; although the lat-
ter may usually be used as two separate circuits,
the two single delays should not be used for stereo
unless they share a common clock frequency gen-
erator. Failure to ensure this will result in varying
phase errors between the two outputs as the relative
frequency drift of the two oscillators produces slight
delay changes.

Delay lines find a multiplicity of uses: matching
terrestrial audio circuits to satellite delayed vision;
resynchronizing audio to synchronizer delayed video
and to video delayed by the many processes used
to achieve composites of real foreground and vir-
tual background pictures. It may also be neces-
sary to delay ancillary e.g. prefader outputs by

matching amounts in order to avoid the confusion of repeated audio. Other delay line uses include public-address systems correcting for early arrival of sound from close loudspeakers; and multimicrophone arrays retiming close microphone signals to coincide with more distant ones. In addition, delays can produce effects of their own such as single echoes, flutter echoes and comb filters. Digital mixing consoles may provide this facility on a per channel basis thus avoiding further analogue-to-digital and digital-to-analogue conversions. Finally, the move from analogue to digital distribution of signals to transmitters requires consideration to be given to synchronism if not inherent in the distribution system.

5.3.6 Equipment bays

The bays discussed here are the full height 19-in. wide type normally found arranged side by side in a continuous run. The number of bays will vary considerably depending on factors such as whether the sound control room houses the communications system for the whole studio, or whether the console contains its own electronics completely. A small facility may have no central technical area and so will require all line sending, receiving, equalizing and testing equipment to be here.

The bays are primarily intended to carry fixed pieces of equipment constructed on a standardized chassis system in a way that is economical in both floor and panel space. Maintenance access is usually provided by doors to the rear of the cabinets (if they are free-standing) and removable panels to the sides. Sometimes for reasons of economy of space, bays form part of a control-room wall and access to the rear is from another room or corridor. Where neither form of rear access is possible, front-access swinging bays may be used, though they are wider than standard bays and may also restrict the maximum equipment depth which can be mounted.

When placing equipment in bays, consideration must be given to mutual interference due to stray fields and to heat dissipation. Following these considerations, it is convenient to group equipment either in similar-use categories or those requiring frequent interconnection. Inevitably there will be some conflicts; for reasons of weight it is convenient to place power supplies toward the bottom of the bays, but heat generation would suggest the top. Some equipment that has front panel controls would most obviously appear near the centre of the stack, but unperforated top and bottom screening would prevent the flow of cooling air up the bay and may well rule

this out. Many equipment designers circumvent this problem by fitting cooling fans to individual pieces of equipment, but the noise penalty can be high. Direct connection of the top of the bay to the sound control room air conditioning extract will usually be a better answer.

It is not possible to give a comprehensive list of what should appear in sound control room bays, but some items will appear in most. Since circuits to and from the sound control room will probably pass through blocks and amplifiers contained in the bays, as will the inputs and outputs of many pieces of equipment, a bay monitoring position will be very valuable, allowing if necessary a second operator to check circuits and systems. A comprehensive jackfield allows flexibility in the use of equipment as well as checking and substitution. Pin matrices may be employed as a space saving and elegant alternative to plugs and sockets for routing communications and other systems, and these too will require bay front panel space at eye height. Now digital signal line termination, distribution and patching will increasingly be found in addition to analogue.

5.3.6.1 The jackfield
The jackfield may in fact contain some other plug-and-socket system or may be based on jacks of one of several different gauges. The jackfield (or patchfield or patch bay, etc.) may be one large panel or distributed as rows between blocks of equipment, or a mixture of both. Whichever way it is constructed, its purpose is to provide access to inputs, outputs and listens of as many circuits and pieces of equipment as may be thought advantageous. It is common practice to add spare amplifiers, attenuators, mixing pads and repeater coils, and tie lines to other jackfields, control rooms and apparatus rooms. Parallel strips (rows of perhaps half a dozen jacks wired together in parallel) are a convenient and very useful way to use up short runs of otherwise uncommitted jacks.

The bay jackfield then, is the central monitoring, testing, and linking position within the sound installation. It is wise to plan to leave occasional rows of jacks free for later expansion.

5.3.6.2 Monitoring position
It is perfectly possible to wire all incoming and outgoing music and communications circuits directly to and from the console; it is, however, good practice and much more convenient for all such circuits to pass through blocks in the bays and from there to line sending or receiving amplifiers, as appropriate. These amplifiers will provide monitoring facilities

which can be wired to the bay jackfield and the monitor selection matrix, giving easy selection to loudspeaker and meter monitoring. A fully comprehensive matrix will also have listens of the inputs and outputs of all fixed or standard items of equipment in the control room, and also all console outputs including auxiliaries, clean feed (mix minus) and multitrack selections. Circuits to and from the studio floor and listens of permanent communications facilities should appear, as should outputs of routeable switches such as switched talkbacks, switched outputs and the like. Finally, a few miscellaneous patchable inputs to the matrix should always be added for connection to temporary arrangements. However, constraints of costs increasingly mitigate in favour of plugged rather than switched monitoring selection.

5.3.6.3 Public address equipment
Not every sound installation in a television studio will include permanent PA equipment but many, especially larger studios used for light entertainment programmes, will. The installed equipment will normally comprise an input selection system, a number of power amplifiers, an output routeing system and a number of loudspeakers. If the design is based on the so-called 100 V line high impedance output method, where loudspeaker cable length is relatively unimportant, then it is normally possible to mount the power amplifiers in the sound control room bays where level control and spare amplifier substitution is easily effected. If, however, the low impedance output method is chosen it may be necessary to have the power amplifiers nearer the loudspeakers in the studio, and probably completely rigged and derigged on each occasion of use.

Two basic loudspeaker types will commonly be found useful in studio work: the column type, which is directional to a degree at all frequencies: and the many designs of cabinet loudspeakers, which are directional at high frequencies but become increasingly omnidirectional as the frequency falls. Column loudspeakers will find their principal use where high music powers are not required but the ability to avoid feedback is, whilst cabinets can often provide a wider frequency response and higher powers but may give rise to feedback problems if some other method of control is not exercised.

The input selection system, which may well be on the console or the bay, can be anything from a row of input jacks to a sizeable switch matrix. It is normal practice to use console auxiliary outputs as sources to PA either directly through the input selector to the amplifiers, level control being exercised from the console, or via an additional mixing console. The additional console may be on the studio floor, where a second operator controls the mix to PA.

The output routeing system may not exist as such. Some 100-V line designs simply use one power amplifier to drive each of several output loops in the studio and loudspeakers are connected as desired; a difficulty may arise when the programme requirement leads to long loudspeaker cable runs in the studio in order to find the required feed. A more flexible, if more expensive, alternative is to install permanent cable runs from individual studio floor sockets to the control room and to feed them via a high level patch-field or pin matrix, the sources to which are the power amplifier outputs. The high levels and multiplicity of mixes required by popular music presentation are usually met by specialized equipment temporarily imported onto the studio floor.

5.3.6.4 Radio microphones
The improvement in and expansion of radio microphone technology parallels the similar developments in personal microphones, and there is an obvious connection. Both have become smaller, more reliable and have improved in technical standards. Television producers have applied consistent pressure to be free of first, the constraints of microphone booms, with their restricting effect on camera shots, and second of any cables which may inhibit the movement of artists. Hence even quite simple programmes will be planned around extensive use of radio microphones, freeing the producer of problems but creating new ones for the sound operator.

The essentials of a radio microphone system are a microphone, often but not always a personal microphone, a small transmitter which may be body worn or hand-held, and a receiver connected to the sound console. The transmitter will have an aerial which may be separate (a free wire or helical type) or may be incorporated in the transmitter body or microphone lead.

The receiver is the key to successful radio microphone operation, and this is why it is beneficial to have the receiver system engineered into the sound equipment bays if at all possible. In any case, many of the better receivers do not lend themselves to totally portable operation due to their size, chassis format and power requirements. Fixed aerials in the studio, with head amplifiers if necessary and low loss cables to rack-mounted diversity receivers offer the best chance of success. In a large studio centre this will often mean having frequency agile transmitters

and receivers and a well thought out band plan to make the best use of available channels and avoid mutual interference problems. In the sound control room a position must be found for the receivers such that the operator can readily check the several indications of operating status normally provided on the receiver front panel. These will always include received radio-frequency signal strength and carrier modulation depth and may also add diversity switching status, transmitter battery condition and receiver muting or 'squelch'. Receiver line outputs will usually be wired directly to the console source jackfield whilst the headphone monitoring outputs may be 'bussed' together for operational convenience.

An alternative approach when it is not desired to provide each studio with its own radio microphone system is to provide a position, perhaps a shelf or a fixed table, to which permanent aerial feeders, output tie lines and mains power distribution are provided so that portable equipment may be rigged as required.

5.3.6.5 Line identification devices
Studio productions which include contributions from outside, broadcast or other outside sources will at times find the need for a means of identifying positively and continuously the intended purpose of a circuit in advance of its actual use. Cartridge and cassette players running continuous loops have largely given way to solid-state systems in which a message is recorded in RAM, usually battery supported. Several channels are often provided and the more comprehensive equipments provide automatic switching at intervals to the studio's tone oscillator to give positive level indication.

5.3.6.6 Telephone balance units
The inclusion in programmes of telephone calls direct from members of the public is only one use of telephone line balancing equipment; production conference circuits, dial-up contribution circuits and stand-by contribution or clean-feed circuits are others.

The problem of using telephone circuits for programme purposes is easy to state and difficult to overcome. The telephone circuit is two wire, and carries signals in both directions simultaneously. If it is desired to broadcast the signal from one direction but not the other, then very accurate balancing of the line is required to separate the two.

For many purposes the restricted bandwidth quality of a telephone circuit is not a disadvantage and broadcasters may wish to use them in place of specially provided music circuits. However, where a two-wire telephone circuit is used in place of a four-wire circuit then the problem arises of how to separate the incoming signal, which is to be broadcast, from the outgoing (usually cue programme) signal which is not. Incomplete separation results in the cue-programme material, which may be the studio end of a two-way interview, returning distorted from the telephone line along with the incoming contribution and being unavoidably added to the programme. In television, where this now distorted sound of a local contributor is accompanied by a clean picture, the effect is very incongruous. When more than one telephone circuit is in simultaneous use, further complications, including feedback, can result.

The traditional answer relying on telephone technology made use of the hybrid transformer, and required very careful setting up and stable line characteristics to produce acceptable results. Later equipment based on this principle, but which continuously and automatically rebalances the line has been more successful, but the most significant improvement came with the digital adaptive filter technique which is fast, accurate and fully automatic and can produce separations of the order of 40 dB routinely. Since only one or two simple switches need to be within reach of the operator, and can normally be located remotely, the balance unit itself may be conveniently mounted in the bay, leaving the console surfaces free.

5.3.7 Vision monitoring

The business of making television programmes proceeds on the basis that of the two of the viewers' senses to be engaged, the visual is the dominant one. Sound, therefore, is made to match picture, rarely the other way around. For this to be possible it is clearly essential that the sound operators can constantly refer to the studio output picture as the programme proceeds, in order that they can match it. Equally important is the ability to select onto a second picture monitor a vision source required for cue purposes – a preview selection. Television sound control rooms will therefore have at least two picture monitors: a 'studio output' or 'transmission' monitor, usually a large colour display; and one or more 'preview' monitors the inputs of which are selected by a switching system under the control of the sound operator and having all the principal vision sources of the programme available to it.

Picture monitors must be carefully placed so that they may be seen with the minimum of head and

eye movement away from other operational controls and indicators; in particular, it must be possible for the operator to glance quickly from script to peak programme (PPM) or volume unit (VU) meters, to picture monitor and back again without refocusing the eyes. For working in stereo, the visual and aural images must coincide. Picture monitor height must be set so that the whole of the picture area may be seen by the seated operator, but as low as reasonably possible so as to limit the eye movement necessary to look from meters to picture. Refinements to the picture monitoring may include a complete bank of smaller monitors each tied to a single vision source in the manner of a production control room. The picture monitors are therefore in a central and dominant position, and make the television sound control room obviously different from any other.

5.3.8 Miscellaneous facilities

Most television sound control rooms will have a variety of other items which may include some of those described here. Talkback and intercom loudspeakers may be built into the console or separately, as in *Figure 5.9*, where they are supported on microphone stands. Discs are required to be instantly available and are, therefore, kept in specially made racks; the tape operator's position will often have its own picture monitor and talkback loudspeaker, and perhaps its own jackfield.

Assignable consoles and those with full automation will have a video display, perhaps a personal computer (PC). A cassette recorder for logging purposes may have been superseded by a VHS recorder with a feed of video with burnt-in time-code. An essential in any control room is an accurate clock, which may be on a loop or may rely on radio code or GPS for accuracy. In some designs the clock is the focus for a number of other indicators, such as rehearsal and transmission lights, and lamp indications of telephone call and studio doors open. Finally, time-code displays of time of day and programme elapsed time have replaced the once ubiquitous stopwatch.

5.4 Studio installations for local radio

5.4.1 Introduction

Although no formal links exist between BBC local radio and independent local radio (ILR), and although their respective *raisons d'etre* are quite different, nevertheless stations in each category are provided with equipment which is remarkably similar in type, scale and operation. One fundamental difference between the two types is that in the case of BBC local radio stations the equipment is standardized, whereas in ILR it is not. Rather than consider these types of station in two separate sections which would involve much duplication, the requirements and typical equipment for any local radio station are described, with significant differences between the BBC and independent stations highlighted. It is helpful first to consider how the two networks came into being, and where their philosophies may be different and where similar.

5.4.1.1 BBC local radio

On 8 November 1967, the BBC opened the first local radio station in the UK as part of an 18-month experiment. Radio Leicester was followed by Radios Sheffield. Merseyside, Nottingham. Stoke-on-Trent. Brighton, Leeds and Durham. and these were equipped with Peto Scott modular hardwired mixing desks to a standard design. Technical areas were the operations room, with a control desk and amplifier and communications bays, a main studio with another desk facing, the operations room through a window, and a separate conventional studio, cubicle suite known grandly as the 'education studio'. This experiment being deemed a success by Government and by those at Broadcasting House, and with the start of commercial radio seen as inevitable, the BBC embarked on a programme of local radio building. A second batch of 12 stations was planned, and as early as 1970 the *BBC Handbook* reported that it was hoped 'to increase the number to 40 by 1974'. At the time of writing the network is all but complete, although many stations have been refurbished, rebuilt, extended and, in one case, transplanted from one side of the country to the other (Radio Durham was closed and replaced by Radio Carlisle – now Radio Cumbria.)

The second group of 12 stations had mark-II installations, essentially similar to mark I with a few extra features and the desk modules removable for maintenance. The electronic building blocks were assembled and wired by BBC staff using BBC bays, amplifiers and other equipment, but for the following wave (up to 20 locations with two or three desks in each) the Capital Projects department realized that it could be both desirable and necessary to use a commercial manufacturer, and Neve were commissioned to build a batch of desks to a BBC specification. This proved to be a passing flirtation, however, and the stations were re-equipped with BBC-built mark-III

gear, Neve having faithfully served its allotted time-span.

The protototype BBC mark-III desks were installed at Radio Carlisle, and subsequently almost every station in the chain was later equipped or re-equipped with them. Since 1990, however, the mark-III has been superseded by the mark-IV desk, built by commercial manufacturers to a BBC specification, but this time fully turnkey, i.e. all fixed technical plant provided and installed by an outside firm. Contracts were initially placed with MBI (for Radio Solent) and Audix. In 1992 there were two sorts of equipment in front-line service, all the mark-II and Neve desks having been replaced, but the mark-IV desk had two forms (MBI and Audix) and could be provided by others also. The predominant BBC mark-III desk, built in-house by BBC Equipment Department (as it then was), varies little from station to station.

BBC local radio's primary task, not significantly changed since its inception, is to provide a speech-based service of local entertainment, education and information, and to involve the local community as much as possible. All versions of local radio equipment were designed with this in mind, and the mixing desks and other studio equipment are straightforward and uncomplicated, capable of being operated (with a little tuition) by part-timers, amateurs, and main-tenance engineers as well as producers, production assistants and journalists – there are no Studio Man-agers in local radio!

5.4.1.2 Independent local radio

In the early 1970s the decision was taken by the Government that BBC radio should enjoy competi-tion from a commercially based rival. It was further decided that this should be on a local rather than on a national basis and so 18 areas were earmarked for new independent local radio stations. These were to be operated by independent companies but regulated by the then ITA (Independent Television Authority) which was in consequence renamed the IBA (Inde-pendent Broadcasting Authority). There were to be 19 stations in the first phase. It was decided that Lon-don would have two stations and that one of these would be a station specializing in news. The other stations were expected to provide a broad range of output of which much would be music-based, but they were also expected to provide local news, cur-rent affairs. discussion programmes and drama. The London news station was also required to set up and operate a national and international news service (Independent Radio News (IRN)) for all the other

ILR stations. This was to be distributed as both an audio and text service upon which the other stations could draw. Initially distribution was by land line, but with the expansion of the ILR network it became more convenient to distribute the service by satellite.

The early requirement that ILR stations should provide a broad range of programming rather than specialize in their output meant that a variety of studio types and equipment was needed. Thus many of the first 19 stations constructed and equipped music studios which could also be used for drama, in addition to one or two self-operated studios which could be used by presenters to play music, mostly from commercial records. The restriction on the amount of music which could then be played from records, 9 hours per day, also encouraged the stations to use their music studios to record local musicians to fill their broadcasting hours as most were on the air 24 hours per day.

When the first ILR stations came on air in 1973, the majority of UK radio listening was to stations in the medium wave (MW) band. In consequence, all the stations were equipped with both MW and very high frequency (VHF) stereo transmitters with the MW transmitter being the primary service for most listeners. However, all the stations were equipped with stereo studio facilities and from the beginning provided virtually all their output in stereo whether it was music. drama or discussion programmes. Under the IBA the stations were not responsible for the pro-vision or maintenance of the transmitters. These were provided by the IBA Engineering Department which was also responsible for setting technical standards for the contractors' studios and equipment and for making both pre-transmission and annual checks on compliance with these standards.

With the passing of the 1990 Broadcasting Act the IBA was dissolved and separate authorities set up to regulate UK commercial radio and television. The Radio Authority (RA) came into existence from 1 January 1991 as the regulator of all Independent Radio licensees. The RA, unlike the IBA, does not concern itself with studio standards but only with standards for transmitters which the radio stations may now install, operate and own if they so wish or employ some other agency to provide and operate. The Engineering, Division of the IBA was set up as a separate transmission company under the name of National Transcommunications Ltd (NTL).

Until 1988, ILR stations enjoyed only limited licence to provide separate programmes on their AM/MW and FM/VHF transmitters. Nowadays sta-tions have to provide separate programmes at all

times and the stations which were originally allocated separate MW and VHF transmitters now provide this. A number of the ILR stations set up after the first 19 were awarded as dual areas with two or more VHF and MW transmitters needed to cover the area. These stations were required to provide some degree of separate programming to these areas, typically local programme services relevant to each area at times of peak listening. This opportunity was used to sell separate advertising to each transmission area and although most stations now provide common core programme material to these areas there has remained a need to provide separate news, traffic, weather forecast and commercials to the individual areas. The continuing development of ILR under the Radio Authority has led both to a large expansion in the number of stations in the UK (to over 180 in 1998) while at the same time there has been considerable consolidation of ownership in the industry. A large number of stations are now owned by one of the eight larger groups and may share certain programme elements. This has led to some quite complex studio equipment requirements in order that the staffing can be held to an economically viable level.

Since January 1991 when the RA became responsible for the regulation of ILR it has been the practice to award licences to new stations for either AM/MW or FM/VHF frequencies but not both as in the past. Stations which were originally awarded a franchise by the IBA had to choose which of their licences to reapply for if by the end of their current licence period they had not established identifiably different programme services on their MW and VHF transmitters.

5.4.1.3 General studio layout

The particular requirements of a radio station will depend on a number of factors. These include the population size, the editorial area to be covered, the programming policy of the radio station and the size of staff and budget to be made available. However, it is likely that at the very least one or two self-operated studios will be needed together with a studio which may be used for discussion or phone-in programmes. For convenience these studios will probably be located next to each other with windows providing visual communications between them (see *Figure 5.10*). Although the station may use an external supplier for its national and international news it is likely to provide local news from within its own resources. This may well require specific news-gathering facilities together with the necessary editing facilities.

Figure 5.10 *Typical LR transmission suite layouts*

If the station is to be supported by commercial revenue it may also require a production studio to allow local commercials to be made in house. Alternatively, one of the on-air studios may also double as a commercial production studio for which additional consideration will need to be given when equipping the studio.

Good acoustic isolation between the studios and between the studios and areas outside will gain maximum flexibility of usage. This good isolation will demand suitable air conditioning if the studios are to remain comfortable to occupy, and the air conditioning ventilation plant will in turn need to operate quietly to prevent intrusive background noise. It is important to consider that where non-professional broadcasters are to be extensively used on air there will often be a lack of microphone technique. This can frequently mean the presence of studio guests who project poorly and for whom a correspondingly higher level of microphone amplifier gain is needed to compensate for the low audio level. The higher gain will necessarily make any deficiencies in the studio background level more obvious to listeners. For this reason it is just as important to invest in good acoustic conditions in local radio as in a network radio studio and with suitable design and choice

of a sensible location for the studios it does not have to be prohibitively expensive. In many cases more time and care spent in identifying a suitable studio site can pay dividends in the cost of the necessary building work.

5.4.2 Audio mixing desks

The audio mixing desk acts as the central point for every studio operation even though it may no longer look like the mixer of previous years. Through it all local and outside sources may be mixed together as required to provide the output of the radio station. In addition communications with other areas both within and outside the radio station may be established to enable the smooth working of programme operations. The mixing desk may be considered as a number of different but integrated sections (described below). *Figures 5.11* and *5.12* show a local radio desk which is typical of many in use. Not all the features described in the text are incorporated in this particular desk, either for reasons of cost, to simplify operation, or because they were not required. It is now common in many radio stations to augment this conventional mixing console surface with one or more PC driven screens. These will present information on the programme content and running order and often interact with the mixer providing 'live assist' and automation facilities.

5.4.2.1 Input channels

The typical studio mixing desk will normally have two main types of input channel. The first will be a monophonic signal input channel which will often accept signals at both microphone and line level. In the case of a channel which accepts both types of input signal there will usually be separate inputs to the channel with a switch to select the microphone or line level input. There should be a microphone input preset gain control to allow a gain variation between 80 and 20 dB. For electrostatic microphones requiring a power supply it is common practice to provide a 48 V phantom supply at the input to the desk, and there is sometimes an accompanying switch to allow this supply to be disabled. The line-input preset gain control should ideally allow a variation from 20 dB of attenuation to 20 dB of gain. The gain control may be a stepped attenuator or a continuously variable potentiometer. The stepped attenuator has the advantage of repeatability while the potentiometer will allow the gain to be adjusted, if necessary, while the microphone or line source is 'live' to air. In the case of a stereo desk there should

be a pan(oramic) control which allows the signal to be moved between the left and right channels by progressively altering the relative signal levels sent to the left and right outputs of the desk.

BBC transmission desks do not normally incorporate a variable equalizer section, but many ILR desks do. This equalization section can vary widely from a highly sophisticated parametric equalizer to a very simple high-pass filter. However, for use in self-operated studios it is probably unwise to allow too much variation in the equalization as the operator or presenter will often select most unwise settings for the controls, particularly when it comes to settings for their own microphone. The use of a high-pass filter to remove studio rumble and microphone popping is generally to be recommended.

In the case of microphone channels it may often be wise to incorporate some form of dynamic range limiting in the input channel by the use of a compressor or limiter section. BBC desks provide a compressor which acts on a microphone subgroup, rather than on individual channels. Care should be taken to avoid excessive amounts of compression if, as is often the case, there is considerable audio processing employed on the final audio output of the radio station.

In a studio which is self-operated it is necessary to mute the monitor and talkback loudspeakers when the fader associated with a local microphone is opened. If this were not done then there would be positive feedback from the monitor loudspeakers into the local microphone(s) causing attendant howlround. This muting facility is operated by a microswitch on the microphone fader when it is moved away from the 'off' position.

Mono channels are used for microphones and for most outside sources, radio car and telephone sources, all of which are described in more detail later.

The second type of input channel employed is for stereo input signals. This will normally accept line level signals only. There should be a preset gain control to allow a variation in gain from -20 to $+20$ dB and a balance control to allow for the correction of a left-to-right signal level inequality on stereo input signals. A useful addition can also be a switch to enable the routing of either the left or right input signal to both left and right outputs; this allows the channel to cater for a mono signal input or for the occasion on which one of the channels of a stereo source fails for some reason. In some cases a channel phase switch may be fitted which when operated inverts the phase of the left channel

Figure 5.11 *Layout of a typical local radio sound desk.*

Figure 5.12 *A typical local radio sound desk.*

with respect to the right channel. This allows for the correction of input material which has suffered an unintended phase reversal between channels. A frequency equalizer section may be provided, but on BBC and most ILR desks is omitted for simplicity.

Within the channel it will be usual to have the means for listening to the input signal before the channel fader is raised and allowed to reach the output of the desk. This prefade listen facility (PLF) allows the preset gain and pan or balance controls to be set by routing the input signal to the monitor system of the desk where it can be auditioned and the level checked on a reference programme meter. Generally the operation of the PFL facility on the channel will cut the selected monitor source and replace it with the prefade signal. In some desks there is an additional means of operating the channel prefade by moving the associated fader against a spring loaded microswitch at the maximum attenuation position.

To facilitate the operation of the local studio equipment it is common to provide a remote control

switch associated with each of the reproducer source channels (grams. tapes, etc.). Operation of this switch will start the equipment and will usually be indicated at the desk by an associated lamp. The fader may also be used for this operation ('fader start') by making use of a microswitch which operates when the slider is moved from its backstop (i.e. from the position of maximum attenuation). This allows a silent start and avoids transmitting any clicks which might be present on the audio output of the source at the moment when it starts. To prevent inadvertent operation of the fader start facility a switch is usually provided to disable it when not required.

The channel may be fitted with an on/off switch, but in self-operated desks this can often cause problems if it is inadvertently operated by the presenter and there is insufficient visual indication. The provision of auxiliary send controls on the channels to allow the setting up of an alternative output mix of the desk input signals can be very useful. In particular this can be used with the microphone channels to

allow the setting up of a signal to feed an electronic reverberation device, thus allowing the sound of the studio microphones to be modified. In BBC-built desks this auxiliary feed is provided as a hardwired postfader mix of all channels except outside and rebroadcast sources. Channels are likely to be permanently dedicated to source categories, and control panels can therefore be specially tailored.

5.4.2.2 Outside source channels

An 'outside' source may be any source of audio originating outside the studio, and therefore requires a feed of cue or clean feed from the studio and communication with it. A channel dedicated to handling outside sources will usually have controls associated with these requirements grouped together on or near it, including switches to select a particular source, to route *cue or clean feed* back to the source for remote presenter or audience, and individual source talkback. Indication should be provided to show if an outside source is in use by another studio. To economize on booked lines, cue circuits are often provided by the public switched telephone network (PSTN), and dialling arrangements (possibly automatic) may be on the desk.

5.4.2.3 Phone-in channels

These will usually be mono channels similarly equipped to outside source channels but dedicated to accepting the output of a *telephone balance unit* (TBU). As in the case of an outside source the TBU must receive a clean feed from the desk.

5.4.2.4 Radio-car channels

At least one channel may be dedicated for use with a radio car. Controls may be available on the channel for selection of base stations or receivers to be used for the incoming radio circuit from the vehicle, as a large editorial area or difficult terrain often necessitates multiple receiver points. Cue or talkback may be returned via a radiotelephone circuit. Both ultra-high frequency (UHF) (full bandwidth) and VHF (speech bandwidth) links are in common use, and many stations now have more than one equipped vehicle.

5.4.2.5 Rebroadcast source channels

Distributed services provided by other agencies are used by both the BBC and ILR. BBC stations can rebroadcast BBC national networks, usually taken off-air, and also line-distributed news services; ILR have various commercial satellite feeds available

should they wish to take advantage of them. All such services may be fed to a rebroadcast source channel on the desk which will have a simple selector switch, but none of the other facilities associated with other outside source channels. This switched input channel may be a convenient location to make other sources available: a return feed from a reverberation unit, or the output of a newsreading booth are examples.

5.4.2.6 Reproducer source channels

For simplicity of operation, reproducer sources are likely to be wired to dedicated channels (or channel inputs where the channel can accommodate more than one input). A typical complement in 1998 might be: one gram, two or three CDs, one DAT or minidisc player, one to three reel-to-reel tapes, and three cartridges and/or a hard disk player. However cartridge players are progressively being replaced by hard disk storage systems. In many cases radio stations are using hard disk storage units to replace most of the individual local audio replay sources.

All channels will usually be similarly equipped, with controls for remote and fader start and often a recording feed selector. Sometimes adjustment is provided for depth of ducking of such sources under the level of the microphones. If a 'direct-to-transmitter' channel is provided (usually a tape channel) the desk and studio can then be used for other work while a prerecorded programme is being transmitted.

5.4.2.7 Split transmission output routing

It was mentioned in an earlier section that many ILR stations have a requirement for playing separate commercials to one or more transmitters which otherwise receive common programme material from the studio. This has led to the need for mixing desks which employ multiple stereo output buses. In its simplest format separate sources are provided for the commercial material. Such sources are brought into the desk on separate channels. During a commercial break the outputs of these channels are routed to separate stereo output buses each of which is feeding a different transmitter circuit. During common programming the separate output buses receive the same material from the other desk channels.

Some BBC stations which require several FM/VHF transmitters to cover their service area, in common with many ILR stations, provide differing programme services on them during parts of the week. This would generally be to provide a more localized news and information service.

5.4.2.8 Monitoring

In self-operated studios (*Figure 5.13*) it is necessary to provide monitor outputs for both loudspeakers and headphones with separate level controls. Headphones are needed to allow the presenter to continue to listen to the output of the desk or the radio station whilst any local microphone fader is open. The particular sources which may be selected for monitoring will vary depending on local requirements but they should include any off-air signal(s) from the local transmitter(s) and the signal(s) leaving the radio station to feed those transmitter(s). Sources regularly used as sustaining feeds or contributions to station output (e.g. a news studio) may also merit their own selector position. Selection of a signal for monitoring would also cause it to be displayed on a programme meter so that its level may be observed and in the UK this is usually by means of a peak programme meter (PPM). For ease of observation of stereo material a meter with a coaxial movement will most easily allow the left and right signals to be measured. While the meters are showing the level of the selected source they are unavailable for monitoring the main output of the desk or the output of the station. It is therefore desirable to have an additional pair of meters which are dedicated to this function. To indicate the degree of phase coherence either a mono meter may be provided, or a further stereo meter showing the sum and difference of the left and right channels, labelled M and S (sum and difference signals).

Figure 5.13 *A typical self-operated local radio studio.*

When the prefade facility is operated at the input channel the monitoring will usually be arranged to indicate the level of the signal on the monitor meters at the same time. This will allow the source signal level to be brought to the correct range with the input channel fine gain control before the fader is opened to put it on air.

Since it is likely that guests will be invited into the studio to participate in programmes it is necessary to provide an additional headphone monitor feed for them. This will normally be switchable between a limited number of sources such as studio and station output. The provision of this separate feed allows the presenter to continue to use his or her own monitor facilities without disturbing the guest who may not be a professional broadcaster and would be disturbed to hear, for example, a source being prefaded whilst on air.

5.4.2.9 Talkback and communications

Talkback and intercom are both means of providing immediate communication between areas, but talkback is concerned with operational programme contributions, whereas intercom is used for other purposes. Thus talkback is integral to the functioning of the desk and will be dealt with here (intercom is described in Section 5.4.7.1), although an intercom panel is very likely to be found on the desk.

It is important that the desk operator (the presenter in *self-operation* mode) should be able to communicate privately with a contribution source, and talkback is provided to satisfy this requirement. Talkback circuits must therefore be independently selectable (one switch per contribution source) and audible by the contributor whether broadcasting or not (routed to headphones and overriding cue or programme feeds). Talkback keys will thus be found on outside-source channels, radio car channels, channels dedicated to outside broadcasts, and for studio guest headphones and other on-air studios. To make the talkback more intelligible to the contributor, the local monitor loudspeakers are normally dimmed when a talkback key is operated.

Reverse talkback (private communication by the contributor to the desk operator) is not usually associated with outside sources, as these are often only a single voice, but may be used with radio car contributions (via a radio communication link separate from the programme circuit, for instance) or particularly from other studios, when it is important prior to programme junctions for exchange of cues, timings, etc. Reverse talkback will normally appear on a small speaker built into the desk, muted when a microphone is live and will cause the main monitors to dim when active. It will also appear on the presenter's headphones, sometimes displacing the programme in one ear only to increase intelligibility,

but more usually, mixed with a reduced level of the monitored source.

Talkback to and from destinations remote from destinations remote from the radio station may also be needed if outside broadcasts are to be undertaken successfully. The means of carrying this talkback may be by a circuit booked from the local telecommunications operator, a radio system or a dialled telephone circuit, but ideally the operation of the system should appear to remain the same to the presenter or operator. This will often require some sort of interfacing circuitry to allow the presentation of the talkback channel to the desk talkback system. The use of voice operated switching (vox) may often be employed to detect the presence of incoming talkback. In the case of telephone company talkback circuits allowance also needs to be made for the considerable transmission loss which may be incurred. It may be necessary to provide 20–30 dB of gain to bring the signal up to line level. In any event it is important that the compensating gain for this circuit may be adjusted independently of any local talkback loudspeaker control so that the talkback signal level may be equalized to that of the radio station internal levels. Also, to avoid spurious noise causing annoyance on such circuits, it is advisable to have some means of switching off the individual talkback circuits, although an indicator light is important so that any incoming talkback can at least be seen if not immediately heard.

5.4.3 Studio audio sources

5.4.3.1 Analogue disc reproducers (grams)
Although the compact disc (CD) has almost completely taken over the role of primary music source in radio stations there is still sometimes a need to be able to play vinyl records. However, this is now rarely done live since the material is normally transferred into a hard disk storage unit or to a CD using a recorder. The main requirements for a broadcast turntable are fast starting speed, robust and durable construction and ease of cueing the record to the start of audio. This last requirement means that the pickup cartridge fitted must be considerably more robust than that fitted to a hi-fi system since it has to withstand the reverse rotation of the record while tracing its surface. This need to be able to backcue usually leads to a much shorter life for the pick-up stylus than would otherwise be enjoyed and a requirement for replacements to be economically priced. It is standard practice to place the pick-up amplifier close to the turntable and many manufacturers of broadcast

turntables mount the amplifier directly under the unit. The output of the preamplifier should be at line level and will have been correctly equalized to compensate for the RIAA (Record Industry Association of America) recording curve. Some benefit can also be gained if the amplifer gain is adjusted to compensate for the higher cutting levels commonly employed on 45 r.p.m. singles. This would take the form of a link between the turntable speed selection and the amplifier. To increase the ease of operating the turntable the local start control is normally wired in parallel to a connector, allowing it to be started from the desk channel either via a momentary pushbutton or by the opening of the associated fader. The turntable run status may likewise be indicated at the desk channel by a lamp.

5.4.3.2 Compact (and mini) disc players
The requirements of broadcast compact-disc and mini-disc (MD) players have parallels with their vinyl disc counterpart. However, the means of cueing to the start point is necessarily different. Generally the presenter will wish to start playing a CD or MD track from the start of the programme material. This will often not be located precisely at the point at which the player will indicate the start of the track. To facilitate cueing many players incorporate an audio level detector which on detection of a digital coding level above a preset value will place the player into pause. Subsequent restarting of the player will then give an immediate audio output. In some players the threshold may be varied to allow a degree of flexibility. As an alternative or in addition to this autocue function the player may incorporate a cue wheel which allows the play point on the disc to be moved backwards and forwards in the pause mode. This causes a short segment of the disc to be played repeatedly and by adjusting the play point the required cue point may be set. The advantage of the cue wheel over the threshold detector is the ability to start playing from a cue point which may be in the middle of a track. In practice this facility is often only required in production studios and in most cases the cue to threshold facility is the only requirement for an on-air studio. Remote start and indicator facilities will be the same as for the analogue disc reproducer.

Since low-cost recorders for both CD and MD formats are now widely available it has made the assembly of material for playback in studios very convenient. However, it should be borne in mind that the MD uses data compression to achieve a comparable playing time to a CD, but on a much smaller disk. It is always safer therefore to use CD

as an archiving format when it is not known what further use may be made of the material in the broadcast process.

5.4.3.3 Analogue tape recorders

The standard tape width used for stereo replay is 1/4 in (6.3 mm). Both National Association of Broadcasters (NAB) and cine spools are in common use. Tape speeds employed are generally 7.5 in/s (19 cm/s) or 15 in/s (38 cm/s). Noise reduction is not in common use for stereo broadcast masters and has generally been avoided as adding complication and the potential for operator error at the time of replay. European broadcasters generally employ the CCIR (Comité Consultatif International Radio) equalization characteristic rather than the NAB standard more usually employed in commercial music recording studios. However, many of the broadcast tape machines can be switched to replay tapes recorded to either characteristic.

If a studio is to be used for on-air purposes only then the suitability of the tape machine for editing is secondary to its other requirements. These are the ability to start quickly and with negligible speed variation and to suppress any spurious signals in the audio output related to the transport start process. In the case of an on-air studio remote start and play indication is all that is needed, but in studios used for production a full set of remote controls for the transport together with indication will be appreciated by operators.

5.4.3.4 Analogue cartridge players

The need for the repeated playing of short audio items in radio studios led to the development of the cartridge player. One of the most common audio items to come within this definition is a commercial which commonly runs for between 10 and 60 s. Since commercial radio was most widely established in North America it was the National Association of Broadcasters (NAB) standard cartridge which became the most common accepted standard throughout the world. There are two standard sizes of NAB cartridge, but of these only the smaller NAB 'A' size cartridge is in common use in broadcast studios.

The NAB A cartridge contains a length of 1/4 in tape, with a lubricant on the back coating, formed into a loop. Standard recording speed is 7.5 in/s (19 cm/s). A stereo cartridge will, when recorded, contain three parallel tracks: the two upper tracks will contain the left and right audio signals, while the lower cue track contains tones which allow the

playing machine to locate significant points on the tape. On commencing recording a short section of 1 kHz tone will have been placed on the cue track. Detection of this tone causes the replaying machine to stop, thus making it ready to play the programme item from the start. Further secondary (150 Hz) and tertiary (4 kHz) cue tones may be used to signify additional cue points. Generally the secondary cue is used to signify the end of the audio. Cartridges are normally sold already loaded with tape and are specified by their playing time. For this reason the maximum playing time of the cartridge will always need to be longer than the duration of the programme material. The secondary cue is used to enable the starting of a subsequent machine where there is a need for automatic sequencing. Many cartridge machine manufacturers also use the secondary cue to trigger an increase in capstan speed to two or three times reproduce speed, at the same time muting the replay amplifier to reduce noise, to reduce re-cue time. The tertiary cue is generally not used. In addition to the three cue tones it is also possible to record data on the cue track between the end of the primary stop cue and the beginning of the secondary cue. This allows the cartridge to be labelled with its number and title so that when replayed on air this data may be recovered and automatically logged for station records. However, this facility has not been used much other than in automation systems.

With the availability of various competitively priced higher quality options such as recordable CD, floppy disk, hard disk, etc. the NAB cartridge has almost become obsolete in the broadcast environment.

5.4.3.5 Digital audio tape

Digital audio tape (DAT) is useful primarily as a medium for programme acquisition and playback in radio stations. However, it is relatively cumbersome for programme preparation and it is usual to copy the recorded material to an intermediate format for editing. The need to copy material to another format to edit can slow down the production process, but it is no worse than the use of analogue cassettes and the end quality is usually higher.

For transmission purposes the major drawback with DAT when it first appeared was the short delay inherent in the time between the starting of the tape and the appearance of audio. This has been overcome in the latest professional machines, but illustrates the lower development level of the medium when compared with analogue tape.

5.4.3.6 Digital audio hard disk storage

The use of a digital hard disk for storing audio as an alternative technique to the analogue or digital cartridge has become commonplace. In such systems the audio storage system itself may be located at a distance from the studio environment as its contents can be played by means of a remote control. It is a medium that initially lent itself particularly well to the storage of commercials and other short items such as jingles and news cuts. However, with the progressive reduction in the cost of hard disk capacity it has become quite normal to store most of a radio station's currently played music library on disk so that the original source CD can be retained safely in the music library. The storage medium is usually a Winchester-type hard disk which is completely sealed from the external environment and typically has a mean time between failure (MTBF) of 10 years or more. To maximize the storage time available per megabyte of disk space it is common practice to compress the previously digitized audio data to a quarter or less of its original linear encoded data rate during recordings. Digital expansion is then provided upon playback before decoding to the original analogue signal. By using several thousand megabytes of disk storage playing times of several hours may be achieved. (For example, a currently common sized disk of 9 Gbytes will store around 80 hours of 15 kHz stereo audio material compressed 4:1 to 256 kb/s) The economics of such storage systems initially meant that the control and playback from the system was shared between a number of studios and between a number of programme channels. However, the falling cost of PCs with large capacity hard disks now means that the use of a single device per studio, with a network connection to allow the transfer of audio between storage units, can be equally affordable. Unlike a cartridge only one recording is needed to satisfy the requirements of all the users if a system is capable of supporting multiple and simultaneous users or allows network transfers. Furthermore, as a commercial schedule is predetermined and usually stored in a computer it is possible to make an electronic transfer of the items to be played, so that the need for the presenter to load a cartridge or key in a track number, together with much of the potential for playing an incorrect item, can be largely eliminated. This also allows the presenter to spend more time on the contents of the programme and less in clerical work, logging the playout, as the system can automatically record the time and date of transmission of each item.

5.4.3.7 Digital cartridge players

In its time the analogue cartridge player served well as a means of playing short audio items. However, it suffered from some inherent defects. Firstly, the track width available for audio recording had to be reduced from that of a normal stereo open-reel machine due to the need to accommodate the third, cue, track. This cue track must not be placed too close to the centre (right) audio track for fear of *cross-talk* allowing the cue tones to be reproduced by the right audio replay amplifier. This reduction in track width increases the tape noise level compared with an open-reel machine which only needs to record two tracks.

The second problem area stems from the design of the cartridge. On an open-reel machine the tape is guided across the replay heads by precision-machined and positioned guides. This ensures that the tape is guided accurately across the record and replay heads particularly with regard to *azimuth*. In an NAB cartridge player the tape guidance can only partly be dependent on the external guides due to the limited part of the tape path which is accessible from outside the cartridge. Also the positioning of the tape with respect to the replay and record heads is dependent on the surface accuracy of the cartridge shell which is made of a plastic pressing. All this means that the resulting azimuth angle of the tape to the heads may vary from cartridge to cartridge and to some degree from machine to machine. The result tends to be an azimuth tolerance which at best will always be wider than with a well-designed open-reel machine. This in turn can often lead to poor phase coherence and stability between the two audio channels. Thirdly the wear and tear on the tape in the cartridge is much higher than for tape contained on open reels as it is permanently running in contact with adjacent layers on the central pack.

For these reasons there was an obvious need for an alternative but similar form of storage which did not suffer from these defects. The rapidly declining cost of computer hardware and of digital encoding and decoding electronics saw a number of alternatives to the analogue NAB cartridge become available. The technique common to all these systems is to use some sort of removable disk which can be quickly loaded and unloaded from its reader drive to emulate the ease of use of the NAB cartridge. Because computer disk drives are made in far greater numbers than cartridge player transports they are very cheap and consequently computer disk-based systems have largely displaced analogue cartridges as radio stations were equipped or re-equipped.

5.4.3.8 Telephone balance units

The telephone balance unit (TBU) contains an electronic *hybrid transformer* which takes advantage of the characteristics of the telephone line. The hybrid transformer attempts to adjust automatically to match itself to the characteristic impedance of the telephone line. In the case of a perfect match it would be possible to turn the two wires of the telephone line, which carry speech in both directions, into a four-wire line with separate incoming and outgoing speech circuits. In practice the TBU is not perfect and between 20 and 30 dB of separation may be achieved between the incoming and outgoing speech in a real system. This separation is needed to reduce the coloration of the studio sound by its connection to the telephone line which has to carry the same studio signal to the incoming caller. In order to prevent feedback while the caller is on air the outgoing signal presented to the TBU must be a clean feed (mix-minus) signal that is, it will contain the composite studio output signal after the subtraction of the incoming signal from the telephone line. The TBU is also necessary to provide a barrier between the telephone line, which carries a d.c. voltage, and the mixing desk. In the UK it is also necessary for the TBU to be approved by the *BABT* (British Approvals Board for Telecommunications) if it is to be connected to the public switched telephone network (PSTN).

5.4.4 Other studio facilities

5.4.4.1 Telephone switching facilities

Many radio stations will wish to allow the audience to participate via the telephone in the radio station's programming, whether for competitions or for listener advice programmes. In either case the need will be for some form of telephone call handling system which will allow the answering of one or more telephone lines by the presenter or production staff. The system should allow calls to be answered either off-air, so that a choice can be made as to which caller to put on air, or on air in which instance the caller will be put directly on air at the moment of being answered.

If the call is to be answered off-air then the caller will need to be provided with a feed of the radio station output via the telephone line. It is important that the caller is requested to listen to this rather than their own radio for two reasons. Firstly, because the radio station is likely to be operating an audio profanity delay system (see below), and secondly because if they are selected to go on air the microphone of the telephone handset may pick up the audio output from their radio and cause acoustic positive feedback. When the caller is selected to go on air it is necessary to connect the telephone line to a two-to-four wire conversion system. This will usually take the form of a TBU (see Section 5.4.3.8).

If it is wished to have more than one caller on air at the same time it will be necessary to use as many additional TBUs as callers, with a separate clean feed for each unit. This will enable each caller to hear not only the studio contribution but also any other callers who may be on air at the same time. In order to enable the smooth handling of callers some mixing desks allow the presenter to talk to the caller, when the line has been extended to the desk via the TBU, by using the desk internal talkback system. In this mode desk talkback is mixed in with the clean feed to the caller while the telephone channel is prefaded.

Because the content of telephone calls cannot be entirely under the control of the radio station it is common practice to introduce an audio delay into the output of the studio or radio station during phone-in programmes. This delay, typically of 7–10 seconds duration, allows the presenter to remove a potentially obscence or libellous caller from air before the offending audio reaches the transmitter. These days the delay is normally provided by means of an electronic device which encodes the audio digitally and then applies the resultant signal to a large shift register. Sufficient memory is provided so that the digital signal will take up to 7 seconds or more to reach the end of the shift register where it is then decoded back to an analogue signal. These electronic units have largely replaced the earlier tape loop systems which were prone to a much higher failure rate and did not provide the near-transparent quality now available.

The method of entering the delayed mode varies depending on the device employed, but there are two common techniques. In the first the digital delay device is used in a so-called 'catch-up' mode. The delay is initially set to zero and the signal is routed through the system. The device gradually introduces more delay into the system during natural pauses in the audio until it has reached the preset maximum. The rate at which the delay period is incremented is often adjustable so that it is possible to adjust the operation for the most pleasing result, but care is needed to ensure that the rate of increase is not too fast otherwise the resulting output can sound unnatural. In the second method of operation the delay system always operates at its maximum duration. To enter the delayed mode the normal transmission path is cut and a prerecorded segment of audio of the same or slightly longer length than the delay system

is played to line. Just before this audio reaches its end the output of the delay system is switched to line and mixed with the tail end of the prerecorded audio. After the prerecorded audio has ended it is switched from line and allowed to recue. Typically the prerecorded audio would be played from a cartridge machine. The output of the studio or station will now experience the required delay.

Should a caller then say something that needs to be censored the following needs to occur: in the case of the catch-up system the delay line will dump all the audio in the shift register and reset the delay to zero (that is the studio or station output will return to real time). It will then repeat the same progressive introduction of delay that was used initially to enter the delay mode. Meanwhile the desk channel will have been faded out and another caller selected. The presenter must not pause during this time otherwise the listeners will hear a gap in the programme. In the case of the fixed delay system the output of the delay is disconnected from the output path and the prerecorded audio is replayed. The audio from the delay system is then switched back to line in the same way as in the initial entry into delay.

To exit from the delay in both cases the sequence is similar. All programme input to the delay system is stopped. The presenter listens to the post-delay signal and waits until the last audio item is heard. The delay system is then switched out of circuit and the station or studio returns to real time at which point the programme may continue in the usual way. To enable the switching operation to be carried out most easily it is usual to provide the studio with three push-buttons which will provide the respective 'delay in', 'censor' and 'delay out' functions when operated. All the switching and equipment operation will then be controlled automatically rendering the operation as simple as possible for the presenter. It is important to remember that during the period in which delay is in operation the presenter must be able to monitor the station output at a point before the delay system and that the callers also hear the same signal whilst they are waiting to go on air. BBC local radio stations do not normally operate profanity delay systems.

5.4.4.2 Indicator lights
A variety of indicator lights may be used in studio areas, depending on the conventions of the individual broadcasting organization.

Microphone live, transmission A red light in the studio or the control room or both acts as a cue and as a warning to contributors that a microphone

is, 'open', and will be repeated outside the door as a warning to visitors. It will be switched automatically from a fader backstop switch on the relevant desk channel, and may also be manually controlled. 'Transmission' lights are rarely used in Local Radio.

Rehearsal/studio powered A blue light was traditionally fitted outside the door of BBC studios, illuminated via the technical power breaker in the studio and thereby indicating when the area was in use. ILR stations, normally leaving their studios powered continuously, have no such requirement and it is now considered largely unnecessary in the BBC also.

Telephone, radio car Because telephone ringing (like other sound sources) is muted when a microphone is live, some visual indication of an incoming call is necessary. Several separate lamps may be provided either for different telephone circuits, for intercom or talkback, or to indicate radio-car calling.

5.4.4.3 Acoustics
Studio acoustics are dealt with elsewhere in this book. The technical considerations of sound transmission through partitions are determined by the laws of physics, and sufficient isolation must be maintained between studios and control rooms for monitoring to be possible without feedback and howl-round. Isolation from other sources of noise will be determined by the budget available, but the acoustic treatment of the studios and control rooms must be taken seriously if the broadcast result is to sound professional.

5.4.4.4 Power
Power to technical equipment throughout the station will often be segregated from the general power distribution. This is to avoid irregularities in the mains which may originate in non-technical apparatus (e.g. lift motors or air-conditioning plant) finding its way into the broadcast signal, parts of which may be highly amplified and subject to interference. Particular care should be taken to provide a low impedance technical earth.

5.4.4.5 Clocks
Time is important to the broadcaster for a number of separate reasons: time of day is frequently broadcast, programme schedules are run to a time-table, and air-time is sold by duration and time of broadcast. It is therefore vital that both the time of day and means of accurately timing are widely available throughout a radio station. Each operational area will have at least one clock as close to the line-of-sight

of presenters and operators as possible; traditionally this would be a 25-cm diameter analogue face with a pulsed sweep second hand for accuracy of cueing, driven by a master clock in the racks room (see Section 5.4.10.2). A smaller clock, similarly driven, is often built into the desk or set into a presenter's box on the studio table. All-digital displays have not found widespread favour because they are more difficult to interpret, but analogue faces using lamps at each second position and hybrid analogue/digital are also in common use. All-electronic clocks have the virtue of being silent, a quality which only the most expensive electromechanical pulsed movements can equal. Furthermore, electronic displays can be driven from a time-code signal, also from a master clock, so providing absolute synchronism, summertime resetting, leap seconds and the like. The advent of low-cost self-contained clocks which are synchronized to one of the radio time code transmissions has enabled their use as an alternative in many smaller radio stations, where the cost of a master/slave clock system cannot be justified.

A digital stopwatch is built into most modern control desks for timing programme segments, but this is not associated with the station clock system. This stopwatch may often be started and stopped by the operation of CD players or other audio sources so that the duration of the items played may be automatically indicated.

5.4.5 Sustaining and regional programming (local networking)

5.4.5.1 BBC
As BBC local radio stations came under the control of BBC administrations natural geographical grouping could be supported and directed by a parallel financial and programme structure. Experiments in programme sharing within regions have become formalized, although each region operates a slightly different system. Typically this is based around a destination-controlled router, each station in the region having a send and a receive stereo circuit to this hub. Any station can originate a programme and any or all stations can elect to take it or not, with the policies of the region and financial pressure determining the actual arrangements. Cartridge machines with local identification jingles may be located in the racks room in each station and triggered by tones or data signals from the studio originating the programme. Thus the programme feed passing through individual stations on its way to that station's transmitters can be interrupted and the local

identification inserted. Hence local programming can be provided at substantially reduced cost, particularly during those times of day when audiences are low.

5.4.5.2 Independent local radio
Since franchises (nowadays licences) were first awarded to ILR stations there has been a tendency for the economically strong stations to buy the less strong stations. In some cases this has taken place on a regional basis and to maximize the economy of scale, much of the programme output of the individual radio stations has been produced from a single studio. In order for the local content of the individual radio stations to be retained use is often made of a number of digital storage systems which are associated with the transmitter feeds serving each of the original individual franchise areas. These systems contain the commercials, station identity liners (using the programme presenter's voice if required) and news and local information relevant to the area served. At appropriate points in the centrally provided programme the storage systems are activated to provide the local content to each of the areas served. This enables the local identity of the original areas to be preserved but significantly reduces the operating cost as one presenter may now be broadcasting to two or more areas.

For similar reasons of economy, stations which may be geographically separated from the programme providing station may receive programmes which they transmit for the part of the day for which it would be uneconomic to employ a 'live' presenter. This could be overnight for example. The main programme may be delivered to the remote station by landline or satellite. Subaudible frequency tones or a data channel may be used to start the storage system playing out the required items and switch the main programme over to the local items. The items that are to be played will have been previously entered and scheduled by the local station staff so that the system may be left to operate unattended if required. If everything works reliably the local audience should not normally be aware that the majority of the programme is being produced outside their region since they will hear the same local station identification and commercials that they might hear when programmes are locally generated.

5.4.6 Station logging systems

5.4.6.1 Transmission logging
For legal and regulatory purposes local radio stations must be able to log the output of their transmissions

onto some form of audio recorder, which for BBC local radio is a change from past practice as a consequence of the 1990 Broadcasting Act. In the past it was common to employ low-speed, multiple-track, open-reel recorders, but the quality of reproduction of these machines is usually severely limited by the very low recording speed. The availability of domestic video cassette recorders with hi-fi FM stereo recording facilities brought with it the possibility of providing a high-quality recording lasting 8 hours at the long-play speed. These machines can also be used in conjunction with a PCM (pulse code modulation) encoder which enables a second stereo digital signal to be recorded as a pseudo-video signal. This combination allows two high-quality stereo signals to be recorded simultaneously, while still leaving the low-speed longitudinal track free for recording a time announcement. This has the advantage for the radio station that the logging system is able to satisfy legal, regulatory and archiving functions at the same time.

5.4.6.2 Music logging

There will usually be a need to record the title and duration of all the commercial music tracks that have been played by a local radio station. An accurate means of providing this information will be required by any organization responsible for the collection of fees due for the use of these music tracks or for the regulation of 'needletime'. In its simplest form this is done by the presenter using a stopwatch and compiling a handwritten log, but poor accuracy together with the illegible handwriting of many presenters has led radio stations to improve on this system. Using bar codes either printed on the record sleeve or on stick-on labels the presenter can enter record and track details into a computer-controlled storage system in a simple operation. Timing information can be obtained from the channel fader end-stop switch, and a complete log of music played can be compiled with very little of the effort of the manual system. Statistical information can be returned at intervals on a floppy disk or as convenient. The BBC 'SABLE' (studio automatic barcode logging equipment) is an example of such a system, and others are in use elsewhere.

If the radio station uses a computer system to compile the playing order of records the presenter may well have a preprinted listing which will only need alteration from the full running length of the track in the case that it was not played in its entirety.

5.4.6.3 Commercial logging

For ILR stations it is necessary to be able to provide a log of the actual times at which each commercial has been played. This allows the advertiser to be provided with a transmission certificate when an invoice is sent out. Most stations write the transmission time onto a preprinted log. With the advent of digital storage systems it has become simple for the information to be logged automatically by the storage system either to a printer or to a computer disk. In the case of stations which take programmes from a remote location and insert local commercials automatically it is important that this logging process is reliable. In the case of a dispute over transmission times the station audio log can be used and it is customary for stations to keep this log for 90 days.

5.4.7 Communications

5.4.7.1 Intercom

It is essential within a broadcasting environment that instant communication is available between operational areas and those most involved with, and responsible for, station output. In local radio this means, in practice, an intercom system with stations in management offices, production office, news room, technical areas (particularly the racks room and engineer's workshop) and studios, where it must be muted when microphones are live or on-air. A PBX telephone system, vital for more general communication, is no substitute for a station intercom.

5.4.7.2 Private automatic branch exchange systems

Most radio stations will require some sort of telephone system for handling business-call traffic. In the very smallest of stations it may be sensible to combine this with the same system which is used to handle calls for on-air use. In most stations it is more likely that a separate system will be employed, although a number of extensions on the system may also be connected to the 'on-air' system. Most of the modern PABX systems use digital circuitry both for the control system and for the call path. The use of a computer to control the operation of the system makes it simple to move extension numbers and to determine the level of facilities and access allowed by extensions. This can be particularly useful on special occasions such as during charity appeals, or at times of emergency when special groups of extensions may need to be set up for programme support.

5.4.8 Outside broadcast facilities

5.4.8.1 Music and control circuits, radio links and integrated services digital network (ISDN)

This section is not intended to deal with the provision and equipping of outside-broadcast vehicles or remote locations, but rather with the facilities required at the radio station to handle outside broadcasts.

The audio signals from most outside broadcasts are likely to be returned to the radio station by means of lines or radio links or a combination of the two. As well as the audio from the outside-broadcast site there will be a need for some sort of communications with the presenter or production staff at the site, and this likewise can be by a mixture of lines and radio links. If the site is outside the range of the radio station's transmitters a feed of cue programme to the outside-broadcast site will also be necessary.

In many cases it will be most convenient, where the broadcast is taking place within the service area of the radio station, if the audio signals and communications can be relayed using the broadcaster's own radio equipment. However, it is likely that signals from the outside broadcast will be received at a site remote from the studio, such as a tall building or mast where the advantages of aerial height may be gained, and then be relayed back to the studio by lines.

In the case of contributions from fixed sites such as other radio stations, concert halls or sports grounds it may be most economically advantageous to use lines for the entire route. If the venue is used frequently it may be sensible to rent a permanent circuit from the site to the radio station. Where a venue is only occasionally used it may be cheaper to rent just a local end between it and the adjacent local exchange and pay for connection only during periods of use. A temporary local end may be installed for a one-off event. In all these cases the nature of the programme and the available budget will determine whether the circuit should be monophonic or stereophonic and of what audio bandwidth. Generally the cue circuit may be of a lower grade than the contribution circuit unless it has to be relayed to a wider audience than the presenter; a dial-up telephone line may be quite satisfactory. Where the programme or cue programme has to travel via a satellite path it is important that the cue programme is derived as a clean feed and is not simply a studio or station output signal, otherwise the contributing presenter will hear a delayed version of his or her voice due to the long satellite path length.

If the radio station and the outside-broadcast site are both connected to the integrated services digital network (ISDN) it is possible to obtain an audio bandwidth and quality greatly in excess of a normal dialled call while using the same digital transmission capacity as a standard voice call. To make use of this capacity the radio station will need to provide a suitable terminal adapter and digital codec which will be compatible with the equipment used at the outside broadcast site. Cost-effective outside broadcasts may then be mounted. ISDN is generally available on request in most areas of the UK.

5.4.8.2 Radio frequencies for outside broadcasts

Radio frequency allocation in the UK is currently regulated by the Radio Communications Agency which in turn is responsible to the Department of Trade and Industry. In addition to frequencies allocated for broadcasting a number of auxiliary broadcast frequencies have been allocated for programme production. These are regulated, in the case of the BBC and radio broadcasters licensed by the RA (Radio Authority), by the Joint Frequency Management Group Ltd (JFMG) which is directly owned by the broadcasters. All broadcasters must apply to the JFMG for spectrum and be licensed before they can make use of radio equipment for programme making. In the case of air-to-ground broadcasts, such as may take place from a traffic spotting plane, the agreement of the Civil Aviation Authority (CAA) must also be obtained.

5.4.9 Newsroom areas

The newsroom in a local radio station and the output that it provides can make a significant contribution to the success of the station in winning and retaining listenership. It plays a major part in rooting the radio station firmly into the community to which it broadcasts. However, it is likely to be considerably more limited in its human resources than its national network competitors and, therefore, must make the most efficient use of those resources. This will mean investing in equipment which will provide the most cost effective operation.

In general the news operation is likely to be dependent on an external agent for the supply of national and international news, while it concentrates on the news stories of local interest. It is likely that from time to time the station may be called upon to provide services to the national broadcasters where a major event occurs in the area and its reporters are first on the scene.

The collection of news material from interviews will require a number of portable tape machines. In general the resulting material is likely to be taken back to the radio station for editing before transmission. Portable recorders using $\frac{1}{4}$ in tape, have largely been superseded by analogue cassette machines, RDAT recorders or MD recorders in most cases due to their smaller size and weight and lower cost. However, these formats generally require dubbing to another medium, such as a hard disk system, for editing. Whatever type of machine is used it is important that it incorporates some form of audio limiter or automatic gain control to optimize the recording levels and prevent tape overload. It is very difficult to carry out an interview, often under less than ideal circumstances and in a hurry, and manually adjust recording levels at the same time.

If the radio station receives an audio service of national and international news it will require some type of recorder to store the audio provided by this service. These machines would typically be activated to record by the presence of audio or coded tones on the incoming circuit. BBC News has a system which can address local radio stations individually or in groups, and is capable of sending material from fixed and mobile sources (London news room, Parliament, political conferences, etc.) to tape or cartridge machines. In addition it is important to provide a means for journalists to phone out to interview a subject and record the resultant interview for later editing. This may mean extending the phone-in system to the newsroom or simply providing one or more individual telephone lines which may be routed via a telephone balance unit.

Within the newsroom area there will be a requirement for a number of editing positions which may also provide the means of transferring from the acquisition medium to the editing medium and finally to the transmission medium. Transmission may be from $\frac{1}{4}$ in tapes or from a cartridge player.

The newsroom may also receive one or more text services and these may make use of hard-copy printers or be directly connected into an electronic storage and retrieval system.

5.4.10 Engineering facilities

5.4.10.1 Test equipment
In most cases the radio station will be totally dependent on its own resources for the alignment and repair of equipment, maintenance only rarely being contracted out. At the least it will require a low distortion test oscillator, a dual-beam oscilloscope, an accurate

noise, distortion and level measuring set, a multimeter, a laboratory power supply and, if analogue tape or disk reporoducers are still used, a wow and flutter measuring set. Together with this test equipment a stock of basic components and significant spares should be held. In order to comply with health and safety requirements a separate workshop area should be set aside with good lighting and a proper workbench on which equipment may be safely maintained without danger to the engineer or other staff.

5.4.10.2 Racks area (apparatus room)
An area will be required in which to install any racks of equipment associated with transmission circuits and other common facilities. A typical apparatus room will house the following equipment.

Station logging recorders These were described in Section 5.4.6.1. They are usually bay mounted in the racks area together with an automatic mechanism to change between machines and an end-of-tape alarm.

Master clock system If a station master clock system is justified it will typically be a high-accuracy crystal clock slaved to a receiver which is in turn tuned to a time-code transmitter (MSF Rugby in the UK). The internal crystal must be stable enough to withstand loss of received clock for a period of several days, to allow for outages. Most master clocks produce both pulsed and time-code outputs to cater for various types of display driver circuits: suitable displays are discussed in Section 5.4.4.5. It is important for the station clock system to be reliable, stable and accurate, but precision greater than about 0.1 s is of no particular value. As digital audio becomes more widely used and synchronizing clock feeds such as the AES/EBU standard are found more frequently, clock precision will become an important issue separate from broadcast time.

Main distribution frame This is simply a permanent termination point for cables to all parts of the radio station to aid installation, modifications, and fault finding at a detailed level. Early frames were of the Post Office tag-block variety, but modern installations use insulation displacement connectors. In both cases, external circuits should be landed on separate tags and *jumpered* together, to allow flexibility and change.

Jackfield This is composed of Post Office 316 or 'B' gauge jacks. It has two primary functions: firstly as a front-line fault-finding aid to establish where a signal is lost or becomes distorted; and secondly as a cross- or over-plugging interface for

non-standard signal path arrangements. The more points in the signal path which appear on the jack-field, the greater the flexibility. All signals should appear at *zero level* unless specifically otherwise labelled.

Receivers for off-air check and rebroadcast sources Rebroadcast receivers will ideally be fixed tuned devices the output signal quality of which is comparable with station-originated output: rack-mounted professional equipment is commercially available for this purpose. Off-air check receivers are not primarily provided for quality monitoring, and hi-fi tuners may be used.

Telephone central apparatus (PABX and phone-in) The racks room offers a suitable and convenient location for the station's PABX central apparatus, although it is wise to separate the telecom-munication contractor's equipment physically from the broadcast plant to avoid disputes in case of inter-ference or other difficulty. Telephone balance units (see Section 5.4.3.8), to enable the phone-in lines to be used successfully for programmes, are often housed in the racks room as they are a facility shared between studio areas.

Line termination equipment Permanent audio circuits incoming to the local radio station are now generally provided using digital circuits although some analogue circuits still remain.

The terminating equipment is usually provided by the broadcaster, with all parameters, possibly includ-ing temperature compensation for analogue circuits, being fixed during installation. A transformer and approved safety barrier, to isolate the broadcaster's plant from the external cables or circuit provider's equipment, is also part of the line termination equip-ment. Typical examples of circuits are those from an outside broadcast radio link receiver site, or local ends to an exchange.

Line sending equipment Although audio is distributed at standard level within the radio station, and distances are usually small enough for there to be no appreciable attenuation, this cannot be assumed to hold good for signals sent to an external line. Line sending amplifiers, capable of providing adequate power output into typical line impedances, are therefore used. Maximum voltages, however, are laid down by the transmission contractors to avoid crosstalk, and these levels are strictly enforced. The line sending amplifier will be isolated from the line itself by a transformer, for protection.

Transmission interface Incoming and outgo-ing permanent lines are customarily terminated on a bay-mounted block known as a tablet, which defines

the limits of responsibility of broadcaster and trans-mission contractor. The circuits on each side of the tablet are joined by a removable U-link on which the circuit may be monitored.

Line limiters and transmission processors Audio processing is almost universally applied to station transmission output; several manufacturers market equipment which can make the signal sound more 'punchy' or just plain louder by more or less subtle compression techniques. In many processors the input signal is split into a number of frequency bands, each band then being separately compressed, after which the signals are recombined. Whether this attracts more listeners is debatable, but what is indisputable is that a compressed signal is easier to hear under adverse listening conditions such as in factories, shops or vehicles. In general the best results are achieved if the audio processor is sited at the transmitter. In this case it is advisable to install a simple limiter before the signal is fed to the transmitter circuit to prevent it being overloaded. A limiter is also likely to be incorporated in the processor or installed following it. This will prevent transmitter overload and consequent distortion or, in the worst case, the triggering of protection circuitry which could shut down the transmitter.

Automatic programme switching equipment In the case of a station which is broadcasting unattended for part of the day, automatic programme switching may be used to cut between different source material; tapes, incoming news, commercials, or sustaining programme. Any switcher may be sited in the racks room, and could be controlled by the master clock or an incoming signal.

5.4.10.3 Power supplies

A radio station is often required to be on the air 24 hours per day, 7 days per week and is often an important resource in providing help and advice dur-ing a local or national emergency. It therefore needs an alternative source of mains power should the local electricity supply fail. In ideal circumstances the entire power supply to the radio station will be main-tained by a generator which will automatically run and provide alternative power in the case of mains failure. For some stations this might seem difficult to justify on economic grounds since it may only be required once every few years. For commercial sta-tions the potential loss of advertising revenue may be sufficient justification but in addition most sta-tions will recognize a need to be able to continue to broadcast through times of local or national emer-gency when the public turn to their radio for help

and guidance. This has led to the installation of generating plant at all BBC local and most independent radio stations. Sufficient stand-by power should be provided to operate a minimum level of service to allow the station to remain on the air; this implies the support of technical plant and enough lighting for the safe continuation of broadcasting. Similar stand-by power arrangements will often be made at transmitter sites as these are usually remote from the studios. Generators ideally should start automatically upon loss of mains power since power cuts may occur at times when there are minimal staff on the station none of whom may be technically able to start a generator.

5.5 Outside broadcast and mobile equipment and operations

5.5.1 Introduction

The principles of audio engineering, like the laws of physics which govern them, are universal, and the operational techniques required to achieve particular styles of sound balance and control are well established. Outside the predictable environment of the studio, however, engineers must practise their skills in conditions which may range from inconvenient to downright unpleasant. Adverse acoustics, dubious power supplies and the need to interface with unfamiliar apparatus are just a few of the problems encountered by technicians 'on the road'.

At their simplest, mobile recordings can be made by radio reporters, etc., using nothing more elaborate than a pocket cassette machine. On the other hand, location audio can equally involve live coverage of huge public events, with over 100 microphones feeding domestic radio and television, foreign broadcasters, record companies and public address systems. Between these extremes lie the many routine assignments of moderate complexity which are the daily stock-in-trade of the mobile sound engineer.

Strictly, this section should examine all these issues. However, it would be superfluous to dwell on procedures so basic that they can be performed by the technically untrained; instead, emphasis will be placed on the equipment and techniques used by expert staff in mainstream professional audio. The topic is enormously varied, but the basic principles will be explained, potential dangers highlighted and advice given on how to overcome some of the more common operational problems. References to recording and broadcasting may be considered interchangeable, unless otherwise indicated.

5.5.2 Working on location

Recording and broadcasting away from the studio are very specialized branches of sound engineering. Working conditions are extremely varied and it is helpful if staff are both physically fit and capable of responding imaginatively to unforeseen demands. Many of the facilities required on location are similar to those found in a studio. However, the ways in which they are implemented may be quite different because of the temporary (and often inhospitable) situations in which the work takes place.

For most purposes, an isolated mixing and monitoring position is required and this is often conveniently provided by means of a fitted control vehicle. This offers a ready-made substitute for a studio mixing area and, if equipped with a sufficiently large sound desk, may prove suitable for many different kinds of programme. If a vehicle is inappropriate or unavailable, an *ad hoc* mixing area must be constructed in some convenient room. On these occasions, which are common in radio, time must be allowed for the extra work involved in installing the equipment. In addition, each individual item must be sensibly transportable and robust enough to withstand the rigours of day-to-day mobile use.

5.5.2.1 Sound-control vehicles
There is much to be gained from installing sound equipment permanently in a vehicle. The practice is almost universal in television outside broadcasts, where an audio mixer of appropriate specification will be provided alongside the production and video facilities. Any programmes demanding sound operations of unusual complexity can be serviced by a separate, more sophisticated, mixing vehicle, linked to the main *scanner* by programme and communication circuits. In the field of radio, a great many outside broadcasts make quite modest demands upon technology, and the advantage of a fitted mobile often lies in the faster operation it affords when production schedules are tight. Indeed, mobile control rooms, of whatever form, can provide standard facilities in a fraction of the time required to set up equipment in a temporary control point, and with a significant long-term reduction in wear and tear. A further benefit is that the listening environment, though not as good as in the best studios, is at least consistent, and balancers will need to make fewer mental allowances for unfamiliar acoustics.

Dedicated sound-control vehicles are as varied as the projects for which they are employed, ranging in size from the radio cars used for news gathering to

massive articulated trucks extensively equipped for multitrack recording and weighing in excess of 20 tonnes. When choosing or designing a mobile, planners must therefore consider not only the technical requirements of the work in hand but also the logistic and economic implications.

Ultimately, the facilities and accommodation provided by any technical vehicle are limited by its physical dimensions. These, in turn, are restricted by statute, the relevant UK legislation being contained in documents published by the Department of Transport. Whilst the allowance is generous, it would be unwise to think that a very large vehicle is necessarily the first choice; comparatively few venues can accommodate a multi-axled pantechnicon and even fewer can readily provide adequate power for the amount of equipment it would be likely to contain. It should also be remembered that graded licences are required by drivers of heavy goods vehicles and that this could result in the need to employ specialists for the purpose rather than allowing an engineer, where appropriate, to take the mobile to the work site. So far as acoustics is concerned, there is also little advantage in having a vehicle of enormous size; width will always be severely restricted, so the permissible internal volume could never be translated into anything approaching an ideal shape for listening purposes (see Section 5.5.6). In practice, therefore, the very largest stand-alone audio vehicles are likely to be those offering particularly complex or specialized facilities, sometimes even incorporating lounge and refreshment areas for staff and clients.

5.5.2.2 Temporary control points and modular equipment

Fitted vehicles, like their static studio counterparts, are generally designed to deal with any programme up to a specified level of complexity; generously sized mixing consoles may be complemented by a range of permanent outboard equipment sufficient to meet any reasonable demand at minimal notice. From an operational point of view this is highly convenient, but if a majority of work is comparatively simple there may be a great deal of technical capacity lying idle most of the time. Such overspecification is a liability if a temporary control point is to be used, since time, space and labour are often at a premium. Rather, equipment should be closely tailored to the job in hand and be available in easily manageable components.

An elegant and economical way to deal with work of varied complexity is to use modular equipment. Engineers will choose from a range of compatible units according to the needs of each programme.

One component, for example, may provide outputs to feed recording machines or transmission circuits and offer monitoring and test facilities; it would be supplemented by other free-standing modules containing suitable numbers of input channels, subgroups, etc., to build up an outside broadcast of near-perfect specification which may be laid out to make the best use of available space and furnishings in the mixing room (*Figure 5.14*).

Figure 5.14 *A temporary outside broadcast control point utilizing modular equipment. Here, two 12-channel mixers are feeding a central group/master section, with a further small unit providing monitor selection, metering and loudspeaker drive.* (Courtesy of Hugh Robjohns)

5.5.2.3 Equipment durability

Where such a choice is available, equipment for use on location should be selected for its durability under adverse mechanical and electrical conditions.

Outdoor operation Any apparatus to be used in the open air must obviously be weatherproof. In the case of control equipment (except, perhaps, at PA mixing positions) this is rare; however, microphones, loudspeakers and other items of stage equipment are sometimes exposed to conditions quite unlike those in a studio and should be chosen and protected accordingly. Microphone noise is a common problem, whether caused physically by wind or electrically by ingress of moisture, particularly into capacitor models. Large windshields are generally more effective than small ones and are to be recommended whenever microphones are rigged in exposed positions. It is also worth recalling that pressure-operated microphones are less prone to wind noise than are pressure-gradient types. Thus, an omnidirectional model may prove satisfactory in conditions where a figure-of-eight or even a cardioid would be unusable. Noise can also result when water penetrates connectors in low-level circuits, especially when phantom power is present. In addition

to protecting plugs and sockets, therefore, it is worth switching off the phantom supply to circuits where it is not required.

Cables and connectors Robust, military-grade connectors are to be preferred, with substantial cable-grip and strain-relief components. Power cables should be generously rated and also armoured to prevent puncture or laceration. For microphone circuits, *star-quad cable* is preferable to twin-core types because of its high immunity to induced interference. Very thin specimens are best avoided except for special purposes, e.g. discreet in-shot rigging for television. All cables should be able to withstand indefinite amounts of handling and remain flexible even at very low temperatures.

Carriage of vulnerable items Inherently delicate equipment needs special attention during transit. Larger items should be transported in padded flight cases, devoid of sharp external corners which could cause damage or injury, and with handles positioned so that weight is sensibly distributed. Smaller articles, e.g. microphones, may remain in their own protective containers or else be collected into convenient sets and carried in a well-cushioned box. Stout plastic utility crates are useful for sundries, especially if they can be pre-sorted for use in specific areas. Canvas holdalls are another handy accessory, suitable for coils of cable, etc.

General safety precautions All reasonable precautions must be taken to protect personnel from injury at outside broadcast sites. Most formal safety measures are little more than applied common sense: cables crossing floors should be taped down or covered with matting; equipment slung overhead must be adequately secured against falling, etc. In addition to statutory requirements, any local directives must be observed, and there are particular obligations to prevent hazards when members of the public are present.

Hazard assessment is an essential part of planning, and nominated persons must take responsibility for health and safety management. Similarly, accidents and injuries must be reported in line with officially approved procedures.

5.5.3 Outside-broadcast mixers

Any attempt to describe outside-broadcast mixers in detail would be highly unproductive. Not only are there many different kinds but, as with all equipment, they are subject to constant evolution as technology progresses. It is feasible, however, to suggest the range of facilities likely to be found in various situations although the list can be neither specific nor exhaustive.

(1) *Small radio vehicle:* 12–20 channels; stereo and mono outputs.
(2) *Typical radio vehicle:* 32–48 channels; stereo and mono outputs; possible options for multi-track recording.
(3) *Single camera TV drama unit:* 6–12 channels; stereo and mono outputs.
(4) *Typical TV scanner:* 24–40 channels; various main, group and individual channel outputs in mono and stereo.
(5) *Multitrack recording vehicle:* 40–60 channels; group outputs and monitor mixer to suit tape machines in use.
(6) *Modular mixers:* 6–16 channels per module; indefinite scope for use in multiple with group and output modules.

In all but the simplest cases, it should be possible to derive clean-feed (mix-minus) outputs, either to produce versions of programme sound devoid of commentary, etc., or to permit multi-way working with other outside broadcasts or studios. The number and nature of clean feeds available will depend on the mixer design, which will in turn have been predetermined by programme expectations.

5.5.3.1 Space limitations

Lack of space is the usual limiting factor in mobile control rooms. Designers of new vehicles are often asked to install mixers with the maximum possible number of channels, and when an old vehicle is refurbished there may be a request for more channels in the space occupied by the previous console. Some mixers are available with unusually narrow modules, thereby permitting an increase in packing density. Alternatively, there may be scope for a trade-off between the number of channels and the sophistication of other facilities. Each case would be examined on its own merits.

A more sophisticated way to save space with analogue mixers is by the use of 'virtual' consoles. Audio signals are processed remotely from the desk proper, and common 'assignable' panels used to adjust all normal parameters (equalization, dynamics, etc.) via digital interfaces. The control surface is thereby much simplified and, since faders may be designated to perform more than one function, a very powerful console can be provided within modest dimensions. To quote just one example, BBC Television's Mobile Sound Control vehicle 3 can accept up to 112 inputs and control them by means

of 88 faders. Undoubtedly, however, the future lies with all-digital consoles, which, in addition to these ergonomic advantages, can offer all the characteristic benefits of digital signal processing. For a variety of reasons, the introduction of digital mixers has been relatively slow but, as their performance and reliability improve (and price and power consumption fall), more and more OB and recording companies are likely to abandon analogue equipment completely when refurbishment becomes due.

5.5.3.2 Interconnection

In spite of many technical innovations, much of the internal routeing of analogue signals in control vehicles is still carried out using traditional jacks and patch cords. In the UK, 1/4-in 'Post Office' components remain common, though 'bantam' systems with 4.4-mm plugs and sockets are becoming widely used, especially where space is unusually tight. Cleanliness is vital for trouble-free operation, and bantam plugs, in particular, require regular maintenance to remove the tarnish characteristic of outside-broadcast use.

External connections will be made via a termination panel (or 'tail board') mounted behind protective doors in the vehicle's coachwork. Both single and multiway audio circuits are normally provided, together with video, aerial and power connectors as appropriate.

5.5.4 Power supplies

Except for truly portable (i.e. hand-held) equipment, and one or two other exceptions, a stable and reliable mains power supply is essential on location. In many cases, a local domestic supply will be entirely adequate, though it is always prudent to check the ratings of fuses or circuit-breakers before making connection to convenient wall sockets. At sites where control vehicles are regularly used, dedicated power outlets may be provided, often via industrial connectors; throughout Europe, CEE (Commission L'Equipement Electrique) plugs and sockets are becoming widely adopted as standard. Power consumption of sound equipment ranges from under 1 kVA, for a small mixer and tape recorder, to more than 20 kVA for a sophisticated vehicle with multitrack recorders, digital processors and air-conditioning plant.

A suitable supply may not necessarily be forthcoming. Some venues have insufficient spare capacity; others have very old wiring systems which may give cause for concern about safety. At worst, there may be no public supply of any kind. In all

these circumstances, planning engineers will need to make alternative arrangements. Sometimes the local electricity company can provide a dedicated feed from a nearby substation, or else a mains supply may be replicated by means of a mobile generator. Where very small loads are involved, alternating current may be derived from car batteries, or similar, by means of an electronic inverter.

5.5.4.1 Supply voltage

Nominal electricity supply voltage varies throughout the world, e.g. 110 V in the USA and Canada and 230 V in Europe. Equipment to be transported abroad must therefore be capable of accepting this wide range of inputs, even if only by means of tapped power transformers. Modern apparatus incorporating switched-mode power supplies may be able to function satisfactorily in all countries without manual adjustment. In all cases, however, it is important that unpredictable voltage fluctuations do not adversely affect performance, least of all to degrade the quality of sound.

5.5.4.2 Voltage regulation

Even when all reasonable precautions have been taken, exceptional problems may arise from time to time. For example, in some areas the available voltage may actually change significantly during the course of a day as a result of varying consumer demand on a long supply line. One remedy here is to employ an automatic voltage regulator (AVR) in the technical power distribution, and many control vehicles are fitted with one as standard. AVRs not only raise the voltage of overstressed supplies, but also reduce unusually high line voltage to the correct nominal value. When several vehicles are co-sited, it is common practice to use one of them as a power distribution point, so it may be possible to feed other mobiles from its regulated supply.

5.5.4.3 Supply frequency

Until comparatively recently, motor-driven equipment such as tape machines used the frequency of the a.c. supply as a speed reference. Thus it was not possible to operate a European-standard machine (designed for 50-Hz operation) in countries such as the USA where 60 Hz is used. The advent of crystal control has enabled production of machines which can be used with equal facility all over the world.

5.5.4.4 Interference on supply lines

Control vehicles will generally accept more than one incoming power cable, thereby allowing separation

of supplies to technical equipment from those to utilities such as lighting and air conditioning. Two main benefits ensue from this arrangement. Firstly, there is a degree of insurance against failure of the technical supply; non-essential services may be abandoned in an emergency and their power source switched over to feed the technical circuits. Secondly, there is a reduced risk of interference from the thermostats, etc., which control these on-board 'domestic' items. Switching heavy electrical loads, especially those with an inductive component, tends to superimpose momentary high voltage transients on to the supply. These, if not suppressed at source, can produce clicks on the output of sound equipment sharing the same power circuit.

It should be remembered that impulsive interference of this kind occurs to some extent on most mains supplies, and may prove troublesome if audio equipment is not sufficiently immune. A broadcast of a church service, for example, could be marred by clicks from the boiler being used to prepare tea for the congregation. In such cases, where there is no possibility of removing the source of interference, a transient filter may be inserted in the power line to the technical equipment; its degree of effectiveness will depend on the severity of the interference and the susceptibility of individual pieces of apparatus. In exceptional cases of mains-borne interference, the use of battery-powered equipment may prove the only effective panacea.

5.5.4.5 Battery operation

Apart from their obvious use in powering portable equipment, batteries provide a vital reserve facility for small-scale live outside broadcasts. Switches to effect changeover from mains to battery operation must be readily accessible, and emergency headphones provided if loudspeaker monitoring is lost in the process. A further important use for batteries is to power the stand-by lighting system in control vehicles. Lead-acid accumulators are normally used, and these are continuously charged during the presence of a mains supply. Battery lighting is invaluable, not just in emergencies but also when inspecting the vehicle at night after external power has been disconnected.

5.5.4.6 Safety precautions and power supply monitoring

Except where supplies are guaranteed to remain segregated (e.g. for technical and air-conditioning circuits), all incoming power feeds to a vehicle or other control area should originate from the same phase of the electricity supply. In the UK, the potential between any two different phases is approximately 400 V, enough to cause great damage if supplies are inadvertently connected together and most certainly a danger to life. Power supply monitoring systems typically include indicators to confirm that only one phase is present, and equipment should not be switched on until this fact has been established. The correct connection of line and neutral legs must also be verified. Facilities should be provided for measuring the voltage of each supply cable and the current drawn from it. Suitably rated miniature circuit breakers will safeguard distribution paths, with a residual current device giving overall protection. Adequate earthing is also important, and qualified engineers should measure the *earth loop impedance* to ensure that it is tolerably low. As a further precaution, the metal bodywork of a vehicle may be separately earthed through a spike driven into the ground nearby. Where portable equipment is installed in a temporary location, the condition of the local mains supply should also be carefully scrutinized. Again, checks should be made to ensure that live and neutral have not become reversed and that the earth return is satisfactory. If there is any doubt about the state of earthing, additional security may be provided by bonding equipment to water pipes, etc.

5.5.5 Monitoring in temporary control rooms

One of the greatest problems encountered on location, especially when dealing with music, is arranging satisfactory conditions for loudspeaker monitoring. Programmes which originate in public auditoria will be heard alongside the finest studio recordings and, by inference, will be judged against them. Yet the outside broadcast balance engineer must often pursue excellence in quite unhelpful and highly makeshift surroundings.

As with a studio, however, three basic criteria affect the potential quality of the listening environment:

(1) the amount of space available,
(2) the nature of the acoustics, and
(3) the degree and nature of background noise.

Venues in frequent use are likely to have areas set aside for recording or broadcast control. These vary in elegance from plain, unfurnished store rooms where portable equipment may be set up, to lavishly appointed cubicles with permanent mixers and full acoustic treatment. Whatever the starting point,

however, efforts should be made to create the best possible environment for quality assessment. In practice, referring to the above, this means:

(1) attempting to place the loudspeakers and the balancer at the corners of an equilateral triangle of suitable size;
(2) arranging available carpets and furnishings to simulate, as far as practicable, a domestic acoustic; and
(3) taking steps to minimize unwanted noises.

Anomalies in monitoring are, by their very nature, likely to result in defective recordings. For example, if the listening environment is excessively reverberant the balance engineer may produce a sound which is acoustically too 'dry', and vice versa. Similarly, monitoring conditions which enhance low frequencies may result in programmes with inadequate bass, and so on.

If it is necessary for the balancer to sit well below the level of the loudspeakers, the more directional high frequency sounds may not be heard to best advantage. In this case it may be helpful to mount the loudspeakers upside down, bringing the tweeters closer to ear level to improve tonal balance and stereo imagery. *Figure 5.28* shows the same principle applied in a fitted vehicle.

Having taken all reasonable steps to optimize the listening conditions, engineers are well advised to reproduce a familiar recording through the monitor system. Audible imperfections may then be attributed to the immediate surroundings and allowances made when the new balance is attempted. A variation on this theme is for the engineer to listen for part of the time on trustworthy headphones. Whilst this does not necessarily give the correct stereo effect, it is nevertheless a means of judging sound quality against a known standard, and this, of course, is the very essence of monitoring.

5.5.6 Monitoring in control vehicles

5.5.6.1 Layout of equipment

For practical reasons, it is usual to mount the sound desk across the width of a mobile, with the loudspeakers as close as possible to the walls. European statute decrees that a vehicle's external width must not exceed 2.5 m, and if the walls are constructed with typical amounts of acoustic isolation treatment the internal dimension may be in the region of 2.3 m. Thus, a pair of loudspeakers each 50 cm in width, 40 cm deep and angled at 60° towards the balancer

can never effectively be more than 1.47 m apart, even though the vehicle is the maximum legal size. This is rather less than the spacing favoured in most studios, but is perfectly acceptable provided the balancer is allowed to sit at a commensurate distance from the loudspeakers. Unfortunately, the depth of many sound desks, together with the need to reserve access to their rear, often forces the engineer's chair a little too far away, resulting in a narrowing of the stereo image (*Figure 5.15*).

Figure 5.15 *Monitoring anomalies caused by positioning a large sound desk across the width of a vehicle. The ideal listening position (A) is unattainable, and the engineer must tolerate the narrower stereo image heard at position B.*

Recognizing this problem, at least one manufacturer has produced loudspeakers of modified cross-section, allowing them to be mounted more snugly against the walls. Alternatively, smaller loudspeakers might be considered; this is not necessarily an inferior option, since the extended bass response of larger units is seldom heard to advantage in a confined space and indeed may exacerbate unpleasant acoustic effects (see *Figure 5.19*). Again, internal geometry may be improved by using a mixer with a steeply rising profile and correspondingly reduced depth. In certain cases this may even allow it to be mounted satisfactorily along the side wall of the vehicle, possibly improving visual contact with other

areas (see *Figure 5.28*). One risk here, however, is that ill-informed designers may mount the loudspeakers too far apart – an arrangement likely to compromise stereo imaging or destroy it completely.

A highly radical solution to monitoring problems was adopted by BBC Radio for their prototype Digital Stereo Vehicle. In this case, the entire listening area was made expandable by use of hydraulic rams, increasing its useful width by some 1.2 m (*Figure 5.16*).

Figure 5.16 *BBC Radio's prototype Digital Stereo Vehicle (DSV), showing sides expanded to improve monitoring acoustics.* (Courtesy of Hugh Robjohns)

5.5.6.2 Acoustic environment

The two parameters of greatest significance are reverberation time and frequency response, the former being environmental and the latter a product of the listening space and whatever monitoring equipment

is installed. Other minor problems may include resonant or rattling cupboard doors, but these can usually be dealt with in a simple pragmatic fashion.

Reverberation time Ideally, the reverberation characteristics of a control area should be similar to those in which the sound will eventually be heard, i.e. an average domestic living room with average furnishings. In practice, this is not difficult to achieve in any reasonable vehicle, although the greater ease with which short wavelengths can be absorbed will generally result in a reduction of reverberation with increasing frequency. The properties of a domestic listener's carpets and curtains, however, will not be dissimilar. *Figure 5.17* shows a plot of the reverberation time inside one type of BBC control vehicle and clearly shows the frequency-dependent nature of the sound decay.

Frequency response Obtaining a linear frequency response inside a vehicle is rather more problematic. Indeed, the size and shape of the average mobile control room could hardly be less helpful in this regard. The worst shortcomings are likely to be caused by standing waves, which develop between parallel surfaces such as opposing walls, floor and ceiling, etc. Standing waves occur at the frequency where the surfaces are a half wavelength apart, and thence at multiples of that frequency. Taking the width of our typical control room as an example, a wall spacing of 2.3 m will result in standing waves at approximately 75, 150, 225 Hz, etc. Listeners midway between these walls will experience alternate nodes and antinodes at these frequencies and thus perceive a very irregular response. Moreover, each

1/3 octave band centre frequency, Hz

Figure 5.17 *Reverberation characteristics inside a sound control vehicle.* (Courtesy of the BBC.)

other set of parallel surfaces will produce a further series of standing waves in its own plane, all these waves becoming superimposed and making the response still worse. And if this were not bad enough, listeners in other positions will hear a completely different pattern of peaks and nulls, and their impression of the sound balance will be quite disparate. Cheap and unsophisticated wall treatments, e.g. carpeting or foam tiles, will usefully inhibit reflections at high frequencies, and thus standing waves can readily be controlled in this part of the spectrum. At the troublesome low frequencies, however, these porous absorbers would have to be of great thickness to be equally effective and their use in an already confined space is out of the question. Remedies available to the acoustic designer therefore include the avoidance of truly parallel boundaries (or giving them irregular, diffusing surfaces) and the use of specialized, wideband absorbers (*Figure 5.18*).

Figure 5.18 *Preparing a new vehicle for preliminary acoustic tests.* (Courtesy of the BBC.)

When the acoustic environment has been optimized by physical means, objective frequency response measurements should be carried out and consideration given to correcting remnant errors by means of permanent electronic equalization in the monitor chain. *Figure 5.19* shows the initial test results for a BBC radio vehicle. The loudspeakers in this case were full-sized studio types, whose extended bass response reacted with the vehicle's acoustics to produce a peak at 50–60 Hz. This anomaly was subsequently corrected by the use of a specially designed filter.

Even when the best acoustic treatment has been applied to a small control room, the perceived sound quality will depend very much on the precise listening position. This fact should be impressed upon visitors to the vehicle, who may otherwise be inclined to criticize the balance without justification.

5.5.6.3 Background noise

Adequate soundproofing is vitally important if the versatility of a control vehicle is not to be compromised. The difference between ambient noise levels in, say, a country churchyard and a city high street is very considerable, and yet the same vehicle may be required to work in both locations. Unfortunately, some of the classic methods of impeding sound transmission are wholly unsuitable for use in the construction of a mobile unit. Massively dense walls, for example, would be far too heavy, and cavity construction likely to squander unacceptably large amounts of space. Compromise is inevitable and has been attempted in a number of ways. *Figure 5.20* details a wall construction used by the BBC, and shows how a sandwich of different materials has been used to create an acoustic barrier.

Equal attention should obviously be given to the floor and ceiling, with care taken to minimize the impact effects of heavy rain or hail. Doors and windows are best double glazed and fitted with airtight seals. *Figure 5.21* shows the test results for a vehicle using all these design features and clearly illustrates the difficulty of providing isolation at low frequencies.

Noise, of course, is not all generated externally, and several sources can usually be identified within the vehicle itself. Equipment cooling fans, tape capstan motors and buzzing transformers will all contribute to the background noise level. The greatest single problem is often the air-conditioning plant, though fitting the compressor outside on elastic mounts and connecting it to the chiller unit by means of flexible pipework will generally keep vibration to an acceptable level. Air circulation from any fan system should be kept at low velocity to prevent turbulence.

5.5.7 Communications

No broadcast operation can run effectively unless there is an adequate system for rapid exchange of information. Parts of this system will be used for engineering purposes, e.g. during rigging, line-up and fault-finding; other elements are required by production departments concerned with organizing the programme and directing staff and artistes. It is not unusual for communications schemes to be far more elaborate than those for actual programme sound. Arrangements at outside-broadcast sites are further complicated because outstations may be far removed from the control centre and each may demand a different range of facilities. Thus, a great deal of time is

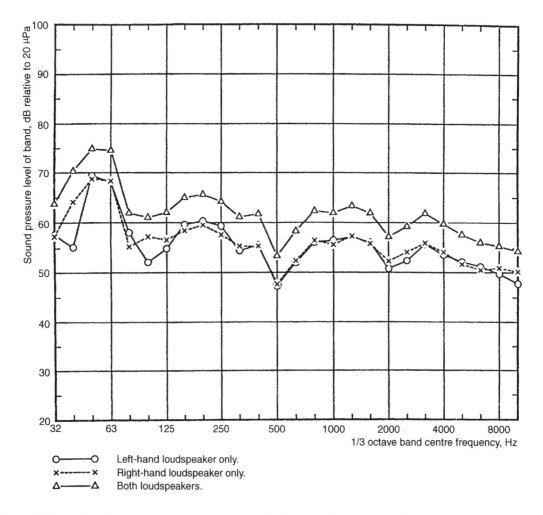

o———o Left-hand loudspeaker only.
x--------x Right-hand loudspeaker only.
△———△ Both loudspeakers.

Figure 5.19 *Unequalized frequency response at operator's position in a sound control vehicle.* (Courtesy of the BBC.)

often expended in setting up individual requirements and then testing the system as a whole. With complexity comes the risk that signals not intended for broadcast may find their way into the programme chain. Everyone should be aware of this possibility and take all sensible precautions against it, e.g. by not leaving headphones near a live microphone. In sound-control areas, especially in the confines of an outside-broadcast vehicle, talkback loudspeakers should be positioned so that their output cannot be confused with programme sound.

5.5.7.1 Talkback

In its many and varied forms, talkback is the fundamental means of communication. For a simple radio programme it may be no more elaborate than a single switched microphone for relaying instructions to

Figure 5.20 *Multilayer wall construction for acoustic insulation of a control vehicle.* (Courtesy of the BBC.)

o———o **Right side**
x······x **Left side**

Average of marked frequencies (dB):
right side 40.1
left side 39.7

Figure 5.21 *Acoustic isolation characteristics of a sound-control vehicle, showing the difficulty of excluding low-frequency noise.* (Courtesy of the BBC.)

an announcer's headphones. More typical though, certainly in television, is a comprehensive network of circuits, all carrying different information, capable of operating simultaneously and with scope for cross-connection. Arguably the most important feature of an outside-broadcast talkback system is its adaptability. Not only must all regularly-used facilities be instantly available, but it should be possible to service any request for unusual feeds, foreseen or otherwise, in a quick and efficient manner. Staff in charge must therefore be thoroughly acquainted with the talkback system, as the effectiveness of its use can often determine the success or failure of a programme, especially a live transmission.

5.5.7.2 Radio talkback
Talkback by cable is adequate for many staff working remote from the control point, but others, e.g.

stage managers, require complete freedom of movement and are generally provided with radio talkback. Typically, a TV outside-broadcast vehicle will carry a transmitter of approximately 25 W and broadcast continuous director's talkback from a roof-mounted aerial, staff replying via portable transceivers. Additional base stations may be installed for use by other departments, e.g. for audio staff moving between areas or setting up complicated arrangements of microphones. Most radio outside broadcasts and location sound recordings do not require such a formal installation but nevertheless make wide use of walkie-talkie sets, especially during the rigging and testing of equipment.

5.5.7.3 Communications matrix
A highly flexible method for setting up and distributing talkback, etc., is to employ a matrix. This may

be imagined as a series of summing amplifiers into which can be fed any combination of signals from a choice of buffered sources. In practice, the inputs and outputs are arranged along the axes of a perforated rectangular panel and selection made by inserting conductive pins into the appropriate cross-points. Resistors inside the colour-coded pins determine the levels of individual signals at the mixing points and thus allow complete control of the balance sent to any destination (*Figure 5.22*). Where demands are modest, a matrix may have perhaps 10 inputs and outputs. Large TV units are likely to require more elaborate devices, possibly with 50 in and 100 out. The largest British outside-broadcast matrix was installed in BBC Television's Mobile Central Control Room; with 100 inputs and 150 outputs it provides 15 000 cross-points and almost unlimited versatility.

5.5.7.4 Telephones

Whether used in isolation or as part of a complex network, telephones are a vital resource at outside broadcasts. One of their simplest applications is to permit conversations between the OB and its related studio centre, using either a line from the public switched telephone network (PSTN) or a circuit specially installed for the purpose. Within the OB site itself, use is often made of 'field' telephones. These provide their own polarizing voltage and a manually operated AC ringer, enabling them to be used at either end of any convenient pair of wires. In permanently fitted vehicles, it is common to find plain telephone handsets with external power supply, ringing and switching arrangements, perhaps allowing the same instrument to be connected to a number of different lines. Larger OB units are equipped with fully automatic telephone exchanges, providing flexible communications between staff working inside the vehicle and those in remote operational positions. Access to the PSTN may again be arranged, via a direct exchange line (DEL), both for general convenience and to provide an emergency liaison facility for live transmissions if more formal control systems (talkback, etc.) should fail. Control vehicles are also increasingly being equipped with cellular telephones, allowing greater freedom of communication where no physical line is provided.

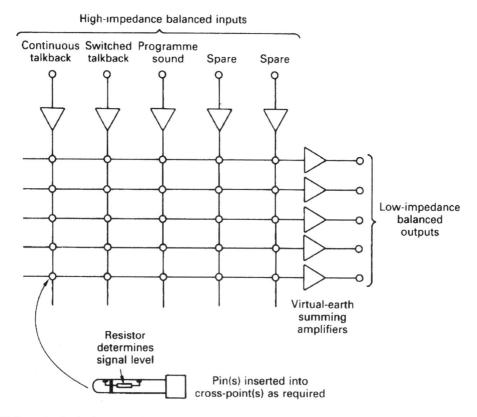

Figure 5.22 *Example of a simple communications matrix.*

5.5.7.5 Cue lights

Even when highly sophisticated communications are possible, use can still be found for devices as simple as cue lights. Whether supplied as separate components or incorporated into an elaborate facilities box, cue lights are frequently used to pass distinctive messages to commentators and performers at outside broadcasts and recording sessions. Return cues may also be possible, either by a complementary light system or audible signals in the control area.

5.5.8 Live transmissions

Outside broadcasts are often relayed to a remote studio centre for recording or live transmission, and this requires various kinds of communications circuits.

5.5.8.1 Types of circuit

Circuits to carry programme material must offer a high standard of performance, not least when music is involved. With modern digital communications, it is relatively simple to provide the basic requirements of a wide, flat frequency response with low noise and distortion. The additional need for accurate phase matching between left and right legs of a stereo pair has also become easier to achieve since the demise of analogue lines and their associated equalization problems. It has always been customary to use the term 'music circuit' even when programmes consist entirely of speech, though nowadays it may be possible to achieve economies by deliberately reducing the circuit specification if the very highest fidelity is not required.

Far less exacting standards are required of circuits used for telephones and talkback between an outside broadcast and the studio centre. In some cases these operate in bi-directional 'two wire' mode, terminated either with telephone instruments or with hybrid devices which transmit and receive signals without significant mutual interference. Alternatively, a so-called 'four wire' may be provided, actually a pair of unidirectional circuits configured so as to carry talkback both ways. The general term 'control circuit' is used to describe these systems used for engineering and production liaison.

5.5.8.2 Circuit implementation and testing

During recent years, the telecommunications revolution has had dramatic effects upon outside broadcast engineering and operations. Long-established conventions and practices have been swept aside as new technology has rendered them obsolete. As an example, temporary analogue music circuits (once universal at live OBs) are now unknown in the UK and their place has been taken by various digital systems, with signals routed either along cables or via terrestrial or satellite radio links.

One of the most significant developments has been the introduction of integrated services digital networks (ISDN). These provide dial-up circuits, similar to conventional telephone lines but exchanging data at 56 or 64 kbit/s, which may be used for a variety of purposes. For example, the same circuit could be used for rapid facsimile transmission, communications between computer terminals or to carry audio signals suitably processed and companded by coder/decoder (CODEC) units. At frequently-used OB sites, the expense of installing permanent ISDN lines is soon offset by the convenience and low cost of their subsequent use. Several different forms of audio coding have been developed, some using several ISDN circuits simultaneously to improve performance and reliability. Each variant has its own unique parameters for optimizing frequency response, noise, distortion and processing delay, so it is essential that compatible equipment is installed at each end of the link. ISDN features fully isolated bi-directional operation, so the CODEC at an OB can provide an output of cue programme and/or talkback from the central studio with no separate control circuit required. Processing delays of several tens of milliseconds (sometimes much more) should, however, be expected and allowed for (see Section 5.5.8.4).

An earlier digital technique still in common use in television is 'sound-in-syncs' (S-I-S). Audio signals are sampled and coded and the resulting data inserted into the line synchronizing period of the video waveform, conveniently marrying sound and vision for distribution purposes. Picture synchronization of domestic receivers would be severely disturbed by the modified signal, so the intrusive data must be removed at some convenient stage and the audio passed on as required, typically in digital stereo and analogue mono forms. Where suitable equipment is provided throughout the broadcast chain, it thus becomes possible for live sound to pass from an OB to the viewer's home entirely in the digital domain. Dual-channel (stereo) S-I-S and the final transmission system both use forms of the NICAM coding system developed by the BBC. In future, as digital distribution of video signals becomes the norm, even S-I-S may become extinct, discrete channels of digital audio being carried in the same bit-stream as the picture information.

On the day of a broadcast, once circuits have been established, it is usually to carry out basic tests

between the OB site and the receiving centre. Audio level, frequency response and distortion performance may be confirmed by sending a variety of tones, and noise measured with all signals muted. The rugged nature of digital circuits means, however, that these tests need be only rudimentary, involving none of the 'fine tuning' of errors associated with analogue lines.

Although many traditional communications techniques have passed into history, it is likely that some analogue circuits will persist in parts of the world for the foreseeable future, so it is important that engineers involved in their maintenance should maintain a grasp of basic principles. It should be appreciated, for example, that the level and frequency response of signals passed along a 'copper' line are substantially affected by the terminal impedances of the equipment at each end. Any tests carried out with the circuit disconnected must therefore utilize oscillators, meters, etc., with characteristics similar to those of the normal apparatus.

5.5.8.3 Line-up and circuit identification
Outside-broadcast line-up can often prove a lengthy business as programmes are routed systematically through the distant studio centre as far as a network control. To avoid undue disturbance or waste of valuable rehearsal time, it is useful if the broadcast can inject tone 'downstream' of its local sound distribution and monitoring. Line-up can then continue indefinitely without disrupting preparations for the programme. Larger control vehicles may be equipped with a discrete communications bay with a nominated engineer supervising all test procedures.

Sometimes there will be several outgoing circuits carrying different versions of the same programme. A sports outside broadcast, for example, may generate a composite sound mix, foreign-language commentaries and a clean-feed (mix-minus) of atmosphere and effects. Here, each outgoing circuit would be provided with its own tone injection system and a means of selecting the appropriate audio source. With increasing complexity comes the risk that circuits may become transposed or confused between the outside broadcast and its various destinations. An effective answer to these problems is to employ what are commonly known as identification ('ident') loops. Typically, these are solid-state memories which constantly repeat a verbal identification of each circuit and a short period of line-up tone. Their use, in lieu of constant tone, also ensures that circuit checks may continue even when outside-broadcast engineers are engaged in other work or their control area is unattended. Various systems are also used for breaking

line-up tone momentarily in a predetermined fashion to help identify the two legs of a stereo contribution.

5.5.8.4 Operational techniques
Many recorded outside broadcasts, especially those of public events, are already treated as if they were live. Much of their content may be impossible to retake, and so engineers, producers and performers are well accustomed to getting things right first time. An actual live transmission, therefore, requires only minor adjustments to operational techniques and very few extra facilities.

Clocks Although programme content is sometimes unpredictable, most broadcasts are given defined start and finish times. Reliable clocks are therefore needed at OB sites and it is becoming common to utilize models which automatically synchronize to dedicated time-reference transmitters, e.g. those at Rugby (UK) or Frankfurt (Germany). Television outside broadcasts often generate a time-code set to time of day for video logging and a readout of this may be displayed by a suitable decoder.

Cue arrangements Immediately before a live transmission, programme staff will need to listen to their network so they can respond promptly and effectively at the moment of handover. The necessary feed of network sound may be provided along a control circuit or else obtained from an off-air receiver. One hazard of modern technology is that appreciable delay may be introduced into music and control circuits by digital encoding and/or satellite links. As a result, an outside-broadcast commentator responding to a network cue on headphones will be most disconcerted to hear his or her own voice returning perhaps several hundreds of milliseconds later. A simple solution, but one which requires a little dexterity, is to switch the commentator's monitoring to a local output at the moment the commentator begins to speak. This is clearly inappropriate if a conversation is required with the remote studio and in this case the destination must send a clean feed (mix-minus) along the associated reverse circuit and the contributor forgo his or her own voice for the duration. This may, in itself, prove troublesome in noisy environments, when commentators are unable to hear their own voices on headphones. It is therefore useful if local sound can be mixed with the clean feed at the OB site to restore a natural and reassuring balance. Some modern self-operated commentary equipment allows contributors to set up headphone feeds to their own satisfaction.

Cue lights from the control area provide an additional or emergency facility in radio, while in

television a floor manager relays information, often by means of cryptic hand signals out of sight of the camera.

Stand-by equipment The immediacy of live broadcasting demands that reasonable precautions are taken against equipment failure. At very least, some spare microphones should be provided, other apparatus being duplicated as may prove expedient or in line with company policy. If possible, battery operation should be available as insurance against loss of the electricity supply (see Section 5.5.4.5).

5.5.9 Shared facilities

When high-grade equipment is properly installed, adequately powered and competently maintained, any self-contained mobile sound operation should prove reliable and predictable. Often, however, engineers must provide interconnections with equipment belonging to other people. Although this can avoid unnecessary duplication of facilities, it sometimes brings the risk of mutual interference or may even compromise safety. Thus, even if the 'foreign' equipment is technically compatible, interfacing should be implemented such that all parties in a complex operation remain substantially independent.

Some of the more common examples of shared facilities are:

(1) direct connection of audio circuits to powered musical instruments;
(2) dividing microphone feeds, on a stage or elsewhere, to supply more than one user (e.g. for foldback, public address, recording and broadcasting); and

(3) distributing a common sound mix to several destinations (e.g. at a major public event).

5.5.9.1 Musical instruments and direct-inject boxes

As an alternative to placing a microphone in front of a musician's amplifier, signals from a guitar or synthesizer, etc., may be obtained by the use of a direct-inject (DI) box (*Figure 5.23*). This normally intercepts the unbalanced jack lead between instrument and amplifier, delivering a balanced output to the mixer, usually at microphone level. Passive DI boxes commonly use a step-down transformer to effect the necessary impedance matching, balancing and attenuation. Active boxes, powered by batteries or the mixer's phantom supply, achieve the same result by means of an amplifier. Both types allow separation of the instrument and mixer earths to avoid hum loops, and some offer additional facilities such as phase reversal, extra attenuation and filtering. Certain stage amplifiers provide their own direct outputs, the signals of which may contain further electronic effects chosen by the musician. Where such an output is available, it may be possible to dispense with a DI box, but this should be done only if hum does not result and electrical safety is not jeopardized.

5.5.9.2 Distributing low-level signals

Splitting microphone and other low-level signals requires care, since all cables must remain screened to prevent noise pick-up. *Figure 5.24* shows the schematic of a typical passive splitting box. There is one direct input/output circuit whose continuous earth provides screening all the way to the microphone and allows phantom powering where necessary. Isolated outputs, at slightly lower level, are

Figure 5.23 *Example of a passive direct-inject box.*

Figure 5.24 *Typical passive microphone splitter.*

taken from the secondary windings of a transformer. On the latter circuits, screens are earthed by their respective mixers but are discontinued inside the box to avoid hum loops. Series resistors (R) are often included in the outputs so that changing the load on one does not seriously affect the others. Larger values of R improve this isolation but reduce the signal level, and vice versa. For most purposes, R may be a few hundred ohms.

Active splitters are also commonly used, their outputs being even more effectively isolated. Their other advantages are that they present microphones with a constant load impedance and may amplify signals and send them to line at a considerably reduced impedance. If situated fairly close to the microphones, therefore, they can immunize signals against noise induction and high-frequency losses on subsequent long cable runs. Dedicated (non-splitting) microphone amplifiers may also be used for the latter purposes, some types being capable of remote gain adjustment from the sound control position.

5.5.9.3 Distributing high level signals
Mixer outputs and similar high-level signals are comparatively easy to handle. If circuits are properly balanced, they are not seriously prone to noise pick-up, even on long cable runs; thus, screening may be dispensed with and hum loops avoided. Simple, unity-gain amplifiers are used to distribute high-level signals. They should offer a high input impedance, to

allow several to be connected across the same circuit without loading, and a low output impedance for minimizing losses due to cable capacitance, etc. At major events, a central bank of distribution amplifiers may be installed to provide 'clean' sound for a large number of users (*Figures 5.25* and *5.26*).

5.5.9.4 Non-standard equipment
Occasionally it will be necessary to interface with non-standard or domestic equipment and difficulties may arise from combinations of unusual signal levels, impedances and connectors. In practice, there are very few problems which cannot be solved by judicious use of transformers, attenuators, etc., and experienced engineers often carry a selection of these useful accessories to help in such eventualities. As a general rule, audio signals are best fed from a low impedance into a higher one, and this should be remembered when selecting adaptors and components for temporary use.

5.5.9.5 Electrical safety
Whatever the complexity of a technical rig, and however tight the schedule, matters of electrical safety must be given top priority. This is certainly true when facilities are shared, since linking different sets of equipment, often widely separated, can lead to unforeseen problems even if only through poor liaison between engineers. The topic is wide-ranging, but a few basic rules should be emphasized.

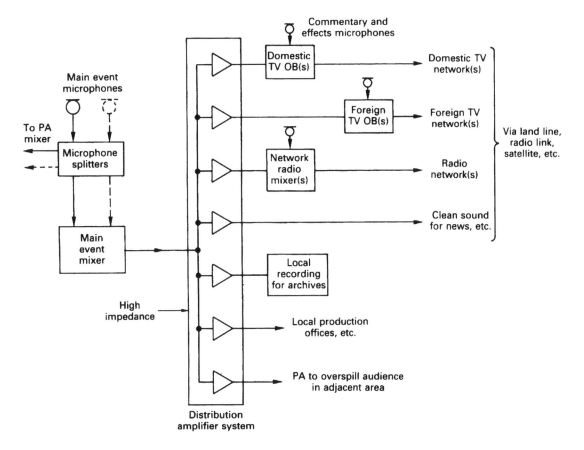

Figure 5.25 *Typical sound distribution at a major outside broadcast. For security, spare circuits would be set up and for stereo all lines would be duplicated.*

Figure 5.26 *Sharing facilities at a major outside broadcast. On the left are passive microphone splitters, on the right a bank of distribution amplifiers. (Courtesy of the BBC.)*

(1) Keep all equipment in first-class condition, regularly inspecting and testing power connections in line with statutory requirements.

(2) Ensure that all exposed metalwork is securely earthed. Never arbitrarily disconnect earth wires as a quick cure for hum loops.

(3) Be aware that equipment in other areas may be powered from a different supply phase (see Section 5.5.4.6). Avoid running power cables between such areas.

(4) Thoroughly test all available safety apparatus, e.g. residual current devices, and observe local regulations.

5.5.10 Specialized vehicles

There are occasions, not least in radio broadcasting, when location work is best served by sound vehicles of a more specialized nature. These may be used to supplement conventional mobiles or else replace them for specific types of programme.

5.5.10.1 Mobile radio studios

Several of these have been designed and operated by the BBC over the years. Recent examples are 11 m in length and provide most of the features of a static talks studio, including an interview area, tape, disc and telephone facilities, TV monitoring and extensive communications. A 30-channel stereo mixer accepts up to 24 outside sources, providing individual talk-backs and clean-feed (mix-minus) as required. At the rear of each vehicle is a dedicated radio links area, with a 10-m pneumatic mast to carry transmitting and receiving aerials. Commentators have the option of sitting in the cab or even on the roof, which has been reinforced and fitted with guard rails for the purpose. Mobile studios are widely used for broadcasting political conferences, major sporting events and state occasions (*Figures 5.27 and 5.28*).

Figure 5.27 *Mobile radio studio at a sporting event. Communications aerials are mounted on the roof, which is also guarded and reinforced for use by commentators.* (Courtesy of the BBC.)

Figure 5.28 *Control area of mobile radio studio. The profile of the desk allows it to be placed along the side of the vehicle without blocking the passageway. The loudspeakers are unusually high, but are mounted upside-down so that the tweeters are close to ear level.* (Courtesy of the BBC.)

5.5.10.2 Studio/recording caravans

During exceptionally busy periods, extra mobile studio facilities may be required. Holding additional fully equipped vehicles in reserve is clearly uneconomic and, given that simpler arrangements are often quite satisfactory, an alternative is to make use of studio caravans. These are trailers, perhaps 7–9 m in length, divided into two or more sections and able to accommodate the basic furnishings of a recording suite. One area will be nominated as the studio proper and equipped with tables, chairs, microphones and headphones, etc. In the other main compartment, portable equipment will be set up to provide mixing, recording and editing facilities appropriate to the work in hand. Studio caravans may lack such luxuries as sound insulation and air-conditioning, etc., but are nevertheless both adequate and cost-effective in the situations where they are likely to be used.

5.5.10.3 'Road show' vehicles

A number of popular travelling radio programmes are broadcast in front of a live audience and hosted from large trailers convertible for use as performance areas. By opening out panels in the side of the coachwork, stages are produced to accommodate disk-jockey equipment and a high-powered PA system. Control and monitoring take place in a separate, isolated compartment, where an engineer mixes the outputs of the presenter's console with audience microphones, etc., as required. Road show vehicles are painted attractively to advertise their network and reflect the style of its programmes.

5.5.10.4 Radio cars

For quick response to newsworthy events, few vehicles can match the effectiveness of a radio car. Despite the blanket description, radio cars can take many forms, ranging from actual estate cars to light commercial vehicles. Programme requirements are often very simple, with perhaps one or two reporters or interviewees inside the car or standing nearby. Audio equipment is correspondingly basic and connected to transmitting and receiving apparatus communicating with the studio centre. If the link equipment is terrestrial, a vehicle will be restricted to a prescribed area, within which raising the aerials on a telescopic mast is likely to improve propagation. When equipped with a satellite terminal, however, a radio car is able to operate virtually anywhere with a clear line of sight to the appropriate area of sky. Some local radio stations mount more complex outside broadcasts using the radio car principle, though

it could be argued that their vehicles are really just small conventional sound mobiles fitted with on-board radio links.

5.5.10.5 Radio link vehicles

Radio links are commonly used to transmit programmes from an outside broadcast to a studio centre, but where an outside-broadcast (OB) site covers a very large area they are also convenient for internal communications. For example, operators carrying sound effects microphones may follow competitors around a golf course, transmitting to a base station and receiving talkback via headphones. Suitably equipped, they may also be able to arrange interviews with celebrities, thus providing a valuable resource to the programme producer. Where complexity demands, all radio links may be conveniently installed in a single vehicle and supervised by specialist engineers.

5.5.10.6 Mobile generators

These provide a power supply where no suitable local source is available (see Section 5.5.4). Smaller types are mounted on trailers; larger versions are built as complete dedicated trucks, able to cater for the substantial requirements of TV lighting, etc., as well as technical equipment. In all cases, generators must deliver a good sinusoidal output, free from harmonics which could cause interference. They should also be effectively muffled to minimize vibration and mechanical noise. The ability to synchronize their output frequency with that of the local supply is often helpful in reducing hum on both sound and vision equipment.

5.5.11 Video facilities

The ability to watch a programme's vision output is fundamental to the television sound engineer. At very least it permits rapid and appropriate response to events on the screen, but more importantly it allows the sound to be matched artistically to the pictures so that a composite illusion is created for the viewer. At outside broadcasts, as in the studio, sound control areas will be equipped with a main transmission monitor, together with one or more preview monitors for watching selected cameras, etc., as required.

Where a separate audio vehicle is employed, individual vision circuits may be extended from the main scanner for each monitor source. Alternatively, the sound engineer may select preview signals remotely using a switching device to generate special control data. With the latter system, only two video circuits are required: one for the transmission picture, and the other for the output of the remote switching matrix.

Even in radio, the ability to see what is happening in front of the microphone can be invaluable, especially if a performance is spontaneous or there has been inadequate time for rehearsal. Where an event is also being televised, it is usually possible to obtain a courtesy feed of a suitable picture; failing this, closed-circuit cameras may be employed.

Joining areas together with video cables is not without potential problems, since the circuits are not balanced and unwanted coupling may occur between the earths at each end of the line. If the ends are remote from one another, a hum loop may result and interfere with vision and/or sound. The choice of remedy will depend on circumstances. Sometimes the receiving end can be terminated by an amplifier or monitor which is not locally grounded; alternatively, a remote closed-circuit camera may be powered via a mains isolating transformer, effectively removing its earth reference without compromising safety. Under certain circumstances, hum on vision circuits may be reduced by inserting a *longitudinal stop coil*.

Picture monitors are notoriously sensitive to spurious magnetic fields, and the close proximity of loudspeakers can often upset their colour purity and scan linearity. In the confined space of a control vehicle, this can prove an acute problem and sometimes the only remedy is to shield the monitors with magnetic screening material.

Acknowledgement

The author is grateful to BBC Radio and Television Outside Broadcasts and the Corporation's Engineering Information and Communications Departments for their valuable assistance.

Bibliography

DEPARTMENT OF TRANSPORT, *Road Vehicles (Construction and Use) Regulations,* current edition, HMSO, London (1986)

5.6 The tapeless studio

We define a 'tapeless studio' as one in which the primary recording and editing medium is not tape.

5.6.1 History of development

In 1985 the only two disk-based digital audio workstations commercially available were the AudioFile from AMS and Direct-to-Disk from New England Digital (NED). Denon also had a system called the DN-052ED, but initially this was only used for in-house mastering for compact disc (CD) and was not made commercially available until a couple of years later. Both the AudioFile and Direct-to-Disk were multichannel systems, both provided eight channels and both were based on custom hardware. But whereas the AudioFile also had a custom user interface, the Direct-to-Disk used a monitor, keyboard and mouse. In fact at first it could only operate as a slave to the Synclavier music sampling/sequencing system and NED marketed this combination as 'the tapeless studio'.

This caused quite a stir since the implication was that the tapeless system was going to replace the multitrack tape machine. Although this generated a certain amount of interest, it put tape machine manufacturers on the defensive as well as those studios which had already invested heavily in digital multitrack tape. Indeed, NED strove to provide more and more channels in order to complete directly in the higher end multichannel music recording market, but the technology at the time proved inadequate and NED consistently encountered difficulties with their larger multichannel systems. The safest bet therefore was to stick with eight channels and find a new approach to marketing the system.

AMS on the other hand, did not make such claims and concentrated straight away on applications where disk-based systems have the advantage such as in audio post production. In this area in particular, the AudioFile did much better than the Direct-to-Disk. This was almost certainly due to several factors. Firstly, the AudioFile was marketed as an adjunct to the multitrack tape machine and was therefore not perceived to be in direct competition with it. Secondly, the AudioFile provided operational control which was more familiar to those who were used to conventional editing methods and thirdly, the system was stand-alone. The Direct-to-Disk, on the other hand, did well as a natural extension to existing Synclavier systems owned by higher-end musicians/studios but the use of the keyboard/mouse and musical keyboard was generally met, in Europe at least, with resistance by the audio post market.

From 1985 to 1988 growth in the number of systems was slow and world sales were relatively small. However, the launch of the AMS and NED systems

encouraged a number of manufacturers to develop their own products (the typical development time being 2–3 years), so that by early 1989 a further 17 systems had been launched onto the market. Among these were three new multichannel systems which were all generally aimed at the audio for video post market. These were again based on customized hardware and had customized user interfaces. One or two flirted with the music market but there was still the perception that it was a choice between the digital audio workstation and multitrack tape. However, there was news of at least two systems which claimed to provide up to 64-channel operation, but these systems obviously ran into difficulties because neither was launched at the time.

In addition, a number of stereo systems came onto the market. These were generally aimed at bringing stereo recording/editing to areas which did not require expensive multichannel systems or time-code slaving, such as CD premastering, pop-music editing and editing for radio. Some of these systems used custom hardware and user interfaces but some were based around the personal computer. This provided a convenient hardware platform and user interface on which to develop a basic system. The best known of these stereo systems was the Digidesign Sound Tools, a Mac-based package which was largely responsible for bringing tapeless technology to the masses.

A couple of cart replacement systems were launched and these used customized hardware and user interfaces. These did not have much success, probably due to their operational methods being unfamiliar to the broadcast market and the features they offered did not fully exploit random access/non-destructive editing. In other words, people could not see how they would improve working methods.

By 1990, a few more systems had come onto the market bringing the total number of systems available to 29. These new systems generally had the same applications as before. Until then, the only types of system which were very application specific had been the cart replacement systems, but 1990 saw the introduction of a few other systems aimed at specific applications such as noise reduction and automatic dialogue replacement for film and TV. In general, the increased performance and reduction in hardware costs saw a growing popularity in personal computers (PCs) with more PC-based stereo systems and a couple of PC development systems coming onto the market. In addition more systems in general, even if they used custom hardware, were starting to use the PC as the user interface.

By mid-1991 the number of systems on the market had more than doubled to 74. Application-specific systems had increased, particularly in the area of cart replacement and station automation, audio postproduction for film, music recording for MIDI sequencers and video spotting. By 1998, the number of systems launched was well in excess of 300. Many earlier systems were no longer in production, a number had changed ownership, and most stereo editing systems had been replaced by multichannel versions. It is now possible to avoid tape completely, from acquisition, through editing to delivery, and whatever the budget or application, there are many digital audio workstations to choose from.

There are high-cost proprietary systems with custom controllers, real-time processing, sophisticated synchronization and machine control features, integrated automated mixing and custom archiving/backup systems, to low-cost card and software packages for the personal computer, which generally rely on a keyboard and mouse for operational control. Largely depending on cost, such packages may offer less sophisticated synchronization and machine control features, non-real-time processing, and no specific archiving system, but will still include many useful time-saving features, and will above all, maintain high-quality digital audio.

For those interested in the highest quality audio, an increasing number of systems now support 'high density' 24-bit/96 kHz audio, whereas for those who do not require this, yet more systems offer optional compression for application such as DVD mastering, interfacing with broadcast delivery systems or for faster transfer of audio via telephone or the Internet.

While the data rates and storage capacity of the fixed hard disk ever increase, and its cost reduces, it remains the most popular medium for tapeless editing systems. The removable magneto optical (MO) disk as well as removable hard disk drives are often available as an alternative, but for multichannel operations their performance still cannot compete with the fixed hard disk.

5.6.2 Software

An increasing number of software packages are now available in both Mac and PC versions, and as disk performance ever improves and the power of the personal computer grows, users can turn their own personal computer into an audio editor simply by purchasing a relatively low-cost software package which relies on the audio I/O of the host platform. Multitrack operation is now the norm rather than the exception, and such software packages will generally allow the editor to use as many tracks as the hardware permits. The advantage of these packages is that they can take advantage of off-the-shelf hardware with readily available spares and peripherals, but the disadvantage is that additional hardware must be sourced if professional audio I/O and synchronization are required.

Relying solely on the host can lead to variable results, particularly in terms of processing speed, so for more guaranteed performance there are various audio engines available, some comprising a card, perhaps with an external interface box, and some a standalone unit containing all required processing, storage and I/O. Most are also bundled with proprietary editing software, although many also support third-party software, particularly in the form of plug-ins. These can cover a comprehensive range of applications including machine control, EDL management, effects processes, and file conversion utilities, and allow the user to customize a platform for specific applications at a relatively low cost. It should be noted, however, that their performance, particularly if used in unusual combinations, cannot always be guaranteed.

5.6.3 Hardware

Although friendlier operating systems such as Windows 95/NT have helped in improving the user interface, many manufacturers have also recognized the need for more intuitive operational control. More support optional dedicated controllers for machine control, editing and mixing, be they proprietary or third-party solutions, and an increasing number of systems are also eliminating the time wasted waiting for the VTR to spool by providing integrated non-linear video, or interfaces to external digital video disk recorders.

For those who prefer not to have to configure a system themselves, most engine and/or software suppliers also offer turnkey packages, which at least ensure that the platform and storage have the correct specifications. Nonetheless, for many, the optimized proprietary system is still an attractive solution. There are proprietary products for a wide variety of applications, ranging from the large integrated recording, editing and mixing systems complete with dedicated console and multi-format surround mixing to the studio-in-a-box style compact desktop system, complete with built-in hard disk and editing mixer

controls. Then there is the new breed of simple tape replacement unit or dubber, most of which allow multiple units to be chained for more tracks.

5.6.4 Basic requirements of a tapeless studio

To define the tapeless studio first requires that a studio itself is defined. Conventionally, this has generally been understood to be a building of sufficient size in which recording, editing, mixing, mastering and/or archiving could take place. Depending on the application, there could be separate rooms (and even specialist personnel) dedicated to each of these functions. For example, a television studio would be quite different from a music recording studio. In television, 'studio' is considered a global term covering a range of activities that are carried out in different physical locations by various people, whereas in a music studio, almost all activities (except acoustic recording) are carried out in the same room by the same person(s). Thus, to define a truly tapeless studio for one type of application may be quite inappropriate for another. However, there are certain functions which cover all requirements of any generally accepted studio application. These functions include, recording, playback, editing, mixing, processing, mastering, compiling, archiving, transferring and conversion.

Depending on the application, a studio will be complete by being able to perform some or all of these functions. The truly tapeless studio will be able to perform them without having to use tape at any stage. However, the extreme ends of the audio manipulation process, i.e. location recording and archiving, can be considered as grey areas since these activities still mostly use tape.

5.6.5 Basic processes

5.6.5.1 Recording
This covers the acquisition of source material. The source material may be recorded 'live', e.g. location recording or musicians in a studio, or may be provided in a prerecorded form, e.g. a sound library or archive material. Depending on the source, recording may require analogue to digital conversion or direct digital transfer. Other recording functions may include the ability to 'drop-in' either manually or automatically for automated dialogue replacement (ADR) or musical purposes. The number of simultaneous record inputs required will depend on how many individual channels are to be recorded simultaneously. This may not be the same as the number

of simultaneous replay channels required. For example, recording a live band may require eight or more simultaneous channels, whereas a sound track consisting of eight channels may be constructed from a number of stereo sources which do not require to be recorded simultaneously. For location recording, the tapeless system must be portable, robust, have its own power source (if none is available) and either have sufficient recording capacity or removable media. For multichannel recording, the tapeless system must provide a sufficiently high number of input channels.

5.6.5.2 Playback
This covers the ability to replay recorded audio in order to establish whether it has been recorded satisfactorily and to gauge the effects of any editing which may have been carried out. For the purpose of this section, the term 'cue' means a segment of audio which has a defined 'in' and 'out' point. Other common terms for cue include segment, clip, take, cut and event. Depending on the application, playback requirements will vary. For example, for cart replacement the playback requirement would be to trigger cues from their beginning, either manually or automatically. For audio post production for video the requirement would be to replay the audio from any point, in synchronization with the picture.

5.6.5.3 Editing
The term 'editing' can be used broadly to describe any change which is made to the source audio and can include mixing or processing of any kind. However, for the purposes of this discussion, the term 'editing' is used to describe the process of marking audio for basic random access functions such as marking 'in' and 'out' points, cutting, pasting, deleting, copying, moving, etc. For applications such as cart replacement, editing need only consist of being able to mark 'in' and 'out' points for a cue. For classical music editing, editing may involve more complex features including the use of customized crossfades.

5.6.5.4 Mixing
For the purposes of this discussion, this term is used to describe the processes of level adjusting, pan control and equalization as well as combining two or more streams of audio.

5.6.5.5 Processing
This covers the ability to change the audio in other ways apart from editing or mixing.

5.6.5.6 Mastering

Mastering requirements will vary according to application. For most applications, masters are stereo recordings which are then encoded in the appropriate format for replay/transmission. For music mastering, this can involve producing a stereo master which must be PQ encoded and then sent to a CD pressing plant for mass production. For TV broadcast or film applications the master may comprise more than two tracks or may require a particular surround format.

5.6.5.7 File transfer

Material may need to be transferred from one system to another, be it for sound library purposes, transfer from source recorders, transfer between different editing/mixing systems, or transfer to delivery systems. For music recording, for example, the artist may wish to record at one studio and mix at another. For post-production, material may be required to be transferred between different editing suites for specialist functions such as ADR, sound effects, music editing or foley. For broadcast, material prepared in the production studio must be transferred to a delivery system, be it in-house or located elsewhere.

5.6.5.8 Compiling

A master may consist of more than one recording. For music purposes a master album will comprise the individual tracks sequentially arranged in the order in which they are to appear on CD (or other final product). For film purposes a master may comprise sequentially recorded scenes and in radio broadcast, a master will comprise a playlist of cues.

5.6.5.9 Conversion

Material may not always be provided in the desired format or on the desired medium. There must therefore be a suitable method of converting material from one format or medium to another.

5.6.5.10 Archiving

Conventionally, the master is often used as the archiving medium. If tape is being replaced, then it is essential that if the material is to be used or reworked at a later date it must be suitably archived. In addition, for post production work especially, it is essential to have access to a sound library and this must be stored on a suitable medium.

5.6.6 Available tapeless technology

5.6.6.1 Recording

Mobile recorders are now available in all shapes and forms, from laptop systems to custom-designed compact units which resemble the more traditional mobile recorder in appearance. The first major leap into digital location sound recording came with the introduction of the portable DAT recorder. The next step was to allow the acquisition media to be directly transferable into a post-production station for editing, hence the introduction of various products using PCMCIA hard disk, which can be removed and inserted directly into a PC to be edited. There are also mobile recorders which use PCMCIA flash memory, and recorders based on Mini-Disc (MD). As an alternative to physical transfer, an increasing number of recorders also allow transfers back to base via telephone or ISDN. Until recently, the digital audio workstation has found it difficult to compete with the multitrack tape machine in terms of providing a high number of input channels, but there are instances where a tape machine with a certain number of channels can be replaced by a digital audio workstation with a smaller number.

This is because with a tape machine, each track (or channel) is generally assigned to a particular instrument or sound type and is output to the mixing desk which must also have the same number of input channels as is used on the tape machine. Each channel on the mixing desk is set up specifically for the sound type routed to it and automation may be available to control the sound dynamically. But if each sound type is not playing continuously there is a certain amount of redundancy in the tape and mixing desk channels. It may even be the case that at any instant, the number of sound types playing simultaneously may never reach the number of channels being used in total (*Figure 5.29*).

With a digital audio workstation, the times when the sound type is actually playing can be defined

Figure 5.29 *Linear tape-recording with redundant space.*

as events and some systems allow any event to be routed to any channel at any time (provided that the total number of sounds being played does not exceed the total number of channels available). In addition, some systems allow individual mix settings for each cue, and these settings will follow the cue no matter which channel it is being played through. Thus, for example, a composition consisting of 12 different sound types, but in which no more than any eight are ever required to play simultaneously, can be successfully arranged and mixed within an eight-channel digital audio workstation (*Figure 5.30*).

Figure 5.30 *Non-linear recording.*

Although the method used in *Figure 5.30* allows 12 sound types to be arranged in an eight-channel system, in operational terms, there are several drawbacks to this solution. Firstly, depending on how frequently a sound type occurs, marking it out as cues can be laborious. Secondly, if a system only has as many editing tracks (i.e. internal software tracks) as there are channels, then it may be confusing to have two or more different sound types occurring sequentially on the same track, especially when a sound type is to be soloed for mixing purposes. Thirdly, depending on how many cues there are for a sound type, assigning a mix setting to each can be laborious.

A number of digital audio workstations have solutions to one or more of these problems. In some instances the marking out process is unnecessary, particularly if cues are taken from a system's sound library and have thus already been defined. However, at least three systems can automatically generate cues by sensing audio above a certain level either during the recording process or after recording has been completed. The system marks the start of a cue on sensing an input level and marks the end on sensing 'silence'. A generic cue name can be given and this is automatically updated numerically when the next cue is generated.

Although there are a number of digital audio workstations which provide more internal editing

tracks than actual channels, in most cases the editing tracks must be assigned to available output channels, so that a 99-track system with eight channels will only allow eight of the 99 tracks to be selected for replay. However, some of these systems do not have this limitation and will automatically allocate any free channel(s) to any track(s) which requires it.

An alternative to using a digital audio workstation for arranging short cues is to use RAM-based samplers. These are generally used for musical purposes, but can also be used to provide sound effects. They can provide high polyphony, allow editing and processing to varying degrees and are useful as sound libraries. The amount of recording capacity is limited by the amount of RAM in the system, so a practical solution is to use the sampler for short sounds and the digital audio workstation for longer ones as shown in *Figure 5.31*. Both the digital audio workstation and the sampler could be synchronized via MIDI using an external sequencer which would trigger cues from both systems by using MIDI events. Some digital audio workstations can support third party sequencing packages which can operate within the digital audio workstation, thus allowing the digital audio workstation itself to trigger sampler cues. In addition, a number of digitial audio workstations integrate both sampling and disk-based recording, allowing both sampler and disk-based cues to be sequenced within the same operating system.

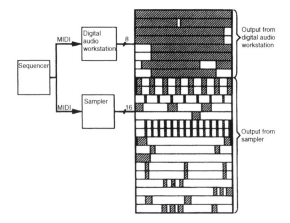

Figure 5.31 *A digital audio workstation synchronized with a sampler/sequencer.*

So, although the digital audio workstation may not always be able to compete with the multitrack tape machine in terms of numbers of simultaneous inputs, digital technology can offer alternative and more flexible solutions in cases where multichannel

playback, rather than recording is required. It should also be noted that a growing number of disk-based alternatives to the multitrack tape machine are available, in the form of modular tape replacement units or dubbers. Typically available in record/playback or playback-only versions, such systems generally allow multiple units to be chained in order to increase the number of tracks. Some use removable media, so that material can be transferred to an editing system, and a growing number support direct playback of various file formats, thus allowing playback or files that have been generated on different makes of editing system.

5.6.6.2 Playback

Most digital audio workstations have as many analogue outputs as they have channels, but in some cases, the number of simultaneous replay outputs may not correspond with the number of channels which are to be replayed simultaneously. For example, some digital audio workstations aimed at production for radio will allow eight channels to be mixed internally to a stereo output. The degree of sophistication of playback features will depend on the application(s) for which a system is designed.

For example, there are two basic types of cart replacement system. The first is similar in design to the cart machine and allows recording and replay from floppy disk, CD or MD. The controls on the unit are simple and allow manual triggering or remote triggering via contact closures. There are usually two types of unit for such a system, namely a record/edit module and a playback-only module. A studio will normally have one record module and several playback modules. The record module allows recording and simple editing for defining 'in' and 'out' points of cues. The playback module will allow cues to be replayed only from their 'in' points.

The second type of cart replacement system will generally use a personal computer for controlling all functions such as recording, editing and replaying cues. Such systems also allow the cues to be sequenced or playlisted in any order of playback. This means that an event time is assigned to each cue, or to a block of sequenced cues, as in the case of an advertisement break for example. In simpler systems, if event times cannot be assigned, cues can often be playlisted by telling the system what the order of playback should be. Cues are then sequentially played, one immediately after the other.

Both types of sequencing are also possible with systems which are aimed at other applications. For audio to picture, cues can be assigned event times

which are displayed in time code. For music purposes, these event times will be displayed in measures/beats and some systems offer a selection of timing displays.

As already mentioned, a number of digital audio workstations support MIDI and can therefore be controlled by external MIDI devices. MIDI events can be used to trigger cues and MIDI time-code (MTC) can be used for synchronizing and locating purposes. Such systems can slave to picture by employing an LTC/MTC converter, however, the resolution of MTC may not be high enough for some applications. Some systems, although aimed primarily at music, will have an LTC interface so that not only can they slave to picture, but can also slave to multitrack tape machines for which LTC is the usual method of synchronization.

Digital audio workstations aimed at audio post production for picture will have at least an LTC interface, if not VITC (vertical interval time code), house sync, pilot tone and others. These will allow the digital audio workstation to slave to picture and the general trend is also to provide an RS422 interface which allows the digital audio workstation to control the transport of the VTR. In cases where a digital video disk recorder is used, this can lead to extremely fast locating and synchronization.

Because it has random access, the digital audio workstation can offer other playback functions which a tape machine cannot. Examples include the ability to loop for atmospheres (without having physically to copy any audio) or instantly return to a user-defined point for ADR or other performances which require repetitive auditioning. Because disk and RAM do not stretch or wear, systems based on these media are also far more reliable than those based on tape in terms of maintaining a correct time-base.

5.6.6.3 Editing

Whatever the editing function required, there will be a tapeless system capable of it. As a general rule, the amount of reliability, polyphony and speed of a system will be reflected in its cost. However, it is not essential to have just one digital audio workstation for all applications within the studio. It may be the case that a low-cost digital audio workstation can be used for preproduction work which can then be transferred to a more sophisticated system for editing, synchronization, mastering or multichannel purposes.

Conversely, operations which normally require multiple systems can often be done with just one digital audio workstation. An example of this can

be found in cart replacement. Conventionally there would be three separate stages, firstly recording and editing, then transferring to cart and finally replaying a playlist from multiple cart machines. However, with a suitable tapeless system, all recording, editing and replay operations can be carried out and automated within the one system without the need for transferal.

As with any market, increased competition is bringing prices down and/or increasing the number of functions offered by each system for the same price. Amongst the huge range of systems available are some whose operation is designed to be familiar to a particular type of editor and because of its non-destructive editing capabilities and flexibility, the digital audio workstation will generally provide overall savings in time, hardware and, in many cases, personnel.

5.6.6.4 Mixing

Many digital audio workstations offer internal mixing, but the features offered will vary greatly from system to system. Some will offer real time digital mixing with automation and surround sound whilst others provide control of the level at the analogue output stage, non real-time equalization and perhaps 'snapshot' automation. With such systems, it must be considered that level changes will not affect the audio if it is transferred to another system in the digital domain. The most frequent complaint about digital audio workstations in terms of mixing is the lack of tactile control – many users find it cumbersome to perform dynamic functions using a mouse and screen. However, some systems provide integrated mixer sections comprising faders, pots, buttons and level meters. Others provide optional tactile controllers or will work with third party controllers.

If internal mixing is not provided, or is insufficient, then an external mixer must be used. If using an analogue console for this purpose, then the audio must be sent from the digital audio workstation to the console via analogue outputs or external DAs (digital-to-analogue converters). However, for those who wish to keep audio in the digital domain, audio can be transferred from the digital audio workstation to a digital mixer via digital interfaces. Depending on which interfaces both systems have in common, there are several types which can be used. The most common is the AES/EBU (Audio Engineering Society/European Broadcast Union) interface which is designed for stereo transfer. For multichannel transfer, multiple AES/EBU interfaces can be used but

MADI (multichannel audio digital interface) would be more suitable or proprietary multichannel interfaces such as Sony and Yamaha.

It should be noted that if two or more digital signals are added together, they may result in a signal which has a larger word size than its component parts. This will cause a problem if, for example, a mix consisting of greater than 16-bit audio is to be directly transferred to a CD mastering device which only accepts 16-bit audio. It is therefore essential that the digital mixer is capable of limiting the resolution of the resultant audio.

5.6.6.5 Processing

Whether processing takes place in the digital or analogue domain, there is no need for tape to be involved. Some processes such as reverberation, dynamics, sound reversal and varispeed can be done in the analogue domain. However, these can all be done in the digital domain as well as processes which can only be done in the digital domain such as time compression/expansion and intelligent recognition functions (as used for noise removal, filtering and other specialist applications). Digital audio workstations are tending to offer more and more internal processing, which makes them increasingly attractive. But some processes, such as reverberation and time compression, involve subjective judgement on the resulting quality of the sound and engineers still often use their favourite external devices for such processes. In the case of digital varispeed (as in the case of pitch changing with samplers), it is essential that the processed signal has the same sampling rate as before if it is to be digitally transferred. Thus if the audio is to be digitally transferred, variable sampling rates are not acceptable.

5.6.6.6 Compiling

A number of digital audio workstations, especially those aimed at CD premastering, will allow songs or projects themselves to be compiled in the desired order of replay. Some will also automatically generate the PQ subcode required for CD. For radio applications, most of the cart replacement systems allow cues to be playlisted and many allow playlists themselves to be scheduled.

5.6.6.7 Mastering

For music mastering, a number of systems integrate recordable CD and these discs are being accepted by CD pressing plants as masters, so that the audio never has to be transferred to tape at any stage. In

the case of cart replacement systems, the system itself is used to provide the master material and if systems are networked, the master will reside on disk within the common central store and be triggered internally by whichever subsystem is being used for playback.

5.6.6.8 File transfer

For transfer from source recorders, manufacturers are increasingly offering interfaces such as AES/EBU and S/DPIF for stereo transfer, or ADAT and TDIF for transfer to/from Alesis or Tascam multichannel digital tape recorders respectively. For file transfer from sound libraries or between different makes of system, there are various file conversion utilities on offer, and for faster bulk transfer, various companies have been cooperating on a one-to-one basis in order to provide some form of file compatibility between their respective systems.

An increasing number of systems support OMF (Open Media Framework), not only for transfer between different audio workstations but also between audio workstations and non-linear video editing systems (which all now include some level of audio editing capabilities). However, while OMF is mainly used as a translator between proprietary formats, on a longer-term basis, it is hoped that the industry will agree on a standard native format such as WAVE or the Broadcast Wave Format (BWF). Already implemented by a number of manufacturers, BWF is a version of WAVE that has been designed to be straightforward to implement. It contains the minimum information considered necessary for all broadcast applications, and accommodates both linear and MPEG audio, as well as higher-level descriptors which can be used to refer to other files containing more complex sets of information. Using standard formats, there are also systems which support cross-platform operation, allowing, for example, projects created on a Mac or Windows-based system to be directly played and edited, with no translation required.

In order to facilitate the physical transfer of material, an increasing number of systems support removable hard disk and optical media. These can be used to transfer files between lower-cost 'prep stations' or simple recorders to larger systems of the same make within a studio, or for transfer of whole projects between different makes of system. They also serve as a useful means of archival, particularly where fast changeovers between projects are required.

While removable media are appropriate for situations where systems are not physically connected, the demand for networking and electronic transfer is increasing. This not only removes the need for transporting media, but also allows a more efficient use of resources, such as access to a shared archiving device or printer. Various types of network have been developed recently, using SCSI, ATM, CDDI and FDDI. Some allow multiple access to centralized storage, others allow one workstation to audition multichannel audio from the local storage of another connected workstation, for example, and it is also now possible to transfer audio via wider networks such as the Internet for non-real-time transfer, or via ISDN in real-time using compression.

5.6.6.9 Conversion

There are cases where the source material in terms of raw audio has a different sampling rate and/or format to the one being used for transfer/mastering. In the music industry for example, the sampling rate of 44.1 kHz dominates because this is the rate used for CD. In broadcast however, there may be several sampling rates being used such as 32 kHz for transmission, 44.1 kHz for source material and 48 kHz for mastering/editing. If the material is to be kept in the digital domain, there is a need for sample rate conversion and/or format conversion. Many digital audio workstations offer sample rate conversion and/or format conversion either as part of the main software or as an option using a software plug-in. There are also now a number of peripheral devices designed specifically to do either job or both.

5.6.6.10 Archiving

Although removable hard disk and optical media can be used for archiving, the most cost-effective medium, especially for longer-term archiving, is still tape. However, transfer to tape has the drawback that it requires normal operation to stop while the process takes place, unless a cut-down off-line workstation is available for transferring material from removable media to tape for example. Another solution supported by some systems with multitasking or parallel processing capabilities is to allow transfer in the background either by direct connection to an archiving device, or via a proprietary network to a remotely connected archiving device. Another time-saver is intelligent archiving, which could, for example, not waste time reloading elements which have been detected as already existing on disk.

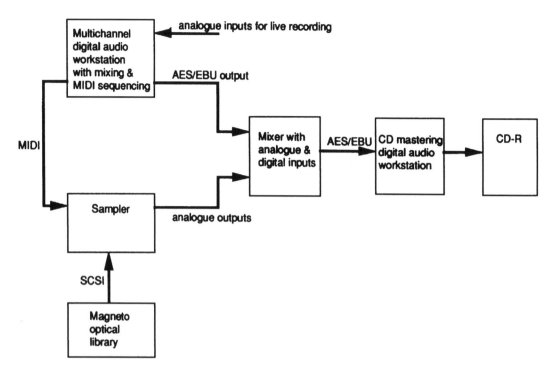

Figure 5.32 *An example of a tapeless studio using analogue and digital equipment.*

5.6.7 Connecting and synchronizing the tapeless studio

5.6.7.1 Basic connections
Now that the essential elements for a tapeless studio have been provided they must be connected in a way which is both practical and problem free. *Figure 5.32* shows an example of a tapeless music recording studio using a combination of analogue and digital techniques.

5.6.7.2 Digital interfacing
Naturally, if the input and output of each system is in the analogue domain connections would be fairly straightforward. In an ideal world, there would be no degradation of the audio were it passed through several stages of analogue-to-digital and digital-to-analogue conversion. However, in reality this is not the case, so where possible, it is desirable to keep the audio in the digital domain and, as already mentioned, a number of digital interface formats are available for direct digital transfer.

5.6.7.3 Synchronization
If digital audio is to be transferred from one system to another in real time, the sampling clock of both systems must be running at the same frequency.

When transfer is directly from one system to another, the sample clock of the receiver usually derives its clock reference from the data provided by the transmitter. In the case of the AES/EBU and SDIF interface, the sample clock is embedded within the data itself.

For multichannel digital audio transfer, MADI is recommended. This differs from the above formats in that it is asynchronous, i.e. the data is transmitted at a constant rate irrespective of what the sampling rate is. It is therefore necessary to have a separate synchronizing clock which can either be generated by the transmitter or a separate device.

In fact it has been suggested that all digital audio systems have a separate synchronization input so that they can be connected to a common synchronization reference. This is to avoid errors in data recovery which may arise when multiple systems are serially connected (each introducing a delay with respect to the original transmitter's sample clock) or where there are multiple, independent sources (such as the inputs to a mixer). The AES has recommended that this digital audio reference signal (DARS) is provided by the AES/EBU format. With systems such as a domestic CD player, there is no input for synchronization, so the output would need to be connected to a synchronizing system which takes

Figure 5.33 *A tapeless studio connected using a digital audio reference signal (DARS).*

its reference from the DARS. An example of a tapeless studio connected using a DARS is shown in *Figure 5.33*.

For audio post to film, some digital audio workstations are now capable of interfacing with film transports so that audio can be edited against film rather than video.

5.6.8 Conclusion

As digital distribution of material increases, be it in the form of CD, MD, DVD, digital broadcasting or Webcasting, so too will the need for more material, faster turnaround times and more access to shared resources. All this makes the use of random access technology for audio and video inevitable. It is not until recently that the true potential of this technology, as a whole, has begun to be unleashed as the major hurdles of limited tracks, insufficient storage, file transfer and long archiving times, have now mostly been overcome. Nonetheless, while tape remains a convenient and cheap acquisition and archiving medium, in practice the truly tapeless studio will perhaps only be as common as the truly paperless office.

Bibliography

AUDIO ENGINEERING SOCIETY, *Proceedings of the 10th International AES Conference*, Berkshire (1991)

RUMSEY, F.J., *Digital Audio Operations*, Focal Press, Oxford (1991)
SYPHA, *A Review of the Digital Audio Workstation Market*, Sypha, London (1991)
SYPHA, *The Tapeless Directory*, 6th edn, Sypha, 216A Gipsy Road, London SE27 9RB (1997). E-mail: sypha@compuserve.com

5.7 Studio air-conditioning systems

5.7.1 Introduction

The term 'air-conditioning' describes the process of treating air in order to control its temperature, humidity and purity. A studio which is air-conditioned must be supplied with the correct quantity of suitably treated air so that the occupants can work safely and comfortably in an otherwise sealed environment. The plant which supplies the conditioned air must be designed and maintained to adjust its output automatically to meet changes in demand brought about by variations in climate and studio use. This flexibility of output must not compromise the efficiency of the air-conditioning system since even moderately sized installations consume relatively high amounts of the overall studio running costs. A well designed air-conditioning system is unobtrusive, the user should be unaware of its presence, i.e. it is not noisy nor too hot or cold. It is draught free and provides a comfortable environment to work in. This section describes the requirements of air-conditioning and how practical systems operate.

5.7.2 Air quality

5.7.2.1 Air pollutants

In order to assess the requirements of an air-conditioning system it is necessary to consider the environment to be treated. A studio for example, which is normally sealed and acoustically treated will contain equipment, furniture and personnel, all of which will contribute to the degradation of air quality. Broadcast equipment and studio lighting consume electrical power and as a by-product give out heat energy. Studio furnishings can emit formaldehyde gas which is, in sufficient concentration, an irritant to eyes and respiratory tracts. The occupants of a studio, in addition to producing heat energy, may also cause considerable pollution. Oxygen is taken from the air and carbon dioxide is exhaled. Perspiration and flatulence are obvious examples of body odours, but water vapour,

cigarette smoke and bad breath can also contribute to the pollution of the air that is expected to be used by all those working within a shared environment. Ionization of the air has been cited as a contributor to poor air quality. Positive ions tend to cause symptoms such as nausea and headache. Negative ions can give a perception of the air being fresh and wholesome.

5.7.2.2 Measuring air quality

Having identified the principal sources of air pollution it is necessary to determine what constitutes an acceptable level of air quality. This is less easy than it sounds. Our perception of air quality is highly subjective. Occupants of a room will tend not to notice a progressive deterioration in air quality unless they leave and then re-enter the room. Even then it is unlikely that two people would agree as to how good or bad the air quality was. Human beings have a highly refined sense of smell, the nose being capable of identifying many hundreds of different and complex odours. However, this sense is not especially objective and can be rendered ineffective by perfume or by an illness such as the common cold. Other pollutants such as carbon monoxide, carbon dioxide and formaldehyde are even more difficult to identify. All three gases would escape detection by humans in the concentrations which are likely to cause ill effect.

There are commercially produced gas sensors available which are capable of measuring air quality. A common design operates on the principle of absorption of oxidizable gases by a semiconductive material, such as zinc dioxide, which is formed into a tube and surrounds a small heating element. The porous lining of the tube absorbs oxidizable gases which causes the conductivity of the material to increase. The change in current flowing through the device is proportional to the level of gases within the air sample and can thus be used to indicate air quality. Experience has shown that, when suitably positioned, such devices are capable of responding to changes in air quality. This has been demonstrated by conducting trials within this type of sensor alongside a conventional carbon dioxide sensor. Whilst sampling a room over a given time period and for different levels of occupancy (and thus differing values of air quality), it has been shown that the output of gas sensors will follow the changes measured in levels of carbon dioxide.

5.7.2.3 Ventilation requirements

Experience has shown that the most critical factor in determining ventilation requirements is the satisfactory dilution of body odours. The minimum ventilation requirement for maintaining carbon dioxide at the recommended maximum allowable concentration of 0.5% by volume, is as little as 0.8 l/s for persons who are seated. However, in order to dilute body odours to acceptable levels, the minimum fresh air ventilation rate per person is normally quoted as 8 l/s, i.e. 10 times the volume flow rate necessary to maintain the carbon dioxide concentration at a safe level. This figure assumes sedentary non-smoking occupants, and would be increased to a minimum of 32 l/s where heavy smokers are present.

5.7.2.4 Summary

Air quality is a very important factor in determining the air conditioning requirements of a studio. Dilution of typical pollutants such as body odours, cigarette smoke and carbon dioxide is essential to maintain a safe, healthy and pleasant working environment. The studio air quality, which may be monitored by proprietary air sensors, varies according to occupancy density and the type of activity performed. Fresh outside air must be continuously introduced into the studio in order to dilute and displace the stale air. This effectively dictates the ventilation requirements for the studio.

5.7.3 Thermal comfort

5.7.3.1 Temperature

One of the most challenging aspects of air-conditioning for any studio is ensuring that all those who use the area perceive it to be comfortable, i.e. that they feel neither too warm nor too cold. This is difficult to achieve in reality and there is unfortunately no simple solution. Individuals sense skin temperature rather than room temperature and, in practice, are relatively intolerant to changes in their environment which promote feelings of warmth or cold in exposed parts of the body.

Through the process of metabolism, humans convert energy in food to heat. Persons seated and at rest will liberate approximately 115 W of heat, emitted as radiation and convection (75%) and evaporation (25%). By involuntary physiological mechanisms, such as shivering and perspiring, we are able to maintain a constant body temperature for a wide range of external conditions. However, thermal comfort tends to be subjective, and may depend on an individual's metabolic rate, activity, clothing and acclimatization. A survey showed that a group of individuals in a common environment, when questioned as to their perception of the temperature, responded quite

differently. Whereas there was no single temperature which everyone described as 'neutral', it was possible to find a point at which no one felt 'uncomfortable'. Between these extremes responses such as 'slightly warm' or 'slightly cool', suggest that, at best, it is only possible to please most of the people for most of the time.

Clothing is obviously a major influence of personal thermal comfort and in practice occupants of a studio should expect to perceive a benefit of several degrees celsius by, say, simply wearing a jacket. The use of clothing to improve personal comfort in a shared environment has limited use however, and clearly cannot be seen as a substitute for accurate temperature control of a studio.

5.7.3.2 Heat losses and gains

Establishing the criteria for creating the ideal comfort zone involves an analysis of the mechanisms of heat exchange that take place between a person and his or her environment. Essentially heat losses or gains may be considered as radiant, sensible and latent. Sensible heat exchanges, which take place through convection, are essentially dry gains or losses, i.e. there is no transfer of moisture. Latent heat entails the transfer of heat via moisture, e.g. heat loss by perspiration.

Radiant heat Radiant heat loss or gain is dependent on the difference between the temperature of the body surfaces, and the mean radiant temperature of the surroundings. The latter is defined as a mean of the surrounding surface temperatures within direct sight of the individual. When these are at any temperature above or below the average surface temperature of occupants, normally taken as 27°C, then radiant heat exchange will take place. Many of the surfaces inside a studio will be at room temperature, but extremes do exist such as incandescent lamps, equipment and the surfaces of external glazing. Where occupants are seated close to such surfaces, radiated heat loss or gain may cause discomfort, but the solution may be to move the individuals concerned rather than adjust the room air temperature, although in practice this frequently may not be feasible.

Convected heat Convective heat exchange is dependent on the velocity and temperature of the surrounding air. When studio air temperatures change the heat flow to or from occupants will depend upon their average surface temperature. As with radiant heat transfer, the temperature difference determines the rate of heat exchange and when both air and body surface temperature are the same there is no transfer of heat.

Air velocity will also contribute to convective heat exchange. By moving air across a studio, convection currents tend to accelerate the mechanisms of heat transfer. There is therefore an optimum air velocity for the comfort zone of any studio, insofar as a high air velocity will cause excessive heat exchange, whilst too low an air speed can promote stagnation and stratification of the air.

The sensible and radiant heat loss or gain experienced by studio occupants is thus dependent on:

(1) the mean radiant temperature of the surroundings,
(2) the ambient air temperature, and
(3) the average air velocity

For studio personnel, engaged in light duties, the typical sensible heat loss to the environment is approximately 100 W for an air temperature of 20°C.

Latent heat In order to keep the lungs moist it is necessary for humans to constantly evaporate moisture into the air which we inhale. This action requires approximately 15 W of latent heat even when the body is at rest. Should the air temperature rise above say 25°C the body will be receiving sensible heat from the air and radiant heat from its surroundings. For even light duties such a heat gain would cause body temperature to rise and so, to maintain equilibrium, the person would begin to perspire. As body glands open to release sweat on to the skin, so latent heat is lost by the body in the process of evaporating the sweat. The mechanism is of course completely self-regulating, and should the temperature of the air continue to rise, then more latent heat is rejected by the body as extra sweat is released. Whilst at low temperatures, sensible heat losses predominate, as the air becomes hotter then latent heat losses become more important. The rate at which the body can lose latent heat is dependent on the humidity of the surrounding air. As the moisture content of the surrounding air increases so it becomes more difficult for the body to evaporate sweat because the air will only hold a finite amount of water.

Thermal indices Consideration of the foregoing mechanisms of heat exchange between humans and their environment leads us to conclude that the thermal comfort of an individual is dependent on:

(1) the ambient air temperature (dry bulb temperature),
(2) the mean radiant temperature,
(3) the average air velocity,
(4) the humidity of the air,

(5) clothing worn, and
(6) activity undertaken.

Clearly it is not just enough to measure and control the dry bulb temperature of the air, although this is possibly the single most important influence on thermal comfort. Instead thermal indices have been devised which combine these variables and give a more reliable indicator of personal thermal comfort.

The four measurements required to assess the thermal environment are dry bulb temperature, wet bulb temperature, mean radiant temperature and air velocity. The terms 'dry bulb' and 'wet bulb' refer to the thermometers used to measure the dry air temperature and humidity. A conventional dry bulb thermometer records ambient air temperature only. A similar thermometer the bulb of which is kept moist, normally by a linen wick and water reservoir will, when in air, measures a lower wet bulb temperature due to the cooling effect caused by water being evaporated from the bulb. The difference between wet and dry bulb temperatures is proportional to the air humidity since the drier the air (low humidity) the more readily moisture is evaporated and thus the lower the wet bulb temperature. Such instruments, sling hygrometers, are commonly used to determine the relative humidity of the air.

The resultant temperature (t_{res}) is the index most widely used in the UK and is given by:

$$t_{res} = \frac{t_r + (t_{ai}\sqrt{10v})}{1 + \sqrt{10v}}$$

where t_{ai} is the inside air temperature (°C), t_r is the mean radiant temperature (°C), and v is the mean air velocity (m/s).

The recommended resultant temperature range for sedentary occupations, such as in studios, is 19–23°C, and providing the relative humidity is maintained at between 40% and 70%, and the air velocity is less than 0.1 m/s, most people will feel comfortable. Where the air velocity is 0.1 m/s, the equation for finding the resultant temperature simplifies to:

$$t_{res} = \tfrac{1}{2}t_r + \tfrac{1}{2}t_{ai}$$

5.7.3.3 Summary
Occupants of a studio are very sensitive to thermal influences. Variations in air temperature, humidity, velocity and radiated heat can all affect how warm or cool an individual feels. Thermal comfort is most commonly achieved by controlling the resultant air temperature of studios within the range 19–23°C.

There are a number of designs of sensor for detecting air temperatures, but the majority operate on the principle that a change in temperature causes the sensor resistance to alter. Thus the current that flows through the sensor is determined by the temperature of the surrounding air, and can be used to control the heating or cooling output.

5.7.4 Humidity

5.7.4.1 Psychrometry
Psychrometry, the study of air and water vapour mixtures, is particularly relevant to air-conditioned spaces, since the moisture content of the air can have adverse effects on occupants and equipment. Accordingly, it is usual to build into air-conditioning systems some means of controlling the humidity of the air supplied to studios.

5.7.4.2 Relative humidity
Air will naturally and readily absorb water from its surroundings. This water is held in suspension as a vapour and will exert a pressure such that the atmospheric pressure in a given room is the sum of the dry air pressure plus the vapour pressure. The vapour pressure is directly related to the mass of moisture in the air.

The mass of moisture that can be held by a sample of air is governed by the air temperature. As air becomes warmer it will hold more moisture in suspension, whilst conversely, as air is cooled so moisture is less readily absorbed. For each temperature there is a point at which the air is unable to absorb any more water, and is thus described as being saturated. The vapour pressure of the moisture in the air at this condition, the dew-point temperature, is referred to as the 'saturation pressure.' It is temperature dependent and relates to the ability of air to absorb moisture.

The term 'relative humidity' (RH) is commonly used to determine the quantity of moisture in air, and is expressed as:

$$RH = \frac{\text{Actual vapour pressure}}{\text{Saturation vapour pressure}} \times 100\%$$

The relative humidity of air is measured by using an instrument known as a sling hygrometer which directly measures the wet and dry bulb temperatures. A scale, supplied with the instrument, is used to determine the relative humidity based upon the wet and dry bulb temperatures.

Consider the following examples of typical readings obtained from a sling hygrometer: dry bulb

temperature = 20°C. Wet bulb temperature = 15°C. From the scale, the percentage relative humidity is 58.5%. This relates to the following pressures:

Actual vapour pressure	13.67 mbar
Saturated vapour pressure	23.37 mbar
At atmospheric pressure	1013.25 mbar

It can be shown that this equates to 8.5 g of moisture per 1000 g of dry air, i.e. 0.85%. So for a typical air sample, a relative humidity of 60% represents air containing less than 1% by mass of water vapour.

For most air-conditioned studios the humidity will be controlled within the range 40–70% RH. Should the humidity of the air be allowed to go outside of this range problems will arise.

High humidity will cause physical discomfort since it becomes more difficult for the body to lose latent heat in a humid atmosphere. So whilst the majority of people would feel comfortable for a resultant air temperature range of 18–21°C and humidity of 40–70%, they would feel unpleasantly warm were the humidity to increase for the same temperature. Another problem associated with high humidity is the condensation that is likely to occur where the humid air meets cool surfaces. This may lead to water damage of furnishings and equipment.

Low humidity (less than 40% RH) can cause a dangerous build up of static electrical charges, particularly where certain synthetic materials are used in the studio furnishings. Simply walking across a carpet containing a high percentage of nylon fibres can cause occupants to build up high static electric charges, caused through the unnaturally dry conditions associated with low RH, which leads to a dramatic increase in the carpet resistance. The shock threshold of 2 kV will increase below 40% RH to approximately 5 kV at 15% RH. Low humidity is also associated with higher than average incidences of infections such as colds and flu amongst occupants of air-conditioned buildings. This is because the nasal passages and respiratory tract become dehydrated as a consequence of breathing in relatively dry air, and so become less efficient at filtering out the invasive organisms responsible for common infections.

5.7.4.3 Summary
Studio air humidity should generally be maintained between 40% and 70% RH. Higher than average humidity can cause condensation and personal discomfort whilst low humidity increases incidences of static electric charge and common colds. Humidity sensing elements may be either of the electrical or electronic type. The former consists of a hygroscopic

element, such as a strand of hair, which changes length according to the humidity of the surrounding air. The element is held under tension by a spring and the movement caused by the change in length operates a switch in a electrical circuit, so sending an output signal. The electronic type of humidistat contains an element in which two electrodes are separated by a hygroscopic material the resistance of which changes with the humidity of surrounding air. The current flow through the element determines the output signal from the device.

5.7.5 Ventilation

5.7.5.1 Mechanical ventilation
In order to satisfy the requirement for air quality control, described in Section 5.7.2, it is essential for studios to be adequately ventilated. Since noise constraints prevent the use of natural ventilation it is necessary to employ a mechanical system. The ventilation plant may also include provision for temperature and humidity control to meet the criteria outlined in Sections 5.7.3 and 5.7.4.

The air is supplied to the studio by a fan through suitably sized ducting and diffusers at a pressure slightly above atmospheric, i.e. at positive pressure. An extract fan removes air from the conditioned space and exhausts a proportion of this from the building, the remainder being recirculated via the supply system. This combination of supply and extract air systems is known as balanced ventilation. By designing a slightly higher flow rate for the supply than the extract, it is possible to maintain a small positive pressure above atmospheric and thus prevent inward leakage of air (and pollutants) from outside of the studio. The supply and extract air flow rates must be balanced in order to maintain the studio pressure within the design limits, since an excessive positive or negative pressure within the studio would prevent the opening of doors.

The flow rates of supply and extract air to the studio will be determined by the designer based on the studio size, and likely number of occupants. It is possible to determine the minimum fresh air supply based on the requirement outlined in Section 5.7.2.3, which recommends 8 l/s per person. A studio suitable for a presenter and up to three other occupants would therefore require a fresh air supply of at least 32 1/s (0.032 m³/h). This calculation assumes that all of the supply air is taken from outside of the building, i.e. there is no recirculation. In practice the air provides more than just ventilation, it is usually temperature and humidity controlled which is an

expensive process. It is common practice therefore to recirculate a percentage of the extract air and make up the remainder of the supply air from outside. The fresh air intake must be at least 10–15% of the total supply air so that the recirculated stale air is adequately diluted.

Since the air supply to the studio is measured in units of volume flow rate it is usual to quote, for a given studio, the number of air changes in every hour:

$$\text{No. studio air changes per hour} = \frac{\text{Volume of air supplied per hour}}{\text{Volume of studio}}$$

5.7.5.2 Air delivery systems

The ventilation system must be designed to deliver the correct volume flow rate of air to the studio with the maximum possible efficiency and at a minimum level of noise. In addition, the air should be distributed through the studio in such a way as to minimize draughts and yet permit thorough scavenging or mixing of the air to ensure that there are no stagnant pockets of untreated air.

Ducted air flow The distribution system must convey the design flow rate of air between air-conditioning plant and studio with the minimum amount of losses and at a low noise level. The flow of air through the ducting causes the pressure in the duct to rise. The air pressure has two components, static pressure and velocity pressure, the sum of which is known as 'total pressure':

$$\text{Total pressure} = \text{Static pressure}$$
$$+ \text{Velocity pressure}$$

Static pressure is due to the force exerted by the mass of air within the duct, whilst velocity pressure is, as its name suggests, caused by the movement of the air pressure through the duct.

Static pressure is measured by attaching a manometer and probe to the side of the ducting. The manometer consists of a U-shaped tube which is partially filled with a fluid, typically water. The changes in static pressure cause the fluid level to rise or fall in the two limbs of the instrument. The manometer may be calibrated to record either inches or millimetres of water gauge pressure. In order to detect velocity pressure it is necessary to insert another probe into the ductwork and direct the open end of this into and parallel to the flow of air. Again by attaching a manometer to the probe, it is possible to record the pressure in inches or millimetres of water gauge, but this time the manometer will sense

both the velocity and static pressures. Thus velocity pressure can be determined by deducting the static pressure reading from the total pressure reading. Electric measuring instruments employing pitot tubes and transducers may also be used to record air pressures within ducts.

The total pressure of air flowing through a duct-work system will decrease as a result of frictional losses and leakage. These changes in pressure are important since the supply and extract fans are sized for a design air flow rate against the effective pressure which the fan develops as it delivers air to the system. The pressure of the air within the duct is a function of the air mass flow rate (\dot{m}), density (ρ) (and hence temperature), velocity (c) and duct area (A), i.e. for air flowing through a duct:

$$\dot{m} = \rho A c \ (\text{kg/s})$$

As air flows through ductwork it will not only decrease its total pressure but, as a result of changes in duct cross sectional area, the air velocity will also change. By rearranging the above equation it will be seen that:

$$c = \frac{\dot{m}}{\rho A} \ (\text{m/s})$$

Given that the mass flow rate and density are practically constant for a length of ducting, a reduction in duct area will thus cause the air velocity to increase. Consequently the velocity pressure of the air will also increase for a reduction in duct area although the static pressure will decrease by a greater amount, giving a nett fall in total pressure.

In practice, air flow through ductwork is most commonly observed by measuring its pressure and velocity using instruments such as manometers and anemometers, respectively. The designer is able to calculate air delivery rates, pressures and velocities for a given system, and engineers may check these figures using the aforementioned instruments. For this reason air distribution systems are fitted with removable rubber bungs at various strategic points to enable air pressure and velocity to be measured conveniently. This is particularly useful when commissioning an air-conditioning system and also when fault finding. Manometers are often fitted permanently to ducts either side of air filters since they are a useful indicator of how dirty the filter membrane has become.

Where it is required to adjust air flow this is most usually by the use of adjustable dampers fitted in the ducts. These devices, normally motor driven, may be set to vary the air flow in the same way as a

valve controls liquid flow. Adjustable dampers of this type should not be confused with fire dampers which are designed to close automatically in the event of fire. They are operated by either a fusible link or an electromagnetic device linked to a smoke detector. The correct siting and operation of these dampers will preserve a building's fire compartments and thus limit the spread of fire and smoke through the air-conditioning system.

Fans Various designs of fan may be used in the supply and extract systems, but the two most common types are centrifugal and axial. The names of these fans describe the direction of air flow through the rotating impeller. The centrifugal design of fan may be further differentiated according to the arrangement of the impeller blading; radial, forward curved and backward curved. Axial fans may be sub-classified according to the position of the motor, i.e. totally enclosed, in which the motor is immersed in the air flow, or bifurcated where the motor is isolated from the air.

The choice of fan for a particular application is made with reference to the fan performance characteristic, which compares power and delivery pressure volume flow. From this data the designer can select a fan that for the delivery required will operate at near to its optimum efficiency. Other considerations influencing fan selection are noise, axial fans being comparatively poor in this respect due to their high blade tip speed, and whether the fan is to be operated at fixed or variable speed.

There are many laws governing the performance of fans, but one of the more fundamental relationships establishes that for a constant impeller diameter, volume flow is directly proportional to impeller speed. Thus if a fan is connected to a variable speed drive system its output may be adjusted across a given range, permitting more efficient operation of the ventilation system for, say, periods of low occupancy.

The final selection process will involve comparing the fan static pressure at the design volume flow rate with the static pressure of the system. 'Static pressure' refers to the pressure that will build up in the duct as the fan accelerates from rest to its operating speed. The static pressure will vary as the square of the volume flow rate for turbulent flow systems. *Figure 5.34* shows a typical fan performance curve superimposed on a system characteristic. The intersection of the two curves identifies the operating point for the fan from which it is possible to determine the volume flow rate delivered by the fan and the static pressure of the system at this output.

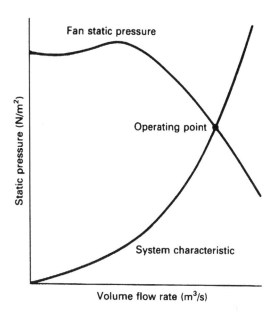

Figure 5.34 *Fan and duct pressure–flow relationship.*

Studio air distribution The air will enter and leave the studio through air terminals. The position of these diffusers will determine the distribution of air through the studio and hence significantly influence the quality of the air-conditioning. Other factors include the air supply velocity, the studio geometry and the difference between room and supply air temperature. Studio occupants, particularly when seated, are very sensitive to the velocity and temperature of air over exposed parts of the body. The system must be designed to limit the air velocity within a given occupation zone to an acceptable level. For most studios this will mean a maximum air velocity of 0.25 m/s reducing in winter to 0.15 m/s. Although there is no recommended minimum velocity, the problems associated with very low air speed are that of stagnation, in which a temperature gradient of more than 3°C between ankle and neck causes noticeable discomfort to occupants.

The establishment of occupation zones within studios is necessary because outside of these zones the air velocity needs to be much greater in order to obtain satisfactory air displacement from the studio, i.e. to meet the required number of design air changes per hour.

The location of the supply and extract outlets will depend on the following considerations:

(1) The studio design and in particular the predicted occupation zone so that the airstream does not

impinge directly on partitions or other fixtures except where a cooling effect is required.

(2) The extracts should be located near to stagnation zones.

(3) There is the minimum amount of 'short circuiting', i.e. air flowing directly from supply to extract.

(4) To provide maximum air entrainment the supply (primary) air should be so directed as to collect the maximum possible amount of room (secondary) air whilst not exceeding the design air velocities within the occupation zone.

The two principal designs of air terminal employ diffusers and perpendicular jets to direct supply air into the air-conditioned space. Diffusers often rely upon a phenomenon known as the 'Coanda effect' where air, discharged parallel to a plane surface, tends to cling to and flow along the wall or ceiling, thus reducing the amount of air movement within the occupation zone. Diffuser design, which may be radial, part-radial, or linear, may also cause the air to swirl, again reducing the amount of air disturbance in the critical studio zone. Perpendicular jets may use louvres, grilles, nozzles, etc., to impart a perpendicular air flow from the terminal. There are over a dozen different designs of diffuser and jet, some adjustable, which can be selected to suit any requirement. The result of such choice is that numerous air flow patterns can be created within the studio environment. The skill of the designer is to select and position the air terminals to give the best possible air distribution for the particular studio arrangement. Some systems use ceiling-mounted supply and extract terminals which direct the air along the ceiling and then down the sides of the studio, leaving the occupied zone relatively draught free. Others systems have an outlet mounted at a high level on a wall which projects air across the ceiling and down the opposite wall, or else from a low-level outlet from which the air is projected up the wall across the ceiling and down the opposite wall.

A major determining factor regarding the air flow pattern is the temperature of the air itself. Cold air is denser than warm air and so depending on the season, and hence the supply air temperature, the air will behave differently when it discharges into a studio the air which is at a different temperature. If the temperature difference between room and supply air is excessive then the air distribution will become turbulent, resulting in draughts. It is recommended that the air supply in winter should not be above about 50°C, whilst in summer this can be as low as 5–10°C.

Air passing through any design of terminal will undergo a loss of kinetic energy as it impinges on the vanes or grill of the device. Unfortunately, some of this energy is converted to produce sound waves and the ventilation noise may be intrusive, particularly during otherwise quiet pauses in a studio broadcast. There are two principal ways of limiting the noise generated in this way: by reducing the air velocity, and by altering the flow of air through the terminal. Both of these factors are taken into consideration at the design stage, although retrospective measures may be taken where, for example, the air terminal has adjustable louvres which may be reset to give a reduced noise level. However, partially closing a terminal will raise noise levels since the air velocity will be increased. In extreme cases the only solution may be to change the air terminal entirely or else reduce the air flow to the studio altogether.

Another source of noise is excessive air turbulence at the terminal caused by poorly designed supply ductwork, which should be arranged so that there are no abrupt changes in air direction immediately adjacent to the terminal inlet. Where such changes are unavoidable then guides may be fitted inside the duct to reduce the air turbulence.

Finally, noise can originate from the air-handling plant itself, and in particular the fan which produces both rotational noise and vortex noise. The former is caused purely by the action of the rotating impeller striking the air and this produces a pulsating pressure point close to the fan inlet or outlet. The effect is made worse by obstructions close to the fan and as far as is possible the air should be steamlined to reduce such noise. Vortex noise is a function of the air velocity which must be kept as low as possible by designing the ductwork with a reasonably large cross-sectional area and a minimum of restrictions. Main supply and ducts should be sized to give air velocities of no more than 7.5 m/s, and terminal branch ductwork less than 3.5 m/s.

Mechanical fan noise can also be heard in studios where there is insufficient attenuation. The problem arises from the slight out of balance forces produced by the fan at speed which can be transmitted through the ductwork system in the form of vibration. Such noise can be successfully reduced by isolating the motor and fan vibration from the ductwork and buildings. This is accomplished by supporting the fan and motor installation upon antivibration mountings and flexible couplings between the fan and duct. The mountings selected for this purpose will only provide noise isolation across a given fan/motor speed range and, therefore, if the fan is run outside

its design speed range, excessive noise levels may be produced.

5.7.5.3 Summary
Ventilation of studios requires a fan and distribution system which are properly matched in terms of air flow, pressure and velocity. Once commissioned, design parameters of a studio's vent plant may be confirmed by recording air velocity and pressure at appropriate points within the system.

The air supplied to a studio must satisfy stringent requirements in order to ventilate the environment safely and effectively. The supply and removal of many litres of air per second, between 10 and 20 studio air changes per hour, must be accomplished without causing unnecessary discomfort to occupants. At the same time air changes must be effective in displacing stale air and providing heating and cooling for the studio without causing excessive temperature stratification.

Last, but by no means least these large volumes of air must be circulated through a network of ducts and diffusers without intruding on the noise-sensitive environment of the studio.

5.7.6 Temperature and humidity control

5.7.6.1 Heating
To raise the temperature of the air-conditioning supply, air is drawn through a heater battery. This is essentially a heat exchanger, the most common design being similar in principle to a car radiator. Water or steam flows through a series of finned tubes placed in the path of the air. The moving air stream is warmed as it crosses these tubes whose fins increase the effective heat transfer. The amount of heat exchanged between the battery and air is dependent on the air flow and the heater output. The temperature and flow rate of water or steam through the heater determine its output and hence a control valve linked to an air sensor positioned downstream of the heater battery will provide automatic control of air temperature.

The water or steam coil heater battery is normally designed so that fluid flow inside the tubes is opposite to the direction of air flow over the tubes, i.e. counterflow heat exchange. This is more effective than the alternative parallel flow arrangement and results in a more compact heater battery.

An alternative design of heater uses electric elements in place of water or steam. These are much less common and are only really justifiable where the heater output is low or space is at a premium.

5.7.6.2 Cooling
Air cooling is accomplished by using a finned heat exchanger, similar in design to a heater battery, but with either chilled water or a refrigerant. Both systems rely on refrigeration plant as a source of cooling and, like the heater battery, the air temperature can be controlled by regulating the water or refrigerant flow through the coils.

The refrigeration plant used to remove heat from the air, either directly or indirectly, operates on the vapour-compression cycle. The refrigerant has a low boiling point such that at atmospheric pressure and ambient temperature it will exist as vapour. By reducing the pressure of the refrigerant to less than atmospheric it is possible to condense the refrigerant to a liquid which will boil at low temperatures and, in so doing, absorb heat from its surroundings.

The vapour-compression cycle consists of four stages:

(1) evaporation,
(2) compression,
(3) condensing, and
(4) expansion.

Evaporation of the liquid refrigerant, which is at low pressure and temperature, takes place as heat is absorbed from around the evaporator. Since heat energy will only 'flow' from hot to cold, then provided the refrigerant is cooler than the surrounding water or air it will absorb sensible and latent heat to produce the desired cooling effect. The volume or mass flow rate and the temperature of the refrigerant can be adjusted to increase or decrease the cooling effect.

Superheated (dry) vapour leaves the evaporator, is drawn into a compressor and discharged as a high pressure, high temperature gas. This compressor is driven by an electric motor whose size determines the effective running cost of the refrigeration plant. The refrigerant is now above ambient temperature and so the vapour is passed to a water or air cooled condenser. This removes the heat absorbed by the refrigerant during evaporation and compression, and condenses the vapour back to a liquid.

Finally, the high-pressure subcooled refrigerant passes through an expansion device which causes a pressure drop. The liquid refrigerant is now cooled enough to re-enter the evaporator and begin the whole cycle once more.

Some designs of air-conditioning plant use a chiller in which the evaporator itself is immersed in

the air stream. These are referred to as 'direct expansion systems'. The relatively warm air flows across the finned coils, causing the refrigerant within to evaporate, and in so doing cools the air. An alternative arrangement is to use the refrigerant evaporator to chill water which is then circulated around the cooling coils of the air-conditioning chiller battery.

5.7.6.3 Humidity control

Depending on the weather and the studio conditions, it may be necessary to either humidify or dehumidify the air supply. Where it is required to humidify the air it is necessary to add water to the supply air before it enters the studio. Traditionally this was accomplished by spraying warm water into the air stream. By atomizing the water the air would readily evaporate the moisture into vapour form and so increase in terms of relative humidity. Unfortunately the warm water provided a haven for the bacteria responsible for legionnaire's disease which could then be all too easily dispersed by the air-conditioning system. For this reason steam humidification has found favour since the *Legionella* bacteria are destroyed by the high-temperature steam, and the moisture is very readily absorbed in this form. Where a steam supply is not readily available, self-contained electric steam boiler/humidifiers provide a compact and efficient solution. By linking the output of the humidifier to a suitably positioned humidistat, fully automatic control may be achieved.

Air dehumidification is possible using chemical absorbants, but this method is not usually employed in studio air-conditioning systems. A simpler method of removing moisture from the supply is to cool the air to below its dew-point (saturation) temperature, which will cause the water vapour to condense out of the air. This is easily achieved by using the cooling coil and then reheating the air to the desired temperature. It can be expensive, however, particularly where large air volumes are involved, and so not all studios are fitted with this capability. A potential hazard of the cooling coil used in this way is that as the water condenses out on the chiller it can collect in the duct and be warm enough for the *Legionella* bacteria to colonize it. Consequently, great care must be taken with the drainage arrangements adjacent to cooling coils to ensure that all the water can readily drain away and that there is little risk of back syphonage from the drain.

5.7.6.4 Psychrometric charts

Psychrometry, the study of air and water mixtures, is obviously important to the air-conditioning engineer.

The psychrometric chart is a complex plot of wet and dry bulb temperature, humidity, enthalpy (heat energy), specific volume and moisture content of air at atmospheric pressure.

The quality of a sample of air from an air-conditioning system can be established simply by measuring the wet and dry bulb air temperatures and transferring this data to the psychrometric chart. It is then possible to determine humidity, enthalpy and moisture content at this condition. It is also possible to predict how these values will change for a given process. For example, if the air is sampled before the cooling coil, it is possible to predict what the air temperature and humidity will be after the coil, and therefore at what temperature it will subsequently enter the heater battery.

The psychrometric chart is useful to the designer and engineer, but it is by no means essential to understanding the basic air-conditioning process described in this section.

5.7.6.5 Summary

Heating and cooling of the air is accomplished by passing air across heated or chilled coils located in the air-conditioning duct. Steam or hot water are used to heat the air and refrigerant, directly or by chilled water, to cool the air. Humidity is increased by spraying steam directly in to the air, whilst the chiller coils are used to dehumidify the air.

5.7.7 Air-conditioning plant

5.7.7.1 Air-conditioning systems

There are essentially four classes of air-conditioning system and the choice of installation will depend on studio size, utilization, location and maintenance and running costs of the proposed air-conditioning plant. Since all four designs are commonly used in studio air conditioning it is appropriate to make a comparison between them. Each system, regardless of the design, will normally be able to control the air temperature, humidity and ventilation.

5.7.7.2 All-air systems

A central air-conditioning plant typically consists of fans, heating and cooling coils, humidifiers, preheaters, reheaters, filters and dampers, interconnected by ductwork to the conditioned space, e.g. the studio. *Figure 5.35* shows a typical installation in which outside (fresh) air is drawn in by the supply air fan, across a preheat coil and mixes with return air from the conditioned space. The percentage of fresh

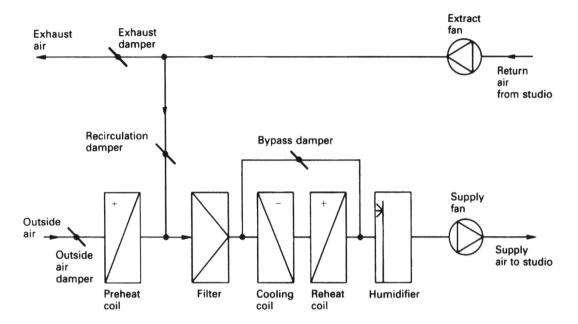

Figure 5.35 *Arrangement of a typical central air-conditioning system.*

air to return air is determined by the system controller which regulates the position of the appropriate dampers. The mixed air is drawn across the filter and through the cooling and reheat coils to give the desired temperature. Finally, air passes across the humidifier before being drawn through the fan and then distributed to the various zones within the building, each of which may have individual reheaters to give fine adjustment to the supply air temperature. Return air is ducted from the studio extracts and, depending upon the damper position, a certain proportion of the air is discharged to outside whilst some is recirculated. At certain times of the year it may be possible to operate the plant with 'free' cooling, i.e. the outside air is cool enough to be able to switch off the chiller and simply use full fresh air supply without recirculation.

The arrangement of the coils, filter, fans and humidifer mean that the air can be treated in one single pass for the following conditioning processes:

(1) heat and humidify,
(2) cool and humidify,
(3) dehumidify and heat, and
(4) dehumidify and cool.

An alternative design replaces the individual zone reheaters with variable air volume (VAV) outlets. These VAV units are fitted with small heaters and motorized dampers to regulate both the temperature and flow of air to the studio. The supply and extract fans for such systems must be fitted with variable inlet vanes so that their delivery may be regulated to meet the demands of the individual VAV units.

In other designs of system the cooling coil and heater are arranged one above the other within a partitioned duct so that some air flows over the heater battery in the top 'deck' and across the chiller on the lower 'deck'. The air streams are then recombined, the air temperature being determined by dampers which control the air mixture. Alternatively, the air streams may be ducted away separately and mixed only at individual zones. This is referred to as a dual-duct system and has the advantage of allowing a rapid response to temperature variations within studios, because the air streams are mixed immediately adjacent to the room. The individual mixing boxes have motorized dampers which are controlled from a zone or studio thermostat, and are designed to deliver a constant air volume. An attenuator fitted in the box will reflect and absorb noise from the air handling unit.

5.7.7.3 Air–water systems

A central plant supplies primary conditioned air to individual induction units located at each studio. These units contain a coil through which either hot or cold water flows. The majority of the studio's thermal load is balanced by the induction unit

which takes primary air from high velocity duct-work and passes it through nozzles causing room (secondary) air to be drawn into the unit across the heating/cooling coil. The two air streams combine and discharge to the studio. The quantity of heating added or removed from the secondary air is controlled by the studio thermostat. The primary air is supplied at the correct humidity and ventilation rate and also contributes to the heating or cooling load of the studio.

5.7.7.4 All-water systems
An air terminal unit situated in the studio receives piped hot and cold water from remotely sited plant. The fan coil unit receives outside air which is then filtered, dehumidified, heated or cooled as appropriate. Air may be recirculated and supplemented by outside fresh air to meet the required ventilation demand.

5.7.7.5 Direct expansion systems
Direct expansion (DX) systems are the simplest type of air conditioning equipment. Two common designs are the window box and the split pack. The window box consists of a fan coil unit mounted on an outside wall, usually beneath a window, in which a self-contained refrigeration system and heater are mounted. Outside air and recirculated air are blown across the refrigeration unit evaporator coil and a heater element. A separate fan directs outside air across the refrigeration system condenser and forces the resultant hot air outside.

'Split-pack' systems are so-called because the refrigeration condenser and compressor, i.e. the condensing unit, is mounted outside the conditioned studio space. The fan coil unit comprising filter, fan and refrigeration evaporator is mounted in the studio and supplies cooled recirculated air only.

5.7.7.6 Summary
Studio air-conditioning systems can vary considerably in terms of their sophistication, ranging from large automated central air systems to the compact and simple split-pack unit. The choice of system for a given studio will depend on the size, complexity and proximity of the studio with respect to the building as a whole. Regardless of design, the air-conditioning plant should satisfy the basic requirement of automatically delivering air to the studio at the correct temperature, humidity and ventilation rate. Sensors and actuators linked via suitable controllers provide the means by which this is achieved. Regular maintenance and testing of the air-handling plant is necessary to ensure that the system performance is compatible with the original commissioning data.

Bibliography

CIBSE, Ventilation and air conditioning (requirements, systems equipment and control). In: *CIBSE Guide*, Vol. B, Sections B2 and B3, Chartered Institute of Building Services Engineers, London (1986)

LUFF, M.G., *Air Conditioning for Students*, Technitrade Journals, London (1980)

MILLER, L.M., *Students Textbook of Heating, Ventilating and Air Conditioning*, Technitrade Journals, London (1976)

6

Distribution of audio signals

David M. Davidson
Chief Engineer, Telephone Systems, Northern Telecom Europe Ltd
6.1 Telephony 6/2
6.2 Telephony transmission systems 6/20

David C. Yates
Broadcast Transmission Training Consultant, BBC Transmission
6.3 Radio propagation 6/35

Bev Marks
Broadcast Consultant
6.4 Transmission developments: NICAM and RDS 6/46

Francis Rumsey
Senior Lecturer, University of Surrey
6.5 Digital audio interfacing, synchronization and networks 6/57

6.1 Telephony

6.1.1 Introduction

The basic objective of telephony is to provide a service to enable two customers at a distance to set up a call via the network so that both may talk to each other. The telephone system must supply the means of connecting two telephones together via signalling, switching and transmission functions. The *signalling* function enables the customer to interface with the telephone system and inform the telephone system which telephone the customer wishes to contact. The signalling system also informs the user of the progress that the telephone system is making in setting up the requested connection and alerting the desired distant party that someone is attempting to make contact. *Switching* is the mechanism to interconnect the individual telephones with each other. *Transmission* is the end-to-end conveyance of the voice signal at the level of loudness and intelligibility that the user originates into the telephone.

To set up a connection through the network from one telephone to the other there must be a means of identifying individual telephone lines: the directory number. Thus every telephone throughout the world must have a unique number. From the directory number of the calling and the called number it must be possible for the exchange equipment to determine a route for the call via the network of telephone exchanges. When a caller lifts the handset this action signals to the exchange that the caller wishes to set up a call. The exchange signals to the caller that it is ready to receive instructions by returning the dial tone. The caller signals the desired destination telephone by dialling the directory number of the distant telephone. The exchange then decodes this directory number and translates it into routing information to set the connection up through the switches within the exchange or exchanges. When the distant telephone's status has been determined, i.e. busy or free, this is signalled back to the caller. If the distant party is free the equipment will apply a ringing signal to the called telephone to alert the telephone. When the called telephone handset is taken off the hook this signals to the exchange that the call has been answered and the ringing should be removed from the line, the line placed into the busy state as far as other callers are concerned and the call connection established for end-to-end voice communication. When the caller or the called party concludes the conversation, either will replace the telephone handset and this will signal to the exchange that the call

has terminated and the equipment used in setting up the call should be released and both telephone lines returned to the idle condition to await the next call.

6.1.2 Electromagnetic telephone

All the early telephones were electromagnetic in operation and had to perform several basic functions.

Hybrid The telephone is essentially a four-wire device. Two wires connect the transmitter and two wires connect the receiver, but the wiring from the public telephone exchange is only two wires so the telephone must separate the transmitter signal from the receiver directing the transmitted signal towards the telephone exchange and separating out the received signal and directing this towards the receiver all at the same time.

Side-tone Some of the transmitted signal needs to be directed back into the receiver to provide the user with side-tone so that the user has confidence that the telephone is operating.

Power feed The transmitter in electromechanical telephones has to be fed direct current (d.c.) from the telephone exchange and the receiver, being electromechanical, needs to be isolated from the d.c. to function correctly.

Signalling When the user lifts the handset the telephone has to present a low d.c. resistance towards the telephone exchange as part of the signalling system.

Impedance matching The a.c. impedance of the telephone has to match the impedance of the telephone line for maximum power transfer of the a.c. speech signal.

Compensation for line variation The telephone has to interface with a wide variation in telephone line resistance without the need for any customizing of the telephone to match the particular line conditions.

Alerting A means must be provided to alert the user that a caller wishes the user to pick up the telephone.

Protection Protection of the telephone and the user from extraneous voltages, lighting, power line crosses and acoustic shock from induced impulse voltages.

6.1.2.1 Speech circuit

To overcome these conflicting requirements the electromechanical telephones were designed with a form of hybrid autotransformer called an anti-side-tone induction coil (ASTIC) as the main speech circuit (*Figure 6.1*). The three-winding transformer is

Telephone speech circuit

Figure 6.1 *Telephone speech circuit.*

Figure 6.2 *Carbon granule transmitter.*

arranged as a balanced bridge, the balance impedance being matched to that of the average line. Speech currents from the transmitter split into two components, i_1 and i_2 flowing in the balance impedance and the line, respectively. The current i_1, in winding L_1 induces a potential in winding L_3 to produce a current i_3. Similarly, current i_2 flowing in winding L_2 produces a current i_4 in winding L_3. The windings are arranged such that the currents i_3 and i_4 are flowing in opposite directions and if the bridge is perfectly balanced cancel each other out. In practice, perfect balance is not achieved since the balance network is only an approximation to the line impedance and the small current that is not fully cancelled out and is flowing in the circuit provides the side-tone.

The transmitters in the electromechanical telephones were all carbon microphones (*Figure 6.2*). The transmitter is constructed in such a way that when the sound impinges on the diaphragm, the diaphragm vibrates causing the carbon granules to be compressed and decompressed. The variation in pressure on the carbon granules causes a variation in resistance between the two carbon electrodes resulting in a variable current on the line (*Figure 6.3*).

The receiver in the electromechanical telephones were moving iron diaphragm devices (*Figure 6.4*). The diaphragm moves under the influence of a varying magnetic field created by currents flowing in the coils. The permanent magnet is made from an alloy of iron, aluminium, nickel and cobalt and has a high permeability and, therefore, it produces a strong and lasting magnetic field. The diaphragm is made of a high permeability but low retentivity material so that it responds readily to the changing fields induced by the currents in the speech coils.

Figure 6.3 *Line current.*

Figure 6.4 *Moving iron receiver.*

6.1.2.2 Basic signalling to the exchange (dial pulse and offhook)

The simplest telephones signalled to the exchange by lifting the receiver so that a mechanically operated contact short circuited the line and current flowed down the line from the exchange through the contact and back to the exchange, thus operating the electromagnetic relay in the exchange. If the exchange was manual then the relay would either light a lamp or move a mechanical device to signal to the telephone operator that a customer was calling. This fundamental method of operation, i.e. short circuiting the line via the telephone circuit with a contact, is still used today to signal to the exchange that a call is being initiated (*Figure 6.5*).

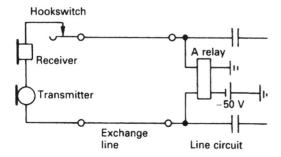

Figure 6.5 *Basic signalling.*

In manual exchanges the caller would request the telephone operator to obtain the desired number and the operator would set up the connection. On automatic exchanges the telephone would be fitted with a mechanical dial. The dial worked on the principle of making and breaking the line current with a contact in response to the numbers selected on the dial. The user would turn the dial until it hit the stop then let the dial return to the normal position. The dial would return to the normal position at a controlled speed regulated by a governor and would open and close the line contact, at 10 pulses/s, only on the return path. One break of the line represents a digit 1, two breaks a digit 2, and so on, the number of breaks corresponding to the digits dialled. A further contact was provided on the dial which always closed when the dial was moved from the off-normal position. This contact was used in the telephone circuit to take the transmitter and receiver out of circuit to ensure a 0 Ω path for transmission of the dialled digits from the telephone and prevent the user from hearing the dialled digits.

6.1.2.3 Basic signalling from the exchange (ringing)

To alert the user that a caller is attempting to make connection the electromechanical telephone is provided with a bell. The bell operates on an a.c. signal varying in frequency depending upon the ringing generator fitted in the exchange. The ringing signal from the exchange is superimposed on the normal d.c. line feed to power the telephone. The ringing signal can be anywhere from 50 V to 130 V at frequencies of 17–50 Hz. Whatever the actual ringing voltage or frequency, the bell must react only to the ringing current, not the d.c. line feed to the telephone and not to the speech signal during conversation. *Figure 6.6* shows a typical bell circuit in a telephone. The 1.8 µF capacitor acts as a d.c. block to prevent polarization of the bell by the d.c. feed from the exchange and the contact is part of the telephone hookswitch. When the user lifts the handset the contact makes, looping the line via the telephone. The telephone circuit is now connected across the line presenting a low impedance to the exchange and a relay in the exchange operates to this condition removing the ringing source from the line.

Figure 6.6 *Typical bell circuit.*

6.1.3 Electronic telephones

Although a large number of electromechanical telephones still exist throughout the world and are still in everyday use they are expensive to manufacture and limited in capability; electronic designs have largely taken over and offer more features and cheaper, more efficient transducers have been incorporated into the designs to provide better quality of speech reproduction along with higher reliability. Development of the electronic telephone at first mirrored the operation of the electromechanical telephone as far as signalling to the exchange was concerned and dialling integrated circuits reproduced faithfully the functions of

the mechanical dial, although the user interface was changed to a push button keypad for digit selection. Ringer circuits also mimic the bell operation, basically to ensure compatibility with the existing exchange equipment.

6.1.3.1 Speech circuit

Integrated-circuit manufacturers offer a vast array of electronic speech circuits all interfacing to different forms of transducers or switchable between electrodynamic or ceramic types. They also offer different forms of balance network and gain settings for send or receive loudness ratings. The essential requirements of any electronic telephone speech circuit is much the same as in the electromechanical telephone:

(1) to provide two- to four-wire conversion,
(2) to extract power from the line to power the transducers and the integrated circuits,
(3) to set the send and receive loudness gains,
(4) to provide side-tone for the user,
(5) to present a matched impedance to the line for maximum power transfer of the speech signal.

The two- to four-wire conversion is implemented by amplifiers and components arranged in a bridge formation (*Figure 6.7*). In *Figure 6.7*, for perfect balance of the bridge:

$$Z_{\text{line}}/Z_{\text{bal}} = R_1/R_2$$

The a.c. signal from the microphone is sent to one diagonal of the bridge. A small percentage of the signal is lost in the balance impedance Z_{bal}, but the majority is sent to line via R_1. Some of the signal from the microphone will be fed back into the earphone because of the imperfect balance. Many integrated circuits have line current sensing circuits built in so that when the telephone is nearer the exchange more current flows in the loop and a different balance impedance is switched in to control the gain on the receive amplifier and also the level of side-tone. To prevent see-sawing at the boundaries between different line current values, hysteresis circuits are used. As all integrated circuits need a supply voltage at the same polarity all the time, a diode bridge is always present in front of the speech circuit to ensure that the polarity of the supply voltage remains the same even if the line is reversed from the exchange.

Various forms of balance network are used depending on the overall transmission plan for the telephone network. Traditionally, 600 Ω (resistive) was the balance impedance, but this was rarely achieved in practice. The impedance that a telephone

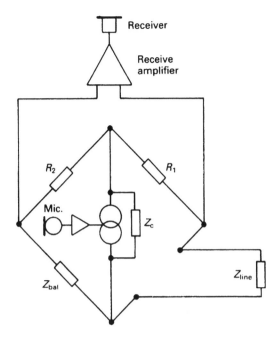

Figure 6.7 *Telephone bridge.*

should present to line is chosen to restrict the range of impedances that a telephone circuit and local cable present to the exchange feed bridge. The better the balance, then the less prone the telephone connection will be to echo on long distance telephone calls and provide a better means of controlling side-tone on short lines. For these reasons a complex balance impedance network (*Figure 6.8*) has been adopted in new designs.

Figure 6.8 *Complex impedance balance network.*

The transmitters in an electronic telephone can be moving coil or electret types. *Figure 6.9* shows the essential features of a moving coil transmitter. It consists of a coil attached to a diaphragm and suspended in a strong radial magnetic field formed by the permanent magnet. When the sound impinges on the diaphragm a voltage is induced in the coils by the movement in the magnetic field. *Figure 6.10*

Figure 6.9 *Moving-coil transmitter.*

Figure 6.10 *Electret microphone.*

Figure 6.11 *Rocking armature receiver.*

shows the essential features of an electret microphone (a term originally described by Heaviside). A thin polymer film permanently retains an electric charge equivalent to a d.c. polarizing voltage of 100 V imparted to the film at the time of manufacture. When the film is vibrated by sound waves there is a variation in capacitance between the ground and the backplate output connection. Since electret transducers have an extremely high impedance, and the output is low, a compensating amplifier is necessary between the electret and the speech circuit. Normally, a field-effect transistor (FET) device is used; it is integrated within the body of the electret microphone.

Receivers in electronic telephones can either be moving-coil devices as described above or rocking armature devices (*Figure 6.11*). The device consists of a rocking armature connected to a conical diaphragm. The armature is held in equilibrium, when no signals are present, by the equal attraction of the two pole pieces of the permanent magnet. When signals are received in the coils wound on the pole pieces the magnetic balance is upset, and the armature will rock in sympathy with the variations in current in the coils. The movement of the armature is

transmitted to the conical diaphragm which vibrates to produce the sound output.

6.1.3.2 Signalling to the exchange (pulse or loop disconnect dialling)

When the customer lifts the telephone handset the contact is made by the hookswitch and the line is offhook with current flowing round the loop from the exchange battery. The exchange relay is energized and a dial tone is returned from the exchange to the caller. The telephone circuit for dial pulse generation consists of a keyboard scanning matrix that scans the rows and columns of the keypad and provides persistency checking, to ensure that a key has been pressed, and also key debouncing circuitry. *Figure 6.12* shows a typical pulse train for the digits 3 and 1. The pulses are arranged to replicate the mechanical dial with a pulse repetition rate of 10 per second. The offhook period at the beginning of the pulse train is necessary to ensure that the exchange has time to react to the calling condition and return dial tone.

The make/break ratio depends on the design of the exchange equipment, but generally is biased towards the break periods, with about 60% break and 40% make. This enables electromechanical relays to operate to the make period within 40 ms, giving a considerable margin to allow for adverse line conditions. As the electromechanical relays generally take longer to release this gives a greater margin to ensure that the relay is fully released before the next pulse arrives. The interdigit pause is the time interval that the mechanical dial had built into it from the space on the dial from the digit-1 hole to the dial stop. This has to be around 300 ms which is the time required for the exchange equipment to seize the next selection device and be ready to receive the next pulse train. The last part of the pulse train is the on-hook period which again must be around 300 ms to ensure that the exchange equipment has cleared down and is ready to accept the next call request.

Digital integrated circuits can readily generate these pulse trains and scan the keypad of the telephone. In addition they can provide features such as last-number redial by storing the last number dialled and can provide memory to store the most frequent numbers dialled. As these devices all tend to be complementary metal oxide semiconductors (CMOS) it is usually necessary to connect the pulsing output to the line via a high voltage drive transistor due to the high back-e.m.f. pulses that appear on the line during pulsing into electromechanical exchanges.

6.1.3.3 Signalling to the exchange (multifrequency or tone dialling)

The dial tone multifrequency (DTMF) was developed in the mid-1950s at the Bell Laboratories in the USA and was introduced as an alternative to pulse dialling offering greater speed of dialling and improved performance. It has since been adopted by the Comité Consultatif International Telephonique et Telegraphique (CCITT) and is now widely used throughout the world for most new telephone designs. DTMF signalling is an in-band signalling system composed of 16 audio frequencies arranged in a 4 × 4 matrix with four low frequencies forming the rows and four high frequencies forming the columns. Digits are represented by two out of eight frequency combinations. For example, the digit 5 is formed by 770 Hz plus 1336 Hz transmitted to the exchange (*Table 6.1*).

The DTMF tones are usually 50 ms in duration with intervals between the tones of 40 ms, thus giving a total cycle time of 90 ms. Thus a string of 10 digits can be transmitted to the exchange

Table 6.1 Dial-tone multifrequency (DTMF) tone assignments

Low frequencies (Hz)	High frequencies (Hz)			
	1209	1336	1447	1633
697	1	2	3	A
770	4	5	6	B
852	7	8	9	C
941	*	0	#	D

in 900 ms, whatever the combination of the digits. This contrasts very well with the loop disconnect signalling at 10 pulses per second and variable digit length, depending on the digits dialled (see *Figure 6.12*).

To send 10 individual digit 5s to the exchange in pulse dialling will take 7800 ms, including the inter-digit pauses, and to send the same digits in DTMF takes only 900 ms, i.e. DTMF is 8 times quicker. The division of frequencies into high and low groups simplifies the detection at the exchange (*Figure 6.13*). The receiver first filters out dial tone and any line noise and then separates the signals into the high-frequency components and into the low-frequency components. It then digitally measures zero crossings over a period of time to produce digit decoding.

Detection of DTMF signals is complicated by the presence of 50/60 Hz power, line noise, dial tones of various frequencies, random noise and voice signals, the latter being more a problem of

Figure 6.12 *Typical pulse trains.*

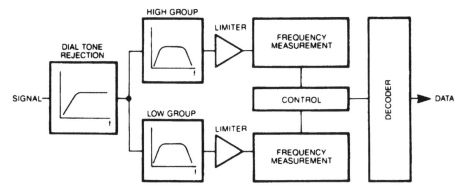

Figure 6.13 *Dial-tone multifrequency (DTMF) receiver.*

voice by unwittingly triggering the DTMF receiver when not required. Dealing with these problems while still retaining speech immunity has presented a design challenge to the communication engineer. *Figure 6.14* shows a typical line wave form, with the levels of significant parameters annotated. Because DTMF receivers are designed for a unique electrical environment the parameters that control their operational performance have developed their own particular vocabulary.

Sensitivity at the A level is the lowest level of signal that is guaranteed to be detected. The range of levels detected is known as the 'dynamic range', which usually extends to about 30 dB above the A level.

DTMF twist is the ratio of the level of the high-frequency DTMF component to that of the low-frequency signal component. Most receivers are designed to handle level differences from +4 dB to −8 dB, with respect to 0.775 V.

Frequency deviation from the nominal sometimes arises because of variations in DTMF senders and frequency shifts induced by analogue line carrier systems. Receivers are therefore expected to detect

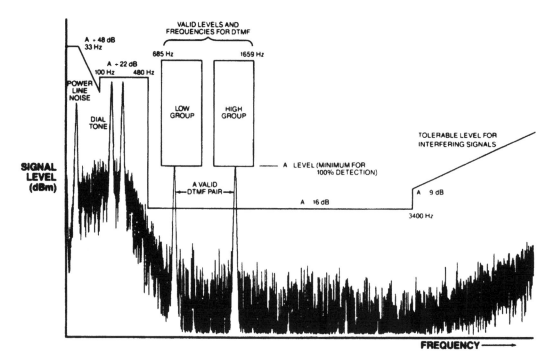

Figure 6.14 *Typical waveform line.*

signals that may drift by as much as 1.5% plus 2 Hz from the nominal throughout the dynamic range.

The signal/noise ratio is the minimum separation between signal and noise that will not degrade signal detection accuracy over strings of successive digits.

Detect/reject times tend to be 40 ms for both; i.e. the signal must persist for 40 ms and, due to all the sources of tones and interruptions on the telephone network, this becomes a critical parameter to ensure reliable detection of tones.

Dial tone immunity is a severe test for the DTMF receiver. By its very nature dial tone is present on the line until the first digit has been successfully detected and decoded. Dial tones also vary from country to country and are of a fairly high level, as *Figure 6.14* shows.

DTMF signalling has enabled an additional set of digits to be generated from additional keys such as * and # on the basic telephone and A, B, C, D if the 1633-Hz column was utilized. Telephone administrations have used these extra signals to enable customers access to additional services such as call diversion, etc. DTMF also enabled the user to gain access to other services such as into banking computers or voice mail machines. These attachments rely on the end-to-end signalling nature of DTMF. It is because DTMF signalling is totally within the 4 kHz speech band that telephony operates world-wide and the DTMF signals can be transmitted through the network as if they are speech.

6.1.4 Signalling from the exchange (ringing)

The ringing signals from the exchange are still the same as in the electromechanical telephone. The duration between bursts of ringing may have altered, but the requirement is still fundamentally to supply enough power to drive an a.c. bell. Early electronic telephones still had a bell fitted and it is still possible to buy telephones with bells fitted. Other forms of electroacoustic devices are also available to provide a more cost-effective implementation of the alerting device. Alerting devices can be miniature loud-speakers or piezo-ceramic disk resonators. These devices are interfaced to the line ringing voltage by an integrated circuit which senses the incoming ringing signal and then amplifiers drive the output device. These devices extract power from the ringing supply and have a built-in hysteresis to prevent unwanted operation by spurious pulses. Facilities are provided for volume and tone warble adjustment. The d.c. fed from the exchange is blocked by the bell capacitor in the telephone circuit to ensure the device only 'sees' the a.c. ringing signal and the ringing device does not trip the ringing supply from the exchange until the telephone has been taken off the hook.

6.1.4.1 Audible tones

Audible tones are provided to advise the user of the progress of the call request through the network. Some tones are provided to give the user other information, e.g. line temporarily out of service. Tones are one frequency or a combination of frequencies and may be cadenced to provide a more distinctive meaning. The particular tones used are dependent on the telephone operating company and vary throughout the world.

Dial tone is provided to inform the user that the exchange has been seized (i.e. has recognized the calling signal) and is now ready to receive the dialled digits. Dial tone is superimposed on the d.c. voltage feed to the telephone and is normally a combination of two frequencies.

Busy tone is returned to the caller when the wanted party is engaged on another call and this has given the tone its alternative name 'engaged tone'. Some exchange systems distinguish between equipment busy, i.e. blocking has occurred in the network, and customer engaged on another call. Two slightly different tones are then provided. In the USA equipment busy or path busy is called 'reorder' tone and the cadencing is twice that of normal busy tone.

The **ringing tone** is returned to the caller when the called party is being rung and persists until either the caller hangs up or the called party answers the call and trips the ringing at the exchange.

The **howler or receiver-off-hook tone** is used to signal to the user that the receiver has been left off the hook. The tone starts at a low intensity with a high audiofrequency and gradually builds up to maximum intensity over a period of time.

6.1.5 Public switched network

The public switchboard telephone network (PSTN) is made up of a hierarchy of telephone exchange equipment. Local exchanges serve the customers' telephone lines and these local exchanges can be directly connected to each other via interexchange circuits (junction circuits) if the traffic demand is high enough. Access to other local exchanges is gained from the next level in hierarchy, the tandem exchange or primary trunk switching centre. These exchanges are interconnected to the next level

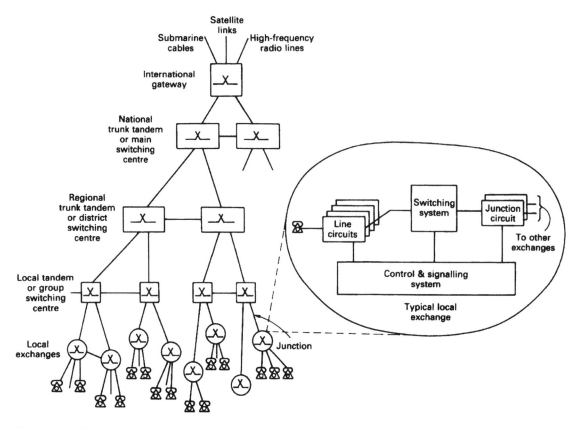

Figure 6.15 *Public switched network.*

of regional switching centres then national switching centres and, finally, to international gateway exchanges (see *Figure 6.15*). All exchanges can be cross-connected with direct junction circuits at any level depending on the traffic demands of the users.

6.1.5.1 Line circuits

A line circuit is the access point into the PSTN for the telephone. A line circuit can be regarded as a number of functional blocks to provide the translation from a two-wire analogue telephone line into the four-wire pulse code modulation (PCM) data stream. These functions are commonly called by the acronym 'BORSHT' standing for 'battery feed, overvoltage protection, ringing, supervision, hybrid and test'. Most of the problems associated with the design of the exchange line circuit stems from the electrically severe environment of the telephone line which is subjected to transients, lightning surges, steady-state longitudinally induced voltages, and power line crosses. Further constraints are placed on new designs of line circuits with the need to interface with existing customer premises equipment to

provide the low frequency, high voltage to drive the bell circuit in old telephones. *Figure 6.16* shows the functional elements of a line circuit.

The **battery feed** can take one of two basic forms: constant current or constant voltage. Constant voltage is the traditional method of feeding power out on the telephone line. The battery voltage is nominally -50 V in the range -44 to -56 V with the positive side of the battery connected to ground. The feed voltage is connected to line via a 200 Ω resistance in each wire. This resistance was traditionally the feed relay monitoring the loop on the line through the telephone. The original choice of 50 V feed was to ensure that sufficient current was available on an 800 Ω telephone line to operate a carbon microphone. The constant voltage is not ideal for modern telephone equipment and has a number of drawbacks. If most of the lines are relatively short, and therefore take between 20–60 mA when the telephone is in use, then the power drain on the battery can be excessive. Under fault conditions, i.e. an earth on one leg of the pair, the feed resistance heats up and will dissipate 13 W. Constant-current feed uses

Figure 6.16 *Functional partitioning of a telephone line circuit.*

active devices in each leg of the pair arranged so that a constant current is fed to the line when the telephone is connected, i.e. off the hook, irrespective of the line length, the telephone impedance, or the battery voltage. Normally the current is limited to some preset value such as 25 mA. When the telephone is idle, i.e. on the hook, there is insufficient battery voltage to drive this current into the line leakage resistance so the control circuitry switches off the feed to the line. This change of line state can be used as a signal to monitor the condition of the distant telephone, i.e. off or on the hook. Thus constant current can reduce the total exchange power requirements and limit the short-circuit power dissipation to that of the current limit. Constant-current operation represents an inherently high impedance to speech signals. Constant-current feed can cause some problems to existing telephone equipment; if the telephone has a regulator fitted the effect of the regulator can be overridden and some form of line length sensing is required to sense the line length and then increase the gain to compensate for the absence of the regulator. The performance of old equipment which relied on drawing high currents on short exchange lines could be seriously impaired.

Protection of the line circuit is required to ensure that lightning strikes or power-line crosses do not cause damage to the exchange equipment. Lightning protection normally takes the form of a pair of gas discharge tubes connected between each line and earth. A more efficient method is to use a single gas discharge tube with a common centre electrode connected to earth. The latter arrangement has the advantage that if a discharge fires one side of the tube the other will automatically trigger, and thus both legs of the telephone line are clamped to

earth. Due to the speed of some lightning pulses it is still possible for the equipment to be subjected to a 1500 V pulse and gas discharge tubes will not protect against power line crosses, so therefore additional protection is required. Semiconductor devices used for switching or active line-feed devices require secondary protection in the form of clamping diodes and varistors with a series resistor between the line and the clamping device to make them effective. This form of protection has the disadvantage of introducing unwanted transmission loss on the line. Furthermore, it does not protect adequately against power crosses and some form of detector which senses either overcurrent or overvoltage, and then disconnects the line from the source, is preferable. The simplest way to provide this is a fusable link, but this method does not give automatic reinstatement of the line when the power source is removed from the line.

Ringing sources are normally centrally generated a.c. supplies between 17 and 50 Hz, depending on the particular telecommunications operator's specification, and will be superimposed on the battery feed to the line. The ringing voltage is applied to the line via a relay in most equipment and the ringing voltage is cadenced with an on period and then an off period, giving the telephone its characteristic sound. The loop detector circuits must stay on line during the ringing phase to detect when the telephone has been taken off the hook and the ringing should be disconnected. Not all ringer circuits are fed to the line by relays. Designs exist using high power semiconductor devices that generate the ringing voltage from a d.c. voltage supply.

Supervision of the telephone loop condition is one of the principal functions of the line circuit.

In electro-mechanical exchanges relays were used and the need for periodic readjustment was a severe disadvantage. The electronic line circuit requires a different approach to responding to loop disconnect dialling pulses, immunity to longitudinally-induced a.c. signals, and leakage currents of a few milliamperes. Many forms of loop detector circuit exist. *Figure 6.17* shows a resistive network. This relies on the principle that most line-feeding arrangements use a pair of balanced feed resistors which are used for current limiting on short lines. By cross-connecting four resistors, and monitoring the voltage at the appropriate junction with an operational amplifier, it is possible to detect when the loop current exceeds a predetermined threshold. It is necessary that all the resistors are well matched, with a close tolerance.

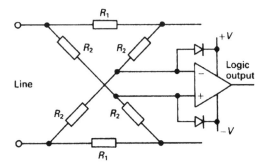

Figure 6.17 *Loop detector. $R_2 \gg R_1$, R_1 may be feed resistors.*

Hybrid or two- or four-wire conversion is necessary in line circuits on digital exchanges due to the unidirectional nature of digital switching on PCM exchanges. It is then necessary to split the telephone line into transmit and receive paths. Transformers can be used for this function and are described in detail elsewhere in this section. In digital telephone exchanges it is now simpler to use only an isolating transformer and operational amplifiers incorporated into the design of the silicon chip. *Figure 6.18* shows a typical circuit based on operational amplifiers. The balance network should match the impedance of the telephone line, but in practice it is a compromise resistive-capacitive network.

Test access is the last remaining function in the line interface and this is normally performed by relays at the front end of the line circuit. The relays enable the line to be disconnected from the line circuit and a test access bus connected so that test equipment can be used to test the telephone line without the line circuit being connected, and to test the line circuit without the line being connected. Modern test equipment is computer controlled and

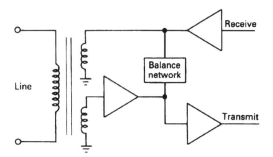

Figure 6.18 *Hybrid two- to four-wire conversion.*

lines are generally tested automatically at night when traffic on the exchange is low. The results of the tests are compared with the previous results and/or against a standard to determine if the line is still maintaining its standard.

6.1.5.2 Signalling

Many different signalling schemes have been derived over the years and, in principle, a similar set of signals is required between exchanges as that required between the customer and the exchange. The equipment requires a set of supervisory signals to control the connection (e.g. answer signal) as well as the address signals. Signals are transmitted over the network either in band, i.e. within the speech band, or out of band.

Decadic interexchange signalling Decadic, or loop disconnect pulsing between exchanges, is similar in principle to loop disconnect signalling between a telephone and the exchange, except in the decadic case it is normally an electromechanical relay that performs the pulsing and not dial contacts. Loop disconnect signalling is restricted to circuits where a direct physical path exists between the exchanges, e.g. a copper pair. The loop resistance must be low enough to permit operation of the distant pulse receiving relay. Thus this type of signalling is limited to short distance junction circuits (*Figure 6.19*). To

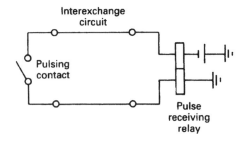

Figure 6.19 *Decadic interexchange signalling.*

Figure 6.20 *Double circuit signalling.*

extend the distance, double current signalling has been employed using a polarized relay as the digit receiving relay. This has the following advantages over the loop disconnect method. The polarized relay is very sensitive and can operate at higher line resistances. Double current signalling (make and break signals are represented by opposite directions of current flow) ensures that line leakage cannot suppress break signals and the line is discharged more quickly by the application of the opposite potential (*Figure 6.20*).

Single-frequency interexchange signalling
On amplified interexchange circuits it is not possible to use d.c. signalling methods as no physical path exists for the signals to go through the amplifiers. With in-band voice-frequency signalling it is necessary to take certain precautions to prevent speech triggering the receivers. Voice immunity is partly obtained by using a voice frequency where the energy is relatively low for most normal speakers. The maximum power from the voice is around 500 Hz and falls off as the frequency increases and the frequency should not be used for other purposes. One common frequency in use is 2280 Hz and a typical functional diagram is shown in *Figure 6.21*. The 2280-Hz tones are generated for seizing the line, dialling, answer and clear-down signals in either direction, as required. During speech the signalling tones cannot be on the line, but the detector must be monitoring the incoming receive channel to wait for the clear signal from the distant end. On frequency division multiplexing (FDM) carrier circuits out-of-band signalling is employed. The channels are spaced at 4-kHz intervals and the speech channels are restricted to a 300-Hz to 3.4 kHz bandwidth, enabling a signalling channel to be associated with each channel. The signalling channel is usually between 3.4 and 4 kHz and uses a signalling frequency of 3825 Hz. This is available in both directions of the call. In this case signalling is not just restricted to the period prior to call set-up, but can go on during the call, as necessary.

Figure 6.21 *Single-frequency switching.*

Multifrequency interexchange signalling
CCITT R2 signalling is an inter-register signalling system analogous to DTMF but designed for in-band interexchange signalling during the call set-up phase. While the exchange equipment is receiving the dial pulses from the caller a register engages a multifrequency sender and can start to send the routing digits to the next exchange in the hierarchy. The signalling system has the ability to request retransmission of digits if the distant register receives incomplete information. Although faster than decadic signalling it is still limited to link-by-link operation and relies on using the speech path to signal between the different exchanges and, therefore, cannot be used to provide signalling while a call is in progress. The signalling system was mainly used between electromechanical exchanges and semi-electronic exchanges prior to digital working.

Digital signalling CCSS No. 7 With the continued digitization of the telecommunications network, from the application of digital transmission between exchanges and then the digitization within the exchanges, the need arose for a signalling system more suited to full digital interworking. In 1981 CCITT first published a standard in the yellow books for Common Channel Signalling System No. 7 (CCSS7) (see Section 6.1.11). In common channel signalling all the call set-up information, dialled digits, call release information, etc., is sent between the exchanges over a dedicated signalling channel. The signalling links need not be on the same bearer as the speech paths and, therefore, may take a totally separate route through the network. Signalling channels can occupy one or more of the 64 Kbit/s channels on a 2 Mbit/s digital link and it is common to have at least two channels on different 2-Mbit/s systems for increased reliability of the overall signalling system. CCSS7 can be regarded as two separate networks, one for the basic speech information and the other a packet switched data network for handling the variable length text messages that are used to signal the call set-up information. Because CCSS7 operates with a much greater bandwidth than earlier signalling systems, more detailed information can be sent across the signalling network providing more facilities to the user such as the calling parties number. CCSS7 is based on the open systems interconnect (OSI) seven-layer model specified by the International Organization for Standardization. CCSS7 is a set of protocols for the first three layers of the OSI model plus some functions in layers 4 to 7. The key elements in CCSS7 are the 2-Mbit/s links to the digital exchanges and the signal transfer point (STP) or packet switch. A digital network may comprise a number of digital exchanges and a number STPs all interconnected. The STPs do not need to be co-located with the digital exchanges, and individual administrations design their networks to provide the best service in the most economical way and with regard to the likely future growth of the network. Within the CCSS7 specification a number of discrete elements have been defined:

(1) the telephony user part (TUP),
(2) the integrated services digital network (ISDN) user part (ISUP), and
(3) the transaction capabilities part.

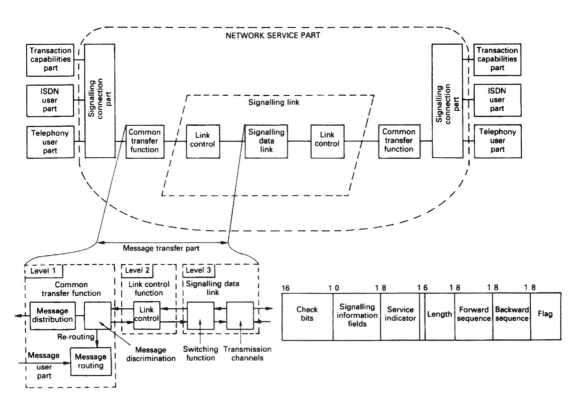

Figure 6.22 *The overall structure of CCSS7.*

Within the network the network-service part is divided into:

(1) the signalling connection part, and
(2) the message transfer part.

Figure 6.22 shows the overall structure of the CCSS7 messaging functions.

6.1.5.3 Switching

There is a great variety of switching systems used throughout the world. The original switching systems were manual operators plugging cords into the switchboard to make the connection and using a series of lights or indicators to monitor the progress of the call before breaking down the connection. Today operators are only used to provide assistance when difficulties arise or to provide a directory enquiry service. Modern switchboards are electronic computer terminals with the call information presented electronically on the screen of a vdu for the operator to interpret.

Electromechanical switching The earliest electromechanical equipment used was the 'Strowger' system, invented by A.B. Strowger in 1889. The mechanism took the place of the manual operator, with the dialled pulses directly controlling the electromagnets (*Figure 6.23*). The pulse trains from the dial lift the vertical pawl to the required level on the bank of contacts and during the interdigit pause the rotary pawl automatically drives the selector round the bank of contacts until the first free outlet is obtained. Groups of switches are connected together, depending on the size of the exchange until the final selector is reached. At the final selector the selector does not automatically hunt round the bank of contacts but waits for the final digit to be dialled and then is driven round the bank of contacts by the dial pulses until the desired number is reached. The ringing tone is then applied. Connections to other exchanges were made at appropriate levels in the switch train. Cross-bar switching is another and later example of electromechanical exchange which was developed to reduce the high level of maintenance being experienced on Stowger switching due to the large number of moving parts which required cleaning, adjustment and contact wear on the wiper contact mechanism. The principle of the cross-bar switch is illustrated in *Figure 6.24*. An electromagnet operates a horizontal cross-bar to which a spring finger is attached. The cross-bar can pull the finger up or down, depending on which magnet is energized, and then a vertical magnet operates a holding bar which traps the finger under a bank of contacts and

the whole assembly is moved by the vertical holding bar until the bank of contacts mate with the other side. The horizontal bar is then released and the vertical bar remains held until the call is released. The horizontal bar is then free to be used on another call.

Figure 6.23 *Strowger Z-motion selector.*

Reed relays were the next family of electromechanical switch. A reed relay is a magnetic spring contact sealed in a glass envelope to provide a dust- and corrosion-free environment for the contact. A number of these contacts are then enclosed in the centre of a coil and, when the coil is energized, the magnetic contact springs close. Arrays of relays are arranged to form a switching matrix. The advantage of the reed relay is its speed of operation, which at around 1 ms is suitable for electronic control. Reed relay exchanges are nearly all under electronic control of some form.

Digital switching Digital switching takes many forms but for voice traffic it concerns the ability to connect individual 64-Kbit/s streams consisting of 8-bit speech samples occurring every 125 µs from one point to the other for one direction of speech, and the same again in the other direction. The analogue signal from the telephone line will have been

Figure 6.24 *The elements of a cross-bar switch.*

previously split into send and receive paths by the hybrid circuit and then the send and receive channels will be converted from analogue to PCM digital bit streams by Coder decoders (CODECs). Digital switching can be space division, which uses a different physical path to transmit the bit stream, or it can be time switching (time division multiplexing). Time-division multiplexing is a multiport switch in which all ports have access to the same physical path onto which transmitted bit streams from individual ports are allocated a unique time slot. The send and receive paths are interconnected by accessing the same time slot. The physical paths may be serial or parallel data highways. Large switches are constructed by using combinations of space–time–space switching, or time–space–time switching.

6.1.6 Business systems

In business applications it is not always practical to have individual telephone lines to each person in the organization and not everyone needs to have

full telephone access to all the world. To cater for the needs of the business environment three basic classes of business communication equipment have evolved: private branch exchanges (PBX), key telephone systems (KTS), and CENTREX. Each has its own particular method of operation and each offers facilities to suit particular business needs.

6.1.6.1 Private branch exchanges
These systems developed from the days when telephone operators were required to connect individual calling lines to one or more people in the organization. Originally they were manual cord-based switchboards where lines appeared on banks of lights (or electromechanical indicators) and underneath each light was a jack socket into which the operator could plug a cord and speak to the caller via a headset. The operator then used another cord to select the desired extension from a bank of jack sockets, and by operating switches could ring the required extension, announce the caller, set up the through-connection and then leave the circuit to attend to another incoming call. The key element was that

a telephone operator controlled the distribution of all incoming calls. Modern PBXs are now all electronic and mainly use digital techniques for switching the calls from one destination to another, but they still retain the operator's position, although the operator's console will have a display, keyboard and even a voice-response unit to play the initial greeting message to the caller.

6.1.6.2 Exchange line circuits

Exchange line circuits provide the interface between the PBX and the public exchange and these circuits are all under processor control. They provide the following basic functions.

A circuit to seize the exchange equipment. This can be loop start which is similar to that of a telephone where the line is short circuited to seize the exchange, or ground start/earth calling where the exchange is seized by applying an earth potential to one of the wires. A ring detector circuit which normally takes the form of an opto-coupler. This senses the incoming ringing signal and passes this onto the PBX call processor to alert the telephone operator. The circuits normally contain the interface between the line and the PBX switching equipment which is normally an isolating transformer and a hybrid two- to four-wire conversion circuit and the CODECs, if it is a digital exchange. Signalling to the exchange is either loop disconnect or DTMF signalling. However, in a PBX it is the line circuit that is sending out the digits not the telephone. The PBX telephone signals to the PBX in the same way as it would to a public exchange and the PBX receives these digits, checks to determine if the call is permitted and the route it should take (public network or private network), and then it sends the digits to the public exchange if permitted by the PBX programming. The digits can be received into the PBX in loop-disconnect form and then retransmitted in DTMF form, or vice versa. Some of the exchange line circuits can be provided with extra functions such as a power failure cut-through facility: if the PBX power fails then a number of lines are automatically enabled to bypass the PBX and go direct to selected telephones. Another common facility, particularly in hotel PBX systems, is meter-pulse detection circuits to detect either the 50 Hz longitudinal meter pulses or the 16/20 kHz meter pulse signals sent by special equipment in the public exchange to advise the PBX of the number of meter pulses being used on a particular call.

6.1.6.3 Line circuits

Line circuits in the PBX perform the same function as in the public exchange but normally can be designed to less demanding requirements as most of the wiring to the telephones is of limited length, i.e. 1 km where public exchange loop wiring is up to 5 or 7 km in length. Most of the wiring is internal and is not subjected to the same environmental problems as external cabling.

6.1.6.4 Tie line trunks

Tie line circuits are used on a PBX to connect two PBXs together over a private circuit, i.e. a direct connection between the two systems provided by the telecommunication administration but which only the PBX can access. These circuits are usually rented and come in a number of forms; direct connections on copper pairs; four-wire amplified circuits; and digital circuits. Each type of circuit has its own form of signalling system for the transmission of call set-up information.

E&M signalling This is the simplest form of signalling and is also used as an interface into multiplex and other transmission equipment because of its simplicity. In E&M signalling the speech pairs are kept separate from the signalling on one pair of wires and the signalling is sent over the other pair.

The contact on the E wire is pulsed to send the dialled digits and at the distant end the relay on the M wire detects the pulses. The sequence is reversed in the opposite direction. The system is limited to short distances due to the resistance that is present in the wiring between the exchanges and the differences in earth potential that can occur between two exchanges.

Signalling on amplified private circuits On amplified private circuits the signalling information must be capable of going through the amplifiers, and to this end a signalling system based on pulses of 2280-Hz tones has been used similar to that used on the public network (see Section 6.1.5.2).

Digital private network signalling (DPNSS) When digital PBXs started to be interconnected with 2 Mbit/s digital private circuits the need for a digital signalling system arose. Digital private network signalling (DPNSS) was produced by a committee of PBX manufacturers in the UK and released as an open standard for any PBX manufacturer to use. It is a common channel signalling system. The signalling system uses a digital bearer comprising 32 channels with channel 0 reserved for frame alignment and synchronization and channel 16 used for signalling. The other 30 channels carry speech. Each

channel runs at 64 Kbit/s to give a link speed of 2.048 Mbit/s. Two such links are required to provide 30 both-way speech channels. When a call is set up on DPNSS a message is sent on channel 16 to the distant PBX with a call request associated with information on who the call is coming from and any service information. The distant PBX responds to this with an acknowledgement message on channel 16 of the other 2-Mbit link in the opposite direction. Channel 16 can carry messages associated with the voice call or can carry virtual messages between the PBXs. These virtual messages can travel between the PBXs at any time to provide sophisticated networking facilities enabling the two PBXs to appear as one large switch to the user. Within the DPNSS specification a range of facilities is specified with sections 1 to 6 specifying basic call set-up and sections 7 to 39 specifying a range of user facilities, e.g. call back when free, add-on conferencing, and centralized operator.

6.1.6.5 Switching
Switching on a PBX uses very similar techniques to public switches but on a smaller scale. With modern digital switching devices it is possible to realize a 1000-port digital switch with just three or four integrated circuits.

6.1.7 Key telephone systems

Key telephone systems originated from the special requirements of business organizations which required greater flexibility in call handling than could be offered by the standard operator answering incoming calls and passing the calls onto the desired party. In key telephone system operation the incoming calls are presented directly to the telephone or group of telephones and the user has the option to answer the call (pick up the line) or let someone else answer the call. In effect each telephone is a prime answering point, like the telephone operator in the PBX operation. Thus the telephones need to be more complex to be able to cater for this extra functionality. Many types of key telephone system are in operation and originally in the USA systems were custom wired on site to suit the individual needs of the customer until American Telephone and Telegraph (ATT) brought out the 1A2 key telephone system which consisted of prewired electromechanical modules and telephones with integral keys and lamps. In the UK the key telephone systems have their roots in the GPO house exchange systems, i.e. large telephones with banks of keys and lamps and multipair cables between the

units. All these old systems have now been replaced with modern microprocessor-controlled designs, but the basic objective is still the same – to provide the business user with more flexible call handling facilities both on the incoming and on the outgoing side.

6.1.7.1 Line circuits and telephones
The exchange line circuits are very similar in operation to that of a PBX and the circuits used to interface to the public exchange are identical. The line circuits to the telephones as well as the telephones are quite unique and each manufacturer has their own implementation. Essentially the line circuit has, in addition to the normal audio circuits, a bidirectional data path to the telephone and in the telephone a data transmitter and receiver. Both ends of this data path are normally under processor control, and in simple terms a call appears on an incoming trunk circuit and the calling message is passed from the trunk to the line circuit data path and a message sent to the telephone. The message is decoded in the telephone and the processor within the telephone will then drive some sort of display, either light-emitting diode (LED) or liquid crystal, to alert the user that a call is waiting. The telephone may or may not ring, depending on the specific application. The message on the data path can be directed to all telephones in the system or only to specific telephones, depending on the way in which the system is configured. When a user answers the call the telephone answering the call sends a message along the data path to the central processor which then removes the calling message from all the other telephones and directs the incoming trunk call to the telephone which answered the call.

6.1.7.2 Switching
Switching on key telephone systems uses very similar technologies to PBX working. Analogue switching is generally by CMOS switches arranged in arrays to provide the size of switch necessary for the application.

6.1.8 Centrex

In Centrex operation the business customer does not have any switching facility on their premises but relies on the public exchange to provide the switching functions. The user is directly connected to the public exchange from every desk in the organization and this may be spread over a wide area and several exchanges. The physical interface to the public

exchange is that of an ordinary telephone and the signalling is normally DTMF. At the public exchange the lines are formed into a closed user group and the user can obtain anyone else in the organization by dialling only four digits in much the same way as a PBX. The public exchange allocates a class of service to each line which determines what facilities the user is permitted, the lowest level being calls within the organization, then local calls, national calls, and finally international calls. The user's numbers are all published, and if the number is known then there are no restrictions on calling individuals in the organization directly without the need for operator intervention. A common business number is also allocated for people who do not know the individual they wish to contact, but want someone in a specific department. In this case the call is answered by an operator. Special signalling circuits are provided between the Centrex and the customer's premises for the operator positions. The advantages of Centrex to the user are that all the maintenance of the switching equipment is taken care of at the local exchange, organization can be spread over multilocations and private networks are not required to interconnect the locations, and a smaller staff is required to maintain the company communications system.

6.1.9 Integrated services digital networks

The principle of integrated services digital networks (ISDN) is to provide voice and non-voice services in the same network. A key element of the service is to provide 64 Kbit/s channels directly to the customer over existing copper cable from the public digital exchange. Thus for voice services this means moving the CODEC function out of the digital exchange line circuits and providing this within the telephone, together with a method of signalling between the customer's equipment and the telephone exchange. Two basic delivery methods have been defined for ISDN: a basic rate service running at 144 Kbit/s yielding 2×64 Kbit/s channels and 16 Kbit/s signalling channel $(2B + D)$ and a primary rate service running at 2.048 Mbit/s yielding 30×64 Kbit/s channels for voice or data and 2×64 Kbit/s D channels for signalling and synchronization.

6.1.10 Numbering plans

CCITT have recommended that a user should not need to dial more than 11 digits on an international call and have prepared a world numbering plan based on a number of zones and within each zone a code has been allocated to each country.

Code Zone

1	North America (including Hawaii and the Caribbean, except Cuba)
2	Africa
3	Europe
4	Europe
5	South America and Cuba
6	Australasia and the South Pacific
7	The area formerly known as the USSR
8	Eastern Asia and the North Pacific
9	Far East and Middle East
0	Spare Code

Within each zone the countries have been allocated a code depending on their size and the telephone traffic density. The country code can be a one-, two- or three-digit code.

The CCITT international numbering plan naturally places restrictions on the way in which individual countries can design a national numbering plan.

(1) The number of digits dialled on an international number cannot exceed 11 and, therefore, national numbering must comply.
(2) Letters cannot be used as part of the national numbering plan as not all countries have letters on dials. Letters, where used, are only as an *aide memoire* to the user as the basic numbering plans are all-figure numbers.
(3) National numbering plans must allocate an international prefix and reserve this for all parts of the country.

National numbering plans were once restricted by the public exchange equipment mainly of the electromechanical type which relied on a specific number of digits to route the call. However, now with computer-controlled exchanges and fast digit translation capabilities it is no longer necessary to have such a rigid numbering scheme. It is still useful to group numbers by exchange area as this gives the user an indication of whether the call is local or long distance. It is also useful in minimizing the number of digits dialled for local calls, as the long-distance access code does not need to be dialled for local calls.

6.1.11 CCITT

CCITT, the International Telegraph and Telephone Consultative Committee, is a committee of the International Telecommunications Union which is an

agency of the United Nations and is based in Geneva. CCITT publish a series of recommendations in alpha numeric order covering numerous aspects of international telecommunications from methods of charging for international calls to maintenance and quality of service standards. The recommendations are published, after being ratified by the plenary assembly, every 4 years in a series of coloured books. The last plenary assembly took place in Melbourne in 1988 and the results were published in what is known as the 'Blue Books' more like a book shelf than book as the amount of technical information is extremely comprehensive.

Bibliography

AT & T, *AT & T Notes on Local Access and Transport Areas*, AT & T, Illinois (1980)

BELL, R., *Private Telecommunications Networks*, Communications Educational Services Ltd, London (1987)

BRITISH STANDARDS INSTITUTION, *BS 6301, Electrical Safety Requirement for Apparatus for Connection to Telecommunications Networks*, Milton Keynes (1989)

BRITISH STANDARDS INSTITUTION, *BS 6317, Simple Telephones for Connection to Public Switched Telephone Networks*, Milton Keynes (1992)

BRITISH STANDARDS INSTITUTION, *BS 6450, Private Branch Exchanges for Connection to Public Switched Telephone Networks*, Milton Keynes (1993)

BRITISH STANDARDS INSTITUTION, *BS 7378, Apparatus for Connection to Public Telecommunications Systems using the Digital Access Signalling System No. 2 (DASS 2) via a 2048 Kbit/s CCITT Recommendation 6.703 Interface*, Milton Keynes (1991)

CCITT, *CCITT Recommendations*, CCITT, Geneva

FLOOD, J.E., *Telecommunications Networks*, P. Peregrinus, London (1975)

GTE, *GTE Automatic electric*, GTE, Northlake, Illinois

GTE, *GTE Interface Manual for Registered Terminal Equipment*, GTE, Northlake, Illinois (1982)

YOUNG, P., *Power of Speech*, Allen & Unwin, London (1983)

6.2 Telephony transmission systems

6.2.1 Introduction

Telecommunication transmission systems have grown to provide communication between two points either in close proximity or at the opposite side of the world. A very high proportion of all the traffic carried on the transmission systems is voice-frequency signals for audio communication in a fairly narrow bandwidth of between 300 and 3400 Hz. This fairly narrow bandwidth out of the 16 Hz to 20 kHz that the human ear can detect and the 100 Hz to 10 kHz that the human voice can generate has been agreed internationally by all the operators of telecommunications systems via the Comité Consultatif International Telephonique et Telegraphique (CCITT) (see Section 6.1.11) recommendations as adequate to provide an acceptable level of intelligibility for the transmission of speech. The rationale behind the adoption of this narrow bandwidth is that most of the energy of the average male voice is contained within the band 125–2000 Hz and that of the average female voice is in the band 300–3400 Hz. The whole structure of telecommunications transmission has been based on the distribution and interconnection of signals within this bandwidth and other forms of information, e.g. data, are usually transmitted via the telephone network using suitable terminal equipment to adapt to the limited bandwidth available on the network. Connection between each telephone can be made over a wide variety of transmission media and via many types of transmission media on a specific telephone call. The transmission media can be overhead open copper wire, twisted-pair copper or aluminium wire, coaxial cable, optical fibre, microwave radio transmission, satellite or undersea cable. The speech signal will become attenuated the further that it travels from its source and will therefore require to be amplified and regenerated to ensure that the distant party can hear the original sound clearly reproduced. The speech signal will also go through many transitions as it moves from one form of transmission medium to another. The longer the connection the more equipment that is used and the greater the cost of the connection. A hierarchical approach has been adopted by all telecommunications operators nationally and internationally. The hierarchical structure contains a number of levels of concentration and switching with progressively more complex means of interconnection to satisfy the many performance criteria that have to be met. Two notable inventions have had a very significant effect on the course of the development of transmission systems. In 1938, Reeves, while studying various methods to improve the quality of speech transmission in the areas of noise, distortion and cross-talk, filed a patent changing the basic concept of audio transmission which had been in use since Bell's invention of the telephone. The idea was to substitute digitally encoded signals for the analogue signals using pulse-code modulation. In 1966 Kao and Hockham proposed the principle of information transmission via a transparent dielectric medium

(e.g. glass fibre) and within 4 years fibres were available with attenuation less than 20 dB/km to make a viable working system. All three were employed by STC's Standard Telecommunication Laboratories now Bell Northern Research.

6.2.2 Network structure

The network structure is made up of the following basic elements:

(1) Internal wiring on the customer's premises which connects individual telephones to each other, or in business premises connect each telephone via a private branch exchange (PBX) and then out to the network.
(2) The distribution network consisting of cables from the customer's premises to distribution points. These are small connection boxes which enable the individual cables from the subscriber's premises to be connected to larger cables, then on to 'pillars', which are larger cross-connection frames. These enable the smaller cables to be connected to the larger multipair cables and then on to much larger cross-connection frames ('cabinets') as the cables are gathered together in ever increasing numbers towards the telephone exchange.
(3) The local exchange (in North America 'class 5 end office') where the customer's access cable is terminated on a main distribution frame (MDF) and then cross-connected to the telephone switching equipment of the local exchange.
(4) Local exchanges are interconnected via junction networks to either other local exchanges or local tandem exchanges (in North America class 5 offices are interconnected via trunk networks to class 4 toll offices).
(5) The next level in the hierarchy is the trunk network (in North America the toll network) made up of secondary trunk switching centres and tertiary trunk switching centres at the top (in North America class 3 primary centres and Class 2 sectional centres interconnected by toll circuits).
(6) The final level in the hierarchy is the international network with international gateway exchanges (in North America class 1 offices).

Where there is an appropriate community of interest with sufficient telephone traffic to justify the expense of the route, individual exchanges are directly interconnected at the same level in the hierarchy. Where this is not the case exchanges are connected at the next highest level.

6.2.3 Transmission media

In telephone transmission many forms of transmission media are used to convey speech:

(1) bare copper wire on insulators at the top of telegraph poles,
(2) copper or aluminium twisted-pair in multiple cables,
(3) coaxial cable,
(4) optical-fibre cable, and
(5) wireless propagation.

Each of these media have their own particular characteristics and these have been exploited, depending on the application, in telephony. Overhead bare copper wire was used in the early days of telephony where a pair of individual wires was distributed from the telephone exchange to the customer, but this form of cabling has all but disappeared.

6.2.3.1 Twisted-pair cable

A 'twisted pair', as the name implies, consists of two individually insulated wires twisted together to prevent cross-talk between pairs in the cable. Copper wire (normally 0.5 mm diameter) is normally preferred for the conducting core, but aluminium is sometimes used for lower cost. Each of the two wires is surrounded by its own colour-coded polyethylene, polyvinal chloride (PVC) or, in early days of multicore cable, paper insulation. Anything from 6 to 4000 twisted pairs can be layed up together to form a multicore cable.

Distribution of twisted-pair cables To enable customers to be given service without delay, every distribution network must contain spare pairs in the cable. For economy the telephone operators require to balance the number of spare pairs in the cable for future usage against the cost of pulling in additional cables. This has resulted in the need for a flexible distribution system from the customer's premises to the local exchange. A typical telephone operator's distribution network is shown in *Figure 6.25*. Groups of customers are connected to a distribution point which is a multiple termination block either in a building or on a telegraph pole. The wiring from the telegraph pole to the customer's premises is strengthened to withstand the strain of its own weight and snow, and is normally called the 'drop wire'.

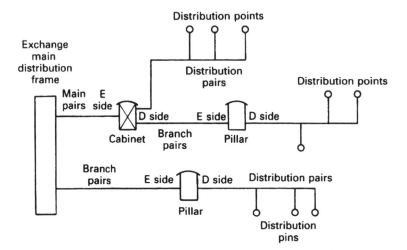

Figure 6.25 *Typical operator's distribution network. MDF, Main distribution frame.*

Distribution points are connected by multipair cables to a street cabinet where a small distribution frame provides flexibility to interconnect the cable pairs and to connect the cable pairs to larger cables and then onto the telephone exchange.

Characteristics of speech in twisted-pair cable

The response of a particular telephone line to carry a speech signal will depend on the basic line characteristics, the frequency of the signal and the length of the line. When a telephone speech signal is transmitted along a cable attenuation occurs of the speech signal a typical attenuation curve for a cable is shown in *Figure 6.26*. The higher frequencies are attenuated more than the lower frequencies and the quality of the speech signal will degrade as the distance increases. For this reason the connection from the local telephone exchange to the customer's premises

is normally limited to no more than about 10 km (7 miles) of copper. Techniques have been used to compensate for the degradation of signal with line length. The attenuation is mainly due to the capacitance in the cable and as the distance increases then so does the capacitance. The effect of the capacitance in the cable is to make the current and voltage of the signal go out of phase resulting in a loss of power at the other end of the cable. In the early days of telephony, inductance was added to the cable and this was termed 'loading'. This enabled the speech signal to be transmitted over much greater distances, but at the expense of a reduction in the available bandwidth as the introduced inductance prevented any signal greater than 3400 Hz being transmitted (see *Figure 6.26*). Loaded cables were very effective on early long-distance trunk circuits, increasing the

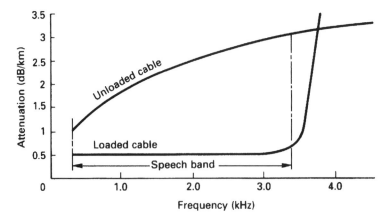

Figure 6.26 *Typical attenuation curve for a cable.*

distance on a heavy duty copper cable line from 20 to 110 km (12 to 70 miles). This technique of loading cables has been superseded by other techniques as the need to make use of the full bandwidth of the twisted-pair cable emerged. Modern telephones now have the characteristic that as the frequency of the sending signal increases then so does the signal transmitted to line to counteract the effect of the attenuation of the higher frequencies in the cable.

6.2.3.2 Coaxial cable
Coaxial cable can be used for transmitting much higher frequency signals. A coaxial cable consists of two conductors: an inner solid copper conductor surrounded by a concentric screen separated by insulation. To minimize dielectric loss the insulation consists mainly of air. The conductors are separated by either polyethylene disks at intervals or an extruded honeycomb of polyethylene. At high frequencies the skin effect causes the current to flow at the perimeter of the inner conductor. This is due to the non-uniform distribution of the magnetic field within the conductor forcing the current to flow near the surface of the conductor. The current tends to flow on the inner surface of the outer conductor. This makes the coaxial cable ideal for high-frequency transmission, as any interfering currents tend to flow on the outside of the outer conductor, providing an efficient screen and having little effect on the main signal being transmitted. Two basic sizes of coaxial cable are in use 1.2/4.4 and 2.6/9.5 mm (inner/outer conductor diameters).

6.2.3.3 Optical-fibre cable
Optical-fibre cable has a number of significant advantages over copper cable:

(1) large bandwidth,
(2) low loss resulting in fewer repeaters on long distances,
(3) electrical isolation,
(4) freedom from electromagnetic interference,
(5) better security (more difficult to tap into than copper), and
(6) failure is open circuit – there can be no short-circuit faults.

Optical-fibre transmission works on the principle that light travels more slowly in an optically dense medium than it does in a less dense medium. Two parallel glass–air interfaces will trap an incident ray of light which has a sufficiently shallow angle of incidence and transmit it over the length of the glass. A glass cylinder behaves in a similar manner except that most of the light travels in a helical manner within the cylinder. Some light will be lost at the surface of an air–glass interface and, therefore, an optical cladding with a much lower index of refraction than the fibre must be used to ensure total reflection within the fibre. The loss of light along the fibre is caused by imperfect reflection due to scratches or absorption. It is measured by a power ratio in decibels and is exponential, i.e. the longer the fibre the more loss in power.

6.2.3.4 Submarine cables
Submarine transmission systems have some special characteristics because of their environment. The repeaters require to be highly reliable as the cost of recovering and repairing a repeater from deep water is extremely high and time consuming, if possible at all. Apart from the physical requirements of armouring the cable and providing tensile strength with steel wire and tape to enable the cable to remain operational under the immense pressure of deep water, the cable system has other unique features. Unlike land-based systems, where power can be made readily available at each repeater station or radio relay station, power cannot be fed into the cable transmission system in mid-ocean. The repeater amplifiers need to be powered and this is done by feeding a high d.c. voltage from each end of the cable, one at a negative potential and the other at a positive potential. The return path is via the sea and the d.c. is fed through a central conductor. Voltages of the order of $+10$ and -10 kV are common, giving a total potential of 20 kV. With low-loss optical-fibre cables up to 180 km can now be reached without the need for any undersea electronics which makes optical-fibre undersea cables a very cost-effective option over radio or satellite links. With these forms of undersea optical-fibre cable the land-based terminal equipment can be upgraded from, say, 565 Mbit/s systems to 2.488 Gbit/s synchronized digital hierarchy (SDH) systems or beyond by changing the terminal equipment and continuing to use the same fibre. This makes island hopping and routes across sea channels very attractive to telecommunications administrations who are not sure of how the demand for additional circuits is going to increase. Optical amplifiers will also play a significant part in future transoceanic cables. The optical amplifier will replace the electronic repeater. The optical amplifier consists of a length of fibre doped with the rare-earth metal eribium in-line with the incoming weak signal and fused onto another length of fibre with a laser pumping a new beam of light into the fibre. The

new beam of light and the weak incoming signal are mixed and the resulting output is an optically amplified signal capable of going on for the next 100 km before requiring amplification again.

6.2.3.5 Wireless microwave propagation

Line-of-sight microwave transmission is fairly common in telecommunications systems. A narrow beam of microwave energy is focused by a transmitting antenna mounted on a parabolic dish reflector towards a corresponding antenna 50–60 km away. Microwave radio transmission was first developed for use in telecommunications in the 1940s for broadband television transmission, but quickly became adapted for high capacity telephone traffic routes providing typically 1800 telephone channels on one link. The microwave antenna consists of a parabolic dish with a launching unit situated at its focal point. Modern microwave transmitters operate in the range 2–23 GHz and cover distances of 50–60 km between aerials. At each hop between aerials the incoming signal is translated to an intermediate frequency, amplified and then stepped up to the main carrier frequency for onward transmission. The incoming and outgoing signals will be at different frequencies, typically 200 MHz apart, to limit the coupling between the receiving aerial and the transmitting aerial on the same structure. As a telephone connection, unlike a television broadcast, requires a bidirectional link, the signals in the reverse direction are arranged to be 200 MHz apart from the signal in the forward direction. Aerials are arranged in a zig-zag manner along a route and this ensures that even under adverse conditions the attenuation of a signal from the first aerial reaching the fourth aerial is very high, being of the order of 50 dB for a shallow angle of 10° (*Figure 6.27*).

Figure 6.27 *Siting of microwave radio stations.*

6.2.3.6 Satellite communications

Satellite systems are a particular form of microwave radio system but, due to the very long distances involved, they present their own particular problems to the communications engineer. The most significant problem to be catered for in the design of any satellite communication system for speech transmission is the delay. The closest an earth station can ever be

to a satellite is 22 300 miles, and that is if the station is on the equator and directly under the satellite, so for most of the world's population the distance is going to be much greater. With electromagnetic waves travelling at 300 000 km/s (186 000 miles/s) that means that a signal will take about a quarter of a second to go up to the satellite and then the same again on the way down. From the telephone user's aspect this means that both parties can start talking without the other realizing that the conversation˙has started, causing collisions and reducing the conversation to an almost simplex working mode.

6.2.4 Quality of telephone transmission

The objective of any telephone transmission system is to ensure that any two users can converse over a telephone network, wherever they may be located, without experiencing any difficulties in understanding each other. To achieve these aims various key parameters have evolved throughout the history of telecommunications to define the performance of a transmission system. Methods have been defined to measure these parameters to enable consistent standards to be set for each element in the transmission system. Because the system includes the user's perception of the ability of the system to convert the sound from the mouth to the telephone transmitter and from the receiver to the ear as well as the transmission through the network, many of the parameters have had to include subjective assessment by users before limits were set on the parameters or the impairments to the transmission of speech.

6.2.4.1 Loudness

Loudness is a key element in speech transmission and one of the most difficult to assess as the normal direct air path in a face-to-face conversation is replaced with the telephone instrument and the transmission network. The problem is how to get enough volume of speech to the listener for the talker to be heard. If too much loss of loudness is present in a transmission path the user at the distant end becomes dissatisfied as listening effort is increased, and with very high loss of loudness the intelligibility of the received sound begins to suffer. Conversely, if the received sound is too loud the listener is again dissatisfied as the telephone is uncomfortable to use. In general, the loudness of a transmission circuit is expressed in terms of transmission loss, independent of frequency, that must be present in a speech path. Values of loudness rating have been measured and defined for various elements in the network.

Send loudness rating The send loudness rating is the loudness loss between the talker's mouth and an electrical interface in the network, measured in decibels.

Receive loudness rating The receive loudness rating is the loudness loss between the listener's ear and an electrical interface in the network, measured in decibels.

Overall loudness rating The overall loudness rating is the loudness loss between the talker's mouth and the listener's ear via a speech path in the network, measured in decibels.

Transmission planning with loudness ratings A typical speech connection consists of a number of elements, local loop cables, local exchange, transmission circuits, multiplexors, etc., and for transmission purposes it is convenient to evaluate the loudness ratings of the individual elements (*Figure 6.28*). The losses between the telephones are called 'circuit loudness ratings' (CLRs). The overall loudness rating (OLR) is given by:

$$OLR = SLR + CLR_1 + CLR_2 + CLR_3$$
$$+ \cdots + CLR_n + RLR$$

where SLR is the send loudness rating and RLR is the receive loudness rating.

6.2.4.2 Side-tone

When a telephone user speaks, various paths exist to transmit the sound from the talker's mouth back to his own ears, via the mechanical path of the handset, the acoustic path around the head, and in the telephone circuit due to mismatch at the hybrid balance impedance between telephone and telephone line. The effect of side-tone is that if it is too loud the talker lowers his or her voice and the listener has difficulty in hearing, or the talker moves the handset away from his or her head, thus reducing the effectiveness of the transmitter. In areas with high ambient noise, then all the room noise is also picked up and transmitted back to the user, which could impair the incoming speech signal.

Side-tone masking rating The side-tone masking rating (STMR) is the loudness loss between the talker's mouth and his or her ear via the electrical side-tone path in the telephone, measured in decibels. For telephone transmission planning purposes the STMR can be calculated using the telephone's send loudness rating and receive loudness rating and the balance return loss between the mean line impedance and the side-tone balance impedance (see *CCITT Recommendation G111* where a suitable algorithm is specified). Subjective tests have shown that an STMR of 7–12 dB is acceptable for normal operation and high values of the order of 20 dB cause the user to think that the connection has gone dead.

Listener side-tone rating Listener side-tone rating is the loudness loss between room noise sources and the talker's ear via the electrical side-tone path in the telephone measured in decibels. To control the effects of high room noise subjective tests have shown that a value of 13 dB for listener side-tone rating is necessary.

6.2.4.3 Signal propagation time (or delay)

Frequently telephone speech paths are set up over very long distances and even with the signals in fibre-optic cable travelling at the speed of light the user can detect the delay and this can cause difficulties in the conversation if the talker has to await a discernable time before receiving a reply from the listener. Delay

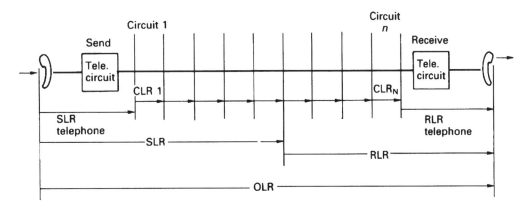

Figure 6.28 *Loudness rating transmission planning. OLR, overall loudness rating; SLR, send loudness rating; RLR, receive loudness rating; CLR, circuit loudness rating.*

can also cause further problems if the signal is reflected back from the distant end due to some mismatch and creates an echo. A delay of 25–300 ms has proved to give acceptable user performance; delays above 400 ms are totally unacceptable for most users. Delay can be introduced from many sources in a speech connection. The following gives some typical sources and the delays introduced.

Coaxial cable	4 µs/km
Radio relay systems	4 µs/km
Frequency division multiplexing	4 µs/km
Optical-fibre cable system	5 µs/km
Satellite link geostationary	260 ms
Frequency division multiplexing channel modulator/demodulator	0.75 ms
Pulse code modulation coder or decoder	0.3 ms
Echo cancellers	1 ms
Digital exchange (all digital)	0.45 ms
Digital exchange (analogue in & out)	1.5 ms
Integrated services digital network basic rate access	1 ms
Mobile network	90 ms

In large European countries the delay over the national network can be as much as 15 ms.

6.2.4.4 Echo

Echo is an undesirable effect that can occur in a telephone speech path when the speech signal in a four-wire transmission path reaches an imperfect line balance hybrid and part of the signal is transmitted back down the transmission path in the receive direction. Another source of echo is the acoustic path between the distant receiver and transmitter. Electrical echo can be eliminated by the use of end-to-end four-wire transmission. If the propagation time for the speech signal is short (1–2 ms) the talker will hear the echo as side-tone, but if the propagation time is longer than 15 ms or so the talker will hear the returned signal as echo.

Talker echo loudness rating Talker echo loudness rating is the loudness loss between the talker's mouth and his or her ear via the echo path along the transmit circuit through the distant four-wire to two-wire hybrid termination and back along the receive path to the talker's ear, measured in decibels. To reduce the level of talker echo various echo control devices are fitted in the transmission path. With the increasing digitalization of the telephone network the overall loss on any connection is reduced and performance of the network is better. However, this also

leads to a reduction in loss in the talker echo path, which is not desirable. Users are increasingly expecting improved performance, but reduction in the loss of the talker echo path will not be seen as a performance improvement.

Listener echo Listener echo is the portion of the talker echo signal that gets retransmitted from the talker end of the connection and arrives at the listener end of the connection delayed from the main signal. The listener echo signal is therefore characterized by additional loss and additional delay from the main signal and if the delay is long the listener reports a 'rain barrel' effect.

Echo control Electrical echo can be controlled by the use of either echo cancellers or echo suppressors. Subjective tests have shown that echo cancellers provide better performance in a transmission network compared with echo suppressors. An echo suppressor is essentially a pair of voice-operated switches which insert a loss of 35 dB or above in the echo return path while one user is talking. If both ends are talking simultaneously then the talker who has the strongest signal controls the echo suppressor (*Figure 6.29*). The problem with echo suppressors is that they can sometimes clip the beginning of words. Echo cancellers work on the principle of correlation of the incoming-signal on the four-wire side of the hybrid with the out-going signal and determine the filters to parameters to be switched in.

6.2.4.5 Distortion

A number of forms of distortion are of interest in telecommunication circuits: non-linear distortion, quantizing distortion, group delay distortion, and attenuation distortion. Not all forms of distortion are significant to the quality and intelligibility of speech due to the very nature of speech signals and some types of distortion are more important in achieving consistent data transmission due to the way data make use of specific frequencies for individual character transmission.

Non-linear distortion Non-linear distortion occurs in systems where the output signal is not a linear function of the input signal, resulting in the creation of harmonics of the input at the output. The major source of non-linear distortion in telephony has been the carbon microphone and subjective tests have shown that linear microphones give an improvement at the optimum listening level, but little difference in other environments. Non-linear distortion is not regarded as a significant factor for speech transmission and has more importance in data transmission.

Figure 6.29 *Echo paths or two- to four-wire transmission circuits.*

Attenuation distortion On a telephone circuit some frequencies will be attenuated more than others. To assess this characteristic the loss/frequency response for each frequency is measured relative to the loss at a reference frequency of 800 or 1000 Hz. The effect of attenuation distortion is that the voice sounds higher or lower at different frequencies. The effect on loudness tends to be greater at the lower end than at the higher frequencies. The effect is more pronounced on intelligibility of the speech signal at the higher frequencies. This is because the 'sharpness' and hence intelligibility of speech is contained in the higher frequencies.

Group delay distortion On a telephone circuit some frequencies will be slower than others at propagating through the network. The main source of this group delay distortion is the presence of filters in a telephone channel. The bandpass filters used in multiplexing equipment increase the delay at the upper and lower frequency extremities of each channel and the effect of loading coils on a long cable route creates a low-pass filter which increases the delay at high frequencies. The effect of group delay distortion on speech at the upper edges of the band is to cause ringing (an oscillation at the affected frequency) and at the lower edges to cause blurring of the speech. However, these effects are only significant in speech circuits with severe delays as the human ear is relatively slow in this respect, requiring 4 ms to react to a change. Group delay distortion is therefore more important to data transmission where the detectors

rely on all the frequencies arriving at the same time to ensure correct decoding of the incoming signal.

Quantizing distortion Quantizing distortion occurs in any telephone circuit where the analogue signal is sampled and each sample is then allocated a discrete code from a coding algorithm. The distortion is a measure of the difference between the original signal with the decoded signal reconstructed into as near the original form as the coding algorithm will allow. In practice the measurement unit has been based on the pulse code modulation 8-bit codec analogue-to-digital and digital-to-analogue conversion and this is 1 quantizing distortion unit (qdu). All other conversions are referenced to this basic process. The effect on speech of 1 qdu is negligible, but it is the accumulation of these throughout the network on a particular call that is important, and CCITT have set a limit of 14 qdu on an international call. The following are some of the sources of quantizing distortion units.

8-bit pulse code modulation codec pair	1 qdu
32 kbit adaptive differential pulse code modulation	3.5 qdu
A-law to μ-law convertor	0.5 qdu

6.2.4.6 Cross-talk

Cross-talk is any sound picked up in a telephone circuit from another telephone circuit. Low levels of cross-talk can be intelligible and cause the user to complain of lack of secrecy. Audible cross-talk at

low levels is dependent on a number of factors in addition to the actual coupling between the channels, e.g. the channel noise, room noise, telephone-set noise, and the state of the conversation (a period of silence or the caller is listening for call progress tones). Cross-talk has numerous possible paths and can originate from line capacitance, mutual inductance or intermodulation products of adjacent carrier channels. Cross-talk limits of 65 dB are usually acceptable and normally inaudible to the user.

6.2.4.7 Noise

Noise will be present on all telephone circuits and results from internal or external sources. Internal noise may be due to bad connections or electromagnetic relays causing fluctuations in voltage, and external noise can be due to power line interference or electromagnetic interference from other equipment. In addition, internally there is thermal or white noise due to electron agitation in electronic components. There are two classifications of noise for measurement purposes: random circuit noise and impulsive noise. The former can be controlled by setting a limit for the circuit and ensuring the random noise power does not exceed this predefined limit. Impulsive noise is more difficult to control and can come from dial pulses in adjacent circuits. These are much more annoying to the user as they appear in short bursts.

6.2.5 Transmission systems

The number of transmission systems in use to convey telephone traffic is enormous and increasing as new techniques are evolved. They can be roughly classified as follows:

(1) two-wire analogue,
(2) two-wire digital,
(3) four-wire analogue, and
(4) four-wire digital.

6.2.5.1 Two-wire circuits

The basic telephone has four transmission wires, two from the transmitter and two from the receiver, and these need to be combined to make connection onto the two wires from the local exchange. This has been achieved with some forms of hybrid termination circuit. Until the introduction of digital local exchanges the two-wire analogue transmission was used up to the trunk network where it was converted with a hybrid into send and receive four-wire paths. With digital local exchanges the two- to four-wire conversion is done at the entry point into the local exchange. Early telephones had a three-winding transformer for the two- to four-wire conversion and side-tone was created by purposely mismatching the hybrid and feeding a portion of the signal back into the receiver. Modern telephones use electronic circuits, but the principle is the same. The impedance of the balance network is therefore at best a compromise impedance match since the characteristic impedance of any two-wire line will vary from one line to the other and there will be a small amount of energy reflected from the mismatch. The degree of accuracy with which the impedance Z_b, of the balance network simulates the characteristic impedance of the line Z_0 is known as the 'balance return loss'. The balance return loss (BRL) is given by:

$$\text{BRL} = 20\log_{10}(Z_0 + Z_b/Z_0 - Z_b)$$

A value of at least 20 dB is necessary for all practical circuits.

6.2.5.2 Four-wire circuits

The standard method of operating long-distance telephone transmission circuits is to maintain separate paths for the transmit and receive direction over the whole distance with the signal amplified at various points along the route (*Figure 6.30*). At each end of the line a hybrid transformer connects the send and receive paths onto the two-wire line and a balance impedance similar to that of the two-wire line. If both ends of the four-wire line were simply connected together a signal could circulate around the complete loop which would result in 'singing' or continuous oscillation. The output signal from the receive amplifier causes equal and opposite voltages to be induced into the secondary of transformer T_1. If the impedances of the two-wire line and the balance network are equal then equal currents will flow in each of the primary windings of T_2 and as these windings are connected in antiphase then no signal will be coupled to the secondary and no input will appear at the send amplifier. The output power from the receive amplifier is divided equally between the two-wire line and the balance impedance and thus will yield a 3 dB loss. In the opposite direction the signal appearing on the two-wire line results in zero current flowing in the balance impedance and the power is split equally between the send and receive amplifiers. The portion presented to the output of the receive amplifier has no effect on the receive amplifier as the receive amplifier is a unidirectional device. The

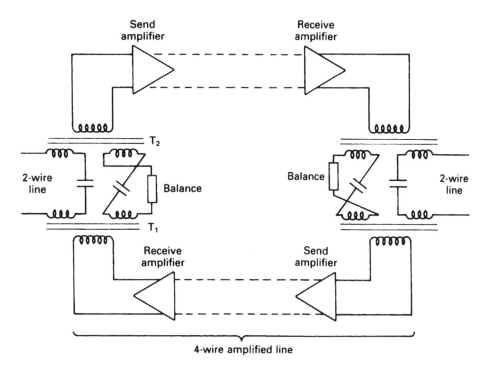

Figure 6.30 *Two- to four-wire conversion for long-distance amplified circuits.*

portion presented to the send amplifier will be transmitted down the line. As with the receive side a 3 dB loss occurs as the input power is divided equally by the hybrid transformer. *Figure 6.30* shows a physical audio channel, but in practice the channel may not be physical pairs but rather a frequency allocation in a high-frequency transmission system or a channel in a digital transmission system.

6.2.5.3 Multiplexing

In the early days of telephony, telephone exchanges were interconnected by one pair of wires and the distance was limited by the gauge of cable and the loading. Later, amplified circuits were used and the existing cables had the loading coils removed. The circuits between the exchanges were still in pair cable and two pairs of wires were allocated to each circuit to form a transmit-and-receive path which could be amplified. As the number of circuits between exchanges increased then the need to make more use of the existing copper cable to carry more traffic became a necessity. However, solid copper cable has a much greater bandwidth than 4000 Hz and the amplifiers on the lines could be designed to cater for a much higher frequency range.

Frequency division multiplexing Frequency division multiplexing (FDM) was the first technique

used to increase the number of speech channels on two pairs of wires between exchanges. In its very simplest form, channel 1 was transmitted unchanged at 0–4000 Hz, channel 2 was translated into the range 4000–8000 Hz, channel 3 into 8000–12 000 Hz, and so on. Adding all the channels together produced a broad-band signal that could be treated as one signal as far as line amplifiers are concerned. The channels could then be separated at the distant end. To achieve the 'lifting' of each individual channel from its baseband of 300–3400 Hz to the higher carrier frequency the base-band frequency is modulated onto a carrier frequency. The modulation wave is a composite of the carrier frequency, the sum of the carrier frequency and the base-band signal, and the difference between the carrier frequency and the base-band frequency. These latter two are known as the 'side-bands'. As each side-band carries the speech signal it is normal practice to suppress the upper side-band with low-pass or band-pass filters and transmit only the lower side-band. The carrier is also suppressed to minimize the total power to be handled by the amplifiers. Practical systems cannot be exactly 3100 Hz (base-band) wide due to the limitations of the cut-off frequency of the band-pass filters so a guard band is left between each band of

frequencies. The guard band is 900 Hz, making the spacing between each carrier 4000 Hz (*Figure 6.31*).

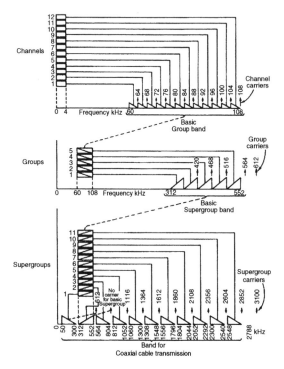

Figure 6.31 *Assembly of channels spaced at 4-kHz intervals in a high-frequency transmission system.*

Time division multiplexing Time division multiplexing (TDM) works on the principle of transmitting the speech signal of 300–3400 Hz over the whole bandwidth of the wide-band transmission channel, but for only a short period of time for each speech so that a number of speech channels can share one wide-band transmission channel. Thus TDM does not divide a transmission line into frequency bands but rather assigns a short time interval to each channel in rotation known as a 'time slot'. The human ear has a reaction time of approximately 4 ms so, if speech is presented as a series of short pulses each less than 4 ms long, the recipient will hear the speech as a continuous sound, the ear not having sufficient time to react to the discontinuities. A sample of the speech is therefore transmitted down the line and if the sample is to be recovered without distortion it is necessary to use a sampling rate at least twice the bandwidth of the basic signal. In simple terms a sample must detect every change in direction of the analogue signal (up or down). Therefore, if the bandwidth of the analogue signal is

limited to 4000 Hz then the sampling rate must be twice the bandwidth frequency to detect all changes in direction of the analogue signal. Thus 8 kHz is used for sampling speech-band signals.

Pulse amplitude modulation The amplitude of the analogue speech signal is sampled every 125 μs (i.e. the 8-kHz rate) and transmitted down the line. This is known as 'pulse amplitude modulation' (PAM) and some early systems used this technique for transmission of telephone signals. However, the amplitude of the signal can be attenuated by the transmission line and will require to be amplified at points along the route to reconstitute the original signal. The transmission line will also pick up noise from all sources, adjacent channels, etc., and this noise will be added to the original signal so that the end result will be a signal of a greater amplitude than desired. Other forms of analogue modulation are possible but these have not been used for telephone transmission, and analogue transmission has generally given way to digital transmission.

6.2.5.4 Digital transmission

Digital transmission has been used to improve the quality of the speech signal transmitted, particularly with respect to the attenuation and distortion that an analogue signal can suffer when transmitted over long distances.

Pulse code modulation In pulse code modulation (PCM) the speech signal is sampled every 125 μs, twice the highest frequency to be transmitted and then the sample is coded into an 8-bit binary code which is transmitted as a series of on/off pulses down the line. It is then decoded and converted back at the other end to an analogue speech signal. This two-step process converts an analogue speech signal into a digital stream of pulses at 64 000 bit/s, i.e. 8 bits × 8000 samples per second. A number of speech signals can be transmitted down the line at the same time by allocating a short period of time for the transmission of the 8-bit code to each speech signal in turn, i.e. a time slot. The time slot is 125 μs, the time required to transmit the 8-bit coded sample of speech.

The transmission line then is presented by *n* speech channels of 64 000 bit/s, where *n* is commonly either 24 or 32, giving a line rate of 1 536 000 or 2 048 000 bit/s.

PCM coding algorithms An analogue speech signal is infinitely variable with time and over a wide frequency range, but most of the intelligence is contained in a narrow band around 1000 Hz and the coding algorithms make use of this principle. The

first seven bits are used to describe the magnitude of the signal sampled and the eighth bit is used to indicate whether the signal is positive or negative. The coder/decoder (Codec) therefore samples the speech signal every 125 μs, compares the voltage level of the signal and its sign (+ or −) with the value in the coder and then transmits the 8-bit binary value for the sample to the line. Two coding algorithms are in common use in telecommunications, the A-law and μ-law. A-law is used in North American and Japanese systems and μ-law is used in European and some other countries.

Differential PCM Voice signals vary relatively slow and seldom change over the full volume amplitude range in under 1 ms. This means that it is possible to convey the same speech signal with fewer bits if the encoding algorithm measures only the difference between the successive samples rather than the full dynamic range of each sample. PCM sends the absolute value of the analogue input but differential PCM sends only the change since the last sample. The differential can be encoded into 4 bits, thus saving on bandwidth or applying voice compression. The disadvantage of this system is that it cannot cope with sharp changes in frequency or volume and may cause distortion in some cases (*Figure 6.32*).

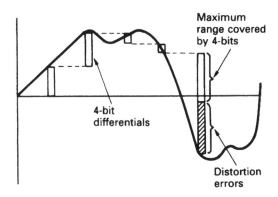

Figure 6.32 *Differential pulse code modulation.*

Adaptive differential PCM By varying the meaning of the 4-bit word then the algorithm can be used to suit different conditions of the signal applied. As differential PCM only sends the change since the last sample it would be difficult for the system to cope with a rapidly changing signal. If the range that the sample covered was increased for loud signals then the sample can still be represented in a 4-bit word. When the volume drops the range covered by the 4 bits reduces; thus the algorithm adapts

to the change in signal. If the sampling rate is maintained at twice the band-width of the speech signal, i.e. 8 kHz, and only 4 bits are used to represent the change then the speech signal can be transmitted at 32 000 bit/s.

Continuously variable slope delta modulation In this form of algorithm for encoding the speech signal the signal that is sent to line relates to the change in the slope of the speech signal. The speech sample is compared with a reference level and if the signal is greater than the reference then the binary digit 1 is sent to line and the reference level is increased. If the speech signal is less than the reference level then binary 0 is sent to line and the reference level is decreased. The encoding algorithm therefore follows the input signal up or down the slope of the input speech signal. A further refinement is added to the algorithm to improve the reproduction of the signal: the steeper the slope (in any direction), represented by successive transmission of all 0 s or all 1 s, the more the algorithm increases the size of the step taken between samples. The distant end reconstructs the changing slope of the signal and smooths out any irregularities in the signal. The sampling rate is dependent on the quality of speech signal required and can vary from 64 000 to 9600 bit/s to give acceptable quality of speech transmission.

Codebook excited linear predictive coding This form of encoding of a speech signal can achieve very low bit rates for the transmission of speech. Codebook excited linear predictive (CELP) was originally developed at AT&T Bell Laboratories. CELP transmission is essentially the analysis of the speech signal in real time. A section of speech signal, e.g. a 20-ms block, is compared with a look-up table and then only the table value is transmitted down the line. At the receiving end the code is passed through the look-up table and the original speech signal is reconstructed. This method is possible with speech signals as speech does not have any dramatic variation over a short time such as 20 ms and various algorithms have been developed to exploit this to reduce speech transmission down to 2400 bit/s with acceptable quality as perceived by the listener. The technique relies on the speed with which digital signal processors can compute the value that matches the speech signal presented.

Time-assigned speech interpolation In a normal telephone conversation one party tends to be speaking while the other party is listening. This can be made use of on expensive long-distance circuits to make maximum use of all the available bandwidth

for as many conversations as possible. Each input circuit is continuously monitored for the presence or absence of a speech signal and if no speech signal is present then the circuit is disconnected and made available for another input that has a speech signal present. Time-assigned speech interpolation (TASI) can give speech compression ratios of up to 3:1 on transatlantic or satellite links. A problem that can be encountered with this sort of system is that clipping of the speech signal can occur if the switching in of channels is not fast enough. The system also needs to have a fast signalling channel dedicated to informing the other end which channel is transmitting the speech signal.

6.2.5.5 Line coding

It is possible to transmit simple binary digits over a pair of twisted wires but difficulties can arise if the data is left in raw binary format. The raw binary format may legitimately contain a series of all 1 s or 0 s and this can charge the line up with the d.c. component present in the data stream and not provide sufficient transitions for the distant end to synchronize and recover the data from the incoming stream. Two broad classifications exist for line coding and block coding where block coding makes use of look-up tables to convert the signal. Alternate mark inversion (AMI) is an example of linear coding where a 0 is sent to line as the absence of a pulse but 1 s are sent alternately as positive- then negative-going pulses. The resulting waveform then has a mean voltage very close to zero. 2B1Q (two (2) level binary input to give a one (1) four-level quaternary output) is an example of block coding used extensively for integrater services digital network (ISDN) transmission on a single pair of wires. In this type of encoding scheme 2 bits are converted into one four-level quaternary output signal, i.e. sequential pairs of bits are translated into four voltage levels +2.5, +0.8, −0.8 and −2.5 V. The net effect of this line code is to limit the frequency components in the transmission line waveform to below 80 kHz from the 160 kHz input.

Thus the input signals in combinations of 2 bits are put through a look-up table (*Figure 6.33*) and the output waveform is as shown in *Figure 6.34*. Thus with this coding method the information transfer rate is the same but the information is transferred in one-quarter of the bandwidth of straight binary coding.

6.2.5.6 Digital hierarchy

Just as in FDM systems where speech channels are multiplexed together to form larger and larger groups for transmission over wide-band channels, digital

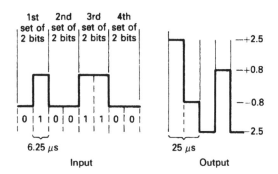

Figure 6.33 *Conversion to a 2B1Q line code.*

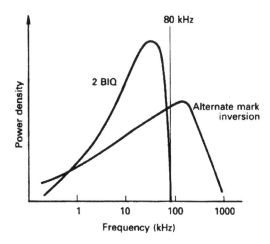

Figure 6.34 *Plot of line code power density versus frequency. AMI, alternate mark inversion.*

systems have grown up and been successively multiplexed together. North American and European systems have been built up on different foundations. In North America the base level in the hierarchy is 64 Kbit/s for on speech channel and is known as DS–0 (digital signal level zero). The next level is 24 channels of 64 Kbit/s plus 8 Kbit/s for control giving 1.544 Mbit/s, known as DS–1, or digital signal level 1. This is often referred to as T1 which was derived from the physical cable terminating equipment associated with the digital transmission system. From this basis a series of different orders of multiplexing was developed to produce the following levels:

		No. of circuits
DS–0	64 Kbit/s	1
DS–1	1.544 Mbit/s	24
DS–1C	3.152 Mbit/s	48
DS–2	6.312 Mbit/s	96
DS–3	44.736 Mbit/s	672
DS–4	274.176 Mbit/s	4032

In the UK and Europe the base level is still the same at 64 Kbit/s, but the next level is based on multiples of 32×64 Kbit/s, i.e. 2.048 Mbit/s, yielding 30 channels for speech and two for signalling and control:

Order	Mbit/s	No. of circuits
1st	2	30
2nd	8	120
3rd	34	480
4th	140	1920

The system are all plesiochronous with bit interleaving: 2, 8 Mbit/s, etc., until the highest level is reached.

Plesiochronous working is historically derived from the way in which digitalization of the network happened and the methods used for synchronizing 2.048 Mbit/s PCM digital systems. The transmission is always synchronous for each 64 Kbit/s channel in the 2.048 Mbit/s stream, since the receiver derives its own bit timing from the incoming signal and the time slot and frame alignments are obtained from predefined formats. However, as the transmission network is made up of a number of elements, not all the 2 Mbit/s links are synchronous with one another, e.g. in point-to-point working, which was the first realization of digital transmission, two multiplexors connected together each receiver derives its timing from the incoming bit steam and these need not be identical in both directions of transmission (*Figure 6.35*).

In the case where a digital multiplexor is connected to a digital exchange the digital exchange runs from a master clock and uses time division switching so the digital transmission speeds must be equal in both directions (*Figure 6.36*). Plesiochronous signals have independent clocking arrangements in which each component signal is out of synchronization with the others for the reasons detailed above, and other phenomena such as jitter (short variations of the digital signal) and wander (slow random phase variations of the digital signal over a period of time). To correct the digital signal, bit stuffing must be resorted to during the multiplexing stage due to the variation on the bit rates. Getting at the individual 64 Kbit/s channels within the higher orders of multiplexing is difficult and requires complex systems. To overcome these problems for future systems synchronous digital hierarchy has been developed.

Synchronous digital hierarchy for broad-band transmission The synchronous digital hierarchy (SDH) had its roots in SONET (synchronous optical network) which was enhanced by CCITT to cover a wide range of digital transmission. The SDH standards are targeted at the multigigabit optical-fibre transmission networks and asynchronous transfer mode (ATM) switching for broadband communication, including high-capacity voice switching and transmission. The basic element of the SDH is the synchronous transmission module (STM). A series of STMs has been defined from the basic unit at 155 Mbit/s (STM-1) to higher order units up to 2.488320 Gbit/s. The CCITT bit rates were chosen in such a way that both the North American and European asynchronous digital hierarchies could be supported and a guarantee given that the existing lower bit rate digital transmission systems would not become obsolete. The STM-1 frame consists of a matrix of nine rows by 270 columns of 8-bit bytes (*Figure 6.37*) with a frame repetition rate of 125 µs. The first 9 columns of the matrix are allocated to section overhead (SOH) which is reserved for management of specific network elements such as multiplexing, demultiplexing, control, protection, alarms

T_x = transmit
R_x = receive
MUX = multiplexor

Figure 6.35 *Point-to-point 2 Mbit/s digital line.* T_x, R_x, *MUX*, F_T, F_o.

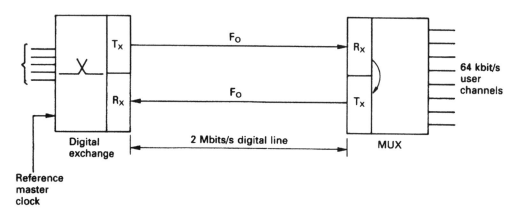

Figure 6.36 *A MUX connection to a digital exchange. R_x, T_x, F_o.*

Figure 6.37 *Synchronous digital hierarchy frame structure.*

or operator use. This leaves 261 columns of the frame for the payload or traffic channels, each being a 64 Kbit/s 8-bit byte of information transfer or speech element. The payload section carries the network user information in virtual containers (VCs) which consists of digital information at plesiochronous rates of 2.048/1.544, 34/45, Mbit/s or asynchronous transfer mode (ATM) cells for broad-band information. The plesiochronous VCs are adapted to the higher container bit rates by means of bit stuffing bits or bytes. This stuffing facility is necessary to adapt the plesiochronous signals to the synchronous network clock. Within each VC is a path overhead (one column of 9 byte) and is transmitted with the associated VC data from end to end throughout the network. This is used for monitoring and management of the path. Higher order transmission rates are obtained by byte interleaving multiplexing of n individual STM-1 frames into an STM-n frame where n is any integer. Thus STM-4 provides for transmission

at 622.080 Mbit/s and STM-16 provides for transmission at 2488.320 Mbit/s transmission. The ability to support existing plesiochronous digital trunks and the emerging ATM broadband networks makes SDH an ideal migration path from the present mixed networks to future ultra-high-speed networks on common optical-fibre transmission media.

Bibliography

CATTERMOLE, K.W., *Principles of Pulse Code Modulation*, Iliffe, London (1969)

CHAPMAN, R.A., *Theory and Problems of Transmission Lines*, McGraw-Hill, New York (1968)

HAMSHER, D.H., *Communication System Engineering Handbook*, McGraw-Hill, New York (1967)

HILLS, M.T. and EVANS, B.G., *Transmission Systems*, Allen & Unwin, London (1973)

MARTIN, J., *Future Developments in Telecommunications*, Prentice Hall, Englewood Cliffs, NJ (1971)

RICHARDS, D.L., *Telecommunications by Speech*, Butterworths, Oxford (1973)
SMITH, S.F., *Telephony and Telegraphy*, Oxford University Press, Oxford (1969)
Transmission Systems for Communications, Bell Telephone Laboratories Inc., USA

6.3 Radio propagation

6.3.1 Introduction

The propagation of radio waves in the broadcast bands at medium frequency (MF), high frequency (HF), and very high frequency (VHF) depends on the frequency being considered and the medium through (or over) which the waves travel. Ground conductivity and the properties of the ionosphere over the transmission path determine the resultant field strengths at MF. The properties of the ionosphere determine the resultant field strengths at HF. The nature of the terrain and the properties of the troposphere over the transmission path determine received signal levels at VHF.

Assessment of received field strength for known transmitter powers and antenna conditions requires the application of free-space theory modified by the inclusion of data obtained from numerous measurements of the properties of the transmission medium. Prediction involves the statistical extension of this data, based on historical observations and resulting trends.

The free-space conditions which will be summarized and modified with idealized theory, are complemented by practical modifications using methods and data noted by the CCIR (the International Radio Consultative Commmmittee which serves the International Telecommunications Union (ITU)).

The Broadcast range of frequencies is as follows.

Medium wave (MF): 526.5–1606.5 kHz.
Very high frequency (VHF Band 2): nominally 88–108 MHz.
Short wave (SW) or high frequency (HF):
 3900–4000 kHz (75 m band)
 5950–6200 kHz (49 m band)
 7100–7300 kHz (41 m band)
 9500–9775 kHz (31 m band)
 11 700–11 975 kHz (25 m band)
 15 100–15 450 kHz (19 m band)
 17 700–17 900 kHz (16 m band)
 21 450–21 750 kHz (13 m band)
 25 600–26 100 kHz (11 m band)

6.3.2 Propagation under free-space conditions

In this context free space implies no ground, no ionosphere and no troposphere. Though this sounds far from reality the results do have a practical application and can be progressively modified to give more practical results.

6.3.2.1 Power density from an isotropic antenna

The isotropic antenna radiates P watts uniformly in all directions. At a distance d metres from the source the energy is spread uniformly over the surface of a sphere of radius d metres.

The power flux density P_d is then given by:

$$P_d = \frac{P}{4\pi d^2} \, (\mathrm{W/m^2}) \qquad (6.1)$$

If an antenna of gain factor G_i with respect to an isotrope is used, then the power density is given by:

$$P_d = \frac{PG_i}{4\pi d^2} \, (\mathrm{W/m^2}) \qquad (6.2)$$

6.3.2.2 Effective isotropically radiated power

The effective isotropically radiated power (EIRP) is the product of power to the actual antenna multiplied by the antenna gain with respect to an isotrope (EIRP $= PG_i$). *Figure 6.38* shows how the power density varies with distance in free space for an EIRP of 1 kW.

6.3.2.3 Field strength

The radiation field strength in volts per metre is related to the power density by the characteristic impedance of free space which, away from the antenna, is $120 \, \pi \Omega/\mathrm{m^2}$. The field strength at distance d metres from an isotropic source radiating P watts is then given by:

$$E = \frac{30\sqrt{P}}{d} \, (\mathrm{V/m}) \qquad (6.3)$$

6.3.2.4 Field from half-wave dipole

A half-wave dipole has a gain with respect to an isotropic source of 1.64 times. Then for a half-wave dipole, equation (6.3) becomes:

$$E = \frac{7\sqrt{P}}{d} \, (\mathrm{V/m}) \qquad (6.4)$$

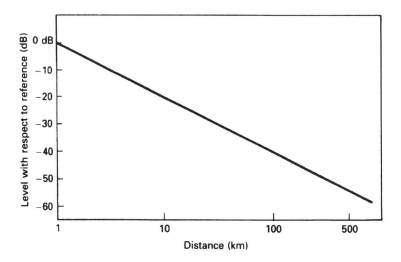

Figure 6.38 *Variation of power density with distance. EIRP = 1 kW; free space; reference (0 dB) at 1 km is approximately 7.96 × 10⁵ W/m.*

The field from an antenna with gain G with respect to a half-wave dipole is:

$$E = \frac{7\sqrt{PG}}{d} \text{ (V/m)} \tag{6.5}$$

6.3.2.5 Reference field strength

The reference field strength is one microvolt per metre (1 μV/m) and other fields are often quoted in decibels with respect to this reference (dB(μV/m)). A typical field for band-2 stereo radio would be 1 mV/m, which is also quoted as a 60 dB field. *Figure 6.39* shows the relationship between dB μV/m field and the field in μV/m.

6.3.2.6 Variation of free-space field with distance

Figure 6.40 shows how the maximum field produced by a half-wave dipole varies with distance for a radiated power of 1 kW.

6.3.2.7 Terminated output voltage from a receiving antenna

If a half-wave dipole is immersed in a field of E V/m then the maximum output voltage E_t produced for a matched termination of 75 Ω is given by:

$$E_t(75) = \frac{E\lambda}{2\pi} \text{ (V)} \tag{6.6}$$

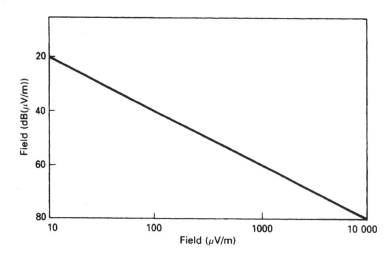

Figure 6.39 *The relationship between field strength and decibel field. The reference field is 1 μV/m corresponding to 0 dB (μV/m).*

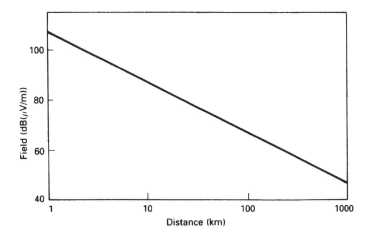

Figure 6.40 *Free-space variation of field with distance for a half-wave dipole of 1 kW. At 1 km the field is approximately 107 dB (μV/m).*

Figure 6.41 shows solutions for equation (6.6) at 100 MHz for various fields. If the impedance is transformed to some other value, say 50 or 300 Ω, then matched terminated voltages are given by:

$$E_t(50) = E_t(75)\sqrt{\left(\frac{50}{75}\right)}(V) \qquad (6.7)$$

$$E_t(300) = 2E_t(75)(V) \qquad (6.8)$$

6.3.3 Field requirements for FM broadcasting

FM broadcasting in band-2 VHF requires a median field strength as listed in *Table 6.2* (CCIR 1990a). These fields are measured at 10 m above local ground level.

Table 6.2 The median field strength required for FM broadcasting in band-2 VHF

	Field strength (dB)	
Area	*Monophonic*	*Stereophonic*
Rural	48	54
Urban	60	66
Large cities	70	74

6.3.4 Polarization used for band-2 FM

Vertical: the electric field is vertical.
Horizontal: the electric field is horizontal.
Slant: the electric field is inclined at 45° clockwise as viewed from the transmitter.

Figure 6.41 *Terminated output voltage (75 Ω) for a 100-MHz half-wave dipole. The terminated voltage is 10 μV for 26.4 dB (μV/m).*

Circular: the electric field is rotating and of constant amplitude.

Elliptical: the electric field is rotating but has an amplitude which varies with angular position.

Clockwise (or right hand)/*anticlockwise* (or left hand): the direction of rotation as seen from the transmitter.

Mixed: a wave with vertical and horizontal components. This includes circular, elliptical and slant.

6.3.5 Field strength prediction at VHF

A step towards a practical modification of the free-space field equations (6.4) and (6.5) is· to include ground effects.

6.3.5.1 Flat smooth earth
The geometry is shown in *Figure 6.42*. For most parts of the service area the angle θ is small, reflection at A takes place with phase inversion and low loss. The net field is the phasor sum of direct and ground reflected waves. The phase difference between these depends on the height of the transmitting antenna h_t, the height of the receiving antenna h_r, the ground distance d between T_x and R_x and the wavelength. For most of the service area the resultant field E_r is given by (IEE, 1989a):

$$E_r = \frac{7\sqrt{(PG)} \times (4\pi h_t h_r)}{\lambda d^2}\text{(V/m)} \qquad (6.9)$$

where PG is the ERP with respect to a half-wave dipole.

Figure 6.43 shows solutions to equation (6.9) for an ERP of 1 kW at 100 MHz, a transmitting antenna height of 100 m, and a receiving antenna height of 10 m.

Figure 6.42 *The effect of flat ground on field strength at VHF.*

6.3.6 Radio horizon

Due to the spherical nature of the earth and the fact that the refractive index of the troposphere (lower atmosphere) produces wave bending normally towards the earth, the effective radius is increased (for a well mixed troposphere) by 4/3. The radio horizon distance is then given by:

$$D_r = 4.12\ \sqrt{H}\ \text{(km)} \qquad (6.10)$$

where D_r is the distance to the horizon (km) and H is the antenna height (m).

6.3.6.1 Improvements to equation (6.9)
Equation (6.9) can be modified to take into account a spherical earth of enhanced radius due to tropospheric refraction (radius of earth increased by 4/3). Effective heights H_t and H_r can be used (EBU, 1982a):

$$H_t = h_t - D_t^2/2kr \qquad (6.11)$$

$$H_r = h_r - D_r^2/2kr \qquad (6.12)$$

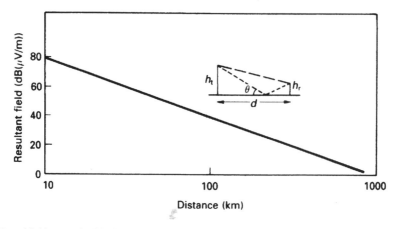

Figure 6.43 *Variation of field strength with distance. Flat earth,* $h_t = 100\ m$, $h_r = 10\ m$, *100 MHz.*

These effective values of antenna heights are used in equation (6.9).

H_t is the effective height of the transmitting antenna, H_r is the effective height of the receiving antenna, D_t and D_r are the distances from T_x and R_x to the reflection point, r is the radius of the earth (approximately 6400 km), and k is, under normal VHF conditions, 4/3.

6.3.7 Propagation into a shadow area behind an obstacle

Figure 6.44 shows a path from transmitting antenna T to receiving antenna R obstructed by an obstacle with peak at O.OR is the path taken by the wave which is diffracted at the edge O. Path TR is blocked. If E_o is the field which would be received in free space with no obstacle and no ground and if E is the field received due to diffraction at edge O, then (EBU, 1982a; CCIR, 1990b):

$$\frac{E}{E_o} = \frac{0.225}{Z}\text{(approx.)} \qquad (6.13)$$

where the parameter Z is:

$$Z = H\sqrt{\left\{\frac{2(d_1 + d_2)}{\lambda d_1 d_2}\right\}} \qquad (6.14)$$

Figure 6.44 *Direct path* $T_x(T)$ *to* $R_x(R)$ *obstructed. OX = H.*

6.3.8 Tropospheric variations affecting VHF band 2

Absolute values of refractive index (n) in the troposphere are close to unity (for example, 1.0004). Refractivity (N) is generally used instead (CCIR, 1990b; IEE, 1989b):

$$N = (n - 1) \times 10 \qquad (6.15)$$

The surface value of N (Ns) is around 300 and the mean gradient per kilometre is around −40 N/km.

These are values for an average atmosphere. Pressure decrease accounts for around −30 N/km, humidity decrease for around −17 N/km and temperature decrease for around +8 N/km. For calculation purposes a modified refractive index m and refractivity M are used enabling propagation to be modelled over a flat earth with atmosphere of refractivity M.

6.3.9 Wave bending

6.3.9.1 Refraction
For a vertical negative gradient of N, waves are bent towards the region of higher refractive index. Refractive index generally decreased with height which gives rise to the 4/3 earth's radius figure already quoted for a well-mixed troposphere.

6.3.9.2 Subrefraction
If the refractive index gradient is less steep than −40 N/km then the wave is bent down less than normal and the radio horizon approaches the geometrical horizon.

6.3.9.3 Superefraction
If the refractive index gradient is steeper than −40 N/km then the wave is bent down more than normal and the radio horizon extends.

6.3.9.4 Temperature inversion
Temperature normally decreases with height but in some circumstances this may not be so. Clear skies and cooling of the surface of the earth around sunset (by radiation) can lead to temperature inversion which may lead to superefraction. If ground temperature falls such that water vapour condenses and fog forms a humidity increase with height may result (at least at low levels) and subrefraction may then result.

6.3.9.5 Air movement
The boundary between two discrete air masses is called a front. A warm mass of air meeting a cold one will rise over the colder air forming a temperature inversion and possible superefraction.

6.3.10 Multipath propagation in band 2

A receiving antenna may be subject to a direct signal from a VHF transmitter and at least one multipath signal arriving by reflection from a distant hillside or other object. The net received field is the phasor sum of the incident components. The problem with FM is that the path difference which determines the

phase difference varies with modulation. If the actual path difference were 9 km and the carrier frequency was 100 MHz then the path difference is 3000 wavelengths. If the deviation is ±100 kHz (rather high, but eases the following calculation) then the path difference (determining the phase difference) varies by ±0.1%. This represents ±3 wavelengths. The resultant phasor now has ·a variation in phase which is dependent on deviation and path difference. The net result is distortion more evident on louder passages where the deviation is large.

The only solution is to use a directional receiving antenna to discriminate against the multipath signal. A mobile installation cannot do this and is susceptible to multipath distortion for band-2 FM broadcasting.

6.3.11 Practical propagation curves

CCIR have produced several curves based on data extending over several years (CCIR, 1990c). An example extracted from this work is illustrated in *Figure 6.45* giving fields for 1 kW ERP. These values are exceeded at 50% of locations for 50% of the time over land. Various heights of transmitting antenna are shown with a receiving antenna height of 10 m as usual. Terrain irregularity from 10 to 50 km from the transmitter is taken as no worse than 50 m. If the irregularity is substantially different then a correction is required.

ITU-P Rec. 370 (VHF and UHF propagation curves for the frequency range 30 MHz to 1000 MHz Broadcast service) is used as the basis for several

implementations for computer use. The ITU version is referred to as 'Rec.370'.

6.3.12 Sporadic E

'Sporadic E' (E_s) refers to the possibility of reflections from the E layer of the ionosphere from sporadic regions of very high ionization giving very high critical frequencies. The E layer is at around 110 km above the surface of the earth and reflections from this height will give single hop ranges out to around 2000 km. More will be said about this when discussing the ionosphere.

6.3.13 Ionospheric effects

6.3.13.1 The ionosphere
The ionosphere (CCIR, 1990c) stretches from approximately 50 km above the surface of the earth. Ionization is due to ultraviolet and X-ray radiation, though high-energy charged particles from the sun also play an important part. The ionosphere contains regions with differing properties according to the frequency of the incident sky-waves. These regions are the D, E and F 'layers'. F is sometimes divided into F_1 and F_2 *Figure 6.46* illustrates the height of these layers ahd gives a much simplified indication of the diurnal variations (over 24 h).

6.3.13.2 Wave bending in the ionosphere
The refractive index of the ionosphere varies with ionization and frequency. For a given frequency the refractive index falls with increasing ionization. For

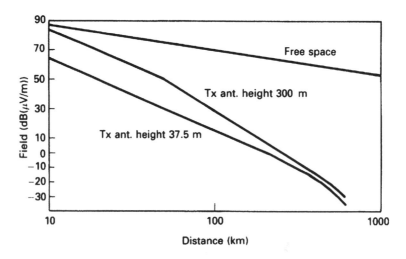

Figure 6.45 *Practical variation of field strength with distance 1 kW ERP terrain irregularity 50 m Rx antenna height 10 m.*

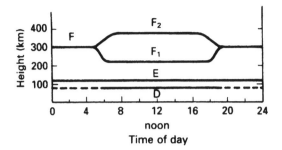

Figure 6.46 *Simplified indication of layer heights and diurnal variations.*

a given ionization the refractive index falls with decreasing frequency. Refraction occurs with the possibility of the wave being returned to earth as if reflection occurred from an effective height. This concept of reflection is used in the definitions which follow.

6.3.13.3 Critical frequency

The critical frequency (F_{crit} or F_o) is the maximum frequency returned to earth from a given layer (E, F, F_1, F_2) at vertical incidence. Typical values are:

$$F_{crit}E = 4 \text{ MHz}$$

$$F_{crit}F_1 = 6 \text{ MHz}$$

$$F_{crit}F_2 = 15 \text{ MHz}$$

6.3.13.4 Maximum usable frequency

The maximum usable frequency (F_{muf} or F_1) is the maximum frequency returned to earth from a given layer at other than vertical incidence. F_{muf} and F_{crit} are often related by:

$$F_{muf} = F_{crit} \sec \theta \qquad (6.16)$$

where θ is the angle of incidence to the 'layer'.

6.3.13.5 Lowest usable frequency

The lowest usable frequency (LUF) describes the lowest frequency for which the required signal/noise ratio is obtained for a given power and antenna gain.

6.3.13.6 Optimum working (traffic) frequency

The optimum working (traffic) frequency (OWF or FOT) is the highest frequency that is liable to propagate, between a given source and destination, via the ionosphere for 90% of the days. If E or F_2 layers control the path then FOT is taken as 0.95 of MUF. F_2 values vary wildly and prediction from many variables including solar activity, season, local time and latitude is necessary.

An MUF over a clearly defined route is indicated by the number of reflections from the relevant layer:

(1) 1E MUF is one reflection via the E layer from source to destination.
(2) $2F_2$ MUF is two reflections via the F_2 layer from source to destination.
(3) Ground range is also included as follows: $1F_2(3000)$MUF is one reflection via the F_2 layer out to 3000 km.

6.3.14 Layer properties

6.3.14.1 Properties of the D layer

The D layer ranges from approximately 50 to 90 km above the surface of the earth. This layer is the main absorbing region of the ionosphere for radio waves. In the HF band the absorption decreases with increasing frequency. During daytime MF skywaves are virtually totally absorbed in the D layer. At sunset ionization falls rapidly and MF skywave absorption also decreases rapidly permitting reflection from the exposed E layer.

6.3.14.2 Properties of the E layer

The E layer lies at around 110 km above the surface of the earth. Reflection of MF signals occurs from the E layer at night time when the D layer de-ionizes. Low HF signals suffering small attenuation in the D layer can also be reflected in the daytime. Ionization levels are greatest at local noon in summer. F_{crit} E peaks in summer.

Sporadic E (E_s) occurs in the E layer when clouds of intense ionization form. These give rise to very high critical frequencies such that reflection at low VHF (and even band 2 at times) can occur resulting in abnormal propagation over large distances.

6.3.14.3 Properties of the F region

This extends upwards from about 150 km and is often divided into two layers called F_1 and F_2. F_1 and F_2 generally combine at night to the single F layer.

6.3.14.4 Properties of the F_1 layer

This occurs around 200 km above the surface of the earth. Ionization is greater than the E layer with resulting higher critical frequencies. F_{crit} F_1 peaks in summer.

6.3.14.5 Properties of the F_2 layer

This occurs around 300 km above the surface of the earth though the actual value varies a great deal.

Ionization is greater than any other layer resulting in the highest critical frequencies. F_{crit} F$_2$ is greater in winter than summer.

6.3.15 Solar cycle variations

6.3.15.1 Solar activity
Ionization levels are greater during more active solar periods. Solar activity is assessed by solar flux measurements at 10 cm wavelength and by deriving 'sunspot' numbers from observed sunspots. The more active the sun the higher the critical frequencies and MUFs.

The 12 month running mean sunspot number R_{12} is generally used. Data to derive R_{12} are centred on a time at least 6 months previously. R_{12} is obtained from (CCIR, 1990e):

$$R_{12} = \frac{1}{12}\left\{ \sum_{n-5}^{n+5} R_k + \frac{1}{2}\left\{ (R_{n+6}) \right. \right.$$

$$\left. \left. + (R_{n-6}) \right\}\right\} \qquad (6.17)$$

where R_k is the mean of the daily sunspot numbers for 1 month. Values are taken for 5 months either side of month n and added to the average for the two values for the months 6 months either side of the month n. The average is then taken over 12 months.

6.3.15.2 Solar cycle (11-year cycle)
Absolute maxima in R_{12} vary with epoch but peaks tend to occur at around 11-year intervals. Activity varies cyclically and variations for cycle 21 and cycle 22 are reproduced in *Figure 6.47* (CCIR, 1990e). The mean of cycles 8 to 21 is drawn to indicate recent activity, for comparison.

6.3.15.3 Solar flares
Intense X-rays from solar flares give rise to enhanced electron density in the D layer resulting in sudden ionospheric disturbances (SID) which produce HF fadeouts (enhanced D layer absorption). Intense bombardment with charged particles later than the visible start of a flare can produce aurora in the auroral zones and associated increased high latitude absorption (polar cap absorption). Magnetic storms may also result where the magnetic field strength varies widely resulting in rapid variations of polarization and subsequent fading.

Figure 6.47 *Sunspot numbers R$_{12}$ for the years 1976–1992. (-----) Observed numbers for cycles 21 and 22. (——) Mean values for cycles 8 to 21 and projected thereafter.*

6.3.16 Diurnal variation of critical frequency

Figure 6.48 shows an idealized example of variation in F_{crit} F$_2$ over 24 h, in winter, for sunspot maxima and minima.

Figure 6.48 *Diurnal variation of F$_{crit}$ F$_2$. Winter, northern middle latitude.*

6.3.17 Single hop ground range

A sky wave launched towards the ionosphere may suffer refraction such that it is returned to earth at a distance dependent on the layer which was responsible for the reflection. This assumes that the frequency is not so high as to ensure no reflection. The maximum range for a single 'hop' occurs for the lowest possible launch angle (take-off angle). The corresponding approximate ranges are:

E layer: 2000 km

F$_1$ layer: 3400 km

F$_2$ layer: 4000 km

6.3.18 Prediction of remote field strength over an ionospheric path

6.3.18.1 Prediction methods

It is possible to either fit empirically derived formulae to measured data or to attempt to produce theoretical models which incorporate all factors affecting received field strength. A combination of the two may be more realistic.

6.3.18.2 The factors to be considered, together with simplified sample values, from transmitter to receiver

Factor	*Sample value*
TX power	300 kW (55 dBW)
Antenna nominal gain	20 dB (ref. dipole)
Free-space path loss	130 dB (6000 km)
	25 m band
Gain reduction due to pointing errors	3 dB
Free-space path field	4.5 mV/m
Ionospheric absorption	5 dB
Intermediate ground reflection loss	2 dB
Polarization coupling loss	3 dB
Other losses	2 dB
Net field strength	1 mV/m approx. (60 dB(μV/m))
Effective receiving antenna gain	-15 dB (simple antenna)

6.3.18.3 Propagation models

Computer-based methods of field strength prediction vary in complexity and reliability (e.g. CCIR 1990f–h). The CCIR methods described in these references rely to a great extent on empirical equations derived from the analysis of large amounts of measured data. It is necessary to predict the state of the ionosphere in terms of ionization, determine all propagation modes that are possible over the route and calculate their relative contributions. Use is made of sunspot numbers, solar flux (10.7 cm), magnetic indices, latitude, solar zenith angle, frequency and other factors. The basic stages of computation are typically those listed below.

6.3.18.4 Stages of computation

(1) Establish circuit conditions: location, TX power, antenna characteristics, solar activity, hour, month.
(2) Determine ionospheric reflection locations, ground constants and magnetic field.
(3) Determine antennae patterns.
(4) Predict ionospheric characteristics.
(5) Determine MUF.
(6) Locate raypaths for frequency used. Consider all possible active modes between Tx and Rx.
(7) Determine relative importance of possible modes.
(8) Calculate effective Tx and Rx antennae gains.
(9) Evaluate freespace attenuation.
(10) Evaluate ionospheric absorption.
(11) Evaluate polarisation coupling loss and ground reflection loss.
(12) Determine overall transmission loss.
(13) Add powers of active modes and calculate field strength.

6.3.18.5 Computer implementations

The ITU recommendation for HF propagation is ITU-P Rec. 533. This is used as the basis of some implementations for computer use. The ITU version is referred to as 'Rec.533'. Other approaches include VOACAP and VOAAREA which are US-sourced versions.

Some of these display the results on a world map projection and an example is shown in *Figure 6.49*. This indicates field strengths in dB (reference a microvolt per metre) with distance from Kuwait City for 500 kW to a particular HF antenna on a bearing of 315 degrees and 1900 UT on 9.00-MHz in June for a (low) smoothed sunspot number of 10.

6.3.19 MF propagation

During daylight hours ionospheric absorption is so high in the D layer that the service area can only be provided by the ground wave. The transmitting antenna is a vertical radiator often a mast or wires hanging down from a mast. The radiator is generally less than a quarter wave tall for practical reasons, and is sometimes in the form of a top loaded T or umbrella arrangement. These antennae launch vertically polarized signals which propagate over the surface of the earth inducing ground currents as they go. These ground currents have associated energy loss depending on the ground conductivity.

6.3.19.1 Propagation over loss-free ground

The field at distance d metres produced by a short monopole radiating P watts over a loss-free ground is given by:

$$E = \frac{9.5\sqrt{P}}{d} \tag{6.18}$$

Distance (km)

Figure 6.49 *Plot of predicted field strength (dB with reference to a microvolt per metre) from Kuwait City. Power is 500 kW, bearing 315 degrees, 1900 UT, 9.00 MHz, June, Sunspot Number 10, antenna broadent type HR 441*

For 1 kW this field is 300 mV/m at 1 km, i.e. 109.5 dB(μV/m).

For any distance D km the field for 1-kW radiated would then be:

$$E = \frac{300}{D} \text{ mV/m} \qquad (6.19)$$

6.3.19.2 Cymo-motive force

The cymo-motive force (CMF) (CCIR, 1990j) of an antenna is the limiting value of the product of field strength and distance from the antenna as the distance is reduced to zero, it has dimensions of voltage. Over a loss free ground, for 1 kW radiated from a short vertical radiator, the CMF is 300 V. Practical values for CMF can be obtained by measuring values of field strength at varying distances multiplying them by the distance and plotting the result against distance. Extrapolation gives the value at zero distance, i.e. the CMF.

6.3.19.3 Propagation over non-perfect ground at MF

CCIR have produced graphs for various values of ground conductivity and permitivity (CCIR, 1990k). Values for the MF band are shown in *Figure 6.50* for a CMF of 300 V and average ground conductivity.

Figure 6.50 *Ground-wave propagation curves. Land; good ground.*

6.3.19.4 Application to MF broadcasting
The curves notes in Section 6.3.19.3 (where *Figure 6.50* is one example) may be used, but the actual CMF value, which will differ from 300 V, is required. When this is known a correction factor is used to modify the values read from the curves. The correction factor is given by:

$$\text{Correction} = \frac{\text{CMF}}{300} \qquad (6.20)$$

In decibels:

$$\text{Correction} = 20\log\left(\frac{\text{CMF}}{300}\right) \qquad (6.20)$$

6.3.20 Ionospheric propagation of MF signals

The conditions considered are those of 'night time', D layer absorption has reached a very low value and reflection occurs normally from the E layer at around 100 km.

6.3.20.1 Cairo curves
The so-called 'Cairo curves' were derived from measurements on north–south and east–west paths in 1937 and 1938. The original curves have since been modified for median values and fields read from the curves require further modification due to polarization coupling losses which are more significant in lower latitudes and north-south directions. One kilowatt of radiated power is assumed.

A rather approximate fit to the north–south Cairo curves is given by (CCIR, 19901):

$$E_c = \frac{231}{(3 + .001d)} - 18 \qquad (6.21)$$

where d is kilometres and E_c is in dB(μV/m).

For general planning use an empirical formula producing a curve midway between the north–south and east–west Cairo curves is given by (CCIR, 19901):

$$E_c = \frac{396}{4 + 0.001d} - 40 \qquad (6.22)$$

where d and E_c are in the same units as above.

6.3.20.2 General method of calculation
The field E is given by:

$E =$ Field for a CMF of 300 V (E_o) at a reference time for a given distance (dB(μV/m)) $+20\log$ (actual CMF/300) (dB) $-$ Hourly loss factor (dB) (6.23)

The hourly loss factor varies with the time of day and is taken at the midpoint position. The reference time for paths less than 2000 km is that at midpath.

The hourly loss factors are approximately (IEE, 1989c):

Hours after sunset	Loss (dB)	Hours after sunrise	Loss (dB)
0	10	−1	2
1	5	0	10
2	2	1	25
4	0	2	Large

6.3.20.3 The field for a CMF of 300 V (E_0)

Charts are produced for (E_0) for various frequencies, latitudes and distances, values extracted from CCIR (1990 m) are shown in *Figure 6.51*. Values read from the chart are used in equation (6.23).

Figure 6.51 *Variation of* E_0 *with distance. Northern middle latitude.*

6.3.20.4 Computer implementations

ITU have produced recommendation ITU-R P.435 for LF/MF skywave propagation. ITU computer implementation is referred to as 'Rec 435' Groundwave predictions can be obtained manually from ITU Recommendation ITU-R P.368. Computer implementations of this recommendation are also available.

References

CCIR, *CCIR Recommendation 412–5*, Vol. X, part 1 Geneva (1990a)

CCIR, *CCIR Report 563–4*, Annexe to Vol. V, Geneva (1990b)

CCIR, *CCIR Recommendation 370–5*, Vol. V, Geneva (1990c)

CCIR, *CCIR Report 725–3*, Vol. VI, Geneva (1990d)

CCIR, *CCIR Recommendation 371–6*, Vol. VI, Geneva (1990e)

CCIR, *CCIR Report 894–2, CCIR*, Vol. VI, Geneva (1990f) (Third CCIR computer-based prediction method.)

CCIR, *CCIR Supplement to Report 252*, Geneva (1990g), (Method for estimating skywave field strength and transmission loss at frequencies between 2 and 30 MHz.)

CCIR, *CCIR Report 340*, Geneva (1990h) CCIR atlas of Ionospheric characteristics

CCIR, *CCIR Recommendation 561–2*, Vol. X, part 1, Geneva (1990j)

CCIR, *CCIR Recommendation 368–6*, Vol. V, Geneva (1990k)

CCIR, *CCIR Report 575–4*, Annexe to Vol. VI, Geneva (1990l)

CCIR, *CCIR Recommendation 435–6*, Vol. VI, Geneva (1990m)

IEE, *IEE Electromagnetic Wave Series No. 30*, Chap. 2, London (1989a)

IEE, *IEE Electromagnetic Wave Series No. 30*, Chap. 9, London (1989b)

IEE, *IEE Electromagnetic Wave Series No. 30*, Chap. 7, London (1989c)

EBU, *EBU Report Tech. 3236–E*, Appendix 2, Brussels (1982a)

ITU-R P. 533–5. In force 1995. HF propagation prediction method

6.4 Transmission developments: NICAM and RDS

6.4.1 Introduction

From an audio standpoint the development of both radio and television receivers has been driven over recent years by the various improvements in audio reproduction which have become available from record players, audio cassette players, hi-fi amplifiers, loudspeakers, and compact disc (CD) players. Transmission standards and, consequently, receiver performance always lag behind such developments because of the massive investments in transmitters, masts and all the other aspects of establishing a reliable transmission system which is needed to support the millions of listeners and viewers who rely upon these services. However, there has been much effort put into improvements of the existing transmission standards in order to attempt to emulate the audio performance of those common audio products.

6.4.1.2 VHF/FM radio

The audio quality of radio transmissions improved significantly in the 1950s when frequency modulation (FM) broadcasting was introduced on the

very high frequency (VHF) band. Both audio bandwidth and signal/noise ratio were improved. However, reception of VHF/FM transmissions does carry a burden of careful tuning and substantial antennas to achieve the best from the signal. Over the years tuning improvements have come from the introduction of receivers with synthesizers. While the improvement was welcome it still meant that users had to be very 'frequency literate' to be sure of tuning to the best signal. Gradually more portable VHF/FM receivers became available and their performance worsened because of the rather inadequate whip antenna which was seldom deployed correctly by the non-technical user. Some improvement was gained by a general change from horizontal to vertical or slant polarization of VHF/FM transmissions. These problems in many ways delayed the hoped for success of VHF/FM radio broadcasting and many broadcasters were looking for an answer to improve matters during the late 1970s and early 1980s. Also in the 1970s programmes began to be broadcast as stereo transmissions and audio listening standards improved as a result of mass produced equipment meeting high standards, colloquially known as 'hi-fi'.

6.4.1.3 UHF television
In the meantime television sound quality took a major step forward with the introduction of ultra high frequency (UHF) transmissions, but it took many years before the reproduced audio quality became as good as that provided by FM tuner amplifiers, mainly because the television receiver had most development work put into the picture quality and audio quality actually lagged behind the transmitted audio signal quality.

6.4.1.4 Mobile radio reception
Mobile reception of VHF/FM radio transmissions had the disadvantage that frequent retuning was necessary and with the inevitable multipath effects the user was often forced to retune when perhaps that particular transmission would be the best signal just a few hundred metres further away. Here was a very real problem which initially the receiver designers could only attempt to solve by improving the radio frequency (RF) sensitivity of car receivers and some amazing advances were made in the mid-1980s; acceptable listening to VHF/FM signals 50 km or more outside the defined service area became commonplace with top end of the range car receivers! Automatic retuning marked a significant improvement in the listening pleasure given by VHF/FM

transmissions when the Radio Data System (RDS) was introduced in the late 1980s.

6.4.1.5 Stereo television sound
Television sound had a more difficult period because it appeared that the UHF band plan would not permit the addition of stereo sound to the transmissions, however the German Institut for Rundfunktechnik (IRT) developed a second sound carrier method which fits into the channel space allocated and this gave stereo performance with broadly the same quality.

Then digital audio found its way into television transmissions by using NICAM 728 sound which is claimed to be as good as CD quality, although the transmitted audio bandwidth of NICAM 728 is somewhat limited by comparison with CD bandwidth.

6.4.1.6 Digital technology impact
By the end of the 1980s digital technology in television and radio receivers had made a considerable impact and both high definition television (HDTV) and digital audio broadcasting (DAB) developments were poised to revolutionize transmission and reception of television and radio broadcasts in the mid- to late-1990s.

This section now concentrates on the significant developments of the last decade: NICAM and RDS. The really clever thing about both of these advances is that they have been 'slotted' into the existing transmission standards in a compatible way so that existing receivers are not affected yet new receivers can take full advantage of their benefits.

6.4.2 NICAM origins

NICAM stands for 'Near-instantaneous companded audio multiplex' and the technology was originally developed for sound signal distribution, not as a transmission standard, but this slightly strange acronym stuck and is now very well known to television users! In the 1970s digital transmission of mono television sound from studio to television transmitter was first implemented in the UK by using a technique known as 'sound in syncs' (SIS). The relatively low sample rate of 31.25 kHz, actually twice television line frequency, needed some companding to achieve an adequate signal to noise ratio and it equated to a 12-bit linear system. The digital signal was cleverly inserted into the line sync pulses and carried to the television transmitter with the normal video distribution to give a significant improvement to the transmitted analogue television sound after it

had been decoded before modulating the television transmitter in the normal way.

Similar improvements in distributing radio sound signals were sought and work in the mid-1980s was aimed at distribution of audio signals in the standardized 2 Mbit/s telecommunication primary multiplex channel. The NICAM 3 standard was developed in the UK by the British Broadcasting Corporation (BBC) to provide digital distribution of six high-quality 15 kHz sound signals each using 338 Kbit/s. The NICAM 3 specification calls for 32 kHz sample rate with a maximum resolution of 14 bit. Simply, to cram it all into the transmission channel, it must be compressed down to 10 bits and this is achieved by looking at blocks of samples and deciding how they fall into one of five ranges which is then transmitted along with the 10 bits to inform the decoder how to reconstitute the full 14 bits before the analogue-to-digital converter converts back to an analogue sound signal.

6.4.2.1 NICAM 728

In order to distinguish NICAM television sound from its distribution forerunners, it is technically known as 'NICAM 728.' This number is derived from the fact that there are 728 bits per frame which results in a digital data rate of 728 Kbit/s. As with NICAM 3 the coding process starts by taking 32 bits per sample with 14-bit accuracy and companded to 10 bits. Then to achieve stereo sound transmission it is necessary to interleave left and right channel bits which gives 352 bits per channel and with 24 bits added for additional data the total is 728 bits per frame. This has to be distributed from studio to television transmitter and is sent in dual SIS format. The digital signal is now modulated onto a 6.552 MHz carrier which sits above the vision carrier; in order to achieve a final spectrum which can fit into the predefined UHF television channel as shown in *Figure 6.52*, differential encoded quadrature phase shift keying is used.

The NICAM 728 specification was first published jointly in 1988 by the BBC, British Radio Equipment Manufacturers Association (BREMA) and the Independent Broadcasting Authority (IBA) (BBC, BREMA, IBA, August 1988). Transmissions in the UK started less than 2 years later. This technology provided the impetus for significant improvements in audio reproduction from television receivers. Whilst it is often advertised as being of CD quality, the sampling rate of 32 kHz means that the audio bandwidth is indeed less, being band limited at 15 kHz. However, most television listening takes place directly listening to the television receiver and the cabinet

Figure 6.52 *Television channel spectrum showing analogue and data components.* (Courtesy of BBC.)

constraints are such that the benefits of improved distortion and signal to noise ratio balance well with this frequency response. Television cabinet design immediately improved and some novel design ideas to include larger bass loudspeakers with internal audio ducting and cross-over filters to drive tweeters have resulted in the listener hearing a much improved reproduced audio output. Most of these improvements would have been impossible without the single chip NICAM decoder which had the effect of permitting this technology to be incorporated into many television receivers at little extra cost to the user.

6.4.3 Radio Data System (RDS) origins

By the early 1970s it was clear to listeners that the air waves (mostly from listening to medium frequency (MF) broadcasts) were becoming very crowded and broadcasters realized that things could only worsen on both MF and VHF/FM bands as time progressed. Some countries were turning towards VHF/FM-only broadcasting because the MF band was so grossly overcrowded. In the UK many new radio stations had been granted permission to broadcast initially on both MF and VHF/FM providing a simultaneous broadcast on both frequencies of a single programme service. As time progressed two separate programmes were produced and transmitted separately increasing the listener choice even further. Pressure on the radio spectrum has continued in the 1990s to provide even more radio services with increased emphasis on smaller community areas.

This can only be done with a considerable change in the way the channels are allocated to the broadcasters. New services will generally

have to use the VHF/FM band since it has some expansion capability, especially at the higher frequency end where additional spectrum between 105 and 108 MHz became available in the UK from 1996. In passing it is worth noting the importance of a very well ordered FM band plan; the use of a frequency at a particular power and location must be carefully worked out to ensure that co-channel and adjacent channel interference is avoided. Of course this calculation even has to take account of transmissions from other countries and international agreements are essential if the listener is to hear anything at all.

6.4.3.1 Crowded broadcast bands

In the UK for example, the BBC uses over 20 different frequencies to provide coverage for one of the national networks, creating a potential difficulty for listeners, particularly mobile ones, in deciding which frequency they should use for the best reception. This difficulty is represented graphically in

Figure 6.53, which shows the potential broadcasts which can be received to the west of London on a modern car radio.

Potential tuning problems, difficulty of knowing what station has been tuned and the possible provision of enhancements to the service led researchers in the mid-1970s to investigate ways of making VHF/FM broadcasts more convenient for the listener. The work was undertaken by members of the European Broadcasting Union (EBU) which encourages uniformity of operations amongst the members and it has overseen the development of RDS as a standardized broadcast system.

6.4.3.2 Added subcarrier

Engineers started collaboration in many broadcasting organizations notably the German IRT, BBC Research and Swedish Telecom Radio, the latter having developed a paging system operating on a 57 kHz subcarrier. This work progressed throughout Europe under the guidance of the EBU including field trials

Figure 6.53 *Diagram of a car-radio dial west of London.* (Courtesy of the BBC.)

in several countries and culminated in the publication of EBU document *Tech 3244: Specifications of the Radio Data System for VHF/FM Sound Broadcasting*, which was issued in March 1984 (EBU, 1984).

The choice of subcarrier frequency and amplitude is critical to the technical success of the system because it is vital that receivers do not convert the subcarrier into in-band interference, regardless of reception conditions. The multiplex signal and added RDS subcarrier is shown in *Figure 6.54*.

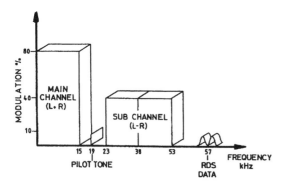

Figure 6.54 *Multiplex FM signal with RDS subcarrier added.* (Courtesy of VG Broadcast.)

This was a very important milestone: the system had been carefully nurtured by the EBU in order to create a large market making it worthwhile for the radio manufacturing industry to incorporate RDS in their radios for use in all countries. Indeed this is a world-wide issue especially since there are now so many excellent portable radios which can be taken on journeys to other countries. With RDS transmissions on the increase it will be necessary for radios to provide the advantages of RDS wherever they are to be used. Also, of course, a European standard means that the radio manufacturer has a potential to sell to about 40% of the world market for high-specification radios.

6.4.3.3 RDS standards
However, things have not stopped there. The RDS standard is reflected in ITU-R *Recommendation B5.643-2* so it is well on the way to becoming a world-wide standard (ITU-R, 1986). Then the Comité Européen de Normalisation Electrotechnique (CENELEC) published EN 50067:1992, which confirmed the earlier published EBU Specification and added certain additional items following field experience of the initial RDS systems. These standards

gave information for broadcasters and radio designers and they were the key to RDS performance. In addition, the EBU published EBU Tech 3260: Guidelines to the implementation of RDS, which gives important background to detailed aspects of RDS implementation, following field experience in several countries (EBU, 1990). More details, including descriptions of baseband coding, data message structure and decoding can be found in a number of excellent books now available (see Bibliography).

During the years 1994 to 1997, the RDS Forum – the association of RDS users – managed the development of a further backwards-compatible upgrading of the RDS standard. This was voted by CENELEC in April 1998 and was published thereafter, numbered EN 50067:1998 and becomes the definitive RDS standard. The EN 50067:1998 dated RDS specification includes several new features and further enhances harmonization with the US National Radio Standards Committee (NRSC) specification for Radio Broadcast Data System (RBDS), which now contains the RDS specification as a core to the NRSC specification, having been completely harmonized (EIA/NAB 1997). This will allow operators using the CENELEC standard to implement, for example, the originally, US-only, feature of Programme Type Name (PTYN). But the major upgrade is in the form of a completely new RDS feature called Open Data Applications (ODA) which allows new data services – not defined in the standard – to be added to RDS transmissions as they are required in the future.

6.4.3.4 RDS logos
Two RDS logos have been developed to help identify products which have RDS features built into them. *Figure 6.55* shows the original RDS logo intended for use on basic RDS products, all the

Figure 6.55 *RDS logo.* (Copyright BBC and EBU)

acceptable forms are detailed in EN 50067: 1998. It will frequently be found on RDS car radios, where panel space is at premium, in the interleaved rings form only. The logo is intended to convey the smooth transition from one transmitter service area to another. The concentric rings represent reducing signal strength going away from the transmitter.

Figure 6.56 shows the enhanced other networks (EON) logo which may be used on RDS radios with an approved minimum EON performance.

Figure 6.56 *RDS EON logo.* (Copyright BBC and EBU)

6.4.3.5 RDS features
EN 50067:1998 refers to over 20 RDS features and the major features are shown in *Table 6.3*. Many of these are in use by broadcasters around the world. Other features are less used but specific features have been implemented for specialist applications in some locations.

The objective for a broadcaster is to use features which will cost relatively little to provide and give the maximum of help to the listener. Secondly, there are features which do not have a direct relation to the broadcast programme and may be used for revenue earning. These are both the so-called 'static' and 'dynamic' features which can be selected by broadcasters to make up their transmissions.

6.4.3.6 Static RDS features
Initially most broadcasters just provide static RDS features. Briefly, the three features known as programme identification (PI), programme service (PS) and alternative frequencies (AF) provide automatic tuning capability and station name capability which is so helpful for a car radio. These features are all derived from data held in an erasable programmable read-only memory (EPROM) in the RDS encoder which will continue cycling around to provide the necessary groups in a predetermined sequence. Hence the description 'static RDS'.

6.4.3.7 Programme Service name
Straight away one can see the advantage that Programme Service (PS) name will give the listener. All those confusing frequency numbers are redundant – all that matters is the station's identity. For example, BBC Radio 1 which can be found around the UK on some 18 different frequencies, these are all indicated as >‿BBC‿RI‿<. Other PS names to be seen in south-east England are shown in *Table 6.4*.

Table 6.3 Major RDS features and abbreviations

Feature	Acronym	Description
Application ID	AID	Application ID for an ODA
Alternative Frequencies	AF	List of alternative frequencies
Clock Time	CT	Clock time and date
Enhanced Other Networks	EON	Information about other services
Open Data Application	ODA	ODAs registered for use with AID
Programme Item Number	PIN	Time and date for start of programme
Programme Identification	PI	Programmed identification code
Programme Service name	PS	Programme Service name, 8 characters only
Programme Type code	PTY	Programme type code, e.g. pop music
Radio Paging	RP	Bleeps and alphanumeric messages
Radio Text	RT	Radio text messages of up to 64 characters
Travel Announcement	TA	Travel Announcement switch: on-air
Transparent Data Channel	TDA	Addressable data services
Traffic Message Channel	TMC	A special ODA for coded traffic messages
Travel Programme	TP	Travel Programme switch: TAs in a service

Table 6.4 Examples of some PS names used in south-east England

Radio station name	PS name
BBC Radio 4	>⎵BBC⎵R4⎵<
KFM (Tonbridge, Sevenoaks)	>KFM⎵1016<
BBC Southern Counties Radio	>BBC⎵SCR⎵<
Classic FM	>Classic⎵<
Southern FM	>SOUTHERN<

6.4.3.8 Programme Identification code

The Programme Identification (PI) code for BBC Radio 1 (C201) is never seen by the listener but it is vitally important to the radio to ensure that all data received with that code is relevant to BBC Radio 1 and does not become confused with any other data received. The structure of this number can convey some vital details to the radio in order to carry out some special auto tuning function. For example, a group of commercial stations, with the same management, transmitting in adjacent areas can use generically linked PI codes. RDS radios including some first generation car radios will build up knowledge about the other local services and load them into the memory 'behind' the chosen station's preset button, so that once out of range of the first choice the radio will retune to an adjacent service. Ideally this response will be put under user control; many second-generation RDS radios offer this level of control to avoid the tendency to jump between programmes just at an interesting point!

6.4.3.9 Alternative Frequencies list

The Alternative Frequencies (AF) list comprises all the available frequencies, geographically surrounding the presently received transmission which also carry the selected programme, so that in whatever direction the radio should be taken it will always have a selection of possible new frequencies to choose from to maintain reception of that service. For example, at the Sutton Coldfield transmitter site in the West Midlands there are seven VHF/FM transmitters: Radio 1, Radio 2, Radio 3, Radio 4, Radio Derby, Radio WM and Classic FM.

The AF list for the BBC Radio 3 transmitter at Sutton Coldfield includes over 20 different alternative frequencies, providing references to all the Radio 3 transmitters surrounding Sutton Coldfield, including: Sutton Coldfield, Rowridge, Blaenavon, Bow Brickhill, Ridge Hill, Llangollen, Pontypool. Holme Moss, Oxford, Long Mountain, Ludlow,

Llanrhaeadr-ym-Monhant, Wenvoe, High Wycombe. Llandinam and Peterborough, respectively.

There might be one or two surprises in this list which has been very carefully compiled to ensure the best possible RDS radio performance. The Sutton Coldfield Radio 3 transmitter frequency is in the list because there are several relay transmitters which simply retransmit the received signal from Sutton Coldfield, so *their* AF list must include the Sutton Coldfield Radio 3 frequency. Some entries are quite far away, but in general the frequencies in the list are those of the transmitters the service areas of which surround the service area of Sutton Coldfield, so that in whatever direction the radio should be taken it will always have a selection of possible new frequencies to choose to maintain reception of that service. Small relay transmitters which are often used to fill in valleys (e.g. High Wycombe) are also included in the lists to ensure smooth reception of the desired service in those previously difficult reception situations with a non-RDS radio.

The AF list for Radio 1, Radio 2 and Radio 4 will be broadly similar to Radio 3 because the adjacent transmitter sites also have these services radiated with the same power.

6.4.3.10 Clock time

In the case of Clock Time (CT) it will be obvious that CT is not really static or there will be a bit of a problem! In fact it is transmitted in a special group on the minute edge, but the RDS encoder has this built into the software and no external command is required to put this group into the data stream. In the UK, the CT feature is synchronized by off-air reception of mean standard frequency (MSF) 60 kHz from Rugby to give the required accuracy considered essential these days. RDS car radios will be able to reset themselves after the car has been serviced and the battery has been disconnected and better still will automatically reset to or from British Summer Time twice a year.

6.4.3.11 Dynamic RDS features

BBC national-network transmitters can provide dynamic RDS features because they are connected to the BBC digital distribution network which carries high quality digital audio to the transmitter and, in this case, more importantly provides a link for RDS update data from the central RDS computer in London to the RDS encoders to change the normal cycle of transmitted groups. The following features

are included in this category.

Travel Programme	(TP)
Travel Announcement	(TA)
Programme Type	(PTY)
Programme Item Number	(PIN)
Radio Text	(RT)
Enhanced Other Networks	(EON)

6.4.3.12 Travel flags

Both TP and TA are simple 'flags' which can be set on or off and they form the core of the RDS Travel Service. TP just signifies: 'this programme carries travel information at some time'. An RDS radio can detect it so that on arriving in a new area – with no prior knowledge of the radio stations there – a search tune could be invoked to find a station with a TP flag set on; the listener then knows he will hear travel information during the programme. The TA flag is more specific and indicates that travel information is being transmitted *now*. Several resultant radio responses are possible. In first generation RDS radios. the volume can be lifted automatically to ensure the announcement is heard; if cassette listening is in progress then a switch back to radio listening will again ensure the information is received. Of course, when the TA flag is switched off the RDS radio will revert to the original status.

6.4.3.13 Programme Type

Programme Type (PTY) allows the broadcaster to label his or her programmes with a code which the RDS radio will be able to use for tuning purposes. These are shown in *Table 6.5*. For example, one normally thinks: "I would like to listen to news now.' Since there is a PTY code for news the radio could be asked to search for a matching programme. This PTY feature has both speech and music based codes which are also standardized with the European radio satellite codes to ensure greater user acceptance. Perhaps this feature could be used in a car to select a programme with news information to be heard when it becomes available, but in the meantime listening to a rock music programme would be preferred.

6.4.3.14 Programme Item number

The Programme Item Number (PIN) feature uniquely identifies each programme during the day and could be used to switch the radio on at exactly the programme start time. Perhaps this feature could be used in a car to select a programme to be heard at a particular time but to be listening to cassette until that time; in this case the RDS radio would 'switch on' exactly as the programme started.

6.4.3.15 RadioText

The RadioText (RT) feature can provide messages of up to 64 characters and would be particularly useful on a tuner to give the phone-in number for a programme or the address to write to for further information. Safety issues associated with moving text in a car have become a significant challenge for Human–Machine Interface designers and this is the subject of several regulatory investigations. Following the introduction of the RT feature in a few car radios, the RDS Forum has declared that RT should only be implemented in a car radio when the car is stationary.

6.4.3.16 RDS scheduling

Most broadcasters, including even smaller radio stations, use scheduling systems to run their programme planning systems, for music selection and it is possible to connect an output from these systems to

Table 6.5 Programme Types specified in EN 50067:1998

Speech based	Music based	Other types
Current Affairs	Country music	Alarm
Documentary	Easy listening music	Alarm Test
Finance	Folk music	Children's programmes
Information	Jazz music	Culture
News	Light classics music	Drama
Phone-in	National music	Education
Social Affairs	Oldies music	Leisure programmes
Sport	Other music	Religion
Weather	Pop music	Science
	Rock music	Travel features
	Serious classics music	Varied

allow RDS features such as PTY, PIN and RT to be provided dynamically when implemented. In the UK a range of broadcasters – BBC, Classic FM and Melody FM – have implemented dynamic RT to provide programme-related information. Each have taken slightly differing views to their services, providing simple text on each programme through to detailed information about the music on-air.

6.4.3.17 RDS features world-wide

As mentioned before the RDS system has been standardized across Europe, but it is interesting to note the differing features that have been chosen by the broadcasters to provide their initial services. In Sweden, for example, they have a network and local radio structure comparable to that in the UK and chose similar features apart from the addition of radio paging (RP), which in the UK is already served elsewhere. This is not entirely surprising since RDS has origins in their earlier 57 kHz paging system. Telediffusion de France also started a RP service when they first implemented RDS service, but now both Sweden and France have decided to close these services in the light of competition from mobile phone services. However, the RP feature is used in several other countries where such infrastructures are less well developed.

Throughout Europe, many countries have well-established RDS implementations from both the large multi-network public broadcasters and from national commercial broadcasters, as well as smaller commercial radio stations.

There are many countries around the world (e.g. Brazil, Hong Kong, Kenya, the USA) where RDS transmissions are now providing a service to listeners or where test transmissions are in progress. The status varies from initial testing to paging service in operation and full public service implementation.

6.4.3.18 Radio design considerations

A vital aspect of the introduction of RDS has been the development work needed by the radio manufacturers across the world. Initially car radio manufacturers were deeply involved, and they continue to lead developments in consumer products, with well over 50 manufacturers worldwide developing RDS car radio products. The enormous advantage that RDS gives with automatic tuning still remains the most compelling feature.

At the least they have to decode PI and AF to achieve automatic tuning and almost all provide station name from the PS. Really, an RDS car radio needs to use these three features fully to give a suitable minimum performance. The advantages of RDS are quite obvious to the car driver, so it was not surprising that the RDS car radio was the first to appear on the market. Although it is perhaps the most technically difficult to design, especially because of the ever changing reception conditions, there is one relatively easy aspect and that is the possibility of using the large car battery as a power source reducing the need for low-current very large scale integrated (VLSI) circuits at the outset.

Before the availability of RDS car radios it was necessary for the designer to provide an automatic stereo to mono blend capability in order to overcome drastic signal strength changes causing undesirable image shifts. Also automatic variable low-pass filtering is employed to reduce hiss. Both these systems are often retained in RDS radios and together with the automatic tuning give considerably improved listening pleasure. Field trials have shown that most unexpected reception of distant transmissions can occur for very brief periods when relatively local transmissions are subject to severe multipath distortion. Manual tuning just cannot achieve such good results.

When it comes to portable RDS radios a designer has to provide a product that uses very little power so it was easier to then develop home tuners. A number of home tuners have been marketed with RDS functionality, but generally the automatic tuning aspects have been less well implemented in favour of providing the RT and PTY features. Several portable and 'walkman style' RDS radios with just PS display functionality are now available.

A number of different implementations are possible for displaying the PS name, some receivers use dot-matrix displays, some use seven-segment displays, some are back-lit for increased visibility. For the user, dot-matrix displays have the added advantage that upper- and lower-case characters can be displayed which assists station-name recognition if the broadcaster is using both. BBC Radio studied display recognition and uses upper case for the national networks but finds that a mixture is best for local radio where the slight limitation of only eight characters is easily overcome, e.g. BBC Radio Gloucestershire is seen as >BBC_Glos<.

The display features of RDS are very important but the radio manufacturer has to use RDS for control functions if automatic tuning is to be implemented; the processing of the incoming data stream demands considerable complexity to decode with enough accuracy to be useful and then act on the

data in a logical way. This needs a great amount of development to achieve sensible responses from the RDS radio in the variable reception conditions that will exist as it is moved about. The RDS car radio manufacturer has another challenge to pack all the extra electronics into the comparatively small standard case, thus large-scale integration has been essential from the start.

6.4.3.19 Enhanced Other Networks feature

The Enhanced Other Networks (EON) feature was standardized early in 1989 and BBC Radio agreed to be the first broadcaster in the EBU to provide EON data. Now every BBC transmission has EON data and other countries including Sweden and Germany have added the feature.

BBC local radio services provide static EON information about all adjacent BBC services. Additionally BBC national network services in England provide dynamic RDS in order to provide information for the RDS travel service which is voiced from BBC local radio stations.

It has already been shown that the AF list will help an RDS radio retune as it is moved around. But the AF list can only inform the radio about the *tuned* service. For example, if you want to retune to another BBC service after driving 150 miles or so listening to BBC Radio 4, then the likelihood of knowing the appropriate frequency for, say, BBC Radio 3 is really quite slim. RDS can come to your aid with the EON feature which provides data to inform second-generation RDS radios which are becoming available, as a background routine, all about the other BBC services. So the EON RDS radio will already 'know' what frequency BBC Radio 3 is available on in that area. This is done by what is known as the 'mapped-frequency method' which simply means that the broadcaster works out beforehand what the best frequencies will be and the EON data is then kept very short to save on RDS radio memory and processing time.

In addition, EON allows the radio to react to status changes on other services and thus provides an 'intelligent' response to various useful radio advances. So that second-generation RDS radios are able to, for example, give the PS name instantly upon pressing any preset button, whereas only some of the original designs offered this possibility by the RDS radio memorising the PS last time it was used. However, the broadcaster might have changed the name in the meantime! For example, regional broadcasts may be simultaneously supplied from a single studio and at certain times of the day form a small network with

a different name requiring a subtle PS name change to keep the listener informed.

EON allows the radio to react to short-term status changes on other services and thus provides an 'intelligent' response to various useful radio advances.

6.4.3.20 Travel service with EON

For the RDS car radio user the most valuable use of EON permits a refined RDS travel service to be offered by national broadcasters, so that whatever programme is being listened to, local travel information from the nearest transmitter may be heard.

The advantages are enormous: no longer does the listener have to find which station transmits travel information, whether in their area, what frequency they use and what time the bulletins are broadcast. All of this is made available to the RDS car radio by the EON data transmitted by the tuned service. In the UK it is now possible to listen to BBC Radio 3 and also have the benefit of travel information from the nearest BBC local radio service, because the EON data from the national network transmitters which are under dynamic control from the central RDS computer can inform an RDS radio about all the other BBC services including information that the TA flag is being switched on for BBC local radio services. In this way the RDS car radio will 'find' a travel message from the nearest BBC local radio station, giving relevant, timely and, hopefully, accurate information about the local travel conditions.

Buried in this process is a vast network of data communications between the travel announcement studio, the national network transmitters, the local radio transmitters and the central RDS computer in London. The BBC pioneered methods of effecting this at low cost to the broadcaster. At first sight it would appear necessary to use new lines to send data to the transmitters to provide update data to the RDS encoders to control the RDS travel flags – the TP and TA flag. Presumably these data lines would have to run from the travel information studio to the transmitters.

Firstly, there was spare data capacity in the digital programme distribution to network transmitters to add the RDS update without any trouble. The whole system has been engineered taking advantage of this fact and the central RDS computer in London (the source of all network radio programmes for the whole UK) sends RDS update data to all the network transmitters.

Secondly, there is the regional data network which links every BBC local radio studio with London and normally carries administration information, in

both directions, such as news wire services and schedule planning information. This data network has additional capacity which can be used to carry the small RDS control data requirement, which at the simplest level is just four states: TP on, TP off, TA on, and TA off. Actually a little more data is sent between the local radio studio and the central RDS computer to obtain a reliable link. In practice, this technique can be used to send commands to the central RDS computer from any location and the RDS travel service uses both local radio studios and other travel information studios to control the RDS flags.

The control is achieved by the use of update data being labelled with a service number so that an RDS encoder 'knows' that it must reflect certain RDS data in what is transmitted, to ensure that the RDS data received off air at the local radio transmitters will provide a command for them to switch their flags, without the need for purchasing additional data circuits. At the local radio transmitter sites there is a Radio 2 RDS monitoring receiver which is engineered to receive Radio 2 off-air and the decoded RDS data carries information which includes the travel flags commands. Thus there is a data path from the person, wishing to switch an RDS travel flag, via an ordinary datacomms link to the central RDS computer then on to the Radio 2 transmitters and over air to the local radio transmitter. It will be seen that the RDS travel flags can be switched on or off without having to rent a single additional data circuit from the telecommunications providers. Indeed, for the whole command path from the person pressing the TA 'on' button to the RDS radio, there are some 16 microprocessors involved including data multiplexers, demultiplexers, encoders and decoders!

'Vectoring' information is provided to an RDS radio by the EON feature to inform it that a particular local radio transmission is: 'now changing one of the travel flags'. It is transmitted just at the time of any status change so that the radio can immediately check if that local radio transmission is of adequate signal strength and if so it can retune to it. It is very important that the listener will previously have had to instruct the radio that this 'interrupt' service is required, otherwise it could be most annoying if not required.

6.4.3.21 Travel information in the future
The use of RDS to convey Traffic and Travel Information (TTI) has long been a desire of RDS system developers and indeed the 1992 edition of the

EN 50067 standard included some coding suggestions, but development work was still incomplete at that time. The basic concept of the RDS-Traffic Message Channel (TMC) was, however, established by that time. In essence the idea is that TTI messages should be generated, using a predetermined Location Database, and densely coded to reduce the data capacity requirements and then be fitted into a single RDS group for transmission. Upon receipt, TMC messages would only be decoded in a special RDS-TMC receiver fitted with a Location Database to facilitate the decryption process and allow the receiver to present the TTI messages in the 'language or icon comprehension' of the user.

The long-term objective is for all travellers to have access to TTI from before leaving to their destination (perhaps by teletext or the Internet), covering all possible road route situations (by FM RDS), including events as they occur, using the ALERT-C protocol, and some status information for urban situations, using another protocol known as ALERT-Plus.

Apart from the RDS standard, which now contains the ODA feature needed for TMC and other 'hooks' for TMC, it requires another series of CEN standards (ENV 12313) to define the ALERT-C and ALERT-Plus protocols, Events Lists and Location Databases. These standards are nearly finalized, but being pre-standards will need to be fully voted into full standards in 2 years' time, following some final implementation activities.

Throughout Europe, development of RDS-TMC has been undertaken with significant funding from the European Commission (EC) over some 10 years and many evaluation field trials have been undertaken. The first RDS-TMC implementations were completed in some countries in 1997 and more began in 1998. These implementations still had to manage the challenge of supplying the Location Databases to the consumer, in smart card or CD-ROM format. This challenge and the RDS-TMC receiver design issues revealed in attempting to offer multiple languages are still none to easy to solve in a reasonable consumer-priced product. Nevertheless the political pressure to overcome these difficulties are considerable, and yet more EC funded demonstration projects are set to establish more user-oriented services across Europe.

6.4.3.22 Is RDS really delightfully simple?
It will be appreciated that RDS is a complex subject for the radio manufacturer and the broadcaster. In a way they each have half an operating system under their control and each half must work perfectly with

the other. Maintaining perfect sympathy between the two halves has been a vital objective of the standards which have been developed by the EBU, ITU-R and CENELEC.

For the listener RDS must be very simple, requiring the complex standards to be distilled into a very easily used product. It is then possible for the listener to rely upon the automatic responses which have become possible in the RDS radio.

First generation RDS radios are quite excellent if all that is required is to listen to a radio service with more than one transmitter. For example, listening to Radio 2 whilst driving all the way from Newcastle to Newquay will cause the RDS radio to unobtrusively retune at least 6 or 7 times on the way and $>_$BBC$_$R2$_<$ is all that will be seen on the display. From a radio user's point of view this really can be a hands-off journey.

Already the introduction of RDS has led to some novel ways of presenting information on the radio status to the user and no doubt more simplification will follow. For example, touch-sensitive displays and bleepers to indicate that the radio cannot give that action requested at present. Like all technological developments the prices are falling progressively and even more innovation in the broadcasting and reception of RDS can be expected.

RDS has been designed for both broadcasters and radio manufacturers to add features from the existing standards when the demand develops; furthermore, it is possible to compatibly add new features which have not so far even been thought about!

References

BBC, BREMA, IBA, *NICAM 728: Specification for Two Additional Digital Sound Channels with System I Television*, London (August 1988)

ITU-R, Recommendation BS.643-1 (10/95) System for automatic tuning and other applications in FM radio receivers for use with the pilot-tone system

CENELEC, *EN 50067: 1998 Specification of the radio data system (RDS) for VHF/FM sound broadcasting in the frequency range from 87.5 to 108.0 Mhz*

EBU, *Tech 3244: Specifications of the Radio Data System, RDS for VHF/FM Sound Broadcasting*, European Broadcasting Union, Geneva (March 1984)

EBU, *Tech 3260: Guidelines for the Implementation of the RDS System*, European Broadcasting Union, Geneva (January 1990)

EIA/NAB, *National Radio Systems Committee, United States RBDS Standard Version 2.0, Specification of the radio broadcast data system (RBDS)* (1997)

Bibliography

BEALE, T. and KOPITZ, D., RDS in Europe, RBDS in the USA – What are the differences and how can receivers cope with both systems? *EBU Technical Review* No. 225 (Spring 1993)

EBU, *EBU Technical Review*, European Broadcasting Union, Geneva (Feb. 1991) (Many articles.)

HOWE, P., *RDS – The Engineering Concept*, BBC Radio (February 1990)

KOPITZ, D., *Radio Data System – From Specification to Practical Reality*, European Broadcasting Union, Geneva (Feb. 1989)

KOPITZ, D. and MARKS, B., *RDS – The Radio Data System*, Artech House, London (1998)

LEWIS, A.R., PARNALL, S.J. and ROBINSON, J., *RDS, the Technical Realisation of a New Broadcast Service*, IBC, London (1988)

MARKS, B., RDS – *Really Delightfully Simple? f.fwd* (3) 59–61 and (4) 67–69 (Mar./Apr. & May/Jun. 1989)

MARKS, B., RDS – The Way Ahead for Radio, *IEE Review*, 127–129 (Apr. 1989)

MARKS, B., The introduction of the radio data system to the United Kingdom. Presented at: BroadcastAsia 90 (1990)

MARKS, B., Dynamic RDS – its features and management. Presented at: BroadcastAsia 92 (1992)

MARKS, B., What is happening to RDS? *Electronics and Wireless World*, **94** (1633), 1096–1100 (Nov. 1988)

MOTHERSOLE, P.L. and WHITE, N.W., *Broadcast Data Systems – Teletext and RDS* p. 148 (1990)

PARNALL, S.J., Decoding RDS *Electronics and Wireless World*, **95** (1636), 148–152 and (1637), 284–287 (Feb./Mar. 1989)

PARNALL, S.J. and RILEY, J., *RDS Developments*, IBC, London (1990)

SAUNDERS, M., RDS. *Electronics Today International*, **18** (1), 12–19 (Jan. 1989)

WRIGHT, S., *The Broadcaster's Guide to RDS*, Focal Press, p. 203, Oxford (1997)

6.5 Digital audio interfacing, synchronization and networks

This section is concerned with the interconnection of digital audio equipment in order that audio signals may be transferred between devices in a digital system. It describes the interface formats which are used most widely, and gives advice concerning the correct use of such interfaces.

6.5.1 Overview

6.5.1.1 Interconnection in the analogue domain
Digital recordings may be copied from one machine to another either via one of the standard digital

interfaces (see below) or via the analogue inputs and outputs. In the case of the latter, replayed audio is converted back to the analogue domain by the replay machine's digital-to-analogue convertors, routed to the recording machine via a conventional audio cable, and then reconverted to the digital domain in the recording machine's analogue-to-digital convertors (*Figure 6.57*). The audio is subject to any gain changes which might be introduced by level differences between outputs and inputs, or by the record gain control of the recorder and the replay gain control of the player. Analogue domain copying is necessary if any analogue processing of the signal is to happen in between player and recorder, such as gain correction, equalization, or the addition of effects such as reverberation. More and more of these operations, though, are now possible in the digital domain.

Figure 6.57 *Connections for analogue-domain copying. The record gain control will affect the level of the copy. A/D, analogue-to-digital; D/A, digital-to-analogue.*

Interconnection in this way is perfectly feasible, although it will involve a slight increase in noise of about 3 dB per generation of copying. If a recording has been copied in this way it is no longer quite correct to say that a recording has remained in the digital domain throughout its production and post-production (a fact indicated by the 'DDD' symbol on compact discs (CDs)), although very few people would notice the difference introduced by a single generation of copying in the analogue domain, provided that levels were carefully matched between recorder and player. Multiple generations of copying via the analogue domain should be avoided.

An analogue-domain copy cannot be said to be a 'perfect copy' or a 'clone' of the original master, since the data values will not be exactly the same (due to slight differences in recording level, differences between convertors, etc.) – for perfect copies it is necessary to make a true digital copy.

6.5.1.2 Interconnection in the digital domain

In order to make a perfect 'clone' of a digital recording it is necessary to copy the binary audio data directly from the player to the recorder, without passing through conversion stages. Professional systems, and some consumer systems, have digital interfaces conforming to one of the standard protocols (see below), and allow for a number of channels of digital audio data to be transferred between devices, with no loss of sound quality. Any number of generations of digital copies may be made without affecting the sound quality of the latest generation, provided that errors have been fully corrected. The digital outputs of a tape-recorder are taken from a point in the signal chain after error correction, which results in the copy being error corrected. Thus the copy does not suffer from any errors which existed in the master, provided that those errors were correctable.

Digital interfaces may also be used for the interconnection of recording systems and other digital audio devices such as mixers and effects units, and this will become more prevalent as more digital signal processing equipment becomes available. It will eventually become common only to use analogue interfaces at the very beginning and end of the signal chain, with all other interconnections being made digitally.

6.5.1.3 Computer networks and digital audio interfaces compared

Real-time audio interfaces are the digital audio equivalent of signal cables, down which digital audio signals for one or more channels are carried in real time from one point to another, possibly with some auxiliary information attached. A real-time audio interface uses a data format dedicated specifically to audio purposes, unlike a computer data network which is not really concerned with what is carried in its data packets. A recording transferred over a digital interface to a second machine may be copied 'perfectly' or cloned, and this process takes place in real time, requiring the operator to put the receiving device into record mode such that it simply stores the incoming stream of audio data. The auxiliary information may or may not be recorded (usually most of it is not).

Real-time interfaces are normally unidirectional, point-to-point connections, and should be distinguished from buses and computer networks which are often bidirectional and which carry data in a packet format. Sources may be connected to destinations using a routing matrix, very much as with analogue signals. Audio data is transmitted in an unbroken stream, there is no handshaking process involved in the data transfer, and erroneous data may not be retransmitted – there is no mechanism

for requesting its retransmission. The data rate of a real-time audio interface is directly related to the sampling rate, wordlength and number of channels of the audio data to be transmitted, thus ensuring that the interface is always capable of serving the specified number of channels. If a channel is unused for some reason its capacity is not normally available for assigning to other purposes (such as higher speed transfer of another channel, for example).

Clearly real-time interfaces are best suited to operational situations in which analogue signal cabling needs to be replaced by a digital equivalent, and where digital audio signals are to be routed from place to place within a studio environment so as to ensure dedicated signal feeds.

In contrast, a computer network typically operates asynchronously and data is transferred in the form of packets. One would expect a number of devices to be interconnected using a single network, and for an addressing structure to be used such that data might be transferred from a certain source to a certain destination. Bus arbitration is used to determine the existence of network traffic, and to avoid conflicts in bus usage. The 'direction' of data flow is determined simply by the source and destination addresses. The speed of the network effectively determines the amount of traffic which may be present on the bus and the rate at which it will be transferred, this bearing little or no relationship to the audio sample rate, the number of channels or the wordlength. The format of data on the network is not necessarily audio specific (although protocols optimised for audio transfer may be used), and one network may carry text data, graphics data and E-mail, all in separate packets, for example.

6.5.1.4 Video-based pulse code modulation adaptors

The digital inputs and outputs of a tape recorder *must* be distinguished from the video or 'composite digital' interfaces of a pulse code modulation (PCM) adaptor or VTR used for digital audio (as found in older systems such as the Sony PCM-1630 and Sony PCM-701). The video or 'composite digital' outputs carry a digital audio signal encoded into a pseudo-video waveform for recording on to a video tape recorder (VTR), and are not the same as the digital outputs of a system. Although it is possible to make copies by simply connecting the video output of a player VTR into the video input of a recorder VTR (as when copying a domestic video tape), the copy will not be error corrected, and is likely to have more errors on it than the original. The correct procedure in this case is to copy the recording by

using the relevant PCM adaptor in the 'dub' or 'copy' mode (see *Figure 6.58*) in which the video signal from the player is first decoded and error corrected before being sent back out to the recorder. Video or 'composite digital' outputs must not be connected to ordinary digital inputs in any circumstances, even if they use the same BNC-type connectors.

Figure 6.58 *When copy mode is used on a PCM adaptor, video from a player is decoded and error corrected before being re-encoded and sent out to a recorder via the video or 'composite digital' output.*

On the Sony PCM-1610 and 1630, the input selector has three positions: 'analogue', 'digital' and 'dubbing'; it is the 'dubbing' position which refers to the above mode.

6.5.2 Digital interfaces

There are a number of types of digital interface, some of which are international standards and others of which are manufacturer specific. They all carry digital audio with at least 16-bit resolution, and will operate at the standard sampling rates of 44.1 and 48 kHz, as well as at 32 kHz if necessary, with a degree of latitude for varispeed. The increased importance of higher sampling rates such as 96 kHz and 88.2 kHz has led to modifications of some standards to enable IFs to operate at these rates. Most of the interface standards are one- or two-channel only, but one of them, known currently as MADI (see Section 6.5.2.6) is a multichannel interface. A number of interface standards are manufacturer specific, in that they are not the subject of Audio Engineering Society (AES) recommendations or International Electrotechnical Commission (IEC) standards, and this refers to such as the various Sony interfaces, TASCAM, ADAT and the Yamaha interfaces. Manufacturer-specific interfaces may also be found on equipment from another manufacturer if it has been considered necessary for the purposes of

signal interchange, especially on devices manufac-
tured prior to the standardisation and proliferation of
the AES/EBU (European Broadcasting Union) inter-
face. The interfaces vary as to how many physical
interconnections are required, since some require one
link per channel plus a synchronization signal, whilst
others carry all the audio information plus synchro-
nization information over one cable.

Making a copy of a recording using any of the
interface standards involves the connection of the
appropriate cables between player and recorder, and
the switching of the recorder's input to 'digital' as
opposed to 'analogue', since this sets it to accept
a signal from the digital input as opposed to the
analogue-to-digital convertor. It is normally neces-
sary for both machines to be operating at the same
sampling rate and may require the recorder to be
switched to 'external sync' mode, so that it can lock
its sampling rate to that of the player. The recorder
should be capable of at least the same quantizing
resolution (number of bits per sample) as the player,
or greater, otherwise audio resolution will be lost. If
there is a difference between the systems it may be
advisable to use a processor in between the machines
which optimally dithers the signal for the new reso-
lution (see Section 6.5.3.3) and may be able to offer
sample rate conversion as well (see Section 6.5.4.4).

6.5.2.1 The AES/EBU interface (IEC 958)
The AES/EBU format allows for two channels of
digital audio (A and B) to be transferred serially
over one balanced interface, using RS422 drivers
and receivers. It has been standardized by the IEC
as IEC 958 (1989) which is currently undergoing
major revision. The hardware interface is shown in
Figure 6.59. The interface allows two channels of
audio to be transferred over distances up to 100 m,
but longer distances may be covered using combina-
tions of appropriate cabling, equalization and termi-
nation. Standard XLR-3 connectors are used, often
labelled 'DI' (for digital in) and 'DO' (for digital
out).

Figure 6.59 *AES/EBU hardware interface. EQ, equalizer.*

Each audio sample is contained within a 'sub-
frame', the format of which is shown in *Figure 6.60*,
and each subframe begins with one of three synchro-
nizing patterns to identify the sample as either A or B
channel, or to mark the start of a new channel status
block, as shown in *Figure 6.61*. These synchronizing
patterns violate the rules of biphase mark coding (see
below), and thus are easily identified by a decoder.
Additional data is carried within the subframe in the
form of 4 bits of auxiliary data (which may either be
used for additional audio resolution or for other pur-
poses such as low-quality speech), a validity bit (V),
a user bit (U), a channel status bit (C) and a parity
bit (P), making 32 bits per subframe and 64 bits per
frame.

Figure 6.60 *AES/EBU subframe format. Sync, synchronization
pattern; Aux, auxiliary data; LSB least significant bit.*

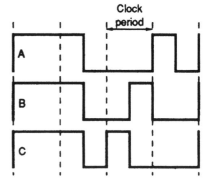

Figure 6.61 *AES/EBU sync patterns. A, start of A-channel
subframe; B, start of B-channel subframe; C, start of new
channel status block.*

One frame (containing two audio samples) is trans-
mitted in the time period of one audio sample, and
thus the data rate varies with the sampling rate. Chan-
nel status bits are aggregated at the receiver to form
a 24-byte word every 192 frames, and each bit of
this word has a specific function relating to interface
operation. Examples of bit usage in this word are
the signalling of sampling rate and pre-emphasis, as
well as the carrying of a sample address 'time-code'

and labelling of source and destination. Bit 1 of the first byte signifies whether the interface is operating according to the professional (set to 1) or consumer (set to 0) specification.

Biphase mark coding, the same channel code as used for SMPTE/EBU time-code, is used in order to ensure that the data is self-clocking, of limited bandwidth, d.c. free, and polarity independent. The interface has to accommodate a wide range of cable types, and a nominal 110 Ω characteristic impedance is recommended. AES 3 (1985) said that up to four receivers with a nominal input impedance of 250 Ω may be connected across a single professional interface cable, but AES 3 (1992) recommends the use of a single receiver per transmitter, having a nominal input impedance of 110 Ω.

6.5.2.2 Sony–Phillips digital interface (SPDIF)

The SPDIF is very similar to the professional AES/EBU interface. When the IEC standardized the two-channel digital audio interface, two requirements existed: one for 'consumer use', and one for 'broadcasting or similar purposes'. A single IEC standard (IEC 958: 1989) has resulted with only subtle differences between consumer and professional implementation. The electrical interface for the consumer version is shown in *Figure 6.62*. The data format of subframes is the same as that used in the professional interface, already described, but the channel status implementation is different.

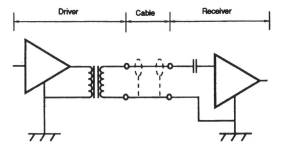

Figure 6.62 *SPDIF hardware interface.*

The second byte of channel status in the consumer interface has been set aside for the indication of 'category codes', these being set to define the type of consumer usage. Current examples of defined categories are (00000000) for the general category; (10000000) for CD; (01000000) for a PCM adaptor (such as a Sony PCM-601) and (11000000) for a digital audio tape (DAT) machine. Once the category has been defined, the receiver is intended to interpret

certain bits of the channel status word in different ways, depending on the category. For example, in CD usage, the four control bits from the CD's 'Q' channel subcode are inserted into the first four control bits of the channel status block (bits 1–4).

The consumer format interface can be found on many items of semiprofessional or consumer digital audio equipment, such as CD players and DAT machines. It usually terminates in an RCA phono connector, although some hi-fi equipment makes use of optical fibre interconnects carrying the same data. Format convertors are available for converting consumer format signals to the professional format, and vice versa.

Modifications to the consumer interface standard are currently being published by the IEC. The advent of DVD players has led to the interface being used in 'data' mode to carry Dolby AC-3 or MPEG audio data for interconnection between players and low bit rate multichannel media decoders. DTS surround data may also be transferred in this way.

6.5.2.3 Sony digital interface (SDIF)

The most common Sony interface is SDIF-2, and is intended for the transfer of one channel of digital audio information per cable, at a resolution of up to 20 bits (although most devices only make use of 16). The interface, as used on most two-channel equipment, is unbalanced and uses 75 Ω coaxial cable terminating in 75 Ω BNC-type connectors, one for each audio channel. TTL-compatible electrical levels (0–5 V) are used. The audio channel connectors are accompanied by a word clock signal on a separate connector, which is a square wave at the sampling frequency used to synchronize the receiver's sample clock. Sony's PCM-3324 and 3348 multitrack machines use SDIF also, but with a differential electrical interface conforming to RS422 standards and using D-type multiway connectors. A single BNC connector carries word clock as before.

In each audio sample period, the equivalent of 32 bits of data is transmitted over each interface, although only the first 29 bits of the word are considered valid, since the last three bit-cell periods are divided into two cells of one-and-a-half times the normal duration, in order to act as a synchronizing pattern. As shown in *Figure 6.63*, audio data are transmitted with the most significant bit (MSB) first, followed by nine control or user bits. The resulting data rate is 1.53 Mbit/s at 48 kHz sampling rate and 1.21 Mbit/s at 44.1 kHz.

A 256-word block is used (a word being the 32 bits described above), the first of which carries

control information in bits 21–28, the subsequent words carrying user information in these bits. In this first word, bits 26 and 27 signal whether pre-emphasis is used or not (00 for no emphasis, and 01 for emphasis to the 50/15 µs standard), bit 28 signifies copy protection status (1 is copy protected), and bit 29 is used to signify the start of a 256 word block, being set to a one at this point.

Figure 6.63 *SDIF-2 data format. MSB, most significant bit; sync, synchronizer.*

The SDIF-2 interface is used mainly for the transfer of audio data from Sony professional digital audio equipment, particularly the PCM-1610 and 1630, but also from semiprofessional equipment which has been modified to give digital inputs and outputs (such as the PCM-701 and various DAT machines). It is also apparent on a number of disk-based workstations for the loading and unloading of audio data.

6.5.2.4 Modular digital multitrack (MDM) interfaces

The recent growth in popularity of modular digital multitrack recorders has led to widespread adoption of eight-channel interfaces by TASCAM and Alesis (ADAT). These are incompatible with each other. The ADAT interface is an optical serial interface in its most common form, whereas the TASCAM interface uses a multiple D-type parallel connector.

6.5.2.5 Yamaha cascade interface

Yamaha digital audio equipment is equipped with a 'cascade' connector to allow for a number of devices to be operated in cascade, such that the two-channel main outputs of a mixer, for example, may be fed into a further mixer to be combined with another group of channels. This interface terminates in an eight-pin DIN-type connector (as shown in *Figure 6.64*) and carries two channels of 24-bit audio data over an RS422-standard differential line. The two channels of data are multiplexed over a single serial link, with a 32-bit word of left-channel data followed by a 32-bit word of right-channel data (the 24 bits of audio are sent the least significant bit (LSB) first, followed

by eight 0s). Word clock alternates between low state for the left channel and high state for the right channel. Inductors of 20 µH are connected between pins 6 and 7 and ground to enable suppression of radio frequency interference. The 'out' socket is only enabled when its pin 8 is connected to ground.

Pin	Function
1	WCLK +
2	GND
3	Audio data -
4	WCLK -
5	Audio data +
6	20 µH coil to GND
7	20 µH coil to GND
8	GND (in), ENABLE (out)

Yamaha professional equipment also makes use of multiple D-type connectors carrying audio data in either the 'Y1' or 'Y2' formats. The latter of these is bidirectional and has a format similar to that described above using two channels multiplexed on a single pin. These connectors carry eight channels each.

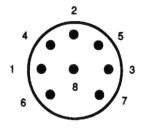

Figure 6.64 *Yamaha interface connector pins.*

6.5.2.6 MADI interface (AES10–1991)

A number of manufacturers combined forces to propose a multichannel serial interface standard known as 'MADI' (the multichannel audio digital interface), which is based on the two-channel AES/EBU data. It has applications in large-scale digital routing systems, and for the interconnection of multitrack audio equipment. MADI naturally uses a higher data rate to carry the increased amount of information. MADI allows for 56 channels of audio to be carried serially over one 75 Ω coaxial cable or optical fibre, such that one sample for each channel is transferred within one audio sample period.

Allowing for maximum commonality with the two-channel AES/EBU data format, MADI allows for subframes containing 20- or 24-bit audio data, together with the additional status bits of the AES/EBU data for each channel. The main difference

in the MADI format is in the first 4 bits of each subframe (which in the two-channel interface break the biphase mark coding law to provide a sync pattern). Channel subframes thus formed are linked sequentially to form a 56-channel frame (*Figure 6.65*).

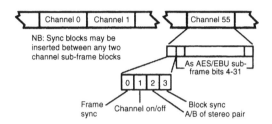

Figure 6.65 *MADI frame format. Sync, synchronized.*

Unlike the two-channel interface, the transmission data rate of MADI is fixed at 125 Mbit/s, irrespective of the sampling rate or number of channels, and the actual data transfer rate is 100 Mbit/s due to the use of a 4–5-bit encoding scheme. In this channel code each 32-bit subframe is broken up into 4-bit words which are then encoded to 5-bit words according to a look-up table, the reason being to maintain a low d.c. content to the code. A synchronization symbol (11000 10001) is inserted at least once per frame, and in cases where the full bandwidth of the link is not being used, additional synchronization symbols are inserted to take up the capacity of the bus.

Communication is expected to be handled entirely by so-called 'TAXI' (transparent asynchronous Xmitter receiver interface) chips, which automatically take care of the insertion of sync symbols. Unlike the two-channel format, it is intended that the link will be asynchronous and accompanied by a separate synchronizing clock (in the form of an AES/EBU signal), as it is not intended that the data be self-clocking. BNC 75-Ω connectors are to be used, and the maximum coaxial cable length is intended to be no more than 50 m (although optical fibre would allow more). The modulation method used is to be NRZI in which a transition from high-to-low or low-to-high represents a binary 1, and no transition represents a 0. Because of the asynchronous nature of the interface, buffers are required at both ends of the link so that the data can be re-clocked from the buffer at the right rate (see *Figure 6.66*). At the receiver, data are clocked out under control of the synchronizing signal.

Figure 6.66 *MADI buffering arrangement. WCLK, word clock.*

6.5.2.7 IEE 1394 or 'fine wire'

The recently developed amplifier desktop bus known as IEE 1394 has growing applications in audio and video, allowing a large number of channels to be transferred at speeds well in excess of real time if required. This technology is likely to replace consumer interfaces such as IEC 958 over the next five years.

6.5.3 Practical considerations

6.5.3.1 Simple two-channel interfacing

For a two-channel copy between one digital system and another (say between two PCM adaptors, or two digital tape machines, or between a disk-based system and a tape machine) it is necessary to use a digital interface such as AES/EBU or SDIF-2. The two possibilities are shown in *Figure 6.67*, and it will be seen that in the case of the AES/EBU link only a single balanced line, terminated in XLR connectors, connects the two devices, whereas in the SDIF-2 link three 75-Ω BNC leads are required, one for word clock (WCLK) and one for each audio channel. This difference results from the fact that the AES/EBU

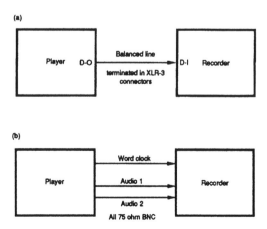

Figure 6.67 *(a) AES/EBU and (b) SDIF-2 interfacing compared. AES/EBU requires a single XLR cable for the clock and two channels of audio. SDIF-2 requires three BNC cables.*

signal is 'self-clocking', whereas the SDIF-2 requires a separate clock. It has been demonstrated in practice, though, that some devices appear to interface better without the WCLK signal in SDIF-2 interfacing, requiring only the two audio signals, and this is a matter for experimentation in each case. (This arises from the fact that the WCLK signal, although at the same sampling rate as the audio data, may not be at the correct *phase* to act as a reliable reference for the latching of received data.) WCLK should be linked between devices wherever possible, and in the case of a synchronized system it is vital that each device is locked to a common word clock or other sync reference.

Signals are normally transferred over the interface with whatever flags are needed to indicate the status of the player tape. For example, a replayed tape may be pre-emphasized and copy protected, and thus flags will be set in the interface which signal this fact to the recorder, using data embedded within the audio data. In occasional cases such flags may not have been set, or may have been set incorrectly, in which case some manual intervention will be required (see Section 6.5.3.3). A machine receiving copy-protected data will normally refuse to record such data, and indicates the copy-protection status on its front panel. This is normally only common in consumer equipment, where digital copying may be considered inappropriate. In some systems the record-gain controls of a recorder will be inactive in the case of a digital copy, since they normally operate on analogue signals, but there are now more examples of systems which have digital record gain controls allowing level correction in the digital domain. Such a digital gain control should be bypassed or set to unity (0 dB) if a perfect digital 'clone' copy is required.

When the interconnection is working correctly, the recorder should indicate the same signal on its meters as the player, the recorder's 'WCLK lock' or 'ext sync' light (or equivalent) should be illuminated (indicating that the recorder is correctly locked to the incoming signal) and sound should be audible from the recorder's outputs when monitoring its inputs.

6.5.3.2 Problems with digital interconnection

First-line troubleshooting If a digital interface appears not to be working (that is no signal is present on the receiving device when connected digitally to a transmitting device), it could be due to one or more of the following conditions.

(1) Sampling rates of transmitter and receiver are different. Set the two to be the same, preferably locked to a common reference.

(2) Receiver is not locked to transmitter sample clock. Ensure that receiver is in external synchronization mode and that a synchronizing signal (common to the transmitter) is present at the receiver's sync input. Alternatively, use an AES/EBU interconnect, which should be self-synchronizing.

(3) Receiver is not switched to accept a digital input.

(4) Received data are in the wrong format. Both transmitter and receiver must operate to the same format.

(5) There may be a consumer-professional conflict in channel status (see below). Use a format convertor (see Section 6.5.3.3) to set the necessary flags.

(6) Cables or connectors may be damaged.

(7) Cable may be too long or of poor quality.

(8) Copy protect or SCMS flag may be set by transmitter. For professional purposes, use a format convertor to set the necessary flags.

(9) Incompatible channel status, user or validity data.

(10) Receiver is not in record mode. Some recorders must be at least in record-pause before they will give an audible and metered output derived from a digital input.

(11) Digital signal may be of poor quality. Check eye-height on scope against specification, and check for possible interference sources. Alternatively make use of an interface error analyser.

Difficulties with AES/EBU interfaces Incompatibilities exist between the AES/EBU interfaces used in professional equipment. Furthermore, the interface specifications allow for electrical interconnections which may result in poor performance. The user may be faced with the need to test digitally interfaced systems to verify their correct operation, especially when installing large networks.

Although the specification allows for the use of microphone cable, the characteristics of microphone cable vary widely. AES/EBU signals are very-high-frequency data signals. They cannot be treated just like audio signals. AES/EBU data signals may suffer when subjected to changes in cable and interface impedance, and such changes may arise if two dissimilar cables are connected together, if the signals is split between multiple receivers, or if the cabling used within any receiver or equipment bay does not match the impedance of the signal cable. The corruption of the data which results may affect sound quality, since errors may arise, although this is only

likely in extreme cases, since for the majority of the time the interface works perfectly well with ordinary cables. The data may also be corrupted if cables of insufficient bandwidth or high attenuation are used, increasing the susceptibility of the AES/EBU signal to noise and external interference. The design of the receiver also plays a major part in the effective operation of the interface, and there are good receivers and bad receivers in existence. Any audio distribution amplifiers (DAs) in the signal chain must be replaced with DAs suitable for handling the bandwidth required.

It has been suggested that the use of 75-Ω coaxial cable with unbalanced 75-Ω interfaces at either end may provide a better transmission line than the standard balanced interface, and a modification to the standard may appear to allow for this possibility. None the less, the better transmission line would be replaced by the problems suffered by an unbalanced interface. Microphone cables may well be found to work successfully over long distances, but this is often only reliable when a single device transmits to a single receiver over a single cable, and depends on the quality of the receiver.

Not only may there be electrical difficulties with interfacing, but also there may be data incompatibilities. These arise principally out of the vague interpretation of how validity, channel status and parity bits should be interpreted by receivers. The validity bit is asserted to indicate an invalid sample (one which has been found to be erroneous by a previous device) and some receivers ignore this bit, whilst others may mute or interpolate. Parity errors will only be signalled when an odd number of bits are in error, and even numbers of errors may be passed without attention. The standard does not incorporate any sophisticated error checking or correction mechanisms (except on channel status data which includes a cyclic redundancy check (CRC)), and thus errors on the interface will affect sound quality. Furthermore, channel status implementation is not fully standardized, and certain receivers may expect bits to be set in a particular fashion, which may or may not match that presented to them.

The AES/EBU receiver may extract a clock signal from the incoming biphase mark data, and it is common for this clock to be noisy (having timing jitter and instability) after it has passed over a lengthy interconnect. Receivers vary as to the quality of their clock-recovery circuits, it being most important to reconstruct a stable, jitter-free clock signal if such a derived clock is to be used to reference other parts of a system such as the digital-to-analogue convertors.

Two examples of dedicated AES/EBU transceiver chips, one from the BBC (called 'AESIC') and one from Motorola, promise low-jitter clock recovery in the face of poor input signals. Further, clock 'de-jitterizers' are becoming available with extremely high specifications, which will match the recent generation of high-quality digital-to-analogue convertors, ensuring good sonic performance.

Testing interfaces New test equipment is appearing, specifically designed for the purpose of testing digitally interfaced equipment. Audio Precision's 'System One Dual Domain DSP' is an example of a set which can perform selected tests on AES interfaces, to check for errors, measure performance, and to set certain data conditions to check the implementation of the standard on different receivers. Some receivers and receiver chips have provision for the indication of a 'near failure' status, which monitors the state of received data. An oscilloscope may be used to view the 'eye pattern' of the received data, to check the minimum difference between high and low voltages of each bit cell (which should be 200 mV), although the scope must be triggered from the transmitter clock output to view the eye pattern correctly. Although the eye-height is a valuable indicator of the state of the signal, it should not be taken as the only criterion for satisfactory performance.

6.5.3.3 Use of interface format processors and convertors

As suggested in the preceding sections, a certain amount of intervention may be required between devices when interconnecting them digitally. This may be in order to modify status flags to suit a purpose, or to correct the status of certain flags to match the signal. It may also be necessary to perform operations on the audio signal, such as de-emphasis and gain correction, or to convert the signal from one interface format to another. One way of doing some of these things is to pass the signal through a full blown digital mixer, but this may not be necessary in most cases, and a mixer may not allow the format conversion or flag modification. For this reason, a number of devices have been introduced which allow various degrees of modification to signals emanating from the standard digital interfaces, and some examples are described below by way of illustration.

The simplest operations of this type involve the conversion of AES/EBU signals to SPDIF consumer format or SDIF-2 format and vice versa, with possible modification of the copy prohibit status to allow copying of protected recordings. Audio & Design in

the UK makes a range of 'Digital ProBoxes' for such purposes.

A more complex device is exemplified by the Audio Digital Technology FC-1, which has facilities for converting any of three input formats (SPDIF, AES/EBU and SDIF-2) to any of three output formats, with the possibility for de-emphasis in the digital domain to either CCITT J17 (used in broadcast) or 50/15 µs characteristics. It has two stereo signal paths which may be mixed together with the option of a mixer panel which can be used to perform level or channel balance correction and cross-fades in the digital domain. It will also dither the digital signal optimally for one of three possible output resolutions (16, 18 or 20 bits), for conversion of signals between systems of different resolutions, ensuring the optimization of sound quality when copying high-resolution signals to lower resolution systems. A number of other functions are also offered, such as d.c. blocking (important for removing d.c. offsets introduced by badly aligned analogue-to-digital convertors), digital high-pass filters, CRC, channel phase and left/right reversal, and output status modification. There is also an oscillator for alignment purposes.

Another example of an advanced device in this vein is the Digital Audio Research DASS 100. It takes the important step of offering sample rate conversion between any of the standard rates (32, 44.1 and 48 kHz, ± 2%), as well as compensating for varispeed inputs of up to ± 10% of these frequencies. There are optional interfaces to virtually every possible format (AES/EBU, SPDIF, SDIF-2, ProDigi, Yamaha DMP-7D and Sony 50 pin SDIF), and inputs may be mixed together with gain correction. Other features include the synchronization of inputs to a reference (see below), the variable delay of signals, the generation of a system sync signal from one of a number of sources, the pre- or de-emphasis of signals in the digital domain, and the generation of audio test waveforms in the digital domain.

6.5.3.4 SCMS in DAT machines

The serial copy management system (SCMS) is to be installed in all consumer DAT machines, and introduces a flag into a first-generation DAT copy of copyright material (only when made digitally) which prevents that DAT tape from being digitally copied subsequently. It does not prevent copying via the analogue outputs. The same restrictions also apply, unfortunately, to recordings made on SCMS DAT machines via their analogue inputs – they may only be copied digitally once.

This system will not be applied to truly professional recorders, since it would be unnecessarily restrictive, but might still be found in 'professionalized' consumer DAT machines. Furthermore, it should only be implemented on the consumer SPDIF port, not on the professional AES/EBU port, if fitted. It is possible that a digital format processor, such as those described in the previous section, could be made to modify the relevant flag and thus allow copying, although of course this would only be relevant to appropriate professional usage and would be illegal for consumers. Professional users intending to purchase consumer DAT machines would be wise to check whether or not SCMS has been implemented on the machine of their choice. SCMS would become obtrusive in professional operations such as editing, where digital copying might extend through more than one generation.

For a more detailed discussion of SCMS and its implications refer to the author's article *The Truth about SCMS* (see Bibliography).

6.5.4 Synchronization of digital audio signals

If digital signals are to be combined in any way, or interconnected between devices, it is important that they are identically timed. In the following sections some strategies and criteria for signal synchronization are discussed.

6.5.4.1 Sync references

A number of reference signals may be used for locking the sampling-rate clock of a digital audio device. In a large system it is vital that all devices remain locked to a common sampling rate reference clock, and there are now AES recommendations for this purpose. Signals will be considered synchronous when they have identical sample rates. In order to achieve this, all devices within a system should be able either to lock to a reference signal in the form of a separate AES/EBU input, or to the clock rate implicit in the AES/EBU programme signal present at the device's digital audio input. Input signal frame edges must lie within ± 25% of the reference signal's frame edge, and output signals within ± 5% (*Figure 6.68*), although tighter accuracy than this is preferable.

The AES/EBU format reference signal may either contain programme or not. If it does not contain programme it may simply contain the sync preamble with the rest of the frame inactive. Two grades

Figure 6.68 *Timing stipulations for synchronous signals.*

of reference are specified: grade 1, having a long-term frequency accuracy of ± 1 ppm (parts per million), and grade 2, having a long-term accuracy of ± 10 ppm. The grade-2 signal conforms to the standard AES sample-frequency recommendation for digital audio equipment, and is intended for use within a single studio which has no immediate technical reason for greater accuracy, whereas grade 1 is a tighter specification and is intended for the synchronization of complete studio centres (as well as single studios if required). Byte 4, bits 0 and 1 of the channel status data of the reference signal indicate the grade in use (00 = default, 01 = grade 1, 10 = grade 2, 11 = reserved). The reference signal may be distributed in a star configuration with a maximum of four spurs per circuit.

Currently, digital audio recorders are provided with a wide range of synchronizing inputs, and most systems may be operated in the external or internal sync modes. In the internal sync mode a system is locked to its own crystal oscillator, which should be accurate within ±10 ppm if it conforms to AES recommended practice (AES5-1984). If in the external sync mode the system should lock to one of its sync inputs, which may either be selectable using a switch, or be selected automatically based on an order of priority depending on the mode of operation (for this the user should refer to the operations manual of the device concerned). Typical sync inputs are word clock (WCLK), which is a square-wave TTL-level signal (0–5 V) at the sampling rate usually available on a BNC-type connector; 'composite video', which is a video reference signal consisting of either normal picture information or just 'black and burst' (a video signal with a blacked-out picture); or AES/EBU, which is a reference signal conforming to the tolerances outlined above. In all cases one machine or source must be considered to be the 'master', supplying the sync reference to the whole system, and the others as 'slaves'.

It is also possible to lock a device to its own AES/EBU digital audio input (if one exists), and

this technique is often known as 'genlock'. It can be used for small systems operating in a daisy-chain, or where an external reference is not available (*Figure 6.69*). Alternatively, a separate WCLK may be daisy-chained between devices in cases where the AES/EBU interface is not available, such as when equipment is interfaced using SDIF-2. A 'sync' light is usually provided on the front panel of a device (or under a cover) to indicate good lock to an external or internal clock. and this may flash or go out if the system cannot lock to the clock concerned, because it has too much jitter, is at too low a level, or conflicts with another clock, or because it is not at a sampling rate which can be accepted by the system.

Figure 6.69 *Genlock synchronization. The slave locks to the master's clock by extracting it from the AES/EBU data stream.*

Video synchronization (composite synchronization) is often used when a system is operated within a video environment, and where a digital recorder is to be referenced to the same synchronizing reference as the video machines in a system. This is useful when synchronizing audio and video transports during recording and replay, using either a synchronizer or a video editor, when time-code is only used initially during the pre-roll, whereafter machines are released to their own video synchronizing reference. It also allows for time-code to be recorded synchronously on the audio machine, since the time-code generator used to stripe the tape can be locked to video synchronizers also.

6.5.4.2 Distribution of synchronization references

It is appropriate to consider digital audio as similar to video when approaching the subject of synchronized distribution, especially in large systems. Consequently, it is advisable to use a central high-quality synchronizing signal generator (the equivalent of a video synchronizer pulse generator (SPG)), the output of which is made available widely about the studio centre, using digital distribution amplifiers (DDAs) to supply different outlets in a 'star' configuration. In video operations this is often called

'house sync', and such a term may become used in digital audio as well. Each digital device in the system may then be connected to this house sync signal, and each should be set to operate in the external sync mode. In the future, as described above, it is most likely that house synchronization will take the form of an AES/EBU-format signal, and this should be adopted wherever possible for the sake of standardization.

Using the technique of central synchronizing-signal distribution (*Figure 6.70*) it becomes possible to treat all devices as slaves to the synchronizing generator. In this case there is no 'master' machine, since the sync generator acts as the 'master'. It is a concept which many analogue recording engineers may find unfamiliar, since analogue machines are typically 'free running', requiring no synchronization reference. It requires that all machines in the system operate at the same sampling rate, unless a sample-rate convertor or signal synchronizer are used (see Section 6.5.4.3).

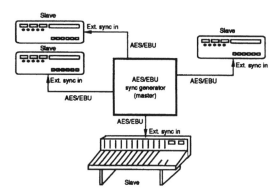

Figure 6.70 *Central sync signal distribution. A high-accuracy master AES/EBU synchronizing generator is used as a reference for all devices in the system. All devices operate in the external synchronized mode.*

Alternatively, in a small studio, it may be uneconomical and impractical to use a separate SPG, and in such cases one device in the studio must be designated as the master. This device would then effectively act as the SPG, operating in the internal sync mode, with all other devices operating in the external sync mode and slaving to it (see *Figure 6.71*). In situations where sources are widely spread apart, perhaps even being fed in from remote sites (as might be the case in broadcast operations) the distribution of a single sync reference to all devices becomes very difficult or impossible. In such cases it may be necessary to synchronize external 'wild' feeds to the house synchronizer before they can be connected to

in-house equipment, and for this purpose a sample rate synchronizer should be used.

Figure 6.71 *Where it is not economically viable to use a separate synchronizing reference, one device must act as the master clock, as shown in (a). In this example the mixer operates in internal synchronized mode and sends a reference signal (either AES/EBU or WCLK) to all other devices (operating in external synchronization). Alternatively, as shown in (b), slaves may lock to the incoming AES/EBU audio signal, which contains a clock.*

6.5.4.3 Use and function of a sample rate synchronizer

A sample rate synchronizer is used to synchronize a non-synchronous digital audio signal to a reference signal. In this context, the synchronizer is used only to lock signals of nominally the *same* sampling frequency, but the sampling frequency of which may differ within the maximum tolerances allowed in the AES recommendation. The audible effect resulting from a signal which drifts with relation to the synchronized reference is usually the appearance of a glitch or click at the difference frequency between the signal and the reference, nominally at a level around 50 dB below the signal. This usually appears when attempting to mix two digital audio signals whose sampling rates differ by a small amount. Some

systems may not operate at all if presented with asynchronous signals.

It may be necessary to correct for timing errors in signals synchronous to the master clock, but which have travelled long distances, since these may have been delayed slightly and thus be out of phase with the reference. This is more properly referred to as 'frame alignment'. Propagation delays are not great: for example, an AES/EBU signal must travel 3.7 km down a cable before it is delayed by one sample period. A 1-bit delay results when a signal has travelled through approximately 58 m of cable (although this depends on the cable). For signals of different nominal sampling frequencies, a sample rate convertor should be used.

A synchronizer's job, when operating with serial data conforming to the AES/EBU standard, is to synchronize signals whose frame edges lie within ±25% of the reference signal's frame edges. This should result in signals at the output which are within ±5% of the reference signal (the definition of synchronous signals), although tighter lock is preferable in order that errors in later stages do not accumulate. The synchronizer is normally to be located in the operational area accepting remote or asynchronous signals, and may eventually be located on a chip within devices themselves. Depending on how far adrift the incoming signal is from the reference, a number of techniques may be used to bring it into lock, but often a simple buffer memory is adequate, with a short-cross-fade being used to reset the buffer when it either overflows or becomes empty. The synchronizer could also perform the useful function of re-clocking noisy remote signals to remove short-term timing errors. The DAR DASS 100 is a commercial example of such a device.

6.5.4.4 Sample rate converters

Since there are a number of possible sampling rates in use in digital audio it will be necessary from time-to-time to interconnect digitally devices whose sampling rates differ. In such cases a sample rate convertor is required, this being a device which takes in audio data at one rate and sends it out at another. This process could either be used between two tape machines which operate at different rates, for converting a 48 kHz recording to 44.1 kHz so that it could be used as a CD master, for example, or it could be used in between a tape machine and a digital mixer, in order that the tape machine could be varispeeded without affecting the input rate to the mixer. The 32 kHz rate is used widely in digital audio broadcasting, and thus the broadcaster will be interested in translating material at the professional rate of 48 kHz to the broadcast rate.

In fully integrated digital systems, where most devices are interconnected digitally, it will become important for one common and fixed sampling rate to be used throughout, locked to a central sampling rate clock. If any devices in the system do not conform to this rate they should be passed through a sample rate convertor in order to allow interchange of audio at the system rate, and a number of modern disk-based devices are now becoming available with internal sample rate convertors which ensure that, no matter what the internal operations of the system, the input and output sampling rates remain constant. This is important in any system which is varispeeded or synchronized to a time-code source, since without sample rate conversion the sampling rate of the digital outputs of the system would be likely to follow any speed variations of the replay.

It is now possible to perform high-quality sample rate conversion over a wide range of input and output rates which do not have to have a simple integer or fractional relationship. Older convertors were only capable of converting between a fixed input and output rate, and will be less useful.

6.5.5 Networks

General purpose data networks are increasingly used for digital audio, not as signal cables but as a means of transferring files between distributed equipment. Not only can networks be used to link a number of local workstations and storage resources, but they can also be used to transfer files over very large distances, even across the world if necessary. This has made it possible for people to work on common material, using shared resources such as storage and backup, and sometimes to avoid the time delay and expense of physically transporting recording media to other sites.

As with storage devices, it is only recently that networks have become fast enough and cheap enough to be viable in this application, since, as already explained, digital audio requires considerable bandwidth for real-time transfer. If networks are to be truly useful for digital audio applications then they really need to be capable of handling the file transfer for a number of channels of audio simultaneously, and preferably at speeds in excess of real time. It may, nevertheless, be acceptable in some circumstances for a network to transfer in non-real-time, especially if one is dealing with an international transfer of audio that would have taken longer or

cost more if one had had to physically transport the recording instead.

6.5.5.1 Network principles

A network carries data either on wire or optical fibre, and is normally shared between a number of devices and users. The sharing is achieved by containing the data in packets of a limited number of bytes (usually between 64 and 1518), each with an address attached, rather like MIDI data. The packets may share a common physical link, which is normally a high-speed serial bus of some kind, being multiplexed in time either using a regular time slot structure or in an asynchronous fashion whereby the time interval between packets may be varied, as shown in *Figure 6.72*. The length of packets also may not be constant, depending on the requirements of different protocols sharing the same network. Packets for a particular file transfer between two devices may not be contiguous and may be transferred erratically, depending on what other traffic is sharing the same physical link.

Figure 6.73 shows some common physical layouts for local area networks (LANs). LANs are networks which operate within a limited area, such as an office building or studio centre, within which it is common for every device to 'see' the same data, each picking off that which is addressed to it and ignoring the rest. In order to place a packet of data on the network, devices must have a means for determining whether the network is busy, and there are various protocols in existence for arbitrating network access. In the 'backbone' configuration devices are connected to spurs off a common serial bus which requires the bus to be 'chained' between each successive device. Here, a break in the chain can mean disconnection for more than one device. The star configuration involves a central hub which distributes the data to each device separately, and this is more reliable because a break in any one link does not affect the others. Bus arbitration in both these cases is normally performed by collision detection which is a relatively crude approach, relying very much on the rules of spoken conversation between people. Devices attempt to place packets on the bus whenever it appears to be quiet, but a collision may take place if another device attempts to transmit before the first one has finished. The network interface of the transmitting device detects the collision by attempting to read the data it has just transmitted and retransmits it after transmitting a brief 'blocking signal' if it has been corrupted by the collision.

A token ring configuration places each device within a 'ring', each having both an 'in' and an 'out' port, with bus arbitration performed using a process of token passing from one device to the next. This works rather like trains running on a single track line, in that a single token is carried by the train using the line, and trains can only use the line if carrying the token. The token is passed to the next train upon leaving the single track sector to show that the line is clear.

Network communication is divided into a number of 'layers', each of which determines an aspect of communication. The ISO seven-layer model for open systems interconnection (OSI) shows how many levels there are at which compatibility between systems needs to exist before seamless interchange of data can be achieved (*Figure 6.74*). It shows that communication begins with the application and filters down through various stages to the layer most people understand – the physical layer, or the piece of wire over which the information is carried. Layers 3, 4 and 5 can be grouped under the broad heading of 'protocol', determining the way in which data packets are formatted and transferred. There is a strong similarity here with the exchange of data on physical media, as discussed earlier, where a range of compatibility layers from the physical to the application determine whether or not one device can read another's disks.

Figure 6.72 *Packets for different destinations (A, B and C) multiplexed onto a common serial bus. (a) Time division multiplexed into a regular time slot structure. (b) Asynchronous transfer showing variable time gaps and packet lengths between transfers for different destinations.*

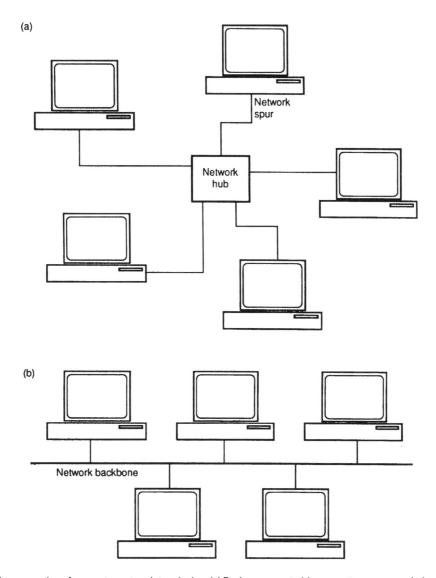

Figure 6.73 *Two examples of computer network topologies. (a) Devices connected by spurs to a common hub, and (b) devices connected to a common 'backbone'.*

6.5.5.2 Network standards

Ethernet, FDDI (Fibre Distributed Data Interface) and ATM (Asynchronous Transfer Mode) are examples of network standards, each of which specifies a number of layers within the OSI model. FDDI, for example, specifies only the first three layers of the OSI model (the physical, data link and network layers).

Ethernet allows a number of different methods of interconnection and runs at a maximum rate of 10 Mbit/s (although there is now a version running at 100 Mbit/s), using collision detection for network access control. Interconnection can be via either thick (10-Base-10) or thin (10-Base-2) coaxial cable, the thin being quite common and normally working in the backbone-type configuration shown in the previous section, using 50 ohm BNC connectors and T-pieces to chain devices on the network (see *Figure 6.75*. Such a configuration requires resistive terminators at the ends of the bus to avoid reflections, as with SCSI connections. Twisted-pair (10-Base-T) connection is also becoming quite popular, and is often configured in the star topology using a central hub. Devices can then be plugged and unplugged without affecting others on the network.

| 7 Application Layer |
| 6 Presentation layer |
| 5 Session layer |
| 4 Transmission layer |
| 3 Network layer |
| 2 Data link layer |
| 1 Physical layer |

Figure 6.74 *The ISO model for Open Systems Interconnection is arranged in seven layers, as shown here.*

FDDI is a high-speed optical fibre network running at 100 Mbit/s, and operating on the token passing principle described above, allowing up to 2 km between stations. There is also a copper version of FDDI called CDDI which runs at the same rate but restricts interconnection distance. ATM is really only a protocol for data communication, which does not specify the physical medium for interconnection. Physical systems already exist which are capable of running at 250 Mbit/s, and plans are in hand for systems running at 2.4 Gbit/s. Both FDDI and ATM have been considered quite suitable for audio and other multimedia purposes, and special protocols have been devised by some manufacturers to optimise the transfer bandwidth for realtime recording and replay purposes.

6.5.5.3 Network protocols

A protocol specifies the format of communication on a network. In other words it determines things like the format of data packets, their header information and addressing structure, and any handshaking and error retrieval schemes, amongst other things. One physical network can handle a wide variety of protocols, and packets conforming to different protocols can co-exist on the same bus.

Some common examples of general purpose network protocols are such as AppleTalk, used for file sharing and printing between Macintosh computers; TCP/IP, used for communications over the Internet (see below); and Novell, used widely in PC networks. These general purpose protocols are not particularly efficient for real-time audio transfer, but they can be used for non-real-time transfer of audio files between workstations. Specially designed protocols may be needed for audio purposes, as described below.

6.5.5.4 Audio network requirements

The principal application of computer networks in audio systems is in the transfer of audio data files between workstations, or between workstations and a central 'server' which stores shared files. The device requesting the transfer is known as the 'client' and the device providing the data is known as the 'server'. When a file is transferred in this way a byte-for-byte copy is reconstructed on the client machine, with the file name and any other header data intact. There are considerable advantages in being able to perform this operation at speeds in excess of real time for operations in which real-time feeds of audio are not the aim. For example, in a news editing environment a user might wish to load up a news story file from a remote disk drive in order to incorporate it into a report, this being needed as fast as the system is capable of transferring it. Alternatively, the editor might need access to remotely stored files, such as sound files on another person's system, in order to work on them separately. In audio post-production for films or video there might be a central store of sound effects, accessible by everyone on the network, or it might be desired to pass on a completed portion of a project to the next stage in the post-production process.

Simple desktop computer networks such as Apple's LocalTalk are very slow by audio standards and are not capable of transferring even a single channel sound file at real-time speeds, whereas Ethernet is fast enough to transfer audio data files slightly faster than real-time, depending on network loading. Approaches using FDDI or ATM over optical fibre are most appropriate for handling large

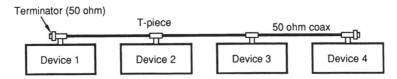

Figure 6.75 *Typical thin Ethernet interconnection arrangement.*

numbers of sound file transfers simultaneously at high speed. Unlike a real-time audio interface, the speed of transfer of a sound file over a network (when using conventional file transfer protocols) depends on how much traffic is currently using it. If there is a lot of traffic then the file may be transferred more slowly than if the network is quiet (very much like motor traffic on roads). The file might be transferred erratically as traffic volume varies, with the file arriving at its destination in 'spurts'. There therefore arises the need for network communication protocols designed specifically for the transfer of real-time data, which serve the function of reserving a proportion of the network bandwidth for a given period of time, as described below.

Without real-time protocols designed as indicated above, the computer network may not be relied upon for transferring audio where an unbroken audio output is to be reconstructed at the destination from the data concerned. The faster the network the more likely it is that one would be able to transfer a file fast enough to feed an unbroken audio output, but this should not be taken for granted. Even the highest speed networks can be filled up with traffic! This may seem unnecessarily careful until one considers an application in which a disk drive elsewhere on the network is being used as the source for replay by a local workstation, as illustrated in *Figure 6.76*. Here it must be possible to ensure guaranteed access to the remote disk at a rate adequate for real-time transfer, otherwise gaps will be heard in the replayed audio.

6.5.5.5 Audio networks

Recent networking solutions in the audio field, such as those proposed by AMS/Neve, Siemens, Sonic Solutions, Avid and others, use high-speed optical fibre technology and FDDI or ATM. Other companies have stuck with the slower Ethernet for the time being, although not attempting to use it for the purpose of transferring large numbers of channels of audio in real time. Although Ethernet is capable of a maximum data rate of 10 Mbit/s, its performance under heavy load with many users is often as low as 2-3 Mbit/s because of the network access method used.

Sonic Solutions, for example, has taken the step of developing its own network protocol called Sonic-Net for transporting audio and other multimedia data over FDDI in a form which ensures optimum use of bandwidth by using large packets, and allows systems to guarantee that real-time file transfers, once started, will be able to complete without breaks (the so-called 'reservationist' approach). It allows over 80 channels of audio to be transferred simultaneously. Phase 2 of SonicNet will allow the network to be used for routing real-time digital audio as well as for transferring files, thereby further blurring the distinction between digital audio interfaces and computer data networks.

Sonic Solutions has gone further through the OSI model than the first three layers specified in the FDDI protocol, allowing greater control over the way in which data is transferred. The alternative to this highly controlled approach is a form of network

Figure 6.76 *In this example of a networked system a remote disk is accessed over the network to provide data for real-time audio playout from a workstation used for on-air broadcasting. Continuity of data flow to the on-air workstation is of paramount importance here.*

'free for all', which assumes there is more network bandwidth than will ever be needed at any one time, allowing applications to grab whatever bandwidth they can for file transfer operations. This is simpler, but less reliable in cases of heavy network load.

An example of a networked audio system is shown in *Figure 6.77*, showing how resources can be distributed around the network. Note that this is based on SonicNet which allows stand alone disk drives to be connected directly to the network using a special interface chassis and card.

ATM allows for either guaranteed bandwidth communications between a source and a destination (needed for AV applications), or for more conventional variable bandwidth communication. It is a switched network, which means that the physical network is made up of a series of interconnected switches that are reconfigured to pass the information from sources to destinations according to the header information attached to each data packet. A network management system handles the negotiation between different devices that are contending for bandwidth, according to current demand.

6.5.5.6 Extending a network
It is common to need to extend a network to a wider area or to more machines. As the number of devices increases so does the traffic, and there comes a point when it is necessary to divide a network into zones, separated by 'repeaters', 'bridges' or 'routers'. Some of these devices allow network traffic to be contained within zones, only communicating between the zones when necessary. This is vital in large interconnected networks because otherwise data placed anywhere on the network would be present at every other point on the network, and overload could quickly occur.

A repeater is a device which links two separate segments of a network so that they can talk to each other, whereas a bridge isolates the two segments in normal use, only transferring data across the bridge when it has a destination address on the other side. A router is very selective in that it examines data packets and decides whether or not to pass them depending on a number of factors. A router can be programmed only to pass certain protocols and only certain source and destination addresses. It therefore acts as something of a network policeman and can be used as a first level of ensuring security of a network from unwanted external access. Routers can also operate between different standards of network, such as between FDDI and Ethernet, and ensure that packets of data are transferred over the most time/cost-effective route.

Figure 6.77 *Example of a networked audio system with distributed storage and processing resources, allowing access to all disk and tape devices by both workstations (based on Sonic Solutions SonicNet).*

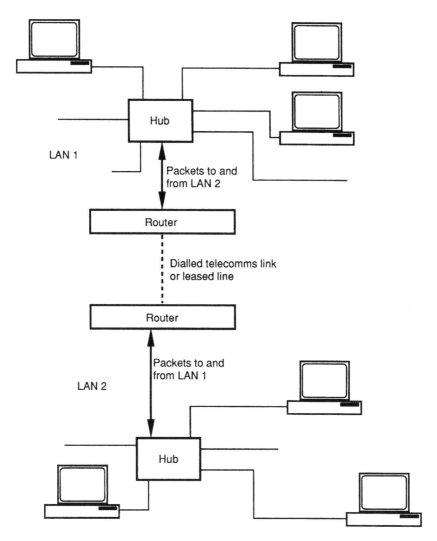

Figure 6.78 *A WAN is formed by linking two or more LANs using a router.*

One could also use some form of router to link a local network to another which was quite some distance away, forming a so-called Wide Area Network (WAN), as shown in *Figure 6.78*. Data can be routed either over dialled data links such as ISDN (see below), in which the time is charged according to usage just like a telephone call, or over leased digital circuits which are rented for a certain period of time from telecomms companies and are always available for data transfer without dialling. The choice would depend on the degree of usage and the relative costs. The data rate handled by such links is usually lower than that provided on the LAN, thereby limiting their usefulness for multichannel audio work, but still making them very useful for transferring

files to distant destinations and for auditioning stereo audio material, possibly using a form of data reduction. Leased lines are often available at bandwidths of between 1 and 2 Mbit/s, whereas ISDN runs at 64 or 128 kbit/s. It is to be expected that lines with greater bandwidth will gradually become available at reasonable cost as time goes by.

6.5.5.7 SCSI assignment matrix
Although not a true network, it is possible to create some of the advantages of a network using a method of SCSI assignment, which allows central storage resources to be accessed from a variety of workstations. This is the approach adopted by SSL in its *SoundNet* system. In SoundNet up to 16 SCSI

Figure 6.79 *Storage resources assigned to workstations using a SCSI assignment matrix (based on SSL SoundNet).*

devices can be connected to a switcher which can assign the devices to different workstations, as shown in *Figure 6.79*. It requires that the workstation processors be located close to the switcher, because SCSI cables have to be short, so a lower speed Ethernet connection is then used to link the workstations for control purposes. The user terminals are then connected using RS422 serial interfaces.

This approach restricts the storage device assignment to a single machine at a time, but storage devices may be reassigned at any time so that a file could be quickly copied between two systems, for example. The Ethernet network can also be extended to the control terminal where an optical disk drive can be located for playing library disks. Ethernet is generally fast enough for transferring at least stereo audio in real-time, provided not too many people try to do it at the same time. Multiple workstation processors can be synchronized together for large-scale projects involving multiple channels, each replaying eight channels of audio from separate disks.

6.5.5.8 ISDN
ISDN, the Integrated Services Digital Network, is a development in the telecommunications field which is gaining wider usage. ISDN is an extension of the digital telephone network to the consumer, providing two 64 kbit/s digital channels ('B' channels) which can be connected to ISDN terminals anywhere in the world simply by dialling. Data of virtually any kind may be transferred over the dialled-up link,

and potential applications for ISDN include audio transfer.

Since the total usable capacity of a single ISDN-2 connection is only 128 kbit/s it is not possible to carry linear PCM data at normal audio resolutions over such a link, but it is possible to carry moderately high quality stereo digital audio at this rate using a data reduction system such as MPEG-Audio, or to achieve higher rates by combining more than one ISDN link to obtain data rates of, say 256 or 384 kbit/s. Multiple ISDN lines must be synchronized together using devices known as inverse multiplexers if the different delays which may arise over different connections are to be compensated. There are also ISDN-30 lines, providing 30 simultaneous 'B' channels of 64 kbit/s (giving roughly 2 Mbit/s), but these are expensive.

It is possible to use ISDN links for non-real-time audio file transfer, and this can be economically viable depending on the importance of the project and the size of files. The cost of an ISDN call is exactly the same as the equivalent duration of normal telephone call, and therefore it can be quite a cheap way of getting information from one place to another.

In the USA, there still remain a lot of circuits which are very similar to ISDN but not identical. These are called 'Switched 56' and carry data at 56 kbit/s rather than 64 kbit/s (the remaining 8 kbit/s that make up the total of 64 kbit/s is used for house-keeping data in Switched 56, whereas with ISDN the house-keeping data is transferred in a 'D' channel

on top of the two 64 kbit data channels). This can create some problems when trying to link ISDN terminals, if there is a Switched 56 bridge somewhere in the way, requiring file transfer to take place at the lower rate.

For more information on ISDN, the reader is referred to an excellent book by John Griffiths (see Bibliography).

6.5.5.9 The Internet

There is now wide interest in the Internet as a means for world-wide communication. At the moment its primary use is not for real-time multimedia information such as audio and video, but there is growing evidence of applications which allow low quality (very low bit rate) audio and video information to be carried in this way. The RealAudio format, for example, developed by Progressive Networks, currently allows 8 bit, 11 kHz sampling rate, mono sound to be transferred in real time over the Internet, giving roughly medium wave radio quality, and with no guarantee of unbroken transfer. People are using the Internet for transferring some sound files in non-real time, but mainly short sounds for the games and music markets because of the limited data rates available to most users. Whole books have been written about the Internet, because it is a very large subject, so it is only proposed to give a brief introduction here.

The Internet is a collection of interlinked networks with bridges and routers in various locations, which originally developed amongst the academic and research community. It is used widely in these fields, but has also become extremely popular for personal and business applications, with many commercial operations now offering to connect users to the Internet for a reasonable fee. The bandwidth (data rate) available on the Internet varies from place to place, and depends on the route over which data is transferred. In this sense there is no easy way to guarantee a certain bandwidth, nor a certain 'time slot', and when there is a lot of traffic it simply takes a long time for data transfers to take place. Many people access the Internet through a commercial service provider, using a telephone line and a modem, and the maximum speed that is possible using such technology is currently 28.8 kbit/s (slow by audio standards). At this rate the error rate can be quite high, and many people operate at lower rates such as 2400, 9600 or 14 400 bit/s. ISDN is another alternative for higher speed communications, and the most intensive users will probably opt for high speed leased lines giving direct access to the Internet. The inter-university network in the UK for example, called Super JANET, runs on leased lines at 10 Mbit/s.

The common protocol for communication on the Internet is called TCP/IP (Transmission Control Protocol/Internet Protocol). At a more detailed level, as part of the TCP/IP structure, there are high level protocols for transferring data in different ways. There is a file transfer protocol (FTP) used for downloading files from remote sites, a simple mail transfer protocol (SMTP) and a post office protocol (POP) for transferring e-mail, and a hypertext transfer protocol (HTTP) used for interlinking sites on the so-called WorldWide Web (WWW). The WWW is a collection of file servers connected to the Internet, each with its own unique IP address (the method by which devices connected to the Internet are identified), upon which may be stored text, graphics, sounds and other data. Applications are available which make it easy to access these WWW servers, with links to applications that can replay any downloaded multimedia information.

Bibliography

AES, AES10-1991 (ANSI S4.43-1991). Serial multichannel audio digital interface (MADI), *Journal of the Audio Engineering Society*, **39**, 369–377 (1991)

AES, AES11-1991 (ANSI 4.44-1991). *Synchronization of digital audio equipment in studio operations*, Audio Engineering Society (1991)

AES, AES3-1992. Serial transmission format for two channel linearly represented digital audio data, *Journal of the Audio Engineering Society*, **40**, 3, March (1992)

ANDERSON, D., High speed networking for professional digital audio, AES UK Digital Audio Interchange Conference, 18–19 May, London, pp. 60–69, Audio Engineering Society (1993)

ANON, *DDP Reference Manual* (1990). Doug Carson and Associates, 300 North Harrison, PO Box 1646, Cushing, OK 74023, USA. Tel: +1 918 225 0346; Fax: +1 918 225 1113

ANSI, ANSI X3.166-1990 *Physical Layer Medium Dependent*, American National Standards Institute (1990)

APPLE COMPUTER, *Audio Interchange File Format AIFF-C* (1991). Apple Computer Inc., Developer Technical Support, 20525 Mariani Ave, MS:75-3T, Cupertino, CA 95014, USA. (Also available as a PostScript file by anonymous ftp from *ftp.sgi.com* in */sgi/aiff-c.9.26.91.ps*)

AUDIO ENGINEERING SOCIETY, *Proceedings of the AES/EBU Interface Conference*, AES British Section (1989)

AUDIO ENGINEERING SOCIETY, *AES 11 Synchronization of Digital Audio Equipment in Studio Operations* (1991)

AUDIO ENGINEERING SOCIETY, *AES 10 Serial Multichannel Digital Audio Interface (MADI)* (1991)

AUDIO ENGINEERING SOCIETY, *AES 3 Serial Transmission Format For Linearly Represented Digital Audio Data* (1992)

AVID, *OMF Interchange Specification Version* 1.0 (1994). OMF Developers' Desk, Avid Technology Inc., Metropolitan Technology Park, One Park West, Tewksbury, MA 01876, USA. Tel: +1 508 640 3400; Fax: +1 508 640 9768. E-mail: *omf-request@avid.com*

BURKHARDTSMAJER, B. *et al.*, 'The ISDN Music TAXI', presented at the 92nd AES Convention, Vienna, Austria, 24–27 March, preprint 3344 (1992)

DIGIDESIGN, *Sound Designer I and II file format specifications*. Digidesign, 1360 Willow Rd, Menlo Park, CA 94025, USA. Tel: +1 415 688 0744; Fax: +1 415 327 3131

DIGIGRAM *MusiFile format specification* (1993). Digigram, Parc de Pré Milliet, 38330 Montbonnot, France. Tel: +33 76 52 47 47; Fax: +33 76 52 18 44

ELECTRONIC Arts, *EA IFF 85 standard for interchange format files* (1985)

GRIFFITHS, J., *ISDN Explained*, Wiley, Chichester (1990)

HARRIS, J. and RUBEN, I. *Bento Specification, Revision 1.0d4.1*. Apple Computer Inc. (1992)

IEC, IEC 1119 *Part 6: Serial Copy Management System*, International Electrotechnical Commission, Geneva (1992)

INTERNATIONAL ELECTROTECHNICAL COMMISSION, *IEC 958 Digital Audio Interface* (1989)

RUMSEY, F., *The Audio Workstation Handbook*, Focal Press, Oxford (1996)

RUMSEY, F. and WATKINSON, J., *The Digital Interface Handbook* (2nd edn) Focal Press, Oxford (1995)

WATKINSON, J., *The Art of Digital Audio*, (Second edition) Focal Press, Oxford (1994)

7

Miscellaneous topics

Steven Dixon
Department of Physics, University of Warwick

Vivian Weeks
Formerly of BBC Design and Equipment Department

John Ratcliff
Proprietor, The Television Training Company

7.1 Ultrasonics

7.1.1 Introduction

'Ultrasonics' is the broad term which describes the study or application of sound waves with frequencies greater than can be detected by the human ear. This definition is somewhat arbitrary, but 16 kHz is usually considered to be the lower frequency limit of the ultrasound spectrum despite the fact that some human beings, particularly young children, are capable of hearing frequencies of up to 20 kHz. The techniques described in this chapter use ultrasound with frequencies up to 10^{13} Hz, at which point the equivalent wavelength is approaching atomic dimensions. Clearly the choice of the frequency used is endemic in the nature of the technique and the medium through which the ultrasound is to be propagated.

The uses of ultrasound are as varied as the frequency spectrum is broad. The pulse echo technique for measuring distances, employed by bats and porpoises, is used in many different spheres ranging from underwater echo sounders and back fat measurements on pigs and cattle, to process control in steel mills. Pulse echo techniques are commonly used for flaw detection and the non-destructive evaluation of materials. The use of novel methods of ultrasound generation and detection such as the electromagnetic acoustic transducer (EMAT), air coupled transducers and lasers have been shown to add versatility to the use of ultrasound for flaw detection and process control.

High intensity ultrasonic waves may be used for welding both metals and plastics in much the same way as conventional techniques, while avoiding the use of external heat and melting. In addition, the direct application of ultrasonic vibrations may be used for immersed cleaning, foam suppression and for the production of emulsions. These processes are often enhanced by the ability of the ultrasonic vibration to induce cavitation in fluids.

The medical applications of ultrasound are becoming increasingly widespread as the techniques become more sophisticated. Ultrasound scanners are able to monitor the development of the unborn baby, the behaviour of the heart and now monitor blood flow, and are proving to be an invaluable diagnostic tool. Ultrasound is also used for the treatment of muscular problems for instance by stimulating blood flow and a concentrated beam of ultrasonic energy may be used to break up gall and kidney stones.

The purpose of this section is to provide an introduction to the properties and generation and detection of ultrasound and to describe some of the more widespread industrial ultrasonic techniques and medical applications of ultrasonics. The mathematical background to the propagation of ultrasound through various media is extremely complicated and often largely empirical; consequently it has been kept to a minimum. The reader is referred to the appropriate texts listed in the Bibliography for a more complete mathematical account.

7.1.2 The nature of ultrasound

Within the bulk of a material, both longitudinal and shear (or transverse) ultrasonic waves may be produced by the appropriate transducers as will be discussed in the following section. The longitudinal wave may be produced in solid liquids or gases and is the most commonly used mode in ultrasonics. The displacement of particles in the media is along the direction of propagation of the wave. Thus, just as in a conventional sound wave, the wave is made up of regions of compression and rarefaction (*Figure 7.1*). However, in solids it is also possible for an ultrasonic wave to have all or part of its amplitude perpendicular to the propagation direction of the wave, i.e. a shear wave (*Figure 7.2*). These waves are unique in that the medium undergoes no compression or rarefaction, and there are no local fluctuations in density as occurs in longitudinal waves.

Figure 7.1 *A longitudinal wave. In the unstrained state, the horizontal and vertical lines will be equally spaced.*

Upon reflection at an interface a longitudinal wave is likely to induce a shear wave in addition to the reflected longitudinal wave (*Figure 7.3*). Similarly, a reflected shear wave will generate a longitudinal

Direction of propagation

Figure 7.2 *A transverse or shear wave.*

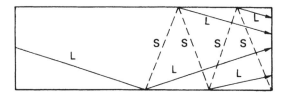

Figure 7.3 *Mode conversion occurring on reflection of both transverse and longitudinal waves. (———) The paths of the longitudinal waves; (- - - - - -) the paths of the shear waves.*

wave. This phenomena is known as 'mode conversion' and will lead to a proliferation of different modes as the wave propagates, along a bar for instance. Clearly, the extent to which mode conversion occurs is dependent on the angle of incidence.

At the surface of a solid the propagation of the ultrasonic wave is somewhat more complicated and a variety of different surface waves may be propagated.

Love waves (*Figure 7.4*) may occur in a thin layer of material lying on a substrate, and are easily visualized as a shear wave propagating along a surface layer with the transverse displacement within the plane of the surface. However, the applications of Love waves are relatively few and they will not be encountered again in this section.

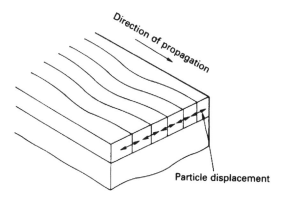

Figure 7.4 *A Love wave propagating in a surface layer.*

If the material is considered to be of infinite thickness (practically, greater than several wavelengths) Rayleigh waves (*Figure 7.5*) may be propagated. Rayleigh waves also propagate along the surface; however, the particle displacement is both in and out of the plane, producing a 'ripple' across the surface. Each particle on the surface of the media traces out an ellipse with the passage of the wave. The amplitude of this motion decreases exponentially with surface depth and at one wavelength below the surface, will be at approximately one-tenth of its value at the surface.

Figure 7.5 *A Rayleigh wave.*

In the case of plates with a thickness of the order of the ultrasound wavelength plate or Lamb waves (*Figure 7.6*) may be propagated. These are best visualized as two sine waves propagating along either side of the plate. If the two sine waves are in phase

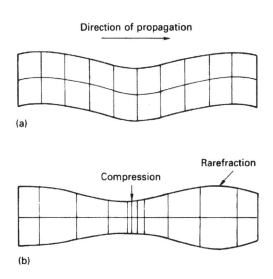

Figure 7.6 *An asymmetrical Lamb wave (a) and a symmetrical Lamb wave (b).*

the net result is an asymmetric Lamb wave. If they are out phase by 180° the net result is a symmetric Lamb wave. The particle displacement in the centre of the plates allows analogies to be drawn between the Lamb waves and bulk waves. The symmetrical wave causes compression and rarefaction of the particles in the centre of the plate analogous to a longitudinal wave, while the asymmetrical mode causes transverse displacement of the central particles analogous with a shear wave. In both types of Lamb wave, the particle displacement at the surface is elliptical.

In an ideal medium Lamb waves are the only types of ultrasonic wave which will show dispersion, i.e. the velocity of the wave is dependent on the frequency or, more correctly, the group and phase velocities are different. In the case of bulk waves and Rayleigh waves, the velocity of the ultrasonic wave in a solid is related to the density of the material and its modulus of rigidity, Young's modulus or Poisson's ratio. In liquids and gasses, the velocities of bulk waves are dictated by the density and the ratio of the specific heats and the bulk compressibility, respectively. The velocities of Lamb waves are affected by the frequency of the wave and the plate thickness and do not conform to a predictable pattern. There are, as ever, exceptions to these rules and in some circumstances Rayleigh waves and bulk waves will display dispersion.

7.1.3 The generation and detection of ultrasound

7.1.3.1 Piezoelectric transducers
The piezoelectric transducer is able to convert electrical energy to sound energy, and vice versa. Operation is based on the fact that when certain materials, referred to generically as 'piezoelectrics', are subjected to an applied electric field they will deform in some way. This deformation can be effected very quickly (the order of picoseconds), and a.c. fields can be used to generate ultrasound of frequencies of up to 500 MHz in some materials. However, for high a.c. frequencies, the period of the electric field becomes greater than the relaxation time of the piezoelectric (the time required for the piezoelectric to relax to its original dimensions) and the material will be unable to respond to the applied voltage. If the piezoelectric is subjected to a particular deformation then the appropriate electric field will be generated in the material. Thus piezoelectric materials are suitable for use as both detectors and generators of ultrasonic waves.

Crystalline quartz is one of the most common naturally occurring piezoelectrics, having the chemical composition SiO_2. A wafer of quartz cut along a particular direction with its plane containing the c axis (i.e. a highly symmetrical axis) of the crystal (this cut is known as the 'X cut') can be used to make a transducer for the generation of longitudinal waves; a force applied normal to each face generates an electric field in the sample in the same direction as the applied force (*Figure 7.7*). A wafer cut with the c axis in the plane of the wafer along another specific direction (the 'Y cut') can be used as a shear wave transducer. In this case if a shearing force is applied to the faces of the wafer an electric field is generated in the sample perpendicular to the faces of the wafer (*Figure 7.8*). Other crystal cuts are also used having different thermal stabilities but slightly lower ultrasonic efficiencies.

Figure 7.7 *A schematic diagram of an X-cut quartz transducer.*

Figure 7.8 *A schematic diagram of a Y-cut quartz transducer, used to generate and detect shear waves.*

More efficient and useful piezoelectric materials than quartz have synthetically produced, including plastics and ceramics, the most popular in use being the group of ceramics containing lead, zirconate and titanate, or PZT. The ceramics are polycrystalline, and before they can be used as a piezoelectric they must be electrically polarized by placing them in a strong electric field. For this reason they are known as 'ferroelectrics' because of the similarity with the magnetization of a ferromagnet. The piezoelectric

nature of a sample can be destroyed by heating the sample to a temperature above its Curie temperature (T_c). Above this temperature the thermal energy becomes greater than the energy of the interaction that keeps the ferroelectric polarized, and thus the sample becomes depolarized. The process is reversible and thus the piezoelectric nature will return below T_c.

Construction Electrical contacts are made on each face of the piezoelectric wafer by sputtering on a thin layer of metal several micrometres thick. Silver or nickel is usually used to which an electrical contact can be soldered. The wafer plus contacts now behaves electrically like a capacitor with some high resistance in parallel. In order to maximize the field in the piezoelectric an LCR resonant circuit is usually constructed, using the piezoelectric as the capacitor. The resistor serves to damp the oscillations in the wafer (*Figure 7.9*).

Figure 7.9 *An LCR circuit. C is the capacitance of the transducer. R is chosen to damp oscillations in the circuit. The resonant frequency is given by* $f_0 = 1/(2\pi)\sqrt{LC}$.

Apart from the electrical resonance the transducer will also have a mechanical resonance, dependent on the physical construction of the transducer. Consider a quartz X-cut wafer, clamped at both its faces in the transducer housing. The wafer faces must move in antiphase (opposite directions at the same time), when an a.c. voltage is applied to them. At resonance, a standing sound wave is generated in the wafer which then passes into the bulk of the sample (if the transducer is acoustically coupled to the sample). The longitudinal wave generated in the piezoelectric wafer consists of a series of compressions and rarefactions of planes in the wafer parallel to the surface planes. The mechanical resonant frequencies of the wafer will have wavelengths λ given by:

$$\lambda = \frac{2t}{(2n+1)} \quad n = 0, 1, 2, 3, \text{ etc.} \quad (7.1)$$

where t is the wafer thickness, and n is the order of the harmonic. The fundamental resonance occurs when $n = 0$ (*Figure 7.10*).

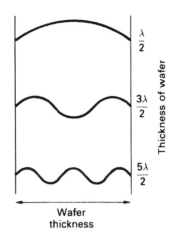

Figure 7.10 *Resonant standing waves generated in a quartz wafer.*

[*Note*: equation 7.1 applies only to this particular case. Mechanical resonance depends on how and where the piezoelectric wafer is held within the transducer.]

Limitations The piezoelectric transducer can be usefully driven at higher frequencies than the fundamental resonant frequency (higher order resonances) becoming less efficient the higher the order. The main limitation on the maximum frequency attainable with a piezoelectric transducer is the wafer thickness. It becomes physically more difficult to produce and manage very thin wafers needed for the higher frequencies. The thickness of the wafer also dictates the maximum voltage which can safely be applied across the wafer before electrical breakdown occurs.

Characterization of piezoelectric transducers The performance and efficiency of piezoelectric transducers are characterized by four constants which govern the bandwidth of the ultrasonic signal and the efficiency at which electrical energy is converted into sound energy.

The bandwidth of a piezoelectric transducer is described in terms of a quality factor Q, given by:

$$Q = \frac{f_{\text{fun}}}{(f_a - f_b)} \quad (7.2)$$

where f_{fun} is the fundamental resonant frequency, and f_a and f_b are the frequencies at which the transducer output amplitude falls to $1/\sqrt{2}$ above and below the resonant frequency, respectively (*Figure 7.11*).

Other factors which are used to classify piezoelectric transducers are the electromechanical coupling

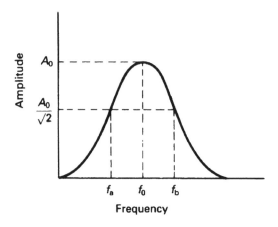

Figure 7.11 *Graph of amplitude versus frequency, yielding the Q factor.*

coefficient k, and the piezoelectric constants d and g.

$$k = \sqrt{} \text{(Fraction of mechanical energy}$$
$$\text{that goes to produce an electric}$$
$$\text{field in the piezo)} \tag{7.3}$$

or

$$k = \sqrt{} \text{(Fraction of electrical energy}$$
$$\text{that goes to produce}$$
$$\text{mechanical energy)} \tag{7.4}$$

Some of the energy put into the piezoelectric is stored in other forms such as elasticity, so k is not strictly a measure of the transducer's overall efficiency. It is quite obvious though that a large value of k is desirable.

$$g = \frac{\text{Electric field in piezo wafer (V/m)}}{\text{Stress applied to piezo wafer faces (Pa)}} \tag{7.5}$$

$$d = \frac{\text{Charge on piezo wafer face (C)}}{\text{Force applied to wafer face (N)}} \tag{7.6}$$

or

$$d = \frac{\text{Change in length of wafer thickness (m)}}{\text{Voltage applied to wafer faces (V)}} \tag{7.7}$$

The implications of the constants g and d are:

Good receiver:	low d,	high g.
Good transmitter:	high d,	low g.

Consequently, if the same transducer is to be used to generate and detect the ultrasound it generated, some compromise between its g and d factors must be achieved.

Using a piezoelectric transducer Transducers can be used in transmission and reflection techniques. The same transducer may be used to generate and detect ultrasonic waves in the sample. Firstly, however, the transducer must be acoustically coupled to the sample so that a maximum amount of acoustic energy is directed into the sample. Sometimes the transducer can be rigidly fixed to the sample, but in most applications the transducer is scanned across, or repositioned on the sample. Thus the couplant is usually viscous in nature (very viscous couplant is required for shear wave transducers as liquids do not propagate shear waves), and chosen so that it has a suitable acoustic impedance. Where getting the sample and transducer wet is not a problem as the whole system may be submerged in water and the water acts as the acoustic couplant).

Diffraction effects Piezoelectric transducers vary in size from approximately 2 mm up to several centimetres in diameter. The fact that the ultrasound is generated over a finite area means that diffraction effects must be taken into account.

Consider a sample with a surface much larger than the face of the transducer which is acoustically coupled to it, and a thickness of many ultrasonic wavelengths. As the ultrasound passes into the bulk of the sample it is diffracted due to the finite width of the transducer face, analogous to the diffraction of light at a small hole (*Figure 7.12*). The near-field diffraction pattern gives rise to a complicated interference pattern extending a distance x into the sample given by:

$$x = \frac{D^2 - \lambda^2}{4\lambda} \tag{7.8}$$

where D is the transducer face diameter, and λ is the wavelength of ultrasound in the sample.

Figure 7.12 *Light diffraction through a hole.*

The far-field has a much simpler structure, the beam half-angle $\theta_{1/2}$ being given by:

$$\theta_{1/2} = \frac{1.22\lambda}{D} \qquad (7.9)$$

Thus higher frequencies and larger transducer faces give rise to less diffuse ultrasonic beams. In detection, however, increasing the transducer diameter decreases the spatial resolution on the sample surface.

Increasing the frequency also means that smaller irregularities in the bulk can be detected, as the higher the frequency then the less the ultrasound diffracts around the irregularity (*Figure 7.13*).

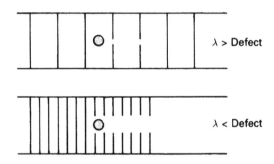

Figure 7.13 *Diffraction effects involving defects.*

7.1.3.2 Electromagnetic acoustic transducers

The electromagnetic acoustic transducer (EMAT) has proved to be a useful transducer for both detection and generation of ultrasound in electrically conducting samples. A constant magnetic field, usually produced by a strong permanent magnet, applied to the sample surface causes the generation of eddy currents on and just under the surface when the surface moves due to ultrasonic waves. The eddy currents can then be detected by a pick-up coil. The reverse process applies when generating ultrasound using EMATS.

A systematic approach to the processes of detection and generation by an EMAT is presented below (see *Figures 7.14* and *7.15*).

Detection The process is as follows.

(1) A constant magnetic field is applied to the sample penetrating the sample surface. The pick-up coil is in the 'middle' of this field (as shown in *Figure 7.14*), wound in a plane perpendicular to the sample surface.

(2) An ultrasonic wave arrives at the sample surface and consequently in-plane or out-of-plane motion of the surface occurs.

(3) As the surface moves so do the electrons within it, dragged by the electrostatic force, because the velocity of the ultrasonic wave is much

Figure 7.14 *An in-plane-motion detecting EMAT.* B_s, *static magnetic field;* B_e, *varying magnetic field due to the eddy current.*

Figure 7.15 *An out-of-plane-motion detecting EMAT.* B_s, *static magnetic field;* B_e, *varying magnetic field due to the eddy currents.*

less than the velocity of free electrons. Thus there is a current moving in a magnetic field, and Fleming's right-hand dynamo rule applies. Consequently, the electrons experience a force perpendicular to the magnetic field and the ultrasonic displacement direction.

(4) The force on the electrons accelerates them and gives rise to an eddy current. This eddy current has its own associated magnetic field (right-hand-grip rule).

(5) The time-varying magnetic field of the eddy currents generates an electromotive force (e.m.f.) in the pick-up coil. The magnitude of the e.m.f. depends on the number of turns in the coil and the dimensions of the coil.

The generated e.m.f. is then usually amplified by electronic circuitry before analysis can be carried out on a normal oscilloscope. Typically, the e.m.f. generated in such a coil is less than 1 mV for a sample that is a good conductor.

The in-plane EMAT is often referred to as the 'shear-wave EMAT', and the out-of-plane EMAT as the 'longitudinal EMAT'. The propagation vector of shear and longitudinal waves need not necessarily be perpendicular to the sample surface. Thus shear and longitudinal waves can give rise to both in-plane and out-of-plane motion simultaneously and be detected by either EMAT. Surface Rayleigh waves give rise to elliptical particle motion (of a point in the surface), and hence surface waves can be detected by either EMAT.

Generation The generation of wideband ultrasonic waves by EMAT requires a sharp pulsed current (of the order of nanoseconds) to be passed through a coil to induce currents in the sample on or just below the surface. For various EMAT designs, the current induced in the sample can be deduced by the methods of imaging, matching the magnetic field boundary conditions at the sample surface (Bleaney and Bleaney, 1976). These currents flow in a static applied magnetic field and thus experience a force. As the electrons are coupled to the lattice by a Coulomb interaction then shock waves propagate as the lattice is 'moved' via the interaction with the electrons. Two types of generators commonly used are the 'shear wave generating' EMAT and the 'meander line' EMAT, used for generating single frequency surface waves (*Figure 7.16*).

Limitations of EMATS The most obvious limitation of an EMAT is that it will only operate on electrical conductors, and gives quite poor results with some low conductivity samples. EMATS as detectors do not have a particularly good instrumental resolution, and quite often their bandwidths and the bandwidths of the amplifiers do not cover a large spectrum of ultrasound. Their instrumental resolution is determined by the width of the pick-up coil. It is desirable to have many turns on a coil to increase the induced e.m.f. and hence the sensitivity, but this tends to make the coil wider. Thus surface waves with wavelengths shorter than the coil width tend to be 'integrated' on traversing the coil. The electronic circuitry used to amplify the e.m.f. will not have a constant gain for all the frequencies of ultrasound detected, as it too has a bandwidth. Consequently, a constant response for all frequencies detected cannot be assumed.

Figure 7.16 *A meander line EMAT.*

The induced e.m.f. in the coil is not a linear function of the displacement of the sample surface. The EMAT signal can be calibrated using interferometric methods if the same sample is used and the distance between the EMAT and the sample is kept constant.

EMATS do not work very well on ferromagnetic materials because the magnetostrictive effect is present. The magnetic field in the sample which gives rise to the e.m.f. in the EMAT coil will produce magnetostriction, and because the field varies with time additional eddy currents will be produced after Le Chatelier's principle. These eddy currents will be in the opposite direction to those which give rise to them. Thus the magnetostrictive effect may reduce the signal received by the EMAT.

The main limitation of the generating EMATS is that they need a large, short pulse of current which must be driven through the coil. The induced current in the sample is proportional to the number of turns in the coil, but as the number of turns increases so does the coil inductance, making it more difficult to drive the high current through the coil.

7.1.3.3 Magnetostrictive transducers

Ferromagnetic metals and compounds possess a property known as 'magnetostriction', where the application of a magnetic field to the sample causes a change in the dimensions of the sample. The degree to which this change in shape occurs depends on the type of sample and the internal structure of the sample (grain shape, alignment, etc.). Nickel, for example, can exhibit an expansion of up to 1 in 10^6,

while the compound Turfinol (Fe, Tb and Py) can exhibit an expansion of up to 1 in 10^2, making it an ideal material for magnetostrictive transducers.

Ferromagnetic materials themselves can be tested directly, and without contact by utilizing their own magnetostrictive properties. This can be done by winding a coil around the sample and passing an a.c. current through the coil (*Figure 7.17*). The magnetostrictive effect is only induced in the area of the sample covering the coil, and only to a depth into the ferromagnet, known as the skin depth δ, given by:

$$\delta = \sqrt{\frac{2}{\omega\sigma\mu_0\mu_r}} \qquad (7.10)$$

Figure 7.17 *Reflection and transmission of ultrasonic waves in a magnetostrictive sample.*

where ω is the angular frequency of the incident electromagnetic wave (the same frequency as the a.c. coil current), σ is the conductivity of the sample, μ_0 is the susceptibility of free space, and μ_r is the relative susceptibility of the sample. For further reading see Bleaney and Bleaney (1976).

As a consequence, the magnetostrictive change in sample dimension only occurs on and close to the surface of the sample, and thus the ultrasonic waves generated travel on or just under the surface. Away from the coil the waves tend to penetrate into the bulk of the sample becoming more like a planar wave. Consequently, the ultrasonic directivity of the sample is much less complicated further away from the coil. The ultrasonic waves travelling through the bulk of the material have an associated change in magnetization which travels through the ferromagnet. The change in magnetization can be detected a distance away from the coil using another coil, thus giving information about the ultrasonic wave in the sample. Using this method it is possible to detect defects as small as 10% of the cross-section of the sample. The sensitivity for the detection of holes and cracks in the surface is highly dependent on the dimensions of the sample and of the defect, and on the orientation of the defect with respect to the ultrasound

propagation vector. Optimizing the method by choosing the most sensitive set-up can detect holes and cracks to within a few per cent of the sample dimension. Of course other effects such as the frequency of ultrasonic waves and diffraction must be taken into account when optimizing performance.

Transducers can also be constructed from ferromagnetic materials. A simple design, for example, is a ferromagnetic rod with a coil wrapped around its circumference, suitably coupled to the sample to be investigated. However, magnetostrictive transducers are only rarely used for defect detection because they are only capable of working at low frequencies. The maximum frequency attainable is limited by the speed at which the magnetic domain wall can move within the ferromagnet.

Consider a ferromagnetic sample, say a rectangular bar. The sample is unmagnetized as a whole, and no magnetic field is applied to the sample. There are regions of magnetization in the sample known as 'magnetic domains'. These domains are aligned in such a way so as to give a net zero magnetization. The most simplified arrangement is that of four magnetic domains as shown in *Figure 7.18*. In general, the size, number and orientation of the domains are governed by physical properties beyond the scope of this book. For a detailed explanation of magnetic domain structure see Bleaney and Bleaney (1976) or Kittel (1986).

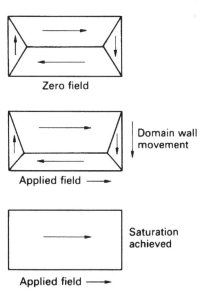

Figure 7.18 *The movement of domain wall across a ferromagnetic sample in response to an applied magnetic field.*

As a magnetic field is applied to the bar, the net magnetization of the ferromagnet changes along the bar so as to oppose the magnetic field applied to it. It does this by moving the magnetic domain wall. The 'speed' at which the domain wall moves thus determines how quickly the magnetization of the sample can change, and thus the maximum frequency of applied magnetic field that can be used usefully to generate ultrasound in the ferromagnet.

There is also the factor of the ferromagnet dimensions to consider, as these determine how far the domain wall must move. Normally, for a manageable and useful size of ferromagnet, the maximum frequency possible is of the order of 1 MHz. The frequency range of such transducers is variable up to its maximum, unlike the piezo-electric transducers that have narrow operational bandwidths.

While being limited to low frequencies (in ultrasonic terms), high powers can be attained. This makes the magnetostrictive generation of ultrasound useful for drilling and welding applications, which are described later. Typically, high power applications require currents of around 100 A. The impedance of the coil is $R + \omega L$, where R is the coil resistance, ω is the a.c. angular frequency, and L the inductance of the coil. Thus it can be seen that at high power applications the current in the coil is the factor that limits the maximum operation frequency.

7.1.3.4 Lasers as generators

All the transducers described so far must be in contact or in very close proximity with the material under test. Lasers may be used to generate and detect ultrasound in a totally remote way, making them ideal for use in hostile environments. As with other ultrasound-generation methods the generating pulse must have a rise time of the order of nanoseconds. To achieve this fast rise time, a Q-switched or a dye-switched laser must be used (see Siegman, 1986). Generation of ultrasound using lasers falls into two distinct extreme cases: surface ablation and the thermoelastic regime. With laser generation, the effectiveness of the method depends on how much of the incident beam energy is absorbed, i.e. on the reflectivity of the sample. The frequency of ultrasound generated depends on the rise time of the laser (the quicker the rise time the higher the generated frequencies), and on how ablative or thermoelastic the technique used is.

Ablative generation In the ablative regime the impact of the laser beam vaporizes a thin surface layer of the sample forming a plasma. This plasma is ejected from the surface of the sample and this gives rise to a net force perpendicular to the surface, over the area of the ablated region (*Figure 7.19*). As this force acts normally to the surface it generates predominantly longitudinal waves. The surface of the metal will usually be damaged to some extent unless a thin film of liquid (oil for example) is applied to the surface, which vaporizes to form the plasma, or if a wavelength of light is used such that it breaks down the air above the sample surface rather than the surface itself. The degree to which a surface is damaged depends on how much energy is absorbed in the ablated region, in a 'shot', and how many shots were taken. (A single shot from a point-focused 800 mJ Q-switched YAG laser would damage the surface of aluminium to a depth of tens of micrometres.) For many samples, ablative damage (without surface fluid) is so slight so as to be unimportant.

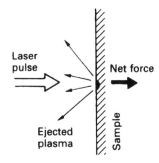

Figure 7.19 *Ablative generation of ultrasound.*

Thermoelastic generation In the thermoelastic regime no plasma is formed and no surface damage occurs, thus it is strictly non-destructive testing. Essentially a volume of the surface equal to the cross-sectional area of the incident laser beam multiplied by the skin depth of the sample is heated up very quickly. The volume of 'hot' sample is heated very quickly, and is free to expand away from the metal having no stress component normal to the surface here (*Figure 7.20*). The 'hot' volume expansion is constrained perpendicular to the surface by its colder surroundings, so there is a net radial hoop of force in the 'hot' volume acting at the interface of the 'hot' and 'cold' regions. Thus the radial force predominantly gives rise to shear waves. Because the laser energy densities are so low, the waves created in the thermoelastic regime tend to be smaller in amplitude than those generated in the ablative regime.

Surface waves and mode conversion Surface waves are generated in both the thermoelastic

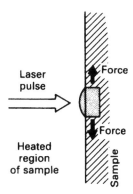

Figure 7.20 *Thermoelastic generation of ultrasonic shear waves.*

and ablative regimes. Because the particle motion associated with a surface wave is elliptical, a force applied either normally or tangentially to a surface could initiate a surface wave.

Quite often in surface wave analysis it is useful if a single frequency surface wave can be produced. This can easily be done by optically splitting the beam into two equal-intensity beams and then recombining them so that an interference pattern is produced on the sample surface. The net result is a convolution of a Rayleigh pulse due to the width of the pattern and a sinusoidal pulse.

Whenever ultrasonic waves are incident on a boundary they can be mode-converted into other types of ultrasonic waves. In both the thermoelastic and ablative regimes some longitudinal and some shear waves are always present. The two main factors for this are that:

(1) a totally ablative or totally thermoelastic regime cannot be achieved; and
(2) there is some mode conversion at the sample surface where the laser beam is incident.

The presence of longitudinal arrivals in the thermoelastic regime has also been explained to some extent to be due to thermal diffusion into the sample.

7.1.3.5 Lasers as detectors
Lasers can also be used to detect small surface displacements caused by the arrival of an ultrasonic wave. The actual detection is done by light-sensitive electronics such as photodiodes. The detector must have a very fast rise time of the order of nanoseconds to detect the surface displacements. Two main techniques for detection are described below.

Interferometric techniques The principle of this technique is that two different light beams of the

same frequency interfere when they are combined. The intensity of the resulting beam is dependent on the phase difference of the two interfering light beams.

In this particular technique the initial laser beam is split into two, one beam travelling a constant fixed distance to a mirror where it is reflected back to a certain point (*Figure 7.21*). The other beam is reflected from the sample and brought to the same final point as the first beam. The beams of light then interfere at this fixed point (or may be made to interfere using some suitable optical arrangement). If a photodetector of some sort is then placed at the point where the two beams meet and interfere then the small changes in optical path length for the two beams can be measured. The low-frequency background vibrations such as sound or movement may be filtered out by employing a feedback system on the 'fixed' mirror so that it moves in response to background vibrations to give no change in the interference pattern at these low frequencies, well below the ultrasonic frequencies of interest. Thus the only observed displacement will be the ultrasonic displacements of the sample surface, normal to the surface.

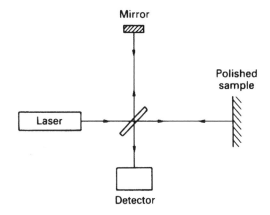

Figure 7.21 *A schematic diagram of a Michelson Interferometer.*

Knife edge detector The knife edge technique is a much simpler, but less sensitive, technique than the interferometer. A laser beam is directed at an angle, slightly off normal to the sample surface, and reflected from the sample surface. The reflected beam is then incident upon a 'split' detector (two separate but identical detectors very close together) (*Figure 7.22*). If the sample surface moves normal to itself or tilts then the position of the reflected beam will move while remaining parallel to its original

(a)

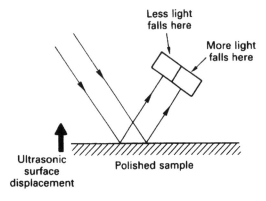

(b)

Figure 7.22 *The knife edge detection technique. (a) The beam falls onto both photodetectors at zero displacement set-up. (b) The amount of light falling on each detector changes the a.c. component, representing a surface displacement.*

direction. This change can be detected as a change in signal on both detectors. The whole system is oriented in such a way that a normal displacement decreases the intensity on one detector while increasing the intensity on the other, giving a measure of the sample surface displacement.

As with the interferometric technique an electronic feedback system can be employed to move both detectors in response to low frequency background vibrations, effectively filtering them out.

Limitations The main drawback with laser generation of ultrasound is that it is not very efficient, as typically laser efficiency itself is less than 1%. This, coupled with the fact that materials such as metals are highly reflective for most of the spectrum, means that high powered lasers are often required.

Laser detection of ultrasound with in-plane motion of the surface is both complicated and difficult, and the resolution of the detector to out-of-plane motion is dependent on the optics used. If a lens is used to focus the laser down to a point on the surface then higher resolution can be achieved for surface waves.

7.1.3.6 Electrostatic transducers
The electrostatic ultrasonic generator works on a capacitance principle. A parallel plate capacitor with an applied voltage has a force acting between the capacitor plates given by:

$$F = -\frac{V^2 \varepsilon_0 \varepsilon_r A}{2d} \qquad (7.11)$$

where V is the voltage on the capacitor plates. A is the cross-sectional area of the plates, and d is the distance between the plates.

This force gives rise to stresses within the dielectric between the plates. By applying an a.c. voltage to a plate, ultrasound can be generated within the dielectric. The surface of a conductor can be used as one plate of the capacitor, with the other being sputtered onto a dielectric on the sample surface. Hence the ultrasound can pass through the dielectric into the sample. Capacitor plates can also be sputtered onto non-conducting dielectric samples for generation or detection of ultrasonic waves within the dielectric.

The variables in the capacitor are given by:

$$V = \frac{Qd}{\varepsilon_0 \varepsilon_r A} \qquad (7.12)$$

where V is the voltage across the plates, Q is the charge on the plates, d is the distance between the plates, and A is the cross-sectional area of the plates. In detecting ultrasound the change in d in the capacitor is to be measured indirectly by measuring the voltage on the plates while keeping the charge on the plates constant. Electrostatic transducers have a wide frequency band of operation.

Limitations The electrostatic detection technique relies on the capacitor plates holding their charge throughout the experiment. The change in voltage across the plates will be small and difficult to measure. The effect being measured is not simply a surface displacement, but a combination of the change in dielectric properties and the change in width across the whole dielectric.

7.1.3.7 Interdigital transducers
This type of transducer is very specialized and is only used for generating single-frequency Rayleigh waves. The transducer consists of two sets of conducting digitae (each like the teeth on a comb) that interlock but do not touch (*Figure 7.23(a)*). These

(a)

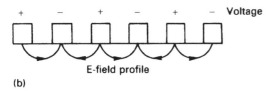

(b)

Figure 7.23 (a) An overhead view of an interdigital transducer and (b) a cross-sectional view showing the electric field profile.

conductive digitae are sputtered onto the sample surface or onto an insulating layer if the sample is conductive. Typical ultrasonic wavelengths are less than 1 mm so that the scale of the digits is quite small. The basic design pulses all the digits on each group with the same voltage at the same time (a positive voltage for one group and a negative voltage for the other). The electric field produced in the sample has a periodic structure as shown (*Figure 7.23(b)*).

The pulsed electric field gives rise to periodic stresses in the simple and thus to periodic surface waves propagating away from the transducer, Refinements can be made to the basic transducer by having a time delay from one digit to the next, introducing phase differences into the individual components of the surface waves.

Limitations The transducers themselves are very small and have to be sputtered onto samples and are thus not only difficult to produce but are not particularly robust. They are not only susceptible to damage by physical contact but also by the current pulsed into them which cannot be very large because the digitae are so thin that thermal expansion produced by the heating effects of a high current could force the digitae off the surface. Interdigital transducers are not very versatile. Once a transducer has been sputtered onto a sample, it cannot be removed and used on another sample.

7.1.4 Industrial applications

Ultrasonic techniques have found widespread use in industry in both product testing and manufacture.

There are several reasons why ultrasound is commonly used.

(1) *A non-destructive probe.* The dimensions and structure of a sample can be assessed non-destructively and, if necessary, without contact between sample and the ultrasound generator and detector.
(2) *Environmentally safe and clean.* Ultrasound does no damage to the environment around the sample, and in general poses no threat to the operator's health, unlike other non-destructive testing techniques such as those which use X-rays and γ-rays.
(3) *Ease of use.* Operators of ultrasonic equipment need very little training and the process can readily be automated for most cases.
(4) *Accurate.* Surface displacements of down to one-fifth of an angstrom (= 0.02 nm) can be measured (over an area the size of the detection spot) using laser interferometry. Thickness measurements of down to 10 µm can be made by looking at the echoes in the bulk from the surfaces, provided that the sample does not absorb ultrasound to the extent that only a few echoes can be observed.
(5) *Overcomes environmental restrictions.* Measurements can be taken automatically and without contact with the sample. Thus there is no need for people or equipment to be within a potentially harmful environment, as occurs with a radioactive environment. Ultrasonic methods can also be employed over a wide range of temperatures and ultrasonic transducers may be positioned where the use of bulkier test equipment would be impossible.

For these reasons, and others which will become clear below, ultrasonics is used in several different ways to measure a wide range of parameters. Listed below are more common applications. There are, however, many more specialized uses of ultrasound which are described in some detail in many of the books given in the Bibliography.

7.1.4.1 Non-destructive testing
Dimensional measurements If the velocity of ultrasound in a material is known, the thickness of a sample of that material may be determined by simply recording the time it takes an ultrasound wave to travel across the sample. The velocity of ultrasonic waves in a sample is an easily measured quantity. It is, however, dependent on the temperature of the sample and on the structure of the sample itself

and these factors must be taken into account when performing thickness measurements.

Thickness measurements can be made by one of two methods, reflection or transmission. In reflection, the bulk waves are reflected from the surface on the other side of the sample, which must be parallel or almost parallel to the transducer side if the same transducer is to be used for generation and detection of the ultrasound. By measuring the time delay between the initial pulse and successive arrivals, or just between successive arrivals, the thickness of the sample can be calculated if the velocity is known. There is a degree of compromise in this method because a good generator is not necessarily a good detector. However, the reflection technique is more convenient than transmission, requiring less equipment, and, more importantly, only access to one side of the item to be measured.

Transmission requires two transducers, one either side of the sample bulk. Unlike reflection it does not require the sides to be almost parallel as the ultrasound is detected before it undergoes its first reflection. The positioning of the receiver with respect to the generator is critical because in certain geometries, the reflected wave may not fall on the detector. However, the transmission technique does permit the use of one transducer optimized to detect and another to generate.

Another dimensional quantity that can be measured is the curvature of a sample using a separate generator and detector of surface waves. This is because even when the sample is isotropic, if its surface is curved then it will be dispersive for Rayleigh waves. Hence the radius of curvature of a sample could also be measured. Sample thickness can also be measured using dispersive Lamb waves. Lamb waves are a surface wave in the limit of a sample thickness comparable with the wavelength of ultrasound propagated within it. Thus for the measurement of thin samples (less than 1 mm) the tendency of Lamb waves to show dispersion is used to calculate sample thickness.

Surface quality A great deal of information can be obtained using Rayleigh wave analysis on a sample surface. Not only defects and flaws are open to investigation, but also other parameters such as curvature of a sample, grain structure or a difference in the macroscopic structure of the surface and the bulk of the material.

(a) *Scratches and cracks*. The presence of scratches or cracks on or just under the surface can be detected using ultrasonic techniques. Rayleigh waves are localized at a surface but do extend into the bulk

of the material. The amplitude of the wave decays exponentially into the sample and falls to a fraction $1/e$ ($e = 2.7183$) of its amplitude at the surface at a depth equal to one wavelength in the material. Because the depth to which a surface wave extends is dependent on frequency, an indication of the depth of the crack or scratch can be obtained. Thus the limiting factor in detecting the depth of the scratch or crack is the highest frequency of Rayleigh waves that can be generated or detected. When using a broadband Rayleigh wave, the same argument applies, i.e. there must be a significant magnitude of frequency component with a wavelength comparable to the size of the structure to be detected.

Scratches and cracks not only give rise to reflected surface waves, but also scatter and mode-convert the surface wave into bulk waves that may propagate into the sample. Thus there is not only a significant loss in the energy of the transmitted surface waves but there are also bulk waves generated that can be detected, indicating the presence of a scratch or crack (*Figure 7.24*).

Figure 7.24 *Detection of ultrasound reflected from a surface breaking crack.*

The theory behind the process of surface wave scattering from surface features is very complicated and can only be performed using very simplified models of surface defects (see Viktorov, 1967). Scratches and cracks are usually complex structures and thus it would be difficult to obtain a mathematical model for their behaviour. Consequently, an empirical approach is often taken in data analysis.

As mentioned earlier, the instrumental limiting factor in detecting a surface defect is the resolution and high-frequency limit of generator and detector. Using laser generation and interferometric detection, the presence of scribed lines can be observed down to a depth of approximately 10 μm. On industrial samples this degree of sensitivity is not required and

piezoelectric transducers are used to detect defects to depths the order of 0.5 mm and deeper.

(b) *Surface structure*. The macroscopic structure (e.g. grain structure in metals) and difference to the bulk structure can be measured using Rayleigh wave analysis. An isotropic media would be non-dispersive for Rayleigh waves. However, materials are rarely isotropic. Anisotropy may produce a directional dependence on the wave velocity or attenuation, or introduce dispersion.

There are three main regimes of wave attenuation with respect to the wave frequency f, and the size of structure in the medium x:

$\lambda \gg x$ scattering factor is proportional to $x^3 f^4$

$\lambda = x$ scattering factor is proportional to $x f^2$

$\lambda < x$ scattering factor is proportional to $1/x$

By analysing how a wave is attenuated after propagating a distance in the medium (as a function of frequency), the size of the structure that attenuates the wave can be calculated. If the surface is of a different structure to the bulk then a Rayleigh wave may become dispersive. This is due to the fact that the depth to which a surface wave penetrates the sample is inversely proportional to the wave frequency. As a result, different frequencies of wave may have different velocities and be attenuated to differing extents.

Bulk defects Bulk defects can be detected in transmission or in reflection. A wave is reflected at an impedance mismatch in the medium of propagation, such as the interface of two different media. The impedance for longitudinal and shear waves is different in the same media and so an indication of defect type may be obtained.

(a) *Cracks, cavities and bonds*. A cavity wall has a large impedance mismatch to both longitudinal and shear waves. Hence the defect can be detected in transmission or reflection. A region of acoustically different material in the bulk will behave like a cavity in the media (to a certain extent) reflecting an amount of the incident energy dependent on the mismatch at the interface. Whether the cavity is large enough to reflect a significant amount of energy depends on transducer size, resolution and frequency (*Figure 7.25*). Transmission detection gives a measurement of the effective cross-sectional size of the cavity (if the frequency of the ultrasound is sufficiently high), while the echo technique is sensitive to the shape and orientation of the cavity (*Figure 7.26*).

Cracks within the bulk may only be detected in certain orientations using an appropriate wave mode. Consider a crack in the media in the form of a plane. If the plane-normal vector lies perpendicular to the ultrasound propagation vector then there will be virtually no effect on the longitudinal and a small effect on the shear wave, perhaps not detectable (*Figure 7.27(a)*).

Now consider the plane normal vector parallel to the propagation vector. If the crack is such that the medium is not coupled across the crack, or coupled poorly, it will behave as a cavity and reflect most of the incident longitudinal and shear waves (*Figure 7.27(b)*). If there is poor coupling across the crack then obviously the shear wave will be most greatly reflected as it has its displacements in the plane of the crack (*Figure 7.27(c)*). The reflectivity of a crack not only depends on its orientation and size (relative to ultrasound wavelength) but also the structure of the crack at its surface; shear waves are

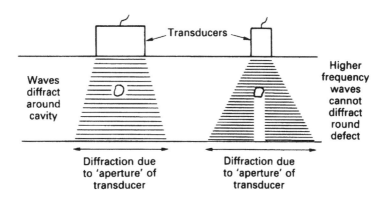

Figure 7.25 *Diffraction effects at bulk defects.*

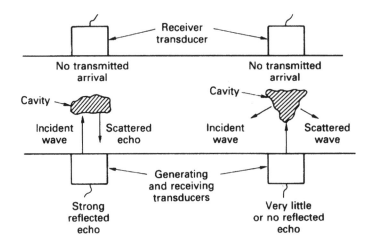

Figure 7.26 *Scattering geometries for transmission and reflection techniques for detecting defects.*

(a)

(b)

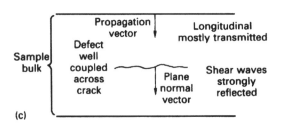

(c)

Figure 7.27 *An ultrasound wave propagating: (a) in the plane of a linear defect; (b) perpendicular to a linear defect; (c) across a crack.*

usually the most sensitive method of detecting cracks within the bulk.

A bond between two media is similar to a simplified planar crack. There will usually be some reflection of ultrasonic waves, even off a well-adhered bond due to impedance mismatch at both the interfaces of the media and adherent. For two media held firmly together but poorly bonded there will be a small reflection from the interface for a longitudinal wave, but a relatively large reflection from a shear wave. This is because (as previously stated) the vibrations of the shear wave are in the plane of the bond, and a rigid connection in the plane of the bond between the bonded media is required in order to propagate the shear wave (*Figure 7.28*). The attenuation of a shear wave propagated through a bond will provide insight into the rigidity of the bond. The thickness of the bond will also affect the transmission and reflection properties of the adherent, and in addition to the general trend there may be certain resonance effects due to the dimensions of the bond compared with the ultrasonic wavelength.

7.1.4.2 Production using ultrasonic techniques

Welding In order to weld two pieces of material together successfully the area of the weld must be hot enough to melt each component locally and the melts should mix, providing a good continuum from one component to the other. The advantage of an ultrasonic weld is that the heating effect is localized to the area of the weld. The principle of operation is simple – a material that can change its dimensions at ultrasonic frequencies by some means (piezoelectric, magnetostrictive) is held fixed at one end, while the other end has the welding 'bit'

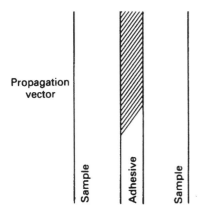

Figure 7.28 *Ultrasound propagation across an adhesive bond. Good adhesives give good acoustic coupling across the layers and thus reflects both longitudinal and shear waves poorly. Poor coupling will always lead to a strong shear wave reflection.*

attached to it and is free to move. The weld bit vibrates up and down at ultrasonic frequencies. When the vibrating bit is pressed onto a softer sample with a lower melting point the vibrations are damped losing energy in the form of heat. Thus the components can be welded together. Where high powers are required such as the welding of metals, a magnetostrictive technique is used as the low-frequency limitations of the ferromagnetic are unimportant, and high power output is available. Piezoelectrics can be used in the welding of materials that have relatively low melting points, such as plastics.

Drilling The method is similar to welding techniques except that in this case a hardened pointed drill bit is used in place of the welding bit. Most drilling applications require high power and thus a magnetostrictive transducer is commonly used.

Cleaning Ultrasonic cleaning works by physically shaking surface contaminants from samples. It is extensively used in the cleaning of devices and components that could easily be damaged and attacked by conventional cleaning techniques such as acid-bath etching. When jet engines are refurbished they are often cleaned ultrasonically. The cleaning process works by ultrasonic waves reaching a sample and being absorbed in the surface. Contaminants are usually not bound into the structure of the components as strongly as the component itself is bound. Consequently forcing vibrations at the sample surface 'tears' contaminant layers off the surface. The ultrasound is coupled to the sample from the transducer via a liquid media that will not attack the sample. Transducers are placed around the ultrasonic baths in arrays, positioned so that the ultrasound is

incident on the appropriate region of the component. For most components alignment is not critical as only a fraction of the ultrasound is absorbed in the actual cleaning process, while most of the ultrasound bounces around the inside of the bath. The cleaning baths can be high power as there is much room to place the transducers around the bath walls; typical powers of jet engine cleaning baths are around 20 kW.

7.1.5 Medical applications of ultrasonics

The use of ultrasonics in medicine has experienced perhaps the most dramatic increase in sophistication in the last two decades of all branches of ultrasonics. Ultrasonic diagnostic probes are now in widespread use. Ultrasonic scanning of the unborn foetus is an almost standard part of every pregnancy in the UK and ultrasonic techniques can now be used to image blood flow in real time.

The use of ultrasound for surgical applications is considerably less widespread. However, gall and kidney stones can be broken up by an intense beam of ultrasonic radiation negating the need for surgery, indeed there are instances when ultrasound has been used to destroy brain tumours. Ultrasonic vibrations are also used to stimulate blood flow to aid the repair of damaged muscle.

The field of medical ultrasound is diverse and ever expanding, containing a wide range of very sophisticated techniques. As a consequence the selection here is devoted to a short review of the more widely employed techniques in diagnosis and, to a lesser extent, treatment.

7.1.5.1 Diagnostic techniques

At their simplest, ultrasonic diagnostic techniques simply measure the time separation between initial pulse and the arrival of a reflection from an interface between portions of the body with different densities. The time difference between echoes provides an indication of the position of the interface within the body and by combining information from several different geometries a more complete image of the local environment may be developed.

Two scanning modes form the basis for most ultrasonic scans of the human body. The simplest scan, known as an 'A-scan', involves normally a cathode ray tube although a paper recorder may be used synchronized to the initial pulse. A low-pass filter is employed to remove the radio frequency signal, so the screen displays a single peak corresponding to a reflection from an interface. An amplifier with a gain

which varies as a function of time may be used to overcome problems caused by the attenuation of the ultrasound pulse as it passes through the body.

Another technique, known as the 'B-scan' technique, involves the generation of many pulses and the detection over a wide area to draw up a composite picture of a section through the portion of the body under examination. The data is displayed on a two-dimensional raster with each pixel corresponding to one echo, the brightness of the pixel being modulated by the amplitude of the echo. White areas represent the most efficient reflectors, black areas represent portions from which there was no reflection.

When operating in real time the B-scan mode will allow scans to study the unborn foetus or to check for gall and kidney stones. When synchronized to the heart beat, for instance, successive B-scans can be used to build up a collection of 'stills' of the heart at different phases of the cardiac cycle.

The relative advantages and limitations of both techniques are quite clear. While the B-scan provides positional and orientational information, the amplitude data which is preserved in the A-scan mode provides valuable information on the build up of fatty-based material on the inner wall of arteries, for example.

Imaging is, however, only one of the uses of ultrasound for diagnosis. Ultrasonic blood flow meters operate in a variety of ways, the most popular technique being pulsed-wave Doppler. An ultrasonic wave is launched into the body, the reflected wave is fast-Fourier transformed and the resulting change in frequency may be related to the velocity of the blood flow by the Doppler equation. This technique offers quantitative information, but is limited to viewing small areas at a time.

Colour flow imaging (CFI) was first developed in 1981 by Eyers and colleagues and is reaching a level of sophistication which is likely to yield a vast increase in application in the near future (Mount, 1992). CFI allows the blood flow to be imaged against a greyscale picture of the local environment collected by conventional B-scan techniques, with the blood flow colour coded according to the direction of flow. By convention blood flowing towards the detector is red, blood flowing away is blue.

CFI also employs a pulsed-wave Doppler technique. However, since several sample volumes are being viewed simultaneously rather than one single sample volume, fast Fourier transforms are too slow. Doppler signals are divided up according to phase into forward and reverse channels. Two consecutive

pulses from each channel are considered. The first is delayed for a period equal to the time between the two transmitted pulses and are fed into a multiplier simultaneously. The output of the multiplier is dependent upon the matching of the two signals and any phase change resulting from flow within the sample will be manifest by a change in the output of the multiplier.

The information gained in this manner is largely qualitative and CFI and pulsed-wave Doppler may be considered as being complementary techniques, the former providing information over a reasonably large area, the latter providing more quantitative detail.

The frequency dependence of the absorption of ultrasound, usually propagated through the heel, has been shown to provide a reasonable indication of the condition of bones and may provide a safe, cheap and quick way of scanning women at risk from osteoporosis (a softening of the bones).

For many medical applications, a steerable beam of ultrasound is required which can be accurately aimed at a particular portion of the body. This is possible using a phased array transducer, which comprises a row of piezoelectric crystals which are pulsed in succession. The time difference between the pulses will dictate the angle at which the waveform made up of the combined signals will leave the transducer (*Figure 7.29*).

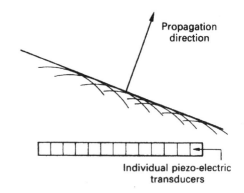

Propagation direction

Individual piezo-electric transducers

Figure 7.29 *A phased array transducer permitting directional control.*

7.1.5.2 Ultrasonics in treatment

The use of ultrasound for medical treatment is more limited than for diagnosis. A focused beam of ultrasound provides a localized source of energy which may be used to stimulate the repair of injured tissue and blood flow by local heating or to break up gall or kidney stones.

When considering the use of ultrasound energy for such application the effects of the local heating on

the surrounding tissue and on the portion of the body through which the ultrasound is propagated must be taken into account. Often several transducers are used to launch ultrasound at a specific region, in the case of a gall stone for instance, from several angles simultaneously, thus reducing the risk of damage to other tissue.

7.1.6 Conclusion

Within the confines of this short section only a brief introduction to the scope of ultrasonic techniques has been possible. Beyond industrial and medical applications ultrasonics has proved to be an invaluable tool in agriculture and even the production and quality testing of food (Javanaud, 1988). Another area that has not been covered is the use of very short wavelength ultrasound for acousto-optics. The reader is referred to a review article by Sliwinski (1990).

References

BLEANEY, B.I. and BLEANEY, B., *Electricity and Magnetism*, 3rd edn, Clarendon Press, Oxford (1976)
JAVANAUD, C., Applications of ultrasound to food systems, *Ultrasonics*, **26**, 117–123 (1988)
KITTEL, C., *Introduction to Solid State Physics*, 6th edn, Wiley, New York (1986)
MOUNT, P., Ultrasound echoes yield pictures of health, *Physics World*, **5**, 58–59 (1992)
SIEGMAN, A.E., *Lasers*, Oxford University Press, Oxford (1986)
SLIWINSKI, A., Acousto-optics and its perspectives in research and applications, *Ultrasonics*, **28**, 196–213 (1990)
VIKTOROV, I.A., *Rayleigh and Lamb Waves, Physical Theory and Applications*, Plenum Press, New York (1967)

Bibliography

Fundamentals and theoretical background
ACHENBACH, J.D., *Wave Propagation in Solids*, North-Holland, Amsterdam (1980)
KRAUTKRAMER, J. and KRAUTKRAMER, H., *Ultrasonic Testing of Materials*, 2nd edn, Springer-Verlag, Berlin (1977)
MALECKI, I., *Physical Foundations of Technical Acoustics*, Pergamon Press, Oxford (1969)
MIKLOWITZ, J., *The Theory of Elastic Waves and Waveguides*, North-Holland, Amsterdam (1980)
ROZENBERG, I.D. (ed.), *Physical Principles of Ultrasonic Technology*, Vol. 2, Plenum Press, New York (1973)
SHUTILOV, V.A., *Fundamental Physics of Ultrasound*, Gordon & Breach, New York (1988)

Industrial applications and techniques
ENSMINGER, D., Ultrasonics, *The Low- and High-Intensity Applications*, M. Dekker, New York (1973)
FREDERICK, J.R., *Ultrasonic Engineering*, Wiley, New York (1965)
OLSON, H.F., *Acoustical Engineering*, Van Nostrand, Princeton, NJ (1957)
SARGENT, M., SCULLY, M.O. and LAMB, W.E., *Laser Physics*, Addison-Weslley, Reading, MA (1974)
SCRUBY, C.B. and DRAIN, L.E., *Laser Ultrasonics, Techniques and Applications*, Adam Hilger, Bristol (1990)
SZILARD, J. (ed.), *Ultrasonic Testing, Non-Conventional Testing Techniques*, Wiley, Chichester (1982)

Medical applications
DUBOVY, J., *Introduction to Biomedical Electronics*, McGraw-Hill, New York, p. 174 (1978)
WELKOWITZ, W, and DEUTSCH, S., *Biomedical Instruments, Theory and Design*, Academic Press, New York, p. 89 (1976)

7.2 International standards for sound systems and equipment

7.2.1 Background

Prior to the early 1960s, preparation of most standards for sound equipment was coordinated individually by National Standards organizations such as the British Standards Institution (BSI). In the UK, the relevant BSI committees are composed of experts nominated by appropriate trade organizations such as ABTT and BREMA together with representatives of large organizations such as the BBC and British Telecom. Usually based on initial work done by one or more of the participating organizations, the final documents originally were published only as British standards and generally reflected established practice in the UK.

For certain subjects this approach was not suitable, such as recordings made on vinyl, tape and film soundtracks where international compatibility of the mechanical features and record/replay characteristics obviously was essential. Therefore, for many years previously, standardization of these subjects together with line and radio transmission practice had been coordinated by one or more of the following bodies: International Standards Organization (ISO). International Electrotechnical Commission (IEC) and International Telecommunication Union (ITU).

About 1960, the IEC commenced work on a multi-part standard: Methods for Specifying and Measuring the Characteristics of Sound System Equipment (now numbered IEC 60268). Apart from recording and

reproducing equipment, which was already covered by separate IEC standards, sound equipment of all types both for professional and household use was to be included. This anticipated the emerging trend towards the use of high-quality household sound equipment in semi-professional situations. Preparation of the standard was to serve two purposes: to promote international trade by ensuring compatibility of equipment in respect both of electrical and mechanical matching of connectors and second, to facilitate realistic comparisons of performance specifications by defining characteristics and specifying preferred methods of measurement.

7.2.2 Preparation of international standards

7.2.2.1 Standards published by the International Electrotechnical Commission (IEC)
For most IEC standards, each part is prepared by a Working Group composed of experts nominated by National Standards organizations, usually delegated from the appropriate committees. The initial draft may be based either on an existing National standard or e.g., proposals for a new technique emanating from a scientific body such as the Audio Engineering Society. After further refinement and sometimes much compromise, the final draft is submitted to all National Standards organizations for agreement. Given a substantial majority of positive votes, the draft is published as an IEC standard.

Most new IEC standards are now harmonized by CENELEC, the European Committee for Electrotechnical Standardization, and are implemented by BSI as BS ENs and carry the IEC number as well (e.g. BS EN 60268-7: 1996, IEC 60268-7: 1996). Where a dual-numbered British Standard series already exists, these standards can also be identified as part of this series by a triple reference (e.g. BS EN 60268-7: 1996, BS 6840-7; 1996, IEC 60268-7: 1996).

Technical Reports type 2 and 3 published by the IEC are now not normally harmonized by CENELEC and are published either as part of a series under dual numbering (e.g. BS 6840-1:1987, IEC 60268-1: 1985) or as a Published Document (PD) or a Draft for Development (DD) if eventual change to a standard is anticipated. If a Technical Report type 2 is, however, harmonized by CENELEC, it should be published as an ENV and will be implemented by BSI as a DD ENV.

National Committees therefore take care to appoint experts to the respective IEC Technical Committee and Subcommittee Project Teams or Working Groups when subjects of significant interest are proposed as new work for the preparation of standards. The concept for the amendment or revision of existing standards is currently changing and will in the near future be handled by Maintenance Teams (MT) and not as New Work. There will therefore be a need for UK experts to be also nominated to serve on these MTs where it is vital to UK interests.

In the past, standards published by IEC were subsequently reviewed by CENELEC for harmonization and resulted in some cases for the need for Common Modifications to be included in the EN. To try to reduce the need for Common Modifications, the CENELEC review now commences in parallel at the IEC CDV stage, where National Committees are asked for comments on the draft and vote for it to proceed for preparation as a Final Draft International Standard (FDIS). The FDIS is then circulated to National Committees for a final yes/no vote for publication. A separate vote at this stage is also undertaken through CENELEC on the draft which is by then identified as an EN.

Although most standards other than those involving safety issues do not have the force of law in European countries, EEC trading practice now requires that all goods such as electronic equipment imported to member countries from any source shall conform to an appropriate EN standard.

Since January 1997, all new IEC publications have been identified by a 60 000 number and this has also been introduced retrospectively for all IEC publications. For example, IEC 27 is now IEC 60027 and orders quoting the earlier number are met with the revised identification. This change is reflected for all references in new IEC standards but will only be implemented in existing IEC publications as the standards are revised in the future. The new numbering system has been employed for most IEC standards mentioned in this section.

7.2.2.2 Standards published by the International Telecommunication Union (ITU)
The International Telecommunication Union (ITU) is one of the specialized agencies of the United Nations Organization. Until 1994 there were two subdivisions of the ITU, namely the International Consultative Committee for Radio Communication (CCIR) and the International Consultative Committee for Telegraphy and Telephony (CCITT). The former was concerned with radio communication (including sound and television broadcasting) and the latter with transmission networks carrying audio, video, telephony

or data signals. Both the CCIR and the CCITT progressed their work through Study Groups composed of delegates from broadcasting organizations and telecommunication authorities, etc. Every four years, groups of new, updated and confirmed Recommendations and Reports were published in two separate series of volumes. Whereas the work of the CCITT was confined to specialized transmission subjects and therefore mainly outside the scope of this publication, CCIR Recommendations and Reports were published in twelve volumes of the 'Green Book' and covered all aspects of sound broadcasting. Many of these are relevant to audio equipment and systems in general and, having been adopted by the IEC, are now incorporated into National Standards.

In 1994, the subdivisions of the ITU, namely CCIR and CCITT, were renamed ITU/R and ITU/T respectively. Their Recommendations and Reports are no longer to be grouped into volumes but are to be published individually with amended references, the document numbers and descriptive titles being unchanged. Thus CCIR Recommendation 468-4:1990 (see Section 7.2.6) has the new reference: Recommendation ITU/R BS 468-4:1994, the document now being one of the BS series of Recommendations.

7.2.3 Overview of IEC 60268: Methods for specifying and measuring the characteristics of sound system equipment

Each part of the standard varies significantly in the amount of detail provided, depending on the need to present theoretical treatment for the derivation of characteristics, etc. Such detail is not usually considered here as the overview is intended only as a guide to the principal characteristics and/or methods of measurement. In all cases, it is essential to consult the published standard to obtain full information.

Application of the standard is generally confined to analogue equipment. However, much digital audio equipment is equipped with analogue interfaces and some characteristics covered in the standard then may be relevant. Methods of measurement for digital amplifiers and similar equipment are given in IEC 61606.

7.2.3.1 IEC 60268-1(2nd edition): 1985/BS 6840: Part 1: 1987 (1993)
Sound System Equipment, Part 1: General
This part of the standard defines uniform methods for specifying characteristics common to all kinds of sound system and parts associated therewith and

also uniform methods of measurement for these characteristics. Descriptions of specialized measurement equipment include:

- Specification of quasi-peak circuit-noise measurement system in accordance with CCIR 468 (see section 7.2.6)
- Specification of acoustic-noise measurement system in accordance with IEC 60651
- Simulated programme-signal using filtered pink-noise source – for testing systems at long-term maximum output power.

7.2.3.2 IEC 60268-2 (2nd edition): 1987/BS 6840 Part 2: 1993
Sound System Equipment, Part 2: Explanation of general terms and calculation methods
General terms relating to characteristics of sound system equipment are defined at length. Items which are less familiar or are often incorrectly expressed include:

- Rated conditions and standard measuring conditions
- Definitions of noise signals – White, Pink, Broadband, Narrow-band
- Expression of voltage and power levels: e.g. power level: 101 g P/Pref.dB(W), i.e. a power level, expressed in dB, is determined by 10×1 g of the ratio of the measured power divided by the reference power. In the above example, the reference power is 1 watt. Similarly, voltage (current) level, expressed in dB is determined by 20×1 g of the ratio of the measured voltage (current) divided by the reference voltage (current). This nomenclature, first defined in IEC 60027-3, *Logarithmic quantities and units*, is normally confined to situations where the test signal is a pure sine wave. The value of the reference quantity, *always enclosed in brackets when the suffix has no other meaning*, is abbreviated as a combination of letters and numbers. The measurement sensing in such tests is normally r.m.s. but other methods, such as peak-to-peak, may be used so long as the reference quantity is appropriately expressed. Generally, the value '1' is omitted.

e.g. 100 mV = −20 dB(V), −18 dB(0.775V),

+40 dB(mV), +100 dB(uV)

Note: Where the signal being measured is not a single frequency but has a broad spectrum (e.g. random noise), the sensing and frequency characteristic of the measurement system must be specified. In

audio practice, this is usually effected by the use of specialized dB suffices. These are covered in detail in Section 7.2.4.

7.2.3.3 IEC 60268-3 (2nd edition):1988/BS 6840: Part 3: 1992

Sound System Equipment, Part 3: Amplifiers
(*Note*: the 3rd edition of Part 3 is currently (1998) being prepared for publication – based on Draft for Public Comment, document 96/211067.) This standard specifies the characteristics which should be included in specifications of amplifiers and the corresponding methods of measurement. It is intended to be used in conjunction with Part 1: General and Part 2: Explanation of general terms and calculation methods. (see Sections 7.2.3.1 and 7.2.3.2).

The range of characteristics specified is comprehensive, appropriate to the most detailed production test schedule or quality appraisal procedure. The 3rd edition of the standard is to incorporate measurements of protection circuits and sustaining time for rated (distortion-limited) output voltage and power. A new section on balance tests appropriate for balanced terminations derived by semiconductor circuits is under consideration.

It should be understood that the measurement methods currently specified for balanced terminations are devised mainly for units incorporating isolating transformers, these usually having precision centre-tapped windings. Installation practices vary in respect of whether the centre-taps are earthed at both input and output, at input or output only or neither. Consequently, the acceptability of figures of input Common-mode Rejection Ratio, CMRR, and output circuit balance obtained with the specified test arrangements is dependent on the intended usage of the equipment.

7.2.3.4 IEC 60268-4 (2nd edition): 1987/BS 6840: Part 4: 1987 (1993)

Sound System Equipment, Part 4: Methods for specifying and measuring the characteristics of microphones The standard applies to sound system microphones of all types. Characteristics specified with appropriate methods of measurement include: electrical impedance, sensitivity, directional response pattern, dynamic range and external influences. For connectors, reference is made to IEC 60574-3: 1983, Audio-visual, video and television equipment and systems. For minimum performance requirements, reference is made to IEC 60581-5:1981 High fidelity audio equipment and systems. For measurement equipment, reference is made to IEC 60651:1979, Sound level meters.

7.2.3.5 IEC 60268-5 (2nd edition): 1989 (1993) (1996)/BS 6840: Part 5: 1995, BS EN 60268-5: 1997

Sound System Equipment, Part 5: Methods for specifying and measuring the characteristics of loudspeakers The standard applies to loudspeakers for sound systems, treated as passive elements, i.e. *not* including amplifiers. A 'loudspeaker' may consist of a single drive unit or a complete system.

7.2.3.6 IEC 60268-6 (1st edition): 1971/BS 6840: Part 6: 1987 (1994)

Sound System Equipment, Part 6: Methods for specifying and measuring the characteristics of auxiliary passive elements Dating from a period when separate units such as passive filters and equalizers, attenuators and transformers were regularly employed for reconfiguring sound systems to meet operational requirements, the standard provides a comprehensive description of characteristics and measurement methods for both balanced and unbalanced input and output circuits. However, some of the measurement methods have been changed in more recently updated parts of IEC 60268 dealing with active units such as amplifiers (see Section 7.2.3.3)

7.2.3.7 IEC 60268-7 (2nd edition): 1996/BS EN 60268-7: 1996

Sound System Equipment, Part 7: Headphones and earphones The standard applies to headphones and earphones intended for use on, or in, the human ear. The possible incorporation of preamplifiers, passive networks and power supplies is covered in the characteristics and test procedures specified.

Test methods for obtaining the frequency response of a headphone or earphone, such that the measured response realistically indicates the fidelity of the sound heard when worn by a user with average hearing, have long been studied by audiologists. The major problem lies in the design of an artificial ear, acting as an acoustic filter between the headphone under test and the microphone of the sound level meter employed to take the measurement. For circum-aural and supra-aural headphones (see Section 2.4.7 earlier), the respective artificial ear has to simulate the damping effect of the pinna (see *Figure 1.49* earlier) and/or the lack of sealing between pinna and headphone. For both headphones and insert earphones, the effect of the auditory canal also has to be simulated as well as the effects of bone conduction through the head. The 2nd edition of the standard has addressed many of these problems.

7.2.3.8 IEC 60268-8 (1st edition): 1973/BS 6840 Part 8: 1988 (1994)
Sound System Equipment, Part 8: Methods for specifying and measuring the characteristics of automatic gain control devices Since this part of the standard was first published, the art and technique of automatic gain control has evolved so that the specified characteristics and tests are no longer adequate for many devices.

7.2.3.9 IEC 60268-9 (1st edition): 1977/BS 6840: Part 9: 1987 (1993)
Sound System Equipment, Part 9: Methods for specifying and measuring the characteristics of artificial reverberation, time delay and frequency shift equipment The standard covers devices intended to give reverberation, time delay and frequency shift to signals of an acoustical origin. Many such devices are now too complex for the basic characteristics and tests specified to be adequate.

7.2.3.10 IEC 60268-10 (2nd edition): 1991/BS 6840: Part 10: 1991
Sound System Equipment, Part 10: Methods for specifying and measuring the characteristics of peak programme level meters While the Peak Programme Level Meter (PPM), is now widely used in sound system equipment of all types, there is very little information published on the theory of its design and operation. Accordingly, in Section 7.2.5 the specification is covered in detail and the References show the evolution of the current IEC and National standards.

7.2.3.11 IEC 60268-11 (2nd edition): 1987/BS 6840: Part 11: 1994
Sound System Equipment, Part 11: Specification for application of connectors for the interconnection of sound system components The standard covers applications, contact arrangements and contact designations for:

- 3.2 mm 'Phono' connectors
- 6.3 mm/1 in, 3.5 mm and 2.5 mm two- and three-contact concentric connectors (jack-plugs)
- All DIN connectors used in domestic audio, vehicle and computer sound connections

7.2.3.12 IEC 60268-12 (2nd edition): 1987/BS EN 60268-12: 1995
Sound System Equipment, Part 12: Specification for application of connectors for broadcast and similar use The standard covers applications, contact arrangements and contact designations for:

- Circular connectors, 3-pin and 5-pin (XLR)
- Concentric B-gauge $\frac{1}{4}$ in jack-plug (PO 316)
- Concentric Bantam 4.4 mm jack-plug
- Loudspeaker connectors for >34 V peak and >4 A peak, with four and eight contacts (Neutrik NL range)

7.2.3.13 IEC 60268-13 (1st edition):1985/BS 6840: Part 13: 1987 (1993) (1998 edition pending)
Sound System Equipment, Part 13: Guide for listening tests on loudspeakers The listening tests specified apply to loudspeakers intended for household systems and environments. The test methods can also be applied to radio and TV receivers.

7.2.3.14 IEC 60268-14 (2nd edition): 1980/BS 6840: Part 14: 1987 (1993)
Sound System Equipment, Part 14: Guide for circular and elliptical loudspeakers outer frame diameter and mounting dimensions The standard is published with a view to minimizing the variety of dimensions in use.

7.2.3.15 IEC 61938: 1996/BS EN 61938: 1997 (revision of IEC 60268-15 (3rd edition) 1996)
Sound System Equipment, Part 15: Preferred matching values for interconnection of sound system components The standard covers:

- Matching of microphones and associated amplifiers
- Power supply for microphones – separate wire or phantom fed
- Matching of all sound sources, recording equipment, loudspeakers and headphones with household system amplifiers (pre-amplifiers).
- Matching of pre-amplifiers and power amplifiers
- Analogue matching of digital audio sources and amplifiers for household use. (*Note*: The rated output voltage for the audio signal is 12 dB below the 'full-scale' digital signal)

7.2.3.16 IEC 60268-16 (2nd edition): 1996/BS 6840: Part 16: 1989 (1996), BS EN 60268-16:1998
Sound System Equipment, Part 16: Guide to the 'RASTI' method for the objective rating of speech intelligibility in auditoria The standard covers three methods:

(1) STI method (Speech transmission index) – fourteen modulation frequencies are applied to seven octave bands
(2) STITEL method – seven modulation frequencies are applied – one to each of seven octave bands

(3) RASTI method – nine modulation frequencies are applied, four contributing to the 500 Hz octave band and five to the 2 kHz octave band

7.2.3.17 IEC 60268-17: 1990/BS 6840: Part 17: 1991

Sound System Equipment, Part 17: Methods for specifying and measuring the characteristics of standard volume indicators The Standard Volume Indicator (VU meter) originated in the USA as a simple passive instrument for monitoring the audio signal level on long transmission circuits at a time when primitive standards of programme quality were the norm. However, its simplicity has led to widespread use where its deficiencies can lead to unacceptable programme control unless a highly skilled operator is employed. In Section 7.2.5.4 the specification is described in detail, leading to the formulation of the current IEC and National standards.

7.2.3.18 IEC 60268-18 (1st edition): 1995/BS 6840: Part 18: 1996

Sound System Equipment, Part 18: Peak programme level meters. Guide for digital audio level meter This is a Technical Report, type 3, offering specifications for characteristics and performance of a peak digital coding level indicator. The intended usage is for measuring the peak level of a digital signal at any point in a digital audio system, the purpose of the report being the establishment of a basic plan for standards development.

In order to preserve some resemblance to the standard PPM (see Section 7.2.5.5), many of the characteristics are expressed in similar terms and values. However, the proposed scale range is 60 dB, with Reference indication (labelled 0 dB) corresponding to full-scale digital level on the respective circuit.

7.2.4 The use of special dB suffixes for the expression of audio signal and noise levels

7.2.4.1 Suffixes generally applied to the expression dB in audio engineering practice

- **dB** This abbreviation for one tenth of a BEL was developed from the concept of Transmission Units originated in the USA in the 1920s. It is based on the concept of comparing any pair of values by means of the logarithm of their ratio. Originally restricted to comparison between two power values, it is now also universally employed for comparison between two voltage values or two current values, taking no regard of impedances.

- **dB(mW)** This expresses a Power Level. i.e. a power measurement compared to a reference of 1 mW (see Section 7.2.3.2)
- **dB(0.775V)** This expresses a Voltage Level, i.e. an r.m.s. voltage measurement compared to a reference of 0.775 V r.m.s. (see Section 7.2.3.2)
- **dBu** This is an informal abbreviation for dB(0.775 V) and therefore must be confined to rms voltage measurements
- **dBm** This is an informal abbreviation for dB(mW) and therefore must be confined to power measurements.
- **dBA** This expresses a sound pressure level where the 'A' signifies two measurement conditions:
(a) The measurement system conforms to BS EN 60651:1994–Precision Sound Level Meter, Type 1, set to 'SLOW' dynamic characteristic
(b) The A-weighting characteristic specified in the standard is employed
The A-weighting characteristic, originally devised to simulate the average response of the ear at low sound pressure levels, is now used for the majority of acoustic noise measurements. However, there are several other weighting characteristics which may be used for special purposes. In these cases, the appropriate letter, B, C, etc. is substituted (See Section 7.2.6).
- **dBq** This expresses a noise level on an audio circuit where the 'q' signifies the following:
(a) The measurement system is a quasi-peak noise meter conforming to CCIR 468–4 (see Section 7.2.6)
(b) The unweighted characteristic specified in the standard is employed.
- **dBqp** As for dBq above, except that the weighted characteristic specified in the standard is employed. The 'p' denotes the use of a psophometric weighting characteristic (see Section 7.2.6).
Note: The use 'q' and 'qp' as part of dB suffixes was included in the CCIR-468 standard to avoid confusion with traditional r.m.s. noise measurements (expressed in dBm) when presenting results of noise measurements using the quasi-peak CCIR instrument. The CCIR-468 standard is concerned with *relative noise levels*, expressed in dBqOp (see Section 7.2.4.2 below), measured on transmission circuits where the programme level is accurately controlled (also defined in CCITT Recommendations J 14 and J 16). However, the acceptance of the standard by the· IEC for use on audio equipment generally required a new expression to replace the use of dBm(or

dBu) for simple noise level measurements; the use of 'dBq' and 'dBqp' for this purpose is a logical step which has been in limited use for some years. This is particularly important considering the large difference in readings that can occur between use of this instrument and others incorporating average or r.m.s sensing and possibly having different frequency characteristics (see Section 7.2.6).

7.2.4.2 Terminology for measurements of relative programme level and relative noise level (equiv. signal/noise ratio) in sound systems

The following dB suffixes have been devised by the CCIR for use with broadcast systems and transmission links as a means of ensuring that correct programme level and circuit noise performance is maintained between broadcasters and transmission authorities.

- **dBr** This expresses the magnitude of a sound programme signal relative to *reference programme level*. As defined in IEC 60268-10, *reference programme level* peaks to *reference indication* on a standard peak programme meter, (see Section 7.2.5). Thus, in the UK a sound programme signal at a 0 dBr point has maximum peaks corresponding to a tone level of +8 dBu (mark 6 on the UK PPM).
- **dBu0** This expresses the magnitude of a line-up tone (single frequency sine wave) relative to the magnitude of the sound programme signal at that point. Thus, at a 0 dBr point, a line-up tone of relative level 0 dBu0 has a level of 0 dBu.
- **dBq0** This expresses the magnitude of the unweighted noise signal, measured in accordance with CCIR 468-4, (see Section 7.2.6) relative to the magnitude of the sound programme signal at that point, e.g. at a −10 dBr point, an unweighted noise signal of relative level −60 dBq0 has a level of −70 dBq. This corresponds to an unweighted signal-to-noise ratio of 60 dB, using the signal line-up level of −10 dBu as the signal reference.
- **dBq0p** This expresses the magnitude of the weighted noise signal, measured in accordance with CCIR 468-4 (see Section 7.2.6) relative to the magnitude of the sound programme signal at that point. e.g. at a +10 dBr point, a weighted noise signal of relative level −70 dBq0p has a level of −60 dBq0p.

7.2.5 Standard programme level meters for the control of audio signal amplitude

7.2.5.1 Introduction

The continuously varying amplitude of audio signals requires constant monitoring to avoid poor signal/noise ratio or overload distortion. Use of primitive early compression techniques could mitigate this problem at the expense of degrading the naturalness of the sound quality. To avoid this while making the most efficient use of the available dynamic range, a monitoring instrument was required which accurately displayed the amplitude of transient programme peaks. Therefore, the development of suitable instruments for programme monitoring mainly has been the concern of broadcasting organizations who initially devised individual solutions to the problem.

International standardization of programme level meters, which commenced around 1970, has now culminated in specifications for the following three instruments:

- IEC standard volume indicator (VU meter)
- IEC peak programme level meter, type 1 (DIN PPM)
- IEC peak programme level meter, types 2a and 2b (UK, EBU PPMs)

The following sections cover: the function of the different instruments, the specified major characteristics and the standard methods of measurement. The References contain the IEC standards and related CCIR/EBU/national standards for each instrument.

The complete specifications presented in the respective standards include many aspects not considered here. The following information is intended only as a guide to the major characteristics and to assist in interpreting a manufacturer's data for commercial instruments.

7.2.5.2 Related standards for programme level meters

Standard volume indicator (VU meter)
Originally published as ASA C16.5.54 (ASA. 1954).

Now superseded by IEC 60268-17:1990/BS 6840: Part 17:1990 (BSI. 1990).

Use of the VU indicator in CCIR practice is described in *CCIR Report 820-1, Relative Values of Sound-programme Signals Established with the VU Meter and with the Peak-programme Meter* (CCIR, 1986).

Peak programme level meter – IEC type 1 (DIN) Originally published as DIN 45406:1965 (DIN, 1965).

Later revised and incorporated in IEC 268–10: 1976.

Re-issued as IEC 60268–10:1991, 2nd edition/BS 6840: Part 10: 1991 (BSI, 1991).

Use of the type 1 PPM within CCIR is covered in *CCIR Report 292:1978.*

Peak programme level meter – IEC type 2a (UK) Originally published as BS 4297:1968 (obsolete).

Revised specification incorporated in IEC 268–10: 1976 with supplement 268–10A:1978. (Combined version published as BS 5428: Part 9:1981.)

Superseded by IEC 60268–10:1991, 2nd edition/BS 6840: Part 10:1991 (BSI, 1991).

Included in the current edition are additional characteristics pertaining to the use of incremental displays. The additional characteristics include:

(1) number of discrete elements in the display,
(2) display resolution,
(3) readability, and
(4) rate of element switching.

The type 2A instrument specified in the current standards corresponds to the majority of 'UK'-type PPMs in use and the published standards now constitute the definitive reference (BSI, 1991).

Peak programme level meter – IEC type 2b (EBU) The instrument specified in BS 4297:1968 formed the basis of the original EBU PPM, published as document EBU Tech. 3205-E:1974. Of particular note is the specification for the 'slow' mode of operation, enabling PPM readings at each end of a transmission circuit to be compared by telephone.

The revised specification prepared for IEC 268–10, 10a:1978 is incorporated in EBU Tech. 3205:D/74/0299/85. This includes revised tests for the 'slow' mode of operation (EBU, 1985).

Use of the EBU PPM is covered in *CCITT Recommendation N15.*

7.2.5.3 Specified characteristics

The performance of programme level meters is defined in terms of the following specified characteristics.

Reference indication The scale marking which denotes the maximum permitted signal level.

Reference input level The level of 1 kHz continuous tone which results in a reading at the reference indication.

Dynamic response The response of a PPM to successively shorter bursts of tone at 5 kHz, each burst spaced by an interval to allow the indicator to reach the no-signal reading. The steady-state amplitude of the tone corresponds to the reference input level.

Integration time The duration of a 5 kHz toneburst the steady-state amplitude corresponding to reference input level, which results in a reading 2 dB below reference indication. (*Note*: it must be emphasized that the integration time is not related to the often misquoted 'rise-time' of the instrument. Also, dynamic tests originally were made using d.c. pulses. The measurements differed from those using tonebursts by a factor of 2.5. Documentation for early PPMs therefore may be misleading.)

Delay time The time interval between the application of continuous 1 kHz tone at the reference input level and the moment when the pointer passes a point 1 dB below the reference indication.

Overswing The transient excess of indication when continuous 1 kHz tone at the reference input level is applied.

Return time The time interval between the removal of a steady-state input signal, at the reference input level unless otherwise stated, and the moment when the pointer passes a specified lower point on the scale.

Overload characteristic The increase in reading when the amplitude of isolated 1.5 ms bursts of 5 kHz tone, corresponding to the reference input level, is raised by 10 dB.

Reference volume (VU only) The level (in dB) with respect to a stated reference, of 1 kHz tone which gives a reading of 0 VU.

Response time (VU only) The time taken for the pointer to reach 90% of the reference indication when a 1 kHz signal, producing the reference indication in the steady state, is applied.

7.2.5.4 Standard volume indicator (VU meter)

General This instrument originated in the USA as a passive moving-coil indicator, driven by a bridge rectifier and equipped with a calibrated input attenuator so that it could be connected directly across any transmission circuit.

Due to its response time of 300 ms, the instrument suffers from a deficient response to transient signals. This necessitates the operator having expert knowledge of sound waveforms when exercising control of the signal amplitude.

The standard instrument must not be confused with other versions such as are often fitted to domestic

tape-recorders. These are usually equipped with a similar scale but may have significantly different dynamic characteristics.

Display characteristics The scale layout of the VU indicator is shown in *Figure 7.30*. The reference indication is 0 VU (100%). (*Note*: some types of sound waveform (e.g. harpsichord) must be controlled to a significantly lower reading.)

All markings at and above 0/100% in red

All markings below 0/100% in black

Figure 7.30 *The scale on a VU indicator.*

Electrical characteristics The standard instrument incorporates a stepped constant impedance attenuator, presenting 7500 Ω at the input terminals; the circuit of the complete unit is shown in *Figure 7.31*. The reference input level is +4 dBu (attenuator at zero-loss setting). (*Note*: the zero-loss setting of the attenuator is labelled '+4 VU'.)

Figure 7.31 *Circuit for a VU indicator.*

The instrument significantly under-reads transients on programme signal, whether these are isolated or in trains as found with some musical instruments. Therefore in CCIR practice it is usually connected so that a 1 kHz tone, at a level 8 dB below that corresponding to the maximum signal, produces a scale reading of 0 VU. Other organizations use different values of sensitivity.

Dynamic characteristics
- Response time: 0.3 s, ±10%.
- 'Fall-back time': 0.3 s nominal.
- Overswing: 1.5%

7.2.5.5 Specification of peak programme level meters (PPMs)

General Unlike the VU indicator, the dynamic performance of which is entirely defined by the ballistics of the moving-coil meter, the performance of the PPM depends primarily on electronic processing of the audio signal. This is essential to provide the quasi-peak response which determines how much of the transient content of the audio waveform is prevented from influencing the reading.

The parameters of the dynamic response are arranged so that those transients exceeding the maximum permitted signal level can be removed without causing significant audible distortion. Failure to control the amplitude of the audio signal so that a proportion of the transients can be removed in this way results in undermodulation of the transmitters and consequent reduction of the service area.

The specified performance of the PPM applies to the combination of electronic processor and display indicator. In the case of the 'UK' PPM, the standard also includes a separate specification for a 'preferred indicator' (*Figure 7.32*). As well as the scale layout, the specification includes calibration tolerances and dynamic performance. The 'preferred indicator' is universally employed in the UK.

Divisions 1–7 are each 4 dB.

All markings, and pointer, to be white on black background

Figure 7.32 *The scale on the preferred meter of a 'UK' PPM.*

The 'DIN' meter usually has a scale length extending from less than −40 dB to +3 dB.

Description of the electronic unit The normal arrangement of the component parts of the electronic unit is shown in *Figure 7.33*. Particular requirements include:

1. Audio input amplifier with specified frequency response
2. Electronic full-wave rectifier
3. Integration circuit
4. Logarithmic d.c. amplifier
5. Linear drive amplifier to match indicator

Figure 7.33 *System diagram for a PPM.*

(1) a full-wave rectifier circuit free both from threshold effects and level-dependent frequency response:
(2) adequate range of logarithmic stage to handle transients; and
(3) output impedance of linear stage suitable for display indicator in respect of speed and damping.

Specified values for characteristics

• *Reference indication*
IEC type 1 (DIN): 0 dB.
IEC type 2a (UK): 6 (see *Figure 7.32*).
IEC type 2b (EBU): +9 dB (see *Figure 7.34*).

All markings, and
pointer, to be white
on black background

Figure 7.34 *Scale for an EBU PPM.*

• *Reference input level*
IEC type 1 (DIN): usually +6 dBu.
IEC type 2a (UK): +8 dBu.
IEC type 2b (EBU): +9 dBu.
• *Dynamic response* For all instruments: five burst durations are specified over the range of 100 to

0.5 ms and the corresponding required deflections have appropriate tolerances applied.
Note: it is important that all burst durations produce the specified readings: deficient performance of the display indicator may be concealed by adjustment for the longer bursts only.
• *Integration time*
IEC type 1 (DIN): 5 ms.
IEC type 2a and 2b (UK, EBU): 10 ms.
• *Delay time* For the type 2 instruments, a minimum value of 75 ms is specified to avoid unacceptable eye fatigue. A maximum value of 150 ms is specified to avoid a perceptible loss of synchronism with the audible sound.
• *Overswing*
IEC type 1 (DIN): 1 dB max.
IEC type 2a and 2b (UK, EBU): 0.5 dB max.
In practice, overswing is a function of the damping applied to the moving-coil indicator. The specified 'preferred indicator' for the type 2 PPM has separate tests applied for speed and overswing.
• *Return time*
IEC type 1 (DIN): range 0 to −20 dB; time 1.4–2.0 s.
IEC type 2a (UK): range 7 to 1; time 2.5–3.1 s.
IEC type 2b (EBU): range +12 to −12 dB: time 2.5–3.1 s.
On multichannel instruments, the return times must be within 0.1 s of each other. It is important that the fallback of the pointer should be 'linear', i.e. the pointer should pass equal numbers of indicated decibels in equal time intervals.
• *Overload characteristic*
All instruments: +5 to +10 dB.
The purpose of the test is to detect excessive clipping or failure of the logarithmic characteristic when large amplitude transients occur.

References

ASA C16.5, *Volume Measurements of Electrical Speech and Program Waves*, American Standards Association, New York (1954)

BSI, BS 6840: Part 10: 1991/IEC *60268-10, Sound System Equipment: Peak-programme Level Meters* (1991)

BSI, BS 6840: Part 17:1991/IEC 60268-17, *Standard Volume Indicators* (1990)

CCIR Report 820-1, *Relative Values of Sound Programme Signal Levels Established with the VU Meter and with a Peak-programme Meter*. In: Green Book, Vol. X11, CCIR, Geneva (1986). (Re-titled: ITU/R Report BS 820-1:1994).

DIN 45406. *Aussteuerungsmesser fur Elektroakustische Breitbandubertragung*, Deutsche Normen, Berlin (1965)

EBU Tech. 3205-D/74/0299/85, *The EBU Standard Peak-programme Meter*, European Broadcasting Union, Brussels (1985)

7.2.6 Measurement of audio-frequency noise signals in professional sound systems and equipment

7.2.6.1 Introduction

Measurement of audio noise signals in sound equipment originally evolved from the use of the precision sound level meter (BSI, 1994), intended for acoustic measurements. This instrument incorporates a root-mean-square (r.m.s.) sensing meter and the A-weighting characteristic (*Figure 35(a)*) simulating the typical frequency response of the human ear at low sound levels.

As the performance of sound equipment has improved, use of the Sound Level Meter (SLM) characteristics has failed to identify audible differences in the level and character of the residual noise signal. This effect may be traced to the following factors.

(1) The A-weighting characteristic progressively attenuates frequencies above 5 kHz. High-quality loudspeakers often have a significant output up to 20 kHz and for a listener who is closely situated, there is negligible absorption of high frequencies such as occurs when naturally originating sounds are heard at a distance or through structures.

(2) The damped r.m.s. response of the SLM circuit does not respond correctly to highly impulsive and sporadic interference such as cross-talk from adjacent programme or data circuits. This is an important aspect for measurements on professional sound systems.

(3) There is no mechanism in the SLM to simulate the relative annoyance to the listener caused by isolated or sporadic noise impulses heard as a background to programme signals emanating from the same loudspeaker(s).

7.2.6.2 Evolution of the CCIR method for audio noise measurement

Development of improved weighting characteristics was undertaken initially for measurements on the long analogue transmission circuits then used extensively in broadcasting systems. These characteristics had the significant difference from A-weighting in that some frequencies above 1 kHz were boosted by as much as 9 dB. However, the relatively narrow bandwidth of the circuits to be measured was reflected in the responses of the weighting networks being severely attenuated above 10 kHz.

In 1970, the general improvements in all types of professional sound equipment including that for transmission, studios and recording, were recognized by the CCIR with the introduction of a weighting curve (CCIR, 1970) with a boost of 12 dB at 6.3 kHz and extending to 12 kHz before any attenuation is introduced (*Figure 7.35(b)*).

As no agreed method initially existed for measurement of impulsive noise, r.m.s. sensing was retained until 1974 when it was replaced by a complex quasi-peak meter characteristic. This was designed to give, in conjunction with the 1970 weighting network, the best correlation with the subjective annoyance factor of all types of noise signal. Specimen noise signals used for the development ranged from pure tones to impulses occurring either singly or in random groups (BBC, 1968). The new meter characteristic was also specified for all unweighted measurements, using a wider pass-band of 22.4 Hz to 22.4 kHz (*Figure 7.35(c)*).

The current instrument, including refined tests and tolerances, was first completely specified in CCIR Recommendation 468–2:1978. There have been two subsequent revisions, CCIR Recommendation 468–3:1982 and the current edition, CCIR Recommendation 468–4:1990 (CCIR, 1990), respectively. Neither of these revisions has introduced any significant modification of the instrument's performance. Test equipment claiming to conform to any of Recommendations 468–2, 468–3 or 468–4 therefore may be assumed to give identical readings. For brevity, in the following texts, the term 'Recommendation 468' has been used to denote any of the editions (Recommendations 468–2, 468–3 or 468–4).

The method has been used in the UK since 1979 for most noise measurements on audio transmission systems by formal agreement between broadcasting and transmission authorities.

Figure 7.35 *Frequency-response characteristics.*

CCIR Recommendation 468 (CCIR, 1990) on noise measurement has been accepted only gradually for sound studios and recording practice. However, the correlation of the noise readings with the subjective annoyance to the listener has proved to be equally valid for measurements on high-fidelity sound equipment. Consequently, in 1985, the method replaced r.m.s./A-weighting in IEC Standard 60268–1 (IEC, 1985) for noise measurements on all types of sound system equipment. The dual-numbered UK standard is BS 6840: Part 1 (1993).

7.2.6.3 CCIR Recommendation 468–4 Measurement of audio-frequency noise voltage level in sound broadcasting

The CCIR publication includes details of passive networks together with test methods and tolerances for the characteristics described below. The following information is intended only as a guide to the characteristics and to assist in correct usage of instruments conforming to the specification.

Weighted characteristic The nominal weighted frequency-response curve is shown in *Figure 7.35(b)*. At 1 kHz, the gain is 0 dB, rising to a maximum of +12.2 dB at 6.3 kHz. Use of the characteristic is specified for all audio systems, regardless of bandwidth.

The skirts of the characteristic are designed to weight subjectively the amplitude of random signals

occurring towards the ends of the audible frequency spectrum. However, the audible effect of continuous tones, particularly at high frequencies, is known to be more severe than that for random signals. Where the presence of spurious tones has to be identified in sound systems, it is usual also to perform unweighted noise measurements using the characteristic described below.

Unweighted characteristic Unweighted noise measurements are made in a passband of 22.4 Hz to 22.4 kHz using the characteristic shown in *Figure 7.35(c)*. The limits of the skirts of the response curve are designed to accommodate band-pass filters having cut-off slopes conforming to the octave filter bands as specified in IEC 60225 (1966), with median frequencies of 31.5 Hz and 16 kHz, respectively. The minimum slope limits also accept simple electronic realization of the characteristic.

Quasi-peak measurement device The measurement device has dynamic characteristics electronically tailored to provide a consistent relationship between the readings and the subjective annoyance of all types of noise signals (see Section 7.2.6.2). The performance is specified in terms of the response both to isolated and to repetitive tone bursts, no constraint being placed on the actual circuit configuration. However, to provide a simple explanation of

the required performance, it is pertinent to describe a possible arrangement as follows:

(1) an input amplifier with adjustable gain, incorporating selectable weighted and unweighted frequency characteristics;
(2) a full-wave rectifier;
(3) an integration circuit, of charge time-constant of 1.6 ms, and a decay time-constant of 390 ms;
(4) an averaging circuit with a time constant of 150 ms;
(5) a drive amplifier for a display indicator, preferably having a logarithmic characteristic; and
(6) a display indicator having a calibrated scale range of at least 20 dB.

Consideration of the time-constant section shows that an isolated short-duration transient may charge the first section to nearly full amplitude. However, transfer of this pulse through the averaging stage results in significant loss of amplitude. The resulting reading for a single transient is therefore 'weighted' to reflect the diminished audible effect.

The presence of a train of transients, each identical to the single transient postulated above, causes the averaging stage to charge to a value closer to the theoretical maximum. The higher reading then reflects the severity of the audible effect and the consequent increase in annoyance factor.

Obviously, the amplitude, duration and repetition interval of transients can be infinitely varied. The time constants have been contrived to give the best overall correlation with the subjective annoyance to the listener.

Performance specification

- *Frequency response* The frequency responses of the weighted and unweighted characteristics are measured in steady state through the complete instrument. Tolerances are applied at each specified frequency.
- *Dynamic response to single tone-bursts (weighted mode)* Single bursts of a 5 kHz tone are applied, having eight burst durations over the range 1–200 ms. Tolerances are applied for the readings obtained at each burst duration.
- *Dynamic response to repetitive tone-bursts (weighted mode)* Three series of 5 ms bursts of 5 kHz tone are applied at repetition rates of 2, 10 and 100 bursts/s, respectively. Tolerances are applied for the reading corresponding to each repetition rate.

- *Overload characteristic* Tests are made with 5 kHz tone-bursts of 0.6 ms duration, the burst amplitude initially adjusted to produce full-scale reading.
- *Reversibility error* Tests are made with repetitive 1 ms d.c. pulses, reversing the polarity between tests.
- *Overswing* A maximum of 0.3 dB momentary excess reading is specified when a 1 kHz tone is applied.
- *Calibration* In both weighted and unweighted modes, applications of a 1 kHz tone at 0.775 V r.m.s. produces a reading of 0 dB. The scale has a calibrated range of at least 20 dB with the reference (0 dB) mark between 2 and 10 dB below full scale.

A convenient arrangement utilizes the 'UK' PPM scale, having equal 4 dB divisions marked 1 to 7, respectively (see Section 7.2.5). Mark 4 is used as the 0 dB indication for noise measurement, corresponding to the indication for an input level of 0 dBu when the unit is switched into PPM mode.

Where the measurement system is incorporated in automatic data-logging equipment, a practical approach is to store noise readings every 100 ms, say, for the duration of the measurement period, e.g. 5 s. The print-out should include both the maximum reading and the average of the readings, comparison of these figures giving a valuable indication of the transient content of the noise signal.

Presentation of results As noted above, the instrument is calibrated with a sine-wave signal so that an applied level of 0 dBu at 1 kHz gives a reading of 0 dB, in both unweighted and weighted modes. However, the quasi-peak response causes unweighted readings for continuous spectrum noise signals to be approximately 4 dB greater than the true r.m.s. values and the difference increases to 11 dB for weighted readings due to the effect of the special weighting network. These discrepancies have necessitated the introduction of new symbols ('q' and 'qp') as part of the dB suffixes to indicate without ambiguity when noise readings have been taken with the instrument (CCIR, 1990).

Noise measurements on sound systems may be made at two classes of monitoring point; either at points where large variations of programme level can occur during normal operation of the controls or at the output where the programme level is fully controlled. At the former points the noise level is

measured, under standard control settings, purely to confirm correct functioning of the equipment. The noise level so measured is expressed either in dBq or dBqp for unweighted or weighted measurements respectively.

At the output point, where the maximum permitted signal level usually corresponds to a tone level of +8 dBu (0 dBr point), the observed value of noise level corresponds to the signal/noise ratio, using the line-up level of 0 dBu as the signal reference. This use of the Rec. 468 meter to measure the 'relative noise level' at a 0 dBr point is denoted by the inclusion of a 0 suffix in the results (see section 7.2.4.2).

Thus unweighted relative noise level readings taken with the Rec. 468 instrument are expressed in dBq0, the use of suffix q (quasi–peak) replacing 'm' or 'u' formerly denoting readings of noise power or r.m.s. noise voltage respectively. Also implicit in the presence of suffix 'q' alone is the use of the unweighted pass-band filter.

Similarly, weighted signal/noise readings are expressed in dBq0p, the introduction of suffix 'p' denoting the use of the specified psophometric weighting network in place of the unweighted characteristic. The use of most dB suffixes in audio practice is detailed in Section 7.2.4.

Note: the method of noise measurement specified in CCIR Recommendation 468 should not be confused with the so-called 'CCIR/ARM' method (AES, 1978) (see Section 7.2.6.4). The current editions of DIN 45405 and CISPR Publication No. 6 specify instruments which are essentially the same as the Recommendation 468 instrument.

7.2.6.4 CCIR/ARM method for audio noise measurement

The use of CCIR Recommendation 468–2 (and later editions (CCIR, 1990)) has not been universally accepted for audio noise signal measurements in all applications because of two practical difficulties:

(1) The cost of the quasi-peak meter characteristic, requiring precision electronics and a high-grade display indicator, cannot be justified where correct registration of impulsive noise is not relevant. This situation applies to individual items of professional equipment, such as sound control-desks or tape machines, as well as to consumer-audio equipment.

(2) The effect of the 12 dB lift in the weighted response at 6.3 kHz combined with the behaviour of the quasi-peak circuit generally result in equipment noise readings which

are about 11 dB higher than those obtained with r.m.s./A-weighting. Commercially, this situation can be unfortunate in a market environment.

In 1978, a proposal was made by Dolby Laboratories Inc. (AES, 1978) modifying the CCIR 468 method to overcome both the problems outlined above. The changes involve the following.

(1) Replacing the quasi-peak meter with an average-reading electronic voltmeter as commonly used by most audio equipment manufacturers and maintenance areas. For continuous-spectrum noise signals of the same bandwidth, these instruments give readings about 5 dB lower than those given by the quasi-peak meter.

(2) Reducing the gain of the CCIR weighting network by 6 dB. This results in the 0 dB gain calibration being effective at 2 kHz, rather than the frequency of 1 kHz used in CCIR 468.

Entitled 'CCIR/ARM' by its inventors, the method has been embraced by many sound-equipment manufacturers since comparative noise readings are still inherently correlated with subjective annoyance due to the use of the CCIR weighting curve. The availability and low cost of suitable test equipment has led to exploration of other possible uses of the method, obviously limited to situations where impulsive noise is not significant. However, because there is no psychoacoustic justification for the changes, the 'CCIR/ARM' method has not been adopted by either the ITU/R or the IEC. The ITU/R deprecates the use of the letters 'CCIR' to indicate the method.

Great care must be exercised when comparing CCIR 468 weighted noise readings with those taken with the CCIR/ARM method. Apart from the basic difference of approximately 11 dB, explained above, CCIR 468 readings are usually taken on professional sound systems at a point where the maximum signal level is +8 dBu (see Section 7.2.6.3). Comparison of unweighted noise readings where a significant level of spurious tone is present also requires caution as the meter characteristics will tend to converge if the random noise signal is swamped by the sine-wave component.

References

AES presentation, _CCIR/ARM: A Practical Noise Measurement Method_, Audio Engineering Society, 60th Convention, Los Angeles (1978)

BBC Research Report No EL-17, *The Assessment of Noise in Audio-frequency Circuits*, British Broadcasting Corporation, London (1968)

BSI, BS EN 60651:1994/IEC 60651:1979(1993), *Precision Sound Level Meter, Type 1* (1994)

CCIR Recommendation 468, *Measurement of Audio-frequency Noise in Sound Broadcasting, in Sound Recording Systems and on Sound Programme Circuits*. In: CCIR Green Book, Vol. X, CCIR, Geneva (1970)

CCIR Recommendation 468–4, *Measurement of Audio-frequency Noise Voltage Level in Sound Broadcasting*. In: Green Book, Vol. X, Part 1, CCIR, Geneva (1990). (Re-titled: ITU/R Recommendation BS 468–4:1994)

IEC 60268–1:1985/BS 6840: Part 1: 1987(1993), *Sound System Equipment, Part 1: General* (1985)

7.3 Digital time signals

7.3.1 The Global Positioning Satellite System (GPS) and timecode

7.3.1.1 The concept

The concept behind the GPS system is that of using time-of-arrival differences in the signals received from a number of satellites in order to compute position. It is not a new idea. Way back in 1713 two mathematicians (William Whiston and Humphrey Ditton) suggested having a fleet of ships spread out at 600-mile intervals across the world's oceans. Each ship would be moored in an exact location, would keep its clock accurate by comparing it with the eclipses of the planet Jupiter's moons, and fire a maroon 6440 feet into the sky at a predetermined time. By observing the flashes from the maroons and counting the time delay before hearing the accompanying explosions, a navigator could calculate the distance from each ship. With the position of the ships being known, the navigator could fix his position. The idea came to nothing, foundering on the logistical nightmare of servicing the vessels, finding enough men to man them for long periods of time at sea, and anchoring them in position. What was needed was unmanned vessels, and satellite fitted with atomic clocks provided the answer. As long as the exact position of each satellite is known, and each satellite transmits identical time signals simultaneously with every other satellite, then the position of the receiver can easily be calculated.

Figures 7.36 to *7.38* illustrate the concept. In *Figure 7.36* signals are being received from two satellites. If the distance from one satellite is r, and the distance from the other is $r + \Delta r$, then the receiver must be at either of points X. If r is unknown, but Δr is known, then the receiver must lie along the curve $y - y$, as *Figure 7.37* shows. Δr is calculated from the time-of-arrival difference between the two transmitted signals. In *Figure 7.38*,

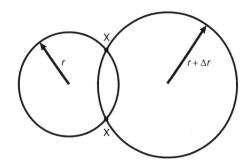

Figure 7.36 *Signals received from two satellites.*

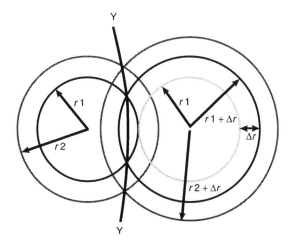

Figure 7.37 *Time-of-arrival difference between two signals.*

signals are being received from three satellites. The differences in the arrival times of the three time signals. r, $\Delta_1 r$, $\Delta_2 r$, enable the receiver's position to be calculated precisely, with no ambiguity. In practice, four satellites are required as the receiver may be either above or below the elevation of one or more of the satellites

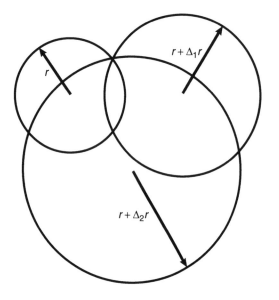

Figure 7.38 *Signals received from three satellites.*

7.3.1.2 Spatial positioning
For GPS to work, the positions of the GPS satellites (SVs or Space Vehicles) must be known. Geostationary satellites cannot be used as the laws of physics determine that for a satellite to be geostationary, it must lie on the plane of the earth's equator. The SVs follow a precisely computed orbit, using an Earth-Centred Inertial coordinate system, in other words, their positions are known relative to the effective centre of rotation of the Earth (or rather, the effective centre of the earth at 12.00 hours UTC on 1 January 2000).

GPS receivers use an Earth-Centred Earth-Fixed (ECEF) coordinate system, in other words their position is computed in terms of latitude and longitude (and height above/below a computed ellipsoid shape, acknowledging that the earth is an oblate spheroid, rather than a sphere).

7.3.1.3 Signal characteristics
The GPS SVs transmit on two common frequencies: L1 = 1575.42 MHz and L2 = 1227.6 MHz. These are 154 times and 120 times the perceived data rate of 10.23 MHz respectively. Interference between

satellites is avoided by the use of Pseudo Random Noise (PRN) scrambling of the data, a technique employed in the NICAM stereo sound transmission system, combined with a Binary Phase Shift Keying (BPSK) of the carrier (again, similar to techniques employed in NICAM).

The L1 frequency link consists of three components, in phase quadrature with each other. One component, the precision P code, is a ranging code of seven days' length, carrying highly accurate time data. The US military reserve the right to make this code unavailable to the general public by substituting another (Y) code details of which information are not in the public domain. For this reason, this code is often referred to as the P(Y) code. The third component is the C/A (coarse/acquisition) code which allows the P(Y) code to be acquired by the receiver. In addition to the P(Y) and C/A codes, the L1 link carries system (NAV) data such as SV ephemerides, system time, SV clock behaviour, status messages and instructions concerning the switching between P(Y) and C/A codes. The L2 frequency link carries either the P(Y) or the C/A codes, selectable by ground command. It can also carry the P(Y) data without the NAV data.

7.3.1.4 Time data
GPS disseminates time data as Universal Coordinated Time (UTC), in a form capable of being used for time synchronization throughout the world. UTC is a composite form of time, having data inputs from both atomic clocks and a time scale derived from the rotation of the earth with respect to the sun (UT1).

Atomic clock time, International Atomic Time (TIA), is highly accurate, being derived from atomic standards held in 50 timing laboratories in different countries, and computed by the Bureau International des Poids et Mesures (BIPM) in Paris to give a statistically definitive time. It is sometimes called a 'Paper Time' since it is not kept by a physical clock.

Irregularities in the earth's orbit, variations in the tilt of the equator and polar wobble result in the length of solar days (and hence of hours, minutes and seconds) varying. These are compensated for to give UT1, which defines the actual orientation of the ECEF coordinates with respect to space and celestial bodies. There are also variations in the rotational speed of the earth and, as a consequence, solar time (UT1) is irregular. Since navigation and both civil and military timekeeping depend on having a uniform time scale, these variations are computed out by the International Earth Rotation Service (IERS) to form UTC, which has seconds of equal length to

TAI. This does not wholly solve the problem since a TIA day is slightly shorter than a UTC day. The difference is in the order of several milliseconds per day, and can add up to a second or so over a year. The IERS determine when to add or subtract leap seconds so that UTC never differs by more than 0.9 second from TAI. The compensation is carried out either in the last minute of the UTC year or in the last minute of June.

The US Naval Laboratory maintains a set of more than twenty caesium atomic clocks and, using astronomical data, generates its own version of UTC, known as UTC(USNO), which is not allowed to vary by more than 1 ms from UTC. Because UTC is periodically corrected, GPS time differs with respect to it. The NAV data contains information for relating GPS time to UTC.

7.3.1.5 Applications
Video cameras and camcorders can be Genlocked to GPS. Timecode generators can jam sync to GPS. Because of its accuracy and stability, GPS time can be used to 'tune' free-running timecode generators, permitting multi-camera shoots to be undertaken with complete confidence without the need for periodic rejamming of the timecode generators. The GPS signal cannot normally be received by a receiver sited indoors, though the author has used GPS as a synchronizing source within the Faraday cage of an aeroplane, when shooting onto video.

7.3.2 The MSF 60 kHz time and date code

7.3.2.1 The code format and transmission
In the UK, the MSF time and date code is broadcast from Rugby (latitude 52° 22′ N, longitude 1° 11′ W). The estimated radiated power is 27 kW from a substantially omnidirectional antenna, giving effective coverage over the UK and northern and western Europe. It can be satisfactorily received in Iceland, Finland and Gibraltar. Carrier frequency is maintained at 60 kHz to within 1 part in 10^{12}. Simple on/off carrier modulation is used, the timing of the pulse edges being governed by the seconds and minutes of UTC, which is always within 1 second of GMT. Every UTC second is marked by a carrier 'off' preceded by at least 500 ms of carrier 'on'. This second marker is accurate to within 1 ms. The first second of each minute (second 00) nominally contains 500 ms of carrier off, though other data may be transmitted by carrier on/off modulation between 25 ms and 330 ms. Seconds 01 to 59 (exceptionally 58 or 60) always begin with at least 100 ms 'off'

and end with at least 700 ms of carrier 'on'. Seconds 01–06 carry information, in signalling bit pairs, about the current minute, hour, day of week, month and year, and about the difference between TIA and UTC, this difference being called DUT code (DUT1). DUT1 is signalled to the nearest 100 ms in the range ±800 ms. A positive number means that GMT is at a higher count than UTC. *Figure 7.39* illustrates the detail of the time and date code for second 00 and seconds 01–60.

7.3.2.2 Allocation of signalling bits
Bits 17B–*51B and bits 01A–*16A inclusive are currently set to '0', as are bits *52B and *59B, their future use being reserved. To identify the minute, bits *53A–*58A are all set permanently at '1', and are always preceded by bit *52A at '0' and followed bit *59A at zero. This sequence 01111110 never appears anywhere else in bits A, so it uniquely identifies the following 00 second minute marker. In minutes shortened or lengthened by the addition of a positive or negative leap second, the second numbers identified as *xx are correspondingly increased or decreased by one, shifting the start of the time and date code by one second relative to the start of the minute. When UK civil time is offset from UTC by one hour for British Summer Time, bit *58B is set to '1'. Bit *53B is set to '1' during the 61 minutes immediately preceding the change, the last being minute 59, when bit *58B changes. *Table 7.1* gives the details. Bits 01B–16B are used to signal the DUT code value. (See *Table 7.2*.)

7.3.2.3 Service continuity
The MSF 60 kHz from Rugby is transmitted continuously except for monthly and annual routine maintenance periods. The monthly maintenance period takes place on the first Tuesday of every month from 10.00 UTC to 14.00 UTC (11.00 BST to 15.00 BST). The annual maintenance period takes place over a two-week period in summer. During this time, the MSF signal is absent during the daytime and restored overnight, though the overnight transmission is not guaranteed. The dates for the annual maintenance breaks can be obtained from the National Physical Laboratory Time & Frequency Service +44 (0) 181 943 6493.

7.3.2.4 Global standard frequency time transmissions (all frequencies in kHz)
- **ATA** 10 000; National Physical Laboratory, Dr K.S. Krishnan Road, New Delhi – 110012, India

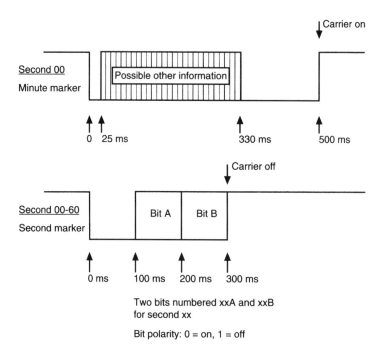

Figure 7.39 *Time and date code.*

- **BPM** 2500, 5000, 10 000, 15 000; Shaanxi Astronomical Observatory, Chinese Academy of Sciences, PO Box 18–Lintong, 710600 Shaanxi, PR China

- **BSF** 5000, 15 000; Telecommunication Laboratories, Ministry of Transportation and Communications, PO Box 71 – Chung-Li, 320 Taiwan, Republic of China

- **CHU** 3330, 7335, 14 670; National Research Council of Canada, Institute for National Measurement Standards – Time Standards, Ottawa, Ontario, K1A OR6, Canada

- **DCF77** 77.5; Physikalisch-Technische Bundesanstalt, Lab. Zeiteinheit, Bundesallee 100, D-38116, Braunschweig, Germany

Table 7.1 Time and date code

Binary coded decimal year (00-99)

80	40	20	10	8	4	2	1
*17A	*18A	*19A	*20A	*21A	*22A	*23A	*24A

BCD month (01-12) / BCD day of month (01-310) / Day of week[a]

10	8	4	2	1	20	10	8	4	2	1	4	2	1
*25A	*26A	*27A	*28A	*29A	*30A	*31A	*32A	*33A	*34A	*35A	*36A	*37A	*38A

BCD hour (00-23) BCD minute (00-59)

20	10	8	4	2	1	40	20	10	8	4	2	1
*39A	*40A	*41A	*42A	*43A	*44A	*45A	*46A	*47A	*48A	*49A	*50A	*51A

Parity bits
Bit *54B, taken with bits *17A–*24A inclusive provides an odd number of 1's
Bit *55B, taken with bits *25A–*35A inclusive provides an odd number of 1's
Bit *56B, taken with bits *36A–*38A inclusive provides an odd number of 1's
Bit *57B, taken with bits *39A–*51A inclusive provides an odd number of 1's

[a]0 = Sunday to 6 = Saturday

Table 7.2 DUT code

DUT 1	Positive	DUT 1	Negative
0 ms	No bits set to '1'	−0 ms	No bits set to '1'
+100 ms	Bit 01B set to '1'	−100 ms	Bit 09B set to '1'
+200 ms	Bits 01B & 02B set to '1'	−200 ms	Bits 09B & 10B set to '1'
+300 ms	Bits 01B to 03B set to '1'	−300 ms	Bits 09B to 11B set to '1'
+400 ms	Bits 01B to 04B set to '1'	−400 ms	Bits 09B to 12B set to '1'
+500 ms	Bits 01B to 05B set to '1'	−500 ms	Bits 09B to 13B set to '1'
+600 ms	Bits 01B to 06B set to '1'	−600 ms	Bits 09B to 14B set to '1'
+700 ms	Bits 01B to 07B set to '1'	−700 ms	Bits 09B to 15B set to '1'
+800 ms	Bits 01B to 08B set to '1'	−800 ms	Bits 09B to 16B set to '1'

- **EBC** 12 008, 6840; Real Instituto y Observatorio de la Armada – 11100 San Fernando, Cadiz, Spain.

- **HBG** 75; Service Horaire HBG, Observatoire Cantonal, CH – 2000 Neuchatel, Switzerland

- **HLA** 5000; Time and Frequency Laboratory, Korea Research Institute of Standards and Science, Yusong PO Box 102, Taejon 305–600, Republic of Korea

- **IAM** 5000; Instituto Superiore delle Poste e delle Telecommunicazioni, Viale Europa 190, 00144 – Rome, Italy

- **JG2AS** 40; **JJY** 5000, 8000, 10 000; Standards and Measurements Division, Communications Research Laboratory, 2–1, Nukui-kitammachi 4-chome, Koganei-shi, Tokyo, Japan

- **LOL1** 5000, 10 000, 15 000; Servicio de Hidrografia Naval Observatorio Naval, Av. Espana 2099, 1107 Buenos Aires, Argentina

- **MSF** 60; Time and Frequency Services, National Physical Laboratory, Middlesex, United Kingdom, TW11 0LW

- **OMA*** 50; Institute of Radio Engineering and Electronics – Academy of Sciences of Czech Republic, Chaberska 57, 182 51 Praha 8 – Kobylisy, Czech Republic

- **PPR** 435, 4244, 8634, 13 105, 17 194.4; **PPE** (inactive); Departemento Servico da hora, Observatorio Nacional (CNPq), Rua General Bruce 586, Sao Cristovao, 20921-030 Rio de Janeiro, Brazil

- **RAB-99** 25; **RBU** 66.66; **RCH** 2500, 5000, 10 000; **RID** 5004, 10 004, 15 004; **RTZ** 50; **RWM** 4996, 9996, 14 996; **RJH-63, RJH-69, RJH-86, RJH-90** 25; Institute of Metrology for Time and Space (IMVP), NPO 'VNIIFTRI' Mendeleevo, Moscow Region 141570, Russia

- **TDF** 162 (Allouis l.f. broadcast transmitter); France Telecom, Centre National d'Etudes des Telecommunications-PAB-STC. Etalons de fréquence et de temps, 196 avenue Henri Ravera, 92220-Bagneux, France

- **VNG** 2500, 5000, 8638, 12 984, 16 000: National Standards Commission, PO Box 282, North Ryde NSW, 2113 Australia

- **WWV** 2500, 5000, 15 000, 20 000; **WWVB** 60; **WWVH** 2500, 5000, 10 000, 15 000 (Hawaii); Time and Frequency Division – 847.00, National Institute of Standards and Technology, 325 Broadway, Boulder, Colorado 80303, USA

- **YVTO** 5000: Direcciòn de Hidrografia y Navigaciòn, Observatorio Cagigal, Apartado Postal No. 6745, Caracas, Venezuela

*This service may no longer be on air.

7.3.2.5 Applications

Video cameras and camcorders can be Genlocked to the MSF signal. Timecode generators can jam sync to MSF. Because of its accuracy and stability, MSF time can be used to 'tune' free-running timecode generators, permitting multi-camera shoots to be undertaken with complete confidence without the need for periodic rejamming of the timecode generators. MSF can normally be received indoors, but can be prone to local interference (in the author's experience, fluorescent lights can cause problems), and reception is sometimes poor in steel-framed buildings. Reception is almost certain to be bad if the receiver's antenna is within a Faraday cage (such as

is sometimes found as part of the fabric of television and radio studios and control rooms.

References

DALY, P., Navstar GPS and GLONASS: global satellite navigation syatems, *Electronics & Communications Engineering Journal*, December, 349–357 (1933)

Document CETM h 099; National Physical Laboratory, UK

Document tafis001.v01 updated 1996-12-18; National Physical Laboratory, UK

Document tafis002.v01 updated 1996-12-18; National Physical Laboratory, UK

Document tafis003.v01 updated 1996-12-18; National Physical Laboratory, UK

Document tafis004.v01 updated 1996-12-18; National Physical Laboratory, UK

Document tafis005.v01 updated 1996-12-18; National Physical Laboratory, UK

GPS Interface Control Document ICD-GPS-200, Navtech Seminars & Navtech Book & Software Store, 2775 S. Quincy St, Suite 610 Arlington VA 22206, USA

HACKMAN, C. and SULLIVAN, D.B., Resource Letter: TFM-1: Time and frequency measurement. *Am. J. Phys.*, **63** (4) April, 306–317 (1995)

KAPLAN (ed.), *Understanding GPS – Principles and Applications*, Artech House, London

Proc. IEEE, **79**, No. 7 July 1991

Time & Frequency User's Manual, US Dept of Commerce: code NSPUE2

SOBEL, D., *Longitude – the true story of a lone genius who solved the greatest scientific problem of his time*, Fourth Estate, London (1996)

7.4 Health and safety

7.4.1 Sources of law

In the UK, health and safety law is derived from three main sources; statute law (Acts of Parliament and Regulations), common law (based on precedents) and contract law (legally binding agreements between two or more parties). Criminal law is concerned with offences against the state. A person found guilty, beyond reasonable doubt, of a crime may be punished by a fine, a prison sentence, or both. The criminal court will not normally concern itself with matters of compensation, though it may order compensation to be paid to an individual to cover personal injury or damage to property. It will not compensate dependents of a deceased person in consequence of that person's death. Civil law deals with disputes between individuals (people and/or companies). A civil case has to be proved on a balance of probabilities, a lesser standard than that required in a criminal case. In a civil court an individual (the plaintiff) sues a defendant for a remedy. In most cases this will be for damages (financial compensation). The UK has two distinct court systems, criminal and civil. At the lowest level of the criminal court system, Magistrates courts deal mainly with less serious criminal offences and refer the more serious up to the Crown courts. County courts deal only with the less important civil matters, the more serious being dealt with by the High Court of Justice, which also hears appeals on points of law from Magistrates and Crown courts.

In the USA the court structure and organization reflect the federal nature of its government. Federal and State courts are roughly parallel in both organization and jurisdiction. Each state has its own, independent, system of courts, adjudicating on most matters. At the same time, a three-tier system of Federal courts adjudicates on federal matters and on cases not appropriate for trial in a state court. At the apex of the tiers is the Supreme Court which primarily reviews decisions of state courts and lower Federal courts on questions of federal law. On the second tier sit thirteen Courts of Appeals (one for each of twelve circuits and one national court, the US Court of Appeals for the Federal Circuit). The Courts of Appeal deal mainly with reviews of decisions made by the courts in the bottom of the federal tiers, the District Courts. Most states have just two tiers of courts, the highest being the Appellate Court (usually referred to as that state's Supreme Court), the other being the trial courts of general jurisdiction

7.4.2 The legislation

In the European Union, health and safety legislation is determined in the first instance by the European Council of Ministers, in effect the decision-making body within the EU. The Council consists of representatives from all governments of the member states. Legislation proposed by the Council of Ministers is considered by the European Parliament consisting of members (MEPs) who represent constituencies in their own member countries. The European Parliament must be consulted on all proposed legislation. The European Commission, a civil service-like body, is empowered to take action against any member state that fails to comply with legislation, and can also make proposals for future legislation to the Council of Ministers. These proposals take the form of Directives. Directives are interpreted and implemented in individual member

states via their own legislative processes. The European Court of Justice is the supreme court in the EU regarding the interpretation of EU law. It enforces EU law through a network of courts and tribunals in all member countries. In the UK, the directives are interpreted and enforced by means of the Health & Safety at Work etc. Act 1974 (HSWA). This is an enabling Act, which allows the Secretary of State for Employment to issue subordinate legislation (Regulations) proposed by the Health and Safety Commission (HSC) after consultation with industry, local authorities and other relevant bodies.

In the USA health and safety legislation is derived from public law enacted by the Senate and House of Representatives assembled in Congress. The law is implemented by the Secretary of Labor, who must report to Congress his or her recommendations, may promulgate, modify or revoke any safety or health standard, and ensure as far as practicable that any rules issued do not differ substantially from an existing consensus standard. The Secretary is advised by a National Advisory Committee on Occupational Health and Safety, comprising twelve people, four of whom are designated by the Secretary of Health, Education and Welfare, the others being appointed by the Secretary of Labor.

7.4.3 UK statute law

Among other things, HSWA places duties of care on an employer with respect not just to his or her employees but to others who may be affected by the employer's work practices, duties in the control of harmful emissions into the atmosphere, duties regarding the safety and suitability of manufactured articles for use at work. It places general duties on every employed person to take reasonable care for the health and safety not just of themselves but of any others who may be affected by their work. There is a duty not to interfere with or misuse anything provided in the interests of health, safety or welfare. There is a duty on the part of the employer to provide, free of charge, any special clothing or equipment required to comply with relevant legislation.

Enforcement of HSWA is the responsibility of the Health and Safety Executive (HSE), who may enter premises at any reasonable time, without giving notice, and who may issue improvement notices, prosecute, or even order an immediate shutdown of the work process. If a company or individual is prosecuted under HSWA legislation it is the responsibility of that person (or company) to prove that they had taken all practicable or reasonable steps to comply with the legislation. This is a new concept in UK law, as previously the onus had been on the prosecution to prove their case.

The duties under HSWA fall into one of three distinct levels:

(1) Absolute requirements
(2) Practicable requirements
(3) Reasonably practicable requirements

Absolute requirements apply where there is a risk of injury or disease if safety requirements are not followed. For example, an employer has an absolute duty to ensure that work equipment is constructed or adapted so as to be suitable for the job of work for which it is intended. Practicable requirements imply that if it is possible to comply with a requirement, then it must be complied with no matter what the cost or sacrifice Reasonably practicable requirements implies that an assessment must be made between the degree of risk involved versus the cost or sacrifice that would have to be made to avert that risk. It does not imply that nothing need be done to make the computation. In practice Risk Assessment is the main vehicle for complying with this requirement.

HSWA allows the Secretary of State to make regulations such as the Health and Safety (First Aid) Regulations 1981. More recently, regulations have been introduced which implement EU directives directly. They include:

- *The Chemicals (Hazard Information and Packaging for Supply) (CHIP 2) Regulations* These cover many aspects of the classification, labelling and packaging of chemicals. There are requirements regarding the provision of data sheets, advertising, labelling for various forms of transportation, child-resistant fastenings and the keeping of records.
- *The Construction (Design and Management) Regulations 1994 (CDM)* These regulations impose certain duties in the design and construction stages of structures (the definition of 'structure' is very wide in the Act). They also impose certain minimum requirements on building sites, on agents and subcontractors, supervisors and designers. Health & Safety files must be kept up to date and available for inspection.
- *The Control of Substances Hazardous to Health (COSHH 2) Regulations 1994* These apply to every workplace and every type of work activity.

They are supported by a number of Approved Codes of Practice (ACOPs), covering more than 40 000 chemicals and materials, together with hazardous substances generated by processes. Employers are required to assess health risks created by work involving hazardous substances and use proper control measures or protective equipment to provide adequate protection.

- *The Electricity at Work Regulations 1989* These apply to all work done with electricity. Even battery-powered equipment and its supplies are covered. Every employer and self-employed person must comply with the regulations in so far as any matters they relate to are under that person's control. The regulations cover systems, work activities and protective equipment. The strength and capability of equipment to perform certain tasks is covered, as are adverse or hazardous environments. Access, earthing, insulation requirements, connections, excess current protection, isolation, competence of equipment users – all have their own regulation. Portable equipment must be tested for integrity on a regular basis, and a suggested testing system is given later.

- *The Health and Safety (Display Screen) Regulations 1992* These implement the EU VDU Screen Directive. All workstations, including recording and post-production suites, whether installed before or after the regulations came into force, have to comply. With certain exemptions, all alphanumeric or graphic display screens are affected, no matter how the screen image is generated. All operators are covered, whether employed, contracted or self-employed. The regulations cover not just the screen itself but also any accessories, peripheral device (e.g. modem, telephone), document holder, printer, chair, desk, worksurface, and the immediate working environment. The regulations do not cover display screens in drivers' cabins, control cabins, on board a means of transport intended for public use, portable systems not for prolonged use, calculators, cash registers or devices having a small data/measurement display screen required for direct use of the equipment (e.g. reverb devices). Whether the VDU screen on a digital sound mixer is covered is a moot point. Under the regulations employers must perform a suitable analysis of the workstations, must plan the daily work activities of users so that periodic interruptions are accommodated, shall provide appropriate eye and eyesight tests, health

and safety training, health and safety information. Additionally, employers must ensure that workstations meet certain minimum requirements regarding the size of the workdesk, the keyboard, the work chair and the quality of the displayed screen image. The work environment shall be adequate in dimension and be designed to provide sufficient space for the user, the lighting shall ensure satisfactory lighting conditions, there shall be sufficient contrast between the screen and the background environment. Sources of light (including daylight) shall not cause glare or distracting reflections. Noise shall be taken into account, as shall room temperature and humidity. All emitted radiation, with the exception of the visible part of the spectrum, shall be negligible. Software provided must be suitable for the task in hand.

- *The Health & Safety (First Aid) Regulations 1981* These apply to nearly all workplaces. Under these regulations employers must provide first aid facilities and inform employees of the first aid arrangements. Self-employed people must provide their own first aid equipment. In addition to the regulations there is an Approved Code of Practice which emphasizes the duties of employers to consider a range of factors, such as number of employees, nature of the work, distribution of the work force, distance from external medical services and shift work arrangements when planning the first aid provision. In the UK there is also an ACOP covering the first aid arrangements for location crews in television and film.

- *The Management of Health and Safety at Work Regulations 1992 (MHSW)* These regulations, which come with an ACOP issued by the HSC, overlap many existing regulations such as COSHH 2 and CHIP 2. All employers are covered, as are the self-employed. Where there is an overlap, the more specific regulation applies. Where MHSW goes beyond the more specific regulations then measures must be taken to comply with MHSW. Under MHSW the employer has a duty to make an assessment of the risks to the health and safety not just of employees but also to those not in their employ who might be affected by their work activities. Appropriate arrangements shall be made for the planning, organization, control, monitoring and review of preventative and protective measures. Where five or more people are employed (even if they are on separate sites with less than five on a

particular site) this assessment must be recorded. Health and safety assistance from a competent person or persons shall be provided, and those persons provided with sufficient resources to do their job. Procedures must be established to deal with emergencies, persons sharing a workplace shall cooperate to comply with the regulations, and employers must ensure that the employer of any employee from an outside organization is provided with appropriate health and safety information. All employees must be provided with appropriate health and safety training and shall be competent to perform their duties. Employees have a duty to inform their employers and co-workers of any perceived serious and immediate danger or of any shortfall in health and safety provision.

- *The Manual Handling Operations Regulations 1992* These implement the EU 'Heavy Loads' directive, and are supported by guidance notes. Employers shall, so far as is reasonably practicable, avoid any manual handling operations that may involve a risk of injury. Where it is not reasonably practicable, the employer shall make an assessment of the risks involved in carrying out the task, take appropriate steps to minimize the risk of injury, and, where reasonably practicable, provide employees with precise information regarding the nature of the load to be lifted.

- *The Noise at Work Regulations 1989* These are accompanied by a number of noise guides issued by the HSE. They give effect to EU Council Directive 86/188/EEC. They bring in the concept of 'daily personal noise exposure'. The regulations introduce the concept of 'Action Levels'. The First Action Level means a daily personal noise exposure of 85 dB(A), the Peak Action Level means a level of peak sound pressure of 200 Pa, the Second Action Level means a daily personal noise exposure of 90 dB(A). Employers shall, when any employee is likely to be exposed to the First Action Level or above, or the Peak Action Level or above, ensure that a competent person makes a noise assessment to identify the employees likely to be affected, provide them with information about the noise. Adequate records have to be kept of these assessments, and risks to hearing from noise should be reduced as far as is reasonably practicable. If the First Action Level is exceeded then ear protection shall be provided.

- *The Personal Protective Equipment at Work Regulations 1992 (PPER)* These regulations replace much of the older, more specific, pieces of legislation. Under PPER, every employer shall ensure that suitable Personal Protective Equipment (PPE) is provided for employees, that items of PPE are compatible one with another, shall ensure that an assessment is carried out before choosing any PPE and that PPE is properly maintained in an efficient state or working order. Lockers shall be provided for PPE when not being used. Employees must be trained in PPE use. All employees and self-employed persons shall make full and proper use of any PPE and make sure that the PPE is returned to its locker after use. Employees are required to report defects in PPE.

- *The Provision and Use of Work Equipment Regulations 1992* These implement EU directive 89/655/EEC and revoke much of the old legislation that had grown up piecemeal over the years. They are intended to simplify and clarify existing laws on the provision and use of work equipment and form a coherent, single set of key health and safety requirements. They come with guidance notes prepared by both the HSC and the HSE. They apply to employers, the self-employed and any person having control, to any extent, over non-domestic premises made available as a place of work (whether or not this is in connection with any form of trade or business, whether for profit or not). In these regulations the term equipment includes any machinery, appliance, apparatus or tool and any assembly of components that are arranged so that they function as a whole to perform a common purpose. Work means any activity involving equipment, including starting, stopping, programming, setting, transporting, modifying, maintaining, cleaning, servicing and repairing. Equipment shall be so constructed or adapted as to be suitable for the purpose for which it is used or provided. Equipment shall only be used for the purpose for which it is intended, and an employer shall bear in mind the working conditions and risks to health and safety of workers. Equipment shall only be used in conditions for which it is suitable. Equipment shall be maintained in an efficient state and working order and be in good repair. Maintenance logs, where provided, shall be kept up to date. Where the use of equipment may involve a specific risk to health and safety the employer shall restrict its use to those authorized to use it, and shall only allow repairs, modifications maintenance or servicing by persons specifically qualified to do so. Adequate health

and safety information and/or instructions shall be provided, including conditions in which the equipment shall be used, foreseeable abnormal situations and action to be taken if those situations occur. Conclusions shall be drawn from experience in using the equipment. Those who use work equipment shall be adequately trained for health and safety. Any equipment shall comply with any enactment that implements any EU directive listed in Schedule 1 that is applicable. Access to dangerous parts shall be prevented by suitable guards or protection devices. No person shall be exposed to any risk from articles falling or ejected, rupture or disintegration of any part of the equipment, equipment catching fire or overheating, unintended or premature discharge of any gas, dust, article, liquid, vapour or other substance, unintended or premature explosion of the equipment or substance used or stored in it. There have to be proper control systems, protection against high or very low temperatures, isolation from sources of energy. The equipment must be mechanically stable, suitably and sufficiently lit and maintainable without exposing the person carrying out the maintenance to risk. Equipment shall be clearly marked for appropriate health and safety reasons, and where appropriate, warning devices shall be fitted for health and safety. The regulations apply not just to new equipment but also to equipment purchased second-hand, hired or leased, the onus being on the employer, not the supplier to ensure compliance.

- _The Reporting of Injuries, Diseases and Dangerous Occurrences Regulations (RIDDOR) 1995_
 These cover the requirement to report certain categories of injury and disease sustained by people while at work, together with certain dangerous occurrences and gas releases. They require a 'responsible person' to notify the relevant authority 'by the quickest practicable means' and subsequently make a written report within ten days on an approved form in any of the following events:

 (1) The death of any person as a result of an accident at work
 (2) Any person at work suffering a specified injury as a result of an accident at work
 (3) Any person not at work suffering an injury as a result of accident at work, where that person is taken to a hospital for treatment
 (4) Where there is a dangerous occurrence

The responsible person must also report as soon as practicable, and in any case within ten days of an accident, using the approved form, any situation where a person is incapacitated for work of any kind which he or she might reasonably be expected to do in the course of their work for more than three consecutive days, because of an injury sustained while working. Should an employee, as a result of an accident sustained while at work, die within one year of that accident, the employer must inform the relevant enforcing authority, in writing, as soon as the death becomes known to him or her.

Under RIDDOR, certain occupational diseases are notifiable. They are listed in Schedule 3, Part 1 of the regulations. In the case of an employee, the presence of the disease has to be confirmed, in writing, by a registered practitioner. For a self-employed person, the diagnosis need not be in writing.

Records have to be kept, by the responsible person, of all reportable injuries, diseases and dangerous occurrences. In the case of reportable injuries to, or dangerous occurrences involving, a person at work the record must contain:

(1) Full name of person concerned
(2) Occupation
(3) Nature of injury

In the case of a person not at work the record must contain:

(1) Full name of person concerned
(2) Status (e.g. visitor, customer, bystander, passenger)
(3) Nature of injury

In addition the record must contain:

(1) Place where accident or dangerous occurrence happened
(2) Brief description of what happened
(3) Date on which the event was first reported to the relevant authority
(4) Method by which the event was reported

In the case of reportable diseases the record must contain:

(1) Date of diagnosis of the disease
(2) Name of person affected
(3) Occupation of person affected
(4) Name or nature of the disease
(5) Date on which the disease was first reported to the relevant authority
(6) Method by which the disease was reported.

Notifiable injuries include:

(1) Any fracture other than to fingers, thumbs or toes
(2) Any amputation
(3) Dislocation of the shoulder, hip, knee or spine
(4) Loss of sight (temporarily or permanently
(5) Chemical or hot metal burn to the eye
(6) Any injury resulting from electric shock or electric burn that leads to unconsciousness, requires resuscitation or admittance to hospital for more than 24 hours.

The following must also be reported:

(1) Any injury leading to hypothermia, heat-induced illness or unconsciousness
(2) Any injury requiring resuscitation or admittance for more than 24 hours
(3) Loss of consciousness caused by asphyxia or exposure to a harmful substance or biological agent
(4) An acute illness requiring medical treatment, or resulting in loss of consciousness, caused by any substance being inhaled, ingested or absorbed through the skin
(5) An acute illness requiring medical treatment where there is reason to believe this resulted from exposure to a biological agent, its toxins, or infected material.

Dangerous occurrences include: events involving lifting machinery, freight containers, pressure systems, overhead power lines, electrical short-circuits, explosives, breathing apparatus, collapse of scaffolding, collapse of buildings or structures, explosion or fire, escape of flammable substances.

- *The Safety Representatives and Safety Committees Regulations 1977* These regulations and the supporting ACOPs and HSE Guidance Notes are designed to promote an effective relationship between employer and workforce in matters of health and safety. They allow for the setting up of safety committees within the workplace and the appointment on those committees of safety representatives appointed by independent trades unions. In addition to representing the employees in consultation with the employer, safety representatives also have the following workplace functions:

(1) To investigate potential hazards and dangerous occurrences and to examine the causes of accidents
(2) To investigate complaints by employees relating to matters of health, safety or welfare
(3) To make representations to the employer on matters arising from the above investigations
(4) To make general representations to the employer on matters of health, safety and welfare
(5) To carry out inspections
(6) To represent employees (of their trade union) with HSE inspectors and any other enforcing authority
(7) To receive information from inspectors
(8) To attend safety committee meetings

An employer must give sufficient time off with pay for the safety representative/s to perform their functions and undergo any reasonable training approved by the HSE.

Any two safety representatives can request an employer to establish a safety committee. The employer must then, after consulting with the safety representatives, establish a safety committee within three months from the date of the request, and publicize the details in the workplace.

- *The Health and Safety (Safety Signs and Signals) Regulations 1996* These bring into force the EU Safety Signs Directive 92/58/EEC. The purpose is to standardize on safety signage throughout the EU. They cover the various means of communicating health and safety information, including use of illuminated signs, hand and acoustic signals (e.g. fire alarms), spoken communication and the marking of pipework carrying dangerous substances. These are in addition to the traditional prohibition and warning signage. Fire safety signage is also included.

 The regulations require employers to provide specific safety signs wherever there is a risk that has not been controlled or avoided by other means. There is no need to provide a sign where it would not help reduce the risk, or where the risk is insignificant. Signs may be needed to regulate road traffic within a workplace. They will be needed to indicate 'safe condition' (e.g. fire escape, first aid provision, assembly point/s, etc.), mandatory requirements (e.g. hard hat areas, need for ear defenders, etc.), to indicate

hazards (e.g. flammable materials, industrial vehicles, etc.) and to post prohibitions (e.g. no access, no smoking, etc.). Hazardous locations (e.g. risk of slipping, falling, low headroom, etc.) need marking, as do stores containing significant quantities of dangerous substances unless the labels on the containers can be clearly seen from the outside. Firefighting equipment and provision (e.g. hose reels, extinguishers, emergency telephones, etc.) must also be signed.

Under the regulations, printed signage on its own is no longer sufficient, the signs must contain symbols that comply with BS 5499: Part 1 1990, and safety signs must be of the correct shape and colour as follows:

Hazard	Triangular, yellow background, black surround, black symbol
Mandatory	Circular, blue background, white surround, white symbol
Prohibition	Circular, white background, red surround, black symbol, with diagonal red bar in foreground from 'top left' to 'bottom right'
Safe condition	Rectangular, green background, white surround, white symbol
Firefighting	Square, red background, white surround, white symbol

Any printed information accompanying the sign must be on a rectangular sign, having the same background colour as the sign and with the printed information the same colour as the symbol.

Fire extinguisher colour and labelling is changed under these new regulations, and, separately, Halon fire extinguishers (the green ones) have been phased out for environmental reasons. All new fire extinguishers are required to be red in colour. They will have a coloured label on them indicating their type, method of use and suitability for different fire types. *Table 7.3* gives the details.

7.4.4 US statute law

This is primarily vested in the Occupational Safety and Health Act of 1970 (Public Law 91–596) (OSHA), as amended by Public Law 101–552, Section 3101, November 1990. The purpose of the act is 'to assure safe and healthy working conditions for working men and women by authorizing enforcement of the standards developed under the Act; by assisting and encouraging the States in their efforts to assure safe and healthful working conditions; by providing for research, information, education and training in the field of occupational safety and health; and for other purposes'. The Act has 34 sections. Its coverage is not confined to mainland US states, but also the District of Columbia, Puerto Rico, the Virgin Islands, American Samoa, Guam, the Trust Territory of the Pacific Islands, Wake Island, the Outer Continental Shelf Lands. Johnson Island and the Canal Zone. It recognizes 'national consensus standards', meaning any standard which has been adopted and promulgated by a nationally recognized standards-producing organization, where there was substantial agreement on its adoption, which was formulated in such a manner that there was opportunity for diverse views to be considered, and that has been designated as a standard by the Secretary of Labor after consulting with appropriate bodies. It also recognizes 'established Federal standards', meaning any current operational health and safety standard established by any agency of the

Table 7.3 Fire extinguisher types, identification and usage

Type	Label colour	Suitable for	Not to be used on
Water	Red	Wood, paper, fabric, textiles, and similar	Burning liquid, electrical, flammable metal fires
CO_2	Black	Burning liquid and electrical fires	Flammable metal fires
Foam	Cream	Burning liquid fires	Electrical and flammable metal fires
Powder	Blue	Burning liquid and electrical fires	Flammable metal fires

USA, or contained in any act of Congress currently in force. Any health and safety standards contained in the Walsh–Healy Act (1936), the Public Law Acts 91–54 (40 USC 333) 1969 and 85–742 (33 USC 941) 1958, and the National Foundation on Arts and Humanities Act (20 USC 951 et seq.) are superseded on the effective date of corresponding standards, promulgated under this Act. The Act is not intended to affect any common law or statutory right of any employee or employer regarding injuries, diseases or death arising out of, or in the course of, employment.

Under the Act, employers have a duty to provide a workplace that is free from recognized hazards that might cause death or serious physical harm to the employees, and to comply with occupational safety and health hazards promulgated under the Act. Each employee is required to comply with occupational safety and health standards, rules, regulations and orders issued under the Act insofar as they are applicable to his or her own actions and conduct. These standards may be promulgated, modified or revoked by the Secretary of Labor after allowing a period of 30 days for written comments, and then a further period of up to 60 days for discussion. An employer may apply for a temporary order granting a variance from the standard, and, under some limited circumstances may be granted a temporary order. A standard, when promulgated, will prescribe the use of labels or other appropriate forms of warning as are necessary to inform employees of all hazards to which they are exposed, any symptoms, appropriate emergency treatment and proper conditions and precautions for safe use.

The Act establishes a National Advisory Committee on Occupational Safety and Health to advise the Secretary, and he or she may also set up additional advisory committees, consisting of not more that 15 people, including at least one person appointed by the Secretary, and containing a balance between employer and employee viewpoints. It permits inspectors to enter a workplace at any reasonable time, without giving notice, to inspect and investigate any person, structure, materials or item/s of equipment. In this event, both employer and employee have a right to have a representative accompany the inspector. Any employee may request such an inspection if he or she believes there is a violation of a safety or health standard that threatens physical harm, or if actual danger exists. Should the inspector believe that a violation exists a citation may be issued, specifying the nature of the violation, the standard, rule, regulation or order violated and the time allowed for the situation to be remedied.

Failure to either act on the citation or to contest it within 15 days may result in a civil penalty of up to $70 000 for wilful or repeated violation, or a civil penalty of up to $7000 for violations not thought to be of a serious nature. If the violation causes the death of an employee, this may result in fines of up to $10 000 and/or six months imprisonment for a first offence, $20 000 and/or one year in prison for a subsequent offence.

In order to develop and establish standards, the Act provides for the setting up of the National Institute for Occupational Safety and Health (NIOSH). NIOSH also undertakes research into standards, training and other related matters. The Act provides for loans to be made available to small businesses to assist with modifications etc. to premises, equipment, facilities or working practices that have to be made to comply with the standards.

Among the many Regulations promulgated under OSHA the following may be relevant:

Part 1902 State Plans for the Development and Enforcement of State Standards
Part 1903 Inspections, Citations and Proposed Penalties
Part 1904 Recording and Reporting Occupational Injuries and Diseases
Part 1910 Occupational Safety and Health Standards
Part 1924 Safety Standards Applicable to Workshops and Rehab. Facilities
Part 1926 Safety and Health Regulations for Construction
Part 1977 Discrimination against Employees under the OSHA Act of 1970
Part 2201 Regulations Implementing the Freedom of Information Act

Among the many Directives promulgated, the following may be of interest:

Std .2 Identification of General Industry Safety and Health Standards
Std 8.1 Interpretation of Standards
Std 1–1.3 Fixed Ladders, Special Requirements, Ladder Safety Devices
Std 1–1.4 Marking of Aisles and Passageways
Std 1–1.5 Protection of Opensided Floors Platforms and Runways; Guardrails
Std 1–1.6 Clearance of Handrails and Railings
Std 1–1.7 Guarding Floor and Wall Opening and Holes
Std 1–1.10 Height of Guardrails in General Industry
Std 1–1.13 Fall Protection in General Industry

Std 1–6.5 Exemption for Religious Reasons from Wearing Hard Hats
Std 1–6.6 Inspection Guidelines for revised PPE Standards (General Industry)
Std 1–8.2 Medical Services and First Aid
Std 1–12.9 General Requirements for all Machines
Std 1–15.1 Clarification of 'Operations Which Involve Construction Work' as Related to Telecommunications Work
Std 1–16.7 Electrical Safety-Related Work Practices – Inspection Procedures and Interpretation Guidelines
Std 3–10.3 Scaffolding, General Requirements Guardrails
Std 3–10.4 Scaffolding

7.4.5 Hazard identification and risk assessment

In the UK, health and safety regulations impose a legal duty on an employer to carry out risk assessment. However, risk assessment should always be regarded as a key management tool in the reduction of injuries and dangerous occurrences in the workplace. It consists of the employer (or someone employed by him or her for the purpose) making a suitable and sufficient assessment of:

(1) Risks to health and safety of employees to which they are exposed while at work
(2) Risks to health and safety of persons not in his or her employment, but which arise in connection with the work activity being carried out

The assessment should usually involve seven separate processes, which should be recorded:

(1) Identify the hazards (a hazard is something with the potential to do harm)
(2) Calculate the risk (a risk is the likelihood that harm will occur)
(3) Calculate the extent (how many people are likely to be affected by a risk)
(4) Analyse the results obtained in steps (1)–(3) above to quantify the totality of the risk in terms of a scale that goes from negligible (no need for action) to extremely serious (take immediate action)
(5) Decide on appropriate control measures (e.g. Codes of Practice, written instructions etc.) together with allocation of responsibilities for implementation
(6) Promulgate *all* information to *all* likely to be affected

(7) Monitor the effectiveness of your assessment and its implementation

Figure 7.40 illustrates the process. *Figures 7.41* and *7.42* illustrate risk assessment forms currently used in the UK television industry. *Figure 7.41* contains a Risk Factor Table. When completed, the factors should be placed in the relevant line/s of the right-hand column of the Risk Analysis Table above it and appropriate action taken as follows:

Very Severe Immediate action required
Severe High priority – attend to as soon as possible
Moderate Include appropriate control measures in the planning
Low Action may be required after further assessment
Negligible Probably acceptable, but be wary of any risk that is calculated as being negligible

7.4.5.1 Working on location
There are special health and safety problems working on location. Work is done in changing environmental conditions. It is often carried on in adverse conditions. People may work alone or in small crews formed together at short notice. There may be limited information available about hazards. Employed persons may be working alongside self-employed freelance crew members. Persons not involved in the work may interfere with, or be affected by, the work. Time is usually limited. There may be no opportunity to retake events.

7.4.5.2 The risk assessment
This should be practical. Be systematic and ensure that all relevant risks are assessed without concentrating on trivial ones. Try looking at 'risk groups' (e.g. all electrical, all concerning working at heights, etc.) instead of going through chronologically. Address what really happens, as opposed to what you or someone else thinks happens. Risks usually increase when actual practice differs from received instructions. Non-routine operations should be looked at carefully. Identify all groups of workers who may be particularly at risk (e.g. newly recruited staff, young or elderly persons, freelances you have not worked with previously).

7.4.5.3 Assigning crew
It is at the crew assignment stage that decisions may be taken that will affect the health and safety of a crew. Size of crew numbers, skill levels and

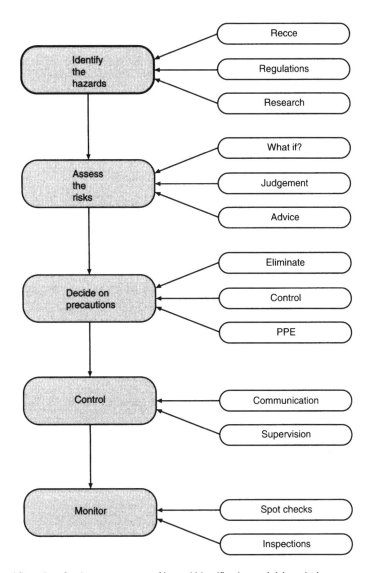

Figure 7.40 *A suggested flow chart for the management of hazard identification and risk analysis.*

experience are factors, as is the ability of crew members to work effectively together. Managers should provide an effective communication channel in order to pass information to the crew in case of changes to circumstances. Make this channel two-way. Pass on safety information in a clear, unambiguous manner. Consider the following checklist:

(1) Who is to do the work? What experience do they have?
(2) Who is to be responsible for what?
(3) Who is to take overall control?
(4) Have health and safety matters been explained fully and clearly?

(5) Have they been put down in writing and given to all concerned?
(6) What emergency procedures might be necessary?
(7) How will any change to a developing story be handled?
(8) Do you know the abilities and limitations (creative, mental and physical) of the crew?

7.4.5.4 Planning
This should be carried out by someone competent in health and safety matters. It is not acceptable for crew to discover new information which could affect

PRODUCTION TITLE:		DATE:	
DIRECTOR:		PRODUCTION MANAGER:	
RISK LEVEL (minor injury to fatality)	HIGH POSSIBILITY	MEDIUM POSSIBILITY	LOW POSSIBILITY
HIGH RISK	DO NOT DO	SERIOUS PRECAUTIONS	GENERAL PRECAUTIONS
MEDIUM RISK	SERIOUS PRECAUTIONS	APPROPRIATE PRECAUTIONS	SMALL PROBLEM
LOW RISK	APPROPRIATE PRECAUTIONS	GENERAL PRECAUTIONS	NO PROBLEM
SITUATION (Write details)	HIGH POSSIBILITY write precautions	MEDIUM POSSIBILITY write precautions	LOW POSSIBILITY write precautions
HIGH RISK			
MEDIUM RISK			
LOW RISK			

Figure 7.41 *A risk analysis form, which incorporates the concept of Risk Factor.*

Does the production involve:	Hazard Present		Severity	Likelihood	Risk Factor
	Yes	No			
1. Audiences					
2. Confined spaces					
3. Electricity or gas					
4. Fire/Flammable materials					
5. Hazardous substances					
6. Heat/Cold/Extreme weather					
7. Lasers/Strobe effects					
8. Machinery/Cranes/Hoists					
9. Night operations					
10. High sound levels/Noise					
11. Non-standard manual handling					
12. Public/Crowds					
13. Radiation					
14. Special effects					
15. People with special needs					
16. Scaffolding/Rostra					
17. Specialised rescue/First Aid					
18. Stunts/Dangerous activities					
19. Vehicles/Speed					
20. Water/Proximity to water					
21. Weapons					
22. Working at heights					
23. Working overseas					
24. Other (specify)					
This list should not be regarded as comprehensive. Use additional sheets if necessary.					

Figure 7.41 (continued)

Risk Factor Table

	Likelihood				
Severity	V Unlikely	Unlikely	Possible	Likely	V Likely
V Severe	Moderate	Severe	Severe	V Severe	V Severe
Severe	Low	Moderate	Severe	Severe	V Severe
Moderate	Low	Low	Moderate	Severe	Severe
Slight	Negligible	Low	Low	Moderate	Severe
Negligible	Negligible	Negligible	Low	Low	Moderate

Risk Assessment Categories

Severity		Likelihood	
Negligible	No visible injury No pain	Very Unlikely	Freak event No known history
Slight	Minor cuts/bruising No lasting effect	Unlikely	Unlikely sequence of events or multiple failure.
Moderate	Deep flesh wound Heavy bruising Degree of disability Possible lost time	Possible	Foreseeable under unusual circumstances. History of occasional occurrence.
Severe	Lost time accident Major injury RIDDOR category	Likely	Known history of regular occurrence. Foreseeable as likely.
Very Severe	Long term disability Amputation or death	Very Likely	Common occurrence. Very foreseeable.

Names/Addresses/Phone Nos of Experts/Contractors used

Figure 7.41 (*continued*)

RISK ASSESSMENT ACTION

Main Risks Identified	Controls to Minimise Risk	Person Responsible

Figure 7.41 (*continued*)

Production: Shooting Date/s:		Production Number: Location:	
	Tick		**Tick**
1 Access/egress blocked/restricted		25 Noise/high sound levels	
2 Alcoholic drinks/hospitality		26 Portable tools above 110 volts	
3 Animals/insects (wild, performing etc.)		27 Practical flame/fire	
4 Special props, tools etc. under the control of presenter/artiste etc.		28 Radiation – sources/equipment etc.	
5 Audience safety/public/crowds/ violence/civil disorder		29 Recording/shooting on location, OBs psc work	
6 Compressed gasses/cryogenics/ low temperature		30 Scaffolds/rostra/decking/platforms/ practical staircase/walkways on set	
7 Confined space/tank/mines/caves/ tunnels		31 Scenery/flats over $10\text{ft} \times 12\text{ft}$/non- standard shape/centre of gravity/flown	
8 Derelict buildings/dangerous structures/ isolation of services/waste control		32 Scenic/set materials – not fire-retardent/ toxicity tested	
9 Diving operations		33 Scenery manual handling difficulties	
10 Explosives/pyrotechnics/fireworks		34 Scenic materials – glass/polystyrene	
11 Fatigue/long hours/physical exertion/ stress		35 Smoking on set (by prior permission)	
12 Fire prevention/evacuation procedures		36 Special 'flying/technical' rigs	
13 First Aid/medical requirements		37 Special needs/children/elderly/disabled	
14 Flammable materials: painting/spraying needed/fuel		38 Special visual effects: rain/snow/fire/ smoke/steam/dry ice/heat	
15 Flying/aircraft/balloons/parachutes		39 Scenery/props storage on premises	
16 Freelance crews, scenic ops		40 Stunts/dangerous activities/hazardous props	
17 Hazardous substances: chemicals/dust/ fumes/poisons/asbestos/battery acid etc./ waste disposal		41 Cranes/camera cables/camera movement/special cable runs/ scanners	
18 Heat/cold/extreme weather/climate		42 Vehicles/motorcycles/speed	
19 Heavy loads on floors/rostra		43 Water/proximity to water/water tanks	
20 Liquid petroleum gas/bottled gasses		44 Weapons/knives/firearms	
21 Lasers/bright lights/strobs effects		45 Working at heights/ladders/tallascopes/ scaffold towers/cherry pickers	
22 Lifting equipment (inc. fork lift trucks)		46 Working on lighting grid/s	
23 Live electrical equipment – no test certificate/not hired from approved supplier		47 Working/storage under seating	
24 Night operations		48 Other: specify	

State overleaf whether risk is high, medium or low.
Specify control measures to reduce risk to acceptable level, and state the resulting risk factor.
Inform persons exposed to risk (in writing) of the control measures to be adopted.

Figure 7.42 *A simple yet comprehensive risk analysis form.*

No.	exposed	Describe risks and state if considered to be High, Medium or Low	Indicate names of experts or contractors to be used. Specify who is responsible for ensuring that these controls are complied with and effective

Figure 7.42 (*continued*)

their health and safety only when they arrive on location. Set up a proper chain of responsibility and ensure that all in that chain are clear about their role. Remember the following:

(1) Knowledge about the location may be scarce. All foreseeable circumstances should be considered.
(2) During the course of a shoot circumstances can change rapidly (especially true of public disturbances and war). The situation may need to be reassessed as the story develops, hence the need for two-way communication.
(3) If crews consist of persons from a number of employers, or contain freelances, the MHSWA Regulations require cooperation and coordination of the work.
(4) Individuals, because of their nature, and that of the work, may be prepared to take risks with both themselves and others. They should be discouraged from doing so.

7.4.5.5 High-risk assignments

It may be necessary to cooperate with other workers to develop and coordinate measures to minimize risk and provide for emergency evacuation. On short-notice assignments arriving on location may be the first opportunity to assess risks. Any precautionary measures worked out in advance should be capable of being adapted to changed circumstances. If crewing on such an assignment remember the following:

(1) Know the way in and out of an area (especially in the dark).
(2) Have an alternative means of escape.
(3) Do not get out of a car between it and a wall if snipers are suspected.
(4) Make sure you have alternative arrangements for communications with fixers.
(5) Arrange for your contingency planning to be coordinated by a key person.
(6) Get to know your PPE, its use, its restriction on movement, its effect on you mentally, and its limitations in terms of effectiveness.

7.4.5.6 Working in areas of public disorder

The decisions regarding crewing and coverage will probably be taken by programme editors. All concerned should be given as much information as possible about the nature of the trouble and its exact location. If possible use only experienced crew. If an inexperienced crew member has to be used, place them with an experienced crew member. Areas with

a history of violence or public disorder should be treated with care even when they appear quiet. Crews should arrive at the outer edge of the disorder and assess the situation before proceeding further. Seek guidance from the authorities (police, army, fire service) to assist the assessment. The decision whether to remain, move further in or withdraw should be made by the senior crew member present, based on their assessment of the situation, and in consultation with the other crew members. That person should keep themselves up to date regarding information about the situation and modify their assessment accordingly. Any crew member concerned about the situation or any specific safety aspect of it should be allowed to withdraw. In that event the senior crew member should re-assess the situation in the light of reduced crew numbers. Useful tips:

(1) Meet at the edge of any disorder.
(2) Make an assessment before doing anything else.
(3) Remain as inconspicuous as possible.
(4) Avoid conspicuous operations with mic boom (or camera).
(5) Withdraw if your presence is exacerbating the situation.
(6) Cease recording if the subject of the shoot starts to behave abnormally.
(7) Never attempt an immediate reconstruction of the event or incident.
(8) Consider the pros and cons of using unmarked cars.
(9) If travelling by vehicle, leave it close to hand and in a secure site. Do not move it into the disturbance unless cleared to do so by the authorities.

Take with you:

(1) The PPE identified by the general risk assessment
(2) A hard hat
(3) A disposable set of overalls
(4) Disposable dust masks
(5) A pair of safety glasses or goggles
(6) Industrial boots or shoes
(7) Disposable surgical gloves

7.5 Portable appliance testing

Although the Electricity at Work Regulations 1989 (EAWR) impose detailed safety requirements, they make no difference between fixed installations and 'transportable' appliances. In practice, however, there

has to be a difference in the management of these two component parts of an electrical system. The fixed installation (sometimes called the 'hard wiring') should either have been installed to IEE 16 Ed. standards (BS 7671) or for older installations, be periodically tested as per BS 7671 and IEE Guidance Note No. 3. For the appliances connected to the fixed installation, there is the distinct hazard of fire and personal shock either from damaged covers exposing live components or from a faulty/damaged/poorly wired flex or connector. EAWR do not specify what tests should be carried out or how frequently testing should be carried out. It is the duty of employer or the self-employed person to devise an effective management system, though to comply with EAWR all decision making and testing should be done by someone with a level of competence beyond that of an untrained non-electrical person.

Any testing system must take into account the construction of the equipment. Class I equipment relies on the integrity of the earthing system to ensure that, in the event of a fault condition, the case will remain close to earth potential. Class II equipment (double insulated) requires no earth, but relies on two separate insulation layers (one can be air) to isolate supply voltages from the equipment user. The environment in which equipment is used will have a bearing on the testing system employed, as will the harshness of use.

In general, a testing regime will combine three levels of action:

(1) Visual checks by the user (after some basic training) each time equipment is used
(2) Formal visual inspection on a regular basis by an appointed and competent person
(3) Combined inspection and test by a competent person or contractor.

User checks should include the following:

(1) Any damage (apart from light scuffing) to the cable sheath
(2) Damage to connector (e.g. cracked plug top, bent pins)

Table 7.4 Recommended inspection intervals for transportable and portable electrical appliances

Equipment/environment	User	Formal visual inspection	Combined inspection and testing
Battery operated (less that 20 V)	No	No	No
Extra low voltage (less than 50 V a.c.)	No	No	No
Personal computers, VDU screens, etc.	No	Yes 2–4 years	No (if Class II) otherwise up to 5 years
Photocopiers and rarely moved fax machines, etc.	No	Yes 2–4 years	No (if Class II) otherwise up to 5 years
Class II equipment not hand held but occasionally moved	No	Yes 2–4 years	No
Hand-held Class II equipment	Yes	Yes 6 months to 1 year	No
Class I equipment	Yes	Yes 6 months to 1 year	Yes 1–2 years
Cables and plugs for Class I equipment and extension leads operating at mains voltages	Yes	Yes 6 months to 4 years depending on type of equipment and harshness of use.	Yes 1–5 years depending on type of equipment it is connected to and harshness of use

(3) No inadequate or taped joins in the mains flex
(4) Outer cable sheath is securely fitted into plug (no internal cable insulation showing)
(5) Equipment not abused (e.g. wet or excessively dirty)
(6) Equipment not obviously damaged (e.g. cracked or bent casing, loose screws)
(7) Signs of overheating (e.g. scorch marks, discolouration)

These checks also apply to extension leads and their associated plugs and sockets. Equipment failing the test should be taken out of service and reported.

Formal visual checks should include all the above, and in addition, removal of the plug cover to check effectiveness of cord grip, tightness of cable terminations, correctly rated and secure fuse, signs of overheating, ingress of water or other contaminant, internal damage, cables correctly terminated (including the earth for Class I equipment). The results of the inspection should be formally recorded. Although not required by EAWR, it helps the management of the testing routine if each item tested is labelled with its latest test date and result. This formal visual test does not involve removing equipment covers. The competent member of staff can be someone with sufficient knowledge and information, who has received appropriate training on what to look for.

Combined inspection and testing is the only way to pick up such faults as loss of earth integrity, deterioration of insulation or internal contamination of equipment. It should be carried out by a competent electrician. It should be carried out:

(1) Whenever there is reason to believe the equipment may be defective, but there is no outward sign of damage
(2) After any repair or modification
(3) At intervals appropriate to the frequency and manner of use of the equipment

As well as the checks above the combined inspection and test should include checks for correct polarity, correct fusing, effectiveness of termination, effectiveness of insulation. Such tests may be made with a Portable Appliance Tester (PAT). The tests may be made on a simple pass/fail basis, or more sophisticated methods may be employed, that require a degree of interpretation. *Table 7.4* illustrates suggested testing routines.

Bibliography

CHANDLER, P., *An A–Z of Health and Safety Law*, 2nd edn, Kogan Page, London (1997)
HSE BOOKS, *CDM Regulations*, HMSO, London
HSE BOOKS, *CDM Regulations, Construction Sheets 39–42*, HMSO, London
HSE BOOKS, *Safe Use of Ladders, Step Ladders & Trestles*, HMSO, London (1984)
HSE BOOKS, *Tower Scaffolds*, HMSO, London (1987)
HSE BOOKS, *Memorandum of Guidance on the Electricity at Work Regulations 1989 HS(R)25*, HMSO, London (1989)
HSE BOOKS, *A Guide to the Health and Safety at Work etc. Act 1974*, 5th edn, HMSO, London (1992)
HSE BOOKS, *Work Equipment – guidance on regulations*, HMSO, London (1992)
HSE BOOKS, *Personal Protective Equipment at Work – guidance on regulations*, HMSO, London (1992)
HSE BOOKS, *Display Screen Equipment at Work – guidance on regulations*, HMSO, London (1992)
HSE BOOKS, *Manual Handling – guidance on regulations*, HMSO, London (1992)
HSE BOOKS, *Workplace Health, Safety and Welfare – approved code of practice* HMSO, London (1992)
HSE BOOKS, *Management of Health and Safety at Work Regulations – approved code of practice*, HMSO, London (1992)
HSE BOOKS, *Managing Crowds Safely*, HMSO, London (1993)
HSE BOOKS, *Electricity at Work: Safe Working Practices HS(G)85*, HMSO, London (1993)
HSE BOOKS, *Maintaining Portable and Transportable Electrical Equipment HS(G)87*, HMSO, London (1994)
HSE BOOKS, *Safety Representatives and Safety Committees*, HMSO, London (1996)
HSE BOOKS, *Camera Operations on Location*, HMSO, London (1997)
HSE BOOKS, *First Aid at Work – approved code of practice and guidance*, HMSO, London (1997)
HSE BOOKS, *The Management of Health and Safety at Work Regulations* (updated), HMSO, London (1997)
HSE BOOKS, *List of Current Health and Safety Legislation 1996*, HMSO, London (1997)
HSE BOOKS, *General COSHH ACOP & Carcinogens ACOP & Biological Agents ACOP*, HMSO, London (1997)
HSE BOOKS, *Young People at Work*, HMSO, London (1997)
HSE BOOKS, *Safety Signs and Signals*, HMSO, London (1997)
MILLWARD, A., *Managing the Safety of Young People in the Workplace*, Croner Publications, London (1997)
STRANKS, J., *Health and Safety Law*, 2nd edn, Pitman Publishing, London (1996)
STRANKS, J., *The Law and Practice of Risk Assessment*, Pitman Publishing, London (1996)
STRANKS, J., *Management Systems for Safety*, Pitman Publishing, London (1996)

Index

Audio Engineer's Reference Book

Second edition

Edited by

Michael Talbot-Smith

Warrington Collegiate

Focal
Press

OXFORD AMSTERDAM BOSTON LONDON NEW YORK PARIS
SAN DIEGO SAN FRANCISCO SINGAPORE SYDNEY TOKYO

Focal Press
An imprint of Elsevier Science
Linacre House, Jordan Hill, Oxford OX2 8DP
225 Wildwood Avenue, Woburn, MA 01801-2041

First published 1994
Paperback edition 1998
Second edition 1999
Reprinted as paperback edition 2001, 2002

British Library Cataloguing in Publication Data
Audio engineer's reference book. – 2nd ed.
 1. Acoustical engineering 2. Sound – Recording and
 reproducing – Equipment and supplies – Handbooks, manuals,
 etc.
 1. Talbot-Smith, Michael
 621.3'893
 ISBN 0 240 516850

Library of Congress Cataloguing in Publication Data
Audio engineer's reference book/edited by Michael Talbot-Smith. –
 2nd ed.
 p. cm.
 Includes bibliographical references and index.
 ISBN 0 240 51528 5
 1. Sound – Recording and reproducing – Equipment and supplies –
 Handbook, manulas, etc. I. Talbot-Smith, Michael.
 TK7881. 4. A926
 621.389'3 – dc21 98-42607
 CIP

For information on all Focal Press publications
visit our website at www.focalpress.com

Typeset by Laser Words, Madras, India

Transferred to digital printing 2005

Printed and bound by Antony Rowe Ltd, Eastbourne